The University of North Carolina Monograph Series
in
PROBABILITY AND STATISTICS

Sponsored by the
Department of Statistics at
The University of North Carolina at Chapel Hill

Number 3

The University of North Carolina Monograph Series
in
PROBABILITY AND STATISTICS

Previously published:

NUMBER 1 RUSSIAN-ENGLISH DICTIONARY OF STATISTICAL TERMS AND EXPRESSIONS AND RUSSIAN READER IN STATISTICS
by Samuel Kotz, with the collaboration of Wassily Hoeffding

NUMBER 2 PROCEEDINGS OF THE SYMPOSIUM ON CONGESTION THEORY
Edited by Walter L. Smith and William E. Wilkinson

Essays in

PROBABILITY AND STATISTICS

Edited by

R. C. BOSE
I. M. CHAKRAVARTI
P. C. MAHALANOBIS
C. R. RAO
K. J. C. SMITH

The University of
North Carolina Press • Chapel Hill

SBN 8078-1109-2
Library of Congress Catalog Card Number 69-63005

General Editor's Preface

The University of North Carolina Monograph Series in Probability and Statistics is sponsored by the Department of Statistics at The University of North Carolina at Chapel Hill and will provide an addition to the publication activities of the Institute of Statistics of the University.

Since the Department was established in Chapel Hill in 1946 by Professor Harold Hotelling, much of the work of the faculty, students, and visitors has been published in the Institute of Statistics Mimeograph Series. It is expected that original research similar in character to that published in the Mimeograph Series in the past will continue to be published as before. There is, however, work of value to research students and senior scholars which deserves publication in more permanent form, and it is this work which will be published in the Monograph Series.

Titles in the Series are approved by an editorial board from the Department of Statistics. Although it is expected that most of the titles will be the work of authors connected with the University, any suitable manuscripts will be considered for inclusion in the Series.

GEORGE E. NICHOLSON, JR.

Preface

The Department of Statistics, The University of North Carolina at Chapel Hill, decided in 1965 to publish a collection of essays by eminent statisticians to be dedicated to the memory of the late Professor Samarendra Nath Roy. At the request of Professor George E. Nicholson, Jr., Chairman of the department, Professor R. C. Bose in the United States and Professor C. R. Rao in India took up the responsibility of contacting eminent statisticians for their contributions to this volume. The response was overwhelming, and the length of the manuscripts of the contributed articles became so large that preparing them for publication was a slow process.

In the initial stages of contacting the statisticians, Professor V. P. Bhapkar, who was visiting the Department of Statistics, University of North Carolina at Chapel Hill, helped Professor Bose. The task of getting the manuscripts checked by editorial readers was handled by Professor I. M. Chakravarti. Copyediting was done entirely by Dr. K. J. C. Smith, who devoted long hours to this task.

The editors wish to thank the contributors for their cooperation and patience.

EDITORS

Samarendra Nath Roy

SAMARENDRA NATH ROY

December 11, 1906—July 23, 1964

Samarendra Nath Roy, the son of Dr. Kali Nath Roy and Suniti Bala Roy, was born December 11, 1906, in Calcutta, India. He died while vacationing at Jasper, Canada. His father was a well-known journalist, being the editor of the Indian nationalist daily, *The Tribune*, published in Lahore, Punjab.

Roy had a uniformly brilliant academic career. He graduated with first-class honors in mathematics in 1928 from the Presidency College, Calcutta, and was ranked first in the university. He received his Master's degree in applied mathematics at the University of Calcutta in 1931 with the theory of relativity as his special subject.

In 1932 Roy started work on cosmological problems in collaboration with Professor N. R. Sen in the Department of Applied Mathematics of the University of Calcutta. In the course of his research work he visited the Indian Statistical Institute in Calcutta several times to use its computational facilities. There he came in contact with Professor P. C. Mahalanobis, who had founded the institute in 1931 with a small group of young, highly talented mathematicians and physicists. Roy was soon persuaded to join the institute, first on a part-time basis in 1934 and a year later as a full-time statistician. He at once became a very active research worker, making outstanding contributions to the field of multivariate analysis, individually and in collaboration with his colleagues. In 1938 he and R. C. Bose worked out the non-null distribution of the studentized D^2-statistic (*Sankhyā*, 1938, pp. 19-38), which Mahalanobis had been using as an important tool in anthropometry. This distribution enables us to compute the power function of Hotelling's T^2, since the studentized D^2-statistic is formally the same as Hotelling's

T^2 statistic. A year later he made another outstanding contribution to multivariate analysis when in a paper published in *Sankhyā* he completely worked out the sampling distributions of p-statistics (which are the latent roots of a certain important matrix connected with samples from two multivariate normal populations). Study of properties of statistical inference procedures based on these latent roots, which serve as key tools in multivariate analysis, became in later years an important and fascinating subject of research to which Roy himself and other illustrious statisticians like R. A. Fisher, H. Hotelling, and T. W. Anderson made significant contributions.

In 1938 Roy was appointed a lecturer in the Department of Applied Mathematics of the University of Calcutta while he continued his research at the Indian Statistical Institute. He transferred to the newly created Department of Statistics in 1941. Here he collaborated with P. C. Mahalanobis and R. C. Bose in building up the new department. From 1946 through 1949 he served as the Assistant Director of the Indian Statistical Institute. During 1947-48 he was also the acting Head of the Department of Statistics of the University of Calcutta.

In spite of a heavy teaching load and administrative duties at the University of Calcutta and the Indian Statistical Institute, his research activities continued to flourish. He derived the non-null distribution of the latent roots (*Sankhyā*, 1942, pp. 15-34), together with the individual distributions of the maximum, minimum, and intermediate roots (*Sankhyā*, 1945, pp. 133-158). At this time he also found a generalization of the analysis of variance to the multivariate case and worked out the distributions of the appropriate test statistics (*Sankhyā*, 1942, pp. 35-50) and published a series of papers on statistical inference in multivariate problems.

In the spring of 1949 Roy came to the United States as a Visiting Professor of Statistics at Columbia University. On his return to India, he was appointed Head of the Department of Statistics at the University of Calcutta. In March 1950 he joined the Department of Statistics of The University of North Carolina at Chapel Hill as a Professor. Here he continued his teaching and research until his death in July 1964.

During this period of fourteen years, Roy, together with his collaborators and students, built up a very impressive record of research. Only a bare mention can be made of the most important of his contributions. One significant contribution was the introduction of a heuristic principle of construction of tests of hypotheses, based on the union and intersection principle (*Ann. Math. Statist.*, 1953, pp. 220-238). This principle was used by him to derive new test procedures as well as to provide new interpretations for existing ones in the case of composite hypotheses. Recently, W. Hoeffding has established optimal properties of Roy's method of test construction, relative to likelihood ratio and other principles of test construction (Wald Memorial Lectures, 1968).

Another important series of his papers was devoted to the construction of simultaneous confidence intervals for meaningful parametric functions arising in the study of multivariate data (*Ann. Math. Statist.*, 1953, pp. 513-536; 1954, pp. 724-761; 1956, pp. 838-858), (*Biometrika*, 1957, pp. 399-410; 1958, pp. 581-586).

He was much interested in developing statistical methods for analyzing experiments where the responses are not necessarily normally distributed (*Biometrika*, 1956, pp. 361-376) and of non-parametric methods especially suited for study of categorical data (*Ann. Math. Statist.*, 1956, pp. 750-757). His recent research interest was in the design and analysis of experiments with more than one factor and more than one response (*Bull. Int. Statist. Inst.*, 1961, pp. 59-72; *Contributions to Statistics: Presented to Professor P. C. Mahalanobis on the Occasion of his 70th Birthday*, 1963, pp. 419-426).

Roy received many academic honors. He was elected Fellow of the National Institute of Sciences in India in 1946, and Fellow of the International Statistical Institute in 1951. He was President of the Statistics Section of the Indian Science Congress in 1948. He was the author of a book, *Some Aspects of Multivariate Analysis* (John Wiley and Sons, New York, 1957), which contained his earlier work on multivariate analysis. At the time of his death he had under preparation two other books and was planning to publish a revised and updated edition of his earlier work. He was an inspiring teacher, and many of his students in India and the United States have achieved distinction and hold responsible positions.

One of the most self-effacing of men, Roy had simple tastes and a warm personality. He was always eager to help his students, friends, and relatives without any regard for his own convenience. His keen sense of humor and his sparkling wit endeared him to all. He was a devoted husband and a loving father and leaves behind him his wife, Mrs. Bani Roy, and four children: Prabir, Subir, Tapan, and Sunanda Roy.

EDITORS

Contents

	General Editor's Preface	vii
	Preface	ix
	Samarendra Nath Roy	xi
1.	ESTIMATION OF COVARIANCE MATRICES WHICH ARE LINEAR COMBINATIONS OR WHOSE INVERSES ARE LINEAR COMBINATIONS OF GIVEN MATRICES, *by T. W. Anderson*	1
2.	ON CONDITIONAL TEST LEVELS IN LARGE SAMPLES, *by R. R. Bahadur and P. J. Bickel*	25
3.	INTERPRETATION AND USE OF A GENERALIZED DISCRIMINANT FUNCTION, *by R. E. Bargmann*	35
4.	ON SUFFICIENCY AND INVARIANCE, *by D. Basu*	61
5.	CATEGORIAL DATA ANALOGS OF SOME MULTIVARIATE TESTS, *by V. P. Bhapkar*	85
6.	ESTIMATING MULTINOMIAL RESPONSE RELATIONS, *by R. D. Bock*	111
7.	SIMULTANEOUS TESTS FOR AVERAGE AND DISPERSION BY COMBINED CONTROL CHARTS, *by P. K. Bose and S. P. Mukherjee*	133
8.	ERROR CORRECTING, ERROR DETECTING AND ERROR LOCATING CODES, *by R. C. Bose*	147
9.	BOUNDS ON ERROR CORRECTING CODES (NON-RANDOM), *by I. M. Chakravarti*	179
10.	NONPARAMETRIC TESTS FOR THE MULTISAMPLE MULTIVARIATE LOCATION PROBLEM, *by S. K. Chatterjee and P. K. Sen*	197
11.	STEP-DOWN MULTIPLE DECISION RULES, *by S. Das Gupta*	229

12. ON THE RELATION BETWEEN UNION INTERSECTION AND LIKELIHOOD RATIO TESTS, *by K. R. Gabriel* 251

13. LINEAR MODELS IN MULTIVARIATE ANALYSIS, *by L. J. Gleser and I. Olkin* 267

14. S. N. ROY'S INTERESTS IN AND CONTRIBUTIONS TO THE ANALYSIS AND DESIGN OF CERTAIN QUANTITATIVE MULTIRESPONSE EXPERIMENTS, *by R. Gnanadesikan* 293

15. AN EXAMPLE OF THE ANALYSIS OF A SERIES OF RESPONSE CURVES AND AN APPLICATION OF MULTIVARIATE MULTIPLE COMPARISONS, *by J. E. Grizzle* 311

16. ON SOME SELECTION AND RANKING PROCEDURES WITH APPLICATIONS TO MULTIVARIATE POPULATIONS, *by S. S. Gupta and W. J. Studden* 327

17. ON CHARACTERIZING DEPENDENCE IN JOINT DISTRIBUTIONS, *by W. J. Hall* 339

18. A GENERAL PURPOSE TEST OF CENSORING OF EXTREME SAMPLE VALUES, *by N. L. Johnson* 377

19. ON THE CONNECTION BETWEEN MULTIPLICITY THEORY AND O. HANNER'S TIME DOMAIN ANALYSIS OF WEAKLY STATIONARY STOCHASTIC PROCESSES, *by G. Kallianpur and V. Mandrekar* 385

20. EXTENSIONS OF FRACTILE GRAPHICAL ANALYSIS TO HIGHER DIMENSIONAL DATA, *by P. C. Mahalanobis* 397

21. A REVIEW OF CONTIGENCY TABLES, *by M. A. Kastenbaum* 407

22. ON A MULTIVARIATE *F* DISTRIBUTION, *by P. R. Krishnaiah and J. V. Armitage* 439

23. TESTING A TABLE OF RANDOM NUMBERS, *by A. Linder* 469

24. ON GAMMAIZATION OF THE VARIANCE RATIO, *by S. K. Mitra and B. M. Mahajan* 479

25. SOME TCHEBYCHEFF TYPE INEQUALITIES FOR MATRIX VALUED RANDOM VARIABLES, *by G. S. Mudholkar* 489

26. m-ASSOCIATE CYCLICAL ASSOCIATION SCHEMES, *by H. K. Nandi and B. Adhikary* 495

27. ON THE NULL-DISTRIBUTION OF THE F-STATISTIC IN A RANDOMIZED PARTIALLY BALANCED INCOMPLETE BLOCK DESIGN WITH TWO ASSOCIATE CLASSES UNDER THE NEYMAN MODEL, *by J. Ogawa, S. Ikeda and M. Ogasawara* 517

28. COMPUTATION OF THE PROBABILITY INTEGRAL OF THE NON-CENTRAL CHI-SQUARE, *by J. Pachares* 549

29. ON THE NON-CENTRAL DISTRIBUTIONS OF THE LARGEST ROOTS OF TWO MATRICES IN MULTIVARIATE ANALYSIS, *by K. C. S. Pillai* 557

30. INFERENCE ON DISCRIMINANT FUNCTION COEFFICIENTS, *by C. R. Rao* 587

31. THE PROBLEM OF THE THREE-WAY ELECTION, *by R. F. Potthoff* 603

32. THE ADMISSIBILITY OF THE LARGEST CHARACTERISTIC ROOT TEST FOR THE NORMAL MULTIVARIATE LINEAR HYPOTHESES, *by S. N. Roy and W. F. Mikhail* 621

33. RANK METHODS FOR COMBINATION OF INDEPENDENT EXPERIMENTS IN MULTIVARIATE ANALYSIS OF VARIANCE. PART ONE: TWO TREATMENT MULTIRESPONSE CASE, *by P. K. Sen* 631

34. PROBABILITIES OF DEVIATIONS, *by J. Sethuraman* 655

35. A NOTE ON EMBEDDING FOR HADAMARD MATRICES, *by S. S. Shrikhande and Bhagwandas* 673

36. OPTIMAL BALANCED 2^m FRACTIONAL FACTORIAL DESIGNS, *by J. N. Srivastava* 689

37. SOME REMARKS ON A DISTRIBUTION OCCURRING IN NEURAL STUDIES, *by W. L. Smith* 707

38. COMPONENT TOLERANCES WHICH ACHIEVE A SPECIFIED SYSTEM TOLERANCE, *by W. A. Thompson, Jr. and R. K. Trask* 733

39. REFLECTIONS ON THE FUTURE OF MATHEMATICAL STATISTICS, *by J. Wolfowitz* 739

Essays in

PROBABILITY AND STATISTICS

CHAPTER 1

Estimation of Covariance Matrices Which Are Linear Combinations or Whose Inverses Are Linear Combinations of Given Matrices

T. W. ANDERSON*, *Columbia University*

1. INTRODUCTION

In many statistical studies the form of the covariance matrix is specified, and the unspecified parameters are estimated. Often the covariance matrix $\mathbf{\Sigma}$ can be written as a linear combination

$$\mathbf{\Sigma} = \sum_{g=0}^{m} \sigma_g \mathbf{G}_g \tag{1.1}$$

of given symmetric matrices $\mathbf{G}_0, \ldots, \mathbf{G}_m$ with the coefficients $\sigma_0, \ldots, \sigma_m$, not specified. We study estimates of these coefficients obtained by the method of maximum likelihood when the p-component observations $\mathbf{X}_1, \ldots, \mathbf{X}_N$ are drawn from the multivariate normal distribution $N(\boldsymbol{\mu}, \mathbf{\Sigma})$ with unspecified mean vector $\boldsymbol{\mu}$ and with covariance matrix $\mathbf{\Sigma}$ of this form.

*This research was sponsored by USAF, School of Aerospace Medicine, Brooks Air Force Base, Texas, under Contract AF 41(609)-2653. Reproduction in whole or in part is permitted for any purpose of the United States Government.

The covariance matrix of a stationary stochastic process (the i-th component of $\boldsymbol{X}_\alpha$ being the value of the α-th realization at the i-th time point) is of the form (1.1) with $m = p-1$ and with σ_g being the covariance between the process g time units apart and $\boldsymbol{G}_g$ being the matrix with 1's g entries above and below the main diagonal and 0's elsewhere ($g = 0, 1, \ldots, p-1$) ; here

$$(1.2) \qquad \boldsymbol{\Sigma} = \begin{bmatrix} \sigma_0 & \sigma_1 & \sigma_2 & \cdots & \sigma_{p-1} \\ \sigma_1 & \sigma_0 & \sigma_1 & \cdots & \sigma_{p-2} \\ \sigma_2 & \sigma_1 & \sigma_0 & \cdots & \sigma_{p-3} \\ \vdots & \vdots & \vdots & & \vdots \\ \sigma_{p-1} & \sigma_{p-2} & \sigma_{p-3} & \cdots & \sigma_0 \end{bmatrix}$$

This model is useful in the statistical analysis of repeated time series.

Another example of this form is the simplest mixed model of the analysis of variance which has a covariance matrix with all variances equal to $\sigma_0+\sigma_1$ and all covariances equal to σ_1 ($\boldsymbol{G}_0 = \boldsymbol{I}$ and $\boldsymbol{G}_1$ having all elements 1). Another case is that of independence, where $\boldsymbol{\Sigma}$ is diagonal, σ_g is the $(g+1)$st element on the main diagonal and $\boldsymbol{G}_g$ is the matrix with 1 in the $(g+1)$st position on the main diagonal and 0's elsewhere ($g = 0, \ldots, p-1$).

A general model of factor analysis defines $\boldsymbol{\Sigma} = \boldsymbol{BMB}'+\boldsymbol{\Delta}$, where $\boldsymbol{B}$ is the matrix of factor loadings, $\boldsymbol{M}$ is the covariance matrix of factor scores and $\boldsymbol{\Delta}$ is the diagonal matrix of variances of errors (and/or specific factors); if $\boldsymbol{B}$ is specified, then $\boldsymbol{\Sigma}$ is in the form of a linear combination of known matrices. (See, for example, Anderson and Rubin (1956)).

The density of $N(\boldsymbol{\mu}, \boldsymbol{\Sigma})$ is

$$(1.3) \qquad \frac{1}{(2\pi)^{\frac{1}{2}p}|\boldsymbol{\Sigma}|^{\frac{1}{2}}} e^{-\frac{1}{2}(\boldsymbol{x}-\boldsymbol{\mu})'\boldsymbol{\Sigma}^{-1}(\boldsymbol{x}-\boldsymbol{\mu})} = \frac{|\boldsymbol{\Psi}|^{\frac{1}{2}}}{(2\pi)^{\frac{1}{2}p}} e^{-\frac{1}{2}(\boldsymbol{x}-\boldsymbol{\mu})'\boldsymbol{\Psi}(\boldsymbol{x}-\boldsymbol{\mu})},$$

where $\Psi = \boldsymbol{\Sigma}^{-1}$. We also study the estimation of parameters when Ψ is given as a linear combination

$$(1.4) \qquad \Psi = \sum_{h=0}^{q} \psi_h \boldsymbol{H}_h$$

of specified symmetric matrices $\boldsymbol{H}_0, \ldots, \boldsymbol{H}_q$.

An example of this model is obtained when a time series $\boldsymbol{x}$ is generated by a circular stochastic difference equation

$$(1.5) \qquad x_t - \mu_t = \rho(x_{t-1} - \mu_{t-1}) + u_t, \quad t = 2, \ldots, p,$$

and $x_1 - \mu_1 = \rho(x_p - \mu_p) + u_1$, where $|\rho| < 1$ and $u_1, \ldots, u_p$, are normally and independently distributed with means 0 and variances σ^2; then the inverse of the covariance matrix of $\boldsymbol{x}$ is of the form (1.4) with $\psi_0 = (1+\rho^2)/\sigma^2$, $\psi_1 = -\rho/\sigma^2$, $\boldsymbol{H}_0 = \boldsymbol{I}$ and $\boldsymbol{H}_1$ having 1's immediately above and below the main diagonal and in the upper right hand and lower left hand corners and 0's elsewhere. For other similar forms useful in time series analysis, see Anderson (1963).

A sufficient set of statistics (but not necessarily minimal) is

$$(1.6) \qquad \bar{\boldsymbol{x}} = \frac{1}{N} \sum_{\alpha=1}^{N} \boldsymbol{x}_\alpha, \quad \boldsymbol{C} = \frac{1}{N} \sum_{\alpha=1}^{N} (\boldsymbol{x}_\alpha - \bar{\boldsymbol{x}})(\boldsymbol{x}_\alpha - \bar{\boldsymbol{x}})'.$$

Since we are interested here in the covariance matrices, we shall make no assumptions about the mean vector $\boldsymbol{\mu}$. Hence, $\bar{\boldsymbol{x}}$ is always used to estimate $\boldsymbol{\mu}$, and the information about $\boldsymbol{\Sigma}$ comes from $\boldsymbol{C}$.

In the case of the covariance matrix having a linear structure we assume that $\boldsymbol{G}_0, \ldots, \boldsymbol{G}_m$ are linearly independent and the values of $\sigma_0, \ldots, \sigma_m$ are such as to make $\sum_{g=0}^{m} \sigma_g \boldsymbol{G}_g$ positive definite. In the case of the inverse matrix having a linear structure we assume $\boldsymbol{H}_0, \ldots, \boldsymbol{H}_q$ are linearly independent and the values of $\psi_0, \ldots, \psi_q$ are such as to make $\sum_{h=0}^{q} \psi_h \boldsymbol{H}_h$ positive definite.

2. COVARIANCE MATRICES WHICH ARE LINEAR COMBINATIONS

2.1. Maximum Likelihood Estimates

The logarithm of the likelihood function L after being maximized with respect to $\boldsymbol{\mu}$ ($\hat{\boldsymbol{\mu}} = \bar{\boldsymbol{x}}$) is proportional to

$$(2/N)\log L = -p\log 2\pi - \log|\boldsymbol{\Sigma}| - \operatorname{tr}\boldsymbol{\Sigma}^{-1}\boldsymbol{C}, \tag{2.1}$$

where tr designates the trace of the matrix. We shall maximize (2.1), using

$$\frac{\partial}{\partial\sigma_g}\boldsymbol{\Sigma} = \boldsymbol{G}_g, \tag{2.2}$$

$$\begin{aligned}\frac{\partial}{\partial\sigma_g}\boldsymbol{\Sigma}^{-1} &= -\boldsymbol{\Sigma}^{-1}\left(\frac{\partial}{\partial\sigma_g}\boldsymbol{\Sigma}\right)\boldsymbol{\Sigma}^{-1} \\ &= -\boldsymbol{\Sigma}^{-1}\boldsymbol{G}_g\boldsymbol{\Sigma}^{-1},\end{aligned} \tag{2.3}$$

$$\begin{aligned}\frac{\partial}{\partial\sigma_g}\log|\boldsymbol{\Sigma}| &= \operatorname{tr}\boldsymbol{\Sigma}^{-1}\frac{\partial}{\partial\sigma_g}\boldsymbol{\Sigma} \\ &= \operatorname{tr}\boldsymbol{\Sigma}^{-1}\boldsymbol{G}_g.\end{aligned} \tag{2.4}$$

Then

$$(2/N)\frac{\partial}{\partial\sigma_g}\log L = -\operatorname{tr}\boldsymbol{\Sigma}^{-1}\boldsymbol{G}_g + \operatorname{tr}\boldsymbol{\Sigma}^{-1}\boldsymbol{G}_g\boldsymbol{\Sigma}^{-1}\boldsymbol{C}, \qquad g = 0, 1, \ldots, m. \tag{2.5}$$

The maximum likelihood estimates are obtained by setting the derivatives equal to 0. The equations are

$$\operatorname{tr}\left(\sum_{i=0}^{m}\hat{\sigma}_i\boldsymbol{G}_i\right)^{-1}\boldsymbol{G}_g = \operatorname{tr}\left(\sum_{i=0}^{m}\hat{\sigma}_i\boldsymbol{G}_i\right)^{-1}\boldsymbol{G}_g\left(\sum_{i=0}^{m}\hat{\sigma}_i\boldsymbol{G}_i\right)^{-1}\boldsymbol{C}, \qquad g = 0, 1, \ldots, m, \tag{2.6}$$

which we can also write as

$$\operatorname{tr}\hat{\boldsymbol{\Sigma}}^{-1}(\hat{\boldsymbol{\Sigma}} - \boldsymbol{C})\hat{\boldsymbol{\Sigma}}^{-1}\boldsymbol{G}_g = 0, \qquad g = 0, 1, \ldots, m, \tag{2.7}$$

where

$$\hat{\boldsymbol{\Sigma}} = \sum_{g=0}^{m}\hat{\sigma}_g\boldsymbol{G}_g. \tag{2.8}$$

If we multiply (2.6) by $\hat{\sigma}_g$ and sum over g, we obtain

$$p = \mathrm{tr}\, \hat{\boldsymbol{\Sigma}}^{-1}\boldsymbol{C}. \tag{2.9}$$

The maximized likelihood is

$$\frac{1}{(2\pi)^{\frac{1}{2}pN} |\hat{\boldsymbol{\Sigma}}|^{\frac{1}{2}N}} e^{-\frac{1}{2}pN}. \tag{2.10}$$

If there is more than one solution to (2.6), we take the solution that makes $\left|\sum_{g=0}^{m} \hat{\sigma}_g \boldsymbol{G}_g\right|$ a minimum. The solution should make $\sum_{g=0}^{m} \hat{\sigma}_g \boldsymbol{G}_g$ positive definite. Since the likelihood function $L \to 0$ for $\boldsymbol{\Sigma}$ approaching a singular matrix or for one or more elements of $\boldsymbol{\Sigma}$ approaching ∞ and/or $-\infty$ (see the proof of Lemma 3.2.2 of Anderson (1958), for example), there is at least one relative maximum in the set of $\sigma_0, \ldots, \sigma_m$ for which $\sum_{g=0}^{m} \sigma_g \boldsymbol{G}_g$ is positive definite. (It will be seen from (2.14) that if $2\boldsymbol{C} - \hat{\boldsymbol{\Sigma}}$ is positive definite, the solution to (2.6) is a relative maximum.)

In general, the equations (2.6) cannot be solved directly, but they can be solved iteratively (by the Newton-Raphson method). Let $s_0, \ldots, s_m$ be an initial approximation to the solution, and let $r_0, \ldots, r_m$ be an adjustment; that is, $\hat{\sigma}_i = s_i + r_i$, $i = 0, \ldots, m$. Then

$$\begin{aligned} \left[\sum_{i=0}^{m} \hat{\sigma}_i \boldsymbol{G}_i\right]^{-1} &= \left[\sum_{i=0}^{m} (s_i + r_i) \boldsymbol{G}_i\right]^{-1} \\ &= \left[\left(\sum_{i=0}^{m} s_i \boldsymbol{G}_i\right)\left(\boldsymbol{I} + \left(\sum_{i=0}^{m} s_i \boldsymbol{G}_i\right)^{-1} \sum_{h=0}^{m} r_h \boldsymbol{G}_h\right)\right]^{-1} \\ &= \left[\boldsymbol{I} - \left(\sum_{i=0}^{m} s_i \boldsymbol{G}_i\right)^{-1} \sum_{h=0}^{m} r_h \boldsymbol{G}_h + \ldots\right]\left(\sum_{i=0}^{m} s_i \boldsymbol{G}_i\right)^{-1}. \end{aligned} \tag{2.11}$$

Then (2.6) expanded into terms linear in $r_0, \ldots, r_m$ is

$$\text{(2.12)}\quad \operatorname{tr}\Big(\sum_{i=0}^{m} s_i \boldsymbol{G}_i\Big)^{-1}\boldsymbol{G}_g - \operatorname{tr}\Big(\sum_{i=0}^{m} s_i \boldsymbol{G}_i\Big)^{-1}\sum_{h=0}^{m} r_h \boldsymbol{G}_h\Big(\sum_{i=0}^{m} s_i \boldsymbol{G}_i\Big)^{-1}\boldsymbol{G}_g$$

$$= \operatorname{tr}\Big(\sum_{i=0}^{m} s_i \boldsymbol{G}_i\Big)^{-1}\boldsymbol{G}_g\Big(\sum_{i=0}^{m} s_i \boldsymbol{G}_i\Big)^{-1}\boldsymbol{C}$$

$$-2\operatorname{tr}\Big(\sum_{i=0}^{m} s_i \boldsymbol{G}_i\Big)^{-1}\boldsymbol{G}_g\Big(\sum_{i=0}^{m} s_i \boldsymbol{G}_i\Big)^{-1}\sum_{h=0}^{m} r_h \boldsymbol{G}_h\Big(\sum_{i=0}^{m} s_i \boldsymbol{G}_i\Big)^{-1}\boldsymbol{C},$$

$$g = 0, 1, \ldots, m.$$

If $\hat{\boldsymbol{\Sigma}}_0 = \sum_{i=0}^{m} s_i \boldsymbol{G}_i$, then (2.12) can be written

$$\text{(2.13)}\quad \sum_{h=0}^{m} r_h \operatorname{tr} \hat{\boldsymbol{\Sigma}}_0^{-1}\boldsymbol{G}_h[2\hat{\boldsymbol{\Sigma}}_0^{-1}\boldsymbol{C}\hat{\boldsymbol{\Sigma}}_0^{-1} - \hat{\boldsymbol{\Sigma}}_0^{-1}]\boldsymbol{G}_g = \operatorname{tr}[\hat{\boldsymbol{\Sigma}}_0^{-1}\boldsymbol{C}\hat{\boldsymbol{\Sigma}}_0^{-1} - \hat{\boldsymbol{\Sigma}}_0^{-1}]\boldsymbol{G}_g,$$

$$g = 0, 1, \ldots, m,$$

which consists of a set of linear equations in $r_0, \ldots, r_m$. These equations can be solved for $r_0, \ldots, r_m$; then $s_0' = s_0 + r_0, \ldots, s_m' = s_m + r_m$ form a new approximation to the maximum likelihood estimates of $\sigma_0, \ldots, \sigma_m$, and $\hat{\boldsymbol{\Sigma}}_1 = \sum_{i=0}^{m} s_i' \boldsymbol{G}_i$ is a new approximation to the maximum likelihood estimate of $\boldsymbol{\Sigma}$. If $\hat{\boldsymbol{\Sigma}}_0$ in (2.13) is replaced by $\hat{\boldsymbol{\Sigma}}_1$, a new adjustment can be found.

The coefficient of r_h on the left hand side of (2.13) is the negative of the second partial derivative of (2.1) evaluated at $\sigma_0 = s_0, \ldots, \sigma_m = s_m$; that is,

$$\text{(2.14)}\qquad -\frac{2}{N}\frac{\partial^2 \log L}{\partial\sigma_h \partial\sigma_g}\bigg|_{\boldsymbol{\Sigma} = \hat{\boldsymbol{\Sigma}}_0} = \operatorname{tr} \hat{\boldsymbol{\Sigma}}_0^{-1}\boldsymbol{G}_h[2\hat{\boldsymbol{\Sigma}}_0^{-1}\boldsymbol{C}\hat{\boldsymbol{\Sigma}}_0^{-1} - \hat{\boldsymbol{\Sigma}}_0^{-1}]\boldsymbol{G}_g,$$

$$h, g = 0, \ldots, m.$$

Unbiased estimates of $\sigma_0, \ldots, \sigma_m$ can be obtained from the fact that

$$\text{(2.15)}\qquad \frac{N}{N-1}\,\mathcal{E}\operatorname{tr}\boldsymbol{C}\boldsymbol{G}_h = \sum_{g=0}^{m} \sigma_g \operatorname{tr}\boldsymbol{G}_g\boldsymbol{G}_h, \qquad h = 0, \ldots, m.$$

The unbiased estimates of $\sigma_0, \ldots, \sigma_m$ are given by the solutions of

$$(2.16) \qquad \sum_{g=0}^{m} s_g \operatorname{tr} \boldsymbol{G}_g \boldsymbol{G}_h = \frac{N}{N-1} \operatorname{tr} \boldsymbol{C}\boldsymbol{G}_h, \qquad h = 0, \ldots, m.$$

These unbiased estimates can be used as initial approximations to form $\hat{\boldsymbol{\Sigma}}_0$. The matrix with elements $\operatorname{tr} \boldsymbol{G}_g \boldsymbol{G}_h$ is positive definite because

$$(2.17) \qquad \sum_{g,h=0}^{m} \operatorname{tr}(\boldsymbol{G}_g \boldsymbol{G}_h) y_g y_h = \operatorname{tr}\left[\left(\sum_{g=0}^{m} y_g \boldsymbol{G}_g\right)\left(\sum_{h=0}^{m} y_h \boldsymbol{G}_h\right)'\right]$$

is positive since $\sum_{g=0}^{m} y_g \boldsymbol{G}_g$ is not $\boldsymbol{0}$ for $(y_0, \ldots, y_m) \neq (0, \ldots, 0)$.

A possible procedure is to compute in turn $\hat{\boldsymbol{\Sigma}}_0 = \sum_{g=0}^{m} s_g \boldsymbol{G}_g$, $\hat{\boldsymbol{\Sigma}}_0^{-1}$, $\hat{\boldsymbol{\Sigma}}_0^{-1}\boldsymbol{C}\hat{\boldsymbol{\Sigma}}_0^{-1}$, $\hat{\boldsymbol{\Sigma}}_0^{-1}\boldsymbol{C}\hat{\boldsymbol{\Sigma}}_0^{-1} - \hat{\boldsymbol{\Sigma}}_0^{-1}$, and $2\hat{\boldsymbol{\Sigma}}_0^{-1}\boldsymbol{C}\hat{\boldsymbol{\Sigma}}_0^{-1} - \hat{\boldsymbol{\Sigma}}_0^{-1}$, and the traces of the matrices involved in (2.13). Then solve the linear equations for $r_0, \ldots, r_m$.

2.2. Asymptotic properties of the maximum likelihood estimates.

The usual asymptotic theory of maximum likelihood estimates applies here as $N \to \infty$. There is a solution $\hat{\sigma}_0, \ldots, \hat{\sigma}_m$ to (2.6) which is consistent, and $\sqrt{N}(\hat{\sigma}_0 - \sigma_0), \ldots, \sqrt{N}(\hat{\sigma}_m - \sigma_m)$ have a limiting normal distribution with means 0 and covariance matrix whose inverse has $\frac{1}{2} \operatorname{tr} \boldsymbol{\Sigma}^{-1}\boldsymbol{G}_h\boldsymbol{\Sigma}^{-1}\boldsymbol{G}_g$ as its h, gth element since the expectation of the matrix of partial second derivatives with respect to the σ_g's for arbitrary $\boldsymbol{\Sigma}$ tends to $\operatorname{tr} \boldsymbol{\Sigma}^{-1}\boldsymbol{G}_h\boldsymbol{\Sigma}^{-1}\boldsymbol{G}_g$ and $\boldsymbol{C}$ and $\bar{\boldsymbol{x}}$ are independent. If the initial approximations $s_0, \ldots, s_m$ are consistent estimates, than $s_0 + r_0, \ldots, s_m + r_m$ are consistent, asymptotically normal, and efficient, and the coefficients in the equations (2.13) are consistent estimates of $\operatorname{tr} \boldsymbol{\Sigma}^{-1}\boldsymbol{G}_h\boldsymbol{\Sigma}^{-1}\boldsymbol{G}_g$.

2.3. Tests of hypotheses.

To test the null hypothesis that the multivariate normal distribution sampled from has a covariance matrix of the form

$\sum_{g=0}^{m} \sigma_g \boldsymbol{G}_g$, where $\boldsymbol{G}_0, \ldots, \boldsymbol{G}_m$ and m are specified, we can use the likelihood ratio test. The likelihood ratio criterion is

$$\frac{|\boldsymbol{C}|^{\frac{1}{2}N}}{\left|\sum_{g=0}^{m} \hat{\sigma}_g \boldsymbol{G}_g\right|^{\frac{1}{2}N}}, \tag{2.18}$$

and the hypothesis is rejected if (2.18) is too small.

To carry out the likelihood ratio test we would like to know the distribution of (2.18) under the null hypothesis. Although the distribution of (2.18) cannot be derived in closed form in all cases, asymptotic theory indicates that -2 times the (natural) logarithm of (2.18) will have a limiting χ^2-distribution with $\frac{1}{2}p(p+1)-(m+1)$ degrees of freedom under the null hypothesis as $N \to \infty$.

One might also be interested in testing hypotheses about specified values of some or all of the coefficients under the assumption that the covariance matrix has this form. For large samples such tests can be based on the asymptotic normal distribution of the maximum likelihood estimates of the coefficients.

3. SOME SPECIAL CASES OF COVARIANCE MATRICES.

3.1. Component matrices which can be simultaneously diagonalized

If the set of component matrices $\boldsymbol{G}_0, \ldots, \boldsymbol{G}_m$ has some special properties, the computation of the maximum likelihood estimates may be easier and the theory may be simpler. A case of interest is when all of $\boldsymbol{G}_0, \ldots, \boldsymbol{G}_m$ are diagonalized by the same matrix. For convenience, we assume $\boldsymbol{G}_0 = \boldsymbol{I}$. (If not, a preliminary transformation can transform $\boldsymbol{G}_0$ to $\boldsymbol{I}$.) Then the transformation shall be orthogonal in order that $\boldsymbol{I}$ be invariant. Suppose there is an orthogonal matrix $\boldsymbol{P}$ such that

$$\boldsymbol{G}_g = \boldsymbol{P}\boldsymbol{\Lambda}_g\boldsymbol{P}', \qquad g = 0, \ldots, m, \tag{3.1}$$

where $\boldsymbol{\Lambda}_g$ is diagonal and $\boldsymbol{\Lambda}_0 = \boldsymbol{I}$. Then the diagonal elements of $\boldsymbol{\Lambda}_g$ are the characteristic roots of $\boldsymbol{G}_g$. A condition for this property is that $\boldsymbol{G}_i\boldsymbol{G}_g = \boldsymbol{G}_g\boldsymbol{G}_i$, $i, g = 0, \ldots, m$. Then

$$\boldsymbol{\Sigma} = \sum_{g=0}^{m} \sigma_g \boldsymbol{G}_g = \boldsymbol{P} \sum_{g=0}^{m} \sigma_g \boldsymbol{\Lambda}_g \boldsymbol{P}'. \tag{3.2}$$

We see that $m+1 \leqslant p$ for $\boldsymbol{\Lambda}_0, \ldots, \boldsymbol{\Lambda}_m$ to be linearly independent.

Let $\boldsymbol{x}_\alpha = \boldsymbol{P}\boldsymbol{z}_\alpha$; that is, $\boldsymbol{z}_\alpha = \boldsymbol{P}'\boldsymbol{x}_\alpha$. Then

$$\mathscr{E}(\boldsymbol{z}_\alpha - \mathscr{E}\boldsymbol{z}_\alpha)(\boldsymbol{z}_\alpha - \mathscr{E}\boldsymbol{z}_\alpha)' = \sum_{g=0}^{m} \sigma_g \boldsymbol{\Lambda}_g, \tag{3.3}$$

which is diagonal. In this form $z_{11}, \ldots, z_{pN}$ are independently normally distributed with variances

$$\mathscr{E}(z_{k\alpha} - \mathscr{E}z_{k\alpha})^2 = \sum_{g=0}^{m} \sigma_g \lambda_{kg}, \tag{3.4}$$

where $z_{k\alpha}$ is the kth component of $\boldsymbol{z}_\alpha$ and λ_{kg} is the kth diagonal element of $\boldsymbol{\Lambda}_g$.

Let $\sum_{\alpha=1}^{N} (z_{k\alpha} - \bar{z}_k)^2/N = v_k$, $\bar{z}_k = \sum_{\alpha=1}^{N} z_{k\alpha}$, $k = 1, \ldots, p$. The logarithm of the reduced likelihood function is proportional to

$$(2/N)\log L = -p\log 2\pi - \sum_{k=1}^{p} \log\left(\sum_{i=0}^{m} \sigma_i \lambda_{ki}\right) - \sum_{k=1}^{p} \frac{v_k}{\sum_{i=0}^{m} \sigma_i \lambda_{ki}}. \tag{3.5}$$

Then

$$(2/N)\frac{\partial}{\partial \sigma_g}\log L = -\sum_{k=1}^{p} \frac{\lambda_{kg}}{\sum_{i=1}^{m} \sigma_i \lambda_{ki}} + \sum_{k=1}^{p} \frac{v_k \lambda_{kg}}{\left(\sum_{i=0}^{m} \sigma_i \lambda_{ki}\right)^2}. \tag{3.6}$$

Setting these derivatives equal to 0, we obtain as equations defining the maximum likelihood estimates

$$\sum_{k=1}^{p} \frac{\lambda_{kg}}{\sum_{i=0}^{m} \hat{\sigma}_i \lambda_{ki}} = \sum_{k=1}^{p} \frac{v_k \lambda_{kg}}{\left(\sum_{i=0}^{m} \hat{\sigma}_i \lambda_{ki}\right)^2}, \qquad g = 0, 1, \ldots, m. \tag{3.7}$$

These equations also follow from (2.6) by making the substitution (3.1); v_k is the kth diagonal element of $\boldsymbol{P'CP}$.

If we let $\hat{\sigma}_g = s_g + r_g$, $g = 0, \ldots, m$, again and expand (3.7) in r_g, the linear terms give

$$\sum_{k=1}^{p} \frac{\lambda_{kg}}{\sum\limits_{i=0}^{m} s_i \lambda_{ki}} \left(1 - \frac{\sum\limits_{h=0}^{m} r_h \lambda_{kh}}{\sum\limits_{i=0}^{m} s_i \lambda_{ki}} \right) \tag{3.8}$$

$$= \sum_{k=1}^{p} \frac{v_k \lambda_{kg}}{\left(\sum\limits_{i=0}^{m} s_i \lambda_{ki} \right)^2} \left(1 - 2 \frac{\sum\limits_{h=0}^{m} r_h \lambda_{kh}}{\sum\limits_{i=0}^{m} s_i \lambda_{ki}} \right), \qquad g = 0, \ldots, m.$$

Let $\hat{v}_k^0 = \sum\limits_{i=0}^{m} s_i \lambda_{ki}$ be the initial approximation to the maximum likelihood estimate of the variance of z_k. The equations (3.8) are equivalent to

$$\sum_{h=0}^{m} r_h \sum_{k=1}^{p} \frac{\lambda_{kh}}{\hat{v}_k^0} \frac{\lambda_{kg}}{\hat{v}_k^0} \frac{2v_k - \hat{v}_k^0}{\hat{v}_k^0} = \sum_{k=1}^{p} \frac{\lambda_{kg}}{\hat{v}_k^0} \frac{v_k - \hat{v}_k^0}{\hat{v}_k^0}, \quad g = 0, \ldots, m, \tag{3.9}$$

which is (2.13) for this case. Unbiased and consistent estimates can be found from

$$\sum_{g=0}^{m} s_g \sum_{k=1}^{p} \lambda_{kg} \lambda_{kh} = \frac{N}{N-1} \sum_{k=1}^{p} v_k \lambda_{kh}, \qquad h = 0, \ldots, m. \tag{3.10}$$

The covariance matrix of the asymptotic distribution of the maximum likelihood estimates has

$$\frac{1}{2} \sum_{k=1}^{p} \frac{\lambda_{kh} \lambda_{kg}}{\left(\sum\limits_{i=0}^{m} \sigma_i \lambda_{ki} \right)^2}, \qquad g, h = 0, \ldots, m, \tag{3.11}$$

as the h, gth element of its inverse. In this case $v_1, \ldots, v_p$ and $\bar{z}_1, \ldots, \bar{z}_p$ form a sufficient set of statistics (not necessarily minimal). Some likelihood ratio criteria use the determinant

$$\left| \sum_{g=0}^{m} \hat{\sigma}_g \boldsymbol{G}_g \right| = \prod_{k=1}^{p} \left(\sum_{g=0}^{m} \hat{\sigma}_g \lambda_{kg} \right). \tag{3.12}$$

3.2. Other special cases

The covariance matrix (1.2) corresponds to a stationary stochastic process; $\boldsymbol{G}_g$ has 1's g entries above and below the main diagonal and 0's elsewhere ($g = 0, 1, \ldots, p-1$). Then

$$
\begin{aligned}
(3.13) \qquad \operatorname{tr} \boldsymbol{G}_g\boldsymbol{G}_h &= 0, & g &\neq h,\\
&= p, & g &= h = 0,\\
&= 2(p-g), & g &= h = 1, \ldots, p-1,
\end{aligned}
$$

$$
\begin{aligned}
(3.14) \qquad \operatorname{tr} \boldsymbol{C}\boldsymbol{G}_h &= \sum_{k=1}^{p} c_{kk}, & h &= 0,\\
&= 2 \sum_{k=1}^{p-h} c_{k,\,k+h}, & h &= 1, \ldots, p-1,
\end{aligned}
$$

where $c_{k,k+h}$ is the k, $(k+h)$th element of $\boldsymbol{C}$. Hence the unbiased estimate of σ_h defined by (2.16) is

$$
(3.15) \qquad s_h = \frac{N}{N-1} \sum_{k=1}^{p-h} c_{k,h+k}/(p-h), \qquad h = 0, \ldots, p-1.
$$

A stationary stochastic process with certain values of the first $p-1$ covariances (subject to conditions of positive definiteness) can be generated by a moving average

$$
(3.16) \qquad x_k = \sum_{j=0}^{p-1} \alpha_j u_{k-j},
$$

where the u_k's are independently distributed according to $N(\nu,1)$. Then

$$
\begin{aligned}
(3.17) \qquad \sigma_h &= \mathscr{E}(x_k - \mathscr{E}x_k)(x_{k+h} - \mathscr{E}x_{k+h})\\
&= \sum_{j=0}^{p-1} \sum_{l=0}^{p-1} \alpha_j\alpha_l \mathscr{E}(u_{k-j} - \nu)(u_{k+h-l} - \nu)\\
&= \sum_{j=0}^{p-1-h} \alpha_j\alpha_{j+h}, \qquad h = 0, \ldots, p-1.
\end{aligned}
$$

For suitable $\sigma_0, \ldots, \sigma_{p-1}$, the above equations can be solved for $\alpha_0, \ldots, \alpha_{p-1}$.

Another special case is where $\boldsymbol{G}_0 = \boldsymbol{I}$ and $\boldsymbol{G}_g = \boldsymbol{g}_g\boldsymbol{g}_g'$, $g = 1, \ldots, m$. Then the equations (2.6) defining the maximum likelihood estimates may be written

$$(3.18)\quad \operatorname{tr}(\hat\sigma_0\boldsymbol{I}+\sum_{i=1}^{m}\hat\sigma_i\boldsymbol{g}_i\boldsymbol{g}_i')^{-1} = \operatorname{tr}(\hat\sigma_0\boldsymbol{I}+\sum_{i=1}^{m}\hat\sigma_i\boldsymbol{g}_i\boldsymbol{g}_i')^{-1}\boldsymbol{C}(\hat\sigma_0\boldsymbol{I}+\sum_{i=1}^{m}\hat\sigma_i\boldsymbol{g}_i\boldsymbol{g}_i')^{-1},$$

$$(3.19)\quad \boldsymbol{g}_g'(\hat\sigma_0\boldsymbol{I}+\sum_{i=1}^{m}\hat\sigma_i\boldsymbol{g}_i\boldsymbol{g}_i')^{-1}\boldsymbol{g}_g = \boldsymbol{g}_g'(\hat\sigma_0\boldsymbol{I}+\sum_{i=1}^{m}\hat\sigma_i\boldsymbol{g}_i\boldsymbol{g}_i')^{-1}\boldsymbol{C}(\hat\sigma_0\boldsymbol{I}+\sum_{i=1}^{m}\hat\sigma_i\boldsymbol{g}_i\boldsymbol{g}_i')^{-1}\boldsymbol{g}_g, \qquad g = 1, \ldots, m.$$

The linear equations (2.13) for the iteration are

$$(3.20)\quad r_0 \operatorname{tr} \hat{\boldsymbol{\Sigma}}_0^{-1}(2\hat{\boldsymbol{\Sigma}}_0^{-1}\boldsymbol{C}\hat{\boldsymbol{\Sigma}}_0^{-1}-\hat{\boldsymbol{\Sigma}}_0^{-1})+\sum_{h=1}^{m} r_h\boldsymbol{g}_h'(2\hat{\boldsymbol{\Sigma}}_0^{-1}\boldsymbol{C}\hat{\boldsymbol{\Sigma}}_0^{-1}-\hat{\boldsymbol{\Sigma}}_0^{-1})\boldsymbol{g}_h = \operatorname{tr}(\hat{\boldsymbol{\Sigma}}_0^{-1}\boldsymbol{C}\hat{\boldsymbol{\Sigma}}_0^{-1}-\hat{\boldsymbol{\Sigma}}_0^{-1}),$$

$$(3.21)\quad r_0\boldsymbol{g}_g'\hat{\boldsymbol{\Sigma}}_0^{-1}(2\hat{\boldsymbol{\Sigma}}_0^{-1}\boldsymbol{C}\hat{\boldsymbol{\Sigma}}_0^{-1}-\hat{\boldsymbol{\Sigma}}_0^{-1})\boldsymbol{g}_g + \sum_{h=1}^{m} r_h\boldsymbol{g}_g'\hat{\boldsymbol{\Sigma}}_0^{-1}\boldsymbol{g}_h\boldsymbol{g}_h'(2\hat{\boldsymbol{\Sigma}}_0^{-1}\boldsymbol{C}\hat{\boldsymbol{\Sigma}}_0^{-1}-\hat{\boldsymbol{\Sigma}}_0^{-1})\boldsymbol{g}_g = \boldsymbol{g}_g'(\hat{\boldsymbol{\Sigma}}_0^{-1}\boldsymbol{C}\hat{\boldsymbol{\Sigma}}_0^{-1}-\hat{\boldsymbol{\Sigma}}_0^{-1})\boldsymbol{g}_g, \qquad g = 1, \ldots, m.$$

Unbiased estimates of $\sigma_0, \ldots, \sigma_m$ are the solutions of

$$(3.22)\quad s_0p+\sum_{g=1}^{m} s_g\boldsymbol{g}_g'\boldsymbol{g}_g = \frac{N}{N-1}\operatorname{tr}\boldsymbol{C},$$

$$(3.23)\quad s_0\boldsymbol{g}_h'\boldsymbol{g}_h+\sum_{g=1}^{m} s_g(\boldsymbol{g}_g'\boldsymbol{g}_h)^2 = \frac{N}{N-1}\boldsymbol{g}_h'\boldsymbol{C}\boldsymbol{g}_h, \qquad h = 1, \ldots, m.$$

This is a factor analysis model where

$$(3.24)\quad \boldsymbol{x} = \sum_{g=1}^{m} \boldsymbol{g}_g y_g+\boldsymbol{u}+\boldsymbol{\mu},$$

and $y_1, \ldots, y_m$ and $\boldsymbol{u}$ are independently distributed according to $N(0, \sigma_1), \ldots, N(0, \sigma_m)$ and $N(\mathbf{0}, \sigma_0 \boldsymbol{I})$, respectively. (See Anderson and Rubin (1956), for example.) The above estimation problem occurs when the vectors of factor loadings $\boldsymbol{g}_1, \ldots, \boldsymbol{g}_m$ are known, the variances $\sigma_1, \ldots, \sigma_m$ of the orthogonal factor scores are unknown, and the error variance σ_0, the same for all components, is unknown.

If the error variances are possibly different, $\mathcal{E}u_k^2 = \tau_k$, $k = 1, \ldots, p$, the covariance matrix is

$$\mathcal{E}(\boldsymbol{x}-\boldsymbol{\mu})(\boldsymbol{x}-\boldsymbol{\mu})' = \sum_{g=1}^{m} \sigma_g \boldsymbol{g}_g \boldsymbol{g}_g' + \sum_{k=1}^{p} \tau_k \boldsymbol{\epsilon}_k \boldsymbol{\epsilon}_k' \tag{3.25}$$

where $\boldsymbol{\epsilon}_k$ is the column vector with kth element 1 and all other elements 0. Let $\boldsymbol{T}$ be the diagonal matrix with τ_k as the kth diagonal element; that is, $\boldsymbol{T} = \sum_{k=1}^{p} \tau_k \boldsymbol{\epsilon}_k \boldsymbol{\epsilon}_k'$; and let $\hat{\boldsymbol{T}} = \sum_{k=1}^{p} \hat{\tau}_k \boldsymbol{\epsilon}_k \boldsymbol{\epsilon}_k'$. Then (2.6) is

$$\boldsymbol{g}_g' \hat{\boldsymbol{\Sigma}}^{-1} \boldsymbol{g}_g = \boldsymbol{g}_g' \hat{\boldsymbol{\Sigma}}^{-1} \boldsymbol{C} \hat{\boldsymbol{\Sigma}}^{-1} \boldsymbol{g}_g, \qquad g = 1, \ldots, m, \tag{3.26}$$

$$(\hat{\boldsymbol{\Sigma}}^{-1})_{kk} = (\hat{\boldsymbol{\Sigma}}^{-1} \boldsymbol{C} \hat{\boldsymbol{\Sigma}}^{-1})_{kk}, \qquad k = 1, \ldots, p, \tag{3.27}$$

where $\hat{\boldsymbol{\Sigma}} = \sum_{i=1}^{m} \hat{\sigma}_i \boldsymbol{g}_i \boldsymbol{g}_i' + \hat{\boldsymbol{T}}$ and $(\boldsymbol{A})_{kk}$ designates the kth diagonal element of $\boldsymbol{A}$. (Here $\sigma_i > 0$, $i = 1, \ldots, m$.) Also (2.13) is

$$\begin{aligned} &\sum_{h=1}^{m} r_h \boldsymbol{g}_g' \hat{\boldsymbol{\Sigma}}_0^{-1} \boldsymbol{g}_h \boldsymbol{g}_h' (2\hat{\boldsymbol{\Sigma}}_0^{-1} \boldsymbol{C} \hat{\boldsymbol{\Sigma}}_0^{-1} - \hat{\boldsymbol{\Sigma}}_0^{-1}) \boldsymbol{g}_g \\ &\quad + \sum_{k=1}^{p} r_k^* \boldsymbol{g}_g' \hat{\boldsymbol{\Sigma}}_0^{-1} \boldsymbol{\epsilon}_k \boldsymbol{\epsilon}_k' (2\hat{\boldsymbol{\Sigma}}_0^{-1} \boldsymbol{C} \hat{\boldsymbol{\Sigma}}_0^{-1} - \hat{\boldsymbol{\Sigma}}_0^{-1}) \boldsymbol{g}_g \\ &\quad = \boldsymbol{g}_g' (\hat{\boldsymbol{\Sigma}}_0^{-1} \boldsymbol{C} \hat{\boldsymbol{\Sigma}}_0^{-1} - \hat{\boldsymbol{\Sigma}}_0^{-1}) \boldsymbol{g}_g \qquad g = 1, \ldots, m, \end{aligned} \tag{3.28}$$

$$\begin{aligned} &\sum_{h=1}^{m} r_h \boldsymbol{g}_h' (2\hat{\boldsymbol{\Sigma}}_0^{-1} \boldsymbol{C} \hat{\boldsymbol{\Sigma}}_0^{-1} - \hat{\boldsymbol{\Sigma}}_0^{-1}) \boldsymbol{\epsilon}_j \boldsymbol{\epsilon}_j' \hat{\boldsymbol{\Sigma}}_0^{-1} \boldsymbol{g}_h \\ &\quad + \sum_{k=1}^{p} r_k^* (\hat{\boldsymbol{\Sigma}}_0^{-1})_{jk} (2\hat{\boldsymbol{\Sigma}}_0^{-1} \boldsymbol{C} \hat{\boldsymbol{\Sigma}}_0^{-1} - \hat{\boldsymbol{\Sigma}}_0^{-1})_{jk} \\ &\quad = (\boldsymbol{\Sigma}_0^{-1} \boldsymbol{C} \boldsymbol{\Sigma}_0^{-1} - \boldsymbol{\Sigma}_0^{-1})_{jj}, \qquad j = 1, \ldots, p, \end{aligned} \tag{3.29}$$

where we have set $\hat{\tau}_k = s_k^* + r_k^*$, $k = 1, \ldots, p$. Unbiased estimates are the solutions of

$$(3.30) \qquad \sum_{g=1}^{m} s_g(\boldsymbol{g}_g'\boldsymbol{g}_h)^2 + \sum_{k=1}^{p} s_k^* g_{kh}^2 = \frac{N}{N-1} \boldsymbol{g}_h' \boldsymbol{C} \boldsymbol{g}_h, \quad h = 1, \ldots, m,$$

$$(3.31) \qquad \sum_{g=1}^{m} s_g g_{jg}^2 + s_j^* = \frac{N}{N-1} c_{jj}, \qquad j = 1, \ldots, p,$$

where g_{kh} is the kth component of $\boldsymbol{g}_h$ and c_{jj} is the jth diagonal element of $\boldsymbol{C}$.

Note that if all the $\boldsymbol{g}_h$ are absent in the last case, $\boldsymbol{\Sigma}$ is diagonal with possibly different diagonal elements and $\hat{\tau}_j = c_{jj}$ is the jth diagonal element of $\hat{\boldsymbol{\Sigma}}$.

One form of Guttman's quasi-simplex (Guttman (1954)) can be expressed in these terms with $\boldsymbol{g}_g$ having all elements 1 except for the first $g-1$ ($g = 1, \ldots, m = p$) which are zero.

The case of oblique factors can also be handled. If $\mathcal{E} y_g y_h = m_{gh}$, $g, h = 1, \ldots, m$, then when different error variances are permitted

$$(3.32) \qquad \mathcal{E}(\boldsymbol{x}-\boldsymbol{\mu})(\boldsymbol{x}-\boldsymbol{\mu})' = \sum_{g,h=1}^{m} m_{gh} \boldsymbol{g}_g \boldsymbol{g}_h' + \sum_{k=1}^{p} \tau_k \boldsymbol{\epsilon}_k \boldsymbol{\epsilon}_k'.$$

Here the symmetric matrix (m_{gh}) must be positive definite.

In the simplest mixed model of the analysis of variance we let

$$(3.33) \qquad x_{k\alpha} = \mu_k + f_\alpha + e_{k\alpha}, \qquad k = 1, \ldots, p, \qquad \alpha = 1, \ldots, N,$$

where $\mu_1, \ldots, \mu_p$ are constants, $f_1, \ldots, f_N$ are independently distributed according to $N(0, \sigma_f^2)$ and the $e_{k\alpha}$'s are distributed independently of the f_α's and of each other, each according to $N(0, \sigma_e^2)$. Here $\bar{\mu} = \sum_{k=1}^{p} \mu_k / p$ is the overall mean; the $(\mu_k - \bar{\mu})$'s are the main effects of the fixed factors; the f_α's are the main effects of the random factors; and the $e_{k\alpha}$'s are the errors. Then $\mathcal{E}(x_{k\alpha} - \mu_k)^2 = \sigma_e^2 + \sigma_f^2$, $\mathcal{E}(x_{k\alpha} - \mu_k)(x_{j\alpha} - \mu_j) = \sigma_f^2$, $k \neq j$, and $\mathcal{E}(x_{k\alpha} - \mu_k)(x_{j\beta} - \mu_j) = 0$, $\alpha \neq \beta$. In terms of the model $\boldsymbol{G}_0 = \boldsymbol{I}$ with $\sigma_0 = \sigma_e^2$ and $\boldsymbol{G}_1 = \boldsymbol{\epsilon}\boldsymbol{\epsilon}'$, where $\boldsymbol{\epsilon}' = (1, 1, \ldots, 1)$, with $\sigma_1 = \sigma_f^2$.

4. COVARIANCE MATRICES WHOSE INVERSES ARE LINEAR COMBINATIONS

4.1. Maximum likelihood estimates

We study the estimation of $\psi_0, \ldots, \psi_q$ when

$$\boldsymbol{\Sigma}^{-1} = \Psi = \sum_{h=0}^{q} \psi_h \boldsymbol{H}_h. \tag{4.1}$$

The logarithm of the (reduced) likelihood function is proportional to

$$\begin{aligned}(2/N)\log L &= -p \log 2\pi + \log|\boldsymbol{\Sigma}^{-1}| - \operatorname{tr} \boldsymbol{\Sigma}^{-1} \boldsymbol{C} \\ &= -p \log 2\pi + \log\left|\sum_{i=0}^{q} \psi_i \boldsymbol{H}_i\right| - \sum_{i=0}^{q} \psi_i \operatorname{tr} \boldsymbol{H}_i \boldsymbol{C}.\end{aligned} \tag{4.2}$$

Since $\boldsymbol{C}$ enters (4.2) only through $\sum_{i=1}^{q} \psi_i \operatorname{tr} \boldsymbol{H}_i \boldsymbol{C}$, we see that $\bar{\boldsymbol{x}}$, $\operatorname{tr} \boldsymbol{H}_0 \boldsymbol{C}, \ldots, \operatorname{tr} \boldsymbol{H}_q \boldsymbol{C}$ form a sufficient set of statistics. The partial derivatives of (4.2) are

$$(2/N) \frac{\partial}{\partial \psi_h} \log L = \operatorname{tr} \boldsymbol{\Sigma} \boldsymbol{H}_h - \operatorname{tr} \boldsymbol{H}_h \boldsymbol{C}, \qquad h = 0, \ldots, q. \tag{4.3}$$

The derivative equations are

$$\operatorname{tr} \left(\sum_{i=0}^{q} \hat{\psi}_i \boldsymbol{H}_i\right)^{-1} \boldsymbol{H}_h = \operatorname{tr} \boldsymbol{C} \boldsymbol{H}_h, \qquad h = 0, \ldots, q. \tag{4.4}$$

If we multiply (4.4) by $\hat{\psi}_h$ and sum over h, we obtain

$$p = \operatorname{tr} \hat{\Psi} \boldsymbol{C} = \operatorname{tr} \hat{\boldsymbol{\Sigma}}^{-1} \boldsymbol{C}. \tag{4.5}$$

The derivative equations can also be written

$$\operatorname{tr} \hat{\boldsymbol{\Sigma}} \boldsymbol{H}_g = \operatorname{tr} \boldsymbol{C} \boldsymbol{H}_g, \qquad g = 0, \ldots, q. \tag{4.6}$$

If we let $\hat{\psi}_i = u_i + t_i$, $i = 0, \ldots, q$, then the linear part of the expansion of (4.4) gives

$$\sum_{g=0}^{q} t_g \operatorname{tr} \left(\sum_{i=0}^{q} u_i \boldsymbol{H}_i \right)^{-1} \boldsymbol{H}_g \left(\sum_{i=0}^{q} u_i \boldsymbol{H}_i \right)^{-1} \boldsymbol{H}_h \tag{4.7}$$

$$= \operatorname{tr} \left(\sum_{i=0}^{q} u_i \boldsymbol{H}_i \right)^{-1} \boldsymbol{H}_h - \operatorname{tr} \boldsymbol{C} \boldsymbol{H}_h, \qquad h = 0, \ldots, q.$$

If $\hat{\boldsymbol{\Psi}}_0 = \sum_{i=0}^{q} u_i \boldsymbol{H}_i$, then (4.7) can be written

$$\sum_{g=0}^{q} t_g \operatorname{tr} \hat{\boldsymbol{\Psi}}_0^{-1} \boldsymbol{H}_g \hat{\boldsymbol{\Psi}}_0^{-1} \boldsymbol{H}_h = \operatorname{tr} (\hat{\boldsymbol{\Psi}}_0^{-1} - \boldsymbol{C}) \boldsymbol{H}_h, \quad h = 0, \ldots, q. \tag{4.8}$$

Consistent estimates can be obtained from

$$\sum_{h=0}^{q} u_h \operatorname{tr} \boldsymbol{H}_h \boldsymbol{H}_g = \operatorname{tr} \boldsymbol{C}^{-1} \boldsymbol{H}_g, \qquad g = 0, \ldots, q. \tag{4.9}$$

4.2. Asymptotic properties of the maximum likelihood estimates

The maximum likelihood estimates are asymptotically normally distributed. Specifically there is a solution $\hat{\psi}_0, \ldots, \hat{\psi}_q$ to (4.4) which is consistent, and $\sqrt{N}(\hat{\psi}_0 - \psi_0), \ldots, \sqrt{N}(\hat{\psi}_q - \psi_q)$ have a limiting normal distribution with means 0 and covariance matrix whose inverse has $\frac{1}{2} \operatorname{tr} \hat{\boldsymbol{\Psi}}^{-1} \boldsymbol{H}_g \hat{\boldsymbol{\Psi}}^{-1} \boldsymbol{H}_h$ as its g, hth element.

4.3. Tests of hypotheses.

The likelihood ratio criterion for testing the null hypothesis that the covariance matrix of the multivariate normal distribution sampled from has the form $\sum_{i=0}^{q} \psi_i \boldsymbol{H}_i$ for specified $\boldsymbol{H}_0, \ldots, \boldsymbol{H}_q$ is

$$|\boldsymbol{C}|^{\frac{1}{2}N} | \sum_{i=0}^{q} \hat{\psi}_i \boldsymbol{H}_i |^{\frac{1}{2}N}. \tag{4.10}$$

When the null hypothesis is true, -2 times the logarithm of (4.10) will have a limiting χ^2-distribution with $\frac{1}{2}p(p+1)-(q+1)$ degrees of freedom as $N \to \infty$.

To test hypotheses of specified values one can use the asymptotic theory of the maximum likelihood estimates, estimating the g, hth element of the inverse of the covariance matrix of the asymptotic normal distribution by $\frac{1}{2}\operatorname{tr} \hat{\boldsymbol{\Psi}}^{-1}\boldsymbol{H}_g\hat{\boldsymbol{\Psi}}^{-1}\boldsymbol{H}_h$.

5. SOME SPECIAL CASES OF INVERSES OF COVARIANCE MATRICES

5.1. Component matrices which can be simultaneously diagonalised

Suppose there is an orthogonal matrix $\boldsymbol{P}$ such that

$$\boldsymbol{H}_g = \boldsymbol{P}\boldsymbol{\Phi}_g\boldsymbol{P}', \qquad g = 0, \ldots, q, \tag{5.1}$$

where $\boldsymbol{\Phi}_g$ is diagonal and $\boldsymbol{\Phi}_0 = \boldsymbol{I}$. Then

$$\boldsymbol{\Sigma}^{-1} = \boldsymbol{\Psi} = \boldsymbol{P}\sum_{i=0}^{q} \psi_i\boldsymbol{\Phi}_i\boldsymbol{P}'. \tag{5.2}$$

Let $\boldsymbol{x}_\alpha = \boldsymbol{P}\boldsymbol{z}_\alpha$. Then

$$\mathcal{E}(\boldsymbol{z}_\alpha - \mathcal{E}\boldsymbol{z}_\alpha)(\boldsymbol{z}_\alpha - \mathcal{E}\boldsymbol{z}_\alpha)' = \left(\sum_{i=0}^{q} \psi_i\boldsymbol{\Phi}_i\right)^{-1}, \tag{5.3}$$

which is diagonal, and $z_{11}, \ldots, z_{pN}$ are independently normally distributed with variances

$$\mathcal{E}(z_{k\alpha} - \mathcal{E}z_{k\alpha})^2 = \frac{1}{\sum_{g=0}^{q} \psi_g\varphi_{kg}}, \tag{5.4}$$

where φ_{kg} is the kth diagonal element of $\boldsymbol{\Phi}_g$.

The equations defining the maximum likelihood estimates are

$$\sum_{k=1}^{q} \frac{\varphi_{kg}}{\sum_{i=0}^{q} \hat{\psi}_i\varphi_{ki}} = \sum_{k=1}^{p} v_k\,\varphi_{kg}, \qquad g = 0, \ldots, q. \tag{5.5}$$

The adjustments $t_0, \ldots, t_q$ to initial approximations $u_0, \ldots, u_q$ are the solutions to

$$\sum_{g=0}^{q}\sum_{k=1}^{p}\left\{\frac{\varphi_{kh}}{\sum\limits_{i=0}^{q} u_i \varphi_{ki}} \frac{\varphi_{kg}}{\sum\limits_{i=0}^{q} u_i \varphi_{ki}}\right\} t_g \tag{5.6}$$

$$= \sum_{k=1}^{p} \frac{\varphi_{kh}}{\sum\limits_{k=1}^{q} u_i \varphi_{ki}} - \sum_{k=1}^{p} v_k \varphi_{kh}, \qquad h = 0, 1, \ldots, q.$$

Consistent estimates can be obtained from

$$\sum_{h=0}^{q} u_h \sum_{k=1}^{p} \varphi_{kh} \varphi_{kg} = \sum_{k=1}^{p} \frac{\varphi_{kg}}{v_k}, \qquad g = 0, \ldots, q. \tag{5.7}$$

The covariance matrix of the asymptotic distribution has

$$\frac{1}{2}\sum_{k=1}^{p} \frac{\varphi_{kg} \varphi_{kh}}{\left(\sum\limits_{i=0}^{q} \psi_i \varphi_{ki}\right)^2} \tag{5.8}$$

as the g, hth element in its inverse.

5.2. Other special cases

Many time series models are of the above form. As an example we consider the model for circular serial correlation coefficients, where $\boldsymbol{H}_0 = \boldsymbol{I}$,

$$\boldsymbol{H}_1 = \frac{1}{2}\begin{bmatrix} 0 & 1 & 0 & \ldots & 0 & 1 \\ 1 & 0 & 1 & \ldots & 0 & 0 \\ 0 & 1 & 0 & \ldots & 0 & 0 \\ \vdots & \vdots & \vdots & & \vdots & \vdots \\ 0 & 0 & 0 & \ldots & 0 & 1 \\ 1 & 0 & 0 & \ldots & 1 & 0 \end{bmatrix}, \tag{5.9}$$

and, in general, $\boldsymbol{H}_j$ has 0 elements except for $\frac{1}{2}$, j entries above and below the main diagonal and $p-j$ entries above and below the main diagonal. The first-order model ($q = 1$) derives from the circular stochastic difference equation ($|\rho| < 1$)

$$(5.10) \qquad \begin{aligned} x_t - \mu_t &= \rho(x_{t-1} - \mu_{t-1}) + u_t, \qquad t = 2, \ldots, p, \\ x_1 - \mu_1 &= \rho(x_p - \mu_p) + u_1, \end{aligned}$$

where $u_1, \ldots, u_p$ are independently distributed, each according to $N(0, \sigma^2)$. Let $\psi_0 = (1+\rho^2)/\sigma^2$ and $\psi_1 = -2\rho/\sigma^2$. Then the special case of our model results.

Using (4.2), (4.5), and $|\mathbf{\Sigma}^{-1}| = (1-\rho^p)^2/\sigma^{2p}$, the equations defining the maximum likelihood estimates may be written $\left(\bar{x}_t = \sum_{\alpha=1}^{N} x_{t\alpha}/N\right)$

$$(5.11) \qquad \begin{aligned} \hat{\sigma}^2 &= \frac{1}{Np} \sum_{t=1}^{p} \sum_{\alpha=1}^{N} (x_{t\alpha} - \bar{x}_t)^2 [(1+\hat{\rho}^2) - 2\hat{\rho} r] \\ &= \frac{\operatorname{tr} \boldsymbol{C}}{p} [(1+\hat{\rho}^2) - 2\hat{\rho} r], \end{aligned}$$

$$(5.12) \qquad r\hat{\rho}^p - \hat{\rho}^{p-1} - \hat{\rho} + r = 0,$$

where

$$(5.13) \qquad \begin{aligned} r &= \frac{\operatorname{tr} \boldsymbol{H}_1 \boldsymbol{C}}{\operatorname{tr} \boldsymbol{H}_0 \boldsymbol{C}} = \frac{\operatorname{tr} \boldsymbol{H}_1 \boldsymbol{C}}{\operatorname{tr} \boldsymbol{C}} \\ &= \frac{\sum_{\alpha=1}^{N} \left[\sum_{t=2}^{p} (x_{t\alpha} - \bar{x}_t)(x_{t-1,\alpha} - \bar{x}_{t-1}) + (x_{1\alpha} - \bar{x}_1)(x_{p\alpha} - \bar{x}_p) \right]}{\sum_{\alpha=1}^{N} \sum_{t=1}^{p} (x_{t\alpha} - \bar{x}_t)^2}. \end{aligned}$$

Here

$$(5.14) \qquad \begin{aligned} |(\hat{\psi}_0 \boldsymbol{H}_0 + \hat{\psi}_1 \boldsymbol{H}_1)^{-1}| &= \frac{\hat{\sigma}^{2p}}{(1-\hat{\rho}^p)^2} \\ &= \frac{\left[\frac{1}{p} (\operatorname{tr} \boldsymbol{C})(1+\hat{\rho}^2 - 2\hat{\rho} r)\right]^p}{(1-\hat{\rho}^p)^2} \\ &= \frac{\left[\frac{1}{Np} \sum_{\alpha=1}^{N} \sum_{t=1}^{p} (x_{t\alpha} - \bar{x}_t)^2 (1+\hat{\rho}^2 - 2\hat{\rho} r)\right]^p}{(1-\hat{\rho}^p)^2}; \end{aligned}$$

r is the circular serial correlation coefficient based on N time series and with the mean for each time point estimated. If $r > 0$, the real roots of (5.12) are a and $1/a$ for some $a > 0$ if p is even and a, $1/a\,(a > 0)$, and -1 if p is odd; if $r = 0$, the real roots are 0 if p is even and 0 and -1 if p is odd; if $r < 0$, the real roots are a and $1/a$ for some $a < 0$ if p is even and a, $1/a\,(a < 0)$, and -1 if p is odd. Since the same process is defined for reciprocal values of ρ (and adjusted σ^2) we take the root between -1 and 1. (The root -1 does not give the absolute maximum of L).

As an initial approximation to the desired root of (5.12), we take r. Then an adjustment is

$$u = -\frac{r^{p-1}(1-r^2)}{pr^{p-2}(1-r^2)+1-r^{p-2}}, \tag{5.15}$$

which will be small if r is small and/or p is large (with high probability).

6. REMARKS

6.1. Estimation of the man vector

The derivation of the maximum likelihood estimates has been on the assumption that $\mathcal{E}x = \boldsymbol{\mu}$, where $\boldsymbol{\mu}$ consists of p components without restriction. In many cases, however, of an assumed structure of the covariance matrix, there would be an assumed structure of the mean vector. For example, in the case of a stationary stochastic process, the mean vector would be assumed to have all components equal. In some other models for time series, the tth component of $\boldsymbol{\mu}$ might be assumed to be a linear function of t (a linear trend).

If the components of the mean vector $\boldsymbol{\mu}$ are assumed to be p functions of less than p real parameters, $\bar{\boldsymbol{x}}$ will not be the maximum likelihood estimate of $\boldsymbol{\mu}$, the reduced likelihood will not be a function of C only, and the equations for the maximum likelihood estimates will be different from the above. For instance, if

$$\boldsymbol{\mu} = \sum_{i=1}^{r} \beta_i \boldsymbol{\gamma}_i, \tag{6.1}$$

where $\boldsymbol{\gamma}_1, \ldots, \boldsymbol{\gamma}_r$ are specified vectors and $\beta_1, \ldots, \beta_r$ are unknown parameters, $2/N$ times the logarithm of the likelihood function is (2.1) plus

$$-\left(\bar{\boldsymbol{x}} - \sum_{i=1}^{r} \beta_i \boldsymbol{\gamma}_i\right)' \boldsymbol{\Sigma}^{-1} \left(\bar{\boldsymbol{x}} - \sum_{i=1}^{r} \beta_i \boldsymbol{\gamma}_i\right). \tag{6.2}$$

If the $\boldsymbol{\gamma}_i$'s are characteristic vectors of $\boldsymbol{G}_0, \ldots, \boldsymbol{G}_m$ (or if they are linear combinations of r such characteristic vectors) then the $\boldsymbol{\gamma}_i$'s are characteristic vectors of $\boldsymbol{\Sigma}$ and $\boldsymbol{\Sigma}^{-1}$, and the maximization of (6.2) with respect of the β_i's does not depend on $\boldsymbol{\Sigma}$. (See Anderson (1948).) In fact, $\hat{\beta}_i = \bar{\boldsymbol{x}}'\boldsymbol{\gamma}_i/\boldsymbol{\gamma}_i'\boldsymbol{\gamma}_i$. (The above conditions force $\boldsymbol{\gamma}_i'\boldsymbol{\gamma}_j = 0$, $i \neq j$, or equivalent vectors which are orthogonal can be used.) Then the preceding theory for the maximum likelihood estimates of $\sigma_0, \ldots, \sigma_m$ holds with $\boldsymbol{C}$ replaced by $(1/N) \sum_{\alpha=1}^{N} (\boldsymbol{x}_\alpha - \hat{\boldsymbol{\mu}})(\boldsymbol{x}_\alpha - \hat{\boldsymbol{\mu}})'$, where $\hat{\boldsymbol{\mu}} = \sum_{i=1}^{r} \hat{\beta}_i \boldsymbol{\gamma}_i$.

The condition that the $\boldsymbol{\gamma}_i$'s are characteristic vectors of the $\boldsymbol{G}_g$'s is very restrictive on both the matrices and vectors. If these conditions do not hold, the maximum likelihood equations will involve both $\hat{\beta}_i$'s and $\hat{\sigma}_g$'s. However, one may for convenience estimate the β_i's by least squares on $\bar{\boldsymbol{x}}$ and replace $\boldsymbol{C}$ by the matrix of sums of squares and products of deviations from the estimated $\boldsymbol{\mu}$.

In general, one can expect that if N is relatively large, the method of estimating $\boldsymbol{\mu}$ using $\bar{\boldsymbol{x}}$ will not have a large effect on estimating $\sigma_0, \ldots, \sigma_m$ on the basis of deviations from the estimated $\boldsymbol{\mu}$.

6.2. Other parametrizations

The model of Section 3 can be written with

$$\boldsymbol{\Sigma} = \sigma_0 \left[\boldsymbol{G}_0 + \sum_{g=1}^{m} \alpha_g \boldsymbol{G}_g\right], \tag{6.3}$$

where $\alpha_g = \sigma_g/\sigma_0$, $g = 1, \ldots, m$. Then

$$\boldsymbol{\Sigma}^{-1} = (1/\sigma_0)\left[\boldsymbol{G}_0 + \sum_{g=1}^{m} \alpha_g \boldsymbol{G}_g\right]^{-1}, \tag{6.4}$$

$$\hat{\sigma}_0 = \frac{1}{p}\operatorname{tr}\left(\boldsymbol{G}_0 + \sum_{g=1}^{m} \hat{\alpha}_g \boldsymbol{G}_g\right)^{-1}\boldsymbol{C} \tag{6.5}$$

which also follows from (2.7) by factoring $\hat{\sigma}_0$ out of $\hat{\boldsymbol{\Sigma}}$, and $\hat{\alpha}_1, \ldots, \hat{\alpha}_m$ minimize

$$p \log \operatorname{tr}\left(\boldsymbol{G}_0 + \sum_{g=1}^{m} \alpha_g \boldsymbol{G}_g\right)^{-1}\boldsymbol{C} + \log\left|\boldsymbol{G}_0 + \sum_{g=1}^{m} \alpha_g \boldsymbol{G}_g\right|. \tag{6.6}$$

If

$$\boldsymbol{\Psi} = \psi_0\left[\boldsymbol{H}_0 + \sum_{g=1}^{q} \omega_g \boldsymbol{H}_g\right], \tag{6.7}$$

where $\omega_g = \psi_g/\psi_0$, $g = 1, \ldots, q$, then

$$\frac{p}{\hat{\psi}_0} = \operatorname{tr} \boldsymbol{H}_0\boldsymbol{C} + \sum_{g=1}^{q} \hat{\omega}_g \operatorname{tr} \boldsymbol{H}_g\boldsymbol{C}, \tag{6.8}$$

which also follows from (4.5), and $\hat{\omega}_1, \ldots, \hat{\omega}_q$ minimize

$$p \log\left(\operatorname{tr} \boldsymbol{H}_0\boldsymbol{C} + \sum_{g=1}^{q} \omega_g \operatorname{tr} \boldsymbol{H}_g\boldsymbol{C}\right) - \log\left|\boldsymbol{H}_0 + \sum_{g=1}^{q} \omega_g \boldsymbol{H}_g\right|. \tag{6.9}$$

The coefficients $\sigma_0, \ldots, \sigma_m$ or $\psi_0, \ldots, \psi_q$ may be specified functions of parameters $\theta_1, \ldots, \theta_r$. Then the derivatives of the logarithm of the likelihood function are

$$\frac{\partial}{\partial\theta_i} \log L = \sum_{g=0}^{m} \frac{\partial}{\partial\theta_g} (\log L) \frac{\partial\sigma_g}{\partial\theta_i}, \qquad i = 1, \ldots, r, \tag{6.10}$$

or

$$\frac{\partial}{\partial\theta_i} \log L = \sum_{g=0}^{q} \frac{\partial}{\partial\psi_g} (\log L) \frac{\partial\psi_g}{\partial\theta_i}, \qquad i = 1, \ldots, r. \tag{6.11}$$

6.3. Statistical analysis of repeated time series

In many areas of investigation several subjects in a study are followed over a period of time. In the simplest cases one measurement may be made on each individual at equally-spaced time points. The succession of measurements is a (univariate) time series.

Much of the study of the statistical analysis of time series is devoted to problems in which only a single time series is available. The present study arose from consideration of problems in which several time series are available. The underlying assumed probability structure is such that some features are the same for every individual. If $x_{t\alpha}$ is the measurement on the αth individual ($\alpha = 1, \ldots, N$) at the tth time point ($t = 1, \ldots, T$), then the column vector $\boldsymbol{x}_\alpha = (x_{1\alpha}, x_{2\alpha}, \ldots, x_{T\alpha})'$ is a time series. In the analysis of a *single* time series $\boldsymbol{x}$ from $N(\boldsymbol{\mu}, \boldsymbol{\Sigma})$ special forms are assumed for $\boldsymbol{\mu}$ and $\boldsymbol{\Sigma}$. For example, every element of $\boldsymbol{\mu}$ may be the same $\boldsymbol{\mu} = m\boldsymbol{\epsilon}$, where $\boldsymbol{\epsilon} = (1, 1, \ldots, 1)'$ or $\boldsymbol{\mu}$ may represent a linear trend $\mu_t = \gamma_0 + \gamma_1 t$. In the simplest cases $\boldsymbol{\Sigma} = \sigma^2 \boldsymbol{I}$. A condition that takes the effect of time into account is a stationary Markov process; the covariance matrix is

$$(6.12)\qquad \boldsymbol{\Sigma} = \sigma^2 \begin{bmatrix} 1 & \rho & \rho^2 & \cdots & \rho^{T-1} \\ \rho & 1 & \rho & \cdots & \rho^{T-2} \\ \rho^2 & \rho & 1 & \cdots & \rho^{T-3} \\ \vdots & \vdots & \vdots & & \vdots \\ \rho^{T-1} & \rho^{T-2} & \rho^{T-3} & \cdots & 1 \end{bmatrix}$$

This can be generated by a first-order stochastic difference equation

$$(6.13)\qquad x_t - \mu_t = \rho(x_{t-1} - \mu_{t-1}) + u_t, \qquad t = 2, \ldots, T,$$

where $u_2, \ldots, u_T$ are independently distributed according to $N[0, (1-\rho^2)\sigma^2]$ and $x_1 - \mu_1$ is distributed according to $N(0, \sigma^2)$.

The exponent of the normal density involves a quadratic form with matrix

$$(6.14)\quad \mathbf{\Sigma}^{-1} = \frac{1}{(1-\rho^2)\sigma^2}\begin{bmatrix} 1 & -\rho & 0 & \cdots & 0 & 0 \\ -\rho & 1+\rho^2 & -\rho & \cdots & 0 & 0 \\ 0 & -\rho & 1+\rho^2 & \cdots & 0 & 0 \\ \vdots & \vdots & \vdots & \vdots & & \vdots \\ 0 & 0 & 0 & \cdots & 1+\rho^2 & -\rho \\ 0 & 0 & 0 & \cdots & -\rho & 1 \end{bmatrix}.$$

The first order difference equation can be generalized to the p-order difference equation

$$(6.15)\qquad \sum_{j=0}^{p} \beta_j(x_{t-j}-\mu_{t-j}) = u_t, \qquad t = p+1, \ldots, T;$$

this is associated with the stationary process with pth order dependence. In these models $\mathbf{\Sigma}^{-1}$ is almost of the form of Section 4 except that some few entries differ from the required pattern. The model of the circular serial correlation coefficient presented in Section 5 is a modification of the stationary model which leads to the desired form of $\mathbf{\Sigma}^{-1}$.

This paper shows how some models for single time series can be used for repeated time series and how inference procedures for single series can be modified to apply to repeated series.

References

Anderson, T. W. (1948). "On the Theory of Testing Serial Correlation," *Skand. Aktuarietidskr.* **31**, 88-116.

Anderson, T. W. (1958). *An Introduction to Multivariate Statistical Analysis.* John Wiley and Sons, New York.

Anderson, T. W. (1963). "Determination of the Order of Dependence in Normally Distributed Time Series," *Proceedings of the Symposium on Time Series Analysis* (M. Rosenblatt, ed.), John Wiley and Sons, New York, pp. 425-446.

Anderson, T. W., and Herman Rubin (1956). "Statistical Inference in Factor Analysis," *Proceedings of the Third Berkeley Symposium on Mathematical Statistics and Probability*, Vol. V, University of California Press, pp. 111-150.

Guttman, L. (1954). "A New Approach to Factor Analysis : the Radex," *Mathematical Thinking in the Social Sciences* (Paul F. Lazarsfeld, ed.), The Free Press, Glencoe, 111., pp. 258-348 ; 430-433.

(*Received Jan. 28, 1965.*)

CHAPTER 2

*On Conditional Test Levels in Large Samples**

R. R. BAHADUR and P. J. BICKEL†, *University of Chicago and Imperial College of Science*

1. INTRODUCTION

Suppose that X is a sample space of points x and that x is distributed according to some unknown one of a given set $\mathcal{P}$ of probability measures P. Let $\mathcal{P}_0$ be a subset of $\mathcal{P}$ and consider the null hypothesis that some P in $\mathcal{P}_0$ holds. Let $x_1, x_2, \ldots$ be a sequence of independent and identically distributed observations on x. For each $n = 1, 2, \ldots$ let T_n be a real valued test statistic based on the first n observations x_i such that large values of T_n are significant. For given data $(x_1, \ldots, x_n)$ let L_n be the level attained by T_n in the given case, and consider the random variable L_n. It is known [1],[2] that when a non-null P holds,

* This research was supported in part by Research Grants NSF GP 3707 and 2593 from the Division of Mathematical, Physical and Engineering Sciences of the National Science Foundation, and in part by the Statistics Branch, Office of Naval Research. Reproduction in whole or in part is permitted for any purpose of the United States Government.

† On leave from the University of California, Berkeley.

L_n cannot tend to zero at an exponential rate faster than exp $[-nJ(P)]$. Here J is a functional on $\mathcal{P}$ defined in terms of the Kullback-Liebler information numbers such that, in typical cases, $0 < J < \infty$ for each non-null P. It is also known [2],[3] that the levels attained by likelihood ratio statistics and certain closely related statistics do tend to zero at the optimal exponential rate $\exp(-nJ)$ in the non-null case.

It is shown here that $\exp(-nJ)$ remains the optimal exponential rate even when L_n is the conditional level attained by T_n, given that a conditioning statistic U_n has the value actually observed. (The conditioning statistic need not be real valued.) This extension of some of the conclusions of [1] and [2] is stated more precisely in Section 2 and the proof is given in Section 3.

As may be seen from Section 3, the proof is only a variation of the proofs in [1] and [2]. It may therefore be worth noting that the present conclusion concerning conditional levels is not immediate from the conclusions in [1] or [2]. The unconditional level attained by T_n is, at least approximately, an average of the various conditional levels corresponding to the potential values of the conditioning statistic U_n. It is not clear that the actual conditional level could not be much smaller than the unconditional level rather frequently. To illustrate this point, suppose for example that the x_i are real valued and normally distributed with mean θ and variance 1, that θ takes only the two values 0 and $\mu > 0$, and that the null hypothesis is that $\theta = 0$. Then $J = \frac{1}{2}\mu^2$ for the non-null distribution. Let $T_n = n^{\frac{1}{2}}\bar{x}_n$; this T_n is an optimal statistic, i.e., with $L_n (= 1 - \Phi(T_n))$ the level attained by T_n, $n^{-1} \log L_n \to -\frac{1}{2}\mu^2$ with probability one when μ holds. In an attempt to make T_n seem even more significant than it is by itself, at least on occasion, let us introduce $U_n = 1$ if $T_n \leqslant n^{\frac{1}{2}}\mu + a_n$ and $U_n = 0$ otherwise, where a_n is a constant. Let L_n^* be the conditional level attained by T_n, given U_n. Let ρ be a constant, $0 < \rho < 1$, and let p_n be the probability that $L_n^* < \rho^n L_n$ when μ holds.

It is readily seen that $p_n > 0$ for each n. The theorem of the following section implies, however, that $p_n \to 0$ as $n \to \infty$ whatever the a_n may be; indeed, with probability one $L_n^* > \rho^n L_n$ for all sufficiently large n when μ holds.

2. STATEMENT OF THE THEOREM

Let X be a space of points x, $\mathcal{B}$ a σ-field of sets of X, and $\mathcal{P}$ a family probability measures P on $\mathcal{B}$. Let $\mathcal{P}_0$ be a proper subset of $\mathcal{P}$.

For any two probability measures P and P_0 on $\mathcal{B}$, let the Kullback-Liebler information number $K(P, P_0)$ be defined as follows. Suppose first that P is absolutely continuous with respect to P_0, i.e., there exists a $\mathcal{B}$-measurable function $f(x)$, $0 \leqslant f < \infty$, such that $P(B) = \int_B f dP_0$ for all B in $\mathcal{B}$. (Such an f will sometimes be denoted by dP/dP_0). In this case let

$$K(P,P_0) = \int_X [\log f(x)]\, dP, \qquad (0 \leqslant K \leqslant \infty). \tag{1}$$

In case P is not absolutely continuous with respect to P_0 let $K = \infty$. Define

$$J(P) = \inf \{K(P,P_0) : P_0 \text{ in } \mathcal{P}_0\}, \qquad (0 \leqslant J \leqslant \infty). \tag{2}$$

Since $K(P,P) = 0$ for all P, $J = 0$ for each P in $\mathcal{P}_0$. As stated earlier, in typical cases $0 < J < \infty$ for P in $\mathcal{P} - \mathcal{P}_0$.

It is assumed that the following integrability assumption holds.

Assumption. If P is a measure in $\mathcal{P} - \mathcal{P}_0$ such that $J(P) < \infty$, there exists a sequence $\{P_j : j = 1, 2, \dots\}$ of (not necessarily distinct) measures in $\mathcal{P}_0$ such that

$$K(P,P_j) \to J(P) \text{ as } j \to \infty \tag{3}$$

and such that

$$\int_X [\log(dP/dP_j)]^2\, dP < \infty \tag{4}$$

for each j.

Now for each $n = 1, 2, \ldots$ let $X^{(n)}$ denote the set of all points $x^{(n)} = (x_1, \ldots, x_n)$ with each x_i in X, and let $\mathcal{B}^{(n)}$ be the σ-field of sets of $X^{(n)}$ determined by $\mathcal{B}$ in the usual way. Let $P^{(n)}$ be the probability measure on $\mathcal{B}^{(n)}$ when $x_1, \ldots, x_n$ are independent and identically distributed according to a probability measure P. We shall sometimes abbreviate $P^{(n)}$ to P when the domain of $P^{(n)}$ is plain from the context.

For each n, let T_n be a real valued $\mathcal{B}^{(n)}$-measurable function on $X^{(n)}$. For each n let Y_n be a space of points y_n, $\mathcal{C}_n$ a σ-field of sets of Y_n, and U_n a function on $X^{(n)}$ into Y_n such that U_n is a measurable transformation of $(X^{(n)}, \mathcal{B}^{(n)})$ into $(Y_n, \mathcal{C}_n)$, i.e., $U_n^{-1}(C_n)$ is in $\mathcal{B}^{(n)}$ for each C_n in $\mathcal{C}_n$.

Suppose that P_0 holds. Let r be a real variable, and for each n let $F_n(r, y_n)$ be the left-continuous conditional distribution function of T_n given $U_n = y_n$. More precisely, $F_n(\cdot, \cdot)$ is a function on $R \times Y_n$ such that (a) for each fixed y_n, F_n is a left-continuous probability distribution function on the real line R, (b) F_n is $\mathcal{C}_n$-measurable in y_n for each fixed r, and (c) for each fixed r, $F_n(r, y_n)$ serves as the conditional probability function $P_0^{(n)}(T_n(x^{(n)}) < r \mid U_n(x^{(n)}) = y_n)$. The existence of such an F_n follows from a theorem of Doob ([5], p. 29). In order to indicate the dependence of F_n on P_0 we shall sometimes write it as $F_n(\cdot, \cdot \mid P_0)$. Now let

$$G_n(r, y_n) = \sup \{1 - F_n(r, y_n \mid P_0) : P_0 \text{ in } \mathcal{P}_0\} \tag{5}$$

and define

$$L_n(x^{(n)}) = G_n(T_n(x^{(n)}), U_n(x^{(n)})). \tag{6}$$

Suppose that in a given case the data consists of n independent observations on x, say $(x_1^0, \ldots, x_n^0)$. Let the values of T_n and U_n in the given sample be r^0 and y_n^0. Then $L_n(x_1^0, \ldots, x_n^0) = G(r^0, y_n^0)$ as defined above is the maximum conditional probability, consistent with the null hypothesis, of T_n being as large as (or larger than) r^0, given $U_n = y_n^0$. In interesting applications, the conditional distribution of T_n given U_n is the same whichever null P_0 holds, so that maximization is redundant. Maximization seems reasonable, however, when arbitrary T_n and U_n are under consideration and $\mathcal{P}_0$ contains more than one measure P_0.

In general, for given n and P_0, $F_n(\cdot, \cdot \mid P_0)$ is not determined quite uniquely by the conditions (a), (b) and (c) above. Moreover, if $\mathcal{P}_0$ is non-denumerable, a particular determination of the family $\{F_n(\cdot, \cdot \mid P_0) : P_0 \text{ in } \mathcal{P}_0\}$ might make G_n defined by (5) non-measurable in y_n. It follows that in general L_n is not uniquely determined and that a given determination of L_n may well be a non-measurable function of $x^{(n)}$. It might be reasonable to require (as a regularity condition on T_n and U_n) the existence of a measurable version of L_n, and to restrict discussion to such a version, but it is not necessary to do so here. The following theorem holds as stated for any given determination of $\{L_n : n = 1, 2, \ldots\}$ with the set E in the statement of the theorem depending possibly on the given determination.

Let $x^{(\infty)} = (x_1, x_2, \ldots, \textit{ad inf})$, let $X^{(\infty)}$ be the set of all points $x^{(\infty)}$, let $\mathcal{B}^{(\infty)}$ be the Kolmogorov σ-field determined by $\mathcal{B}$, and let $P^{(\infty)}$ denote the probability measure on $\mathcal{B}^{(\infty)}$ when the x_i are independently distributed according to P. For each $n = 1, 2, \ldots$ regard $x^{(n)} = (x_1, \ldots, x_n)$ as a function of $x^{(\infty)}$; then $L_n = L_n(x^{(n)})$ is a real valued function on $X^{(\infty)}$, $0 \leqslant L_n \leqslant 1$.

Theorem. *Corresponding to each P in $\mathcal{P}$ there exists a $\mathcal{B}^{(\infty)}$-measurable set $E \subset X^{(\infty)}$ with $P^{(\infty)}(E) = 1$ such that, for each $x^{(\infty)}$ in E,*

$$\liminf_{n\to\infty} \left\{\frac{1}{n} \log L_n\right\} \geqslant -J(P). \tag{7}$$

The theorem shows that in the null case (i.e., if some P in $\mathcal{P}_0$ holds) $n^{-1} \log L_n \to 0$ with probability one. The following stronger conclusion is available from Proposition 4 of Section 3 : If $\{a_n : n = 1, 2, \ldots\}$ is any sequence with $a_n > 0$ and $\sum_n a_n < \infty$, then in the null case $L_n \geqslant a_n$ for all sufficiently large n with probability one. These conclusions are not very useful but perhaps they are not entirely obvious since T_n and U_n are arbitrary statistics.

In the non-null case the theorem is an extension and improvement of Theorem 1 of [2]. To be specific, if we take U_n to be a trivial statistic for each n, e.g. $U_n \equiv 0$, the theorem yields Theorem 1 of [2] under an assumption weaker than those required in [2].

3. PROOF OF THE THEOREM

Let R denote the real line of points r, and $\mathcal{R}$ the class of Borel sets of R. Let Q be a probability measure on $\mathcal{R}$, and let $F(r) = Q\{(-\infty, r)\}$. The following proposition is stated without proof in [2].

Proposition 1. *$Q\{r : 1-F(r) < a\} \leqslant a$ for $0 \leqslant a \leqslant 1$.*

Proof. The inequality holds trivially if $a = 0$ or $a = 1$. Suppose then that $0 < a < 1$ and let $b = \sup\{r : F(r) \leqslant 1-a\}$. Then $F(b) \leqslant 1-a \leqslant F(b+0)$, and $1-F(r) < a$ if and only if $r > b$. Hence $Q\{r : 1-F(r) < a\} = Q\{(b, \infty)\} = 1-F(b+0) \leqslant a$.

Let Y be a set of points y, and $\mathcal{C}$ a σ-field of sets of Y. Let $F(r, y)$ be a real valued function on $R \times Y$ such that F is left-continuous in r for each y and $\mathcal{C}$-measurable in y for each r. Let $\mathcal{R} \times \mathcal{C}$ be the σ-field generated by all sets $B \times C$ with B in $\mathcal{R}$ and C in $\mathcal{C}$.

Proposition 2. *F is an $\mathcal{R} \times \mathcal{C}$-measurable function of (r, y).*

Proof. For any real r and any positive integer k let $m_k(r)$ be the integer such that $2^{-k} m \leqslant r < 2^{-k}(m+1)$. Let $F_k(r, y) \equiv F(2^{-k} m_k(r), y)$. Since $2^{-k} m_k(r) \uparrow r$ as $k \uparrow \infty$, and since F is left-continuous, $F_k(r, y) \to F(r, y)$ for each (r, y) as $k \to \infty$. It will therefore suffice to show that each F_k is $\mathcal{R} \times \mathcal{C}$-measurable. Consider a fixed k, and for any integer j let $g_j(r) = 1$ or 0 according as $m_k(r) = j$ or $\neq j$. Then g_j is a Borel measurable function of r, so $g_j(r) \cdot F(2^{-k} j, y)$ is $\mathcal{R} \times \mathcal{C}$-measurable. Since $F_k(r, y) = \sum_{j=-\infty}^{+\infty} g_j(r) \cdot F(2^{-k} j, y)$, F_k is $\mathcal{R} \times \mathcal{C}$-measurable. The authors are indebted to P. Billingsley for this proof.

Now choose and fix a probability measure P_0 on $\mathcal{B}$. For each n let $F_n(r, y_n)$ be the conditional distribution function of T_n given $U_n = y_n$ when P_0 holds (cf. Section 2). Let

$$M_n(x^{(n)}) = 1 - F_n(T_n(x^{(n)}), U_n(x^{(n)})). \tag{8}$$

Note that M_n is the conditional level attained by T_n in testing the simple hypothesis that P_0 holds.

Proposition 3. *M_n is $\mathcal{B}^{(n)}$-measurable in $x^{(n)}$.*

Proof. Let $V_n = (T_n, U_n)$. Since T_n and U_n are measurable transformations of $(X^{(n)}, \mathcal{B}^{(n)})$ into $(R, \mathcal{R})$ and $(Y_n, \mathcal{C}_n)$ respectively, it follows that V_n is a measurable transformation of $(X^{(n)}, \mathcal{B}^{(n)})$ into $(R \times Y_n, \mathcal{R} \times \mathcal{C}_n)$. According to Proposition 2 (with $Y = Y_n, \mathcal{C} = \mathcal{C}_n$, and $F = F_n$), $F_n(r, y_n)$ is a measurable transformation of $(R \times Y_n, \mathcal{R} \times \mathcal{C}_n)$ into $(R, \mathcal{R})$. Hence $F_n(V_n)$ is a measurable transformation of $(X^{(n)}, \mathcal{B}^{(n)})$ into $(R, \mathcal{R})$, so $M_n = 1 - F_n(V_n)$ is $\mathcal{B}^{(n)}$-measurable.

Proposition 4. *$P_0(M_n < a) \leqslant a$ for $0 \leqslant a \leqslant 1$.*

Proof. For each y_n in Y_n let $Q_n(\cdot \mid y_n)$ be the probability measure on $\mathcal{R}$ corresponding to the distribution function $F_n(\cdot, y_n)$. Then, for any A in $\mathcal{R}$, $Q_n(A \mid y_n)$ is a version of P_0 (T_n in $A \mid U_n = y_n$). It follows (cf. [4]) hence that, for any B in $\mathcal{R} \times \mathcal{C}_n$, $Q_n(\{r : (r, y_n) \text{ in } B\} \mid y_n)$ is a version of $P_0(V_n \text{ in } B \mid U_n = y_n)$. In particular, in view of (8), $Q_n(\{r : 1 - F_n(r, y_n) < a \mid y_n)$ is a version of $P_0(M_n < a \mid U_n = y_n)$; this version is $\leqslant a$ for every y_n, by Proposition 1. Thus $P_0(M_n < a \mid U_n) \leqslant a$ with probability one; hence $P_0(M_n < a) \leqslant a$.

Now choose and fix a measure P on $\mathcal{B}$ such that P admits a density function with respect to the previously chosen P_0. Assume that $dP/dP_0 = f(x)$ satisfies $\int_X [\log f]^2 \, dP < \infty$. Let K be defined by (1), $0 \leqslant K < \infty$. For each n let $f^{(n)}(x^{(n)}) = \prod_{i=1}^{n} f(x_i)$, and let

$$\lambda_n = e^{n(K+\epsilon)} \tag{9}$$

where $\epsilon > 0$ is a constant.

Proposition 5. $\sum_{n=1}^{\infty} P(f^{(n)} \geqslant \lambda_n) < \infty$.

Proof. Since $K < \infty$, $P(0 < f(x) < \infty) = 1$. Let $z_i = \log f(x_i) - K$ for $i = 1, 2, \ldots$ and let $S_n = \sum_{i=1}^{n} z_i$. Then $P(f^{(n)} \geqslant \lambda_n) \leqslant P(|S_n| \geqslant n\epsilon)$ by (9), and Proposition 5 follows from a theorem of Hsu and Robbins [6] (cf. also [7], Theorem 1). It can be shown that the present assumption that $(\log f)^2$ is P-integrable is necessary for the validity of Proposition 5 for every ϵ.

For each n let

$$\mu_n = e^{-(nK+\delta)} \tag{10}$$

where $\delta > 0$ is a constant.

Proposition 6. $\sum_{n=1}^{\infty} P(M_n < \mu_n) < \infty$.

Proof. Given δ, choose ϵ so that $0 < \epsilon < \delta$ and let λ_n be given by (9). Then for each n,

$$\begin{aligned} P(M_n < \mu_n) &= P(M_n < \mu_n, f^{(n)} < \lambda_n) \\ &\quad + P(M_n < \mu_n, f^{(n)} \geqslant \lambda_n) \\ &\leqslant P(M_n < \mu_n, f^{(n)} < \lambda_n) \\ &\quad + P(f^{(n)} \geqslant \lambda_n). \end{aligned} \tag{11}$$

We observe next that $f^{(n)} = dP^{(n)}/dP_0^{(n)}$. Since M_n and $f^{(n)}$ are both random variables defined on the space $X^{(n)}$, it follows that

$$\begin{aligned} P^{(n)}(M_n < \mu_n, f^{(n)} < \lambda_n) &\leqslant \lambda_n P_0^{(n)}(M_n < \mu_n, f^{(n)} < \lambda_n) \\ &\leqslant \lambda_n P_0^{(n)}(M_n < \mu_n). \end{aligned} \tag{12}$$

According to Proposition 4, $P_0^{(n)}(M_n < \mu_n) \leqslant \mu_n$. It therefore follows from (9), (10), (11) and (12) that

$$P(M_n < \mu_n) \leqslant P(f^{(n)} \geqslant \lambda_n) + \exp(-n(\delta - \epsilon)) \tag{13}$$

for every n. Proposition 6 now follows from (13) and Proposition 5.

It is important to note that the proof of Proposition 6 involves an implicit application of the Neyman-Pearson lemma; the critical region $W_n = \{M_n < \mu_n\}$ for testing P_0 against P is compared with an optimal region $\hat{W}_n = \{f^{(n)} \geqslant \lambda_n\}$ of approximately the same size or a larger size, and an upper bound for the power $P(W_n)$ is thus obtained. This connection with the Neyman-Pearson lemma is explicit in [1], and can be made explicit here also by a modification of the preceding arguments.

Proof of the theorem. Choose and fix a P in $\mathscr{P}$. Since (7) holds trivially if $J(P) = \infty$, suppose that $J(P) < \infty$. Let $\{P_j\}$ be a sequence in $\mathscr{P}_0$ such that (3) and (4) are satisfied. (In case the given P is in $\mathscr{P}_0$, $P_j = P$ for all j will do.) Let $M_{nj} = 1 - F_n(T_n, U_n \mid P_j)$. It follows from Propositions 3 and 6 with $P_0 = P_j$ and $\delta = (2k)^{-1}$ that for each j and $k = 1, 2, \ldots$ there exists a $\mathscr{B}^{(\infty)}$-measurable set E_{jk}, with $P^{(\infty)}(E_{jk}) = 1$, such that $x^{(\infty)}$ in E_{jk} implies $n^{-1} \log M_{nj} \geqslant -K(P, P_j) - k^{-1}$ for all sufficiently large n. Let $E = \bigcap_{j,k} E_{jk}$. Then E is $\mathscr{B}^{(\infty)}$-measurable and $P^{(\infty)}(E) = 1$. Since $L_n \geqslant \sup\{M_{nj} : j = 1, 2, \ldots\}$ for each n [cf. (5), (6), (8)], $x^{(\infty)}$ in E implies $\liminf \{n^{-1} \log L_n\} \geqslant -K(P, P_j) - k^{-1}$ for all j and k; hence (7) holds, by (3). This completes the proof.

4. CONCLUDING REMARKS

The integrability assumption of Section 2 is essential to the preceding proof, but it is not known at present whether it is essential to the theorem itself. It may be noted in this connection that there are weaker versions of the theorem which do not require any assumptions whatsoever. In particular, if L_n is $\mathscr{B}^{(n)}$-measurable for each n, then for any probability measure P on $\mathscr{B}$, $0 < \rho < \exp(-J(P))$ implies

$$P(L_n < \rho^n) \to 0 \qquad \text{as } n \to \infty, \tag{14}$$

and there exists a $\mathscr{B}^{(\infty)}$-measurable set E with $P^{(\infty)}(E) = 1$ such that

$$\limsup_{n \to \infty} \{n^{-1} \log L_n\} \geqslant -J(P) \tag{15}$$

for each $x^{(\infty)}$ in E. To verify this we may suppose $J(P) < \infty$. Choose a $\gamma > 0$ and let P_0 be a measure in $\mathscr{P}_0$ such that $K(P, P_0) < J(P) + \gamma$. Since $K < \infty$, dP/dP_0 exists, say f. We now argue as in Propositions 1–6, except that Proposition 5 becomes $P(f^{(n)} \geqslant \lambda_n) = o(1)$ as $n \to \infty$, by the law of large numbers; consequently, proposition 6 becomes $P(M_n < \mu_n) = o(1)$, by (13). Since $L_n \geqslant M_n$, $P(L_n < \mu_n) = o(1)$. Since $\mu_n = \exp(-n[K+\delta]) > \exp(-n[J+\gamma+\delta])$, and since γ and δ are arbitrary, it follows that (14) holds whenever $0 < \rho < \exp(-J)$. Now choose and fix such a ρ. Then there exist positive integers $n_1 < n_2 < \ldots$ such that $\sum_{i=1}^{\infty} P(L_{n_i} < \rho^{n_i}) < \infty$. Consequently, with probability one the left hand side of (15) is not less than $\log \rho$. Since ρ is arbitrary, (15) holds with probability one. These remarks are an elaboration of similar remarks in [2].

References

[1] Bahadur, R. R. (1960). "Asymptotic Efficiency of Tests and Estimates", *Sankhyā*, **22**, 229–252.

[2] Bahadur, R. R. (1965). "An Optimal Property of the Likelihood Ratio Statistic", *Proc. Fifth Berkeley Symp. Math. Statist. Prob.*, **1**, 13-26. Univ. of California Press.

[3] Bahadur, R. R., and Bickel, P. J. (1966). "An Optimality Property of Bayes' Test Statistics". (To appear).

[4] Bahadur, R. R., and Bickel, P. J. (1968). "Substitution in conditional expectation", *Ann. Math. Statist.*, **39**, 377–378.

[5] Doob, J. L. (1953). *Stochastic Processes*, John Wiley and Sons, New York.

[6] Hsu, P. L., and Robbins, H. (1947). "Complete Convergence and the Law of Large Numbers", *Proc. Nat. Acad. Sci.*, **33**, 25–31.

[7] Katz, M. L. (1963). "The Probability in the Tail of a Convolution", *Ann. Math. Statist.*, **34**, 312–318.

(*Received Feb. 28, 1966.*)

CHAPTER 3

Interpretation and Use of a Generalized Discriminant Function

ROLF E. BARGMANN[1], *University of Georgia*

SUMMARY

The linear combinations of response variables introduced by S. N. Roy for the purpose of constructing Union-Intersection tests can be regarded as estimates of a generalized discriminant function. In the present report the parametric function is defined in terms of non-centrality parameters. Following a section on properties, meaning, and interval estimation of these parameters, the report defines estimates of the generalized discriminant function. Four illustrations are given : The standard interpretation for two groups, a well known demonstration for k groups, its meaning in cross classification designs, and its reduction to H. Hotelling's most predictable criterion for a multivariate regression model. After introduction of correlations between artificial and observable variables, the use of the generalized discriminant function is illustrated for the selection of response variables, and a numerical example is supplied.

1. INTRODUCTION

In his development of the Union-Intersection Method for the testing of multivariate hypotheses, *S. N. Roy* [11] introduced the concept of ideal linear combinations of response variables.

[1] This research was supported, in part, by the HEW Cooperative Research Project No. 1132, US Office of Education, and Virginia Polytechnic Institute, 1962.

These are readily recognized as discriminant functions. More precisely, the linear combinations utilized by *S. N. Roy* are maximum-likelihood *estimates* of a discriminant function which should, of course, be defined in terms of *parameters* rather than sample quantities. An attempt will be made in this paper, to define the generalized discriminant function in terms of non-centrality parameters, and to describe its meaning and use in certain standard situations.

2. NONCENTRALITY PARAMETERS

In terms of the *non-central chi-square distribution*, the non-centrality parameter, γ^2, will be defined as follows : Let the non-central p.d.f. (with n d.f.) be

$$f(x) = \sum_{m=0}^{\infty} P(m, \gamma^2/2) g_{n+2m}(x) \tag{2.1}$$

where

$$P(m, \gamma^2/2) = \frac{(\gamma^2/2)^m}{m\,!} e^{-\gamma^2/2} \tag{2.1a}$$

(i.e., a *Poisson* term with parameter $\gamma^2/2$), and

$$g_{n+2m}(x) = \left[2^{(n+2m)/2}\,\Gamma\left(\frac{n+2m}{2}\right)\right]^{-1} x^{(n+2m-2)/2}\,e^{-x/2} \tag{2.1b}$$

(i.e., a *central chi-square* density function with $n+2m$ d.f.).

This choice of γ^2 has certain notational advantages. For example, in the general linear model

$$\boldsymbol{E}(\boldsymbol{y}) = \boldsymbol{A}\boldsymbol{\varphi} \tag{2.2}$$

and the hypothesis

$$H_0 : \boldsymbol{C}\boldsymbol{\varphi} = \boldsymbol{0}, \quad \text{alt.} : \boldsymbol{C}\boldsymbol{\varphi} = \boldsymbol{\eta} \neq \boldsymbol{0}, \tag{2.3}$$

the "*sum of squares*" (actually, *quadratic form*) due to tho hypothesis,

$$SSH = \hat{\boldsymbol{\varphi}}'\boldsymbol{C}'\,[\boldsymbol{C}(\boldsymbol{A}'\boldsymbol{A})^{-1}\,\boldsymbol{C}']^{-1}\,\boldsymbol{C}\hat{\boldsymbol{\varphi}} \tag{2.4}$$

divided by σ^2, is distributed as non-central chi-square with n_h degrees of freedom (*d.f. due to hypothesis*, or rank of matrix C), and with non-centrality parameter

$$\gamma^2 = \boldsymbol{\varphi}'\boldsymbol{C}' [\boldsymbol{C}(\boldsymbol{A}'\boldsymbol{A})^{-1} \boldsymbol{C}']^{-1} \boldsymbol{C}\boldsymbol{\varphi}/\sigma^2. \tag{2.5}$$

In other words, γ^2 is obtainable at once by replacement of the function, $\boldsymbol{C}\hat{\boldsymbol{\varphi}}$, (unique, if $\boldsymbol{C}\boldsymbol{\varphi}$ is estimable) by its expected value under a given alternative, in the expression for the test statistic SSH. The proof of this is straightforward, and has been presented in many texts. In the present notation, it may be found in [9].

As shown in standard textbooks (e.g.,[8], [11]), the test statistic

$$F = \frac{SSH/n_h}{SSE/n_e} \tag{2.6}$$

has, under the alternative, $\boldsymbol{C}\boldsymbol{\varphi} = \boldsymbol{\eta}$, the non-central F distribution with non-centrality parameter γ^2 as defined in (2.5). In this notation, SSE denotes the *sum of squares* due to error,

$$SSE = \boldsymbol{y}'\boldsymbol{y} - \hat{\boldsymbol{\varphi}}'\boldsymbol{A}'\boldsymbol{y}, \tag{2.7}$$

and n_e denotes the degrees of freedom due to error. If the model is singular, $\boldsymbol{A}'\boldsymbol{A}$ may be replaced by $\boldsymbol{A}_1'\boldsymbol{A}_1$, where $\boldsymbol{A}_1$ is a basis of $\boldsymbol{A}$ (see [11]) or $(\boldsymbol{A}'\boldsymbol{A})^{-1}$ may be considered as any *conditional* or *generalized inverse* defined as

$$(\boldsymbol{A}'\boldsymbol{A})(\boldsymbol{A}'\boldsymbol{A})^{-1} (\boldsymbol{A}'\boldsymbol{A}) = (\boldsymbol{A}'\boldsymbol{A}). \tag{2.8}$$

Since, in the singular case, the normal equations

$$\boldsymbol{A}'\boldsymbol{A}\hat{\boldsymbol{\varphi}} = \boldsymbol{A}'\boldsymbol{y} \tag{2.9}$$

are consistent, and since the linear combination $\boldsymbol{C}\hat{\boldsymbol{\varphi}}$ is assumed to be unique ($\boldsymbol{C}\boldsymbol{\varphi}$ *estimable*), the expressions (2.4), (2.5), and (2.7) will be unique, regardless of the choice of a conditional inverse.

The non-centrality parameter plays the role of (the square of) a *standardized distance function*, invariant under change of origin and unit of measurement. Let us assume just two groups of sizes n_1 and n_2. Then, the sum of squares between groups is

$$SSH = (\bar{y}_1 - \bar{y}_2)^2/(1/n_1 + 1/n_2) \tag{2.10}$$

where $\bar{y}_1$ and $\bar{y}_2$ denote the sample means of the two groups (SSH is simply the t^2 statistic without s^2 in the denominator). This expression enables us to write the non-centrality parameter

$$\gamma^2 = (\mu_1 - \mu_2)^2/\sigma^2(1/n_1 + 1/n_2) \tag{2.11}$$

which is readily recognized as the square of the standardized distance between two populations. It should be noted here that the sample sizes enter into this non-centrality parameter for obvious reasons : A given difference, $(\mu_1 - \mu_2)/\sigma$ would result in a greater value of the power function if the sample were large than if it were small. If we wish to describe population distances, without reference to samples to be taken, we usually replace the n_i by q_i, the "quota" probabilities, i.e., the probability that an experimental unit comes from population i.

As another illustration, let us consider the case of a one-way classification analysis of variance, assuming equal group sizes for simplicity. Let each group have n observations. Then

$$SSH = n \sum_{i=1}^{k} (\bar{y}_i - \bar{\bar{y}})^2 \tag{2.12}$$

where $\bar{y}_i$ is the mean of the observations in the i-th group, and $\bar{\bar{y}}$ is the grand mean. Consequently,

$$\gamma^2 = n \sum_{i=1}^{k} (\mu_i - \mu)^2/\sigma^2 \tag{2.13}$$

This expression states that the standardized distance of each group mean from the grand mean is evaluated $[(\mu_i - \mu)\sqrt{n}/\sigma]$, squared, and the squares are added. It is these dimension-free standardized indices which, as for instance the correlation coefficient, allow comparison and study of relationships, even when units of measurement are different. The non-centrality parameters are, in this sense, statements of a standardized distance (squared) between groups, if under the null hypothesis such groups are assumed to have equal location parameters.

We shall propose the use of this non-centrality parameter as a single index of discrepancy. We shall assume that a null hypothesis, $\boldsymbol{C\varphi} = \mathbf{0}$ (equality of effects, absence of regression, etc.) defines, indirectly, some kind of grouping or categorization in such a way that, under the null hypothesis, the location parameters of such categories are identical. Then the non-centrality parameter can be regarded as an index of departure from this central case and, for the discriminant function considered in the following section, we seek that linear combination of response variables which produces a maximum departure from the central case.

The non-centrality parameter has some further uses which, for the convenience of the reader, will be stated explicitly in the sequel. Taking expectations of (2.4) we see that

$$\begin{aligned}(2.14\text{a})\quad E(SSH) &= \operatorname{tr}\{E[\boldsymbol{C}(\boldsymbol{A}'\boldsymbol{A})^{-1}\,\boldsymbol{C}']^{-1}\,\boldsymbol{C}\hat{\boldsymbol{\varphi}}\hat{\boldsymbol{\varphi}}'\boldsymbol{C}'\} \\ &= \operatorname{tr}[\boldsymbol{C}(\boldsymbol{A}'\boldsymbol{A})^{-1}\,\boldsymbol{C}']^{-1}\,[\operatorname{var}(\boldsymbol{C}\hat{\boldsymbol{\varphi}})+E(\boldsymbol{C}\hat{\boldsymbol{\varphi}})E(\hat{\boldsymbol{\varphi}}'\boldsymbol{C}')] \\ &= \operatorname{tr}[\boldsymbol{C}(\boldsymbol{A}'\boldsymbol{A})^{-1}\,\boldsymbol{C}']^{-1}\,[\sigma^2\,\boldsymbol{C}(\boldsymbol{A}'\boldsymbol{A})^{-1}\,\boldsymbol{C}'+\boldsymbol{C}\boldsymbol{\varphi}\boldsymbol{\varphi}'\boldsymbol{C}'] \\ &= \operatorname{tr}\{\sigma^2\,\boldsymbol{I}+[\boldsymbol{C}(\boldsymbol{A}'\boldsymbol{A})^{-1}\,\boldsymbol{C}']^{-1}\,\boldsymbol{C}\boldsymbol{\varphi}\boldsymbol{\varphi}'\boldsymbol{C}'\} \\ &= n_h\,\sigma^2+\boldsymbol{\varphi}'\boldsymbol{C}'[\boldsymbol{C}(\boldsymbol{A}'\boldsymbol{A})^{-1}\,\boldsymbol{C}']^{-1}\,\boldsymbol{C}\boldsymbol{\varphi} \\ &= \sigma^2(n_h+\gamma^2).\end{aligned}$$

Hence, the expected mean square for the hypothesis (MSH) is

$$(2.14\text{b})\qquad E(MSH) = \sigma^2(1+\gamma^2/n_h).$$

Since the last term of (2.14a) divided by n_h is sometimes casually referred to as the variance component (of the effect on which the hypothesis is made), because of its loose analogy with a (true) variance component in a random model, we may use (2.14b) for the purpose of establishing approximate confidence bounds for the ratio of the *variance component* (actually *non-centrality component*) of an effect divided by the *variance component* of error.

This is rather simple if one approximates the non-central F distribution by an improved variance-stabilizing transformation (successive elimination of bias in the variance-stabilizing transformation as proposed by Hotelling [7] for multiple correlations). Defining

$W = 1+mF/n$ (F distribution has (m,n) d.f.)

$a = \sqrt{m+n-2)/(n-2)}$, the transformation is

$$Z = \cosh^{-1}(W/a) \tag{2.15}$$

which is approximately normal with

$$E(Z) = \zeta - \coth\zeta/(n-4) \quad \text{and} \tag{2.15a}$$

$$\text{var}(Z) = 2/(n-4), \tag{2.15b}$$

where

$$\zeta = \cosh^{-1}[\gamma^2/\sqrt{(m+n-2)(n-2)}+a]. \tag{2.15c}$$

These facts can be used to obtain, approximately, confidence bounds on the non-centrality parameter γ^2 (for details, see [4] and a condensed version in the appendix to this paper.)

3. THE GENERALIZED DISCRIMINANT FUNCTION

We now consider the multivariate linear model

$$E(\boldsymbol{Y}) = \boldsymbol{A\Phi} \tag{3.1}$$

where $\boldsymbol{Y}$ denotes the matrix of N experimental units, each measured on p random variables. $\boldsymbol{\Phi}$ denotes the matrix of parameters, with p columns, each referring to one of the response variables. The matrix $\boldsymbol{A}$ is the model or design matrix with the same meaning as in the univariate model. Hence (3.1) is merely an array of p columns, each satisfying (2.2) where each column represents one response variable. The generalized homoscedasticity condition is that elements in different rows of $\boldsymbol{Y}$ are uncorrelated, and that the variance-covariance matrix of all the elements in the same row is the $p\times p$ matrix $\boldsymbol{\Sigma}$, which is assumed to be the same for all rows. We assume that division into groups is indicated by a function $\boldsymbol{C\Phi}$, such that the null hypothesis $\boldsymbol{C\Phi} = \boldsymbol{0}$ represents the case where the location parameters of all these groups are identical.

We wish to find a linear combination of the response variables

$$z = \alpha_1 y_1 + \alpha_2 y_2 + \dots + \alpha_p y_p, \qquad \text{or}$$

$$\mathbf{z} = \mathbf{Y}\boldsymbol{\alpha} \tag{3.2}$$

such that the non-centrality parameter based upon the single variable $\mathbf{z}$ and the grouping C is a maximum. Then

$$E(\mathbf{z}) = \mathbf{A}\boldsymbol{\Phi}\boldsymbol{\alpha} \qquad \text{and} \tag{3.3}$$

$$\text{var}(\mathbf{z}) = (\boldsymbol{\alpha}'\boldsymbol{\Sigma}\boldsymbol{\alpha})\mathbf{I} \tag{3.4}$$

In this notation, $\mathbf{I}$ is an $N \times N$ identity matrix, and $(\boldsymbol{\alpha}'\boldsymbol{\Sigma}\boldsymbol{\alpha})$ is a scalar which takes the place of the usual σ^2 in univariate analysis. The operator var(vector) denotes the variance-covariance matrix of all elements of the vector. Now, the ease of multivariate extensions of univariate analysis is due to the fact that all computational algorithms for obtaining test statistics are the same, with the following generalization :

(3.5) *SSH is extended into a matrix* $\mathbf{H}$ whose *diagonal elements* are the *sums of squares of hypothesis* for *each* response variable separately. The *non-diagonal elements* are corresponding *sums of cross products* involving *pairs* of response variables.

(3.6) *SSE is extended into a matrix* $\mathbf{E}$ whose *diagonal elements* are the *sums of squares of errors,* for *each* response variable separately. The *non-diagonal elements* are corresponding *sums of cross products* involving *pairs* of response variables.

These two rules follow at once from S. N. Roy's [11] presentation, and have been stated repeatedly by many authors. Somehow the rules, in this simplicity, have not entered the usual textbooks, where the traditional "*Wilks-Lawley*" (see, e.g., [1]) formulation is usually preferred. With the generalizations (3.5) and (3.6) the calculations are the same as those in univariate analysis. Merely the distribution functions involved may require special tables [5], or series of chi-square distributions

[1, 9]. The numerical work is, of course, greatly multiplied, and electronic computers are useful especially for the determination of the matrix $\boldsymbol{E}$ (sums of squares and products due to error). The further calculations (determinants, triangularization, characteristic roots and vectors) require considerably less time by contrast.

The formal definition of the matrix $\boldsymbol{H}$ is, of course,

(3.5a) $$\boldsymbol{H} = \hat{\boldsymbol{\Phi}}'\boldsymbol{C}'[\boldsymbol{C}(\boldsymbol{A}'\boldsymbol{A})^{-1}\boldsymbol{C}']^{-1}\boldsymbol{C}\hat{\boldsymbol{\Phi}} \quad \text{(in analogy to (2.4))}$$

and that of $\boldsymbol{E}$ is

(3.6a) $$\boldsymbol{E} = \boldsymbol{Y}'\boldsymbol{Y} - \hat{\boldsymbol{\Phi}}'\boldsymbol{A}'\boldsymbol{Y} \quad \text{(in analogy to (2.7)),}$$

where $\hat{\boldsymbol{\Phi}}$ is the solution of the equations

(3.7) $$\boldsymbol{A}'\boldsymbol{A}\,\hat{\boldsymbol{\Phi}} = \boldsymbol{A}'\boldsymbol{Y} \quad \text{(in analogy to (2.9))}$$

In order to obtain a compact notation for the non-centrality parameter for z as defined in (3.2), we define as the *parametric analogue* of the matrix $\boldsymbol{H}$ (from (3.5) and (3.5a)) a matrix $\boldsymbol{\Gamma}$, obtainable by replacing $\boldsymbol{C}\hat{\boldsymbol{\Phi}}$ in (3.5a) by its expected value $\boldsymbol{C}\boldsymbol{\Phi}$. It will be noted that the diagonal elements of $\boldsymbol{\Gamma}$ each divided by σ_i^2 represent the non-centrality parameters γ^2 for each response variable.

Thus

(3.8) $\boldsymbol{\Gamma} = \boldsymbol{H}$ from (3.5a) with $\hat{\boldsymbol{\Phi}}$ replaced by $\boldsymbol{\Phi}$.

With this notation it follows at once that γ^2, the non-centrality parameter for the linear combination z, is

(3.9) $$\gamma^2 = (\boldsymbol{\alpha}'\boldsymbol{\Gamma}\boldsymbol{\alpha})/(\boldsymbol{\alpha}'\boldsymbol{\Sigma}\boldsymbol{\alpha}).$$

Since

(3.10) $$\partial\gamma^2/\partial\boldsymbol{\alpha} = [2(\boldsymbol{\alpha}'\boldsymbol{\Sigma}\boldsymbol{\alpha})\boldsymbol{\Gamma}\boldsymbol{\alpha} - 2(\boldsymbol{\alpha}'\boldsymbol{\Gamma}\boldsymbol{\alpha})\boldsymbol{\Sigma}\boldsymbol{\alpha}]/(\boldsymbol{\alpha}'\boldsymbol{\Sigma}\boldsymbol{\alpha})^2,$$

we obtain, as a necessary condition for γ^2 to be a maximum.

(3.11) $$\boldsymbol{\Gamma}\hat{\boldsymbol{\alpha}} - \hat{\gamma}^2\boldsymbol{\Sigma}\hat{\boldsymbol{\alpha}} = 0$$

or

(3.11a) $$(\boldsymbol{\Sigma}^{-1}\boldsymbol{\Gamma} - \hat{\gamma}^2\boldsymbol{I})\hat{\boldsymbol{\alpha}} = 0.$$

This shows that the largest value which γ^2 can attain (denoted by $\hat{\gamma}^2$ in (3.11) and (3.11a)) is the largest characteristic root of $\mathbf{\Sigma}^{-1}\mathbf{\Gamma}$. The maximizing linear combination, i.e. the generalized discriminant function, is $z = \hat{\boldsymbol{\alpha}}'\boldsymbol{y}$, where $\hat{\boldsymbol{\alpha}}$ is the eigenvector associated with the largest characteristic root of $\mathbf{\Sigma}^{-1}\,\mathbf{\Gamma}$. The elements of $\hat{\boldsymbol{\alpha}}$ are relative quantities only, i.e. equation (3.11a) remains satisfied if each element of $\hat{\boldsymbol{\alpha}}$ is multiplied by the same constant.

In the (quite unusual) case that $\mathbf{\Sigma}$ and $\mathbf{\Gamma}$ are defined by some a priori mathematical argument, the discriminant function can be directly obtained from (3.11a) by the standard and simple methods of obtaining the largest root and associated vector of a matrix. It is, of course, useful for this and the sample calculations indicated below, to represent the (*at least positive-semidefinite*) matrix $\mathbf{\Gamma}$ as a product $\boldsymbol{TT}'$, where $\boldsymbol{T}$ is often rectangular (number of rows exceeding number of columns) and may be obtained, sometimes by inspection, and in other instances by computing the triangular or quasi-triangular matrix $\boldsymbol{T}$ used by *S. N. Roy* [11], which is readily available as a by-product of inverting or conditionally inverting a matrix by the *Gauss-Doolittle* or Square-Root method. Then the matrix $\boldsymbol{T}'\mathbf{\Sigma}^{-1}\,\boldsymbol{T}$ is symmetric and, frequently, of small order. It has the same characteristic roots (except for zero roots) as $\mathbf{\Sigma}^{-1}\,\mathbf{\Gamma}$ and the eigenvector associated with the largest root of $\boldsymbol{T}'\mathbf{\Sigma}^{-1}\boldsymbol{T}$, call it $\boldsymbol{\beta}$, is related to the discriminant function $\hat{\boldsymbol{\alpha}}$ by the relation $\hat{\boldsymbol{\alpha}} = \mathbf{\Sigma}^{-1}\,\boldsymbol{T}\boldsymbol{\beta}$. The largest root and associated vector of $\boldsymbol{T}'\mathbf{\Sigma}^{-1}\boldsymbol{T}$ can, of course, be obtained readily by a gradient method, even if the matrix is large.

In usual situations, a discriminant function or its estimate is desired for the purpose of finding some *ideal weighted total score* of response variables, to be applied to future experimental units in order to facilitate their classification into one of the groups. For the determination of the discriminant function one has what is commonly known as a *calibration sample*, i.e., a sample of observations from different groups based upon experimental units which are *ideal* representatives of each group. For example, the groups could be especially successful students in science, mathematics, and business administration, and the calibration sample consists of their scores on a battery of tests of scholastic aptitude which they took at the time of college entrance. The estimate of the discriminant function could then be used for the purpose of assigning future students to one of the groups in accordance with the proximity of their z score (weighted total score of the scholastic aptitude tests, with the discriminant function as weights) to the corresponding mean z-score of the ideal representatives of each group. The calibration sample is usually large enough to permit application of large sample methods of approximation. The relative sample sizes of each group in the calibration sample are usually fixed in such a way that they reflect the quotas to be assigned to each group in the future.

Maximum-likelihood estimation of the discriminant function seems indicated and poses a theoretical (though not practical) problem, here as in other aspects of multivariate estimation. Since $(1/N)\boldsymbol{E}$ (N is the number of experimental units) is a maximum-likelihood estimate of $\boldsymbol{\Sigma}$, and $\boldsymbol{H}$ is a maximum-likelihood estimate of $\boldsymbol{\Gamma}$, the set of all characteristic roots and all eigenvectors of $\boldsymbol{E}^{-1}\boldsymbol{H}$ is a maximum-likelihood estimate of the set of all characteristic roots and associated eigenvectors of $\boldsymbol{\Sigma}^{-1}\boldsymbol{\Gamma}$. From this it does *not* follow that the *largest* root of $\boldsymbol{E}^{-1}\boldsymbol{H}$ is necessarily the maximum-likelihood estimate of the *largest* root of $\boldsymbol{\Sigma}^{-1}\boldsymbol{\Gamma}$ and the correspondence of eigenvectors poses the same theoretical problem. The formulation of the generalized discriminant function makes it clear that we are not interested in just some stationary value of γ^2, but only in obtaining the maximum. Pending further research into this evasive problem we can now state that if, perchance, the *largest* root and associated vector of $\boldsymbol{E}^{-1}\boldsymbol{H}$ is the maximum-likelihood estimate of the *second largest* root and vector of $\boldsymbol{\Sigma}^{-1}\boldsymbol{\Gamma}$, and if the calibration sample is reasonably large, then the two roots in the population must be very close indeed, so that either the first or second eigenvector of $\boldsymbol{E}^{-1}\boldsymbol{H}$, or any linear combination of the two, would discriminate almost all well. In most practical applications encountered by this author, the largest root of $\boldsymbol{E}^{-1}\boldsymbol{H}$ stands out rather sharply. The difference between the second and third root has never been found anywhere nearly as large as that between the first and second.

With these qualifications we will define, as the *maximum-likelihood estimate of the discriminant function*, based on a calibration sample, the *eigenvector associated with the largest characteristic root of* $\boldsymbol{E}^{-1}\boldsymbol{H}$, where $\boldsymbol{H}$ and $\boldsymbol{E}$ have been defined in (3.5), (3.5a), (3.6), and (3.6a). By this definition.

$$\boldsymbol{E}^{-1}\boldsymbol{H}\boldsymbol{a} = c^2\boldsymbol{a} \tag{3.12}$$

where c^2 denotes the largest root of $\boldsymbol{E}^{-1}\boldsymbol{H}$, the *estimate of the discriminant function* is recognized as the same *linear combination* which S. N. Roy postulated in the demonstration of the *Union-Intersection test* of the general linear hypothesis. In the next section we shall present a few illustrations in order to interpret such a function.

4. INTERPRETATION OF THE GENERAL DISCRIMINANT FUNCTION

Illustration 4.1 : Two groups

Let a calibration sample consist of n_1 experimental units with sample mean vector $\boldsymbol{y}_1' = [\bar{y}_{11}, \bar{y}_{12}, \ldots, \bar{y}_{1p}]$, i.e., the means in the first group for each of the p response variables, and, correspondingly, with n_2 experimental units from the second group, with sample mean vector $\bar{\boldsymbol{y}}_2'$. The sum of squares for the hypothesis of equality of means in the two groups is, for each response variable,

$$SSH_i = [n_1n_2/(n_1+n_2)](\bar{y}_{1i}-\bar{y}_{2i})^2, \tag{4.1}$$

hence the matrix of sums of squares and products for all response variables is simply

$$\boldsymbol{H} = [n_1n_2/(n_1+n_2)](\bar{\boldsymbol{y}}_1-\bar{\boldsymbol{y}}_2)(\bar{\boldsymbol{y}}_1'-\bar{\boldsymbol{y}}_2') \tag{4.2}$$

The error sum of squares for each response variable is the 'pooled" or "within" sum of squares

$$(4.3) \qquad SSE_i = \sum_{j=1}^{n_1} y_{1ij}^2 + \sum_{j=1}^{n_2} y_{2ij}^2 - n_1 \bar{y}_{1i}^2 - n_2 \bar{y}_{2i}^2 = e_{ii}$$

The corresponding products for a pair of response variables are

$$(4.4) \qquad e_{ik} = \sum_{j=1}^{n_1} y_{1ij}\, y_{1kj} + \sum_{j=1}^{n_2} y_{2ij}\, y_{2kj} - n_1 \bar{y}_{1i}\, \bar{y}_{1k} - n_2 \bar{y}_{2i}\, \bar{y}_{2k}$$

These can be combined into a matrix $\boldsymbol{E}$ (usually called $\boldsymbol{W}$ in this case). Let $\boldsymbol{E} = \boldsymbol{W}$; then

$$(4.5) \qquad \boldsymbol{E}^{-1} \boldsymbol{H} = [n_1 n_2/(n_1+n_2)] \boldsymbol{W}^{-1}(\bar{\boldsymbol{y}}_1 - \bar{\boldsymbol{y}}_2)(\bar{\boldsymbol{y}}_1' - \bar{\boldsymbol{y}}_2').$$

Since a scalar factor does not change the eigenvectors of a matrix, we may obtain the discriminant function from the relation

$$(4.6) \qquad \boldsymbol{W}^{-1} (\bar{\boldsymbol{y}}_1 - \bar{\boldsymbol{y}}_2)(\bar{\boldsymbol{y}}_1' - \bar{\boldsymbol{y}}_2')\boldsymbol{a} = c\boldsymbol{a},$$

where c is the largest root of the matrix on the left. Since, except for zero roots, the roots of $\boldsymbol{AB}$ equal those of $\boldsymbol{BA}$, c is simple $(\bar{\boldsymbol{y}}_1' - \bar{\boldsymbol{y}}_2')\boldsymbol{W}^{-1} (\bar{\boldsymbol{y}}_1 - \bar{\boldsymbol{y}}_2)$. In accordance with the argument following (3.11a) we obtain

$$(4.7) \qquad \boldsymbol{a} = k\boldsymbol{W}^{-1} (\bar{\boldsymbol{y}}_1 - \bar{\boldsymbol{y}}_2),$$

where k is an arbitrary scalar constant. This is the familiar discriminant function for two groups (see, e.g. [10]).

Illustration 4.2 : *k* groups

The hypothesis of equality of group means, in one variable, is tested by

$$(4.8) \qquad SSH_i = SS\ (\text{between}) = \sum_{j=1}^{k} n_j (\bar{y}_{ij} - \bar{\bar{y}}_i)^2,$$

where $\bar{y}_{ij}$ denotes the mean of the jth group on the ith response variable, $\bar{\bar{y}}_i$, is the grand mean on the ith response variable, and

n_j is the number of experimental units in the jth group. This generalizes at once into the matrix

$$(4.9) \qquad \boldsymbol{H} = \boldsymbol{B} = \sum_{j=1}^{k} n_j(\bar{\boldsymbol{y}}_j - \bar{\bar{\boldsymbol{y}}})(\boldsymbol{y}'_j - \bar{\bar{\boldsymbol{y}}}')$$

where the vector $\boldsymbol{y}_j$ is a column vector with p elements which are the means of the jth group on response variables 1 to p, and $\bar{\bar{\boldsymbol{y}}}$ is a column vector whose p elements are the grand means on the response variables. $\boldsymbol{y}'_j$ and $\bar{\bar{\boldsymbol{y}}}'$ are the same vectors written as row vectors, so that each term of the sum in (4.9) is a matrix of rank one.

The matrix $\boldsymbol{E} = \boldsymbol{W}$ has the same elements as those indicated in (4.3) and (4.4), except that there are k sums instead of 2 (one for each group), and k (instead of 2) mean correction terms. Thus the discriminant function is given by the relation

$$(4.10) \qquad \boldsymbol{W}^{-1} \boldsymbol{Ba} = c\boldsymbol{a},$$

where c represents the largest characteristic root of the matrix $\boldsymbol{W}^{-1}\boldsymbol{B}$. Since $\boldsymbol{B}$ is frequently of low rank, it is useful to apply the computational procedure indicated in the paragraph following (3.11a).

It will be noted that c, the characteristic root of $\boldsymbol{W}^{-1}\boldsymbol{B}$, is the Union-Intersection test statistic for the multivariate test of the hypothesis $\boldsymbol{\mu}_1 = \boldsymbol{\mu}_2 = \ldots = \boldsymbol{\mu}_k$, and that the discriminant function $\boldsymbol{a}$ is the linear combination which maximizes the F-ratio as used by *S. N. Roy*.

Illustration 4.3 : Two-way classification

The following hypothetical example may illustrate the possible need of a discriminant function when the grouping is based upon more than one criterion. Suppose that incoming draftees are to be assigned to one of three groups :

Leadership training (A_1)

Technical duties (A_2)

General duties (A_3)

on the basis of the scores in a battery of tests consisting of 10 parts. The educational background influences, of course, performance on these tests. We assume 4 levels :

High school not completed (B_1)

High school completed (B_2)

Some college (B_3)

College graduates (B_4).

A group of, say, 2,000 incoming candidates takes the tests, and individuals are assigned, by some intuitive set of criteria, to one of the three A-groups. At the end of one year, definitely successful and properly placed candidates is each of the three A-groups are identified. We suppose that there are 1,000 of them. They will constitute the *calibration sample* the composition of which may look as follows :

	B_1	B_2	B_3	B_4	
A_1	10	90	60	40	200
A_2	40	110	100	50	300
A_3	150	300	40	10	500
	200	500	200	100	1000

The original records of these, properly placed, candidates on the test battery can now be studied and subjected to a two-way multivariate analysis of variance. A desirable discriminant function would be a weighted total score of the ten tests which maximizes the non-centrality between the A-groups. It is to be noted, however, that since such a design is usually irregular, the *adjusted* estimates of A-effects must be employed. To use unadjusted effect estimates would seem inefficient since, for example, members of group B_4 (*college graduates*) may attain higher scores than members of group B_1 (*unfinished high school*), and yet not be suitable for group A_1 (*leadership training*).

The computation of the adjusted sums of squares and products follows the conventional algorithms for *irregular two-way designs* (see e.g., [3]). Let $n_{i.}$ denote the marginal total counts on the right ($i = 1, 2, \ldots, r$), and $n_{.j}$ the marginal total counts at the bottom ($j = 1, 2, \ldots, c$), and n_{ij} be the entries in the incidence table. Construct a matrix $\boldsymbol{P}$ (symmetric, but singular) whose diagonal elements are

(4.11a) $$p_{ii} = n_{i.} - \sum_{j=1}^{c} \frac{n_{ij}^2}{n_{.j}}$$

and whose off-diagonal elements are

(4.11b) $$p_{ik} = - \sum_{j=1}^{c} \frac{n_{ij}\, n_{kj}}{n_{.j}}$$

We can obtain a conditional inverse of $\boldsymbol{P}$, by one of several methods (e.g., introducing constraints, see [8]). For each response variable, obtain a vector $\boldsymbol{R}_i$ whose elements are the *totals of all observations* (scores) *on the ith test over all subjects in each row*. Thus the vector has r elements. Pack these vectors, one for each response variable, horizontally into a matrix $\boldsymbol{R}$ with r rows and p columns. Analogously, obtain vectors of column totals $\boldsymbol{C}_i$, each with c elements, and pack these column vectors, horizontally, into a matrix $\boldsymbol{C}$ with c rows and p columns. Obtain a matrix of adjusted row totals $\boldsymbol{Q}$, of order r by p, whose elements are

(4.12a) $$q_{ik} = r_{ik} - \sum_{j=1}^{c} \frac{n_{ij}\, c_{jk}}{n_{.j}} \qquad (i = 1, 2, \ldots, r \quad k = 1, 2, \ldots, p).$$

Then the matrix $\boldsymbol{H}$ (generalization of *SS rows, adjusted*) is

(4.13) $$\boldsymbol{H} = \boldsymbol{Q}'\boldsymbol{P}^{(-1)}\boldsymbol{Q}$$

where $\boldsymbol{P}^{(-1)}$ denotes a conditional inverse of $\boldsymbol{P}$. As usual, the matrix $\boldsymbol{E}$ is conveniently obtained by computing the adjusted sums-of-squares and products within each cell, and adding corresponding terms over all cells.

Then

(4.14) $$\boldsymbol{E}^{-1}\boldsymbol{H}\boldsymbol{a} = c\boldsymbol{a}$$

produces the desired discriminant function, $\boldsymbol{a}$, and c, the largest characteristic root of $\boldsymbol{E}^{-1}\boldsymbol{H}$, is recognized as S. N. Roy's test statistic for testing equality of row effects (multivariate) in an irregular cross-classification design. Since $\boldsymbol{H}$ is usually of small rank it is convenient to obtain c and $\boldsymbol{a}$ utilizing the computational steps outlined in the paragraph following (3.11a).

It is interesting to note (for more details, see [2]) that the same discriminant function can be obtained if, from each observation vector, one subtracts the *unadjusted column mean vector* (a row in the matrix $\boldsymbol{C}$ defined between (4.11) and (4.12) divided by $n_{.j}$), and obtains the discriminant function for a one-way classification design, based upon these adjusted scores, by the method outlined in illustration 4.2.

Illustration 4.4 : Multiple Regression

Let there be q *concomitant* (fixed) variables, and p random variables and, for simplicity, let us assume that each random variable has expected overall mean equal to zero. Then the *univariate multiple regression model* is

$$E(y_i) = \beta_1 x_{1i} + \beta_2 x_{2i} + \dots + \beta_q x_{qi} \qquad (i = 1, 2, \dots, n)$$

or, in matrix notation

$$E(\boldsymbol{y}) = \boldsymbol{X\beta} \tag{4.15}$$

The corresponding *multivariate multiple-regression model* is obtained by replacement of the random vector $\boldsymbol{y}$ by a random matrix $\boldsymbol{Y}$ (with p columns), and the parameter vector $\boldsymbol{\beta}$ by the parameter matrix $\boldsymbol{B}$ (with q rows and p columns). Then

$$E(\boldsymbol{Y}) = \boldsymbol{XB}. \tag{4.16}$$

In extending the concept of grouping let us try to obtain the discriminant function (linear combination of response variables) associated with the central case in which regression is absent. The matrix $\boldsymbol{H}$ is thus the matrix of sums of squares and products for regression

$$\boldsymbol{H} = \hat{\boldsymbol{B}}'\boldsymbol{X}'\boldsymbol{X}\hat{\boldsymbol{B}}, \tag{4.17}$$

where

(4.17a) $$\hat{\boldsymbol{B}} = (\boldsymbol{X}'\boldsymbol{X})^{-1}\,\boldsymbol{X}'\boldsymbol{Y}.$$

The error matrix is

(4.18) $$\boldsymbol{E} = \boldsymbol{Y}'\boldsymbol{Y} - \boldsymbol{Y}'\boldsymbol{X}(\boldsymbol{X}'\boldsymbol{X})^{-1}\,\boldsymbol{X}'\boldsymbol{Y}.$$

To obtain an interpretation of the generalized discriminant function in this instance we will consider a *formally similar* case in which the x-variables are *random* variables. Then, except for a constant multiplier (n in this case), $\boldsymbol{X}'\boldsymbol{X}$ represents the sample variance-covariance matrix of the x-set, $\boldsymbol{Y}'\boldsymbol{Y}$ that of the y-set, and $\boldsymbol{X}'\boldsymbol{Y}$ the matrix of sample covariances between members of the x-set and members of the y-set. They will be denoted by $\boldsymbol{S}_{xx}$, $\boldsymbol{S}_{yy}$, and $\boldsymbol{S}_{xy}$, respectively. With this notation,

(4.19) $$\boldsymbol{E}^{-1}\,\boldsymbol{H} = [\boldsymbol{S}_{yy} - \boldsymbol{S}'_{xy}\,\boldsymbol{S}^{-1}_{xx}\,\boldsymbol{S}_{xy}]^{-1}\,\boldsymbol{S}'_{xy}\,\boldsymbol{S}^{-1}_{xx}\,\boldsymbol{S}_{xy}$$

Premultiplying the expression in square brackets by $\boldsymbol{S}^{-1}_{yy}$ one obtains

(4.20) $$\begin{aligned}\boldsymbol{E}^{-1}\,\boldsymbol{H} &= [\boldsymbol{I} - \boldsymbol{S}^{-1}_{yy}\,\boldsymbol{S}'_{xy}\,\boldsymbol{S}^{-1}_{xx}\,\boldsymbol{S}_{xy}]^{-1}\,\boldsymbol{S}^{-1}_{yy}\,\boldsymbol{S}'_{xy}\,\boldsymbol{S}^{-1}_{xx}\,\boldsymbol{S}_{xy}\\ &= [\boldsymbol{I} - \boldsymbol{Q}]^{-1}\,\boldsymbol{Q}, \quad \text{say.}\end{aligned}$$

The matrix $\boldsymbol{Q}$ will be recognized as the matrix *whose characteristic roots are the squares of the canonical correlations* (see [6]). The eigenvector associated with the largest canonical correlation, i.e.,

(4.21) $$\boldsymbol{Q}\boldsymbol{b} = r^2\,\boldsymbol{b}$$

is used in the construction of *Hotelling's most predictable criterion* [6]. The linear combination $\boldsymbol{b}'\boldsymbol{y}$ is that linear combination of y-variables which can be predicted with greatest precision from the x-set. We will show that $\boldsymbol{b}$ equals the discriminant function. From (4.21) it follows that

(4.22a) $$\begin{aligned}[\boldsymbol{I} - \boldsymbol{Q}]\boldsymbol{b} &= (1 - r^2)\boldsymbol{b}\\ [\boldsymbol{I} - \boldsymbol{Q}]^{-1}\,\boldsymbol{b} &= [(1/(1 - r^2)]\boldsymbol{b}.\end{aligned}$$

By (4.21)

(4.22b) $$\boldsymbol{b} = (1/r^2)\boldsymbol{Q}\boldsymbol{b}, \qquad \text{hence}$$

(4.22c) $$[\boldsymbol{I} - \boldsymbol{Q}]^{-1}\,\boldsymbol{Q}\boldsymbol{b} = [r^2/(1 - r^2)]\boldsymbol{b}.$$

Comparison with (4.20) shows that $\boldsymbol{b}$ is identical with the discriminant function. The associated characteristic root, c, is $r^2/(1-r^2)$, where r^2 denotes the square of the largest canonical correlation.

5. USE OF THE DISCRIMINANT FUNCTION FOR VARIABLE SELECTION

The discriminant function is a linear combination of response variables, and chosen in such a way that a maximum separation of groups is attained. Consequently, the discriminant function can be regarded as an *artificial variable.* It may be of interest to interpret this artificial variable, this *best discriminator,* in terms of *its relationship to the original and observable response variables.* The correlation of the *artificial* variable versus each *observable* one seems to be a useful index for the purpose of describing the best discriminator. For example, if it is large (in absolute value) against only some of the response variables, and low against the others, we may conclude that the former response variables contribute essentially to the discrimination between the defined groups. The correlation vector may thus be used to order or select response variables in terms of their contributions to the best discriminator. It should be noted that, in order to use a once established discriminant function for future classification, it is necessary to measure, for each of the new experimental units, *all* the original response variables. If this is impractical, e.g., if the number of variables is very large, it would be preferable to use only those response variables which correlate most strongly with the best discriminator. A new discriminant function may then be established on the basis of the reduced set of variables.

In the special case of just two groups one would, intuitively, select those variables which show the largest standardized mean difference. As shown below, the vector of correlations of the best discriminator versus the original response variables is, indeed, proportional to the vector of standardized mean differences in the case of two groups.

Let $\hat{\boldsymbol{\alpha}}'\boldsymbol{y}$ denote the discriminant function as defined in (3.11a). Then the vector of correlations of this single variable versus the original response variables, $\boldsymbol{y}$, is

$$\boldsymbol{\rho} = [\text{diag}\,(\boldsymbol{\Sigma})]^{-1/2}\,\boldsymbol{\Sigma}\hat{\boldsymbol{\alpha}}/\sqrt{\hat{\boldsymbol{\alpha}}'\boldsymbol{\Sigma}\hat{\boldsymbol{\alpha}}} \tag{5.1}$$

where $diag(\boldsymbol{\Sigma})$ denotes the diagonal matrix with principal elements equal to the diagonal elements of $\boldsymbol{\Sigma}$. The maximum-likelihood estimate of this correlation vector (assuming that means were also estimated by maximum-likelihood) is

$$\boldsymbol{r} = [\text{diag}\,(\boldsymbol{E})]^{-1/2}\,\boldsymbol{Ea}/\sqrt{\boldsymbol{a}'\boldsymbol{Ea}}, \tag{5.2}$$

with $\boldsymbol{E}$ defined by (3.6a) and $\boldsymbol{a}$ by (3.12).

If we set

$$\boldsymbol{Ea} = \boldsymbol{u}, \tag{5.3}$$

then it follows from (3.12) that

$$\boldsymbol{HE}^{-1}\,\boldsymbol{u} = c^2\,\boldsymbol{u} \tag{5.4}$$

and from (5.2) that the vector with elements $u_i/\sqrt{e_{ii}}$ is proportional to the vector of correlations. Formally,

$$\boldsymbol{r} = [\text{diag}\,(\boldsymbol{E})]^{-1/2}\,\boldsymbol{u}/\sqrt{\boldsymbol{u}'\boldsymbol{E}^{-1}\,\boldsymbol{u}} \quad . \tag{5.5}$$

It will be noted that $\boldsymbol{u}$ is the eigenvector associated with the largest root of $\boldsymbol{HE}^{-1}$, i.e., the transpose of $\boldsymbol{E}^{-1}\boldsymbol{H}$, since both $\boldsymbol{E}$ and $\boldsymbol{H}$ are symmetrical. In the special case of two groups, the vector $\boldsymbol{u}$ is proportional to $(\bar{\boldsymbol{y}}_1-\bar{\boldsymbol{y}}_2)$, which can be established at once if one multiplies (4.2) by $\boldsymbol{E}^{-1}(\bar{\boldsymbol{y}}_1-\bar{\boldsymbol{y}}_2)$.

Thus, it follows from (5.5) that, for the special case of two groups, the vector of correlations of the discriminant function versus all response variables is proportional to a vector $\boldsymbol{u}$ with elements $(\bar{y}_{1i}-\bar{y}_{2i})/\sqrt{e_{ii}}$ $(i = 1, 2, \ldots, p)$, the standardized mean difference, as asserted in the beginning of this section.

Numerical Example 5.1

The following data was extracted from a study on mentally retarded children [12] (for more detailed analysis, see [2]). The experimental units were 72 students from public schools, and 108 students from special schools. Twelve variables have been selected for illustration. They are

y_1	Binet Mental Age
y_2	Primary Mental Abilities (PMA) Mental Age
y_3	*Achievement*, Paragraph meaning
y_4	*Achievement*, Word meaning
y_5	*Achievement*, Spelling
y_6	*Achievement*, Arithmetic reasoning
y_7	*Achievement*, Arithmetic computation
y_8	*Gain*, Paragraph meaning (*grade equiv. per year*)
y_9	*Gain*, Word meaning
y_{10}	*Gain*, Spelling
y_{11}	*Gain*, Arithmetic reasoning
y_{12}	*Gain*, Arithmetic computation

TABLE 5.1 : Means and differences

Variable	Public School	Special School	Difference
1	110.0556	106.2685	3.7870
2	112.2847	106.7685	5.5162
3	3.3292	2.5917	.7375
4	3.2569	2.5768	.6801
5	3.6389	2.7954	.8435
6	3.5139	2.8426	.6713
7	3.8569	3.4898	.3671
8	.2944	.3463	−.0518
9	.2625	.3944	−.1319
10	.2806	.3231	−.0426
11	.2833	.4426	−.1592
12	.3583	.3981	−.0398

The discriminant function (in this case simply the solution of the equation system using Table 5.2 as coefficients and the *Difference* column in Table. 5.1 as the right hand side) is, with an arbitrary multiplier,

$$z = \cdot 36y_1 - \cdot 71y_2 - 40\cdot 78y_3 + 36\cdot 34y_4 - 25\cdot 71y_5 - 25\cdot 86y_6 + 20\cdot 87y_7 \\ - 12\cdot 13y_8 + 58\cdot 15y_9 + 3\cdot 53y_{10} + 19\cdot 12y_{11} + 10\cdot 90y_{12}.$$

The *correlation vector* (obtained by (5.2)) is

$$\boldsymbol{r}' = [\cdot 30, \cdot 43, \cdot 70, \cdot 66, \cdot 69, \cdot 65, \cdot 32, -\cdot 08, -\cdot 28, -\cdot 08, -\cdot 32, -\cdot 07]$$

It is thus seen that variables y_3, y_5, y_4, and y_6 (*achievement scores* except arithmetic computation) have the largest correlations with the function of the test scores which discriminates best between students in public and special schools. The *weights* of the discriminant function, even if standardized (divided by the square root of diagonal elements in Table 5.2), would, of course, not permit selection of variables.

6. CONCLUDING REMARKS

Since, except in trivial cases, equation (3.12) will have more than one characteristic root, the question arises whether the eigenvector associated with the smaller roots could be used (in conjunction with the correlation interpretation (5.5)) for further interpretation of discrimination between groups. We suggest that such a formalistic procedure would not be particularly helpful. The linear combination which produces maximum non-centrality between groups has an unequivocal meaning, and may be useful for variable selection and interpretation, but *it is and remains an artificial variable, a mathematical construct.* The corresponding linear combinations associated with the other stationary values of (3.12) could be interpreted as *second best* only in the sense that they would be the best discriminators if, from each variable, *the contribution to the first discriminant function were eliminated or partialed out.* It seems artificial, if interpretation is what we seek, to eliminate effects due to a *mathematical construct.*

Rather, if hierarchial selection is desired, it would be more appropriate to find the best discriminator of the full set of response variables and, based upon a study of the correlations (5.5), to eliminate those variables which were identified as the strongest contributors to the discriminant function. We may then proceed to compute the best discriminant function for the remaining set

TABLE 5.2: Sums of squares and products within

	1	2	3	4	5	6	7	8	9	10	11	12
1	31175.00	23126.00	967.90	1011.90	977.00	1567.10	1638.70	172.90	48.90	—1.40	25.40	10.00
2		32364.00	1055.00	1016.10	1023.10	1582.20	1658.90	53.30	7.70	29.60	—149.20	91.40
3			220.03	181.20	181.27	157.37	120.36	—15.14	19.46	14.62	5.23	14.63
4				209.09	196.06	153.00	114.73	18.87	—0.48	13.94	2.63	9.35
5					300.48	170.97	145.57	23.34	21.29	8.84	12.25	8.21
6						213.15	188.88	15.50	15.76	8.04	—8.89	6.88
7							257.46	13.58	16.16	5.16	9.33	—6.44
8								74.77	10.27	9.95	8.83	1.26
9									43.93	11.57	9.28	8.03
10										52.84	5.33	2.20
11											49.98	7.22
12												58.89

of response variables and select from them the second best determining set. In this manner we would eliminate, successively, *physical variables*, rather than *mathematical constructs*. If we follow this procedure we gain the additional advantage that we are not restricted by the number of characteristic roots in (3.12). We may discontinue the process as soon as we decide that the majority of the response variables have been assigned to the hierarchal strata. In using such a set of discriminant functions for future classification, we need to observe only the first set of variables in the first stage, and postpone the taking of additional measurements until a more precise classification is desired.

APPENDIX

Algorithm for Estimation of Confidence Intervals on Non-Centrality Parameters

Let γ_L^2 denote the lower limit and γ_U^2 the upper limit of the confidence interval on γ^2. Steps 1 through 8 below apply to both upper and lower bounds but, in steps 9 through 18, only the formulas appropriate for lower bounds are given, those for the upper bound follow by analogy.

1. Compute the statistic F from the data.
2. Compute $w = mF/n+1$
3. Compute $a = \sqrt{(m+n-2)/(n-2)}$
4. Compute w/a
5. Find $z = \cosh^{-1}(w/a)$
6. Find $\Phi^{-1}(1-\alpha/2)$, where $\alpha = 1-$(conf. coeff.) and Φ^{-1} denotes the percentage point of the standard normal distribution. Compute $c = \sqrt{2/(n-4)}\,\Phi^{-1}(1-\alpha/2)$.
7. Compute $z_L = z-c$ and $z_U = z+c$.
   Look up $\cosh z_L$ and $\cosh z_U$.
8. Compute $k = \sqrt{(m+n-2)(n-2)}$ and compute $1/k$
9. Compute $\gamma_{L_0}^2 = k[\cosh(z_L)-a]$

10. Compute $b = \gamma^2_{L_0}/k$

11. Compute $v = \gamma^2_{L_0}/\sqrt{(m+n-2)}$ (Record $\sqrt{(m+n-2)}$ for later reference).

12. Compute $s = \sqrt{v^2+m+2\gamma^2_{L_0}}$ and compute s^3.

13. Compute $u = [v+\sqrt{(m+n-2)}]/s$

14. Compute $y = z_L+u/(n-4)$, and look up $\sinh(y)$ and $\cosh(y)$,

15. Compute $d = a+b-\cosh(y)$

16. Compute $e = [(n-2)\sinh(y)]/[(n-4)s^3\sqrt{(m+n-2)}]$

17. Compute $f = d/(1/k+e)$

18. Compute $\gamma^2_{L_1} = \gamma^2_{L_0}-f$.

This is a Newton iteration. Steps 10 to 18 are repeated, using a new value $\gamma^2_{L_{i+1}}$ $(i = 0, 1, 2, \ldots)$ in place of $\gamma^2_{L_0}$, until the desired accuracy for γ^2_L is obtained (usually, until the quantity d in step 15 equals zero to the desired number of places). For the upper bound of the confidence interval use z_U instead of z_L in step 9 to find $\gamma^2_{U_0}$. Then use $\gamma^2_{U_i}$ in place of $\gamma^2_{L_i}$ $(i = 0, 1, 2, \ldots)$ in steps 10 to 18.

Numerical example (One-way analysis of variance, 5 groups, 11 observations per group)

Source	*SS*	*d.f.*	*MS*
Between	178·0728	4	44·5182
Within	198·3636	50	3·9673
Total	376·4364	54	

$(m = 4, \quad n = 50)$

Lower limit. Iteration 1

1. $F = 11{\cdot}2213$

2. $w = 1{\cdot}8977$

3. $a = 1{\cdot}0408$

4. $w/a = 1{\cdot}8233$

5. $z = 1{\cdot}2083$

6. Let $\alpha = {\cdot}10$, then $\phi^{-1}({\cdot}95) = 1{\cdot}6449$, $c = {\cdot}3430$

7. $z_L = {\cdot}8653$, $z_U = 1{\cdot}5513$

   $\cosh(z_L) = 1{\cdot}3983$, $\cosh(z_U) = 2{\cdot}4648$

8. $k = 49{\cdot}96$, $1/k = {\cdot}0200$

9. $\gamma^2_{L_0} = 17{\cdot}8607$

10. $b = {\cdot}3575$

11. $v = 2{\cdot}4768$, $\sqrt{m+n-2} = 7{\cdot}2111$

12. $s = 6{\cdot}7717$, $s^3 = 310{\cdot}52$

13. $u = 1{\cdot}4306$

14. $y = {\cdot}8964$, $\sinh(y) = 1{\cdot}0214$, $\cosh(y) = 1{\cdot}4296$

15. $d = -{\cdot}0311$

16. $e = {\cdot}0005$

17. $f = -1{\cdot}5171$

18. $\gamma^2_{L_1} = 19{\cdot}3778$

After the second iteration, d was zero to 4 places, showing the usual fast convergence of the Newton method especially if, as in the present case, a good starting value is easily obtained (the 0 values were obtained from the unimproved variance stabilizing transformation).

The upper value, again after two iterations, stabilized at 73·8445. The 90% confidence interval on γ^2 is thus

$$[19{\cdot}38 < \gamma^2 < 73{\cdot}84].$$

We went into so much detail in presenting the non-centrality parameter in order to convince the reader of the importance, meaning, and ease of accessibility of this parameter.

References

[1] Anderson, T. W. *An Introduction to Multivariate Statistical Analysis,* John Wiley and Sons, New York, 1958.

[2] Bargmann, R. E. "Representative Ordering and Selection of Variables" (2 vols), U.S. Office of Education, H. E. W., Coop. Research Project No. 1132, 1962.

[3] Graybill, F. A. *An Introduction to Linear Statistical Models,* McGraw Hill, New York, 1961.

[4] Hayslett, H. T., and R. E. Bargmann, "Confidence Interval Estimation, A Summary of Operations," Tech. Report No. 5, Res. spons. by Nat. Inst. of Health, Va. Polyt. Inst., Blacksburg, Va., 1961.

[5] Heck, D. L. "Charts of Some Upper Percentage Points of the Distribution of the Largest Characteristic Root," *Ann. Math. Statist.,* **31** (1960), 625–642.

[6] Hotelling, H. "Relations Between Two Sets of Variates," *Biometrika,* **28,** (1936), 321–377.

[7] Hotelling, H. "New Light on the Correlation Coefficient and its Transforms," *J. Roy. Statist. Soc.,* Ser. B, **15** (1953), 193–232.

[8] Kempthorne, O. *The Design and Analysis of Experiments,* John Wiley and Sons, New York, 1952.

[9] Posten, H. O., and R. E. Bargmann. "Power of the Likelihood-Ratio Test of the General Linear Hypothesis in Multivariate Analysis," *Biometrika,* **51** (1964), 467–480.

[10] Rao C. R. *Advanced Statistical Methods in Biometric Research,* John Wiley and Sons, New York, 1952.

[11] Roy, S. N. *Some Aspects of Multivariate Analysis,* John Wiley and Sons, New York, 1957.

[12] Thurstone, T. G. "An Evaluation of Educating Mentally Handicapped Children in Special Classes and Regular Classes." U.S. Office of Education, H.E.W., Coop. Research Project No. 168(6452), 1960.

(*Received Jan. 25, 1966. Revised Jan. 30. 1967.*)

CHAPTER 4

On Sufficiency and Invariance

D. BASU*, *University of Chicago and Indian Statistical Institute*

1. SUMMARY

Let $(\mathscr{X}, \mathscr{A}, \mathscr{P})$ be a given statistical model and let $\mathscr{G}$ be the class of all one-to-one, bimeasurable maps g of $(\mathscr{X}, \mathscr{A})$ onto itself such that g is measure-preserving for each $P \epsilon \mathscr{P}$, i.e. $Pg^{-1} = P$ for all P. Let us suppose that there exists a least (minimal) sufficient sub-field $\mathscr{L}$. Then, for each $L \epsilon \mathscr{L}$, it is true that $g^{-1}L$ is $\mathscr{P}$-equivalent to L for each $g \epsilon \mathscr{G}$, i.e., the least sufficient sub-field is almost $\mathscr{G}$-invariant. It is demonstrated that, in many familiar statistical models, the least sufficient sub-field and the sub-field of all almost $\mathscr{G}$-invariant sets are indeed $\mathscr{P}$-equivalent. The problem of data reduction in the presence of nuisance parameters has been discussed very briefly. It is shown that in many situations the principle of invariance is strong enough to lead us to the standard reductions. For instance, given n independent observations on a normal variable with unknown mean

* This research was supported by Research Grant No. NSF GP-3707 from the Devision of Mathematical, Physical and Engineering Sciences of the National Science Foundation, and by the Statistics Branch, Office of Naval Research.

(the nuisance parameter) and unknown variance, it is shown how the principle of invariance alone can reduce the data to the sample variance.

2. DEFINITIONS AND PRELIMINARIES

(a) The basic probability *model* is denoted by $(\mathscr{X}, \mathscr{A}, \mathscr{P})$, where $\mathscr{X} = \{x\}$ is the sample space, $\mathscr{A} = \{A\}$ the σ-field of events and $\mathscr{P} = \{P\}$ the family of probability measures.

(b) By *set* we mean a typical member of $\mathscr{A}$. By *function* (usually denoted by f) we mean a measurable mapping of $(\mathscr{A}, \mathscr{A})$ into the real line.

(c) A set A is *$\mathscr{P}$-null* if $P(A) = 0$ for all $P \epsilon \mathscr{P}$. Two sets A_1 and A_2 are *$\mathscr{P}$-equivalent* if their symmetric difference is $\mathscr{P}$-null. Two functions f_1 and f_2 are $\mathscr{P}$-equivalent if the set of points where they differ is $\mathscr{P}$-null. The relation symbol $\sim$ stands for $\mathscr{P}$-equivalence.

(d) By *sub-field* we mean a sub-σ-field of $\mathscr{A}$. A *statistic* is a measurable mapping of $(\mathscr{X}, \mathscr{A})$ into any measurable space. We identify a statistic with the sub-field it induces (see pp. 36–39 of [6]).

(e) By the *$\mathscr{P}$-completion* $\bar{\mathscr{A}}_0$ of a sub-field $\mathscr{A}_0$ we mean the least sub-field that contains $\mathscr{A}_0$ and all $\mathscr{P}$-null sets. Observe that $\bar{\mathscr{A}}_0$ may also be characterized as the class of all sets that are $\mathscr{P}$-equivalent to some member of $\mathscr{A}_0$. Two sub-fields are $\mathscr{P}$-equivalent if they have identical $\mathscr{P}$-completions.

(f) By *transformation* (usually denoted by g) we mean a one-to-one, bimeasurable mapping g of $(\mathscr{X}, \mathscr{A})$ onto itself such that the family

$$\mathscr{P} g^{-1} = \{Pg^{-1} \mid P \epsilon \mathscr{P}\}$$

of induced probability measures is the same as the family $\mathscr{P}$. A transformation g is called *model-preserving* if

$$Pg^{-1} \equiv P, \text{ for all } P \epsilon \mathscr{P}.$$

Observe that, if g is any transformation, then so also is g^n for each integral (positive or negative) n and that the identity map is always model-preserving. Also observe that any transformation carries $\mathcal{P}$-null sets into $\mathcal{P}$-null sets.

(g) Given a transformation g, the sub-field $\mathcal{A}(g)$ of *g-invariant* sets is defined as

$$\mathcal{A}(g) = \{A \mid g^{-1}A = A\}.$$

The $\mathcal{P}$-completion $\overline{\mathcal{A}}(g)$ of $\mathcal{A}(g)$ is then the class of all *essentially g-invariant* sets, i.e., sets that are $\mathcal{P}$-equivalent to some g-invariant set.

(h) The set A is *almost g-invariant* if $g^{-1}A \sim A$. It is easy to demonstrate that every almost g-invariant set is also essentially g-invariant and vice versa (see Lemma 1 for a sharper result). Thus, $\overline{\mathcal{A}}(g)$ is also the class of all almost g-invariant sets.

(i) Given a class $\mathcal{G}$ of transformations g, the three sub-fields of α) *$\mathcal{G}$-invariant*, β) *essentially $\mathcal{G}$-invariant* and γ) *almost $\mathcal{G}$-invariant* sets are defined as follows :

α) $\mathcal{A}(\mathcal{G}) = \bigcap \mathcal{A}(g)$, ($\mathcal{G}$-invariant)

β) $\overline{\mathcal{A}}(\mathcal{G}) = \mathcal{P}$-completion of $\mathcal{A}(\mathcal{G})$, (essentially $\mathcal{G}$-invariant)

and

γ) $\tilde{\mathcal{A}}(\mathcal{G}) = \bigcap \overline{\mathcal{A}}(g)$, (almost $\mathcal{G}$-invariant).

Observe that $\mathcal{A}(\mathcal{G}) \subset \overline{\mathcal{A}}(\mathcal{G}) \subset \tilde{\mathcal{A}}(\mathcal{G})$. With some assumptions on $\mathcal{G}$, one can prove (see Theorem 4 on p. 225 in [6]) the equality of $\overline{\mathcal{A}}(\mathcal{G})$ and $\tilde{\mathcal{A}}(\mathcal{G})$. That $\overline{\mathcal{A}}(\mathcal{G})$ can be a very small sub-field compared to $\tilde{\mathcal{A}}(\mathcal{G})$ is shown in example 1.

(j) A function f is α) *$\mathcal{G}$-invariant*, β) *essentially $\mathcal{G}$-invariant*, or γ) *almost $\mathcal{G}$-invariant*, according as

α) $f = f(g)$, for all $g \in \mathcal{G}$,

β) $f \sim$ some $\mathcal{G}$-invariant function,

or

γ) $f \sim f(g)$, for all $g \in \mathcal{G}$.

Observe that f satisfies the definitions α), β) or γ) above if and only if f is measurable with respect to the corresponding sub-field defined in (i).

(k) When the class $\mathcal{G}$ of transformations g happens to be a group (with respect to the operation of composition of transformations), the sub-field $\mathcal{A}(\mathcal{G})$ of $\mathcal{G}$-invariant sets is easily recognized as follows. For each $x \in \mathcal{X}$, define the *orbit* O_x as

$$O_x = \{x' \mid x' = gx \text{ for some } g \in \mathcal{G}\}.$$

The orbits define a partition of $\mathcal{X}$, and the sub-field $\mathcal{A}(\mathcal{G})$ is the class of all (measurable) sets that are the unions of orbits. The sub-field of essentially $\mathcal{G}$-invariant sets is then the $\mathcal{P}$-completion of $\mathcal{A}(\mathcal{G})$. Our main concern, in this paper, is the sub-field $\tilde{\mathcal{A}}(\mathcal{G})$ and this is not so easily understood in terms of the orbits—unless $\mathcal{G}$ has a structure simple enough to ensure the equality of $\bar{\mathcal{A}}(\mathcal{G})$ and $\tilde{\mathcal{A}}(\mathcal{G})$ (see lemma 1 and example 1).

3. A MATHEMATICAL INTRODUCTION

Let $(\mathcal{X}, \mathcal{A}, \mathcal{P})$ be a given probability model and let g be a fixed model-preserving transformation [definition 2(f)]. We remark that the identity map is trivially model-preserving. In many instances of statistical interest there exist fairly wide classes of such transformations. However, it is not difficult to construct examples where no non-trivial transformation is model-preserving. See for instance example 5.

For each bounded function f [definition 2(b)], define the associated sequence $\{f_n\}$ of (uniformly bounded) functions as follows :

$$f_n(x) = [f(x)+f(gx)+\ldots+f(g^n x)]/(n+1), \qquad n = 1, 2, 3, \ldots.$$

Since g is measure-preserving for each $P \in \mathcal{P}$, the pointwise ergodic theorem tells us that the set N_f where $\{f_n(x)\}$ fails to converge is $\mathcal{P}$-null [definition 2(c)].

If we define f^* as

$$f^*(x) = \begin{cases} \lim f_n(x) & \text{when } x \notin N_f, \\ 0 & \text{otherwise} \end{cases}$$

then, it is easily seen that

i) f^* is $\mathcal{A}(g)$-measurable [definition 2(g)], i.e., it is g-invariant [definition 2(i)], and that

ii) for all $B \epsilon \mathcal{A}(g)$ and $P \epsilon \mathcal{P}$

$$\int_B f dP = \int_B f^* dP.$$

This is another way of saying that f^* is the conditional expectation of f, given the sub-field $\mathcal{A}(g)$ of g-invariant sets. Since the definition of f^* does not involve P, we have the following theorem (lemma 2 in [4]).

Theorem 1. *If the transformation g preserves the model $(\mathcal{X}, \mathcal{A}, \mathcal{P})$, then the sub-field $\mathcal{A}(g)$ is sufficient.*

[*Remark* : Note that in the proof of theorem 1 we have not used the assumptions that g is a one-to-one map and that it is bimeasurable. The proof remains valid for any measurable mapping of $(\mathcal{X}, \mathcal{A})$ into itself that is measure-preserving for each $P \epsilon \mathcal{P}$. A similar remark will hold true for a number of other results to be stated later. However, in a study of the statistical theory of invariance (see [6]) it seems appropriate to restrict our attention to one-to-one, bimeasurable maps of $(\mathcal{X}, \mathcal{A})$ onto itself that preserve the model either wholly or partially.]

Now, given a class $\mathcal{G}$ of model-preserving transformations g, what can we say about the sufficiency of the sub-field

$$\mathcal{A}(\mathcal{G}) = \bigcap \mathcal{A}(g)$$

of $\mathcal{G}$-invariant [definition 2(i)] sets ? The intersection of two sufficient sub-fields is not necessarily sufficient. However, it is known (see Theorem 4 and Corollary 2 of [3]) that the intersection of the $\mathcal{P}$-completions [definition 2(e)] of a countable number of sufficient sub-fields is sufficient. Using this result, we have the following theorem (theorem 2 in [4]).

Theorem 2. *If the class $\mathcal{G}$ of model-preserving transformations g is countable, then the sub-field*

$$\tilde{\mathcal{A}}(\mathcal{G}) = \bigcap \bar{\mathcal{A}}(g)$$

of almost $\mathcal{G}$-invariant sets [definition 2(i)(γ)] *is sufficient.*

Given a countable class $\mathcal{G}$ of transformations, consider the larger class $\mathcal{G}^*$ of transformations of the type $\alpha_1\alpha_2 \ldots \alpha_n$, where each α_i is such that either α_i or α_i^{-1} belongs to $\mathcal{G}$, and n is an arbitrary positive integer. The following properties of $\mathcal{G}^*$ are easy to check :

a) $\mathcal{G}^*$ is a group (the group operation being composition of transformations),

b) $\mathcal{G}^*$ is countable,

c) $\mathcal{A}(\mathcal{G}^*) = \mathcal{A}(\mathcal{G})$ and $\tilde{\mathcal{A}}(\mathcal{G}^*) = \tilde{\mathcal{A}}(\mathcal{G})$.

Now, let A be an arbitrary almost $\mathcal{G}^*$-invariant set, i.e., $g^{-1}A \sim A$ (equivalently, $gA \sim A$) for all $g \in \mathcal{G}^*$. Consider the set

$$B = \bigcap gA$$

where the intersection is taken over all $g \in \mathcal{G}^*$. Since $\mathcal{G}^*$ is a group, the set B must be $\mathcal{G}^*$-invariant. Again, since $\mathcal{G}^*$ is countable, and each gA is $\mathcal{P}$-equivalent to A, we have $B \sim A$. We have thus established the following lemma.

Lemma 1. *For any countable class $\mathcal{G}$ of transformations* [*definition* 2(f)], *every almost $\mathcal{G}$-invariant set is essentially $\mathcal{G}$-invariant, i.e., $\tilde{\mathcal{A}}(\mathcal{G}) = \bar{\mathcal{A}}(\mathcal{G})$.*

Theorem 2, together with lemma 1 and the observation that a sub-field that is $\mathcal{P}$-equivalent to a sufficient sub-field is itself sufficient, leads to the following theorem.

Theorem 3. *If $\mathcal{G}$ is a countable class of model-preserving transformations, then the sub-field $\mathcal{A}(\mathcal{G})$ of $\mathcal{G}$-invariant sets is sufficient.*

[*Remark* : Note that in Theorem 3 we have used the one-to-oneness and bimeasurability of our transformations.]

Before proceeding further, let us consider an example which shows that Theorem 3 is no longer true if we drop the condition that $\mathcal{G}$ is countable.

Example 1. Let $\mathscr{P}$ be a family of continuous distributions on the real line $\mathscr{X}$ where $\mathscr{A}$ is the σ-field of Borel-sets. Let $\mathscr{G} = \{g\}$ be the class of all one-to-one maps of $\mathscr{X}$ onto $\mathscr{X}$ such that $\{x \mid gx \neq x\}$ is finite. Clearly, $\mathscr{G}$ is a group and every member of it is model-preserving. It is easy to check that the sub-field $\mathscr{A}(\mathscr{G})$ of $\mathscr{G}$-invariant sets consists of ϕ and $\mathscr{X}$ only, and hence is not sufficient. The sub-field $\widetilde{\mathscr{A}}(\mathscr{G})$ of almost $\mathscr{G}$-invariant sets is the same as $\mathscr{A}$ and is sufficient.

Consider another example.

Example 2. Let $\mathscr{X}$ be the n-dimensional Euclidean space and $\mathscr{P}$ the class of all probability measures (on the σ-field $\mathscr{A}$ of all Borel-sets) that are symmetric (in the co-ordinates). If $\mathscr{G} = \{g\}$ is the group of all permutations (of the co-ordinates) then $\mathscr{A}(\mathscr{G})$ is the sub-field of all sets that are symmetric (in the co-ordinates) and is sufficient. Since the empty set is the only $\mathscr{P}$-null set, the two sub-fields $\mathscr{A}(\mathscr{G})$ and $\widetilde{\mathscr{A}}(\mathscr{G})$ are the same here.

Given a probability model $(\mathscr{X}, \mathscr{A}, \mathscr{P})$ we ask ourselves the following questions.

1. How wide is the group $\mathscr{G}$ of all model-preserving transformations ?
2. Is the sub-field $\widetilde{\mathscr{A}}(\mathscr{G})$ of almost $\mathscr{G}$-invariant sets sufficient?
3. If a least (minimal) sufficient sub-field $\mathscr{L}$ exists, then is it true that $\mathscr{L} \sim \widetilde{\mathscr{A}}(\mathscr{G})$?
4. What is $\widetilde{\mathscr{A}}(\mathscr{G})$, when $\mathscr{G}$ is the class of all transformations that partially preserve the model in a given manner ?

4. STATISTICAL MOTIVATION

Suppose we have two different systems of measurement co-ordinates for the outcome of a statistical experiment, so that, if a typical outcome is recorded as x under the first system, the same outcome is recorded as gx under the second system. Let us suppose further that the two statistical variables x and gx

have the same domain (sample space) $\mathscr{X}$, the same family $\mathscr{A}$ of events, and the same class of probability measure $\mathscr{P}$. Furthermore, if $P \epsilon \mathscr{P}$ holds for x, then the same probability measure P also holds for gx. The second system of measurement co-ordinates may then be represented mathematically as a model-preserving transformation g of the statistical model $(\mathscr{X}, \mathscr{A}, \mathscr{P})$. The principle of invariance then stipulates that the decision rule should be invariant with respect to the transformation g—and this irrespective of the actual decision problem. If the choice of a new system of measurement co-ordinates leaves the problem (whatever it is) entirely unaffected, then so also should be every reasonable inference procedure.

Thus, if g is model-preserving, the principle of invariance leads to the invariance reduction of the model $(\mathscr{X}, \mathscr{A}, \mathscr{P})$ to the simpler model $(\mathscr{X}, \mathscr{A}(g), \mathscr{P})$, where $\mathscr{A}(g)$ is the sub-field of g-invariant sets. To put it differently, the principle of invariance requires every decision function to be g-invariant or $\mathscr{A}(g)$-measureable. Consider now the class $\mathscr{G}$ of all model-preserving transformations g. Must we, following the principle of invariance, insist that every decision rule be $\mathscr{G}$-invariant (i.e. g-invariant for every $g \epsilon \mathscr{G}$) ? We have already noted in example 1 that, when $\mathscr{P}$ consists of a family of non-atomic measures, the class $\mathscr{G}$ is large enough to reduce the sub-field $\mathscr{A}(\mathscr{G})$ of $\mathscr{G}$-invariant sets to the trivial one (consisting of only the empty set and the whole space). Obviously, we cannot (must not) reduce $\mathscr{A}$ all the way down to $\mathscr{A}(\mathscr{G})$ or even to $\bar{\mathscr{A}}(\mathscr{G})$—the sub-field of essentially $\mathscr{G}$-invariant sets. A logical compromise (with the principle) would be to reduce $\mathscr{A}$ to the sub-field $\tilde{\mathscr{A}}(\mathscr{G})$ of all almost $\mathscr{G}$-invariant sets—that is to insist upon the decision function to be almost $\mathscr{G}$-invariant.

The principle of sufficiency is another reduction principle of the omnibus type. If $\mathscr{S}$ be a sufficient sub-field, then this principle tells us not to use a decision function that is not $\mathscr{S}$-measurable. Suppose there exists a least (minimal) sufficient sub-field $\mathscr{L}$. Following the principle of sufficiency, we reduce the model $(\mathscr{X}, \mathscr{A}, \mathscr{P})$ to the model $(\mathscr{X}, \mathscr{L}, \mathscr{P})$.

Which of the two reductions (invariance or sufficiency) is more extensive ? In other words, what is the relation between $\tilde{\mathscr{A}}(\mathscr{G})$ and $\mathscr{L}$?

Theorem 1 tells us that $\mathscr{A}(g)$ is sufficient (for each g in $\mathscr{G}$). Since $\mathscr{L}$ is the least sufficient sub-field we have (from the definition of $\mathscr{L}$)

$$\bar{\mathscr{L}} \subset \bar{\mathscr{A}}(g) \text{ for all } g \in \mathscr{G}.$$

Theorem 4 follows at once.

Theorem 4. $$\bar{\mathscr{L}} \subset \bigcap \bar{\mathscr{A}}(g) = \tilde{\mathscr{A}}(\mathscr{G}).$$

[*Remark* : Theorem 4 does not establish the sufficiency of $\tilde{\mathscr{A}}(\mathscr{G})$. [See example 1 in [3].]

We thus observe that the invariance reduction (in terms of the group $\mathscr{G}$ of all model-preserving transformations) of a model can never be more extensive than its maximal sufficiency reduction (if one such reduction is available). The principal question raised in this paper is, "When is $\tilde{\mathscr{A}}(\mathscr{G})$ equal to $\bar{\mathscr{L}}$?" We shall show later that in many familiar situations the sub-fields $\tilde{\mathscr{A}}(\mathscr{G})$ and $\mathscr{L}$ are essentially equal. This raises the question about the nature of $\tilde{\mathscr{A}}(\mathscr{G})$, where $\underline{\mathscr{G}}$ is the class of all transformations that preserve the model partially in some well-defined manner. This question is discussed in a later section. In the next two sections we give two alternative approaches to Theorem 4.

5. WHEN A BOUNDEDLY COMPLETE SUFFICIENT SUB-FIELD EXISTS

Let us suppose that the sub-field $\mathscr{L}$ is sufficient and boundedly complete. We need the following lemma.

Lemma 2. *If z is a bounded $\mathscr{A}$-measurable function such that*

$$E(z\,|\,P) \equiv 0 \text{ for all } P \in \mathscr{P},$$

then, for all bounded $\mathscr{L}$-measurable functions f, it is true that

$$E(zf\,|\,P) \equiv 0 \quad \text{for all } P \in \mathscr{P}.$$

The proof of this well-known result is omitted. Now, let S be an arbitrary member of $\mathcal{L}$, and let g be an arbitrary model-preserving transformation. Let $S_0 = g^{-1}S$. Since g is model-preserving we have

$$P(S) \equiv Pg^{-1}S \equiv P(S_0) \quad \text{for all} \quad P \epsilon \mathcal{P}.$$

Writing I_S for the indicator of S, and noting that $I_S - I_{S_0}$ and I_S satisfy the conditions for z and f in Lemma 2, we at once have

$$E[(I_S - I_{S_0})I_S \mid P] \equiv 0, \quad \text{for all} \quad P \epsilon \mathcal{P},$$

or $$P(S) \equiv P(SS_0), \quad \text{for all} \quad P \epsilon \mathcal{P}.$$

Hence, $$\begin{aligned} P(S \Delta S_0) &\equiv P(S) + P(S_0) - 2P(SS_0) \\ &\equiv 2[P(S) - P(SS_0)] \\ &\equiv 0, \quad \text{for all } P \epsilon \mathcal{P}. \end{aligned}$$

That is, for all $S \epsilon \mathcal{L}$, the two sets S and $g^{-1}\ S$ are $\mathcal{P}$-equivalent. In other words,

$$\mathcal{L} \subset \bar{\mathcal{A}}(g),$$

and since g is an arbitrary element of $\mathcal{G}$ (the class of all model-preserving transformations) we have the following theorem.

Theorem 4 (a). *If $\mathcal{L}$ is a boundedly complete sufficient sub-field then* $\mathcal{L} \subset \tilde{\mathcal{A}}(\mathcal{G})$.

[*Remark* : Since bounded completeness of $\mathcal{L}$ implies that $\mathcal{L}$ is the least sufficient sub-field, Theorem 4(a) is nothing but a special case of Theorem 4. The proof of Theorem 4(a) is simple and amenable to a generalization to be discussed later.]

6. THE DOMINATED CASE

We make a slight digression to state a useful lemma. Let T be a measurable map of $(\mathcal{X}, \mathcal{A})$ into $(\mathcal{Z}, \mathcal{B})$. Let P and Q be two probability measures on $\mathcal{A}$ and let PT^{-1} and QT^{-1} be the corresponding measures on $\mathcal{B}$. Suppose that Q dominates P and let $f = dP/dQ$ be the Radon-Nikodym derivative defined on $\mathscr{X}$. It is then clear that QT^{-1} dominates PT^{-1}. Let $h = (dPT^{-1})/(dQT^{-1})$. The function hT on $\mathscr{X}$ (defined as $hT(x) = h(Tx)$) is $T^{-1}(\mathscr{B})$-measurable and satisfies the following relation.

Lemma 3. $$hT = E(f \mid T^{-1}(\mathscr{B}), Q),$$

i.e., hT is the conditional expectation of f, given $T^{-1}(\mathscr{B})$ and Q.

The proof of this well-known lemma consists of checking the identity

$$\int_B f dQ \equiv \int_B hT dQ, \quad \text{for all } B \epsilon T^{-1}(\mathscr{B}).$$

Corollary. *If $T^{-1}(\mathscr{B})$ is Q-equivalent to $\mathscr{A}$, then f and hT are Q-equivalent.*

Now, returning to our problem, let $\mathscr{G}$ be the class of all transformations g that preserves the model $(\mathscr{X}, \mathscr{A}, \mathscr{P})$. Let us suppose that $\mathscr{P}$ is dominated by some σ-finite measure. It follows that there exists a countable collection $P_1, P_2, \ldots$, of elements in $\mathscr{P}$ such that the convex combination

$$Q = \Sigma c_i P_i, \quad c_i > 0, \quad \Sigma c_i = 1,$$

dominates the family $\mathscr{P}$. Let $f_P = dP/dQ$ be a fixed version of the Radon-Nikodym derivative of P with respect to Q. The factorization theorem for sufficient statistics asserts that a subfield $\mathscr{A}_0$ is sufficient if and only if f_P is $\overline{\mathscr{A}_0}$-measurable for every $P \epsilon \mathscr{P}$, where $\overline{\mathscr{A}_0}$ is the $\mathscr{P}$-completion of $\mathscr{A}_0$. We now prove that f_P is $\widetilde{\mathscr{A}}(\mathscr{G})$-measurable for every P. (The $\mathscr{P}$-completion of $\widetilde{\mathscr{A}}(\mathscr{G})$ is itself.)

Since $Pg^{-1} = P$ for all $P \epsilon \mathscr{P}$, it follows that $Qg^{-1} = Q$, From Lemma 3 we have

$$f_P g = E(f_P \mid g^{-1}(\mathscr{A}), Q).$$

The assumption that g is one-to-one and bimeasurable implies that $g^{-1}(\mathscr{A}) = \mathscr{A}$. Hence $f_P g = f_P$ a.e.w. [Q]. Since Q dominates $\mathscr{P}$, it follows that f_P is almost g-invariant, i.e., is $\overline{\mathscr{A}(g)}$ measurable. Since, in the above argument, g is an arbitrary element of $\mathcal{G}$, we have the following theorem. (See problem 19 on p. 253 in [6].)

Theorem 4(b). *If $\mathcal{P}$ is dominated, then $\tilde{\mathcal{A}}(\mathcal{G})$ is sufficient.*

[*Remark* : Since, in the dominated set-up, the least sufficient sub-field $\mathcal{L}$ always exists and since, in this set-up, any sub-field containing $\mathcal{L}$ is necessarily sufficient, it is clear that Theorem 4(b) is nothing but an immediate corollary to Theorem 4. Also note that in the present proof (of Theorem 4(b)) we had to draw upon our supposition that g is one-to-one and bimeasurable. This section has been written only with the object of drawing attention to some aspects of the problem.]

7. EXAMPLES

Example 3 : Let y be a real random variable having a uniform distribution over the unit interval. For each c in $[0, 1)$ define the transformation g_c as

$$g_c y = y + c \pmod 1$$

In this example, $\mathcal{P}$ consists of a single measure and each g_c is model preserving (measure-preserving). If $\mathcal{G}_0 = \{g_c \mid 0 < c < 1\}$, then the only $\mathcal{G}_0$-invariant sets are the empty set and the whole of the unit interval. Here, $\mathcal{A}(\mathcal{G}_0)$ is sufficient.

[*Remark* : In this case, there are a very large number of measure-preserving transformations that are not one-to-one maps of $[0, 1]$ onto itself. For example, let $a_n(y)$ be the nth digit in the decimal representation of y and let

$$gy = \sum_{k=1}^{\infty} \frac{a_{n_k}(y)}{10^k}$$

where $\{n_k\}$ is a fixed increasing sequence of natural numbers.]

Again, if x has a fixed continuous distribution on the real line with cumulative distribution function F, then the class $\mathcal{G}_0$ of transformations g_c defined as

$$g_c x = F^{-1}[F(x) + c \pmod 1)], \quad 0 < c < 1$$

are all model-preserving for x. [In case F is not a strictly increasing function of x, we define $F^{-1}(y) = \inf \{x \mid F(x) = y\}$.]

Thus, for any fixed continuous distribution on the real line, there always exists a large class of measure-preserving transformations.

Example 4. Let $\mathscr{X} = [0, \infty)$ and let x have a uniform distribution over the interval $[0, \theta)$, where θ is an unknown positive integer. Here $\mathscr{P}$ consists of a countable infinity of probability measures. For each c in $[0, 1)$, define the transformation g_c as

$$g_c x = [x] + \{x - [x] + c \pmod 1\},$$

where $[x]$ is the integer-part of x.

Here, each g_c is model-preserving. The minimal (least) sufficient statistic $[x]$ is also the maximal-invariant with respect to the group $\mathscr{G}_0 = \{g_c\}$ of model-preserving transformations defined above. Thus, if $\mathscr{L}$ is the sub-field (least sufficient) generated by $[x]$ and $\mathscr{A}(\mathscr{G}_0)$ is the sub-field of $\mathscr{G}_0$-invariant sets, we have

$$\mathscr{L} = \mathscr{A}(\mathscr{G}_0). \tag{a}$$

Now, if $\mathscr{G}$ is the class of all model-preserving transformations, then (as we have seen in Example 1) the sub-field $\mathscr{A}(\mathscr{G})$ will reduce to the level of triviality. (It will consist of only the empty set and the whole set $\mathscr{X}$.) However, from Theorem 4 we have

$$\bar{\mathscr{L}} \subset \tilde{\mathscr{A}}(\mathscr{G}). \tag{b}$$

Since the group $\mathscr{G}_0$ has a *decent* structure, we can apply Stein's theorem (theorem 4 on p. 225 in [6]) to prove that

$$\mathscr{A}(\mathscr{G}_0) \sim \tilde{\mathscr{A}}(\mathscr{G}_0). \tag{c}$$

Since $\mathscr{G}_0 \subset \mathscr{G}$, we at once have

$$\tilde{\mathscr{A}}(\mathscr{G}) \subset \tilde{\mathscr{A}}(\mathscr{G}_0). \tag{d}$$

Putting the relations (a), (c), and (d) together we have

$$\tilde{\mathscr{A}}(\mathscr{G}) \subset \bar{\mathscr{L}} \tag{e}$$

Putting (b) and (e) together we finally have

$$\bar{\mathscr{L}} = \tilde{\mathscr{A}}(\mathscr{G}),$$

i.e., the least-sufficient sub-field (rather, the $\mathscr{P}$-completion of any version of the least sufficient sub-field) and the sub-field of almost $\mathscr{G}$-invariant sets are identical.

The chain of arguments, detailed as above, is of a general nature and will be used repeatedly in the sequel.

Example 5. Let x be a normal variable with unit variance and mean equal to either μ_1 or μ_2. Does there exist a non-trivial transformation of $\mathscr{X}$ (the real line) into itself that preserves each of the two measures ? That the answer is "no" is seen as follows. Let $\mathscr{G}$ be the class of all model-preserving transformations. In view of theorem 4, $\tilde{\mathscr{A}}(\mathscr{G})$ contains the least sufficient sub-field $\mathscr{L}$. But, in this example, the likelihood ratio (which is the least sufficient statistic) is

$$\exp\left[(\mu_1-\mu_2)x - \frac{1}{2}(\mu_1^2-\mu_2^2)\right],$$

and this is a one-to-one measurable function of x. Thus, every set is almost $\mathscr{G}$-invariant. And this implies that every g in $\mathscr{G}$ must be equivalent to the identity map.

Example 6. Let $x_1, x_2, \ldots, x_n$ be n independent observations on a normal variable with known mean and unknown standard deviation σ. Without loss of generality we may assume the mean to be zero. Let $\mathscr{G}_0$ be the group of all linear orthogonal transformations of the n-dimentional Euclidean space onto itself. Clearly, every member of $\mathscr{G}_0$ is model-preserving. That the class $\mathscr{G}$ of all model-preserving transformations is much wider than $\mathscr{G}_0$ is seen as follows. Let

$$y_i = \phi_i(x_i)\,|x_i|, \quad i = 1, 2, \ldots, n,$$

where $\phi_1, \phi_2, \ldots, \phi_n$ are arbitrary skew-symmetric (i.e., $\phi(x) = -\phi(-x)$ for all x) functions on the real line that take only the two values -1 and $+1$. It is easily checked that, whatever the value of σ, the two vectors $(x_1, x_2, \ldots, x_n)$ and $(y_1, y_2, \ldots, y_n)$ are identically distributed, i.e., the above transformation (though non-linear) is model-preserving. However, the sub-group $\mathcal{G}_0$ is large enough to lead us to Σx_1^2—which is the least sufficient statistic—as the maximal invariant. In view of the decent structure of the sub-group $\mathcal{G}_0$, the arguments given for example 4 are again available to prove the equality of $\bar{\mathcal{L}}$ and $\tilde{\mathcal{A}}(\mathcal{G})$.

8. TRANSFORMATIONS OF A SET OF NORMAL VARIABLES

This section is devoted to a study of the special case (model) of n independent normal variables $x_1, x_2, \ldots, x_n$ with equal unknown means μ and equal unknown standard deviations σ. Hence Σx_i and Σx_i^2 jointly constitute the least sufficient statistic. If $\bar{\mathcal{L}}$ is the $\mathcal{P}$-completion of the sub-field $\mathcal{L}$ induced by $(\Sigma x_i, \Sigma x_i^2)$, then we know from Theorem 4, that

$$\bar{\mathcal{L}} \subset \tilde{\mathcal{A}}(\mathcal{G})$$

where $\mathcal{G}$ is the class of all the model-preserving transformations of $\boldsymbol{x} = (x_1, x_2, \ldots, x_n)$ to $\boldsymbol{y} = (y_1, y_2, \ldots, y_n)$. For any model-preserving transformation from $\boldsymbol{x}$ to $\boldsymbol{y}$, we, therefore, have

$$\Sigma x_i \sim \Sigma y_i$$

and

$$\Sigma x_i^2 \sim \Sigma y_i^2$$

i.e., the statistic $(\Sigma x_i, \Sigma x_i^2)$ is almost $\mathcal{G}$-invariant. If we can demonstrate the existence of a '*decent*' sub-group $\mathcal{G}_0$ of $\mathcal{G}$ for which the statistic $(\Sigma x_i, \Sigma x_i^2)$ is the maximal invariant, then (following the method of proof indicated in examples 4 and 6) we can show that $\bar{\mathcal{L}}$ is indeed equal to $\tilde{\mathcal{A}}(\mathcal{G})$.

Let $\mathcal{G}_0$ be the sub-group of all linear model-preserving transformations. Do there exist non-trivial linear model-preserving transformations (i.e., linear transformations that are not a permutation of the co-ordinates) ? That the answer is "yes" is seen as follows. Let $\mathscr{M} = \{\boldsymbol{M}\}$ be the family of all orthogonal $n \times n$ matrices with the initial row as

$$\left(\frac{1}{\sqrt{n}}, \frac{1}{\sqrt{n}}, \ldots, \frac{1}{\sqrt{n}}\right).$$

Now, if $\boldsymbol{x} = (x_1, x_2, \ldots, x_n)$ are independent $N(\mu, \sigma)$'s and we define

$$\boldsymbol{y}' = \boldsymbol{M}\boldsymbol{x}'$$

(where $\boldsymbol{x}'$ is the corresponding column vector), then the y_i's are independent normal variables with equal standard deviation σ and with means as follows :

$$E(y_1) = \sqrt{n}\,\mu, \quad E(y_i) = 0, \qquad\qquad i = 2, 3, \ldots, n.$$

Thus, for an arbitrary pair of members $\boldsymbol{M}_1$ and $\boldsymbol{M}_2$ in $\mathscr{M}$, we note that $\boldsymbol{M}_1\boldsymbol{x}'$ and $\boldsymbol{M}_2\boldsymbol{x}'$ are identically distributed (whatever the values of μ and σ). Therefore, the two vectors $\boldsymbol{x}'$ and $\boldsymbol{M}_2^{-1}\boldsymbol{M}_1\boldsymbol{x}'$ are identically distributed for each μ and σ. In other words, the linear transformation defined by the matrix

$$\boldsymbol{M}_2^{-1}\,\boldsymbol{M}_1 \qquad (=\boldsymbol{M}_2'\boldsymbol{M}_1,$$

since $\boldsymbol{M}_2$ is orthogonal) is model-preserving for each pair $\boldsymbol{M}_1$, $\boldsymbol{M}_2$ of members from $\mathscr{M}$.

[*Remark* : Later on we shall have some use for this way of generating members of $\mathscr{G}_0$.]

For example, the 4×4 matrix

$$\begin{pmatrix} 1/2 & 1/2 & 1/2 & -1/2 \\ 1/2 & 1/2 & -1/2 & 1/2 \\ 1/2 & -1/2 & 1/2 & 1/2 \\ -1/2 & 1/2 & 1/2 & 1/2 \end{pmatrix}$$

defines a member of $\mathscr{G}_0$ for $n = 4$.

Let H be a typical $n \times n$ matrix that defines a model-preserving linear transformation. We are going to make a brief digression about the nature of H. From the requirement that each co-ordinate of

$$\boldsymbol{y}' = H\boldsymbol{x}'$$

has mean μ and standard deviation σ, we at once have that the elements in each row of H must add up to unity and that their squares also must add up to unity. From the mutual independence of the y_i's, it follows that the row vectors of H must be mutually orthogonal. Thus, H must be an orthogonal matrix with unit row sums. Now, for each model-preserving H, its inverse

$$H^{-1} \ (= H', \text{ since } H \text{ is orthogonal})$$

is necessarily also model-preserving and, thus, the columns of H must also add up to unity, etc.

[It is easily checked that the sub-group $\mathcal{G}_0 = \{H\}$ of linear model-preserving transformations of the n-space onto itself may also be characterized as the class of all linear transformations that preserve both the sum and the sum of squares of the co-ordinates. The author came to learn that G. W. Haggastrom of the University of Chicago had come upon these matrices from this point of view and had a brief discussion on them in an unpublished work of his. We call such matrices by the name Haggstrom-matrix.]

Going back to our problem, we have to demonstrate that the statistic $T = (\Sigma x_i, \Sigma x_i^2)$ is a maximal invariant with respect to the sub-group $\mathcal{G}_0 = \{H\}$ of model-preserving linear transformations. For this we have only to prove that if $\boldsymbol{a} = (a_1, a_2, \ldots, a_n)$ and $\boldsymbol{b} = (b_1, b_2, \ldots, b_n)$ are any two points in n-space such that $T(\mathrm{a}) = T(\mathrm{b})$ then there exists a Haggstrom-matrix H such that

$$\boldsymbol{b}' = H\boldsymbol{a}'.$$

Let M_1 and M_2 be two arbitrary members of $\mathcal{M}$—the class of all orthogonal $n \times n$ matrices with the leading row as $(1/\sqrt{n}, \ldots,$

$1/\sqrt{n}$). If $\boldsymbol{\alpha} = M_1\boldsymbol{a}'$ and $\boldsymbol{\beta} = M_2\boldsymbol{b}'$, then we have, from $T(\boldsymbol{a}) = T(\boldsymbol{b})$, that $\alpha_1 = \beta_1$ that and that

$$\sum_2^n \alpha_i^2 = \sum_2^n \beta_i^2.$$

It follows that there exists an $(n-1)\times(n-1)$ orthogonal matrix that will transform $(\alpha_2, \alpha_3, \ldots, \alpha_n)$ to $(\beta_2, \beta_3, \ldots, \beta_n)$. And this, in turn, implies that there exists an $n\times n$ orthogonal matrix K, with the first row as $(1, 0, 0, \ldots, 0)$, such that

$$\boldsymbol{\beta}' = K\boldsymbol{\alpha}'.$$

Thus, the transformation $M_2^{-1}KM_1$ takes $\boldsymbol{a}$ into $\boldsymbol{b}$. Now, note that, since K is an orthogonal matrix with the initial row as $(1, 0, 0, \ldots, 0)$, the matrix KM_1 is orthogonal with its initial row as $(1/\sqrt{n}, 1/\sqrt{n}, \ldots, 1/\sqrt{n})$, i.e. $KM_1 \in \mathscr{M}$. In other words, there exists a matrix of the form $M_2^{-1}M$ (with M_2 and M belonging to $\mathscr{M}$) which transforms $\boldsymbol{a}$ into $\boldsymbol{b}$. We have already noted that all such matrices belong to $\mathscr{G}_0$, and this completes our proof that $(\Sigma x_i, \Sigma x_i^2)$ is a maximal invariant with respect to $\mathscr{G}_0$.

In example 6 of the previous section, we considered the particular case of the foregoing problem where μ is known. Consider now the other particular case where σ is known and μ is the only unknown parameter. In this case Σx_i is the least sufficient statistic. Now, it is no longer possible to produce a sub-group $\mathscr{G}_0$ of linear model-preserving transformations such that Σx_i is the maximal invarinant with respect to $\mathscr{G}_0$. This is because every linear model-preserving transformation must necessarily be orthogonal and would thus preserve Σx_i^2 also. We proceed as follows.

Suppose (without loss of generality) that $\sigma = 1$. Let $\boldsymbol{y}' = M\boldsymbol{x}'$ where M is a fixed member of the class $\mathscr{M}$ of orthogonal $n\times n$ matrices with the initial row as $(1/\sqrt{n}, \ldots, 1/\sqrt{n})$. Observe that the y_i's are mutually independent and that $y_2, y_3, \ldots, y_n$ are standard normal variables. Let F stand for the cumulative distribution function of a standard normal variable. Let c_2,

$c_3, \ldots, c_n$ be arbitrary constants in [0, 1). If we define $z_1 = y_1$ and

$$z_i = F^{-1}[F(y_i)+c_i \pmod{1}], \qquad i = 2, 3, \ldots, n,$$

then $\boldsymbol{z} = (z_1, z_2, \ldots, z_n)$ has the same distribution as that of $\boldsymbol{y}$ and so it follows that $\boldsymbol{x}'$ and $M^{-1}\boldsymbol{z}'$ are identically distributed. Observe that we have described above a model-preserving transformation corresponding to each $(n-1)$-vector $(c_2, c_3, \ldots, c_n)$ with the c_i's in [0, 1). It is easly checked that Σx_i is a maximal invariant with respect to the above class of model-preserving transformations.

9. PARAMETER-PRESERVING TRANSFORMATIONS

Let $\gamma = \gamma(P)$ be the parameter of interest. That is, the experimenter is interested only in the characteristic $\gamma(P)$ of the measure P (that actually holds) and considers all other details about P to be irrelevent (nuisance parameters). We define a γ-preserving transformation as follows.

Definition. The transformation [see Definition 2(f)] g is γ-preserving if $\gamma(Pg^{-1}) \equiv \gamma(P)$, for all $P \in \mathscr{P}$.

If g is γ-preserving then so also is g^{-1}. The composition of any two γ-preserving transformations is also γ-preserving. Let $\mathscr{G}_\gamma$ be the group of all γ-preserving transformations.

In the particular case where $\gamma(P) = P$, the γ-preserving transformations are what we have so far been calling model-preserving. If $\mathscr{G}$ is the class of all model-preserving transformations, then note that $\mathscr{G} \subset \mathscr{G}_\gamma$ for every γ.

If γ is the parameter of interest, then the principle of invariance leads to the reduction of $\mathscr{A}$ to the sub-field $\widetilde{\mathscr{A}}(\mathscr{G}_\gamma)$. This section is devoted to a study of the sub-field $\widetilde{\mathscr{A}}(\mathscr{G}_\gamma)$.

Since $\mathscr{G} \subset \mathscr{G}_\gamma$ we have

$$\widetilde{\mathscr{A}}(\mathscr{G}_\gamma) \subset \widetilde{\mathscr{A}}(\mathscr{G}).$$

Let us suppose that the least sufficient sub-field $\mathscr{L}$ exists. We have discussed a number of examples where $\widetilde{\mathscr{A}}(\mathscr{G})$ and $\overline{\mathscr{L}}$ are identical. [The author believes that the above identification is true under very general conditions.] In all such situations we therefore have

$$\tilde{\mathscr{A}}(\mathscr{G}_\gamma) \subset \bar{\mathscr{L}}.$$

That is, when the interest of the experimenter is concentrated on some particular characteristic $\gamma(P)$ or P, the principle of invariance will usually reduce the data more extensively than the principle of sufficiency.

Theorems 4, 4(a) and 4(b) tell us that

$$\bar{\mathscr{L}} \subset \tilde{\mathscr{A}}(\mathscr{G}).$$

The following theorem gives us a similar lower bound for $\tilde{\mathscr{A}}(\mathscr{G}_\gamma)$.

A set A is called γ-oriented if, for every pair $P_1, P_2 \epsilon \mathscr{P}$ such that $\gamma(P_1) = \gamma(P_2)$, it is true that $P_1(A) = P_2(A)$. In other words, a γ-oriented set is one whose probability depends on P through $\gamma(P)$. A sub-field is γ-oriented if every member of the sub-field is. The following theorem* is a direct generalization of Theorem 4(a).

Theorem 5. *Lrt $\mathscr{G}_\gamma$ be the class of all γ-preserving transformations and let $\mathscr{B}$ be a sub-field that is*

i) *γ-oriented and*

ii) *contained in the boundedly complete sufficient sub-field $\mathscr{L}$ (which exists).*

Then,

$$\mathscr{B} \subset \tilde{\mathscr{A}}(\mathscr{G}_\gamma).$$

Proof : Let $g \epsilon \mathscr{G}_\gamma$ and $B \epsilon \mathscr{B}$ and let $B_0 = g^{-1} B$.

Then

$$P(B_0) = P(g^{-1}B) \qquad (\because \quad B_0 = g^{-1}B)$$

$$= Pg^{-1}(B).$$

* The author wishes to thank Professor W. J. Hall of the University of North Carolina for certain comments that eventually led to this theorem.

Since g is γ-preserving, we have $\gamma(Pg^{-1}) = \gamma(P)$. And since B is γ-oriented we have

$$Pg^{-1}(B) = P(B).$$

Therefore,

$$P(B_0) = P(B) \quad \text{or} \quad R(I_B - I_{B_0} \mid P) \equiv 0.$$

The rest of the proof is the same as in Theorem 4(a).

In the next section we repeatedly use the above theorem.

10. SOME TYPICAL INVARIANCE REDUCTIONS

Let $x_1, x_2, \ldots, x_n$ be n independent and identical normal variables with unknown μ and σ. Suppose the parameter of interest is σ. Let $\mathcal{G}_\sigma$ be the class of all σ-preserving transformations. If $\bar{x}$ is the mean and s, the standard deviation of the n observations, then the pair $(\bar{x}, s)$ is a complete sufficient statistic. Also s is σ-oriented. From Theorem 5, we then have that s is almost $\mathcal{G}_\sigma$-invariant. Thus, the principle of invariance cannot reduce the data beyond s. That s is indeed the exact (upto $\mathcal{P}$-equivalence) attainable limit of invariance reduction is shown as follows.

Let $\{H\}$ be the class of all $n \times n$ Haggstrom matrices (see section 8), i.e., each H is an orthogonal matrix with unit row and column sums. Let c be an arbitrary real number and let $\boldsymbol{i}$ stand for the n-vector $(1, 1, \ldots, 1)$. Consider all linear transformations of the type

$$\boldsymbol{y}' = H\boldsymbol{x}' + (c\boldsymbol{i})'.$$

If $\mathcal{G}_\sigma^*$ is this class of transformations, then it is easily verified that $\mathcal{G}_\sigma^*$ is a sub-group of $\mathcal{G}_\sigma$. That $\Sigma(x_i - x)^2$ is a maximal invariant with respect to the sub-group $\mathcal{G}_\sigma^*$ of σ-preserving transformations is seen as follows.

Let $\boldsymbol{a} = (a_1, a_2, \ldots, a_n)$ and $\boldsymbol{b} = (b_1, b_2, \ldots, b_n)$ be two arbitrary n-vectors such that

$$\Sigma(a_i - \bar{a})^2 = \Sigma(b_i - \bar{b})^2.$$

Let $c = \bar{b} - \bar{a}$. Then the two vectors $\boldsymbol{a} + c\boldsymbol{i}$ and $\boldsymbol{b}$ have equal sums and sums of squares (of co-ordinates). Hence there exists a

Haggstrom matrix H that transforms $\boldsymbol{a}+c\boldsymbol{i}$ into $\boldsymbol{b}$. In other words, the transformation

$$\boldsymbol{y}' = H\boldsymbol{x}'+(c\boldsymbol{i})'$$

maps $\boldsymbol{a}$ into $\boldsymbol{b}$.

If $\mathcal{B}(s)$ is the sub-field generated by s, then we have just proved that $\mathcal{B}(s) = \mathcal{A}(\mathcal{G}_\sigma^*)$. The proof of the $\mathcal{P}$-equivalence of

$$\mathcal{B}(s) \quad \text{and} \quad \widetilde{\mathcal{A}}(\mathcal{G}_\sigma)$$

will follow the familiar pattern set up earlier for example 4 in section 7.

Now, suppose that our parameter of interest is $\gamma = \mu/\sigma$. The statistic $\bar{x}/s$ is γ-oriented and hence (Theorem 5) is almost $\mathcal{G}_\gamma$-invariant, where $\mathcal{G}_\gamma$ is the group of all γ-preserving transformations.

Consider now the sub-group $\mathcal{G}_\gamma^*$ of all linear transformations of the type

$$\boldsymbol{y}' = cH\boldsymbol{x}'$$

where $c > 0$ and H is a Haggstrom matrix.

It is easily checked that the maximal invariant with respect to the sub-group $\mathcal{G}_\gamma^*$ is the statistic $\bar{x}/s$ and hence, the sub-field $\mathcal{B}(\bar{x}/s)$ generated by $\bar{x}/s$ is $\mathcal{P}$-equivalent to the sub-field $\widetilde{\mathcal{A}}(\mathcal{G}_\gamma)$ of all almost $\mathcal{G}_\gamma$-invariant sets.

The case where our parameter of interest is $|\mu|/\sigma$ is very similar. Once again we observe that the statistic $|\bar{x}|/s$ is oriented towards $|\mu|/\sigma$. Hence, making use of Theorem 5 and observing that $|\bar{x}|/s$ is the maximal invariant with respect to the sub-group of linear parameter-preserving transformations of the form

$$\boldsymbol{y}' = cH\boldsymbol{x}' \ (c \neq 0,\ H \text{ a Haggstrom matrix}),$$

we are able to show that the invariance reduction of the data is to the statistic $|\bar{x}|/s$.

11. SOME FINAL REMARKS

a) In statistical literature, the principle of invariance has been used in a rather half-hearted manner (see, for example, [5] and [6]). We do not find any consideration given to the present project of reducing the data with the help of the whole class $\mathcal{G}_\gamma$ of γ-preserving transformations. In fact, the question of how extensive the class $\mathcal{G}_\gamma$ can be has escaped general attention. One usually works in the framework of a relatively small and simple sub-group of $\mathcal{G}_\gamma$ and invokes simultaneously the two different principles of sufficiency and invariance for the purpose of arriving at a *satisfactory* data reduction. The main object of the present article is to investigate how far the principle of invariance by itself can take us.

b) The main limitation of the principle of sufficiency is that it does not recognize nuisance parameters. Several attempts have been made to generalize the idea of sufficiency so that one gets an effective data reduction in the presence of nuisance parameters. Not much success has, however, been achieved in this direction.

c) On the other hand, the invariance principle usually falls to pieces when faced with a discrete model. The Bernoulli experimental set-up is one of the rare discrete models that the invariance principle can tackle. If $x_1, x_2, \ldots, x_n$ are n independent zero-one variables with probabilities θ and $1-\theta$, then all permutations (of co-ordinates) are model-preserving. And they reduce the data directly to the least sufficient statistic $r = x_1 + \ldots + x_n$. Take, however, the following simple example. Let x and y be independent zero-one variables, where

$$P(x=0) = \theta, \quad 0 < \theta < 1,$$

and

$$P(y=0) = 1/3.$$

Now, the identity-map is the only available model-preserving transformation. The principle of sufficiency reduces the data immediately to x,

d) The object of this article was not to make a critical evaluation of the twin principles of data reduction. Yet, the author finds it hard to refrain from observing that both the principles of sufficiency and invariance are extremely sensitive to changes in the model. For example, the spectacular data reductions we have achieved in the many examples considered here become totally unavailable if the basic normality assumption is changed ever so slightly.

References

[1] Bahadur, R. R. "Sufficiency and Statistical Decision Functions," *Ann. Math. Statist.*, **25** (1954), 423–462.

[2] Basu, D. "Sufficiency and Model-Preserving Transformations," Inst. of Statistics, Mimeo Series No. 420, Univ. of North Carolina, Chapel Hill, 1965.

[3] Burkholder, D. L. "Sufficiency in the Undominated Case," *Ann. Math. Statist.*, **32** (1961), 1191–1200.

[4] Farrell, R. H. "Representation of Invariant Measures," *Illinois J. of Math.*, **6** (1962), 447–467.

[5] Hall, W. J., R. A. Wijsman, and J. K. Gosh. "The Relationship between Sufficiency and Invariance," *Ann. Math. Statist.*, **36** (1965), 575–614.

[6] Lehmann, E. L. *Testing Statistical Hypotheses*, John Wiley and Sons, New York, 1959.

(*Received Sept. 17, 1965.*)

CHAPTER 5

Categorical Data Analogs of Some Multivariate Tests

V. P. BHAPKAR*, *The University of North Carolina at Chapel Hill and University of Poona*

1. INTRODUCTION

Problems of association between several continuous variables are generally handled under the assumption of joint normality of the variables concerned. Similarly in the general multifactor multiresponse situation with observations on several characters of each experimental unit for various factor-combinations, data are generally analyzed under the assumption of normality; the situation is referred to as one involving analysis of variance (ANOVA) or multivariate analysis of variance (MANOVA) according as whether we have observations on a single character or several characters, respectively, of each experimental unit. These methods are fairly well-developed now and form a part of the classical statistical theory.

* This research was supported by the National Institutes of Health Institute of General Medical Sciences Grant No. GM-12868-01.

The position is not so satisfactory when the assumption of normality is discarded. This is particularly true when the experimental data are categorical in nature, i.e., are given in the form of frequencies in cells determined by a finite multi-way cross-classification with predefined categories along each way of classification. Even though analysis of categorical data goes back to the pioneering work of Karl Pearson (1900) and has been developed at subsequent stages by, among others, Fisher (1922), Cramer (1946) and, in particular, Neyman (1949), a lot remained to be done.

Barnard (1947) and Pearson (1947) pointed out by considering the simple 2×2 table that the same 2×2 table could be the outcome of different sampling schemes which makes it necessary to assume appropriate probability models, which may lead to different statistical procedures with obviously different interpretations appropriate to each experimental situation. This line of thought was developed extensively by Roy and his students; refer for example to Mitra (1955), Roy and Mitra (1956), Bhapkar (1959), Roy and Bhapkar (1960). Formulations of some of these categorical data problems as analogs of ANOVA, and MANOVA and *Normal association* problems were offered first by Roy and Mitra (1956) and later on by Roy and Bhapkar (1960); statistical procedures to handle these were given in Roy and Mitra (1956) and Bhapkar (1961) respectively. The present paper is in the same line and develops further some such methods. It deals exclusively with suitable formulations of hypotheses and appropriate tests for multi-response situations.

Section 2 gives notation and preliminaries, Section 3 includes some further results and Section 4 gives specific test criteria for some hypotheses; the test statistic mentioned is usually the Pearson$-\chi^2$ statistic if the maximum likelihood estimates are easy to obtain, and otherwise is the Neyman$-\chi_1^2$ statistic computed from results in Section 3.

2. NOTATION AND PRELIMINARIES

Suppose that s independent random samples of experimental units are taken from s populations, n_{0j} is the size of the sample from the jth population and n_{ij} is the observed frequency in the ith category of the jth sample, $i = 1, 2, \ldots, r$; $j = 1, 2, \ldots . s$. We assume that p_{ij} is the probability that an experimental unit drawn at random from the jth population belongs to the ith category and that either the sampling is with replacement or, if without replacement, sampling fractions are negligible, so that the probability distribution of the observed frequencies is given by

$$\phi = \prod_{j=1}^{s} \left(\frac{n_{0j}\,!}{\prod_{i=1}^{r} n_{ij}\,!} \prod_{i=1}^{r} p_{ij}^{n_{ij}} \right), \tag{2.1}$$

where $\sum_i n_{ij} = n_{0j}$, a given integer, and $\sum_i p_{ij} = 1$ for each $j = 1, 2, \ldots, s$; zero in place of a suffix will indicate sum over that suffix. Let $N = \sum_j n_{0j}$, $q_{ij} = n_{ij}/n_{0j}$, $Q_j = n_{0j}/N$, $\mathbf{p}' = [p_{11}, \ldots, p_{(r-1)1} ; \ldots; p_{1s}, \ldots, p_{(r-1)s}]$ and $q' = [q_{11}, \ldots, q_{(r-1)1}; \ldots; q_{1s}, \ldots, q_{(r-1)s}]$.

According as the marginal frequencies along any dimension or way of classification are held fixed or left free by the experimental scheme, that dimension will be said to be a *factor* or a *response*. Thus, in the above model (2.1), i refers to response categories while j refers to factor categories; n_{i0} is a random variable while n_{0j} is a fixed integer. i may be a multiple subscript, say, $(i_1, i_2 \ldots, i_k)$ with $i_\alpha = 1, 2, \ldots, r_\alpha$, $\alpha = 1, 2, \ldots, k$ so that $r = r_1 r_2 \ldots r_k$; similarly, j also might be a multiple subscript, say $j_1, j_2, \ldots, j_l$ with $j_\beta = 1, 2, \ldots, s_\beta$, $\beta = 1, 2, \ldots, l$ but with the distinction that all combinations may not be selected for the experiment. This will be called a k-response (or k-variate) and l-factor problem where the i_α refer to a category of the αth response while j_β to that of the βth factor.

If a set of real values (scores) is associated with the categories along any way of classification (factor or response), that way of classification will be said to be structured. These may be, for example, the mid-points of class-intervals for a response (or factor) or the values themselves if the response (or factor) is discrete, or may be any scores assigned on some other considerations even for a way of classification without any implied ranking, to start with, for its categories.

Suppose that we have to test the hypothesis

$$H_0 : F_m(\mathbf{p}) = 0, \quad m = 1, 2, \ldots = t(t \leqslant rs\text{-}s) \tag{2.2}$$

where F's are t independent given functions of $\mathbf{p}$. It is assumed that F's possess continuous partial derivatives up to the second order and that the rank of the $t \times (rs-s)$ matrix $[\partial F_m(\mathbf{p})/\partial p_{ij}]$ is t. It is assumed that there is at least one solution such that $p_{ij} > 0$ for all i, j. It is then well known (e.g., refer to [11]) that H_0 can be tested in various ways by using either the χ^2, χ_1^2 statistics, or the likelihood-ratio statistic λ defined, respectively, by

$$\chi^2 = \sum_{j=1}^{s} \sum_{i=1}^{r} (n_{ij} - n_{0j}\,\hat{p}_{ij})^2 / n_{0j}\,\hat{p}_{ij}$$

$$\chi_1^2 = \sum_{j=1}^{s} \sum_{i=1}^{r} (n_{ij} - n_{0j}\,\hat{p}_{ij})^2 / n_{ij} \tag{2.3}$$

$$-2 \log\lambda = 2 \sum_{j=1}^{s} \sum_{i=1}^{r} n_{ij} \{\log n_{ij} - \log n_{0j}\,\hat{p}_{ij}\},$$

where the $\hat{p}$'s are any B.A.N. [11] estimates of the p's obtained subject to constraints (2.2). Neyman (1949) has shown that, in particular, estimates minimizing χ^2 or χ_1^2 or those maximizing ϕ are B.A.N. estimates. If the p's are subject to constraint (2.2), the equations giving these estimates, are, in general, complicated to solve; the minimum—χ_1^2 estimates, though, can be obtained by solving only linear equations whenever the constraints (2.2) are linear in p's. If the functions F_m are not linear, Neyman (1949) has proposed the technique of *linearization* to reduce the problem to the linear case whereby minimum—χ_1^2 estimates are obtained subject to constraints

$$F^*_m(\boldsymbol{p}) \equiv F_m(\boldsymbol{q})+\sum_{j=1}^{s}\sum_{i=1}^{r-1}\left[\frac{\partial F_m(\boldsymbol{p})}{\partial p_{ij}}\right]_{\boldsymbol{p}=\boldsymbol{q}}(p_{ij}-q_{ij}) = 0,$$
$$m = 1, 2, \ldots, t.$$

Neyman has proved that each of the statistics in (2.3), using any system of B.A.N. estimates (using linearization if necessary), has a limiting chi-square distribution with t degrees of freedom as $N \to \infty$ with Q's fixed if H_0 holds; he has also proved that these tests are asymptotically equivalent in the sense that the probability of any two of them contradicting each other tends to 0 as $N \to \infty$ irrespective of whether H_0 is true or false. The author (1965) has shown recently that the χ_1^2 statistic, whenever it is defined, is identical to Wald's statistic (1943) as adapted to the categorical situation and, hence, possesses the same asymptotic optimality properties as those possessed by Wald's statistic and the likelihood-ratio statistic as shown by Wald for the case of sampling from one population (i.e., for $s = 1$) and conjectured for the general case (i.e., for $s \geqslant 2$).

3. SOME FURTHER RESULTS

Theorem 3.1. *Let H_0 be the hypothesis specified by t independent constraints*

$$(3.1) \qquad F_m(\boldsymbol{p}) = \sum_{j=1}^{s}\sum_{i=1}^{r} f_{mij}\, p_{ij} + f_m \equiv \sum_{j=1}^{s}\sum_{i=1}^{r-1} f^*_{mij}\, p_{ij}+f^*_m = 0$$
$$m = 1, \ldots, t,$$

where $f^*_{mij} = f_{mij}-f_{mrj}$ *and* $f^*_m = f_m+\sum_{i=j}^{s} f_{mrj}$; *we assume that these are independent of the basic constraints* $\sum_{i=1}^{r} p_{ij} = 1$. *Let,*

$\boldsymbol{c}' = (c_1 \ldots c_t)$ *with* $c_m = F_m(q)$

$\boldsymbol{G} = [g_{mm'}] \qquad m,m' = 1, 2, \ldots, t$

with

$$g_{mm'} = \sum_{j=1}^{s} n_{0j}^{-1}\left\{\sum_{i=1}^{r} f_{mij}\, f_{m'ij}\, q_{ij}-\left(\sum_{i=1}^{r} f_{mij}\, q_{ij}\right)\left(\sum_{i=1}^{r} f_{m'ij}\, q_{ij}\right)\right\}$$

$$= \sum_{j=1}^{s} n_{0j}^{-1}\left\{\sum_{i=1}^{r-1} f^*_{mij}\, f^*_{m'ij}\, q_{ij}-\left(\sum_{i=1}^{r-1} f^*_{mij}\, q_{ij}\right)\left(\sum_{i=1}^{r-1} f^*_{m'ij}\, q_{ij}\right)\right\}.$$

Then if H_0 holds, the statistic

$$\mathbf{c}'\, \boldsymbol{G}^{-1}\, \mathbf{c} \tag{3.2}$$

has a limiting chi-square distribution with t d.f. as $N \to \infty$ with Q's remaining fixed.

Proof: It can be shown easily (e.g., refer to [2]) that (3.2) is the χ_1^2-statistic to test H_0 and the theorem then follows from Neyman's results.

It may be noted that $\boldsymbol{G}$ is the matrix obtained after replacing $\boldsymbol{p}$ by $\boldsymbol{q}$ in the covariance matrix of $\mathbf{c}$ and hence will be referred to as the *sample covariance matrix* of $\mathbf{c}$; it is nonsingular almost everywhere in view of the assumed conditions.

Theorem 3.2. *Let a linear hypothesis be defined by*

$$\sum_{i=1}^{r} a_{\beta i}\, p_{ij} = d_{j1}\theta_{\beta 1} + d_{j2}\theta_{\beta 2} + \ldots + d_{ju}\,\theta_{\beta u}, \quad \beta = 1, 2, \ldots, k, \tag{3.3}$$

where a's and d's are known constants, θ's are unknown parameters, $\boldsymbol{D} = [d_{j\gamma}]_{s\times u}$ with rank $\boldsymbol{D} = v < s$, and the linear functions on the left in (3.3) *are linearly independent of $\sum_i p_{ij}$ ($\equiv 1$, of course). Suppose*

$$\alpha_{\beta j} = \sum_i a_{\beta i}\, q_{ij}, \quad \boldsymbol{\alpha}_j' = [\alpha_{1j}, \ldots, \alpha_{kj}], \quad \boldsymbol{\theta}_\gamma' = [\theta_{1\gamma}, \ldots, \theta_{k\gamma}]$$

(3.4) *and*

$$\boldsymbol{\Lambda}_j = [\lambda_{\beta\beta' j}]_{k\times k}, \qquad \beta, \beta' = 1, 2, \ldots, k$$

with

$$\lambda_{\beta\beta' j} = n_{0j}^{-1}(\Sigma_i a_{\beta i}\, a_{\beta' i}\, q_{ij} - \alpha_{\beta j}\, \alpha_{\beta' j}).$$

Then the χ_1^2-statistic to test (3.3) *is equal to the minimum value of*

$$S^2 = \sum_{j=1}^{s} (\boldsymbol{\alpha}_j - d_{j1}\boldsymbol{\theta}_1 - \ldots - d_{ju}\boldsymbol{\theta}_u)'\boldsymbol{\Lambda}_j^{-1}(\boldsymbol{\alpha}_j - \ldots - d_{ju}\boldsymbol{\theta}_u) \tag{3.5}$$

with respect to the **θ***'s and has $k(s-v)$ degrees of freedom.*

Proof. Let $\boldsymbol{B} = [b_{\delta j}]$ be a $(s-v)\times s$ matrix of rank $s-v$ such that $\boldsymbol{BD} = \mathbf{0}$, i.e.,

$$\sum_j b_{\delta j}\, d_{j\gamma} = 0, \quad \delta = 1, 2, \ldots, s-v; \ \gamma = 1, 2, \ldots, u.$$

(3.3), then, implies that

$$\sum_j b_{\delta j} \sum_i a_{\beta i}\, p_{ij} = \sum_j b_{\delta j} \sum_\gamma d_{j\gamma}\, \theta_{\beta\gamma} = \sum_\gamma \theta_{\beta\gamma} \sum_j b_{\delta j}\, d_{\gamma j} = 0.$$

Conversely, $\sum_j b_{\delta j} \sum_i a_{\beta i}\, p_{ij} = 0$ implies that the vector $(\sum_i a_i\, p_{ij},$ $j = 1, 2, \ldots, s)$ belong to the vector space orthogonal to that generated by the rows of $\boldsymbol{B}$ and, hence, belongs to the vector space generated by the columns of $\boldsymbol{D}$; thus there exist θ's such that $\Sigma_i a_{\beta i}\, p_{ij} = \Sigma_\gamma \theta_{\beta\gamma} d_{j\gamma}$ which means (3.3) holds. Thus (3.3) is equivalent to $k(s-v)$ independent linear constraints

$$(3.6) \quad \sum_j \sum_i b_{\delta j} a_{\beta i}\, p_{ij} = 0; \quad \beta = 1, 2, \ldots, k; \qquad \delta = 1, 2, \ldots, s-v.$$

The χ_1^2-statistic with $k(s-v)$ degrees of freedom to test (3.6) can be derived, then, from Theorem 3.1. Let

$$c_{\delta\beta} = \sum_j \sum_i b_{\delta j} a_{\beta i}\, q_{ij} = \sum_j b_{\delta j} \alpha_{\beta j}$$

$$\boldsymbol{c}_\delta' = [c_{\delta 1}, c_{\delta 2}, \ldots, c_{\delta k}], \quad \boldsymbol{c}' = [\boldsymbol{c}_1', \ldots, \boldsymbol{c}_{s-v}'],$$

so that

$$\boldsymbol{c}_\delta = \sum_j b_{\delta j} \boldsymbol{\alpha}_j$$

and, hence,

$$(3.7) \qquad \boldsymbol{c} = \boldsymbol{B}^* \boldsymbol{\alpha},$$

where

$$\boldsymbol{B}^* = \begin{bmatrix} b_{11} \boldsymbol{I}_k & \ldots & b_{1s} \boldsymbol{I}_k \\ & \ldots & \\ b_{s-v,1} \boldsymbol{I}_k & \ldots & b_{s-v,s} \boldsymbol{I}_k \end{bmatrix} = \boldsymbol{B} \otimes \boldsymbol{I}_k,$$

the Kronecker product (sometimes also called the Direct product) of $\boldsymbol{B}$ and $\boldsymbol{I}_k$. Also

$$\text{Cov}(\boldsymbol{c}) = \boldsymbol{B}^* \,\text{Cov}(\boldsymbol{\alpha}) \boldsymbol{B}^{*\prime} = \boldsymbol{B}^* \boldsymbol{\Sigma} \boldsymbol{B}^{*\prime},$$

where

$$\boldsymbol{\Sigma} = \begin{bmatrix} \boldsymbol{\Sigma}_1 & 0 & \ldots & 0 \\ 0 & \boldsymbol{\Sigma}_2 & \ldots & 0 \\ 0 & 0 & \ldots & \boldsymbol{\Sigma}_s \end{bmatrix},$$

and $\boldsymbol{\Sigma}_j$ is the covariance matrix of $\boldsymbol{\alpha}_j$. Note that replacing $\boldsymbol{p}$ by $\boldsymbol{q}$ in $\boldsymbol{\Sigma}_j$ gives $\boldsymbol{\Lambda}_j$ defined by (3.4). Thus, the sample covariance matrix of $\boldsymbol{c}$ is $\boldsymbol{B}^* \boldsymbol{\Lambda} \boldsymbol{B}^{*\prime}$, where

$$\boldsymbol{\Lambda} = \begin{bmatrix} \boldsymbol{\Lambda}_1 & 0 & \ldots & 0 \\ 0 & \boldsymbol{\Lambda}_2 & \ldots & 0 \\ 0 & 0 & \ldots & \boldsymbol{\Lambda}_s \end{bmatrix},$$

and the χ_1^2-statistic, in view of (3.2), is

$$\boldsymbol{\alpha}'\boldsymbol{B}^{*\prime}(\boldsymbol{B}^* \boldsymbol{\Lambda} \boldsymbol{B}^{*\prime})^{-1} \boldsymbol{B}^* \boldsymbol{\alpha}. \tag{3.8}$$

On the other hand minimum value of S^2 with respect to $\boldsymbol{\theta}$'s is seen to be

$$\sum_j \boldsymbol{\alpha}_j' \boldsymbol{\Lambda}_j^{-1} \boldsymbol{\alpha}_j - \sum_{\gamma=1}^{u} \hat{\boldsymbol{\theta}}_\gamma' \boldsymbol{\Psi}_\gamma, \tag{3.9}$$

where

$$\boldsymbol{\Psi}_\gamma = \Sigma\, d_{j\gamma} \boldsymbol{\Lambda}_j^{-1} \boldsymbol{\alpha}_j,$$

and $\hat{\boldsymbol{\theta}}' = [\hat{\boldsymbol{\theta}}_1', \ldots, \hat{\boldsymbol{\theta}}_u']$ is a solution of

$$\boldsymbol{\Psi}_\gamma = \sum_{\varepsilon=1}^{u} \boldsymbol{\Phi}_{\gamma\varepsilon} \hat{\boldsymbol{\theta}}_\varepsilon,$$

where

$$\boldsymbol{\Phi}_{\gamma\varepsilon} = \sum_j d_{j\gamma} d_{j\varepsilon} \boldsymbol{\Lambda}_j^{-1}.$$

If we let

$$\boldsymbol{\Psi}' = [\boldsymbol{\Psi}_1', \ldots, \boldsymbol{\Psi}_u'], \qquad \boldsymbol{D}^* = \boldsymbol{D} \otimes \boldsymbol{I}_k$$

and

$$\boldsymbol{\Phi} = \begin{bmatrix} \boldsymbol{\Phi}_{11}, & \ldots, & \boldsymbol{\Phi}_{1u} \\ & \ldots & \\ \boldsymbol{\Phi}_{u1} & \ldots, & \boldsymbol{\Phi}_{uu} \end{bmatrix},$$

then it follows that

$$\boldsymbol{\Psi} = \boldsymbol{D}^{*\prime}\boldsymbol{\Lambda}^{-1} \boldsymbol{\alpha}, \quad \boldsymbol{\Phi} = \boldsymbol{D}^{*\prime}\boldsymbol{\Lambda}^{-1} \boldsymbol{D}^*$$

(3.10) and

$$\boldsymbol{\Psi} = \boldsymbol{\Phi}\, \hat{\boldsymbol{\theta}};$$

hence (3.9) reduces to

$$\boldsymbol{\alpha}'\boldsymbol{\Lambda}^{-1} \boldsymbol{\alpha} - \hat{\boldsymbol{\theta}}' \boldsymbol{D}^{*\prime}\boldsymbol{\Lambda}^{-1} \boldsymbol{\alpha}. \tag{3.11}$$

We have to show that (3.8) is equal to (3.11), taking into account (3.10) and the fact that $\boldsymbol{BD} = \mathbf{0}$, i.e., $\boldsymbol{B}^*\boldsymbol{D}^* = \mathbf{0}$.

Now $\boldsymbol{D}$ is a $s\times u$ matrix of rank v so that $\boldsymbol{D}^*$ is a $ks\times ku$ matrix of rank kv; there exist then a $ks\times kv$ matrix $\boldsymbol{M}$ and a $kv\times ku$ matrix $\boldsymbol{E}$ both of rank kv, such that

$$\boldsymbol{D}^* = \boldsymbol{ME}.$$

From (3.10) then we have

$$\boldsymbol{E}'\boldsymbol{M}'\boldsymbol{\Lambda}^{-1}\boldsymbol{\alpha} = \boldsymbol{E}'\boldsymbol{M}'\boldsymbol{\Lambda}^{-1}\boldsymbol{ME}\hat{\boldsymbol{\theta}},$$

which gives

$$\boldsymbol{\eta} = (\boldsymbol{M}'\boldsymbol{\Lambda}^{-1}\boldsymbol{M})^{-1}\boldsymbol{M}'\boldsymbol{\Lambda}^{-1}\boldsymbol{\alpha},$$

where

$$\boldsymbol{\eta} = \boldsymbol{E}\hat{\boldsymbol{\theta}}.$$

The second term in (3.11) is then $\boldsymbol{\eta}'\boldsymbol{M}'\boldsymbol{\Lambda}^{-1}\boldsymbol{\alpha}$.

The theorem follows if we show that

$$(3.12) \qquad \boldsymbol{B}^{*\prime}(\boldsymbol{B}^*\boldsymbol{\Lambda}\boldsymbol{B}^{*\prime})^{-1}\boldsymbol{B}^* = \boldsymbol{\Lambda}^{-1}-\boldsymbol{\Lambda}^{-1}\boldsymbol{M}(\boldsymbol{M}'\boldsymbol{\Lambda}^{-1}\boldsymbol{M})^{-1}\boldsymbol{M}'\boldsymbol{\Lambda}^{-1},$$

noting that $\boldsymbol{B}^*\boldsymbol{M} = \mathbf{0}$ since $\boldsymbol{B}^*\boldsymbol{D}^* = \mathbf{0}$. Let $\boldsymbol{R}$, $\boldsymbol{S}$, $\boldsymbol{T}$ be non-singular matrices such that

$$\boldsymbol{R\Lambda R}' = \boldsymbol{I},\ \boldsymbol{S}(\boldsymbol{B}^*\boldsymbol{\Lambda}\boldsymbol{B}^{*\prime})\boldsymbol{S}' = \boldsymbol{I},\ \boldsymbol{TM}'\boldsymbol{\Lambda}^{-1}\boldsymbol{MT}' = \boldsymbol{I}$$

so that

$$\boldsymbol{\Lambda}^{-1} = \boldsymbol{R}'\boldsymbol{R},\ \boldsymbol{B}^*\boldsymbol{\Lambda}\boldsymbol{B}^{*\prime} = (\boldsymbol{S}'\boldsymbol{S})^{-1} \text{ and } \boldsymbol{M}'\boldsymbol{\Lambda}^{-1}\boldsymbol{M} = (\boldsymbol{T}'\boldsymbol{T})^{-1}.$$

Then

$$\begin{bmatrix} \boldsymbol{S} & \mathbf{0} \\ \mathbf{0} & \boldsymbol{T} \end{bmatrix} \begin{bmatrix} \boldsymbol{B}^*\boldsymbol{R}^{-1} \\ \boldsymbol{M}'\boldsymbol{R}' \end{bmatrix} [\boldsymbol{R}'^{-1}\boldsymbol{B}^{*\prime} \,|\, \boldsymbol{RM}] \begin{bmatrix} \boldsymbol{S}' & \mathbf{0} \\ \mathbf{0} & \boldsymbol{T}' \end{bmatrix} = \boldsymbol{I}$$

which implies

$$[\boldsymbol{R}'^{-1}\boldsymbol{B}^{*\prime} \,|\, \boldsymbol{RM}] \begin{bmatrix} \boldsymbol{S}' & \mathbf{0} \\ \mathbf{0} & \boldsymbol{T}' \end{bmatrix} \begin{bmatrix} \boldsymbol{S} & \mathbf{0} \\ \mathbf{0} & \boldsymbol{T} \end{bmatrix} \begin{bmatrix} \boldsymbol{B}^*\boldsymbol{R}^{-1} \\ \boldsymbol{M}'\boldsymbol{R}' \end{bmatrix} = \boldsymbol{I},$$

that is,

$$\boldsymbol{R}'^{-1}\boldsymbol{B}^{*\prime}\boldsymbol{S}'\boldsymbol{S}\boldsymbol{B}^*\boldsymbol{R}^{-1}+\boldsymbol{RMT}'\boldsymbol{TM}'\boldsymbol{R}^{-1} = \boldsymbol{I},$$

which leads to (3.12). Q.E.D.

Note that α's are the natural unbiased estimates of the quanties on the left in (3.3) while $\mathbf{\Lambda}_j$ is the *sample covariance matrix* of $\boldsymbol{\alpha}_j$.

4. APPLICATIONS

A. Two-dimensional tables with both dimensions responses

Here we have a single population with i a double subscript $(i_1 i_2)$, $i_\alpha = 1, 2, \ldots, r_\alpha$; $\alpha = 1, 2$ and $r = r_1 r_2$. The probabilities $p_{i_1 i_2}$ are subject to the constraint $\sum_{i_1, i_2} p_{i_1 i_2} = 1$.

(i) *Hypothesis of independence* : This hypothesis is expressed by the condition

(4.A.1) $$H_1 : p_{i_1 i_2} = p_{i_1 0}\, p_{0 i_2},$$

where, of course, $p_{i_1 0} = \sum_{i_2} p_{i_1 i_2}$ and $p_{0 i_2} = \sum_{i_1} p_{i_1 i_2}$. The well known χ^2-statistic to test H_1 is

(4.A.2) $$\sum_{i_1, i_2} \left(n_{i_1 i_2} - \frac{n_{i_1 0}\, n_{0 i_2}}{N} \right)^2 \Bigg/ \left(\frac{n_{i_1 0}\, n_{0 i_2}}{N} \right), \text{ d.f.} = (r_1 - 1)(r_2 - 1).$$

H_1 has been already seen [15] to be an obvious analog of the hypothesis of independence (i.e., no correlation) in the bivariate normal analysis.

(ii) *Hypothesis of equality of two marginal distributions* :

Assuming $r_1 = r_2$, this hypothesis is expressed by the condition

(4.A.3) $$H_2 : p_{i_1 0} = p_{0 i_1};$$

this may be seen to be an analog of the hypothesis $\mu_1 = \mu_2$ and $\sigma_{11} = \sigma_{22}$, in the usual notation, in the normal analysis. Let

$$h_{i_1}(\boldsymbol{q}) = q_{i_1 0} - q_{0 i_1} \qquad i_1 = 1, 2, \ldots, r_1 - 1,$$

and

$$N\,\boldsymbol{G} = \left[\delta_{i_1 i_2}\left(q_{i_1 0}+q_{0 i_1}\right)-q_{i_1 i_2}-q_{i_2 i_1}-h_{i_1}(\boldsymbol{q})h_{i_2}(\boldsymbol{q})\right]$$

$$i_1, i_2 = 1, 2, \ldots, r_1-1,$$

where $\delta_{i_1 i_2} = 1$ if $i_1 = i_2$ and 0 otherwise. Then the χ_1^2-statistic is seen to be

(4.A.4) $\boldsymbol{h}'(\boldsymbol{q})\boldsymbol{G}^{-1}\,\boldsymbol{h}(\boldsymbol{q})$, d.f. $= r_1-1$.

The same expression had been obtained by Sathe (1962) for Wald'r critesion and the two statistics must be identical as observed by the author (1965). The statistic (4.A.4) differs from the one proposed by Stuart (1955) in that the latter deletes the last term in the rectangular brackets for $\boldsymbol{G}$; our statistic should be preferred since $N\boldsymbol{G}$ is a consistent estimator of the covariance matrix of $\sqrt{N}\boldsymbol{h}(\boldsymbol{q})$ even when H_2 is false, while the matrix used by Stuart is consistent only when H_2 holds.

(iii) *Hypothesis of equality of "means" of two variables :*

Let us now suppose that the two responses are "structured", i.e., have an implied ranking, and we have a system of scores $\{a_{i_1}\}$ associated with the categories of the first response with a similar system $\{b_{i_2}\}$ for the second response. Then the above hypothesis may be expressed in the form

(4.A.5) $H_3 : \sum_{i_1} a_{i_1} p_{i_1 0} = \sum_{i_2} b_{i_2} p_{0 i_2}.$

Let

$$c = \sum_{i_1} a_{i_1} q_{i_1 0} - \sum_{i_2} b_{i_2} q_{0 i_2}$$

and

$$Ng = \sum_{i_1} a_{i_1}^2 q_{i_1 0} + \sum_{i_2} b_{i_2}^2 q_{0 i_2} - 2 \sum_{i_1} \sum_{i_2} a_{i_1} b_{i_2} q_{i_1 i_2} - c^2;$$

then the χ_1^2-statistic is seen to be

(4.A.6) c^2/g d.f. $= 1$.

H_3 may be considered as an analog of the hypothesis of equality of means of two possibly correlated variables in the normal analysis. Note that in the special case $r_1 = r_2$ with $a_{i_1} = b_{i_2}$, H_2 implies H_3 and thus H_3 is a weaker hypothesis than H_2 and may be of interest if H_2 does not hold.

B. Three-dimensional tables with all dimensions responses

Here again we have a single population, but now i is a triple subscript $(i_1 i_2 i_3)$, $i_\alpha = 1, 2, \ldots, r_\alpha$, $\alpha = 1, 2, 3$ and $r = r_1 r_2 r_3$, the probabilities $p_{i_1 i_2 i_3}$ are subject to the constraint $\sum_{i_1 i_2 i_3} p_{i_1 i_2 i_3} = 1$.

(i) *Hypothesis of complete independence* :

$$H_4 : p_{i_1 i_2 i_3} = p_{i_1 00}\, p_{0 i_2 0}\, p_{00 i_3} \tag{4.B.1}$$

(ii) *Hypothesis of independence of the first response with the last two* :

$$H_5 : p_{i_1 i_2 i_3} = p_{i_1 00}\, p_{0 i_2 i_3} \tag{4.B.2}$$

(iii) *Hypothesis of independence of the first two responses given the third response* :

$$H_6 : p_{i_1 i_2 i_3} = \frac{p_{i_1 0 i_3}\, p_{0 i_2 i_3}}{p_{00 i_3}} \tag{4.B.3}$$

It has been pointed out by Roy and Mitra (1956) that H_5, H_6 can be considered analogs of the hypotheses of no multiple correlation and no partial correlation, respectively, in the normal analysis while H_4 is that of the hypothesis of zero correlations. The χ^2-statistics are known to be

$$N^2 \sum_{i_1, i_2, i_3} \left(n_{i_1 i_2 i_3} - \frac{n_{i_1 00}\, n_{0 i_2 0}\, n_{00 i_3}}{N^2} \right)^2 \Big/ n_{i_1 00}\, n_{0 i_2 0}\, n_{00 i_3},$$

$$\text{d.f.} = r_1 r_2 r_3 - r_1 - r_2 - r_3 + 2$$

(4.B.4) $$N \sum_{i_1, i_2, i_3} \left(n_{i_1 i_2 i_3} - \frac{n_{i_1 00}\, n_{0 i_2 i_3}}{N} \right)^2 \Big/ n_{i_1 00}\, n_{0 i_2 i_3},$$

d.f. $= (r_1-1)(r_2 r_3-1)$

$$\sum_{i_1, i_2, i_3} \left(n_{i_1 i_2 i_3} - \frac{n_{i_1 0 i_3}\, n_{0 i_2 i_3}}{n_{00 i_3}} \right)^2 \Big/ \left(\frac{n_{i_1 0 i_3}\, n_{0 i_2 i_3}}{n_{00 i_3}} \right),$$

d.f. $= (r_1-1)(r_2-1)r_3$

for testing H_4, H_5 and H_6 respectively.

It may be pointed out here that the hypothesis of pairwise independence of three responses

$$H_7 : p_{i_1 i_2 0} = p_{i_1 00} p_{0 i_2 0},\ p_{0 i_2 i_3} = p_{0 i_2 0} p_{00 i_3},\ p_{i_1 0 i_3} = p_{i_1 00} p_{00 i_3},$$

that of pairwise independence of the first two (separately) with the third, viz.,

$$H_8 : p_{i_1 0 i_3} = p_{i_1 00} p_{00 i_3},\ p_{0 i_2 i_3} = p_{0 i_2 0} p_{00 i_3},$$

and the hypothesis of "no interaction" between three responses (see, for example, [4] or [9]) are hypotheses which have no formal analogs in the normal analysis where pairwise independence between two sets of variables is equivalent to the complete independence of the two sets of variables.

(iv) *Hypothesis of equality of three marginal distributions* :

(4.B.5) $$H_9 : p_{i_1 00} = p_{0 i_1 0} = p_{00 i_1},$$

assuming, of course, that $r_1 = r_2 = r_3$. Let

$$q_1' = \left[q_{100}, q_{200}, \ldots, q_{(r_1-1)00} \right],$$

$$q_2' = \left[q_{010}, q_{020}, \ldots, q_{0(r_1-1)0} \right],$$

$$q_3' = \left[q_{001}, q_{002}, \ldots, q_{00(r_1-1)} \right], \quad q' = [q_1', q_2', q_3'],$$

$$\boldsymbol{Q}_{11} = \text{diagonal}\left(q_{i_1 00},\ i_1 = 1, 2, \ldots, r_1-1\right) \text{ etc.}$$

$$\boldsymbol{Q}_{12} = \left[q_{i_1 i_2 0}\right] \quad i_1, i_2 = 1, 2, \ldots, r_1-1 \text{ etc.}$$

$$\text{(4.B.6)} \quad N\boldsymbol{\Lambda} = [\boldsymbol{Q}_{\alpha\beta} - \boldsymbol{q}_\alpha \boldsymbol{q}_\beta'] \qquad \alpha, \beta = 1, 2, 3,$$

$$\boldsymbol{\Lambda}^{-1} = \boldsymbol{M} \equiv [\boldsymbol{M}_{\alpha\beta}] \qquad \alpha, \beta = 1, 2, 3,$$

$$\boldsymbol{M}_\beta = \Sigma_\alpha \boldsymbol{M}_{\alpha\beta}, \qquad \boldsymbol{M}_0 = \Sigma_\beta \boldsymbol{M}_\beta,$$

and finally

$$\boldsymbol{m} = \Sigma_\beta \boldsymbol{M}_\beta \boldsymbol{q}_\beta .$$

Note that $\boldsymbol{\Lambda}$ is the *sample covariance matrix* of $\boldsymbol{q}$ and hence is nonsingular almost everywhere, excluding, of course, the degenerate case where some variables (or rather the associated probabilities) are linear functions of the remaining variables (i.e., their corresponding probabilities). By applying Theorem 3.2 it can be shown that the χ_1^2-statistic to test H_9 is given by

$$\text{(4.B.7)} \qquad \sum_{\alpha,\beta} \boldsymbol{q}_\alpha' \boldsymbol{M}_{\alpha\beta} \boldsymbol{q}_\beta - \boldsymbol{m}' \boldsymbol{M}_0^{-1} \boldsymbol{m}, \qquad \text{d.f.} = 2(r_1-1).$$

The method can be immediately extended to the case of k variables; the statistic then has $(k-1)(r_1-1)$ degrees of freedom. The case $k = 2$ leads to the statistic (4.A.4). Cochran (1950) has considered the k-variate problem only for the special case $r_1 = r_2 = \ldots = r_k = 2$; even for this special case the statistic (4.B.7) is expected to be more efficient.

H_9 is easily seen to be an analog of the hypothesis $\mu_1 = \mu_2 = \mu_3$ and $\sigma_{11} = \sigma_{22} = \sigma_{33}$ in the normal analysis.

(v) *Hypothesis of equality of "means" of three variables* :

Assume now that the responses are "structured" with $\{a_{i_1}\}$, $\{b_{i_2}\}$ and $\{c_{i_3}\}$ as the scores associated with the respective categories. Then the above hypothesis is expressed by

(4.B.8) $$H_{10} : \sum_{i_1} a_{i1}\, p_{i_100} = \sum_{i_2} b_{i_2}\, p_{0i_20} = \sum_{i_3} c_{i_3}\, p_{00i_3}.$$

Let

$$\gamma_1 = \sum_{i_1} a_{i_1}\, q_{i_100},\ \gamma_2 = \sum_{i_2} b_{i_2}\, q_{0i_20},\ \gamma_3 = \sum_{i_3} c_{i_3}\, q_{00i_3}$$

$$N\lambda_{11} = \Big(\sum_{i_1} a_{i_1}^2\, q_{i_100}\Big) - \gamma_1^2 \text{ etc.}$$

(4.B.9) $$N\lambda_{12} = \Big(\sum_{i_1, i_2} a_{i_1}\, b_{i_2}\, q_{i_1i_20}\Big) - \gamma_1\gamma_2 \text{ etc.}$$

$$\mathbf{\Lambda} = [\lambda_{\alpha\beta}],\ [m_{\alpha\beta}] \equiv \boldsymbol{M} = \mathbf{\Lambda}^{-1},$$

and finally

$$m_\beta = \sum_\alpha m_{\alpha\beta},\quad m_0 = \sum_\beta m_\beta,\quad m = \sum_\beta m_\beta\gamma_\beta,$$

with $\alpha, \beta = 1, 2, 3$. By applying Theorem 3.2 it can be shown that the χ_1^2-statistic is

(4.B.10) $$\sum_{\alpha,\beta} m_{\alpha\beta}\gamma_\alpha\gamma_\beta - (m^2/m_0), \qquad \text{d.f.} = 2.$$

This can be immediately extended to the case of k variables where the statistic would have $k-1$ degrees of freedom.

H_{10} is seen to be an analog of the hypothesis of equality of means of three (possibly correlated) variables in the normal analysis.

C. Three-dimensional tables with two responses and one factor

With the basic set up (2.1) we have now s populations indicated by the subscript j; also i is a double subscript $(i_1 i_2)$, $i_\alpha = 1, 2, \ldots, r_\alpha$, $\alpha = 1, 2$ so that $r = r_1 r_2$ and the porbabilities $p_{i_1i_2j}$ are subject to the constraints $\sum_{i_1, i_2} p_{i_1i_2j} = 1$.

(i) *Hypothesis of independence of the two responses* :

(4.C.1) $$H_{11} : p_{i_1i_2j} = p_{i_10j}\, p_{0i_2j}\,.$$

This implies that H_1 holds for each of the s populations and it is known that we have a χ^2-statistic

$$\text{(4.C.2)} \qquad \sum_j \sum_{i_1, i_2} \left(n_{i_1 i_2 j} - \frac{n_{i_1 0 j}\, n_{0 i_2 j}}{n_{00j}} \right)^2 \Bigg/ \left(\frac{n_{i_1 0 j}\, n_{0 i_2 j}}{n_{00j}} \right)$$

$$\text{d.f.} = (r_1-1)(r_2-1)s$$

which follows immediately from (4.A.2). H_{11} is seen to be an obvious analog of the hypothesis of independence of two variables in each of s (bivariate) normal populations.

(ii) *Hypothesis of "no interaction" between the two responses and one factor* :

This essentially means that the nature of association between the two responses is the same over all factor categories, i.e., for all populations; the formulation depends on what measure of association is chosen for comparison.

$$\text{(4.C.3)} \qquad \begin{aligned} &H_{12} : \frac{p_{i_1 i_2 j}\, p_{r_1 r_2 j}}{p_{i_1 r_2 j}\, p_{r_1 i_2 j}} \quad \text{is independent of } j. \\ &H_{13} : \frac{p_{i_2 i_2 j}}{p_{i_1 0 j}\, p_{0 i_2 j}} \quad \text{is independent of } j. \end{aligned}$$

The formulation H_{12} is due to Goodman (1964) while H_{13} is due to Bhapkar and Koch (1965). These hypotheses are nonlinear and the Wald statistics can be obtained by the "linearization" technique. For details the reader is referred to [9] and [4].

This hypothesis can be seen to be the analog of the hypothesis of equality of s correlations given samples from s bivariate normal populations. Note that the hypothesis H_{11} is a very special case of the hypotheses H_{12}, H_{13}.

(iii) *Hypothesis of homogeneity of marginal distributions* :

$$\text{(4.C.4)} \qquad H_{14} : \begin{aligned} &p_{i_1 0 j} \quad \text{is independent of } j \\ &p_{0 i_2 j} \quad \text{also is independent of } j. \end{aligned}$$

This is seen to be an analog of the hypothesis that $\mu_{\alpha j}$, $\sigma_{\alpha\alpha j}$, $\alpha = 1, 2$, are independent of j, using the usual notation, in the normal analysis. Let

$$\boldsymbol{q}'_{1j} = \left[q_{10j}, q_{20j}, \ldots, q_{(r_1-1)0j}\right],$$

$$\boldsymbol{q}'_{2j} = \left[q_{01j}, q_{02j}, \ldots, q_{0(r_2-1)j}\right], \quad \boldsymbol{q}'_j = [\boldsymbol{q}'_{1j}, \boldsymbol{q}'_{2j}],$$

$$\boldsymbol{Q}_{11j} = \text{diagonal}\left(q_{i_1 0j}, \quad i_1 = 1, 2, \ldots, r_1-1\right) \text{ etc.}$$

(4.C.5) $$\boldsymbol{Q}_{12j} = \left[q_{i_1 i_2 j}\right] \quad i_1 = 1, \ldots, r_1-1 \text{ and } i_2 = 1, \ldots, r_2-1,$$

$$n_{00j}\,\boldsymbol{\Lambda}_j = [\boldsymbol{Q}_{\alpha\beta j} - \boldsymbol{q}_{\alpha j}\,\boldsymbol{q}'_{\beta j}] \qquad \alpha, \beta = 1, 2,$$

$$\boldsymbol{\Lambda}_j^{-1} = \boldsymbol{M}_j, \quad \boldsymbol{M} = \sum_j \boldsymbol{M}_j \quad \text{and} \quad \boldsymbol{t} = \sum_j \boldsymbol{M}_j \boldsymbol{q}_j.$$

Note that $\boldsymbol{\Lambda}_j$ is the *sample covariance matrix* of $\boldsymbol{q}_j$ and is non-singular almost everywhere (excluding the degenerate case). Theorem 3.2 immediately gives the χ_1^2-statistic

(4.C.6) $$\sum_j \boldsymbol{q}'_j \boldsymbol{M}_j \boldsymbol{q}_j - \boldsymbol{t}'\boldsymbol{M}^{-1}\boldsymbol{t}, \qquad \text{d.f.} = (r_1+r_2-2)(s-1)$$

to test H_{14}. The extension to the general k-response case is quite obvious giving a statistic of the same form with $\{\sum_\alpha (r_\alpha - 1)\}$ $(s-1)$ degrees of freedom.

(iv) *Hypothesis of equality of 'means'* :

Suppose now that the two responses are *structured* with $\{a_{i_1}\}$ and $\{b_{i_2}\}$ as the scores. Consider

(4.C.7) $$H_{15}: \quad \begin{array}{ll} \sum_{i_1} a_{i_1} p_{i_1 0j} & \text{is independent of } j \\ \sum_{i_2} b_{i_2} p_{0 i_2 j} & \text{also is independent of } j; \end{array}$$

this is an obvious analog of the hypothesis of equality of mean-vectors of several populations in MANOVA. Let

$$\gamma_{1j} = \sum_{i_1} a_{i_1} q_{i_1 0 j}, \quad \gamma_{2j} = \sum_{i_2} b_{i_2} q_{0 i_2 j}, \quad \boldsymbol{\gamma}'_j = [\gamma_{1j}, \gamma_{2j}],$$

$$n_{00j}\,\lambda_{11j} = \Big(\sum_{i_1} a^2_{i_1} q_{i_1 0 j} \Big) - \gamma^2_{1j} \quad \text{etc.},$$

$$n_{00j}\,\lambda_{12j} = \Big(\sum_{i_1, i_2} a_{i_1} b_{i_2} q_{i_1 i_2 j} \Big) - \gamma_{1j}\,\gamma_{2j},$$

$$(4.C.8) \qquad \boldsymbol{\Lambda}_j = [\lambda_{\alpha\beta j}], \quad \alpha, \beta = 1, 2, \qquad \boldsymbol{M}_j = \boldsymbol{\Lambda}^{-1},$$

and finally

$$\boldsymbol{M} = \sum_j \boldsymbol{M}_j \qquad \text{and} \qquad \boldsymbol{t} = \sum_j \boldsymbol{M}_j \boldsymbol{\gamma}_j.$$

The Theorem 3.2 gives

$$(4.C.9) \qquad \sum_j \boldsymbol{\gamma}'_j \boldsymbol{M}_j \boldsymbol{\gamma}_j - \boldsymbol{t}' \boldsymbol{M}^{-1} \boldsymbol{t} \qquad \text{d.f.} = 2(s-1),$$

as the χ^2_1-statistic to test H_{15}. For the k-response problem we have a statistic of the same type but with $k(s-1)$ degrees of freedom.

Note that H_{14} implies H_{15} and the weaker hypothesis H_{15} may be of interest when H_{14} does not hold.

(v) *Hypothesis of linearity of regression* :

Assume now that the factor is also *structured* with d_j as the score associated with the jth factor-category (or the jth level of the factor). If H_{15} does not hold, we may test the hypothesis

$$(4.C.10) \qquad H_{16}: \quad \begin{aligned} \sum_{i_1} a_{i_1} p_{i_1 0 j} &= \xi^{(1)} + \eta^{(1)} d_j \\ \sum_{i_2} b_{i_2} p_{0 i_2 j} &= \xi^{(2)} + \eta^{(2)} d_j, \end{aligned}$$

where η's are in the nature of regression coefficients, η's and ξ's are unknown. This is an analog of the hypothesis : $\boldsymbol{\mu}_j = \boldsymbol{\xi} + \boldsymbol{\eta} d_j$ in MANOVA. In the notation of (4.C.8) and with

$$\boldsymbol{W} = \sum_j \boldsymbol{M}_j \boldsymbol{\gamma}_j d_j, \quad \boldsymbol{R} = \sum_j \boldsymbol{M}_j d_j, \quad \boldsymbol{S} = \sum_j \boldsymbol{M}_j d_j^2,$$

the χ_1^2-statistic is seen to be

$$(4.C.11) \quad \sum_j \boldsymbol{\gamma}_j' \boldsymbol{M}_j \boldsymbol{\gamma}_j - [\boldsymbol{t}', \boldsymbol{W}'] \begin{bmatrix} \boldsymbol{M} & \boldsymbol{R} \\ \boldsymbol{R} & \boldsymbol{S} \end{bmatrix}^{-1} \begin{bmatrix} \boldsymbol{t} \\ \boldsymbol{W} \end{bmatrix} \quad \text{d.f.} = 2(s-2).$$

Again the k-response extension is immediate where the statistic would have $k(s-2)$ degrees of freedom.

It may be pointed out here that the hypotheses of the type H_{14}, H_{15} and H_{16} were proposed earlier by Roy and Bhapkar (1960). The test of the hypothesis

$$\boldsymbol{\eta} = \boldsymbol{0}$$

with $\boldsymbol{\eta}$ given in H_{16}, or in other words, of the hypothesis H_{15} with H_{16} as the model is provided by the statistic

$$(4.C.9)-(4.C.11) \qquad \text{d.f.} = 2;$$

a statistic of this type would have k degrees of freedom in the general k-response case.

D. Four-dimensional tables with two responses and two factors

In the basic set up (2.1) now we have i a double subscript $(i_1 i_2)$, $i_\alpha = 1, 2, \ldots, r_\alpha$, $\alpha = 1, 2$, so that $r = r_1 r_2$ as before and j also a double subscript $(j_1 j_2)$, $j_\beta = 1, 2, \ldots, s_\beta$, $\beta = 1,2$ with not all combinations selected necessarily for the experiment; let s be the number of $(j_1 j_2)$ combinations selected so that $s \leqslant s_1 s_2$. For convenience we shall denote by j_1 the categories of the *block-factor* (i.e., the *blocks*) and by j_2 the *treatments* (i.e., the categories of the *treatment-factor*).

(i) *Hypothesis of no 'treatment-effect' on the marginal distributions* :

$$(4.D.1) \qquad H_{17}: \begin{array}{ll} p_{i_1 0 j_1 j_2} & \text{is independent of } j_2 \\ p_{0 i_2 j_1 j_2} & \text{is also independent of } j_2. \end{array}$$

This may be seen to be an analog of the hypothesis that $\mu_{\alpha j_1 j_2}$, $\sigma_{\alpha j_1 j_2}$ $\alpha = 1, 2$, are independent of j_2 in MANOVA. In the notation of (4.C.5) with j a double subscript $(j_1 j_2)$ let further

$$\text{(4.D.2)} \qquad \boldsymbol{M}_{j_1} = \sum_{j_2} \boldsymbol{M}_{j_1 j_2}, \qquad \boldsymbol{t}_{j_1} = \sum_{j_2} \boldsymbol{M}_{j_1 j_2} \boldsymbol{q}_{j_1 j_2},$$

the summation being over those j_2 only which occur in conjunction with j_1. From Theorem 3.2 we have the χ_1^2-statistic

$$\text{(4.D.3)} \qquad \sum_{j_1, j_2} \boldsymbol{q}'_{j_1 j_2} \boldsymbol{M}_{j_1 j_2} \boldsymbol{q}_{j_1 j_2} - \sum_{j_1} \boldsymbol{t}'_{j_1} \boldsymbol{M}_{j_1}^{-1} \boldsymbol{t}_{j_1},$$

$$\text{d.f.} = (r_1 + r_2 - 2)(s - s_1)$$

to test H_{17}. The method can be immediately generalized to the case of k responses giving a statistic of the same type with $\{\sum_{\alpha} (r_\alpha - 1)\}(s - s_1)$ degrees of freedom.

(ii) *Hypothesis of no 'treatment effect' on the 'means'* :

If the two responses are *structured* with $\{a_{i_1}\}$ and $\{b_{i_2}\}$ as the associated scores, consider

$$\text{(4.D.4)} \qquad H_{18}: \quad \begin{array}{ll} \sum_{i_1} a_{i_1} p_{i_1 0 j_1 j_2} & \text{is independent of } j_2 \\[2ex] \sum_{i_2} b_{i_2} p_{0 i_2 j_1 j_2} & \text{is also independent of } j_2. \end{array}$$

This is the analog of the usual hypothesis of no *treatment-effects* in MANOVA. In the notation of (4.C.8) but with j a double subscript $(j_1 j_2)$ let further

$$\text{(4.D.5)} \qquad \boldsymbol{M}_{j_1} = \sum_{j_2} \boldsymbol{M}_{j_1 j_2} \qquad \boldsymbol{t}_{j_1} = \sum_{j_2} \boldsymbol{M}_{j_1 j_2} \boldsymbol{\Upsilon}_{j_1 j_2}.$$

For testing H_{18} we have the χ_1^2-statistic

$$\text{(4.D.6)} \qquad \sum_{j_1 j_2} \boldsymbol{\Upsilon}'_{j_1 j_2} \boldsymbol{M}_{j_1 j_2} \boldsymbol{\Upsilon}_{j_1 j_2} - \sum_{j_1} \boldsymbol{t}'_{j_1} \boldsymbol{M}_{j_1}^{-1} \boldsymbol{t}_{j_1}, \qquad \text{d.f.} = 2(s - s_1).$$

For the k-response situation the statistic would have $k(s-s_1)$ degrees of freedom. Here again we note that H_{17} implies H_{18} and the weaker hypothesis H_{18} would be of interest when H_{17} does not hold.

(iii) *Hypothesis of linearity of regression on treatment level* :

We assume now that the second factor is *structured* and d_{j_2} is the score associated with the category j_2; in other words, d_{j_2} is the *level* of the j_2th treatment. If H_{18} does not hold, we may consider

$$\text{(4.D.7)} \qquad H_{19}: \begin{array}{l} \sum\limits_{i_1} a_{i_1} p_{i_1 0 j_1 j_2} = \xi^{(1)}_{j_1} + \eta^{(1)}_{j_1} d_{j_2} \\ \sum\limits_{i_2} b_{i_2} p_{0 i_2 j_1 j_2} = \xi^{(2)}_{j_1} + \eta^{(2)}_{j_1} d_{j_2}, \end{array}$$

where ξ's and η's are unknown. This hypothesis is the analog of the hypothesis that

$$\boldsymbol{\mu}_{j_1 j_2} = \boldsymbol{\xi}_{j_1} + \boldsymbol{\eta}_{j_1} d_{j_2}$$

in MANOVA. In the notation of (4.C.8) with $j = (j_1 j_2)$ and of (4.D.5) let further

$$\text{(4.D.8)} \quad \boldsymbol{w}_{j_1} = \sum_{j_2} \boldsymbol{M}_{j_1 j_2} \boldsymbol{\gamma}_{j_1 j_2} d_{j_2}, \quad \boldsymbol{R}_{j_1} = \sum_{j_2} \boldsymbol{M}_{j_1 j_2} d_{j_2}, \quad \boldsymbol{S}_{j_1} = \sum_{j_2} \boldsymbol{M}_{j_1 j_2} d^2_{j_2}.$$

Then for testing H_{19} from theorem 3.2 we have the χ_1^2-statistic

$$\text{(4.D.9)} \quad \sum_{d_1, d_2} \boldsymbol{\gamma}'_{j_1 j_2} \boldsymbol{M}_{j_1 j_2} \boldsymbol{\gamma}_{j_1 j_2} - \sum_{j_1} \left[\boldsymbol{t}'_{j_1}, \boldsymbol{w}'_{j_1} \right] \begin{bmatrix} \boldsymbol{M}_{j_1} & \boldsymbol{R}_{j_1} \\ \boldsymbol{R}_{j_1} & \boldsymbol{S}_{j_1} \end{bmatrix} \begin{bmatrix} \boldsymbol{t}_{j_1} \\ \boldsymbol{w}_{j_1} \end{bmatrix}$$

$$\text{d.f.} = 2(s-2s_1)$$

For the k-response problem, the statistic would have $k(s-2s_1)$ degrees of freedom.

It may be mentioned here that the hypothses H_{17}, H_{18}, H_{19} were offered as analogs earlier by Roy and Bhapkar (1960). To test the hypothesis

$$\boldsymbol{\eta}_{j_1} = \boldsymbol{0},$$

with η's given by (4.D.7), i.e., the hypothesis H_{18} with H_{19} as the model, we have the statistic

$$(4.D.6)-(4.D.9) \qquad \text{d.f.} = 2s_1;$$

a statistic of the same type would have ks_1 degrees of freedom in the k-response case.

Similarly we can test the hypotheses $\eta_{j_1} = \eta$ (known or unknown as the case may be) and/or $\xi_{j_1} = \xi$ under the model H_{19}; these details are omitted.

(iv) *Hypotheses of "no interaction"* :

This raises a number of problems indeed depending on whether we are interested in the nature of association between the two responses over the categories of the two factors or in the effects of factors on the marginal distributions of the responses. We are then faced with defining *interactions* of different orders and testing hypotheses about these. Hence this is omitted in further discussion and the reader is referred to Bhapkar and Koch (1965) for further details.

5. REMARKS ON HIGHER DIMENSIONAL TABLES

Most of these problems can be handled in much the same way as the simpler problems discussed earlier. Thus, as mentioned earlier, the hypothesis of equality of marginal distributions of several responses, or of equality of *means* of several responses can be tested by methods discussed in 4.B; the hypothesis of homogeneity of marginal distributions or the one of equality of *means* etc., for several populations and the general multi-response problem can be tested by methods in 4.C; the various hypotheses of no *treatment effect* or of the linearity of the treatment-effect in the general multi-response two-factor situation (with one factor in the nature of *blocks*) can be handled by methods in 4.D. Problems of this nature for the general k-response l-factor situation present no further difficulties and can be handled in precisely the same way. It is when we are considering problems of interactions of various types and of various orders that further difficulties arise; some of these are discussed in [4] and [5].

Problems of associations similar to those in H_4, H_5 and H_6 can be handled in a fairly straightforward manner for the general k-response single population situation as follows :

(i) *Hypothesis of complete independence of k responses* :

$$(5.1) \qquad H_{20} : p_{i_1 i_2 \cdots i_k} = p_{i_1 00\cdots 00}\, p_{0 i_2 0\cdots 00} \cdots p_{000\cdots 0 i_k}.$$

It is easy to check, that, as in H_4, we immediately get maximum likelihood estimate $\hat{p}_{i_1 0\cdots 0} = n_{i_1 0\cdots 0}/N$ etc. giving the χ^2-statistic

$$(5.2) \qquad N^{k-1} \sum_{i_1,\ldots,i_k} \left(n_{i_1\cdots i_k} - \frac{n_{i_1 0\cdots 0} \cdots n_{00\cdots i_k}}{N^{k-1}} \right)^2 \Big/ \left(n_{i_1 0\cdots 0} \cdots n_{0\cdots 0 i_k} \right)$$

$$\text{d.f.} = r_1 r_2 \ldots r_k - (r_1 + r_2 + \ldots + r_k) + (k-1)$$

(ii) *Hypothesis of independence of two sets of responses* :

$$(5.3) \qquad H_{21} : p_{i_1 i_2 \cdots i_k} = p_{i_1 i_2 \cdots i_{k_1} 0\cdots 0}\, p_{00\cdots 0 i_{k_1+1} \cdots i_k}$$

where we assume, without loss of generality, that the first k_1 responses form the first set. As in H_5, we get immediately the maximum likelihood estimates $\hat{p}_{i_1 \ldots i_{k_1} 0 \ldots 0} = n_{i_1 \ldots i_{k_1} 0 \ldots 0}/N$ etc. giving the χ^2-statistic

$$(5.4) \qquad N \sum_{i_1, \ldots, i_k} \left(n_{i_1 i_2 \cdots i_k} - \frac{n_{i_1 \ldots i_{k_1} 0 \ldots 0}\, n_{0 \ldots 0 i_{k_1+1} \ldots i_k}}{N} \right)^2 \Big/$$

$$\left(n_{i_1 \ldots i_{k_1} 0 \ldots 0}\, n_{0 \ldots 0 i_{k_1+1} \ldots i_k} \right) \text{d.f.} = \left(r_1 \ldots r_{k_1} - 1 \right) \left(r_{k_1+1} \ldots r_k - 1 \right).$$

Actually the statistic follows directly from (4.A.2) noting that $(i_1 \ldots i_{k_1})$ can be regarded as one subscript i_1^* and $(i_{k_1+1} \ldots i_k)$ as i_2^*.

(iii) *Hypothesis of independence of several sets of responses* :

$$(5.5) \qquad H_{22} : p_{i_1 i_2 \cdots i_k} = p_{i_1 \cdots i_{k_1} 0 \cdots 0}\, p_{0 \cdots 0 i_{k_1+1} \cdots i_{k_1+k_2} 0 \cdots 0} \cdots p_{0 \cdots 0 i_{k_1+ \cdots + k_{t-1}+1} \cdots i_k}$$

where we assume that there are t sets of responses, the first containing the first k_1 responses and so on. The statistic for H_{22} follows immediately from (5.2) regarding $i_1^* = (i_1 \ldots i_{k_1})$ and so on so that $r_1^* = r_1 \ldots r_{k_1}$ etc., and replacing k in (5.2) by t with $r_1^* \ldots r_t^* - (r_1^* + \ldots + r_t^*) + (t-1)$, i.e., $r_1 r_2 \ldots r_k - \left(r_1 r_2 \ldots r_{k_1} + \ldots + r_{k_1 + \ldots + k_{t-1}+1} \cdots r_k \right) + t - 1$ degrees of freedom.

(iv) *Hypothesis of independence of two sets of responses given a third set* :

$$(5.6) \qquad H_{23} : p_{i_1 i_2 \cdots i_k} = \frac{p_{i_1 \cdots i_{k_1} 0 \cdots 0 i_{k_1+k_2+1} \cdots i_k}\, p_{0 \cdots 0 i_{k_1+1} \cdots i_k}}{p_{0 \cdots 0 i_{k_1+k_2+1} \cdots i_k}};$$

here we want to test the conditional independence of the first two sets given the third set of responses. The χ^2-statistic is immediately seen to be

$$(5.7) \qquad \sum_{i_1, \ldots, i_k} \frac{\left(n_{i_1 \cdots i_k} - \dfrac{n_{i_1 \cdots i_{k_1} 0 \cdots 0 i_{k_1+k_2+1} \cdots i_k} n_{0 \cdots 0 i_{k_1+1} \cdots i_k}}{n_{0 \cdots 0 i_{k_1+k_2+1} \cdots i_k}} \right)^2}{\left(\dfrac{n_{i_1 \cdots i_{k_1} 0 \cdots 0 i_{k_1+k_2+1} \cdots i_k}\, n_{00 \cdots 0 i_{k_1+1} \cdots i_k}}{n_{0 \cdots 0 i_{k_1+k_2+1} \cdots i_k}} \right)}$$

$$\text{d.f.} = \left(r_1 \cdots r_{k_1} - 1 \right) \left(r_{k_1+1} \cdots r_{k_1+k_2} - 1 \right) r_{k_1+k_{2+1}} \cdots r_k .$$

Note that H_{20} is an analog of the hypothesis that the correlation matrix is the unit matrix, while H_{21}, H_{22} and H_{23} are analogs of the hypothesis of no canonical correlations, Wilks' hypothesis of independence of sets of variates and the hypothesis of no partial canonical correlations, respectively, in the normal multivariate analysis.

References

[1] Barnard, G. A. (1947). "Significance Tests for 2×2 Tables," *Biometrika*, **34**, 123–138.

[2] Bhapkar, V. P. (1961). "Some Tests for Categorical Data," *Ann. Math. Statist.*, **32**, 72–83.

[3] Bhapkar, V. P. (1966). "A note on the Equivalence of Two Test Criteria for Hypotheses in Categorical Data," *J. Amer. Statist. Assoc.*, **61**, 228–235.

[4] Bhapkar, V. P. and Koch, Gary G. (1965). "On the Hypothesis of No Interaction in Three-Dimensional Contingency Tables," Institute of Statistics, Mimeo Series No. 440, University of North Carolina.

[5] Bhapkar, V. P. and Koch, Gray G. (1965). "The Hypotheses of No Interaction in Four-Dimensional Contingency Tables," Institute of Statistics, Mimeo Series No. 447, University of North Carolina.

[6] Cochran, W. G. (1950). "The Comparison of Percentages in Matched Samples," *Biometrika*, **37**, 256–266.

[7] Cramér, H. (1946). *Mathematical Methods of Statistics*, Princeton University Press, Princeton.

[8] Fisher, R. A. (1922). "On the Interpretation of Chi-Square from Contingency Tables and the Calculation of *p*," *J. Roy. Statist. Soc.*, **85**, 87–94.

[9] Goodman, L. A. (1964). "Simple Methods for Analyzing Three-Factor Interaction in Contingency Tables," *J. Amer. Statist. Assoc.*, **59**, 319–352.

[10] Mitra, S. K. (1955). "Contributions to the Statistical Analysis of Categorical Data," Institute of Statistics, Mimeo Series No. 142, University of North Carolina.

[11] Neyman, J. (1949). "Contribution to the Theory of the χ^2 Test," *Proc. Berkeley Symposium on Mathematical Statistics and Probability*, University of California Press, Berkeley, pp. 239–273.

[12] Pearson, E. S. (1947). "The Choice of Statistical Tests Illustrated on the Interpretation of Data Classed in a 2×2 Tables," *Biometrika*, **34**, 139–167.

[13] Pearson, K. (1900). "On the Criterion that a Given System of Deviations From the Probable in the Case of a Correlated System of Variables Is Such That It Can Be Reasonably Supposed To Have Arisen From Random Sampling," *Philosophical Magazine*, Series 5, **50**, 157–172.

[14] Roy, S. N. and Bhapkar, V. P. (1960). "Some Nonparametric Analogs of Normal ANOVA, MANOVA, and of Studies in Normal Association," *Contributions to Probability and Statistics*, Stanford University Press Stanford, 371–387.

[15] Roy, S. N., and Mitra, S. K. (1956). "An Introduction to Some Nonparametric Generalizations of Analysis of Variance and Multivariate Analysis," *Biometrika*, **43**, 361–376.

[16] Sathe, Y. S. (1962). "Studies of Some Problems in Nonparametric Inference" Institute of Statistics, Mimeo Series No. 325, University of North Carolina.

[17] Stuart, A. (1955). "A Test for Homogeneity of the Marginal Distributions in a Two-Way Classification," *Biometrika*, **42**, 412–416.

[18] Wald, A. (1943). "Tests of Statistical Hypotheses concerning Several Parameters When the Number of Observations is Large," *Trans. Amer. Math. Soc.*, **54**, 426–482.

(*Received Oct. 30, 1965.*)

CHAPTER 6

Estimating Multinomial Response Relations

R. DARRELL BOCK*, *The University of Chicago*

1. INTRODUCTION AND SUMMARY

A method for estimating a multinomial response relation is developed as a generalization of the logit analysis of bio-assay. A definition of a multivariate logit is introduced for this purpose. The method applies generally in the analysis of multi-factor multi-response data when the response is qualitative but not ranked, as in analysis of contingency tables, testing group differences in proportions, item analysis, etc. Maximum likelihood estimation using numerical methods implemented by computer is employed. Examples are presented demonstrating the relationship of the method to other solutions for contingency tables.

In this paper a widely used method for estimating binomial response relations in quantal data is generalized to the multinomial case. The solution obtained parallels in many respects the nonparametric generalization of analysis of variance and multivariate analysis conceived by S. N. Roy (1957, Chapter 15).

*Preparation of this paper was supported in part by NSF Grant GS-1025. Computer time was donated by Jacob L. Gewirtz.

It departs from Roy's formulation only to the extent of expressing the response probabilities of the multinomial law as logistic transforms of parameters of the ultimate model. The logistic transformation is introduced in order to avoid computational difficulties because of inadmissible estimates, and in the hope of simplifying the relationship between the response probabilities and the independent variables.

The proposed solution has the merit of providing a unified treatment of a number of special cases in the analysis of qualitative data. These include 1) fitting the dosage response curve in conventional bioassay; 2) predicting a polychotomous response in bioassay when the categories are purely nominal and have no implied ranking; 3) comparing response relationships in two or more populations; 4) testing equality of proportions; 5) analyzing complex contingency tables; 6) determining item characteristics in psychological tests; and 7) representing the distribution of dichotomous item responses. The case of ordered response categories, which requires a different type of model, has already been considered by Ashford (1959) and Bock and Jones (1968).

The calculations required in the solution are easy to program for machine computation and, except in unusual circumstances discussed below, always give admissible estimates. The model required in the solution may be communicated to the computer by the same symbolism which has been devised for programming of univariate and multivariate analysis of variance (Bock, 1963 and 1965).

To introduce the multinomial generalizations, we review briefly the well-known results for estimation of a binomial response relation with the aid of the logistic response law (see Anscombe, 1956).

2. THE BINOMIAL CASE

In bioassay, and also in certain psychological applications (Bock and Jones, 1967; Maxwell, 1959), the following type of experiment is common. The experimenter arranges n conditions under which the responses of the subjects are observed. These conditions differ in physically identifiable or measurable ways which are hypothesized to influence probability of response. In simple studies, the conditions might represent a level of treatment, but in more complex studies they might represent subclasses of a factorial design of multiple treatments. In any case, let us suppose that the experimenter observes N_j subjects under condition j and scores dichotomously each subject according to the presence or absence of some response. Let the number of subjects who show the response be r_{j1} and the numbers who fail to show the response be $r_{j2} = N_j - r_{j1}$. Let the corresponding response proportions be $p_{j1} = r_{j1}/N_j$ and $p_{j2} = r_{j2}/N_j$. The experimenter's problem is to predict these proportions by estimating response probabilities P_{j1} and $P_{j2} = 1 - P_{j1}$, expressed as functions of the physical (independent) variables which determine the experimental conditions.

In biological and psychological applications it has proved fruitful to assume that the relationship between the response probabilities and the physical variables will be simplified if the probabilities are first transformed by an inverse normal or logistic transformation (see Finney, 1952; Berkson, 1955; Bock and Jones, 1968). In many cases the relationship after transformation is described by a low-degree polynomial (often linear) in the physical variables. Ease of computation makes the logistic transformation particularly convenient when estimating the parameters of such a relationship. The procedure is as follows :

In the binomial case the logistic response law is defined by

$$P_{j1} = \frac{e^{Z_j}}{1+e^{Z_j}}$$

or

$$\log(P_{j1}/P_{j2}) = Z_j,$$

where

$$Z_j = \beta_1 x_{j1} + \beta_2 x_{j2} + \ldots + \beta_q x_{jq}.$$

The x_{jk} are values of the physical variables, or terms of a polynomial in the physical variable, which describe the jth experimental condition.

With random sampling of subjects for each experimental condition, the number of responses r_{j1} is binomially distributed with parameters N_j and P_{j1}. In this case, the parameters $\beta_1, \beta_2, \ldots, \beta_q$ of the response relationship have simple sufficient statistics and the likelihood equations may be immediately written down :

$$\beta_k : \sum_j^n (r_{j1} - N_j P_{j1}) x_{jk} = 0$$

These equations are readily solved by the Newton-Raphson procedure. The second derivatives are :

$$\beta_k, \beta_l : \; - \sum_j^n N_j P_{j1} P_{j2} x_{jk} x_{jl}$$

Thus, from provisional estimates of the parameters, say $\hat{\beta}_k^{(i)}$, improved estimates may be obtained by adding the corrections $(\delta_k^{(i)})$, which are the solution of the equations :

$$\sum_k^q (\delta_k^{(i)}) \sum_j^n N_j \hat{P}_{j1}^{(i)} \hat{P}_{j2}^{(i)} x_{jk} x_{jl} = \sum_j^n (r_{j1} - N_j \hat{P}_{j1}^{(i)}) x_{jl},$$

where $l = 1, 2, \ldots, q$, and $\hat{P}_{j1}^{(i)}$ and $\hat{P}_{j2}^{(i)}$ are obtained from logits computed with $\hat{\beta}_k^{(i)}$.

In computer applications of this solution with the $\hat{\beta}_k^{(i)}$ initially set to zero, the Newton-Raphson process usually converges in five or six iterations to finite values of the parameters. It is possible, however, to encounter samples in which no finite values of the parameters maximize the likelihood. Berkson (1955) gives some examples of this problem for the case of three equally spaced doses. When two of the three sample response proportions are 0 or 1 in this situation, the maximum likelihood solution fails in these cases. Berkson (1955, 1960) has cited these examples in arguing that the estimates of β_k are not one-to-one functions of the minimal sufficient statistics $\sum_j^n r_{j1} x_k$, and, therefore, are not sufficient. Silverstone (1957), on the other hand, shows that certain degenerate response curves of the two-parameter logistic family can be defined which formally extend the domain of response probabilities to 0 and 1 and preserve sufficiency. Some provision for including these limiting forms could be incorporated in the computer program if estimated response probabilities are desired even when some of the estimated β_k becomes infinite.

The asymmetry of the foregoing solution with respect to the response categories can be avoided by formulating a model separately for each category and reparameterizing in the form of contrasts between the two models. Thus, we may define the logistic response law as :

$$P_{j1} = \frac{e^{Z_{j1}}}{e^{Z_{j1}} + e^{Z_{j2}}}$$

$$P_{j2} = \frac{e^{Z_{j2}}}{e^{Z_{j1}} + e^{Z_{j2}}}$$

where $$Z_{j1} = \beta_1 x_{j1} + \beta_2 x_{j2} + \ldots + \beta_q x_{jq},$$

and $$Z_{j2} = -Z_{j1}.$$

With this form of response law, the likelihood equations take the form,

$$\beta_k : \sum_j^n [(r_{j1} - r_{j2}) - N_j(P_{j1} - P_{j2})] x_{jk} = 0.$$

The second derivatives are

$$\beta_k, \beta_l : -4 \sum_j^n N_j P_{j1} P_{j2} x_{jk} x_{jl}$$

Since $(r_{j1} - r_{j2}) - N_j(P_{j1} - P_{j2}) = 2(r_{j1} - N_j P_{j1})$, the estimates obtained with this version of the law differ from the previous estimates only by the factor 1/2, which may be absorbed in the parameters. It is this version of the logistic response law which most conveniently generalizes to the multinomial case.

3. THE MULTINOMIAL CASE

3.1. The response law

Now let the number of categories in the response classification be $m \geqslant 2$. Let the number of subjects under experimental condition j ($j = 1, 2, \ldots, n$) who fall in category h of the classification be r_{jh} where $\sum_{h}^{m} r_{jh} = N_j$. Assume random sampling of subjects, so that the probability of the response frequencies r_{jh} is given by the product multinomial :

$$\prod_{j}^{n} \frac{N_j\,!}{r_{j1}\,!\,r_{j2}\,!\ldots r_{jm}\,!} P_{j1}^{r_{j1}} P_{j2}^{r_{j2}} \ldots P_{jm}^{r_{jm}}$$

To express the response probabilities as functions of the experimental variables, let us introduce the multivariate logits $Z_{j1}, Z_{j2}, \ldots, Z_{jm}$, and generalize the logistic response law as follows :

$$\text{(1)} \qquad \begin{aligned} P_{j1} &= e^{Z_{j1}}/D_j \\ P_{j2} &= e^{Z_{j2}}/D_j \\ &\vdots \\ P_{jm} &= e^{Z_{jm}}/D_j, \end{aligned}$$

where

$$D_j = e^{Z_{j1}} + e^{Z_{j2}} + \ldots + e^{Z_{jm}}.$$

This generalization of the binomial logit is implicit in the work of Luce (1959) and has been proposed more recently by Mantel (1966).

In setting up a linear model connecting the logits with the physical variables, we must provide for the possibility that a structure of the categories is implied in the response classification. Most typically, the structure will arise because the ultimate categories are the result of a crossing or nesting of several mutually exclusive classifications of the same subjects. A model for the logits sufficiently general to include both a structured response classification and multiple experimental variables is as follows:

$$\begin{bmatrix} Z_{11} & Z_{12} & \cdots & Z_{1m} \\ Z_{21} & Z_{22} & \cdots & Z_{2m} \\ \cdot & \cdot & & \cdot \\ Z_{n1} & Z_{n2} & \cdots & Z_{nm} \end{bmatrix} = \begin{bmatrix} x_{11} & x_{12} & \cdots & x_{1q} \\ x_{21} & x_{22} & \cdots & x_{2q} \\ \cdot & \cdot & & \cdot \\ x_{n1} & x_{n2} & \cdots & x_{nq} \end{bmatrix}$$

$$\cdot \begin{bmatrix} \beta_{11} & \beta_{12} & \cdots & \beta_{1t} \\ \beta_{21} & \beta_{22} & \cdots & \beta_{2t} \\ \cdot & \cdot & & \cdot \\ \beta_{q1} & \beta_{q2} & \cdots & \beta_{qt} \end{bmatrix} \cdot \begin{bmatrix} a_{11} & a_{12} & \cdots & a_{1m} \\ a_{21} & a_{22} & \cdots & a_{2m} \\ \cdot & \cdot & & \cdot \\ a_{t1} & a_{t2} & \cdots & a_{tm} \end{bmatrix}$$

In matrix notation,

$$\underset{n\times m}{\boldsymbol{Z}} = \underset{n\times q}{\boldsymbol{X}} \cdot \underset{q\times t}{\boldsymbol{B}} \cdot \underset{t\times m}{\boldsymbol{A}} .$$

The matrix $\boldsymbol{X}$ is the familiar model matrix which appears in univariate and multivariate linear models for continuous normal variables. In the present model, it accounts for variation in the response probabilities over the experimental conditions. In many cases, $\boldsymbol{X}$ is the model matrix for crossed or nested designs and is not of full rank. Suppose the rank of $\boldsymbol{X}$ is $r \leqslant q$. Then it will be necessary to reparameterize the model by expressing the model matrix as, say,

$$\boldsymbol{X} = \underset{n\times r}{\boldsymbol{K}} \cdot \underset{r\times q}{\boldsymbol{L}} ,$$

where the rows of $\boldsymbol{L}$ are coefficients of linearly estimable functions of the parameters in the columns of $\boldsymbol{B}$. That is, $\boldsymbol{L}$ must be of rank r and must satisfy the usual condition for linear estimability:

$$\text{rank} \begin{bmatrix} \boldsymbol{X} \\ \boldsymbol{L} \end{bmatrix} = \text{rank}\,[\boldsymbol{X}]$$

On these conditions,

$$\boldsymbol{K} = \boldsymbol{X}\boldsymbol{L}'(\boldsymbol{L}\boldsymbol{L}')^{-1}$$

(See Bock, 1963, for examples of the construction of $\boldsymbol{K}$). Since the elements in $\boldsymbol{X}$ are physically measurable or identifiable variables of the experimental design, it is appropriate to refer to $\boldsymbol{X}$ as the *physical* part of the model. The matrix $\boldsymbol{K}$ is a column basis of $\boldsymbol{X}$.

The role of the matrix $\boldsymbol{A}$ in the model is to account for variation of the response probabilities across categories of the response classification. Like $\boldsymbol{X}$, it will frequently be of less than full rank. Suppose Rank $(\boldsymbol{A}) = s \leqslant t$. Then it will also be necessary to reparameterize the model by factoring $\boldsymbol{A}$ as, say,

$$\boldsymbol{A} = \underset{t\times s}{\boldsymbol{S}} \cdot \underset{s\times m}{\boldsymbol{T}},$$

where rank $\boldsymbol{S} = s$ and the columns of $\boldsymbol{S}$ are linearly dependent on those of $\boldsymbol{A}$. As before, $\boldsymbol{T}$ may be obtained from,

$$\boldsymbol{T} = (\boldsymbol{S}'\boldsymbol{S})^{-1}\boldsymbol{S}'\boldsymbol{A}.$$

We may refer to $\boldsymbol{A}$ as the *response* part of the model. The matrix $\boldsymbol{T}$ is a row basis of $\boldsymbol{A}$. It is the analogue of the "post-factor" in Roy's (1957) formulation of multivariate analysis of variance, or the matrix of coefficients of latent variables in Bock and Bargmann's (1966) models for covariance structures. Some discussion of its construction is given in Example 1 of Section 3.

When $\boldsymbol{X}$ and $\boldsymbol{A}$ are of deficient rank, reparameterization of the model is introduced so that the matrix of second derivatives of the log likelihood will be non-singular. Non-singularity of this matrix is a necessary condition for obtaining maximum likelihood estimates of the parameters of the model by the method of the following section.

3.2. Estimation

The complete model reparameterized to full rank is

$$\begin{aligned} \boldsymbol{Z} &= \boldsymbol{K}(\boldsymbol{L}\boldsymbol{B}\boldsymbol{S})\boldsymbol{T} \\ &= \underset{n\times r}{\boldsymbol{K}} \cdot \underset{r\times s}{\boldsymbol{\Gamma}} \cdot \underset{s\times m}{\boldsymbol{T}} \qquad (2) \end{aligned}$$

Now it is the $r \times s$ matrix of parameters $\mathbf{\Gamma}$ which is to be estimated. Taking over the terminology of linear statistical models, we will refer to the elements of $\mathbf{\Gamma}$ as *effects*. As in the binomial case, maximum likelihood estimation of these parameters is especially convenient.

Let K_{jk} be the j, k element of $\boldsymbol{K}$, T_{hi} the h, i element of $\boldsymbol{T}$, and γ_{kh} the k, h element of $\mathbf{\Gamma}$.

Then the first derivative of the log likelihood with respect to

$$\gamma_{kh}, k = 1, 2, \ldots, r, \quad h = 1, 2, \ldots, s,$$

is

$$\frac{\partial \log L}{\partial \gamma_{kh}} = \sum_j^n \left\{ \frac{r_{j1}}{P_{j1}} \cdot \frac{\partial P_{j1}}{\partial \gamma_{kh}} + \frac{r_{j2}}{P_{j2}} \cdot \frac{\partial P_{j2}}{\partial \gamma_{kh}} + \ldots + \frac{r_{jm}}{P_{jm}} \cdot \frac{\partial P_{jm}}{\partial \gamma_{kh}} \right\}.$$

From (1)

$$\frac{\partial P_{j1}}{\partial \gamma_{kh}} = P_{j1}\,[T_{h1}(1-P_{j1}) - T_{h2}P_{j2} - \ldots - T_{km}P_{jm}]K_{jk},$$

$$\frac{\partial P_{j2}}{\partial \gamma_{kh}} = P_{j2}\,[-T_{h1}P_{j1} + T_{h2}(1-P_{j2}) - \ldots - T_{hm}P_{jm}]K_{jk},$$

$$\vdots$$

$$\frac{\partial P_{jm}}{\partial \gamma_{kh}} = P_{jm}\,[-T_{h1}P_{j1} - T_{h2}P_{j2} - \ldots + T_{hm}(1-P_{jm})]K_{jk}.$$

Thus, the likelihood equations are, say,

$$\varphi_{\gamma_{kh}} = \sum_j^n \{T_{h1}(r_{j1} - N_jP_{j1}) + T_{h2}(r_{j2} - N_jP_{j2}) + \ldots$$
$$+ T_{hm}(r_{jm} - N_jP_{jm})\}K_{jk} = 0.$$

Similarly, the second derivatives of the log likelihood are

$$\varphi_{\gamma_{kh}\gamma_{il}} = -\sum_j^n N_j\,\{P_{j1}[T_{l1}(1-P_{j1}) - T_{l2}P_{j2} - \ldots - T_{lm}P_{jm}]T_{h1}$$
$$+ P_{j2}\,[-T_{l1}P_{j1} + T_{l2}(1-P_{j2}) - \ldots - T_{lm}P_{jm}]T_{h2} + \ldots$$
$$+ P_{jm}[-T_{l1}P_{j1} - T_{l2}P_{j2} - \ldots - T_{lm}(1-P_{jm})]T_{hm}\}K_{jk}K_{ji}.$$

For purposes of a computer solution of the likelihood equations, it is helpful to express these derivatives in matrix form.

The Newton-Raphson process can then be programmed interpretively using systems of matrix operations such as that of Bargmann (1965) or Bock and Peterson (1967).

For experimental condition j, define the vectors of response frequencies and response probabilities:

$$\boldsymbol{r}_j = \begin{bmatrix} r_{j1} \\ r_{j2} \\ \vdots \\ r_{jm} \end{bmatrix} \text{ and } \boldsymbol{P}_j = \begin{bmatrix} P_{j1} \\ P_{j2} \\ \vdots \\ P_{jm} \end{bmatrix}$$

The $rs \times 1$ vector of first derivatives may then be expressed as

$$\boldsymbol{\varphi}(\boldsymbol{\Gamma}) = \sum_{j}^{n} \underset{s\times m}{\boldsymbol{T}} \underset{m\times 1}{(\boldsymbol{r}_j - N_j\boldsymbol{P}_j)} \otimes \underset{r\times 1}{\boldsymbol{K}_j}, \tag{3}$$

where $\otimes$ denotes the Kronecker product and $\boldsymbol{K}_j$ is the jth row of $\boldsymbol{K}$ written as a column. Successive elements in this expression represent derivatives taken with respect to γ_{kh} with the second subscript varying first.

Define also the $m\times m$ matrix

$$\boldsymbol{W}_j = \begin{bmatrix} P_{j1}(1-P_{j1}) & -P_{j1}P_{j2} & \cdots & -P_{j1}P_{jm} \\ -P_{j2}P_{j1} & P_{j2}(1-P_{j2}) & \cdots & -P_{j2}P_{jm} \\ \vdots & \vdots & & \vdots \\ -P_{jm}P_{j1} & -P_{jm}P_{j2} & \cdots & P_{jm}(1-P_{jm}) \end{bmatrix}.$$

The $rs\times rs$ matrix of second derivatives may then be expressed as, say,

$$\boldsymbol{\psi}(\boldsymbol{\Gamma}) = -\sum_{j}^{n} N_j \underset{s\times s}{\boldsymbol{T}\boldsymbol{W}_j\boldsymbol{T}'} \otimes \underset{r\times r}{\boldsymbol{K}_j\boldsymbol{K}_j'}. \tag{4}$$

For the solution of the likelihood equations to correspond to a maximum of the likelihood, the matrix of second derivatives must be negative definite. Checking on definiteness, we note first that, since $\boldsymbol{W}_j$ is proportional to the variance-covariance matrix of the multinomial frequencies, it is positive semi-definite with one and only one zero root. The characteristic vector corresponding to the zero root is, say, $\boldsymbol{a}$ in which each component is unity. The matrix $\boldsymbol{T}\boldsymbol{W}_j\boldsymbol{T}'$ is therefore positive-definite, provided $s < m$, $\boldsymbol{T}$ is of full rank, and $\boldsymbol{T}$ does not contain the vector $\boldsymbol{a}$ as a row.

The matrix $\boldsymbol{K}_j\boldsymbol{K}_j'$ is, of course, positive-semidefinite and of rank one. The Kronecker-product of $\boldsymbol{T}\boldsymbol{W}_j\boldsymbol{T}'$ and $\boldsymbol{K}_j\boldsymbol{K}_j'$, of which the diagonal matrix of roots is the Kronecker-product of the matrices of roots of the separate matrices (Anderson, 1958, p. 348), is positive-semidefinite of rank s. Provided $n \geqslant r$ and $\boldsymbol{K}$ is of full rank, the sum over j of these Kronecker-products will be positive-definite. The matrix of second derivatives is therefore the negative of a positive definite matrix for all finite $\boldsymbol{\Gamma}$. Thus, the likelihood surface is a paraboloid with a unique maximum. A Newton-Raphson solution of the likelihood equations will therefore converge to the maximum likelihood estimates from any finite initial trial values.

3.3. The Newton-Raphson procedure

The ith iteration of the Newton-Raphson solution of the likelihood equations is carried out as follows : Let $\hat{\boldsymbol{\Gamma}}_i$ be the trial value of the estimated parametric functions at this stage of the solution. Then from (2) the trial logits are

$$[\hat{Z}_{jh}^{(i)}] \quad \underset{n\times m}{\hat{\boldsymbol{Z}}_i} = \boldsymbol{K}\hat{\boldsymbol{\Gamma}}_i\boldsymbol{T},$$

and the trial probabilities are

$$\underset{n\times m}{\hat{\boldsymbol{P}}_i} = \underset{n\times n}{\boldsymbol{D}_i^{-1}} \underset{n\times m}{\left[e^{\hat{Z}_{jh}^{(i)}}\right]},$$

where D_i is a diagonal matrix whose elements are the sum of elements in the rows of $\left[e^{\hat{Z}_{jh}^{(i)}}\right]$. The elements in the rows of the matrix P_i are the components of the vector P_j required for the calculation of the first and second derivatives using (3) and (4).

Then the adjustments, say $\delta_{kh}^{(i)}$, to the trial values $\Gamma_{kh}^{(i)}$, represented as elements of an $rs \times 1$ vector in which the subscript h varies first, are

$$\boldsymbol{\delta}_i = -\boldsymbol{\Psi}^{-1}(\hat{\boldsymbol{\Gamma}}_i)\boldsymbol{\varphi}(\hat{\boldsymbol{\Gamma}}_i),$$

where $-\boldsymbol{\Psi}^{-1}(\hat{\boldsymbol{\Gamma}}_i)$ is the negative of the inverse of the matrix of second derivations evaluated at $\hat{\boldsymbol{\Gamma}}_i$. These corrections are added to the corresponding elements of $\hat{\boldsymbol{\Gamma}}_i$ to obtain improved estimates. The process may be repeated until the corrections vanish. Experience with this solution indicates that the choice of trial values is not critical. The solutions converge rapidly (in five or six iterations) even when all initial trial values are set equal to zero. Other properties of these solutions are discussed in connection with the examples of Section 4.

3.4. Tests of goodness-of-fit

The chi-square approximation for the likelihood ratio statistic provides a convenient test of the goodness-of-fit of the model:

$$\chi^2 = -2 \sum_j^n \sum_k^m r_{jk} \log_e N_j \hat{P}_{jk} / r_{jk}.$$

The $\hat{P}_{jk}$ are the expected response probabilities computed from the maximum likelihood estimates of the parameters in the hypothesized model. These probabilities are computed anew at each stage of the Newton-Raphson solution and their final values may be used in computing this chi-square. Terms for which $\mu_{jk} = 0$ are set to zero in this sum.

The number of degrees of freedom for this chi-square is the difference between the number of parameters fitted when the observed proportions estimate directly the population proportions and the number fitted in the model:

$$\text{d.f.} = n(m-1)-rs.$$

This test may also be construed as a test of a composite hypothesis including all parameters excluded from the model. It is a test of parameters of a conditional distribution, given the estimated values of parameters in the model. Since the use of the logistic response law corresponds to a reparameterization of a multiplicative model for the response probabilities, the hypothesis that certain parameters of the logistic model are zero is equivalent to the hypothesis that the ratio of certain parameters in the multiplicative model is unity. (See Goodman, 1964; Bhapkar, 1965). Some of these tests of hypothesis are discussed in connection with the examples of Section 4.

The Pearsonian chi-square is, of course, an alternative to the likelihood ratio chi-square. In moderate size samples, the two chi-squares usually differ only slightly and there is little basis for choosing between them. Fisher suggests, however, that the L.R. chi-square is more appropriate when some of the expected values are small, but the point has not been investigated in any detail.

4. NUMERICAL EXAMPLES

4.1. The computer program

The method of Section 3 has been programmed in FORTRAN IV with the aid of the matrix subroutines of Bock and Peterson (1967). The program accepts as input the parameters m, n, r, and s, the observed frequencies, including the totals N_j, and the bases $\boldsymbol{K}$ and $\boldsymbol{T}$. (A computer routine for constructing these bases is given in Bock, 1965.) The program routinely performs six iterations of the Newton-Raphson procedure. All problems so far attempted have converged from zero initial trial values in six or fewer iterations. The time required for these calculations in the three problems described in this section was 21, 25, and 30 seconds, respectively (IBM 7094). Two complete solutions were performed in the first problem and four in the second. The time for the last two problems also includes compilation time.

As output, the program prints the final values of the first derivatives, the estimated parameters, logits, and proportions, the large sample variance-covariance matrix of the estimated parameters, i.e., $-\psi^{-1}(\hat{\Gamma})$, and the likelihood ratio and Pearsonian chi-square for the test of goodness-of-fit.

4.2. A 3×3 contingency table

The model may be specialized to one experimental condition and a cross-classification of subjects according to their responses. Multi-way contingency tables provide familiar examples of this type of data. In this case the estimation of a multinomial response relationship is not involved, but association among the ways of classification may be studied.

The rank of the physical part of the model is $r = 1$, and its basis is any scalar constant, say, $\boldsymbol{K} = [1]$. The rank of the response part of the model, on the other hand, depends on the number of categories and ways-of-classification in the contingency table, and on how the response probabilities are assumed to be determined. If the ways of categorizing the subjects are independent, the response probability for each subcategory is the product of the probabilities of main categories to which it belongs. Under a logrithmic transformation (and shift of origin), this multiplicative model becomes a simple additive main-category model for the logits. As a basis for this model, any of these bases commonly used in the analysis of variance of a comparable factorial design may be chosen. There is but one point of difference. In the logistic response law (1), the response probabilities are obviously invariant under change of location of the logits. Thus, the parameter corresponding to the general mean in conventional linear models is indeterminant, and the equiangular vector (all components unity) corresponding to the mean must be excluded from the basis. By constructing all remaining vectors of the basis in the form of contrasts (in which the components sum to zero), we may conveniently fix the origin so that logits sum to zero. Aside from this requirement, any basis of the linear logarithmic model is suitable. Any linear transformation implied in a change of basis is absorbed in the reparameterization (2).

Since the parametric functions corresponding to the logistic main-category model are in one-to-one correspondence with the parameters of the multiplicative model (under the restriction that the multiplicative parameters for each way of classification sum to unity), the maximum likelihood estimates of the response probabilities are the same as those obtained when the multiplicative model is assumed. Thus the logistic analysis gives the same estimated response probabilities as the solutions for contingency tables given by Mood (1950) and Roy (1957). This is illustrated in the following examples.

C. R. Rao (1965, p. 343) presents the following data for a cross-classification of school children according to speech defect (A_1, A_2, A_3) and physical defect (B_1, B_2, B_3) :

Classification

	B_1			B_2			B_3			N
	A_1	A_2	A_3	A_1	A_2	A_3	A_1	A_2	A_3	
Frequencies	45	26	12	32	50	21	9	10	17	217

A main category model for estimating the simple contrasts of log parameters corresponding to A_1 vs. A_3, A_2 vs. A_3, A_2 vs A_3, B_1 vs. B_3, and B_2 vs. B_3 was fitted by the method of Section 2. The basis employed was,

$$\boldsymbol{T}_c = 1/3 \begin{bmatrix} 2 & 2 & 2 & -1 & -1 & -1 & -1 & -1 & -1 \\ -1 & -1 & -1 & 2 & 2 & 2 & -1 & -1 & -1 \\ 2 & -1 & -1 & 2 & -1 & -1 & 2 & -1 & -1 \\ -1 & 2 & -1 & -1 & 2 & -1 & -1 & 2 & -1 \end{bmatrix}$$

The following estimated contrasts and standard errors (in parentheses) were obtained :

$A_1 - A_3$: ·98485 (·21049)
$A_2 - A_3$: 1·20075 (·20486)
$B_1 - B_3$: ·48242 (·17985)
$B_2 - B_3$: ·59232 (·17784)

These estimates are of limited value as estimates of main-category effects, however, because the model appears not to fit the data (i.e., the classifications are not independent). The likelihood ratio chi-square for goodness-of-fit is 30·44478 on 4 degrees of freedom ($p < \cdot 0005$). This result agrees with Rao's to six figures. However, the corresponding Pearsonian chi-square of 32·88427, differs somewhat from Rao's figure of 34·8828, which appears to be in error. The estimated proportions and expected frequencies agree exactly with those computed by Rao from the marginal frequencies.

The significant association among the categories may be studied further by including interactive contrasts in the model. For example, the difference of the contrasts A_1 vs. A_3 and B_1 vs. B_2 may be estimated by adjoining to the basis of vector whose elements are the products of elements in the first and third rows in the matrix $\boldsymbol{T}_c$ above. For purposes of testing the hypothesis that the interactive contrast is null, the reduction in chi-square due to the inclusion of this additional vector may be used as an asymptotically independent chi-square on one degree of freedom (Roy, 1957, p. 132). This hypothesis is equivalent to the hypothesis that the ratio of response probabilities, P_1P_9/P_3P_7, equals unity.

To demonstrate the invariance of the solution under choice of basis, Helment contrasts of the main-class effects were also estimated for these data.

The basis used was

$$\boldsymbol{T}_h = \begin{bmatrix} 1/3 & (2 & 2 & 2 & -1 & -1 & -1 & -1 & -1 & -1) \\ 1/2 & (0 & 0 & 0 & 1 & 1 & 1 & -1 & -1 & -1) \\ 1/3 & (2 & -1 & -1 & 2 & -1 & -1 & 2 & -1 & -1) \\ 1/2 & (0 & 1 & -1 & 0 & 1 & -1 & 0 & 1 & -1) \end{bmatrix}.$$

The resulting chi-square statistics, estimated logits and proportions check identically with those of the previous solution. The reader may verify for himself that the following estimated

Helment contrasts are the appropriate linear combinations of the simple contrasts :

$$A_1 - \frac{A_2+A_3}{2}: \quad \cdot 38448$$

$$A_2 - A_3: \quad 1\cdot 20074$$

$$B_1 - \frac{B_2+B_3}{2}: \quad \cdot 21126$$

$$B_2 - B_3: \quad \cdot 54232$$

4.3. A 2^3 contingency table

The following data appear in a query to Snedecor in *Biometrics*, **14** (1958), 560–562.

A_1				A_2				N
B_1		B_2		B_1		B_2		
C_1	C_2	C_1	C_2	C_1	C_2	C_1	C_2	
79	73	62	168	177	81	121	75	836

These data were subjected to the logistic analysis with bases of rank, 3, 4, 5, and 6 for the response part of the model. These bases consisted of as many rows from the conventional matrix of single degree of freedom contrasts for a 2^3 factorial design (see Cochran and Cox, 1957, p. 156). Displayed in Table 1 are the differences between the goodness-of-fit chi-squares for models of successive rank. For comparison, corresponding chi-squares calculated by Lancaster's method are reproduced from the query. Lancaster's method employs linear contrasts of the row frequencies. Note that the logistic solution concentrates more of the chi-square in the lower order interactions—a generally desirable result from point of view of parsimony.

TABLE 1

Analysis of interactions in a 2^3 *contingency table*

Interaction	d.f.	L.R. Chi-Square	Pearsonian Chi-Square	Lancaster's Solution
AB	1	24.23	38.26	24.10
AC	1	69.54	66.81	68.30
BC	1	20.00	20.12	31.80
ABC	1	6.82	6.80	7.80
Total	4	120.59	131.99	132.00

4.4. Two experimental groups and responses to four test items

Although it is common practice in psychological testing to use the sum of item responses, scored 0 or 1, as a variable for purposes of statistical prediction or classification, useful information may be lost as a result. Actually, the sequence of 0's and 1's of each subject's response to n items assigns him to one of the ultimate categories of a 2^n contingency table. If n is not too large, all of the information may be utilized by directly estimating the response probabilities of the 2^n categories. This has been discussed by Solomon (1961), and his data serve to illustrate how the response probabilities may be estimated by fitting a multinomial response relation (Table 2).

TABLE 2

Response of students in high and low IQ groups to four items of an attitude inventory (from Solomon, 1961)

Item Responses *ABCD*	Observed Low *IQ* Group	Frequencies High *IQ* Group	Expected Low *IQ* Group	Frequencies* High *IQ* Group
1111	62	122	68.6	120.9
1110	70	68	62.6	66.5
1101	31	33	29.7	33.7
1100	41	25	43.2	27.0
1011	283	329	276.3	322.4
1010	253	247	260.6	256.2
1001	200	172	201.4	179.1
1000	305	217	302.7	207.4
0111	14	20	11.6	21.5
0110	11	10	14.3	11.0
0101	11	11	8.2	10.0
0100	14	9	16.0	7.5
0011	31	56	33.5	62.2
0010	46	55	42.6	46.2
0001	37	64	39.7	57.4
0000	82	53	80.2	62.2
Total	1491	1491	1491.2	1491.2

*Rank 2, rank 10 model; see text.

In setting up the model for the analysis, several alternatives must be considered. If the response probabilities of the two

IQ groups differ, the basis for the physical part of the model should be rank 2 : for example,

$$K = \begin{bmatrix} 1 & 1 \\ 1 & -1 \end{bmatrix}.$$

If the groups do not differ, a rank 1 model consisting of the first column of this matrix will suffice.

For the response part of the model, a rank 4 main-category model, rank 10 first-order interaction model, or a rank 14 second-order interaction model are alternatives. If a rank 2 physical part and a rank 15 model should be required, the estimated response probabilities would be equal to the observed relative frequencies and no gain in precision would result from fitting the model. The possible bases for the response part are given by the appropriate single degree of freedom contrasts for a 2^4 factorial design.

For the data of Table 2, the goodness-of-fit chi-square for several alternative models are shown in Table 3.

TABLE 3

Goodness-of-fit tests for various models

Rank of Physical part	Rank of Response part	L.R. Chi-square	Degrees of freedom	Probability
1	4	188.3615	26	$<.0005$
2	4	130.4839	22	$<.0005$
1	10	74.4613	20	$<.0005$
2	10	11.5628	10	$.40 < p < .30$

The rank 2, rank 10 model appears to fit the data well. The estimated parameters assuming this model are shown in Table 4. These estimates may be interpreted as follows : the Low+High effect reflects the general response probability of the items. Item *A* has high response probability and *B* low, so that the response patterns beginning with 10 are especially frequent. Patterns with a *C* response also tend to be more probable, and this tendency is increased by the interactive effect of a joint occurrence of a *C* response with an *A* or *B* response.

TABLE 4

Estimated parameters (and standard errors) for the rank 2, rank 10 model

Response Effect	Physical Effect Low + High		Low − High	
A	.7542	(.03133)	.0433	(.06266)
B	−.7636	(.03195)	.0233	(.06390)
C	.1483	(.02984)	−.1949	(.05967)
D	−.0120	(.02913)	−.2825	(.05827)
AB	−.0316	(.03116)	−.1033	(.06232)
AC	.1240	(.02468)	−.0061	(.04936)
AD	.0289	(.02476)	.0907	(.04952)
BC	.1515	(.02550)	−.0427	(.05101)
BD	.0500	(.02427)	−.0839	(.04853)
CD	.1053	(.01894)	.0222	(.03789)

The estimated contrasts between the two *IQ* groups indicate that the difference between the groups is due almost entirely to differences in their response probabilities for individual items. The very small values of the interaction contrasts suggest little difference in pair-wise association between items from one group to the other. Notice that the *C* and *D* items are responsible for most of the difference contributed by individual items. This illustrates the familiar fact that items which are middling in response probability are better discriminators than items such as *A* and *B*, which are extreme.

The maximum-likelihood estimates of the response probabilities assuming the rank 2, rank 10 model, can be computed from the expected frequencies shown in Table 2. These estimated probabilities can be used, as Solomon (1961) shows, to investigate the value of the items for classifying subjects in the *IQ* groups. They would be somewhat more accurate for this purpose than the observed proportions.

This use of the logistic model is an alternative to Solomon's use of Bahadur's (1961) representation for estimation of dichotomous response probabilities. Like Bahadur's representation it is related ultimately to Lazarsfeld's (1961) latent structure analysis for dichotomously scored items.

References

Anderson, T. W. (1958). *An Introduction to Multivariate Statistical Analysis*, John Wiley and Sons, New York.

Anscombe, F. J. (1956). On Estimating Binomial Response Relations, *Biometrika*, **43**, 461–464.

Ashford, J. R. (1959). "An Approach to the Analysis of Data for Semi-Quantal Responses in Biological Assay," *Biometrics*, **15**, 573–581.

Bahadur, R. R. (1961). "A Representation of the Joint Distribution of Responses to *n* Dichotomous Items", In Herbert Solomon (Ed.), *Studies in item analysis and prediction*, Stanford University Press, Stanford, pp. 158–176.

Bargmann, R. E. (1965). "A Statistician's Instructions to the Computer: A Report on a Statistical Computer Language," *In Proceedings of the IBM Scientific Computing Symposium on Statistics, October 21–23, 1963.* IBM Data Processing Division, White Plains, New York, pp. 301–333.

Bhapkar, V. P. and Koch, G. (1965). "On the Hypothesis of 'No Interaction' in Three-Dimensional Contingence Tables," Institute of Statistics, University of North Carolina, Mimeo Series No. 440.

Berkson, J. (1955). "Maximum Likelihood and Minimum χ^2 Estimation of the Logistic Function", *J. Amer. Statist. Assoc.*, **50**, 130–162.

Berkson, J. (1960). "Problems Recently Discussed Regarding Estimating the Logistic Curve", *Bull. Inst. Internat. Statist.*, **37**, 207–11.

Bock, R. D. (1963). "Programming Univariate and Multivariate Analysis of Variance, *Technometrics*, **5**, 95–117.

Bock, R. D. (1965). "A Computer Program for Univariate and Multivariate Analysis of Variance", In *Proceedings of the IBM Scientific Computing Symposium on Statistics, October 21–23, 1963.* IBM Data Processing Division, White Plains, New York, pp. 69–111.

Bock, R. D. and Bargmann, R. E. (1966). "Analysis of Covariance Structures", *Psychometrika*, **31**, 507–534.

Bock, R. D. and Jones, L. V. (1967). *The measurement and prediction of judgment and choice*, Holden Day, San Francisco (in press).

Bock, R. D. and Peterson, A. (1947). "Matrix Operations Subroutines for Statistical Computation", Research Memorandum No. 7, Statistical Laboratory, Department of Education, The University of Chicago.

Cochran, W. G. and Cox, Gertrude M. (1957). *Experimental Designs* (2nd Ed.). John Wiley and Sons, New York.

Finney, D. J. (1952). *Probit Analysis: A Statistical Treatment of the Sigmoid Response Curve* (2nd ed.). Cambridge University Press, London.

Goodman, L. A. (1964). "Simple Methods for Analyzing Three-Factor Interaction in Contingency Tables. *J. Amer. Statist. Assoc.*, **59**, 319–52.

Lazarsfeld, P. F. (1961). "The Algebra of Dichotomous Systems," In Herbert Solomon (Ed.), *Studies in Item Analysis and Prediction.* Stanford University Press, Stanford.

Luce, R. D. (1959). *Individual Choice Behavior.* John Wiley and Sons, New York.

Mantel, N. (1966). "Models for Complex Contingency Tables and Polychotomous Response Curves," *Biometrics,* **22**, 83–110.

Maxwell, A. E. (1959). "Maximum Likelihood Estimates of Item Parameters Using the Logistic Function," *Psychometrika* **24**, 221–228.

Mood, A. M. (1950). *Introduction to the Theory of Statistics.* McGraw-Hill, New York.

Rao, C. R. (1965). *Linear Statistical Inference and its Applications.* John Wiley and Sons, New York.

Roy, S. N. (1957). *Some Aspects of Multivariate Analysis.* John Wiley and Sons, New York.

Silverstone, H. (1957). "Estimating the Logistic Curve," *J. Amer. Statist. Assoc.*, **52**, 567–577.

Solomon, H. (1961). "Classification Procedures Based on Dichotomous Response Vectors," In Herbert Solomon (Ed.), *Studies in Item Analysis and Prediction.* Stanford University Press, Stanford, pp. 177–186.

(*Received Dec. 30, 1965. Revised Aug. 2, 1966.*)

CHAPTER 7

Simultaneous Tests for Average and Dispersion by Combined Control Charts

P. K. BOSE AND S. P. MUKHERJEE, *University of Calcutta*

1. INTRODUCTION

A usual process control plan involves the computation of two statistics, one describing central tendency and a second measuring dispersion of the measured quality characteristic (x) and observing whether values of both obtained in rational subgroups inspected from the process lie within their respective control limits. Assuming x to have a Normal distribution, the standard values of m and σ required for determining the control limits are say m_0 and σ_0. An assignable cause of variation will result in a population $N(m, \sigma)$. When a small sample has been obtained from this population, the problem is to test whether this population differs significantly from the standard population i.e. to test the 2-parameter simple hypothesis, $H_0 : m = m_0$, $\sigma = \sigma_0$.

The two control charts (say for $\bar{x}$ and R or for $\bar{x}$ and s) must be interpreted simultaneously, since the probability of $\bar{x}$ lying outside its limits depends not only on how m compares with m_0 but also on the values of σ and σ_0. And for this it is reasonable to set the limits on the two charts in such a way that the probability of a sample point going beyond the U.C.L. or below the L.C.L. on either the $\bar{x}$ or the s (or R) chart is the same, say, γ when H_0 is true. A lack of control will be indicated or H_0 will be rejected whenever a sample measure exceeds any of the limits on either chart. Either statistic will lie within its control limits with probability $1-2\gamma$; assuming the measures of central tendency and dispersion used to be independent, the probability that both statistics will lie within their respective control limits is $(1-2\gamma)^2$ and thus, when H_0 is true, the probability of a sample measure going outside of control limits on either chart is $1-(1-2\gamma)^2$. Thus we may set limits by considering $\alpha = 1-(1-2\gamma)^2$.

$$\text{or} \qquad 4\gamma^2-4\gamma+\alpha = 0$$

$$\text{or} \qquad \gamma = \frac{4\pm\sqrt{16-16\alpha}}{8} = \frac{1\pm\sqrt{1-\alpha}}{2},$$

so as to give a preassigned size α to the joint test. (Obviously, the negative sign alone can be considered.) In the case of statistics not distributed independently, the joint distribution must be used to obtain α, given γ.

Walsh (1952) considered three joint tests for $H_0 : m = m_0$, $\sigma = \sigma_0$, viz., $\bar{x}$ and s, t and s, and t and $s(m)$ where

$$s = \sqrt{\frac{1}{n-1}\,\Sigma(x_i-\bar{x})^2} \text{ and } s(m) = \sqrt{\frac{1}{n}\,\Sigma\,(x_i-m_0)^2}$$

By computing powers, he concluded that each of the three tests is found to have regions where its O.C.'s are poor. No one test has uniformly better O.C. than the others. On the whole, tests based on the t-statistic and s appeared to be inferior to the other two, which are roughly equivalent.

The importance of range as a measure of dispersion has always been recognized in Quality Control practice for the facility in its computation. However it remains to investigate power properties of the joint test involving $\bar{x}$ and R. The present paper attempts to study the O.C. surface of this test and to compare the OC of this test with those obtained earlier.

2. OPERATING CHARACTERISTIC OF THE $\bar{x}$ AND R CHARTS USED SIMULTANEOUSLY

Control limits for $\bar{x}$ and R charts used jointly to give a probability α of falsely rejecting the state of control $m = m_0$, $\sigma = \sigma_0$ are to be set at $\bar{\bar{x}} \pm A_1 \bar{R}$ and ($D_3 \bar{R}$ and $D_4 \bar{R}$) when process standards are not given and at $m_0 \pm A\sigma_0$ and ($D_1\sigma_0$ and $D_2\sigma_0$) if standards are provided. Constants A, D_1 and D_2 are defined from the relations :

$$\gamma = \Pr\{\bar{x} \leqslant m_0 - A\sigma_0 / m_0, \sigma_0\}$$

$$= \Pr\{\bar{x} \geqslant m_0 + A\sigma_0 / m_0, \sigma_0\}$$

$$= \Pr\{R \leqslant D_1\sigma_0 / \sigma_0\} = \Pr\{R \geqslant D_2\sigma_0 / \sigma_0\},$$

$$A_1 = A/d_2, \quad D_3 = D_1/d_2 \quad \text{and} \quad D_4 = D_2/d_2$$

where $E(R) = d_2\sigma$.

Tables I(a) and I(b) give values of the factors A, D_1, D_2, A_1, D_3 and D_4 for samples of sizes 3, 4 and 5 and corresponding to probabilities ·005, ·01, ·02, ·05 and ·10 of type I errors using the joint test. These were obtained by linear interpolation in tables of the incomplete integrals for the distributions of the standardised normal deviate and of the standardised range.

For a sample drawn from a population $N(m, \sigma)$ the operating characteristic of the $\bar{x}$ and R charts is

$$\beta = \Pr\{m_0 - A\sigma_0 \leqslant \bar{x} \leqslant m_0 + A\sigma_0 / m, \sigma\} \times$$
$$\Pr\{D_1\sigma_0 \leqslant R \leqslant D_2\sigma_0 / \sigma\}$$

since $\bar{x}$ and R in random samples from a Normal population are distributed independently (Daly, 1946). It is convenient to define the class of alternatives (m, σ) in terms of the parameters k and l where

$$k = \frac{m-m_0}{\sigma_0/\sqrt{n}} \quad \text{and} \quad l = \sigma_0/\sigma.$$

In terms of these values the O.C. of the joint test using ($\bar{x}$ and R) charts is

$$\beta = \int_{-(\sqrt{n}A+k)l}^{(\sqrt{n}A-k)l} \frac{1}{\sqrt{2\pi}} e^{-t^2/2} \, dt \times \int_{D_1 l}^{D_2 l} f(w)dw,$$

w being the standardized range R/σ. The O.C. function is easily found to be the same for k and $-k$. To visualize the probabilities, a three-dimensional O.C. surface is necessary. Values of O.C. for $n = 3, 4$ and 5 for certain values of k and l and corresponding to $\alpha = \cdot 005, \cdot 01, \cdot 02$ and $\cdot 05$ are presented in Table 2. The choice of k and l values was made to facilitate comparison with results obtained by Walsh.

On the plane for each value of σ_0/σ, the curve is symmetrical and roughly bell-shaped. Thus for a given value of σ, the probability of accepting a state of control diminishes as k increases or as m departs from m_0. The curves on planes for large values of l are platykurtic not curving down much below $k = 4{\cdot}0$. Thus for small values of σ (compared to σ_0) the probability of accepting the process continues to remain appreciable even for moderately large shift in process mean. For $\sigma \geqslant 4\,\sigma_0$, however, the test has a very large power to detect lack of control.

On the plane for each value of $k \leqslant 3{\cdot}0$, the curves are positively skew. Thus when m remains within 3σ-limits on the chart for $\bar{x}$, the probability of accepting the process gets smaller as σ becomes larger and then it declines very slowly. For any positive shift in the process variability ($\sigma > \sigma_0$) the probability of accepting the process is much less compared to a negative shift ($\sigma < \sigma_0$) to an equal extent.

It is found that O.C. values for the same deviations (k, l) vary remarkably for different sizes of the joint test. A decrease in the sample size naturally leads to a rise in O.C. values and appreciably so for larger values of l $(\sigma < \sigma_0)$ and moderate deviations in process mean $(k \leqslant 2{\cdot}0)$. The test is almost ineffective for σ near about $2\sigma_0$.

When σ remains at σ_0 the joint test using $\bar{x}$ and R charts is slightly worse than the single test using $\bar{x}$ for detecting deviations in m, the largest inefficiency being observed near $k = 2$. It is remarkable, however, that when m is m_0, the joint test is uniformly more powerful than the single test using R for the class of situations $l < 1$ i.e., $\sigma > \sigma_0$. Thus when there is no deviation in process mean, the joint control chart is superior to the single R-chart for detecting positive shifts in process standard deviation.

3. COMPARISON OF O. C. FUNCTIONS

When σ remains at σ_0 the $(\bar{x}—R)$ charts have exactly the same O.C. as the $(\bar{x}—s)$ charts for all sizes of the joint test. With no change in the process mean, the $(\bar{x}—R)$ charts are almost as effective as $(\bar{x}—s)$ charts, the fall in power diminishing with increase in the size of the test and the largest fall being observed at $\sigma = 2\sigma_0$. Reduction in efficiency for $n = 5$ is 3% at the most. However with the class of alternatives $\sigma < \sigma_0$, the $(\bar{x}—R)$ charts are far more powerful than the $(\bar{x}—s)$ charts, at least for moderate deviations in m and smaller test sizes, since for larger deviations in mean and more stringent tests (with smaller α) even the $(\bar{x}—s)$ charts have high probabilities for rejecting the hypothesis of control. The $(\bar{x}—s)$ test is almost insensitive to moderate deviations in process mean under such situations.

Sometimes, however, we have to consider an increase in the variability of a manufacturing process and pay attention to situations where $\sigma > \sigma_0$. It is found that the $(\bar{x}—R)$ charts are decidedly less effective than the $(\bar{x}—s)$ charts for $\sigma = 2\sigma_0$.

As σ increases further both the joint tests have very small probability of accepting the process and O.C. values for the two tests compare very favourably. It is remarkable to note that under the class of situations $\sigma > \sigma_0$, the $\bar{x}-R$ charts have nowhere a better performance than the $(\bar{x}-S)$ charts.

In Walsh's set, t and s and $\bar{x}$ and $s(m)$ are not statistically independent and much effort is to be put up to obtain their O.C. values. Anyway, tests based on $\bar{x}$ and $s(m)$ are almost equivalent to $(\bar{x}-s)$ and $(\bar{x}-R)$ charts. But (t and s) give mostly unfavourable values. Comparing the O.C. values for $n = 5$ and $\alpha = \cdot 01$ it is found that the (t, s) joint test has a very poor performance so long as $\sigma = \sigma_0$. It is better than the present joint test using ($\bar{x}$ and R) for large k and small l and for some moderate k with large l. For all other situations the $(\bar{x}-R)$ charts are superior to the $(t-s)$ tests.

4. COMPARISON WITH THE SINGLE CHART FOR EXTREME VALUES

It was as early as 1949 that Howell proposed a single control chart for the smallest (s) and the largest (L) observations to describe both central tendency and dispersion. He gave 3σ-control limits as $m \pm A_4\sigma$ (standards given) and $\bar{\bar{x}} \pm A_3\bar{R}$ (no standards given) for sample sizes $n = 2$ to 10.

On this chart probability limits $m \pm A_4\sigma$ or $\bar{\bar{x}} \pm A_3\bar{R}$ may be given from the relations

$$\Pr\{m_0 - A_4\sigma_0 < S, \;\; L < m_0 + A_4\sigma_0 / m_0, \sigma_0\} = 1 - \alpha$$

or

$$1-\alpha, = \left[\int_{-A_4}^{A_4'} \frac{1}{\sqrt{2\pi}} e^{-t^2/2} dt\right]^n$$

and

$$A_3 = A_4 / d_2.$$

The O.C. of this chart for $k = \dfrac{m - m_0}{\sigma_0}$ and $l = \sigma_0/\sigma$ is

$$\Pr\{m_0 - A_4\sigma_0 < S,\ L < m_0 + A_4\sigma_0 / m_0 + k\sigma_0,\ \sigma_0/l\}.$$

Thus

$$\text{O.C. } (\beta) = \left[\int_{-l(A_4+k)}^{l(A_4-k)} \frac{1}{\sqrt{2\pi}} e^{-t^2/2} dt\right]^n$$

Values of probability limits A_4 and A_3 for $n = 3$, 4 and 5 and $\alpha = \cdot 01$, $\cdot 05$ and $\cdot 10$ are given in Table 3. Table 4 gives O.C. values of this chart compared with those for $(\bar{x}-R)$ charts for $n = 5$ and $\alpha = \cdot 05$ for several values for k and l.

It is found that when $l = 1$, i.e., when σ does not change from its initial setting, this test is uniformly better than the $(\bar{x}-R)$ or $(\bar{x}-s)$ charts. For $l < 1$, this test though worse than the other tests for lower values of k, is much better than the $(\bar{x}-s)$ test for larger deviations k, since for such situations, the $(\bar{x}-s)$ test is almost inefficient for $\alpha = \cdot 02$ or $\cdot 05$. For $k = 0$, i.e., for m remaining at its initial setting m_0 the test is very poor for situations $\sigma < \sigma_0$. This test, however, is almost insensitive to changes in process mean $k \leqslant 1{\cdot}25$ below $\sigma = \sigma_0/3$.

From the point of computational facility, the joint test based on range and midrange might have been considered. But there does not exist any distribution with a limited first and a continuous second derivative for which range and midrange are statistically independent (Frechet, 1954). Although the joint distribution of these two statistics in random samples from a normal population has been derived (Pillai, 1950), the moments and incomplete probability integrals have yet to be calculated from the infinite series given.

5. CONCLUDING REMARKS

Detection of lack of control in a manufacturing process involves a test for the two parameter simple statistical hypothesis $H_0 : m = m_0$, $\sigma = \sigma_0$, where m_0 and σ_0 are standards used for setting up control limits for relevant charts. It has been shown (Lehmann, 1952) under certain regularity conditions that unbiased tests of H_0 do not exist. Tests of minimum bias and other types of minimax tests have been given under suitable conditions of monotonicity. But they cannot conveniently be applied for routine quality control. Tests using type C critical regions of Neyman and Pearson, and later developed by Isaacson, are only locally most powerful. The simultaneous use of control charts for $\bar{x}$ and R or for $\bar{x}$ and s may provide useful tests for ready application. Both these joint tests have some "weak points" with small power. On the whole $\bar{x}$ and R charts may be used for situations $\sigma < \sigma_0$ while $\bar{x}$ and s charts are preferable for situations $\sigma > \sigma_0$. The single chart for extreme values can be definitely recommended for detecting deviations in process mean alone, σ being equal to σ_0.

Acknowledgment

The authors would like to thank Sri S. Ganguly for his assistance in computing the O.C. values.

TABLE 1

Factors for obtaining control limits on the control chart for $\bar{x}$ and R to be used simultaneously

(a) Standards given

α	γ	$n = 3$			$n = 4$			$n = 5$		
		A	D_1	D_2	A	D_1	D_2	A	D_1	D_2
.005	.0012, 521	1.7453	.0631	4.9740	1.51145	.2126	5.2370	1.3519	.3877	5.4120
.01	.0025, 063	1.6202	.0930	4.7016	1.4031	.2681	4.9656	1.2550	.4639	5.1492
.02	.0050, 253	1.4862	.1327	4.4229	1.28705	.3428	4.6910	1.1512	.5556	4.8839
.05	.0126, 603	1.2912	.2136	4.0131	1.11825	.4690	4.2980	1.0002	.7078	4.5014
.10	.0256, 584	1.1251	.3067	3.6688	.9744	.5996	3.9719	.8715	.8056	4.1847

(b) Standards not given

α	γ	A_1	D_3	D_4	A_1	D_3	D_4	A_1	D_3	D_4
.005	.0012, 521	1.0311	.0377	2.9386	.7341	.1033	2.5436	.5812	.1667	2.3266
.01	.0025, 063	.9572	.0549	2.7777	.6815	.1302	2.4118	.5395	.1994	2.2136
.02	.0050, 253	.8780	.0784	2.6130	.6251	.1665	2.2784	.4949	.2389	2.0996
.05	.0126, 603	.7628	.1262	2.3709	.5431	.2278	2.0875	.4300	.3043	1.9352
.10	.0256, 584	.6647	.1812	2.1675	.4733	.2912	1.9292	.3747	.3463	1.7990

TABLE 2

Operating characteristic for the joint test based on control charts for $\bar{x}$ and R

$K=\frac{m-m_0}{\sigma_0/\sqrt{n}}$	α	$n = 5$						
		$l = .125$	$l = .250$	$l = .500$	$l = 1.00$	$l = 2.00$	$l = 4.00$	$l = 8.00$
0	.005	.00315	.06911	.59986	.99500	.98210	.80870	.18260
	.01	.00244	.05532	.53571	.99000	.96543	.68370	.06713
	.02	.00184	.04315	.46699	.98001	.93477	.51580	.01448
	.05	.00117	.02806	.36565	.95004	.85489	.26500	.00060
1.0	.005	.00313	.06735	.56718	.97600	.98207	.80870	.18260
	.01	.00242	.05388	.50306	.95967	.96528	.68370	.06713
	.02	.00186	.04199	.43525	.93266	.93400	.51580	.01448
	.05	.00116	.02727	.33708	.86876	.84918	.26500	.00060
2.0	.005	.00306	.06234	.47577	.84477	.96210	.80868	.18260
	.01	.00237	.04977	.41384	.78603	.91388	.68327	.06713
	.02	.00179	.03870	.35050	.70994	.81762	.51023	.01448
	.05	.00113	.02505	.26312	.57856	.58313	.21948	.00058
3.0	.005	.00298	.05479	.34730	.50798	.50922	.43425	.10474
	.01	.00228	.04359	.29330	.42112	.33723	.14997	.00408
	.02	.00172	.03378	.24054	.33183	.18446	.02287	0
	.05	.00109	.02245	.17223	.22586	.05425	.00030	0
4.0	.005	.00280	.04572	.21556	.16393	.02492	.00004	0
	.01	.00216	.03619	.17549	.11574	.00820	0	0
	.02	.00163	.02792	.13829	.07622	.00204	0	0
	.05	.00103	.01843	.09337	.03795	.00018	0	0

TABLE 2—*contd.*

$K \backslash l$	α	$n = 4$						
		.125	.25	.50	1.00	2.00	4.00	8.00
0	.005	.00975	.11538	.65350	.99491	.99050	.93180	.62550
	.01	.00780	.09497	.59144	.99000	.98132	.87300	.42743
	.02	.00610	.07650	.52400	.98001	.96236	.76672	.21172
	.05	.00414	.05243	.42184	.95002	.91079	.54595	.03985
1	.005	.00968	.11245	.61790	.97590	.99047	.93180	.62550
	.01	.00774	.09249	.55540	.95966	.98118	.87300	.42743
	.02	.00605	.07445	.48838	.93265	.96157	.76672	.21172
	.05	.00410	.05096	.38888	.86875	.90471	.54595	.03985
2	.005	.00946	.10408	.51831	.84469	.97033	.93178	.62550
	.01	.00757	.08543	.45689	.78603	.92893	.87245	.42743
	.02	.00591	.06862	.39329	.70994	.84176	.75844	.21172
	.05	.00401	.04681	.30355	.57855	.62126	.45217	.03869
3	.005	.00912	.09148	.37835	.50793	.51358	.50035	.35880
	.01	.00729	.07483	.32382	.42112	.34278	.19149	.02595
	.02	.00570	.05989	.26990	.33183	.18990	.03400	.00007
	.05	.00386	.04194	.19870	.22585	.05780	.00062	0
4	.005	.00865	.07633	.23484	.16391	.02513	.00004	0
	.01	.00692	.06213	.19375	.11574	.00833	0	0
	.02	.00540	.04949	.15517	.07622	.00210	0	0
	.05	.00366	.03444	.10772	.03795	.00019	0	0

TABLE 2—*contd.*

K \ l	α	n = 3						
		.125	.25	.50	1.00	2.00	4.00	8.00
0	.005	.02977	.19065	.70714	.99500	.99540	.98250	.93210
	.01	.02492	.16339	.65432	.99076	.99034	.96244	.85848
	.02	.02034	.13100	.59113	.98001	.98062	.92515	.73321
	.05	.01234	.08478	.44189	.93735	.95080	.81778	.44828
1	.005	.29552	.18581	.66862	.97600	.99537	.98250	.93210
	.01	.02473	.15913	.61444	.96040	.99019	.96244	.85848
	.02	.02019	.13313	.55095	.93266	.97982	.92515	.73321
	.05	.01225	.08241	.40736	.85716	.94445	.81778	.44828
2	.005	.02890	.17198	.56086	.84477	.97513	.98248	.93210
	.01	.02418	.14694	.50546	.78663	.93746	.96184	.85848
	.02	.01974	.12289	.44368	.70994	.85773	.91517	.73321
	.05	.01197	.07569	.31798	.57084	.64855	.67731	.43518
3	.005	.02784	.15116	.40941	.50798	.51612	.52757	.53467
	.01	.02329	.12874	.35824	.42144	.34593	.21111	.05212
	.02	.01901	.10725	.30448	.33183	.19351	.04102	.00024
	.05	.01152	.06782	.20814	.22284	.06034	.00093	0
4	.005	.02643	.12612	.25412	.16393	.02526	.00005	0
	.01	.02210	.10690	.21435	.11582	.00841	0	0
	.02	.01803	.08863	.17505	.07622	.00214	0	0
	.05	.01093	.05568	.11283	.03744	.00020	0	0

TABLE 3

Values of A_3 and A_4

A_4 (Standards given)

α \ n	3	4	5
.10	2.1141	2.2263	2.3110
.05	2.3878	2.4909	2.5688
.01	2.9342	3.0222	3.0891

A_3 (Standards not given)

α \ n	3	4	5
.10	1.2490	1.0813	0.9935
.05	1.4107	1.2098	1.1043
.01	1.7335	1.4679	1.3280

TABLE 4

O.C. of control charts for extreme values and of (x, R) charts

$n = 5 \quad \alpha = \cdot 05$

Chart	K \ l	.25	.50	1.00	1.50	2.00	4.00
Extreme Values	0	.025242	.329429	.950030	.999500	1.000000	1.000000
$\bar{x}, R$		.028910	.365066	.950014	.943649	.854893	.265000
Extreme Values	.5	.024411	.301716	.902537	.995010	.999915	1.000000
$\bar{x}, R$		.027911	.330222	.845944	.900283	.010814	.264999
Extreme Values	1.0	.022046	.231509	.739779	.954357	.995753	1.000000
$\bar{x}, R$		.025116	.241950	.487502	.472426	.427723	.132669
Extreme Values	1.5	.018597	.147869	.463359	.755627	.921305	1.000000
$\bar{x}, R$		.021064	.141750	.128535	.044228	.010860	.000001
Extreme Values	2.0	.014650	.078150	.187258	.334286	.505443	.944285
$\bar{x}, R$		.016460	.065242	.012369	.000377	.000003	0
Extreme Values	2.5	.010789	.033798	.040842	.046343	.052601	.083563
$\bar{x}, R$		.011983	.023210	.000389	0	0	0
Extreme Values	3.0	.007416	.011842	.004107	.001161	.000277	.000003
$\bar{x}, R$		.008125	.006294	.000004	0	0	0

References

Daly, J. F. (1946). "On the Use of the Sample Range in an Analogue of Student's *t*-Test," *Ann. Math. Statist.*, 17, 71–74.

Frechet, M. (1954). *Studies in Mathematics and Mechanics Presented to R. Von Mises*, Academic Press, New York.

Howell, John M. (1949). "Control Charts for Largest and Smallest Values," *Ann. Math. Statist.*, 20, 305–309.

Isaacson, Stanley, L. (1951). "On the Theory of Unbised Tests of Simple Statistical Hypotheses Specifying the values of Two or More Parameters," *Ann. Math. Statist.*, 22, 217–234.

Lehman, E. L. (1952). "Testing Multiparameter Hypotheses," *Ann. Math. Statist.*, 23, 541–552.

Neymann, J. and Pearson, E. S. (1938). "Contributions to the Theory of Testing Statistical Hypotheses," *Statistical Research Memoirs*, 2, 36–57.

Pillai, K. C. S. (1950). "On the Distributions of Midrange and Semi-Range in Samples from a Normal Population," *Ann. Math. Statist.*, 21, 100–105.

Walsh, John E. (1949). "On the Range-Midrange Test and Some Tests with Bounded Significance Levels," *Ann. Math. Statist.*, 20, 257–267.

Walsh, John E. (1952). "Operating Characteristics for Tests of the Stability of a Normal Population," *J. Amer. Statist. Assoc.*, 47, 191–201.

(Received Feb. 2, 1966. Revised May 18, 1967.)

CHAPTER 8

Error Correcting, Error Detecting and Error Locating Codes

R. C. BOSE*,
The University of North Carolina at Chapel Hill

1. INTRODUCTION

Consider a channel which is capable of transmitting any one of q distinct symbols. Such a channel is called a q-ary channel. The special case $q = 2$ is of particular importance. In this case the channel is called binary. Similarly if $q = 3$, we have a ternary channel. The symbols successively presented to the channel for transmission constitute the 'input' and the symbols received constitute the 'output'. Due to the presence of noise a transmitted symbol may be received as one of the other $q-1$ symbols. When this happens we say that there is an error in transmitting the symbol.

In this paper we shall confine ourselves to the case when q is a prime or a prime power, say $q = p^h$ where p is a prime, and $h \geqslant 1$ is any integer. The symbols can then be put in a (1,1)

*This research was supported by the National Science Foundation Grant No. GP-3792 and the Air Force Office of Scientific Research Grant No. AF-AFOSR-760-65.

correspondence with the elements of the Galois field $GF(q)$. For the binary case the field $GF(2)$ contains only two symbols 0 and 1. Consider a set C of $v < q^n$ distinct n-vectors with elements belonging to $GF(q)$. Given a set of v distinct messages we can set up a (1,1) correspondence between the messages and the n-vectors belonging to C. The elements of C may be called code vectors or code words. Thus each message corresponds to a unique code vector (word). To transmit a message over the channel the n elements of the code vector corresponding to the message are presented in succession to the channel. The output is than an n-vector (not necessarily a vector of C) which belongs to the vector space V_n of all n-vectors with elements belonging to $GF(q)$. A decoder is obtained by setting up a decision rule, which specifies a unique vector of C, corresponding to any vector of V_n such that if this vector of V_n is received as an output, it is read as the corresponding vector of C. The code is called a group code if the set C of code words forms a group under vector addition. If C is a vector space (a subspace of V_n), then the code is said to be a *linear code*. Of course a linear code is always a group code. By a code C, we shall mean a code, for which the set of code words is C. The number n is called the length of the code[2].

2. THE HAMMING DISTANCE

Let $\boldsymbol{x}' = (x_1, x_2, \ldots, x_n)$ be any vector of V_n, the vector space of all vectors with elements belonging to $GF(q)$. Then the number of non-zero elements in $\boldsymbol{x}'$ is defined as the weight $w(\boldsymbol{x}')$ of $\boldsymbol{x}'$. Given two vectors

$$\boldsymbol{x}' = (x_1, x_2, \ldots, x_n), \ \boldsymbol{y}' = (y_1, y_2, \ldots, y_n),$$

both belonging to V_n, the Hamming [9] distance $d(\boldsymbol{x}', \boldsymbol{y}')$ between $\boldsymbol{x}'$ and $\boldsymbol{y}'$ is defined as the number of coordinates in which $\boldsymbol{x}'$ and $\boldsymbol{y}'$ disagree. Clearly

$$d(\boldsymbol{x}', \boldsymbol{y}') = w(\boldsymbol{x}' - \boldsymbol{y}') = w(\boldsymbol{y}' - \boldsymbol{x}').$$

[2] Here each code word is considered to be of the same length n. When this is not the case one has variable length codes,

It is readily seen that the Hamming distance satisfies the condition of a metric, i.e.,

(i) $d(\boldsymbol{x}', \boldsymbol{y}') = 0$, if and only if $\boldsymbol{x}' = \boldsymbol{y}'$

(ii) $d(\boldsymbol{x}', \boldsymbol{y}') = d(\boldsymbol{y}', \boldsymbol{x}')$

(iii) $d(\boldsymbol{x}', \boldsymbol{y}')+d(\boldsymbol{y}', \boldsymbol{z}') \geqslant d(\boldsymbol{x}', \boldsymbol{z}')$.

Let $\boldsymbol{g}_1$ and $\boldsymbol{g}_2$ be any words of a group code. Then $\boldsymbol{g}_1-\boldsymbol{g}_2$ is also a code word. Hence the distance between two code words is the weight of some code word. Also $\mathbf{0}$ is a code word. If $\boldsymbol{g}$ is an arbitrary code word then $w(\boldsymbol{g}) = d(\boldsymbol{g}, \mathbf{0})$. Hence

Theorem 2.1. *If d is the minimum distance between the words of a group code, then d is also the minimum weight of the code words.*

3. THE GENERATING MATRIX AND THE PARITY CHECK MATRIX OF A LINEAR CODE

Consider a linear code C. Then the set of code words is a vector space V_k of rank k. Any set of basis vectors of V_k, may be regarded as the set of row vectors of a $k \times n$ matrix

$$(3.1) \qquad \boldsymbol{G} = \begin{bmatrix} g_{11} & g_{12} & \cdots & g_{1n} \\ g_{21} & g_{22} & \cdots & g_{2n} \\ \cdots & \cdots & \cdots & \cdots \\ g_{k1} & g_{k2} & \cdots & g_{kn} \end{bmatrix}.$$

Every other code vector is a linear combination of the rows of $\boldsymbol{G}$. The matrix $\boldsymbol{G}$ is called the generating matrix of the code. Let $\boldsymbol{c}' = (c_1, c_2, \ldots, c_k)$ be any k-vector with elements from $GF(q)$, then $\boldsymbol{c}'\boldsymbol{G}$ is a code word. Since each of $c_1, c_2, \ldots, c_k$ can be taken in q ways, the total number of code words is q^k. Such a code is called a linear code.

Let V_r be the null space of V_k. Then

$$(3.2) \qquad \text{Rank } V_r = n-k = r \text{ (say)}.$$

The number r is defined to be the redundancy of the linear code and k is called the number of information places. Let the row vectors of

$$(3.3) \qquad H = \begin{bmatrix} h_{11} & h_{12} & \dots & h_{1n} \\ h_{21} & h_{22} & \dots & h_{2n} \\ \dots & \dots & \dots & \dots \\ h_{r1} & h_{r2} & \dots & h_{rn} \end{bmatrix}$$

from a basis of V_r. Then $\boldsymbol{H}$ is defined to be the parity check matrix of the linear code. If $\boldsymbol{H}'$ denotes the transpose of $\boldsymbol{H}$, then

$$(3.4) \qquad \boldsymbol{G}\boldsymbol{H}' = \boldsymbol{0},$$

where $\boldsymbol{0}$ is the $k \times r$ null matrix.

The code words can be regarded as the set of independent solutions of the homogeneous linear equations

$$(3.5) \qquad \begin{aligned} h_{11}g_1+h_{12}g_2+\dots+h_{1n}g_n &= 0 \\ h_{21}g_1+h_{22}g_2+\dots+h_{2n}g_n &= 0 \\ \dots \quad \dots \quad \dots \quad \dots &= \dots \\ h_{r1}g_1+h_{r2}g_2+\dots+h_{rn}g_n &= 0 \end{aligned}$$

for the variables $g_1, g_2, \dots, g_n$. The equations (3.5) are called parity check equations. The rows of $\boldsymbol{G}$ are a set of independent solutions of the parity check equations.

Theorem 3.1. *$\boldsymbol{g}'$ is a code word if and only if $\boldsymbol{g}'\boldsymbol{H} = \boldsymbol{0}$ i.e., $\boldsymbol{H}\boldsymbol{g} = \boldsymbol{0}$.*

Theorem 3.2. *Let $\boldsymbol{g}'$ be a word of weight w, belonging to the linear code C, with parity check matrix $\boldsymbol{H}$. Let the i_1th, i_2th, ..., i_wth coordinates of $\boldsymbol{g}'$ be non-zero (all other coordinates being zero). Then there is a linear dependence relation, with non-zero coefficients among the i_1th, i_2th, ..., i_wth column vectors of $\boldsymbol{H}$ and conversely.*

Let $\boldsymbol{H} = (\boldsymbol{h}_1, \boldsymbol{h}_2, \dots, \boldsymbol{h}_n)$, and $\boldsymbol{g}' = (g_1, g_2, \dots, g_n)$. Then

$$\boldsymbol{H}\boldsymbol{g} = g_1\boldsymbol{h}_1+g_2\boldsymbol{h}_2+\dots+g_n\boldsymbol{h}_n = \boldsymbol{0}$$

Now $g_{i_1}, g_{i_2}, \ldots, g_{i_w}$ are non-zero, and the other g's are zero. Hence

$$g_{i_1}\boldsymbol{h}_{i_1}+g_{i_2}\boldsymbol{h}_{i_2}+\ldots+g_{i_w}\boldsymbol{h}_{i_w} = \boldsymbol{0}, \tag{3.6}$$

which proves the first part of the theorem. Conversely if (3.6) holds with non-zero coefficients, then from Theorem 3.1 there exists a code word whose i_1th, i_2th, ..., i_wth coordinates are $g_{i_1}, g_{i_2}, \ldots, g_{i_w}$ and the other coordinates are all zero.

Corollary. *Let C be a linear code with parity check matrix $\boldsymbol{H}$: (i) If no m of the columns of $\boldsymbol{H}$ are dependent then each word of C has weight $\geqslant m+1$. (ii) Conversely if each word of C has weight $\geqslant m+1$, then any m columns of $\boldsymbol{H}$ must be independent.*

(i). Suppose there is a word of C, with weight $m-\alpha$, $\alpha \geqslant 0$. Then there is at least one set of $m-\alpha$ columns of $\boldsymbol{H}$ which are dependent. A set of m columns of $\boldsymbol{H}$ containing these is also dependent. This is a contradiction.

(ii). If a set of m columns of $\boldsymbol{H}$ is dependent, then there is a linear relation among these m columns in which there are $m-\alpha$, $\alpha \geqslant 0$, non-null coefficients. Hence there is a word of weight $m-\alpha$, $\alpha \geqslant 0$. This is a contradiction.

4. EQUIVALENT CODES

If $\boldsymbol{G}$ is the generating matrix of a linear code C, and $\boldsymbol{G}^*$ is obtained from $\boldsymbol{G}$ by column permutations, then $\boldsymbol{G}^*$ generates a linear code C^* defined to be equivalent to C.

The generator matrix $\boldsymbol{G}$ of a linear code C is not unique. If $\boldsymbol{G}_0$ can be obtained from $\boldsymbol{G}$ by elementary row operations (i.e., row multiplication and row addition) then $\boldsymbol{G}_0$ also generates C. If $\boldsymbol{G}^*$ is obtained from $\boldsymbol{G}_0$ by column interchanges, then $\boldsymbol{G}^*$ generates an equivalent code C^*. There is a (1,1) correspondence between the words of C and C^* such that corresponding words have the same weight.

It is readily proved that given an (n, k) linear code C, we can find an equivalent code C^*, for which the generating matrix is

(4.1) $$G^* = [I_k, P],$$

where I_k is the $k \times k$ unit matrix, and P is a $k \times r$ matrix.

Every word of C^* is of the form $c'G^*$ where $c' = (c_1, c_2, \ldots, c_k)$. But $c'G^* = (c_1, c_2, \ldots, c_k;\ c_1p_{11}+c_2p_{21}+\ldots+c_kp_{k1}, \ldots, c_1p_{1r}+c_2\,p_{2r}+\ldots+c_k\,p_{kr})$.

Hence the first k coordinates of any word of C^* can be arbitrarily chosen, then the $(k+1)$th, ..., nth coordinates are certain linear combinations of these. A code of this type is called a systematic code. The first k coordinates of each word are called information symbols and the last r coordinates the check symbols. We thus have

Theorem 4.1. *Every linear code is equivalent to a systematic code.*

Let G^* be given by (4.1). Now

(4.2) $$[I_k, P]\begin{bmatrix} -P \\ I_r \end{bmatrix} = -P+P = 0.$$

Hence if we put

(4.3) $$H^* = [-P', I_r],$$

then the vector space generated by H^* is the null space of the vector space generated by G^*. Hence H^* given by (4.3) is the parity check matrix of the systematic code generated by G^* given by (4.1), and conversely.

5. SYNDROMES AND COSETS

Consider an (n, k) linear code C, with generator matrix G and parity check matrix H. Given any n-vector v', whether belonging to C or not, the syndrome of v' is defined to be the row vector

$$s' = v'H'.$$

From Theorem (3.1), v' belongs to C if and only if its syndrome is zero. Note that the syndrome of any n-vector is an r-vector.

Since the set of code words C, forms a subgroup of the group of all n-vectors, we can form the cosets of C in the usual manner.

Let

$$\mu = q^k - 1, \quad \nu = q^r - 1.$$

We form a table in which the elements of C are written in the first row, the null element being in the initial place.

TABLE I

C	$e_0' = g_0' = 0$	g_1'	g_2'	$\cdots$	g_μ'
C_1	e_1'	$g_1' + e_1'$	$g_2' + e_1'$	$\cdots$	$g_\mu' + e_1'$
C_2	e_2'	$g_1' + e_2'$	$g_2' + e_2'$	$\cdots$	$g_\mu' + e_2'$
	$\cdots$	$\cdots$	$\cdots$	$\cdots$	$\cdots$
C	e_ν'	$g_1' + e_\nu'$	$g_2' + e_\nu'$	$\cdots$	$g_\mu' + e_\nu'$

Let e_1' be any n-vector not belonging to C. Then the coset C_1 is obtained by adding e_1' to the elements of C. The element $g_i' + e_1'$ of C_1 is written in the row corresponding to C_1, below g_i'. Now if e_2' is any n-vector not belonging to C or C_1 we form the coset C_2 in an analogous manner. Proceeding in this manner we get $\nu + 1 = q^r$ cosets counting C itself as one coset. Each n-vector with elements from $GF(q)$ belongs to one and only one coset.

The elements in the first column of Table I are called coset leaders. In forming the coset C_i instead of e_i' we might use and other element of C_i say $e_i' + g_j'$ as the coset leader. This will not change the coset C_i. Only the elements of C_i will now appear in a different order,

$$e_i' + g_j', \; e_i' + g_1' + g_j', \; \ldots, \; e_i' + g_\mu' + g_j'.$$

It is clear that two n-vectors belong to the same coset if and only if their difference belongs to C.

Theorem 5.1. *Two n-vectors belong to the same coset if and only if their syndromes are equal.*

Let v_1' and v_2' be two n-vectors with the same syndrome. Then $v_1'H' = v_2'H'$. Hence $(v_1' - v_2')H' = 0$. Therefore $v_1' - v_2'$ belongs to C, which shows that v_1' and v_2' belong to the same coset.

Conversely if $\boldsymbol{v}_1'$ and $\boldsymbol{v}_2'$ belong to the same coset then $\boldsymbol{v}_1'-\boldsymbol{v}_2' = \boldsymbol{g}'$ where $\boldsymbol{g}'$ belongs to C. Hence

$$(\boldsymbol{v}_1'-\boldsymbol{v}_2')\boldsymbol{H}' = \boldsymbol{g}'\boldsymbol{H}' = \boldsymbol{0}.$$

Therefore $\boldsymbol{v}_1'\boldsymbol{H}' = \boldsymbol{v}_2'\boldsymbol{H}'$, i.e. $\boldsymbol{v}_1'$ and $\boldsymbol{v}_2'$ have the same syndrome.

6. USE OF SYNDROMES FOR ERROR DETECTION AND ERROR CORRECTION

If the code word $\boldsymbol{g}'$ is transmitted and the received vector is $\boldsymbol{v}'$, then the error vector is defined to be

$$e' = \boldsymbol{v}'-\boldsymbol{g}', \tag{6.1}$$

i.e. Received vector $\boldsymbol{v}'$ = Transmitted vector $\boldsymbol{g}'$ + Error vector $\boldsymbol{e}'$.

If there is no transmission error $\boldsymbol{v}' = \boldsymbol{g}'$, and the error vector $\boldsymbol{e}'$ is null. If however w of the coordinates of $\boldsymbol{g}'$ have been wrongly transmitted, then $\boldsymbol{v}'$ and $\boldsymbol{g}'$ disagree in w coordinates. Hence the weight of $\boldsymbol{e}'$ is w. We say that w errors have occurred in transmitting $\boldsymbol{g}'$.

Theorem 6.1. *If the minimum weight of the words of a linear code C is $2t+d+1$, ($t \geqslant 0$, $d \geqslant 0$), then any t or a lesser number of errors can be corrected, and if the number of errors lies between $t+1$ and $t+d$, they can be detected.*

We shall first show that if $\boldsymbol{e}_1'$ and $\boldsymbol{e}_2'$ are any two n-vectors such that $w(\boldsymbol{e}_1')+w(\boldsymbol{e}') \leqslant 2t+d$, then the syndromes of $\boldsymbol{e}_1'$ and $\boldsymbol{e}_2'$ are different. If possible let the syndromes be equal. Then $\boldsymbol{e}_1'\boldsymbol{H}' = \boldsymbol{e}_2'\boldsymbol{H}'$ or $(\boldsymbol{e}_1'-\boldsymbol{e}_2')\boldsymbol{H}' = \boldsymbol{0}$. Hence $\boldsymbol{e}_1'-\boldsymbol{e}_2'$ is a code word. Hence

$$2t+d+1 \leqslant w(\boldsymbol{e}_1'-\boldsymbol{e}_2') \leqslant w(\boldsymbol{e}_1')+w(-\boldsymbol{e}_2') = w(\boldsymbol{e}_1')+w(\boldsymbol{e}_2') \leqslant 2t+d$$

which is a contradiction.

Let Ω_1 be the set of all n-vectors of weight t or less. Also let Ω_2 be the set of all n-vectors whose weight is not less than $t+1$, and does not exceed $t+d$. Then the syndromes of any two vectors belonging to Ω_1 are different from each other. Let S_1 be the set of these syndromes. Then there is a (1,1) correspondence between the vectors of Ω_1 and S_1, such that a vector of

S_1, is the syndrome of the corresponding vector of Ω_1. Note that the null vector is contained in Ω_1, and corresponds to the null vector in S_1.

Again the syndrome of any vector belonging to Ω_1 is different from the syndrome of any vector belonging to Ω_2. In particular the syndrome of any vector belonging to Ω_2 is non-null.

We now set up the following decision rule for decoding: Let $\boldsymbol{v}'$ be the received vector. If the syndrome of $\boldsymbol{v}'$ belongs to S_1, we conclude that the error vector is the corresponding vector of Ω_1. The transmitted vector is then obtained by subtracting this error vector from the received vector. If the syndrome of $\boldsymbol{v}'$ does not belong to S_1 we conclude that the received vector is different from the transmitted word. Thus an error is detected but we do not attempt to correct it.

We have now to show that this decision rule will correct up to t errors and detect up to $t+d$ errors in the transmission of any word. Suppose the transmitted word is $\boldsymbol{g}'$ and the error-vector is $\boldsymbol{e}'$. Then from (6.1),

$$\begin{aligned}\text{Syndrome } \boldsymbol{e}' &= \boldsymbol{e}'\boldsymbol{H}' \\ &= (\boldsymbol{v}'-\boldsymbol{g}')\boldsymbol{H}' \\ &= \boldsymbol{v}'\boldsymbol{H}' \\ &= \text{Syndrome } \boldsymbol{v}'.\end{aligned}$$

If t or a lesser number of errors have occurred $w(\boldsymbol{e}') \leqslant t$. Hence the syndrome of $\boldsymbol{v}'$ belongs to S_1. There is only one member of Ω_1, viz., $\boldsymbol{e}'$ which has the same syndrome as $\boldsymbol{v}'$. Hence our decision rule will correctly pick up the error vector, and then the transmitted word is correctly determined as $\boldsymbol{v}'-\boldsymbol{e}' = \boldsymbol{g}'$.

If between $t+1$ and $t+d$ errors have occurred, then $t+1 \leqslant w(\boldsymbol{e}') \leqslant t+d$. In this case the syndrome of $\boldsymbol{v}'$ will be non-null without belonging to S_1. Hence our decision rule will correctly indicate that errors have occurred in transmitting, but we will not be able to correct them.

If more than $t+d$ errors have occurred, then the syndrome of $\boldsymbol{v}'$ could belong to S_1. If this happens our decision rule would lead to a wrong conclusion.

Corollary. *If the minimum weight of the words of a linear code C is $2t+1$ any t or a lesser number of errors can be corrected. If the minimum weight is $d+1$, errors up to d in number can be detected.*

7. ONE ERROR DETECTING LINEAR CODES

Taking $t = 0$, $d = 1$ in Theorem 6.1, we see that for a one error detecting linear code the minimum weight of each code must be two. Let us take for $\boldsymbol{H}$, the parity check matrix, a single row vector, with non-zero elements from $GF(q)$. Then no column of $\boldsymbol{H}$ is dependent. From the corollary to Theorem 3.2, each word of the corresponding code has weight at least 2. Hence the code must be one error detecting. Thus if

$$\boldsymbol{H} = (h_1, h_2, \ldots, h_n), \quad h_i \neq 0 \text{ for } i = 1, 2, \ldots, n$$

then $\boldsymbol{g}' = (g_1, g_2, \ldots, g_n)$ is a code word if and only if

$$g_1h_1+g_2h_2+\ldots+g_nh_n = 0.$$

We can therefore construct a one error detecting $(n, n-1)$ code for any n. If $\boldsymbol{v}'$ is the received vector, we decide that there has been a transmission error if its syndrome

$$v_1h_1+v_2h_2+\ldots+v_nh_n,$$

is non-null, and that there has been no error if the syndrome is null. In case the error vector is non-null and belongs to the code C, the syndrome of the received word will be zero, and we shall wrongly decide that it has been correctly transmitted. In other cases error will be detected.

8. THE FUNCTION $n_m(r, q)$ AND THE PACKING PROBLEM

Let $m = 2t+d$, $t \geqslant 0$, $d \geqslant 0$. We have shown in Theorem 6.1 that if the minimum weight of the words of C is $m+1$, then we can correct any t or less errors, and detect up to $t+d$ errors.

From the corollary to Theorem 3.2 it follows that one way of obtaining C is to find an $r \times n$ matrix $\boldsymbol{H}$, which has the property (P_m), that no m columns of $\boldsymbol{H}$ are dependent. Then C would be the code with parity check mtrix $\boldsymbol{H}$. One might ask the following question :

For a given r, what is the maximum value of n, for which there exists an $r \times n$ matrix $\boldsymbol{H}$, with elements from $GF(q)$, possessing the property (P_m), that no m columns of $\boldsymbol{H}$ are dependent ? We shall denote this maximum value by $n_m(r, q)$.

The case $m = 1$ is trivial, since any non-null r-vector can be taken as a column of $\boldsymbol{H}$, and repeated as many times as we choose. Hence n does not have a finite maximum. In what follows we shall suppose $m \geqslant 2$.

If $m \geqslant 2$, and $\boldsymbol{H}$ is an $r \times n$ matrix with the property (P_m), then no two columns of $\boldsymbol{H}$ are dependent. The elements of a column vector $\boldsymbol{H}$ may be regarded as the coordinates of a point of the finite projective space $PG(r-1, q)$, distinct columns representing distinct points. Hence alternatively $n_m(r, q)$ is the maximum number of points we can choose in $PG(r-1, q)$ so that no m are dependent. The problem of finding such a set of points in $PG(r-1, q)$ may be called the packing problem.

Lemma 8.1. $n_m(r, q) \geqslant r+1$.

This is obvious since we can choose for columns of $\boldsymbol{H}$, the r unit vectors, and the vector all of whose columns are unity.

Lemma 8.2. *For a given prime power q and a given $m \geqslant 2$, $n_m(r, q)$ is a monotonically increasing function of r such that*

$$n_m(r+1, q) \geqslant 1+n_m(r, q) \tag{8.1}$$

There exists an $r \times n_m(r, q)$ matrix $\boldsymbol{H}$, no m columns of which are dependent. Add an $(r+1)$th null row to $\boldsymbol{H}$, and finally a last column for which the first r elements are zero and the $(r+1)$th element is 1. This extended matrix still has the property (P_m), which proves our result.

Theorem 8.1. *If $\boldsymbol{H}$ is an $r \times n_m(r, q)$ matrix, with elements from $GF(q)$, having the property (P_m), then rank $\boldsymbol{H} = r$.*

Rank $\boldsymbol{H} \leqslant \min[r, n_m(r, q)]$. Hence from Lemma 8.1 rank $\boldsymbol{H} \leqslant r$. Suppose then rank $\boldsymbol{H} = r_1 < r$. Then we can choose r_1 independent rows of $\boldsymbol{H}$, such that the remaining $r - r_1$ rows are dependent on these. The submatrix $\boldsymbol{H}_1$ of $\boldsymbol{H}$, consisting of these r_1 rows has the property (P_m), that no m columns are dependent. Hence $n_m(r_1, q)$ is not less than $n_m(r, q)$. However from Lemma 8.2, $n_m(r, q) \geqslant (r - r_1) + n_m(r_1, q)$. We thus have a contradiction. If follows that rank $\boldsymbol{H} = r$.

The following bounds for $n_m(r, q)$ are known [1], [8], [9], [12], [13]. If $n = n_m(r, q)$ then

$$\text{(i)}\quad 1 + \binom{n}{1}(q-1) + \binom{n}{2}(q-1)^2 + \ldots + \binom{n}{m-1}(q-1)^{m-1} \geqslant q^r,$$

(Gilbert, Varshamov)

$$\text{(ii) (a)}\quad q^r \geqslant 1 + \binom{n}{1}(q-1) + \binom{n}{2}(q-1)^2 + \ldots$$

$$+ \binom{n}{t}(q-1)^t \text{ if } m = 2t, \qquad \text{(Rao, Hamming)}$$

$$\text{(b)}\quad q^r \geqslant 1 + \binom{n}{1}(q-1) + \binom{n}{2}(q-1)^2 + \ldots$$

$$+ \binom{n}{t}(q-1)^t + \binom{n-1}{t}(q-1)^{t+1} \text{ if } m = 2t+1. \text{ (Rao)}$$

Theorem 8.2. *The maximum value of n, for which there exists an $(n, n-r)$ linear code with given redundancy r and such that each word has weight at least $m+1$, is $n_m(r, q)$.*

We can find an $r \times n_m(r, q)$ matrix $\boldsymbol{H}$, with elements belonging to $GF(q)$, such that no m columns are dependent. From Theorem 8.1 its rank is r. Let C be the linear code with has $\boldsymbol{H}$ for its parity check matrix, then C is an (n, k) code where $k = n - r$.

From the corollary to Theorem 3.2, each word of C has weight at least $m+1$.

Suppose there exists a linear code (n_1, n_1-r), $n_1 < n_m(r, q)$ with redundancy r, such that each word has weight at least $m+1$. Then its parity check matrix $\boldsymbol{H}_1$ is an $r \times n_1$ matrix, with the property that no m columns of $\boldsymbol{H}_1$ are dependent. Hence $n_m(r, q) \geqslant n_1$, which is a contradiction.

Theorem 8.3. *For any $c < k$, the existence of an (n, k) linear code, for which each word has weight at least $m+1$, implies the existence of an $(n-c, k-c)$ linear code for which each word has weight at least $m+1$.*

Let C be an (n, k) linear code for which each word has weight at least $m+1$. We can find an equivalent code C^*, for which the generator matrix $\boldsymbol{G}^*$ is in the cannonical form

$$\boldsymbol{G}^* = [\boldsymbol{I}_k, \boldsymbol{P}],$$

where P is an $(n-k) \times k$ matrix. Let $\boldsymbol{G}_1^*$ be the matrix obtained from $\boldsymbol{G}^*$ by dropping the last c rows. Then each word of the code generated by $\boldsymbol{G}_1^*$, belongs to C^*, and must therefore have weight at least $m+1$. Note that the $(k-c+1)$th, $(k-c+2)$th, ..., kth columns of $\boldsymbol{G}_1^*$ are null. Let $\boldsymbol{G}_2^*$ be the $(k-c) \times (n-c)$ matrix obtained from $\boldsymbol{G}_1^*$ dropping these columns. Then $\boldsymbol{G}_1^*$ generates an $(n-c, k-c)$ linear code, each word of which has weight at least $m+1$.

Corollary. *There exists an $[n_m(r, q)-c, n_m(r, q)-r-c]$ linear code, for which each word has weight at least $m+1$, for any c, $0 \leqslant c < n_m(r, q)-r$.*

This corollary follows at once from Theorems 8.2 and 8.3.

9. THE FUNCTION $k_m(n, q)$

Let $k_m(n, q)$ denote the maximum number of information places for a linear code of given length n, with symbols from $GF(q)$, and for which each world has weight at least $m+1$.

Theorem 9.1. *If* $n_m(r, q) \geqslant n > n_m(r-1, q)$, *then*

$$k_m(n, q) = n-r.$$

From the corollary to Theorem 8.3, there exists a linear code

$$[n_m(r, q)-c,\ n_m(r, q)-r-c],$$

for which each word has weight at least $m+1$. Taking $c = n_m(r, q)-n$, we get the existence of an $(n, n-r)$ linear code for which each word has weight at least $m+1$. Hence

$$k_m(n, q) \geqslant n-r.$$

If possible suppose

$$k_m(n, q) = n-r+\theta, \qquad \theta \geqslant 1.$$

Then there exists a linear code $(n, n-r+\theta)$, with redundancy $r^* = r-\theta$, for which each word has length at least $m+1$. Hence from Theorem 8.2

$$n_m(r-\theta, q) \geqslant n.$$

From Lemma 8.2,

$$n_m(r-1, q) \geqslant n.$$

This contradicts the hypothesis.

Corollary 1. *For a fixed m,* $k_m(n, q)$ *is a montonically increasing function of n, but it may stay the same for two consecutive values of n.*

Corollary 2. *If* $n_m(r, q) \geqslant n \geqslant n\ (r-1, q)$, *then the minimum redundancy for a code of given length n, and for which each word has weight at least* $m+1$, *is r.*

10. ONE ERROR CORRECTING (OR TWO ERROR DETECTING) HAMMING CODES

In Theorem 6.1 put $2t+d+1 = 3$, then either $t = 1$, $d = 0$ or $t = 0$, $d = 2$. We thus see that if each word of a linear code C has weight at least 3, then we can either use it to correct a single error or we can use it to detect up to two errors (without attempting any correction). The parity check matrix $\boldsymbol{H}$ of such a code must have the property (P_2), viz. no two columns are dependent. If $\boldsymbol{H}$ is an $r \times n$ matrix, then the columns of $\boldsymbol{H}$ may be regarded as points of $PG(r-1, q)$. The columns corresponding to any two district points are independent. Thus the maximum value of n for given r and q, viz. $n_2(r, q)$, is given by

$$(10.1) \qquad n_2(r, q) = \frac{q^r-1}{q-1},$$

which is the number of distinct points in $PG(r-1, q)$. Thus if we take an $r \times n_2(r, q)$ matrix $\boldsymbol{H}$, whose columns represent all the distinct points of $PG(r-1, q)$ and form the code for which H is the parity check matrix, then we obtain a one error correcting (or two error detecting) $\left(\frac{q^r-1}{q-1}, k\right)$ linear code, where $k = \frac{q^r-1}{q-1} - r$. Since

$$(10.2) \qquad n_2(r-1, q) = \frac{q^{r-1}-1}{q-1}.$$

we have :

Theorem 10.1. *For any given n, we can obtain a one error correcting (or two error detecting) q-ary code, with redundancy r given by*

$$(10.3) \qquad \frac{q^r-1}{q-1} \geqslant n > \frac{q^{r-1}-1}{q-1}$$

This is the minimum redundancy possible.

The proof follows from Corollary 2 to Theorem 9.1.

Example. Let $q = 3$, $n = 10$. Then

$$\frac{3^3-1}{3-1} \geqslant 10 > \frac{3^2-1}{3-1}.$$

Hence the minimum redundancy is 3, and we can get a (10,7) ternary code by taking for the columns of the parity check matrix

$\boldsymbol{H}$, the coordinates of any 10 district points of $PG(2,3)$. Thus we may take

$$\boldsymbol{H} = \begin{bmatrix} 1 & 0 & 0 & 0 & 1 & 1 & 0 & 2 & 1 & 1 \\ 0 & 1 & 0 & 1 & 0 & 1 & 1 & 0 & 2 & 1 \\ 0 & 0 & 1 & 1 & 1 & 0 & 2 & 1 & 0 & 1 \end{bmatrix}$$

To use the code for single error correction, we form the syndrome of the received vector $\boldsymbol{v}'$. If the error vector is $\boldsymbol{e}'$ we have shown that

$$\boldsymbol{v}'\boldsymbol{H}' = \boldsymbol{e}'\boldsymbol{H}'.$$

If $\boldsymbol{e}' = (0,0, \ldots, e_i, 0, \ldots, 0)$. Then $\boldsymbol{v}'\boldsymbol{H}' = e_i\boldsymbol{h}_i'$, where $\boldsymbol{h}_i'$ is the ith row of $\boldsymbol{H}'$. Hence the decoding rule is : Form the syndrome of the received vector. If it is $e_i\boldsymbol{h}_i'$, conclude that the error $(0,0, \ldots e_i, 0, \ldots 0)$ has occurred.

In the example under consideration suppose

$$\boldsymbol{g}' = (1, 2, 2, 0, 1, 1, 2, 0, 1, 2),$$

was transmitted (it is readily verified that this is a code word) and suppose

$$\boldsymbol{v}' = (1, 2, 2, 0, 1, 1, , 2, 1, 2),$$

was received. Now

$$\boldsymbol{v}'\boldsymbol{H}' = (1, 0, 2) = 2\boldsymbol{h}_8'$$

where $\boldsymbol{h}_8'$ is the 8th row of $\boldsymbol{H}'$. Hence we conclude that

$$\boldsymbol{e}' = (0, 0, 0, 0, 0, 0, 0, 2, 0, 0).$$

Then $\boldsymbol{g}' = \boldsymbol{v}' - \boldsymbol{e}'$ is correctly reconstructed.

11. ONE ERROR CORRECTING AND TWO ERROR DETECTING HAMMING CODES

In Theorem 6.1 put $2t+d+1 = 4$, then either $t = 1$, $d = 1$ or $t = 0$, $d = 3$. This shows that if each word of a linear code C, has weight at least 4, then we can use it for correcting one error, and detecting two errors (or alternatively for detecting up to 3 errors without attempting any correction). The parity check matrix of such a code must have the property (P_3), that no three columns are dependent. As has been shown before the problem of finding for any given n, a code with the desired property, and minimum redundancy depends on the solution of the following packing problem : To find in $PG(r-1, q)$, the maximum number of points, so that no three are collinear. A complete answer to this problem is known when $q = 2$, and r is arbitrary, or when $r \leqslant 3$, and q is an arbitrary prime power.

(a) First let us consider the case $q = 2$. Consider the finite projective space $PG(r-1,2)$. The coordinates of any point on a hyperplane Σ i.e. a linear subspace of dimension $r-2$, satisfy a linear equation

$$a_1x_1+a_2x_2+\ldots+a_rx_r = 0 \tag{11.1}$$

where the a_i's are fixed constants (not all zero) belonging to $GF(2)$. Let S be the set of all points not lying on Σ. Any two distinct points of S, lie on a unique line, which meets Σ in a point. Since each line has exactly three points, it follows that no two points of S are collinear. The number of points in Σ is $2^{r-1}-1$, and the whole space has 2^r-1 points. Hence S contains exactly 2^{r-1} points.

Again from the Rao-Hamming bound on $n_m(r, q)$ given in Section 8, $n_3(r, 2) \leqslant 2^{r-1}$. Hence

$$n_3(r, 2) = 2^{r-1} \tag{11.2}$$

For simplicity the equation of the hyperplane Σ may be taken to be

$$x_1+x_2+\ldots+x_n = 0. \tag{11.3}$$

Then S consists of all points with an odd number of non-zero coordinates. Let $\boldsymbol{H}$ be the $r\times 2^{r-1}$ matrix whose columns represent the points of S. Then the $(2^{r-1}, 2^{r-1}-r)$ binary linear code which has $\boldsymbol{H}$ for its parity check matrix, has the property that each word has weight at least 4, and can be used for correcting one error and detecting two errors. These codes were first obtained by Hammng [9]. We can now state :

Theorem 11.1. *For any given n, we can obtain a one error correcting and two error detecting binary code, with redundancy r given by*

$$2^{r-1}-1 \geqslant n > 2^{r-2}-1. \tag{11.4}$$

This is the minimum redundancy possible.

(b) For odd $q > 2$, $r = 3$ it is known that [1],

$$n_3(3, q) = q+1. \tag{11.5}$$

If we take the set of $q+1$ points lying on a non-degenerate conic in the plane $PG(2, q)$, then no three will be collinear. In particular the equation of the conic may be taken as

$$x_1x_3 = x_2^2 \tag{11.6}$$

If the columns of a $3\times(q+1)$ matrix $\boldsymbol{H}$, represent the coordinates of the points lying on (11.6), then for the $(q+1, q-2)$ q-ary linear code which has $\boldsymbol{H}$ for its parity check matrix, each word will have weight at least four.

(c) Again when $q > 2$, $r = 4$, it is known [1], [11] that

$$n_3(4, q) = q^2+1.$$

If we take the set of q^2+1 points lying on a non-degenerate unruled quadric in $PG(3, q)$, then no three are collinear. The equation of the quadric may be taken as

$$a_{11}x_1^2+a_{12}x_1x_2+a_{22}x_2^2 = x_3x_4,$$

where $a_{11}t^2+a_{12}t+a_{22}$ is irreducible over $GF(q)$.

We can now use these points to construct a (q^2+1, q^2-3) q-ary linear code, for which each word has weight 4, and which may therefore be used for correcting one error and detecting two errors.

12. SOME TERNARY LINEAR CODES

Let $q = 3$. It can be proved by geometrical considerations [2], [3] that the set of 12 points of $PG(5, 3)$, whose coordinates are given by the columns of

$$(12.1) \qquad \boldsymbol{H} = \begin{bmatrix} 0 & 1 & 1 & 1 & 1 & 1 & 1 & 0 & 0 & 0 & 0 & 0 \\ 1 & 0 & 1 & 1 & 2 & 2 & 0 & 1 & 0 & 0 & 0 & 0 \\ 1 & 1 & 0 & 2 & 2 & 1 & 0 & 0 & 1 & 0 & 0 & 0 \\ 1 & 1 & 2 & 0 & 1 & 2 & 0 & 0 & 0 & 1 & 0 & 0 \\ 1 & 2 & 2 & 1 & 0 & 1 & 0 & 0 & 0 & 0 & 1 & 0 \\ 1 & 2 & 1 & 2 & 1 & 2 & 0 & 0 & 0 & 0 & 0 & 1 \end{bmatrix},$$

has the property that no 5 are dependent. From the Rao-Hamming bound given in Section 8, $n_5(6, 3) \leqslant 12$. Hence $n_5(6, 3) = 12$. From Section 4, the generating matrix of a ternary linear code C, with $\boldsymbol{H}$ for its parity check matrix can be written as

$$(12.2) \qquad \boldsymbol{G} = \begin{bmatrix} 1 & 0 & 0 & 0 & 0 & 0 & 0 & 2 & 2 & 2 & 2 & 2 \\ 0 & 1 & 0 & 0 & 0 & 0 & 2 & 0 & 2 & 2 & 1 & 1 \\ 0 & 0 & 1 & 0 & 0 & 0 & 2 & 2 & 0 & 1 & 1 & 2 \\ 0 & 0 & 0 & 1 & 0 & 0 & 2 & 2 & 1 & 0 & 2 & 1 \\ 0 & 0 & 0 & 0 & 1 & 0 & 2 & 1 & 1 & 2 & 0 & 2 \\ 0 & 0 & 0 & 0 & 0 & 1 & 2 & 1 & 2 & 1 & 2 & 1 \end{bmatrix},$$

Since $\boldsymbol{H}$ has property (P_5), the minimum weight of the words of the linear code C, generated by $\boldsymbol{G}$ is 6. As a matter of fact it can be verified by actual computation, that all the words have weight 6, 9 or 12. Putting $2t+d+1 = 6$ in Theorem 6.1, we have the following solutions (*i*) $t = 2$, $d = 1$, (*ii*) $t = 1$, $d = 3$, (*iii*) $t = 0$, $d = 5$. Hence the (12, 6) linear code C, can be used either as a 2 error correcting and 3 error detecting code, or as a one error correcting and 4 error detecting code, or as a five error detecting code.

Let $\boldsymbol{H}_1$ be the matrix obtained from $\boldsymbol{H}$ by dropping the last row and the last column. Thus

$$(12.3) \qquad \boldsymbol{H}_1 = \begin{bmatrix} 0 & 1 & 1 & 1 & 1 & 1 & 1 & 0 & 0 & 0 & 0 \\ 1 & 0 & 1 & 1 & 2 & 2 & 0 & 1 & 0 & 0 & 0 \\ 1 & 1 & 0 & 2 & 2 & 1 & 0 & 0 & 1 & 0 & 0 \\ 1 & 1 & 2 & 0 & 1 & 2 & 0 & 0 & 0 & 1 & 0 \\ 1 & 2 & 2 & 1 & 0 & 1 & 0 & 0 & 0 & 0 & 1 \end{bmatrix}$$

It is readily seen that no four columns of $\boldsymbol{H}_1$ are dependent. In fact if any four columns of $\boldsymbol{H}_1$ are dependent, then the corresponding 4 columns of $\boldsymbol{H}$ and the last column would be dependent contradicting the property (P_5) of $\boldsymbol{H}$. Also from the bound given in Section 8, $n_4(5, 3) \leqslant 11$. Hence $n_4(5, 3) = 11$. If we construct the (11, 6) ternary linear code C_1 with $\boldsymbol{H}_1$ as the parity check matrix, then each word of C_1 has weight at least 5. Hence C_1 can be used as a two error detecting code, or as a one error correcting, three error detecting code or as a four error detecting code.

Let the points corresponding to all 11 columns of $\boldsymbol{H}_1$ be denoted by $P_0, P_1, P_2, \ldots, P_{10}$. In $PG(4,3)$, each line has 4 points. Hence the line P_0P_i has two other points say Q_i and Q_i^*. We shall show that the 20 points

$$(12.4) \qquad P_1, P_2, \ldots, P_{10}, \quad Q_1, Q_2, \ldots, Q_{10}$$

have the property that no three are collinear. Three of the points P, say P_i, P_j, P_k cannot be collinear, as in this case P_0, P_i, P_j, P_k would be coplanar. Again $P_iP_jQ_k$ cannot be collinear, since P lies in the plane determined by P_0 and the line $P_iP_jQ_k$. This would make P_0, P_i, P_j, P_k coplanar. Other cases can be similarly disposed of. This shows that $n_3(5,3) \geqslant 20$.

On the other hand it is known [6], that $n_3(5,3) \leqslant 26$. The exact value of $n_3(5,3)$ is not known. If we take for the coordinates of Q_i the sum of the columns corresponding to P_0 and P_i, then the matrix H whose column represent the 20 points (12.6) can be written as

$$(12.4)\quad \boldsymbol{H}_2 = \begin{bmatrix} 1&1&1&1&1&1&0&0&0&0&1&1&1&1&1&1&0&0&0&0 \\ 0&1&1&2&2&0&1&0&0&0&1&2&2&0&0&1&2&1&1&1 \\ 1&0&2&2&1&0&0&1&0&0&2&1&0&0&2&1&1&2&1&1 \\ 1&2&0&1&2&0&0&0&1&0&2&0&1&2&0&1&1&1&2&1 \\ 2&2&1&0&1&0&0&0&0&1&0&0&2&1&2&1&1&1&1&2 \end{bmatrix}$$

The (20, 15) ternary linear code C_2, with parity check matrix $\boldsymbol{H}_2$ has the property that each word has weight at least 4. It can be used either as a one error detecting and two error correcting code, or as a three error correcting code.

13. THE BOSE-CHAUDHURI HOCQUENGHEM CODES [4], [5], [10]

Let V_s be the vector space of all s-vectors with elements from $GF(q)$. We can institute a correspondence between the vector*

$$\alpha = (a_0, a_1, \ldots, a_{s-1}),$$

of V_s, and the element

$$a_0 + a_1 x + a_2 x^2 + \ldots + a_{s-1} x^{s-1},$$

of the $GF(q^s)$, where x is a primitive element of $GF(q^s)$. This is a (1,1) correspondence in which the null vector α_0 of V_s corresponds to the null element of $GF(q^s)$. The sum of any two vectors of V_s corresponds to the sum of the corresponding elements of $GF(q^s)$. We can therefore identify a vector α of V_s, with the corresponding element of $GF(q^s)$. This in effect defines a multiplication of the vectors of V_s and converts it into a field. Thus if

$$\alpha = (a_0, a_1, \ldots, a_s), \quad \beta = (b_0, b_1, \ldots, b_s),$$

are any two elements of V_s, then we can identify α and β with the element $a_0 + a_1 x + \ldots + a_{s-1} x^{s-1}$, $b_0 + b_1 + \ldots + b_s x^{s-1}$ of $GF(q^s)$.

Now x satisfies a certain minimum equation $\phi(x) = 0$ where $\phi(x)$ is a polynomial of the sth degree with coefficients from $GF(q)$. We can form the product of the elements α and β of $GF(q^s)$. Thus let

$$\alpha\beta = \gamma = c_0+c_1x+\ldots+c_{s-1}x^{s-1}.$$

Then the product of the vectors α and β is $\gamma = (c_0, c_1, \ldots, c_{s-1})$.

Each element of $GF(q^s)$ can then be regarded as an s-vector with elements from $GF(q)$.

Let α be a non-zero element of $GF(q^s)$, and let $c > 0$, and $2 \leqslant m \leqslant q-2$, be integers.

Consider the matrix

$$\boldsymbol{H}' = \begin{bmatrix} 1 & 1 & 1 & \ldots & 1 \\ \alpha^c & \alpha^{c+1} & \alpha^{c+2} & \ldots & \alpha^{c+m-1} \\ (\alpha^c)^2 & (\alpha^{c+1})^2 & (\alpha^{c+2})^2 & \ldots & (\alpha^{c+m-1})^2 \\ \ldots & \ldots & \ldots & \ldots & \ldots \\ (\alpha^c)^{n-1} & (\alpha^{c+1})^{n-1} & (\alpha^{c+2})^{n-1} & \ldots & (\alpha^{c+m-1})^{n-1} \end{bmatrix},$$

where we shall suppose that $1, \alpha, \alpha^2, \ldots, \alpha^{n-1}$ are all distinct. Then $\boldsymbol{H}'$ can either be regarded as an $n\times m$ matrix with elements from $GF(q^s)$ or as $n\times ms$ matrix with elements from $GF(q)$. In this later case the element 1 of $GF(q^s)$ is to be regarded as the vector $(1, 0, 0, \ldots, 0)$ of V_s. When we form the transpose of $\boldsymbol{H}'$ i.e.,

$$\boldsymbol{H} = \begin{bmatrix} 1 & \alpha^c & (\alpha^c)^2 & \ldots & (\alpha^c)^{m-1} \\ 1 & \alpha^{c+1} & (\alpha^{c+1})^2 & \ldots & (\alpha^{c+1})^{n-1} \\ 1 & \alpha^{c+2} & (\alpha^{c+2})^2 & \ldots & (\alpha^{c+2})^{n-1} \\ \ldots & \ldots & \ldots & \ldots & \ldots \\ 1 & \alpha^{c+m-1} & (\alpha^{c+m-1})^2 & \ldots & (\alpha^{c+m-1})^{n-1} \end{bmatrix},$$

then $\boldsymbol{H}$ is an $m\times n$ matrix with elements from $GF(q^s)$ or an $ms\times n$ matrix with elements from $GF(q)$. Now elements of $GF(q^s)$ must be regarded as column s-vectors with elements from $GF(q)$.

We shall show that **H** when regarded in the first way has the property (P_m) that no m columns of **H** are dependent over $GF(q^s)$, and hence over $GF(q)$. From this it would follow that when **H** is regarded as a matrix with elements from $GF(q)$, then no m columns would be dependent over $GF(q)$.

Let **M** be the matrix formed by taking any distinct m columns of **H**. Then

$$\boldsymbol{M} = \begin{bmatrix} (\alpha^c)^{j_1} & (\alpha^c)^{j_2} & \dots & (\alpha^c)^{j_m} \\ (\alpha^{c+1})^{j_1} & (\alpha^{c+1})^{j_2} & \dots & (\alpha^{c+1})^{j_m} \\ \dots & \dots & \dots & \dots \\ (\alpha^{c+m-1})^{j_1} & (\alpha^{c+m-1})^{j_2} & \dots & (\alpha^{c+m-1})^{j_m} \end{bmatrix}.$$

where $0 \leqslant j_1 < j_2 < \dots < j_m \leqslant n-1$.

$$\det \ \boldsymbol{M} = \alpha^{c(j_1+j_2+\dots+j_m)} \begin{vmatrix} 1 & 1 & \dots & 1 \\ \alpha^{j_1} & \alpha^{j_2} & \dots & \alpha^{j_m} \\ \dots & \dots & \dots & \dots \\ (\alpha^{j_1})^{m-1} & (\alpha^{j_2})^{m-1} & \dots & (\alpha^{j_m})^{m-1} \end{vmatrix}$$

$$= \alpha^{c(j_1+j_2+\dots+j_m)} \qquad \Pi\, (\alpha^{j_u} - \alpha^{j_v}),$$

where $1 \leqslant u \leqslant v \leqslant m$. But by hypothesis $1, \alpha, \dots, \alpha^{n-1}$ are all distinct. Hence det $\boldsymbol{M} \neq 0$. This shows that the columns of **M** are independent and proves the required result.

Now let **H** be regarded as an $ms \times n$ matrix over $GF(q)$ which has the property (P_m) that no m columns are dependent. Its rank is $r \leqslant ms$. If we now construct the $(n, n-r)$ q-ary linear code C with parity check matrix **H**, then each word will have weight at least $m+1$.

It can happen that many rows of $\boldsymbol{H}$ (or columns of $\boldsymbol{H}'$) are dependent on others and so can be dropped without changing the code C. This will now be illustrated by considering certain examples and special cases.

(a) Let $q = 2$, $s = 6$. We then extend $GF(2)$ to $GF(2^6)$. Let α be a primitive element of $GF(2^6)$. Let us take $c = 1$, $m = 6$, and $n = 63$. We note that $1, \alpha, \alpha^2, \ldots, \alpha^{62}$ are all distinct since α is a primitive root. Then

$$\boldsymbol{H}' = \begin{bmatrix} 1 & 1 & 1 & 1 & 1 & 1 \\ \alpha & \alpha^2 & \alpha^3 & \alpha^4 & \alpha^5 & \alpha^6 \\ \alpha^2 & (\alpha^2)^2 & (\alpha^3)^2 & (\alpha^4)^2 & (\alpha^5)^2 & (\alpha^6)^2 \\ \cdots & \cdots & \cdots & \cdots & \cdots & \cdots \\ \alpha^{61} & (\alpha^2)^{61} & (\alpha^3)^{61} & (\alpha^4)^{61} & (\alpha^5)^{61} & (\alpha^6)^2 \\ \alpha^{62} & (\alpha^2)^{62} & (\alpha^3)^{62} & (\alpha^4)^{62} & (\alpha^5)^{62} & (\alpha^6)^{62} \end{bmatrix}$$

Now $x \to x^2$ is an automorphism of the field $GF(2^6)$. We also note that if c is an element of $GF(2)$, then $c^2 = c$. Hence to any linear relation with coefficients from $GF(2)$, between the elements of column 1 of $\boldsymbol{H}'$, there corresponds the same relation between the elements of the columns 2 and 4 of $\boldsymbol{H}'$. Hence if we drop the columns 2 and 4 of $\boldsymbol{H}'$, then the code C for which $\boldsymbol{H}$ is the parity check matrix will not change. Also the rank of $\boldsymbol{H}$ will not change. In the same way we see that we can drop the column 6. The matrix $\boldsymbol{H}'$ has now been reduced to the form,

$$\boldsymbol{H}_1' = \begin{bmatrix} 1 & 1 & 1 \\ \alpha & \alpha^3 & \alpha^5 \\ \alpha^2 & (\alpha^2)^3 & (\alpha^2)^5 \\ \cdots & \cdots & \cdots \\ \alpha^{61} & (\alpha^{61})^3 & (\alpha^{61})^5 \\ \alpha^{62} & (\alpha^{62})^3 & (\alpha^{62})^5 \end{bmatrix}$$

Regarded as a matrix over $GF(2)$ it is of order (63×18). Since $m = 6$, the (63,45) binary linear code with $\boldsymbol{H}_1$ as parity check matrix has words of weight at least 7 and can be used as a 3 error correcting code.

(b) Now let $q = 2$ and let $s \geqslant 2$ be any positive integer. Let $GF(q)$ be extended to $GF(q^s)$. Let $m = 2t$, $c = 1$, and let α be a primitive element of $GF(q^s)$. Then reasoning as before it is easy to see that if we obtain $\boldsymbol{H}_1'$ from $\boldsymbol{H}'$ by dropping the 2nd, 4th, ..., $(2t)$th columns of $\boldsymbol{H}'$, then the rank of $\boldsymbol{H}_1'$ will be the same as that of $\boldsymbol{H}'$. Hence this rank (when $\boldsymbol{H}_1'$ is regarded as a matrix over $GF(q)$), will be $R \leqslant st$. Hence by following the method explained we shall obtain a $(2^s-1,\ 2^s-1-R)$, t error correcting binary code where $R \leqslant st$.

The estimate st is only an upper bound for the rank of $\boldsymbol{H}'$. The actual rank may be less than this. This is illustrated by the example which follows.

(c) Let $q = 2$, $s = 6$ as in (a). Let $c = 1$, $m = 10$, and as before let α be a primitive element of $GF(q^6)$. We can as explained before drop the even numbered columns of $\boldsymbol{H}'$ and obtain

$$\boldsymbol{H}_1' = \begin{bmatrix} 1 & 1 & 1 & 1 & 1 \\ \alpha & \alpha^3 & \alpha^5 & \alpha^7 & \alpha^9 \\ \alpha^2 & (\alpha^3)^2 & (\alpha^5)^2 & (\alpha^7)^2 & (\alpha^9)^2 \\ \cdots & \cdots & \cdots & \cdots & \cdots \\ \alpha^{62} & (\alpha^3)^{62} & (\alpha^5{}_9{}^{62} & (\alpha^7)^{62} & (\alpha^9)^{62} \end{bmatrix},$$

such that

$$\text{Rank } \boldsymbol{H}' = \text{Rank } \boldsymbol{H}_1' \leqslant 30.$$

Now $(\alpha^9)^7 = \alpha^{63} = 1$. Thus α^9 and its powers constitute a subfield of order 2^3 of $GF(2)$ and α^9 satisfies a third degree equation with coefficients from $GF(2)$. Hence the elements of the last column of $\boldsymbol{H}_1'$ (when regarded as a matrix over $GF(q^s)$)

can be expressed as a linear combination of 1, α^9, α^{18} with coefficients from $GF(2)$. When $\boldsymbol{H}_1'$ is regarded as a matrix over $GF(2)$ then only three of the six columns corresponding to

$$\begin{bmatrix} 1 \\ \alpha^9 \\ (\alpha^9)^2 \\ \cdots \\ (\alpha^9)^{62} \end{bmatrix},$$

are independent. Hence rank $\boldsymbol{H}_1' = 27$. Thus the code for which $\boldsymbol{H}_1$ is the parity check matrix is a (63,36), 5 error correcting binary code.

(d) We can now see how the rank of $\boldsymbol{H}'$ can be obtained in the general case. Consider the factors of $X^{q^s-1}-1$, irreducible over $GF(q)$. Thus let

$$X^{q^s-1}-1 = \phi_1(X)\phi_2(X)\ldots\phi_u(X),$$

where $\phi_i(X)$ is a polynomial in X, with coefficients from $GF(q)$ and irreducible over $GF(q)$. If β is an element of $GF(q^s)$, then β is a root of one and only one polynomial out of $\phi_1(X)$, $\phi_2(X)$, ..., $\phi_u(X)$. On the other hand if β_i is a root of $\phi_i(X)$, then the other roots are β_i^q, $\beta_i^{q^2}$, ... Thus if v is the smallest integer such that $\beta_i^{q^v} = \beta_i$, then the degree of $\phi_i(x)$ is v, and β_i must belong to a subfield of order q^v of $GF(q^s)$. We will say that the index of β_i is v.

Now if among the elements of the set

$$\alpha^c,\ \alpha^{c+1},\ \alpha^{c+2},\ \ldots,\ \alpha^{c+m-1}, \tag{13.1}$$

more than one are the roots of the same polynomial $\phi_i(X)$, then we can immediately drop from $\boldsymbol{H}'$ columns corresponding to all but one. For example in (a), α, α^2, α^4 are the roots of the same irreducible factor of $X^{2^6}-1$ and we therefore can drop the

the columns corresponding to α^2 and α^4. Let α^u be an element of the set (13.1) the column corresponding to which has been retained. When $\boldsymbol{H}_1'$ is now considered over $GF(q)$, this column will become a matrix with s columns. However if v_u is the index of α^u, then out of these s columns only v_u will be independent. This gives us the following rule for the rank of $\boldsymbol{H}'$.

Consider the set of distinct factors of $X^{p^{s-1}}-1$, irreducible over $GF(q)$, whose roots occur one or more times in (13.1). Then the rank of $\boldsymbol{H}'$ is the sum of the degrees of these factors, the degree of each factor counting only once, even if it has more than one root in the set (13.1).

We shall conclude this section with a few more examples :

(3) Let $q = 2$, $s = 6$, $c = 1$, $m = 4$. Let α be the cube of a primitive element of $GF(2^6)$. We can take $n = 21$ since, 1, α, $\alpha^2, \ldots, \alpha^{20}$ are all different but $\alpha^{21} = 1$. The set $\alpha^c, \ldots, \alpha^{c+m-1}$ is now

$$\alpha, \alpha^2, \alpha^3. \alpha^4. \tag{13.2}$$

Now $\alpha, \alpha^2, \alpha^4, \alpha^8, \alpha^{16}, \alpha^{32} = \alpha^{11}$ are the roots of $\phi_1(X)$ a polynomial of the sixth degree ($\alpha^{64} = \alpha^{22} = 1$). Thus $\phi_1(X)$ has roots among (13.2). Again $\alpha^3, \alpha^6, \alpha^{12}$ are the roots of $\phi_2(X)$ a third degree polynomial ($\alpha^{24} = \alpha^3$). Hence the rank of $\boldsymbol{H}'$ is 9, the sum of the degrees of $\phi_1(X)$ and $\phi_2(X)$, and the general method described will lead to a 2 error correcting (21,12) binary code.

(f) Let $q = 3$, $s = 3$. Let $GF(3)$ be extended to $GF(3^3)$. Let $c = 12$, $m = 3$. Let α be a primitive element of $GF(3^3)$ and let $n = 26$. Now consider the set

$$\alpha^{12}, \alpha^{13}, \alpha^{14}.$$

$\alpha^{12}, \alpha^{10}, \alpha^4$ are the roots of a third degree polynomial $\phi_1(X)$, $\alpha^{14}, \alpha^{16}, \alpha^{22}$ are the roots of another third degree polynomial $\phi_2(X)$ and α^{13} is the root of a linear polynomial X-2. Hence we can obtain a (26,19) ternary code correcting one error and detecting two errors.

14. ERROR LOCATING CODES

Elspas and Wolf [14], [15] have recently introduced a new class of codes called error locating codes, with properties intermediate between error detecting codes and error correcting codes. Consider the case of a q-ary channel; where q is a prime power. In an error locating $(n, n-r)$ q-ary code each word is supposed to be divided into N subwords each of length n_0. Thus $n = nN_0$. If errors belonging to a certain class of patterns E_d occur within sub-words, and if the sub-words within which the errors occur belong to a certain class of patterns E_t, then we can detect the presence of transmission errors, and can locate the erroneous sub-words, but cannot actually pin point the errors. For example E_d may be the class of patterns consisting of d or a lesser number of errors in a sub-words, and E_t may be the class of patterns consisting of t or a lesser number of erroneous sub-words. Then it is required to find an $(n, n-r)$ linear q-ary code, such that if errors occur in not more than t sub-words, and consist of not more than d wrongly transmitted symbols in any sub-word, then it should be possible to detect the presence of transmission errors, and locate the erroneous sub-words. We shall now prove the following theorem due to Wolf.

Thereom 14.1. *Let C_0 be a q-ary (n_0, n_0-r_0) linear code which detects the class of error-patterns E_d. Let $Q = q^{r_0}$. Let C^* be a $(N, N-R)$, Q-ary linear code for which the transmission symbols are elements of $GF(Q)$ and which is capable of correcting errors belonging to a class of patterns E_t. Then we can construct an $(n, n-r)$ linear q-ary code, with $n = n_0N$, and $r = r_0R$, such that if errors belonging to E_d occur within a pattern of sub-words belonging to E_t, then the errors can be detected and erroneous sub-words located.*

Let $\boldsymbol{H}_0$ be the parity check matrix of C_0, where $\boldsymbol{H}_0$ is of order $r_0 \times n_0$. For example if $q = 2$, $n_0 = 7$, $r_0 = 3$, and E_d is the class of patterns consisting of two or fewer errors in any word of length 7, then we may take $\boldsymbol{H}_0$ as

$$(14.1)\qquad H_0 = \begin{bmatrix} 1 & 0 & 0 & 1 & 1 & 1 & 0 \\ 0 & 1 & 0 & 0 & 1 & 1 & 1 \\ 0 & 0 & 1 & 1 & 1 & 0 & 1 \end{bmatrix}$$

The columns of H_0 can be regarded as elements of $GF(q^{r_0})$ or $GF(Q)$. Thus in the example if α is a primitive element of $GF(2^3)$, satisfying the equation $\alpha^3+\alpha^2+1 = 0$, we can write

$$(14.3)\qquad H_0 = [1\ \alpha\ \alpha^2\ \alpha^3\ \alpha^4\ \alpha^5\ \alpha^6]$$

In the general case H_0 is a row-vector of length n_0 with elements from $GF(Q)$.

Let H^* be the parity check matrix of C^*, the order of H^* being $R \times N$ (when regarded as a matrix over $GF(Q)$. Let γ_{ij} be element in the ith row and jth column of H^*. Let be the Kronecker product of H^* and H_0 (regarded as a row vector over $GF(Q)$). Thus

$$(14.3)\qquad H = H^* \otimes H_0 = \begin{bmatrix} \gamma_{11}H & \gamma_{12}H & \cdots & \gamma_{1N}H \\ \gamma_{21}H & \gamma_{22}H & \cdots & \gamma_{2N}H \\ \cdots & \cdots & \cdots & \cdots \\ \gamma_{R1}H & \gamma_{R2}H & \cdots & \gamma_{RN}H \end{bmatrix}$$

When H is regarded as a matrix over $GF(Q)$, it is of order $R \times n_0 N$. However each element of H can be regarded as a column vector of length r_0, with elements from $GF(q)$. Thus when H is regarded as a matrix over $GF(q)$ it is of order $r_0 R \times n_0 N$ or $r \times n$. Let C be the code (with symbols from $GF(q)$, which has H (regarded in the second way) for its parity check matrix. Then C is the required $(n, n-r)$ error locating q-ary linear code.

Let us now consider the error-correcting capabilities of C. First consider the situation where errors occur only in the jth sub-word, the errors belonging to the class E_d. Then the error vector can be divided into N sub-blocks. All the sub-blocks are null except the jth which is say $(e_1, e_2, \ldots, e_{n_0})$, this vector belonging to E_d. Let

$$H_0 = (h_1, h_2, \ldots, h_{n_0}).$$

Then the resulting syndrome will contain the components

$$S_i = (e_1 h_1 + e_2 h_2 + \ldots + e_{n_0} h_{n_0}) \gamma_{ij} = a_{ij} \gamma_{ij},$$

where a_{ij} is a non-zero element of $GF(Q)$. If errors occur within several sub-words say $j_1, j_2, \ldots, j_v$ and if the errors within each sub-word are contained in E_d, whereas the pattern of sub-words in which the errors occur belongs to E_t, the resulting syndrome will contain components

$$S_i = a_{j_1} \gamma_{ij_1} + a_{j_2} \gamma_{ij_2} + \ldots + a_{j_v} \gamma_{ij_v}.$$

Note that $a_{j_1}, a_{j_2}, \ldots, a_{j_v}$ are non-zero elements of $GF(Q)$ and do not depend on i. Now if in the code C^* the error vector has as its j_1th j_2th, ..., j_vth coordinates the elements $a_{j_1}, a_{j_2}, \ldots, a_{j_v}$, and the other coordinates are zero, then the resulting syndrome will have exactly the components S_i. Since C^* corrects all patterns belonging to E_t, it is clear that the syndromes resulting from all permissible errors in the error locating code C, are all different. This proves our theorem.

To continue our example let C^* be the (63,55) two error correcting Bose-Chaudhuri octic code ($Q = 2^3$). Let θ be a primitive element of $GF(2^6)$. We can take θ as a root of the equation $\theta^6 + \theta + 1 = 0$ [7, page 262]. Then the elements of the subfield $GF(2^3)$ of $GF(2^6)$ are θ^{9i} ($i = 0,1,2,3,4,5,6$). Let $\theta^9 = \alpha$, then $\alpha^3 + \alpha^2 + 1 = 0$ and α is a primitive root of $GF(2^3)$. Using the relation $\theta^2 = \alpha^3\theta + \alpha$, we can express each element of $GF(2^3)$ in the form $\beta\theta + \delta$ where β and δ belong to $GF(2^3)$ so that elements of $GF(2^6)$ can be regarded as 2-vectors over $GF(2^3)$. Now we can take for the parity check matrix of C^* the matrix

$$(14.4) \quad \boldsymbol{H}^* = \begin{bmatrix} 1 & \theta & \theta^2 & \theta^3 & \cdots & \theta^{62} \\ 1 & \theta^2 & (\theta^2)^2 & (\theta^2)^3 & \cdots & (\theta^2)^{62} \\ 1 & \theta^3 & (\theta^3)^2 & (\theta^3)^3 & \cdots & (\theta^3)^{62} \\ 1 & \theta^4 & (\theta^4)^2 & (\theta^4)^2 & \cdots & (\theta^4)^{62} \end{bmatrix},$$

where $\boldsymbol{H}^*$ is of order 8×63, when regarded as a matrix over $GF(2^3)$, Hence using the method explained we first form the Kronecker product $\boldsymbol{H} = \boldsymbol{H}^* \otimes \boldsymbol{H}_0$. This is of order 24×441 over $GF(2)$, and then construct the code C which has $\boldsymbol{H}$ as a parity check matrix. We thus obtain a (441,417) binary linear code in which each word is to be divided into 63 sub-words of length 7. If then there are not more than two errors in not more than two sub-words, then we can detect them and locate the erroneous sub-words.

References

[1] Bose, R. C. (1947). "Mathematical Theory of Symmetrical Factorial Designs," *Sankhyā*, **8**, 107–166.

[2] Bose, R. C. (1961). "On Some Connections Between the Design of Experiments and Information Theory," *Bull. Inter. Stat. Inst.*, **38**, pt. 4, 257–271.

[3] Bose, R. C. (1961). "Some Ternary Error Correcting Codes and Fractionally Replicated Designs," *Colloque Inter. du C.M.R.S. le Plan d'Experiences*, No 110, 21–32. Editions du C.M.R.S.

[4] Bose, R. C. and Ray-Chaudhuri, D. K. (1960). "On a Class of Error Correcting Binary Group Codes," *Information and control*, **3**, 68–79.

[5] Bose, R. C. and Ray-Chauhuri, D. K. (1960). "Further Results on Error Correcting Group Codes," *Information and control*, **3**, 279–290.

[6] Bose, R. C. and Srivastava, J. N. (1964). "On a Bound Useful in the Theory of Factorial Designs and Error Correcting Codes," *Ann. Math. Statist.*, **35**, 780–794.

[7] Carmichael, R. D. (1956). *Introduction to the Theory of Groups of Finite Order*, Dover, New York.

[8] Gilbert, E. N. (1952). "A Comparison of Signalling Alphabets," *Bell System Tech. J.*, **31**, 504–522.

[9] Hamming, R. W. (1950). "Error Detecting and Error Correcting Codes," *Bell System Tech. J.*, **29**, 147–160.

[10] Hocquengham, A. (1959). "Codes Correctuers d'Erreurs," *Chiffres*, **2**, 147–156.

[11] Qvist, B. (1952). "Some Remarks Concerning Curves of the Second Degree in a Finite Plane," *Ann. Acad. Sci. Fenn. Ser. A.I.*, **134**.

[12] Rao, C. R. (1947). "Factorial Experiments Derivable from Combinatorial Arrangements of Arrays," *Supp. J. Roy. Stat. Soc.*, **9**, 128–139.

[13] Varshamov, R. R. (1957). "Estimate of the Number of Signals in Error Correcting Codes," *Dokaldy A.N.S.S.R.*, **117**, no. 5, 739–741.

[14] Wolf, J. K. (1965). "On an Extended Class of Error Locating Codes. *Information and control*, **8**, 163–169.

[15] Wolf, J. K. and Elspas, B. (1963). "Error Locating Codes—A New Concept in Error Control," *IEEE Trans. Inform. Theory* ***IT*-9**, 20–28.

(*Received Jan. 1, 1966.*)

CHAPTER 9

Bounds on Error Correcting Codes (Non-Random)

I. M. CHAKRAVARTI*,

The University of North Carolina at Chapel Hill

SUMMARY

This is a collection of results on bounds on different parameters of error correcting codes (non-random). The relationship between different criteria of optimality for codes is discussed. The well-known bounds due to Plotkin, Hamming and Rao, Varsharmov and Gilbert and Johnson are described. The asymptotic expressions of these bounds are also given.

1. INTRODUCTION

Let us recall a few definitions and results. A *block code* is a code that uses sequences of n channel symbols or n-vectors or n-tuples. A *q-nary channel* transmits sequences formed from a set of q distinct symbols. The set of all q-nary n-vectors has q^n elements. Only a selected subset of n-vectors is transmitted. The subset is called a *code* and its elements—the n-vectors—are called the *code-words* or *code points*. The *Hamming distance* $\delta(u, v)$ between two code-words u and v is the number of positions

* This research was supported by United States Army Research Office-Durham, Contract No. DA-31-124-AROD-254.

in which they differ. The *Hamming weight* $w(v)$ of a code-word v is the number of non-zero components in it. A code has *minimum distance* d if the distance between any two of its code-words is at least d. If $d = 2t+1$, a code with minimum distance d can correct at least t errors made during transmission through the channel.

If the q symbols are taken as the q elements of a field $GF(q)$, then the q^n n-vectors form a vector space V_n. In this case, q is necessarily a prime number or a power of a prime number. If the n-vectors of a code form a subspace of the n-dimensional vector space, then it is called a *linear code* or a *group code*. The *minimum distance* for a linear code is equal to the minimum weight of its non-zero vectors.

A matrix G whose row-vectors form a basis of a linear code V is called a *generator matrix* for V. If V has dimension, k, $\boldsymbol{G}$ is a $k \times n$ matrix with rank k and V has, then, q^k *code-words*. Such a code is called an (n, k) linear code.

Consider the set V^* of q-nary n-vectors which are orthogonal to every one of the vectors in V. Then V^* is a subspace and thus defines a *linear code*. Let $\boldsymbol{H}$ define a generator matrix for V^*. Then $\boldsymbol{H}$ is of the form $(n-k) \times n$ and it has rank $n-k$. $\boldsymbol{H}$ is called a *parity-check* matrix for V. Similarly, $\boldsymbol{G}$ is a parity check matrix for V^*. It also follows that,

$$\boldsymbol{G}\boldsymbol{H}^T = \boldsymbol{0}. \tag{1.1}$$

For an (n, k) linear code V, n is the *word length*, k is the number of *information places* and $n-k = r$ is called the *redundancy* or number of *check digits*.

Any generator matrix $\boldsymbol{G}$ can be put in the combinatorially equivalent form

$$\boldsymbol{G} = [\boldsymbol{I}_k, \vdots \boldsymbol{G}_1], \tag{1.2}$$

by elementary row operations and permutation of columns. Two *combinatorially equivalent* matrices generate the same code.

Given a generator matrix $\boldsymbol{G} = [\boldsymbol{I}_k \vdots \boldsymbol{G}_1]$, the corresponding parity check matrix $\boldsymbol{H}$ is given by

(1.3) $$\boldsymbol{H} = [-\boldsymbol{G}_1^T : \boldsymbol{I}_r],$$

where $r = n-k$.

Here we quote a theorem [4] without proof, connecting the weight of a code-word and the linear dependence relation between columns of the parity check matrix $\boldsymbol{H}$.

Theorem 1.1. *Let V be a linear code with parity check matrix $\boldsymbol{H}$. Then for each code word of weight w there is a linear dependence relation between w columns of $\boldsymbol{H}$.*

Corollary : *A linear code with parity check matrix $\boldsymbol{H}$ has minimum weight (and hence minimum distance) at least d, if and only if every subset of $d-1$ columns of $\boldsymbol{H}$ is linearly independent. Such a matrix is said to possess the P_{d-1} property.*

2. CRITERIA OF OPTIMALITY

A linear code that for some t has all patterns of weight t or less and no others as coset leaders is called a *perfect* code [17]. A code which for some t has all patterns of weight t or less and none of weight greater than $(t+1)$ as coset leaders is called *quasi-perfect*.

Consider a *binary symmetric channel*. Suppose there exists an (n, k) binary linear code which is *quasi-perfect*. Let q be the probability that the received symbol is the same as the transmitted one and $p = 1-q$ is the probability that the received symbol is other than the transmitted one. Then the probability of correct decoding is given by

(2.1) $$\text{Prob (correct decoding)} = \sum_{i=0}^{n} f_i p^i q^{n-i},$$

where f_i is the frequency of coset leaders of weight i and

$$\sum_{i=0}^{n} f_i = 2^{n-k}$$

If $p < q$, $p^i q^{n-i}$ decreases with increasing i and hence the probability of correct decoding is increased whenever one f_i is increased and another f_{i+j} $(j > 0)$ is decreased. For a quasi-perfect code, f_i for $i = 0, 1, 2, \ldots, t$, is equal to the number of

n-vectors of weight i and is thus as large as possible. The terms f_{t+2} and beyond are all zero and f_{t+1} accounts for the remaining cosets. Then the probability of correct decoding P is given by

$$(2.2) \qquad P = \sum_{i=0}^{t} \binom{n}{i} p^i q^{n-i} + f_{t+1} p^{t+1} q^{n-t-1},$$

where $$f_{t+1} = 2^{n-k} - 1 - \binom{n}{1} - \ldots - \binom{n}{t} > 0.$$

It is easy to see that P is as large as possible. In the case where quasi-perfect codes do not exist, P provides an upper bound on the probability of correct decoding for any (n, k) linear group code.

The codes that have one information symbol repeated $(2t+1)$ times correct all combinations of t or fewer errors and no patterns of more than t errors. These trivial codes, the Golay (23,12) code [12] and the Hamming codes [14] are the only known *perfect* binary codes. Certain codes found by omitting columns from Hamming codes and Bose-Chaudhuri double-error-correcting codes are quasi-perfect [13]. Quasi-perfect codes have been called optimal codes [17], since they maximize the probability of correct decoding.

Other criteria of optimality of non-random codes are discussed in [9]. A brief account is given here.

For a given value of q, a linear code involves three parameters, n, k and d. For fixed n and k, a linear code which maximizes d is called a *maximum-minimum distance* (or *max-mini* for short) code. For fixed n and d, a linear code which maximizes k, is a *maximum size* code. For fixed k and d, a linear code which minimizes n is called a *minimum redundancy* code. These definitions of optimality are related but not equivalent.

Let V_r denote the vector space of all r-vectors whose elements belong to $GF(q)$. Let $n_d(r, q)$ denote the maximum number of r-vectors chosen from V_r, such that any d distinct vectors are independent. This number $n_d(r, q)$ is also the maximum number of points that can be chosen in the finite projective space $PG\ (r-1, q)$, such that no d of the points lie on a flat space of dimension $d-2$ or less. Existence of a set of $n = n_{d-1}(r, q)$ q-nary r-vectors such that no $d-1$ of them are linearly dependent, implies the existence of a parity-check matrix $\boldsymbol{H}$ of the form $r \times n$, whose columns are these r-vectors. $\boldsymbol{H}$ in its turn determines an (n, k) linear code having minimum distance d.

Given d and q, $n_d(r, q)$ is a monotonically increasing function of r.

Let $k_d(n, q)$ denote the maximum k for a linear code, given n, d and q. This implies the existence of a subspace of rank $k = k_d(n, q)$ in the vector-space V_n of all q-nary n-vectors.

The following theorem [6] establishes a relationship between $n_{d-1}(r, q)$, $n_{d-1}(r-1, q)$ and $k_d(n, q)$.

Theorem 2.1. *If* $n_{d-1}(r, q) \geqslant n > n_{d-1}(r, q)$, *then* $k_d(n, q) = n-r$.

In the following discussion, these two well-known results (*see* for instance [4]) are used.

(2.3) $$n_2(r, q) = (q^r-1)/(q-1).$$

(2.4) $$n_3(r, 2) = 2^{r-1}.$$

Since $k_3(8,2) = 4 = k_4(8,2)$, we note that $k_d(n, q)$ is not a strictly increasing function of d for all values of n. So, although, a code with parameters $n = 8$, $k = 4$ and $d = 3$ is of maximum size, it is not of maximum-minimum distance.

Again, $k_3(7,2) = 4 = k_3(8,2)$ shows that $k_d(n, q)$ is not a strictly increasing function of n. Hence a code with parameters $n = 8$, $k = 4$ and $d = 3$ is of maximum size but not of minimum redundancy.

Let $d_k(n, q)$ be the d for a maximum-minimum distance code with given n, k and q. Since $d_k(n, q)$ is not a strictly increasing function of n, a max-mini distance code is not necessarily of minimum redundancy.

Again $d_3(7, 2) = 4 = d_2(7, 2)$ shows that $d_k(n, q)$ is not a strictly decreasing function of k. So a code with $n = 7$, $k = 2$,

$q = 2$ and $d = 4$, is of max-mini distance, but not of maximum size.

Let $N_d(k, q)$ denote the smallest possible n for a code with given d, k and q. We show following [9] that $N_d(k, q)$ is a strictly increasing function of k. Let V be a minimum redundancy code for given d, q and k, that is, $n = N_d(k, q)$.

Let V_i be a subspace of V consisting of all vectors in V, whose ith coordinate is zero. We may choose i so that the ith coordinate is not always zero in V, and then V_i is an $(n, k-1)$ code with minimum weight d. Omitting the ith coordinate of every vector in V_i, one obtains an $(n-1, k-1)$ code with minimum weight d. Hence,

$$N_d(k-1, q) < n = N_d(k, q).$$

Since $N_d(k, q)$ is a strictly increasing function of k, a minimum redundancy code must be of maximum size.

Let V be a minimum redundancy code for given d, q and k, that is $n = N_d(k, q)$. Then choosing i so that the ith coordinate is not always zero in V, we omit the ith coordinate. This gives us an $(n-1, k)$ code with weight at least $d-1$. Hence

$$N_{d-1}(k, q) \leqslant n-1 < N_d(k, q).$$

Thus $N_d(k, q)$ is a strictly increasing function of d and hence, a minimum redundancy code is of max-mini distance.

Finally, the following example shows that a code can be both of max-mini distance and maximum size without being of minimum redundancy. Consider the binary code V consisting of the four vectors,

$$\boldsymbol{a}_0' = (000\ 000\ 000\ 0000)$$
$$\boldsymbol{a}_1' = (111\ 111\ 110\ 0000)$$
$$\boldsymbol{a}_2' = (000\ 011\ 111\ 1110)$$
$$\boldsymbol{a}_3' = (111\ 100\ 001\ 1110)$$

Here $n = 13$, $k = 2$, and $d = 8$. Since the last coordinate of all the vectors is zero, it can be omitted without affecting k or d. Hence V is not of minimum redundancy. But since codes with parameters $n = 13$, $k = 2$, $d = 9$ or $n = 13$, $k = 3$, $d = 8$, do not exist, V is of max-mini distance and maximum size. These results can be stated in the form of a theorem [9].

Theorem 2.2. *If a code is of minimum redundancy then it is of maximum size, and of max-mini distance. The other possible implications between these three properties are not universally valid.*

Definitions of optimality in terms of the probability of error detection or correction do not seem to be very closely related to the three optimalities defined in terms of redundancy, size or minimum distance. For instance, a quasi-perfect code maximizes the probability of correct decoding for a given n and k but it does not necessarily have the max mini distance for that n and k. Conversely, a max-mini distance or minimum redundancy code may not maximize the probability of correct decoding.

On the other hand, several classes of codes which have been constructed with the maximum size as a criterion of optimality turned out to be quasi-perfect and hence optimal in the probability sense. The Hamming codes with $d = 3$, the unreduced $d = 4$ codes, Golay codes (23,12) and the Bose-Chaudhuri $d = 5$ codes are examples in point.

3. BOUNDS ON PARAMETERS OF CODES

3.1. Bounds on $n_d(r, q)$

We first quote a few results on bounds on $n_d(r, q)$ without proof. A full treatment is given in a paper [3]. The symbol $m_d(r, q)$ has also synonymously been used for $n_d(r, q)$ [5]. It has been shown [4] that,

(3.1) $$n_3(3, q) = q+1 \text{ when } q \text{ is odd.}$$

(3.2) $$n_3(3, q) = q+2 \text{ when } q \text{ is even,}$$

(3.3) $$n_3(r, 2) = 2^{r-1} \text{ for } r > 3,$$

(3.4) $$n_3(4, q) = q^2+1 \text{ for } q \text{ odd.}$$

For q even $(q > 2)$, it is shown [19] that,

(3.5) $$n_3(4, q) = q^2+1.$$

For $r > 3$ and $q > 2$, it is known [21] that,

(3.6) $$n_3(r+1, q) < q^{r-1}+1.$$

In [20], it has been proved that

(3.7) $$n_3(5, q) \leqslant q^3-q^2+8q-14$$

for q odd and $q \geqslant 7$; also

(3.8) $$n_3(r+1, q) < q^{r+1}-q^{r-2}+8q^{r-3}-6\left(\sum_{i=0}^{r-4} q^i\right)-8$$

for $r > 4$, q odd ($q \geqslant 7$).

Further some improvements on the inequalities are available in [2]. These are

(3.9) $$n_3(r+1, 7) \leqslant 7^{n-1}-\sum_{i=1}^{r-3} 7^i \qquad \text{for } r > 4,$$

(3.10) $$n_3(5, 5) \leqslant 124,$$

(3.11) $$n_3(r+1, 5) < 5^{r-1}-10\left(\sum_{i=0}^{r-5} 5^i\right)-1 \quad \text{for } r > 4.$$

It has been proved in [8], that for $q > 2$, $r \geqslant 4$, $n_3(r, q)$ cannot exceed the positive root of the equation

(3.12) $$x^2(q^2-q-1)-x\{(q^2-2q-1)+N_r(q-2)\}-2N_r = 0,$$

where $N_r = (q^r-1)/(q-1)$. This provides the following improved inequalities.

(3.13) $$n_3(r+1, 3) \leqslant (3^{r+1}+23)/10 \quad \text{for } r \geqslant 4,$$

(3.14) $$n_3(5, q) \leqslant (q^3-1) \quad \text{for } q \text{ even and } q > 2,$$

(3.15) $$n_3(r+1, q) < \frac{q^2-2q-1+\left(\sum_{i=0}^{r} q^i\right)(q-2)}{q^2-q-1}+1$$

for q even ($q > 2$).

Finally, in [3] it has been shown that

(3.16) $$n_3(r+1, q) \leqslant q^{r-1}-2\left(\sum_{i=1}^{r-4} q^i\right)-1$$

for $r > 4$, $q > 2$ and q even.

3.2. Plotkin bound

We first prove a lemma on the weight of code words.

Lemma 3.1. *The sum of the weights of the code words of an (n, k) linear code with symbols taken from $GF(q)$ is $nq^{k-1}(q-1)$.*

Proof: Let $\boldsymbol{G}$ be the generator matrix of an (n, k) linear code,

$$\boldsymbol{G} = \begin{bmatrix} g_{11} & g_{12} & \cdots & g_{1n} \\ g_{21} & g_{22} & \cdots & g_{2n} \\ \cdot & \cdot & \cdots & \cdot \\ g_{k1} & g_{k2} & \cdots & g_{kn} \end{bmatrix} \tag{3.17}$$

$\boldsymbol{G}$ has rank k and we can choose the columns of $\boldsymbol{G}$ so that none of them is a null vector. The q^k code words are the q^k linear combinations of the k row-vectors of $\boldsymbol{G}$. Consider the linear equation.

$$x_1 g_{1i} + x_2 g_{2i} + \ldots + x_k\, g_{ki} = 0. \tag{3.18}$$

There are q^{k-1} solutions $(x_1, x_2, \ldots, x_k)$ to this equation, x_i's being elements of $GF(q)$. Hence if the q code words are written in the form of a $q^k \times n$ matrix, in each column there will be q^{k-1} zeros and $q^k - q^{k-1} = q^{k-1}(q-1)$ non-zero elements of $GF(q)$. Hence the sum of the weights of the code words is equal to $nq^{k-1}(q-1)$.

This provides an upper bound on the minimum weight (equivalently, minimum distance) for a linear (n, k) code.

Theorem 3.1. *The minimum weight d of a code word in an (n, k) linear code is at most $nq^{k-1}(q-1)/(q^k-1)$.*

Proof: Since there are (q^k-1) code words with non-zero weight and since the minimum weight can be at most equal to the average weight, it follows that,

$$d \leqslant nq^{k-1}(q-1)/(q^k-1). \tag{3.19}$$

Consider an $(r \times n)$ parity check matrix $\boldsymbol{H}$ which determines an (n, k) linear code having minimum distance d and $k = n-r$. This implies that no $(d-1)$ columns of $\boldsymbol{H}$ are dependent. If any c columns of $\boldsymbol{H}$ are deleted the resulting matrix still retains the P_{d-1} property and hence determines an $(n-c,\ k-c)$, $(c < k)$, linear code having minimum distance d. Hence we have the following lemma.

Lemma 3.2. *If there exists an (n, k) linear code with minimum distance d and if c is a positive integer less than k, then there exists an $(n-c,\ k-c)$ linear code with minimum distance d.*

Let $B(n, d)$ be the maximum possible q^k in a linear code for a given n with minimum distance at least d. By Lemma 3.2, the existence of a linear code with parameters n, k and d, where $q^k = B(n, d)$, implies the existence of a linear code with parameters $n-c$, $k-c$ and d, if $c < k$. Hence $B(n-c, d) \geqslant q^{k-c}$ and

$$q^k = B(n, d) \leqslant q^c B(n-c, d). \tag{3.20}$$

From (3.19), we get

$$q^{k-1}\,(qd+n-nq) \leqslant d. \tag{3.21}$$

From (3.20) and (3.21) it follows that

$$q^k = B(n, d) \leqslant \frac{q^c\, qd}{qd-(q-1)(n-c)}, \tag{3.22}$$

where $c < k$, provided $c > n - \dfrac{qd}{q-1}$.

Assuming that c is sufficiently large so that the denominator is positive, one should use that value of c which minimizes the right hand side. Considering the ratio

$$\frac{q^c\, qd}{qd-(q-1)(n-c)} \Big/ \frac{q^{c+1}\, qd}{qd-(q-1)(n-c-1)}$$

$$= \frac{1}{q} + \frac{q-1}{q\{qd-(q-1)(n-c)\}},$$

it is seen that it is greater than 1, provided that

$$qd-(q-1)(n-c) < 1,$$

and in this case c should be increased to $c+1$. Otherwise, one should not.

The inequality $qd-(q-1)(n-c) < 1$, can be written as $c < n-(qd-1)/(q-1)$. Hence to optimize, one takes for c the smallest integer greater than or equal to $n-(qd-1)/(q-1)$. This is the Plotkin bound. This implies that if $n > \dfrac{qd-1}{q-1}$,

$$q^k \leqslant q^{n-\frac{qd-1}{q-1}}\, qd. \tag{3.23}$$

Taking logarithms to the base q, one has

$$k \leqslant n-\frac{qd-1}{q-1}+1+\log_q d. \tag{3.24}$$

This can be stated as follows :

Theorem 3.2. *If $n > (qd-1)/(q-1)$, the number of check symbols required to achieve a minimum weight d is at least*

$$\frac{qd}{q-1}-\frac{1}{q-1}-1-\log_q d.$$

If d is very large, the last three terms in the expression are negligible.

The Plotkin bound holds also for non-linear codes [18].

3.3. The Varsharmov-Gilbert bound

We try to construct a parity check matrix $\boldsymbol{H}$ whose columns are q-nary r-vectors having the property P_{d-1}, that is every $d-1$ columns or fewer are linearly independent. Any non-null q-nary r-vector may be chosen as the first column of $\boldsymbol{H}$. Then the second column of $\boldsymbol{H}$ may be any non-zero q-nary r-vector other than the multiples of the first. The third column may be any q-nary r-vector which is not a linear combination of the first two. In general, the ith column is chosen as any q-nary r-vector that is not a linear combination of any $d-2$ or fewer preceding columns. This method of construction assures that no linear combination of $d-1$ or fewer columns will be the null vector. As long as the set of all linear combinations of $d-2$ or few columns does not include all q-nary r-vectors, another column can be added. In the worst possible case, all these linear combinations might be distinct. There are q^r-1 non-zero q-nary r-vectors. Hence if,

$$(3.25)\quad \binom{n-1}{1}(q-1)+\binom{n-1}{2}(q-1)^2+\ldots+\binom{n-1}{d-2}(q-1)^{d-2} < q^r-1,$$

there exists a linear code with n digits and at most r parity check digits (hence with at least $k = n-r$ information symbols) with minimum distance d. The code is the null space of the matrix $\boldsymbol{H}$ of the form $r\times n$. Hence the following theorem ([22], [11]) is proved.

Theorem 3.3. *It is possible to construct a code of length n and minimum distance d with r parity check symbols where r is the smallest integer satisfying* (3.25).

Writing

$$(3.26)\qquad S = \sum_{i=0}^{d-2}\binom{n-1}{i}(q-1)^i,$$

it is easy to show that,

$$(3.27)\qquad S < \binom{n-1}{n-d+1}(q-1)^{d-2}\,\frac{\lambda(q-1)}{\lambda(q-1)-\mu},$$

provided $\lambda(q-1) > \mu$, where $\lambda = 1-\dfrac{d-2}{n-1}$ and $\mu = \dfrac{d-2}{n-1}$.

Using Stirling's approximation for the factorials, one can then show that the right hand side of (3.27) is asymptotically equal to

$$(3.28)\quad -(n-1)\left\{\left(1-\frac{d-2}{n-1}\right)\log\left(1-\frac{d-2}{n-1}\right)+\frac{d-2}{n-1}\log\frac{d-2}{n-1}\right\} +(d-2)\log(q-1),$$

as $d\rightarrow\infty$, $n\rightarrow\infty$ such that $0 < b_1 < \dfrac{d-2}{n} < b_2 < 1$, where b_1 and b_2 are two preassigned constants. Hence, asymptotically, if

$$(3.29)\quad \left(1-\frac{k}{n}\right) \geqslant \left(1-\frac{1}{n}\right)H\left(\frac{n+1-d}{n-1}\right)+\left(\frac{d-2}{n}\right)\log(q-1),$$

or more simply if,

$$(3.30) \qquad \left(1-\frac{k}{n}\right) \geqslant H\left(\frac{n-d}{n}\right)+\frac{d}{n}\log(q-1),$$

then there exists a code of word length n with minimum distance at least d and with at least k information places for large n and d. For $q = 2$, (3.30) simplifies to

$$(3.31) \qquad 1-\frac{k}{n} \geqslant H\left(\frac{n-d}{n}\right).$$

In the above expressions $H(x) = -x\log x-(1-x)\log(1-x)$.

3.4. The Hamming-Rao bound

Suppose there exists an (n, k) linear code with minimum weight d. There are two cases to be considered :

(i) $d = 2t+1$ and (ii) $d = 2t+2$.

Case (*i*), $d = 2t+1$: In this case one can correct with certainty all patterns of t or fewer errors. Hence all q-nary n-vectors of weight t or less must be coset leaders.

The number of q-nary n-vectors with weight t or less is given by

$$1+\binom{n}{1}(q-1)+\binom{n}{2}(q-1)^2+\ldots+\binom{n}{t}(q-1)^t.$$

This must be less than or equal to the number of cosets q^r where $r = n-k$. Hence,

$$(3.32) \qquad 1+\binom{n}{1}(q-1)+\ldots+\binom{n}{t}(q-1)^t \leqslant q^r.$$

Case (*ii*), $d = 2t+2$: In this case, one can correct with certainty any pattern of t errors and detect with certainty any pattern of $(t+1)$ errors. As in case (*i*), the number of cosets whose leaders are n-vectors of weight t or less is given by

$$1+\binom{n}{1}(q-1)+\ldots+\binom{n}{t}(q-1)^t.$$

All n-vectors of weights $(t+1)$ must lie in other cosets. Consider vectors whose last coordinate is non-zero. The number of such vectors of weight $(t+1)$ is $\binom{n-1}{t}(q-1)^{t+1}$.

The distance between any two such vectors is at most $(2t+1)$. No two of these can be in the same coset because otherwise there will exist two codewords with distance between them equal to $2t+1$ or less. But this contradicts the hypothesis that $d = 2t+2$. Hence,

$$(3.33) \quad 1+\binom{n}{1}(q-1)+\ldots+\binom{n}{t}(q-1)^t+\binom{n-1}{t}(q-1)^{t+1} \leqslant q^r.$$

The Hamming-Rao bounds apply to non-linear codes as well [14].

The left hand side of (3.32) is asymptotically equal to

$$q^{nH\left(\frac{t}{n}\right)+t\log(q-1)}.$$

Thus for large n and d, an (n, k) linear code with minimum distance d does not exist if

$$(3.34) \qquad k/n > 1-H\left(\frac{d-1}{2n}\right)-\frac{d-1}{2n}\log(q-1).$$

For binary linear codes (3.34) takes the form

$$(3.35) \qquad \frac{k}{n} > 1-H\left(\frac{d-1}{2n}\right).$$

$\frac{k}{n}$ is called the transmission rate. For a binary symmetric channel $1-H(p)$ is the channel capacity where p is the probability of a transmitted symbol being received as the other one.

3.5. The Johnson bound

A new upper bound on the size of non-linear binary error-correcting codes and its asymptotic form have been derived in [15] and [16]. The problem is stated as follows : Find the maximum subset of vertices of an n-dimensional unit cube such that any two vertices of this subset are at least Hamming distance d apart, that is, their coordinate vectors differ in at least d places, where $d = 2t+1$ for a t-error correcting code.

Let $A(n, d)$ be the maximum number of rows in a matrix of n columns with entries of zeros and ones, with the property that two row vectors differ in at least d places.

Let $R(m, r, \lambda)$ be the maximum number of vectors each having r ones and $(m-r)$ zeros, with the property that the inner product of any two of the vectors is $\leqslant \lambda$.

Then for any two binary n-vectors with weights r_i and r_j having inner product λ_{ij} and the Hamming distance d_{ij},

(3.36) $$r_i + r_j = 2\lambda_{ij} + d_{ij}.$$

Let E_n be the set of 2^n vertices of the n-dimensional unit cube to be divided into disjoint subsets as follows :

(3.37) $$E_n = S_0 + S_1 + \ldots + S_{d-1},$$

where S_0 is the desired maximum subset of vertices at least Hamming distance d apart from each other and S_k is the subset of vertices at distance k from S_0.

Let $|X|$ denote the number of points in a point set X.

Then,

(3.38) $$A(n, d) = |S_0|$$

and

(3.39) $$2^n = |S_0| + |S_1| + \ldots + |S_{d-1}|$$

For $r \leqslant t$, we have

(3.40) $$|S_r| = \binom{n}{r} |S_0|.$$

Then the Hamming bound for a non-linear code can be easily obtained as

(3.41) $$2^n > |S_0| \sum_{i=0}^{t} \binom{n}{i}.$$

An improvement on the Hamming bound is given in [15] by considering the sets of S_k for $k > t$. This can be stated as follows :

(3.42) $$2^n \geqslant |S_0| \left\{ \sum_{i=0}^{t} \binom{n}{i} + \frac{\binom{n}{t+1} - \binom{d}{t} R(n, d, t)}{\left[\frac{n}{t+1}\right]} \right\},$$

where $\left[\frac{n}{t+1}\right]$ is the largest integer contained in $\frac{n}{t+1}$. A further improvement on (3.42) has been obtained and an asymptotic expression derived in [16]. This expression involves a complicated function $g(F)$ where F is defined as $F = n/t$, which has been tabulated in [16] for several values of F. For an (n, k) linear code Johnson's bound can be stated as

$$1-\frac{k}{n} \geqslant H\left(\frac{1+g(F)}{F}\right). \tag{3.43}$$

The following graph taken from [16], shows the different asymptotic upper and lower bounds for binary linear codes.

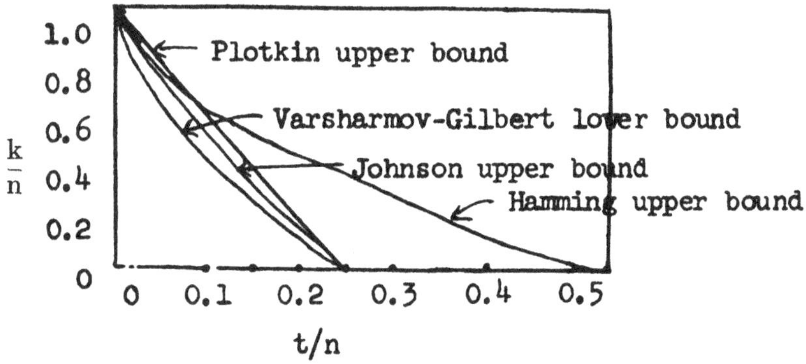

Figure 1. Comparison of asymptotic bounds.

Improved lower bounds on the size of non-linear binary codes are given in [1].

3.6. Discussion on bounds

Among the known codes, the Hamming codes and the equivalent highest rate Reed-Muller codes and Bose-Chaudhuri codes meet the Plotkin bound. Hence these codes have the maximum-minimum distance. For large n their rates all approach 0 or 1. In the middle rate range, for both the Reed-Muller and the best available estimate for the Bose-Chaudhuri codes with any fixed non-zero rate, the ratio d/n approaches 0 as n approaches infinity. There is no known coding system for which it has been proved that d/n remains finite as n approaches infinity with the transmission rate k/n held fixed, though the Varsharmov-Gilbert bound shows that such codes exist. For the binary symmetric channel, Elias' error-free coding [10] is known to be capable of achieving a probability of error which tends to zero, maintaining a transmission rate k/n which is bounded away from zero with large n. But the rate for Elias' code is much below channel capacity.

According to Shannon's fundamental theorem for noisy channels, for any rate k/n less than the channel capacity $1-H(p)$ there exist codes for which the probability of error is arbitrarily small. But no non-random coding is known to achieve this. For a more detailed account see [17].

References

[1] Bambah, R. P., Joshi, D. D. and Luthar, I. S. (1961). "Some Lower Bounds on the Number of Code Points in a Minimum Distance Binary Code," *Inf. and Control*, **4**, 313–323.

[2] Barlotti, A. (1957). "Una limitzione superiore per il numero di punti appartenenti a una k-calotta ($C(k, 0)$ di uno spazio lineare finito," *Boll. Un. Mat. Ital.*, (3), **12**, 67–70.

[3] Barlotti, A. [1966]. "Bounds for k-Caps in $PG(r, q)$ Useful in the Theory of Error-Correcting Codes," Univ. of North Carolina, Inst. of Statistics, Mimeo Series No. 484.2.

[4] Bose, R. C. (1947). "Mathematical Theory of the Symmetrical Factorial Designs," *Sankhyā*, **8**, 107–166.

[5] Bose, R. C. (1961). "On Some Connections Between the Design of Experiments and Information Theory," *Bull. Inter. Statist. Inst.*, **38**, 257–271.

[6] Bose, R. C. and Ray-Chaudhuri, D. K. (1960). "On a Class of Error-Correcting Group Codes," *Inf. and Control*, **3**, 68–79.

[7] Bose, R. C. and Ray-Chaudhuri, D. K. (1960). "Further Results on Error-Correcting Binary Group Codes," *Inf. and Control*, **3**, 279–290.

[8] Bose, R. C. and Srivastava, J. N. (1964). "On a Bound Useful in the Theory of Factorial Designs and Error-Correcting Codes," *Ann. Math. Statist.*, **35**, 408–414.

[9] Burton, R. C. (1964). "An Application of Convex Sets to the Construction of Error-Correcting Codes and Factorial Designs," Univ. of North Carolina, Inst. of Statistics, Mimeo Series No. 393.

[10] Elias, P. (1954). "Error-free Coding," *IRE Trans.*, **PGIT-4**, 29–37.

[11] Gilbert, E. N. (1952). "A Comparison of Signalling Alphabets," *Bell System Tech. J.*, **31**, 504–522.

[12] Golay, M. J. E. (1949). "Notes on Digital Coding," *Proc. IRE*, **37**, 657.

[13] Gorenstein, D., Peterson, W. W. and Zierler, N. (1960). "Two-Error-Correcting Bose-Chaudhuri Codes are Quasi-Perfect," *Inf. and Control*, **3**, 291–494.

[14] Hamming, R. W. (1950). "Error Detecting and Error Correcting Codes," *Bell System Tech. J.*, **29**, 147–160.

[15] Johnson, S. M. (1962). "A New Upper Bound for Error-Correcting Codes," *IRE Trans.*, **IT-8**, 203–207.

[16] Johnson, S. M. (1963). "Improved Asymptotic Bounds for Error-Correcting Codes," *IEEE Trans.*, **IT-9**, 198–205.

[17] Peterson, W. W. (1961). *Error-Correcting Codes*, The M.I.T. Press and John Wiley and Sons, New York.

[18] Plotkin, M. (1960). "Binary Codes with Specific Minimum Distance," *IRE Trans.*, **IT-6**, 445–450.

[19] Quist, B. (1952). "Some Remarks Concerning Curves of the Second Degree in a Finite Plane," *Ann. Acad. Sci. Fenn. Ser. A.*, **I**, No. 134.

[20] Segre, B. (1959). "Le geometrie di Galois," *Ann. Mat. Pura Appl.* (4), **48**, 1–97.

[21] Tallini, G. (1956). "Sulle *k*-calotte di uno spazio lineare finito," *Ann. Mat. Pura Appl.* (4), **42**, 119–164.

[22] Varsharmov, R. R. (1957). "Estimate of the Number of Signals in Error-Correcting Codes," *Doklady A.N.S.S.S.R.*, **117**, No. 5, 739–741.

Note added in proof : This paper was prepared in August 1965 and presented at the Summer School on Combinatorial Methods in Coding and Information Theory held in Royan, France, on August 26-September 8, 1965, under the auspices of NATO. For recent results in this area, one may refer to "Algebraic Coding Theory" by E. W. Berlekamp, published by McGraw Hill Co., 1968.

CHAPTER 10

*Nonparametric Tests for the Multisample Multivariate Location Problem**

SHOUTIR KISHORE CHATTERJEE AND

PRANAB KUMAR SEN[1], *University of Calcutta*

SUMMARY

In this paper, the univariate nonparametric analysis of variance tests by Kruskal and Wallis and by Brown and Mood have been extended to the general case of $p(\geqslant 1)$ variates and $c(\geqslant 2)$ samples. This has been accomplished through a conditional approach which makes the tests distribution-free under the null hypothesis. Various properties of the proposed tests have been studied and their asymptotic powers compared.

1. INTRODUCTION

During the past three decades the pace of development of nonparametric inference procedures has been tremendous. But, from the point of view of applications, this development has been confined mostly to univariate problems in the case of single

* Part of the work was completed with the partial support of the National Science Foundation Grant, GP-2593, at the Statistical Laboratory, University of California, Berkeley, while the second author was there.

[1] Now at The University of North Carolina at Chapel Hill, U.S.A.

as well as several samples and bivariate problems in the case of single sample. Multivariate problems have received comparatively much less attention; so that, at present there are few nonparametric contenders to standard methods of parametric multivariate analysis based on the assumption of multinormal parent distributions. Among the few multivariate nonparametric tests available, mention may be made of the bivariate sign-tests by Hodges (1955), Blumen (1958) and Bennett (1962, 1964); for a comparative study of these tests, the reader is referred to Chatterjee (1966). These tests actually relate to the single sample case. In the several sample case, a permutation test based on Hotelling's T^2-statistic was suggested by Wald and Wolfowitz (1944); but the test suffers from such shortcomings as are common to all tests based on permutations of values. In course of a series of lectures at the Calcutta University in 1962, S. N. Roy referred to a two sample bivariate location test developed by Roy, Bhapkar and Sathe; but the test is based on a step-down procedure in which the roles of the two variates do not appear to be symmetric.

In an earlier paper [7], the present authors considered two nonparametric tests for the bivariate two sample location problem, these being the generalizations of two well-known univariate two sample location tests, namely, Wilcoxon-Mann-Whitney rank-sum test and Mood's median test. Later on, the same principle has been used to formulate certain two sample nonparametric tests for testing the identity of association of two bivariate distributions. The object of the present investigation is to generalize the results of [7] to the general case of $p(\geqslant 1)$ variates and $c(\geqslant 2)$ samples, and develop nonparametric procedures for testing the identity of locations of several multivariate distributions. From one point of view, these results generalize the tests by Kruskal and Wallis (1952) and Brown and Mood (1950) to the multivariate case, and from another, they are the nonparametric analogues of the parametric tests by Roy (1942) and others [*see* Anderson (1958, Chapter 8)] for testing the equality of mean vectors of several multinormal distributions.

2. THE PROBLEM

Let $\boldsymbol{X}_\alpha^{(k)} = (X_{1\alpha}^{(k)}, \ldots, X_{p\alpha}^{(k)})$, $\alpha = 1, \ldots, n_k$ be n_k independent and identically distributed (vector valued) random variables (i.i.d.r.v.) having $p(\geqslant 1)$ variate continuous cumulative distribution functions (cdf) $F_k(\boldsymbol{x})$, $k = 1, \ldots, c$, and let the c samples be distributed independently. Let Ω be the set of all c-tuples of p variate continuous cdf's, and it is asumed that $\boldsymbol{F} = (F_1, \ldots, F_c)$ belongs to Ω. Later, some mild restrictions will have to be imposed on Ω, and these will be stated as occasions arise. On the basis of these c independent samples we desire to test the null hypothesis

$$H_0 : F_1(\boldsymbol{x}) = \ldots = F_c(\boldsymbol{x}), \text{ a. e.} \tag{2.1}$$

In testing this null hypothesis, we are particularly interested in detecting those alternatives which imply any difference in locations among the c distributions; the phrase *difference in locations* will be interpreted differently for the two tests to be considered here. For the rank-sum tests this will mean that for at least one of the p variates, the c sample observations are not all stochastically equal; the stochastic equality of two random variables X and Y being defined by the equality

$$P\{X > Y\} + \tfrac{1}{2}P\{X = Y\} = \tfrac{1}{2}. \tag{2.2}$$

Precisely, this means that for some $i(= 1, \ldots, p)$ there is at least one pair of (k, q) such that

$$P\{X_{i\alpha}^{(k)} > X_{i\beta}^{(q)}\} + \tfrac{1}{2}P\{X_{i\alpha}^{(k)} = X_{i\beta}^{(q)}\} \neq \tfrac{1}{2}, \ (k \neq q = 1, \ldots, c) \tag{2.3}$$

For the median test, let us denote by $\mu_i^{(k)}$ the median (assumed to be uniquely defined) of the random variable $X_{i\alpha}^{(k)}$, for $i = 1, \ldots, p$; $k = 1, \ldots, c$. Then the difference in location means that the following $pc(c-1)$ differences

$$\mu_i^{(k)} - \mu_i^{(q)}, \quad i = 1, \ldots, p; \quad k \neq q = 1, \ldots, c \tag{2.4}$$

are not all zero. There will be a special class of alternatives that will be of interest for both the rank-sum and median tests. This will be the class of *translation type* alternatives, and may be sketched as follows:

$$(2.5)\quad F_k(\boldsymbol{x}) = F(\boldsymbol{x}+\boldsymbol{\theta}^{(k)}),\ \boldsymbol{\theta}^{(k)} = (\theta_1^{(k)}, \dots, \theta_p^{(k)}),\ k = 1, \dots, c,$$

where $\theta_i^{(k)}$, $i = 1, \dots, p$; $k = 1, \dots, c$ are all real constants. Thus the c cdf's may be regarded to differ only by shifts in the locations. The proposed tests are valid for the types of alternatives considered above.

3. PRELIMINARY NOTIONS

Let us rank the N observations $X_{i\alpha}^{(k)}$, $\alpha = 1, \dots, n_k$, $k = 1, \dots, c$ on the ith variate in an increasing order of magnitude and let the rank of $X_{i\alpha}^{(k)}$ so obtained be denoted by $I_{i\alpha}^{(k)}$. Since the cdf's are all continuous, we can take the absence of ties for granted, in probability. The observation vector $\boldsymbol{X}_\alpha^{(k)}$ thus gives rise to the rank-vector

$$\boldsymbol{I}_\alpha^{(k)} = (I_{1\alpha}^{(k)}, \dots, I_{p\alpha}^{(k)}),\quad \alpha = 1, \dots, n_k,\ k = 1, \dots, c.$$

The N rank vectors corresponding to the N observations can be represented by the rank matrix (of order $p\times N$)

$$(3.1)\quad \boldsymbol{I}_N = \begin{pmatrix} I_{11}^{(1)} & \dots & I_{1n_1}^{(1)} & \dots & I_{11}^{(c)} & \dots & I_{1n_c}^{(c)} \\ \cdot & \dots & \cdot & \dots & \cdot & \dots & \cdot \\ I_{p1}^{(1)} & \dots & I_{pn_1}^{(1)} & \dots & I_{p1}^{(c)} & \dots & I_{pn_c}^{(c)} \end{pmatrix}$$

Each row of this random matrix is a permutation of the numbers $1, \dots, N$. Therefore the matrix $\boldsymbol{I}_N$ can have $(N!)^p$ possible realizations. We shall say that two rank matrices of the form (3.1) represent the same collection of rank vectors if one can be obtained from the other by a rearrangement of the columns. Specifically, the collection corresponding to (3.1) can be represented by permuting the columns so that the first row becomes $(1, \dots, N)$. Let the ith row of the matrix so obtained be $(\lambda_{i1}, \dots, \lambda_{iN})$, $i = 2, \dots, p$. If we write

$$(3.2)\qquad \lambda_{1\alpha} = \alpha,\quad \alpha = 1, \dots, N,$$

the collection of rank vectors corresponding to (3.1) can be represented by the *collection matrix*

$$\mathbf{\Lambda}_N = \begin{pmatrix} \lambda_{11} & \dots & \lambda_{1N} \\ \dots & \dots & \dots \\ \lambda_{p1} & \dots & \lambda_{pN} \end{pmatrix} \tag{3.3}$$

The matrix $\mathbf{\Lambda}_N$ being based on the random matrix $\boldsymbol{I}_N$ is itself random. As each row of $\mathbf{\Lambda}_N$, other than the first, can realize all the permutations of the numbers $1, \dots, N$, $\mathbf{\Lambda}_N$ can have $(N!)^{p-1}$ possible realizations. We shall denote this set of possible realizations of $\mathbf{\Lambda}_N$ by $\mathscr{L}_N$. Typically, we shall write $\boldsymbol{L}_N$ for a realization of $\mathscr{L}_N$ where

$$\boldsymbol{L}_N = \begin{pmatrix} l_{11} & \dots & l_{1N} \\ \dots & \dots & \dots \\ l_{p1} & \dots & l_{pN} \end{pmatrix} \tag{3.4}$$

$$l_{1\alpha} = \alpha, \qquad \alpha = 1, \dots, N. \tag{3.5}$$

The probability distribution of $\mathbf{\Lambda}_N$ over $\mathscr{L}_N$ would, of course, depend on the distributions $F_1, \dots, F_c$. However, given a particular realization $\boldsymbol{L}_N$ of $\mathbf{\Lambda}_N$, the conditional distribution of $\boldsymbol{I}_N$ over the $N!$ permutations of the columns of $\boldsymbol{L}_N$ would be uniform under H_0, whatever the common parent distribution may be. In this paper, we shall propose different functions of the elements of $\boldsymbol{I}_N$ as statistics for testing H_0. *From the observation made above, it follows that the conditional distribution of any such statistic given* $\mathbf{\Lambda}_N = \boldsymbol{L}_N$ *would be distribution free when* H_0 *is true.*

4. MULTIVARIATE MULTISAMPLE RANK-SUM TESTS

On the basis of the rank matrix $\boldsymbol{I}_N$ given by (3.1), let us find the mean ranks

$$\bar{I}_i^{(k)} = (1/n_k) \sum_{\alpha=1}^{n_k} I_{i\alpha}, \; i = 1, \dots, p; \; k = 1, \dots, c, \tag{4.1.1}$$

for the p variates in each of the c samples. We shall formulate a test for H_0 on the basis of these mean ranks. For this we shall derive first the expressions for the first and second order moments of the conditional joint distribution (under H_0) of the pc random variables defined in (4.1.1), and given $\mathbf{\Lambda}_N = \mathbf{L}_N$, where $\mathbf{\Lambda}_N$ is given by (3.2) and (3.3), and $\mathbf{L}_N \in \mathcal{L}_N$ is a particular realization of $\mathbf{\Lambda}_N$ given by (3.4) and (3.5).

Let the random variables $Z_\alpha^{(k)}$, $\alpha = 1, \ldots, N$; $k = 1, \ldots, c$, be defined as

$$(4.1.2) \qquad Z_\alpha^{(k)} = \begin{cases} 1, \text{ if } \alpha = I_{11}^{(k)}, \ldots, I_{ln_k}^{(k)} \\ 0, \text{ otherwise.} \end{cases}$$

We can then write

$$(4.1.3) \qquad \bar{I}_i^{(k)} = (1/n_k) \sum_{\alpha=1}^{n_k} Z_\alpha^{(k)} l_{i\alpha} \text{ for } i = 1, \ldots, p; k = 1, \ldots, c.$$

When H_0 is true

$$\begin{pmatrix} Z_1^{(1)} & \cdots & Z_N^{(1)} \\ \cdot & \cdots & \cdot \\ Z_1^{(c)} & \cdots & E_N^{(c)} \end{pmatrix}$$

would be a matrix obtained by permuting randomly the columns of

$$\begin{pmatrix} \overbrace{1 \ldots 1}^{n_1} & \overbrace{0 \ldots 0}^{n_2} & \ldots & \overbrace{0 \ldots 0}^{n_c} \\ 0 \ldots 0 & 1 \ldots 1 & \ldots & 0 \ldots 0 \\ \cdot \ldots \cdot & \cdot \ldots \cdot & \ldots & \cdot \ldots \cdot \\ 0 \ldots 0 & 0 \ldots 0 & \ldots & 1 \ldots 1 \end{pmatrix}$$

and this will be true independently of $\mathbf{\Lambda}_N$. Hence we obtain

$$(4.1.4) \qquad E(Z_\alpha^{(k)} \mid H_0) = n_k/N,$$

$$(4.1.5) \qquad \text{Cov}(Z_\alpha^{(k)}, Z_\beta^{(q)} \mid H_0) = [n_k(\delta_{\alpha\beta}N-1)/N(N-1)][\delta_{kq}-n_q/N];$$

for α, $\beta = 1, \ldots, N$; k, $q = 1, \ldots, c$, where $\delta'_{\alpha\beta}$s and δ'_{kq}s are Kronecker deltas. Hence it is easy to show that

$$(4.1.6) \qquad E(\bar{I}_i^{(k)} \mid \mathbf{L}_N, H_0) = (N+1)/2, i = 1, \ldots, p, k = 1, \ldots, c;$$

$$(4.1.7) \qquad \text{Cov}(\bar{I}_i^{(k)}, \bar{I}_j^{(q)} \mid \mathbf{L}_N, H_0) = \frac{(N+1)(\delta_{kq}N-n_k)}{12n_k} r_{ij}(\mathbf{L}_N),$$

for $i, j = 1, \ldots, p$; $k, q = 1, \ldots, c$, and where

$$(4.1.8)\quad r_{ij}(\boldsymbol{L}_N) = 12 \sum_{\alpha=1}^{N} \left(\ell_{i\alpha} - \frac{N+1}{2}\right) \left(\ell_{j\alpha} - \frac{N+1}{2}\right) \Big/ N(N^2-1),$$

for $i, j = 1, \ldots, p$, (it may be noted that $r_{ii}(\boldsymbol{L}_N) = 1$, for all $i = 1, \ldots, p$).

Now under the null hypothesis, the apportionment of the numbers $1, \ldots, N$ to the c sets $(I_{i1}^{(k)}, \ldots, I_{in_k}^{(k)})$, $k = 1, \ldots, c$ is likely to be equitable for each $i\ (= 1, \ldots, p)$, and hence we would expect that for each $i\ (= 1, \ldots, p)$, the mean ranks $\bar{I}_i^{(k)}$, $k = 1, \ldots, c$ would be close to each other and as a result to $(N+1)/2$. Since, only $p(c-1)$ of these mean ranks are linearly independent, it seems that we may base our test for H_0 on the set of $p(c-1)$ contrasts

$$(4.1.9)\qquad \left(\bar{I}_i^{(k)} - \frac{N+1}{2}\right),\ i = 1, \ldots, p;\ k = 1, \ldots, c-1.$$

Again, for practical convenience, it would be necessary to formulate the test on the basis of a single function of the elements in (4.1.9), and for this we are to choose a function that would reflect the numerical largeness of any of these contrasts. A positive definite quadratic form in these contrasts seems to be the most natural answer. Now, if we write

$$(4.1.10)\qquad \boldsymbol{R}(\boldsymbol{L}_N) = (r_{ij}(\boldsymbol{L}_N))\ i, j = 1, \ldots, p,$$

by (4.1.6), (4.1.7) and (4.1.8), the conditional dispersion matrix (under H_0) of the random variables in (4.1.9) (taken in that order), is readily deduced to be

$$(4.1.11)\qquad \frac{N+1}{12} \left(\frac{N}{n_k}\,\delta_{kq} - 1\right)_{k,\,q=1,\,\ldots,\,c-1} \otimes \boldsymbol{R}(\boldsymbol{L}_N),$$

where $\otimes$ stands for the Kronecker product [*see* Anderson (1958, p. 347)]. Also, it may be easily verified that

$$(4.1.12)\qquad \left(\frac{N}{n_k}\delta_{kq} - 1\right)^{-1}_{k,\,q=1,\ldots,\,c-1} = \left(\frac{n_k}{N}\,\delta_{kq} + \frac{n_k\, n_q}{n_c N}\right)_{k,\,q=1,\ldots,\,c-1}$$

If now $\boldsymbol{R}(\boldsymbol{L}_N)$ is assumed to be positive definite and its inverse matrix is denoted by

$$\boldsymbol{R}^{-1}(\boldsymbol{L}_N) = (r^{ij}(\boldsymbol{L}_N)), \tag{4.1.13}$$

then the inverse of the dispersion matrix in (4.1.11) would be given by

$$\frac{12}{N+1}(\delta_{kq}n_k/N + n_k n_q/n_c N)_{k,q=1,\ldots,c-1} \otimes \boldsymbol{R}^{-1}(\boldsymbol{L}_N). \tag{4.1.14}$$

Now we are in a position to formulate the test statistic. Whenever, the collection matrix $\boldsymbol{\Lambda}_N$ of the pooled sample is such that $\boldsymbol{R}(\boldsymbol{\Lambda}_N)$ is positive definite, we take as our test statistic

$$W_N = \frac{12}{N+1}\sum_{k=1}^{c}\frac{n_k}{N}\sum_{i=1}^{p}\sum_{j=1}^{p} r^{ij}(\boldsymbol{\Lambda}_N) \left[\bar{I}_i^{(k)} - \frac{N+1}{2}\right]\left[\bar{I}_j^{(k)} - \frac{N+1}{2}\right], \tag{4.1.15}$$

which is a symmetric expression in $\{\bar{I}_i^{(k)}, i = 1, , p; k = 1, \ldots, c\}$. When $\boldsymbol{R}(\boldsymbol{\Lambda}_N)$ is not positive definite and is of rank $p' < p$, we may choose a subset of p' variables for which the rank correlation matrix would be positive definite, and write down for W_N an expression similar to (4.1.15) but involving only the p' variables chosen. However, as we shall see in Section 4.2, under some mild restriction on Ω, $\boldsymbol{R}(\boldsymbol{\Lambda}_N)$ will be positive definite, in probability.

From the remark made at the end of section 3, it follows that the conditional distribution of W_N given $\boldsymbol{\Lambda}_N = \boldsymbol{L}_N$ would be the same under H_0, whatever the cdf's $F_1 = \ldots = F_c = F$ may be. From this conditionally known distribution of W_N it is possible to construct the test function $\phi(W_N)$:

$$\phi(W_N) = \begin{cases} 1, & \text{if } W_N > W_{N,\varepsilon}(\boldsymbol{\Lambda}_N), \\ A_{N,\varepsilon}(\boldsymbol{\Lambda}_N), & \text{if } W_N = W_{N,\varepsilon}(\boldsymbol{\Lambda}_N), \\ 0, & \text{otherwise,} \end{cases} \tag{4.1.16}$$

where the constants $W_{N,\epsilon}(\mathbf{\Lambda}_N)$ and $A_{N,\epsilon}(\mathbf{\Lambda}_N)$ (which may depend on $\mathbf{\Lambda}_N$,) are so chosen that

$$E\{\phi(W_N) \mid \mathbf{L}_N, H_0\} = \epsilon \tag{4.1.17}$$

The last equation implies

$$E\{\phi(W_N) \mid H_0\} = \epsilon. \tag{4.1.18}$$

Thus we can always construct a size ϵ test for H_0. In practice, the use of this exact test is forbidden (unless n and p are very small) because of the prohibitive amount of labor involved in the numerical evaluation of the permutation distribution of W_N. Therefore, we have to consider approximations to the exact permutation distribution of W_N that will be satisfactory at least for large samples. We discuss this in the following sub-section.

4.2. Large sample permutation test

Here we shall assume that N is adequately large, and for each N, there is a set $n_1, \ldots, n_c$ such that

$$\sum_{k=1}^{c} n_k = N, \ \lim_{N=\infty} n_k/N = \nu_k, \quad k = 1, \ldots, c; \tag{4.2.1}$$

where $\nu_1, \ldots, \nu_c$ are c fixed numbers lying in the open interval (0,1) and adding to unity.

Theorem 4.2.1. *If* $\{\mathbf{L}_N : \mathbf{L}_N \epsilon\, \mathcal{L}_N\}$ *is a sequence such that the corresponding sequence of matrices* $\{\mathbf{R}(\mathbf{L}_N)\}$, *defined by* (4.1.8) *and* (4.1.10), *has a positive definite limit, then the sequence of conditional distributions of* W_N *(defined by* (4.1.15),) *given* $\mathbf{\Lambda}_N = \mathbf{L}_N$, *converges to a chi-square distribution with* $p(c-1)$ *degrees of freedom.*

Proof. By virtue of (4.1.6), (4.1.7) and (4.1.15), it seems sufficient to show that the set of $p(c-1)$ contrasts in (4.1.9) has jointly (under the conditional probability measure induced by the N equally likely permutations of the columns of $\mathbf{L}_N$) asymptotically a multinormal distribution. For this it suffices to show that any nontrivial linear compound of the form

$$N^{\frac{1}{2}} \sum_{k=1}^{c} \sum_{i=1}^{p} t_i^{(k)}\, \bar{T}_i^{(k)} \quad \left(\sum_{k=1}^{c} t_i^{(k)}\, n_k = 0,\ i = 1, \ldots, p \right) \tag{4.2.2}$$

is asymptotically normal with zero mean and a nondegenerate variance. Using (4.1.2), (4.1.3) and some simple algebraic manipulations, we can rewrite (4.2.2) in the form

$$T_N = N^{\frac{1}{2}} \sum_{\alpha=1}^{N} \left(\sum_{i=1}^{p} t_{i\alpha}\, l_{i\beta_\alpha} \right), \tag{4.2.3}$$

where we define $l_{i\alpha}$ as in (3.4) and (3.5), and write

$$I_{1\alpha} = \beta_\alpha, \quad \alpha = 1, \ldots, N; \tag{4.2.4}$$

and where n_k of the $t_{i\alpha}$ have the common value $(t_i^{(k)}/n_k)$, for $k = 1, \ldots, c$, $i = 1, \ldots, p$. Thus writing

$$\sum_{i=1}^{p} t_{i\alpha}\, l_{i\alpha'} = b_{\alpha\alpha'}^{(N)}, \ \alpha, \alpha' = 1, \ldots, N \tag{4.2.5}$$

we have

$$N^{-\frac{1}{2}}\, T_N = \sum_{\alpha=1}^{N} b_{\alpha\beta_\alpha}^{(N)}$$

By (4.2.4), $(\beta_1, \ldots, \beta_N)$ represent a random permutation of the numbers $(1, \ldots, N)$. Therefore, we can use a combinatorial central limit theorem by Hoeffding (1951), and require only to show that on subtituting

$$d_{\alpha\alpha'}^{(N)} = \sum_{i=1}^{p} \left(t_{i\alpha} - \frac{1}{N} \sum_{\alpha=1}^{N} t_{i\alpha} \right) \left(l_{i\alpha'} - \frac{N+1}{2} \right), \tag{4.2.6}$$

the following condition is satisfied

$$\lim_{N=\infty} \frac{\max\limits_{1 \leqslant \alpha, \alpha' \leqslant N} \{d_{\alpha\alpha'}^{(N)}\}^2}{\frac{1}{N} \sum\limits_{\alpha=1}^{N} \sum\limits_{\alpha'=1}^{N} \{d_{\alpha\alpha'}^{(N)}\}^2} = 0, \tag{4.2.7}$$

Now, by (4.1.8) and (4.2.6), we have

$$\frac{1}{N} \sum_{\alpha, \alpha'=1}^{N} \{d_{\alpha\alpha'}^{(N)}\}^2 \tag{4.2.8}$$

$$= \frac{N^2-1}{12} \sum_{i,j=1}^{p} r_{ij}(\boldsymbol{L}_N) \sum_{\alpha=1}^{N} \left(t_{i\alpha} - \frac{1}{N} \sum_{\alpha=1}^{N} t_{i\alpha} \right) \left(t_{j\alpha'} - \frac{1}{N} \sum_{\alpha=1}^{N} t_{j\alpha} \right).$$

Hence, by (4.2.1) we get after some simple adjustments that

$$(4.2.9)\qquad \lim_{N=\infty} \sum_{\alpha,\alpha'=1}^{N} \{d_{\alpha\alpha'}^{(N)}\}^2/N^2$$

$$= \frac{1}{12}\sum_{k=1}^{c} \nu_k \sum_{i,j=1}^{p}\left(\nu_k t_i^{(k)} - \sum_{q=1}^{c} t_i^{(q)}\right)\left(\nu_k t_j^{(k)} - \sum_{q=1}^{c} t_j^{(q)}\right) r_{ij}(\boldsymbol{L}_N).$$

As the matrix $\boldsymbol{R}(\boldsymbol{L}_N)$ is positive definite (by assumption, for N adequately large), the right hand side of (4.2.9) will be positive unless

$$\nu_k t_i^{(k)} - \sum_{q=1}^{c} t_i^{(q)} = 0 \text{ for all } i = 1, \ldots, p,\ k = 1, \ldots, c,$$

which holds only in the trivial case $t_i^{(k)} = 0$ for $i = 1, \ldots, p$, $k = 1, \ldots, c$. Thus, as $N \to \infty$

$$(4.2.10)\qquad \frac{1}{N}\sum_{\alpha,\alpha'=1}^{N} \{d_{\alpha\alpha'}^{(N)}\}^2 = O(N).$$

Again, from (4.2.6), as $\boldsymbol{L}_N \epsilon \mathcal{L}_N$,

$$|d_{\alpha\alpha'}^{(N)}| \leqslant \frac{N-1}{2}\sum_{i=1}^{b}\left| t_{i\alpha} - \frac{1}{N}\sum_{\alpha=1}^{N} t_{i\alpha}\right| \leqslant \frac{N-1}{2}\sum_{i=1}^{p}\left\{\max_{1\leqslant k\leqslant c}\left|\frac{t_i^{(k)}}{n_k} - \frac{1}{N}\sum_{k=1}^{c} t_i^{(k)}\right|\right\} \text{ for } \alpha, \alpha' = 1, \ldots, N,$$

and hence,

$$(4.2.11)\qquad \max_{1\leqslant \alpha,\alpha'\leqslant N} |d_{\alpha\alpha'}^{(N)}| \leqslant \frac{N-1}{2N}\sum_{i=1}^{P}\left\{\max_{1\leqslant k\leqslant c}\left|\frac{N}{n_k} t_i^{(k)} - \sum_{k=1}^{c} t_i^{(k)}\right|\right\} = O(1)$$

as by (4.2.1), the right hand side of (4.2.11) converges to a finite quantity as $N \to \infty$. (4.2.10) and (4.2.11) together imply that the sets of numbers (4.2.5) satisfy the condition (4.2.10). The rest of the proof of the theorem follows by routine methods and hence is omitted.

We need to prove a lemma before the main theorem of this section is proved. We shall write

$$\overline{F}(x_1, \ldots, x_p) = \sum_{k=1}^{c} \nu_k F_k(x_1, \ldots, x_p) \tag{4.2.12}$$

and denote by $(\eta_1, \ldots, \eta_p)$ a set of random variables following the distribution law (4.2.12) $\overline{\psi}_{ij}$ will stand for the grade correlation coefficient between η_i and η_j, and

$$\overline{\Psi} = (\overline{\psi}_{ij})_{i,j=1,\cdots,p}. \tag{4.2.13}$$

will stand for the grade correlation matrix of $(\eta_1, \ldots, \eta_p)$.

Here as well as later, we shall use the following notations for the univariate marginal cdf's associated with the distributions F_k, $k = 1, \ldots, c$ and $\overline{F}$. Let $F_{k[i]}(x_i)$ and $\overline{F}_{[i]}(x_i)$ stand for the marginal cdf of the ith variate $X_i^{(k)}$ and η_i respectively, for $i = 1, \ldots, p$ and $k = 1, \ldots, c$. In terms of these notations $\overline{\psi}_{ij}$ is explicitly given by

$$\psi_{ij} = 3 \int_{E^p} \{2\overline{F}_{[i]}(x_i) - 1\}\{2\overline{F}_{[j]}(x_j) - 1\} d\overline{F}(x_1, \ldots, x_p), \quad i, j = 1, \ldots, p, \tag{4.2.14}$$

where E^p stands for the p-dimensional real (Euclidean) space.

Lemma 4.2.2. *As $N \to \infty$, $\boldsymbol{R}(\Lambda_N)$ converges in probability to $\overline{\Psi}$.*

The proof follows as a simple extension of a similar lemma (in the particular case of $p = c = 2$) by the authors ([7], pp. 29–31) and hence is not reproduced here.

For the further development of the theory, we shall have to assume that Ω satisfies a mild restriction. This we state below :

Condition A. *Ω is such that for all points $(F_1, \ldots, F_c) \epsilon \Omega$ and for all ν_k, $k = 1, \ldots, c$, $0 < \nu_k < 1$, $\sum_{k=1}^{c} \nu_k = 1$, the grade correlation matrix $\overline{\Psi}$ defined by* (4.2.13), (4.2.14) *is positive definite.*

Throughout the rest of Section 4.2, condition A will be assumed. This assumption, however, is not very restrictive, because $\overline{\Psi}$ is positive definite, unless one or more of the random variables $\{2\overline{F}_{[i]}(\eta_i)-1\}$, $i = 1, \ldots, p$ can be expressed linearly in terms of the others with probability one. This, by our assumption of continuity of $F_1, \ldots, F_c$ for all points in Ω implies that one or more of the random variables $\eta_1, \ldots, \eta_p$ can almost surely be expressed as functions of the remaining variables, the function being monotonic in each of the arguments involved. So, if this form of singularity is excluded, condition A will always be satisfied. We now prove the main theorem of this subsection.

Theorem 4.2.3. *Under condition A on* Ω, *the conditional null distribution of* W_N *given* $\boldsymbol{\Lambda}_N$ *converges, in probability, to chi-square distribution with* $p(c-1)$ *degrees of freedom (d.f.).*

The proof of this theorem follows readily from theorem 4.2.1. and lemma 4.2.2. For the intended brevity of the paper, the details are omitted.

Corollary. For any $\epsilon : 0<\epsilon< 1$, let $W_{N,\epsilon}(\boldsymbol{L}_N)$ and $A_{N,\epsilon}(\boldsymbol{L}_N)$ be defined as in (4.1.16) and (4.1.17) and let $\chi^2_p(c-1)$, ϵ be the upper $100\epsilon\%$ point of a χ^2 distribution with $p(c-1)$d.f. Then as $N \rightarrow \infty$,

$$W_{N,\epsilon}(\boldsymbol{\Lambda}_N) \xrightarrow{P} \chi^2_{p(c-1),\epsilon} \text{ and } A_{N,\epsilon}(\boldsymbol{\Lambda}_N) \xrightarrow{P} 0.$$

The proof is omitted (*see* [13, p. 171]).

Remark. In the theorem and corollary above, $\boldsymbol{\Lambda}_N$ corresponds to $\boldsymbol{F} = (F_1, \ldots, F_c)$, where $\boldsymbol{F}\epsilon\Omega$, but H_0 may or may not be true. Further, it is not essential for the above theorem and corollary that $\boldsymbol{\Lambda}_N$ should be the collection matrix corresponding to samples taken from a fixed point $\boldsymbol{F}\epsilon\Omega$. If $\boldsymbol{\Lambda}_N$ is any sequence of random matrices, $\boldsymbol{\Lambda}_N$ assuming values of $\mathscr{L}_N$, such that $\boldsymbol{R}(\boldsymbol{\Lambda}_N)$ is positive definite, in probability, then the theorem and corollary would hold.

Again, by an adaptation of the method of proof of corollary 3 to theorem 3.2.2. of [7], and using (4.2.12), we obtain the following theorem on the unconditional distribution of W_N (under H_0).

Theorem 4.2.4. *Under* H_0, *the statistic* W_N, *defined by* (4.1.15), *has asymptotically a chi-square distribution with* $p(c-1)$ *d.f.*

Theorem 4.2.4. enables us to suggest an asymptotically distribution free test based on the same statistic W_N, defined by (4.1.15). This test consists in rejecting the null hypothesis when W_N exceeds the value $\chi^2_{p(c-1), \varepsilon}$. This test will be termed hereafter as the *asymptotically distribution-free* rank-sum test.

4.3. Consistency of the tests

As in the preceding section, we define by $F_{[i]}(x_i)$ the marginal cdf of the ith variate η_i, $i = 1, \ldots, p$, where $(\eta_1, \ldots, \eta_p)$ has the cdf $\overline{F}$, defined by (4.2.12). Also let

$$d_i^{(k)} = \int_{E^p} \{2\overline{F}_{[i]}(x_i) - 1\} dF_k(x_1, \ldots, x_p), \quad i = 1, \ldots, p; \quad k = 1, \ldots, c. \tag{4.3.1}$$

Theorem 4.3.1. *The exact test* (4.1.16) *and the asymptotically distribution-free test for* H_0, *are both consistent against the set of alternatives that* $d_i^{(k)}(k = 1, \ldots, c;\ i = 1, \ldots, p)$ *are not all zero. Thus, for translation type of alternatives in* (2.5), *these tests are consistent against the set of alternatives that* $\boldsymbol{\theta}^{(1)}, \ldots, \boldsymbol{\theta}^{(c)}$ *are not all null vectors.*

The proof of this theorem is straight-forward and is omitted. By virtue of the consistency, for any fixed alternative (deviating from the null hypothesis (2.1),) the power of the tests will be asymptotically equal to unity. In the next subsection we shall consider the usual Pitman's type of shift alternatives and study the power proporties of the tests.

4.4. Asymptotic power of the tests

Let us consider the sequence of alternatives $\{H_N\}$, where H_N is represented by $(F_{1N}, \ldots, F_{cN})$:

$$F_{kN}(x) = F(x + N^{-1/2}\, \boldsymbol{\theta}^{(k)}), \ k = 1, \ldots, c; \tag{4.4.1}$$

where $\boldsymbol{\theta}^{(k)} = (\theta_1^{(k)}, \ldots, \theta_p^{(k)})$, $k = 1, \ldots, c$, are c vectors not all of which are identical.

Just as in Sections 4.2 and 4.3 we define the marginal cdf of the ith variate associated with $F(x_1, \ldots, x_p)$ by $F_{[i]}(x_i)$, for $i = 1, \ldots, p$, and write

$$\psi_{ij} = 3 \int_{E^p} \{2F_{[i]}(x_i)-1\}\{2F_{[j]}(x_j)-1\}dF(x_1, \ldots, x_p), \quad j = 1, \ldots, p; \tag{4.4.2}$$

$$\mathbf{\Psi} = (\psi_{ij}), \mathbf{\Psi}^{-1} = (\psi^{ij}) = (\psi_{ij})^{-1}.$$

The grade correlation matrix $\mathbf{\Psi}$ and its reciprocal $\mathbf{\Psi}^{-1}$ will be positive definite by virtue of condition A, stated in subsection 4.2. Further, to simplify the expression for the asymptotic power of the tests, we shall assume that the marginal cdf $F_{[i]}(x)$ is absolutely continuous (for all $i = 1, \ldots, p$) and write $f_{[i]}(x)$ as the corresponding density function (assumed to be continuous), for $i = 1, \ldots, p$. Further, the integrals

$$h_i = \int_{-\infty}^{\infty} \{f_{[i]}(x)\}^2\, dx, \quad i = 1, \ldots, p; \tag{4.4.3}$$

will be assumed to exist and the limiting relations

$$\lim_{N=\infty} \int_{-\infty}^{\infty} f_{[i]}(x+0(1/\sqrt{N}))dF_{[i]}(x) = h_i, i = 1, \ldots, p \tag{4.4.4}$$

will be assumed to hold. Finally, (4.2.1) will be implicit through out this subsection. For each N, $(X^{(k)}_{1(N)}, \ldots, X^{(k)}_{p(N)})$ will denote a set of random variables following the distribution F_{kN}, defined by (4.4.1), there being n_k observations from this distribution, for $k = 1, \ldots, c$. For notational simplicity, we shall keep the subscript N understood, and denote the N observations by

$$(X^{(k)}_{1\alpha}, \ldots, X^{(k)}_{p\alpha}), \quad \alpha = 1, \ldots, n_k; \quad k = 1, \ldots, c. \tag{4.4.5}$$

Also, we define the mean ranks $\bar{T}^{(k)}_i$, $i = 1, \ldots, p$, $k = 1, \ldots, c$ as in (4.1.1) and the statistic W_N as in (4.1.15).

Theorem 4.4.1. *Under* (4.4.1) *through* (4.4.5), W_N *has asymptotically a noncentral chi-square distribution with* $(p(c-1)$ *d.f. and the noncentrality parameter*

$$\Delta_R = 12 \sum_{k=1}^{c} \nu_k \sum_{i=1}^{p} \sum_{j=1}^{p} \psi^{ij}(\theta^{(k)}_i-\bar{\theta}_i)(\theta^{(k)}_j-\bar{\theta}_j)h_i h_j, \tag{4.4.6}$$

where $\bar{\theta}_i = \sum_{k=1}^{c} \nu_k\, \theta^{(k)}_i$, *for* $i = 1, \ldots, p$.

Proof. We shall prove first that $\sqrt{N}\left(\bar{I}_i^{(k)}-\frac{N+1}{2}\right), i=1,\ldots,p;$ $k=1,\ldots,c$, have asymptotically a multinormal distribution. For this, let us define the usual sign-function $s(x)$ to be $+1\ (-1)$ according as $x > (<) 0$, and let it be 0, otherwise. Also write

$$U_{iN}^{(k,\,q)} = \frac{1}{n_k n_q} \sum_{\alpha=1}^{n_k} \sum_{\beta=1}^{n_q} s(X_{i\alpha}^{(k)} - X_{i\beta}^{(q)}) \tag{4.4.7}$$

for $k \neq q = 1, \ldots, c$; $i = 1, \ldots, p$. Then we have

$$\sqrt{N}\left(\bar{I}_i^{(k)} - \frac{N+1}{2}\right) = \frac{1}{2} \sum_{\substack{q=1 \\ \neq k}}^{c} \frac{n_q}{N} \sqrt{N}\, U_{iN}^{(k,\,q)}. \tag{4.4.8}$$

Now, by a well-known technique used in studying the asymptotic distributions of U-statistics (*see*, Hoeffding (1948), and Andrews (1954),) and following some simple but essentially lengthy steps, it can be shown that $\{\sqrt{N}U_{iN}^{(k,\,q)},\ k \neq q = 1, \ldots, c;\ i = 1, \ldots, p,\}$ has asymptotically a multinormal distribution which is essentially singular and is of rank $p(c-1)$ when $\mathbf{\Psi}$ is positive definite. Hence, using (4.4.8) and some routine calculations, it can be shown that under (4.4.1) through (4.4.5) the vector

$$N^{1/2}\left(\bar{I}_1^{(1)} - \frac{N+1}{2}, \ldots, \bar{I}_p^{(1)} - \frac{N+1}{2}, \ldots, \bar{I}_1^{(c-1)} - \frac{N+1}{2}, \ldots, \bar{I}_p^{(c-1)} - \frac{N+1}{2}\right)$$

will have asymptotically a multinormal distribution with mean vector

$$(m_1^{(1)}, \ldots, m_p^{(1)}, \ldots, m_1^{(c-1)}, \ldots, m_p^{(c-1)}),$$

and dispersion matrix $\frac{1}{12}\left(\frac{1}{\nu_k}\delta_{kq} - 1\right)_{k,q=1,\ldots,c-1} \otimes \mathbf{\Psi}$, where

$$m_i^{(k)} = -h_i(\theta_i^{(k)} - \bar{\theta}_i) \text{ for } i = 1, \ldots, p;\ k = 1, \ldots, c. \tag{4.4.9}$$

Hence, by simple reasonings, we may conclude that

$$\frac{12}{N} \sum_{k=1}^{c-1} \sum_{q=1}^{c-1} \sum_{i=1}^{p} \sum_{j=1}^{p} \left(\nu_k \delta_{kq} + \frac{\nu_k \nu_q}{\nu_c}\right) \psi^{ij} \left(\bar{I}_i^{(k)} - \frac{N+1}{2}\right)\left(\bar{I}_j^{(q)} - \frac{N+1}{2}\right) \tag{4.4.10}$$

will be asymptotically distributed as a non-central χ^2 with d.f. $p(c-1)$ and non-centrality parameter

$$\Delta_R = 12 \sum_{k=1}^{c-1} \sum_{q=1}^{c-1} \sum_{i=1}^{p} \sum_{j=1}^{p} \left(\nu_k \delta_{kq} + \frac{\nu_k \nu_q}{\nu_c} \right) \psi^{ij} m_i^{(k)} m_j^{(q)}$$

$$(4.4.11) \quad = 12 \sum_{i=1}^{p} \sum_{j=1}^{p} \psi^{ij} \left\{ \sum_{k=1}^{c-1} \nu_k m_i^{(k)} m_j^{(k)} + \frac{1}{\nu_c} \sum_{k=1}^{c-1} \sum_{q=1}^{c-1} \nu_k \nu_q m_i^{(k)} m_j^{(q)} \right\}$$

$$= 12 \sum_{k=1}^{c} \nu_k \sum_{i=1}^{p} \sum_{j=1}^{p} \psi^{ij} (\theta_i^{(k)} - \bar{\theta}_i)(\theta_j^{(k)} - \bar{\theta}_j) h_i h_j,$$

Again, it follows from lemma 4.2.2, (4.2.14), (4.4.1) and (4.4.2) that $\boldsymbol{R}(\boldsymbol{\Lambda}_N) \xrightarrow{P} \boldsymbol{\Psi}$. Hence the statistic W_N, given by (4.1.15) is seen to be asymptotically equivalent, in probability, to the statistic (4.4.10), and the theorem follows.

With the help of theorem 4.4.1, corollary to theorem 4.2.3 and a result due to Hoeffding (1952, p. 171), we readily arrive at the following.

Theorem 4.4.2. *Both the permutation test and the asymptotically distribution-free test (based on the statistic W_N, defined by* (4.1.15),) *are asymptotically power-equivalent and have asymptotically (under the sequence of alternatives in* (4.4.1),) *a noncentral chi-square distribution with $p(c-1)$ d.f. and the noncentrality parameter Δ_R, defined by* (4.4.6).

5. MULTIVARIATE MULTISAMPLE MEDIAN TESTS

We shall now generalize the well-known univariate median test by Brown and Mood (cf. [16]) to the $p(\geqslant 1)$ variate case, following the same conditional approach as in section 3. We adopt the same notations as in the previous sections. Among the N values of $\{I_{i\alpha}^{(k)},\ \alpha = 1, \ldots, n_k;\ k = 1, \ldots, c\}$ there are exactly $a = [N/2]$ ($[s]$ being the largest integer contained in s) values which do not exceed a, for $i = 1, \ldots, p$. Let us then define a system of 2^p-mutually exclusive and exhaustive cells $\left\{ J_{r_1 \cdots r_p} : r_i = 0, 1, i = 1, \ldots, p \right\}$ by the convention that if for any cell $J_{r_1 \cdots r_p}$ $r_i = 1$ (or 0) then for any observation belonging to it $I_i^{(k)} >$ (or $\leqslant$) a, for $i = 1, \ldots, p$. Let among the N random rank p-tuplets $C_{r_1 \cdots r_p}$ observations belong to the cell $J_{r_1 \cdots r_p}$, for all $\left(r_{1, \ldots, r_p}\right)$. Then $\{C_{r_1 \cdots r_p}\}$ is a set of random variables, depending on $\mathbf{\Lambda}_N$, and we may note that

$$\sum_{(S_{j\alpha})} C_{r_1 \cdots r_p} = a \text{ (if } \alpha = 0) \text{ or } N - a \text{ (if } \alpha = 1), \tag{5.1.1}$$

where the summation $S_{j\alpha}$ extends over all $r_1, \ldots, r_{j-1}, r_{j+1}, \ldots, r_p$ (for a given $r_j = \alpha$), $j = i, \ldots, p$. Thus we have a 2^p-contingency table for the pooled sample of size N, and we term this as the *basic table*. Now, corresponding to the kth sample of size n_k, we have, by reference to the same system of cells $\left\{J_{r_1 \cdots r_p}\right\}$, another 2^p-table, whose entries are denoted by

$$n_{k(r_1 \cdots r_p)}, \; r_i = 0, 1, \; i = 1, \ldots, p; \tag{5.1.2}$$

this table will be termed as *table k*, for $k = 1, \ldots, c$. Obviously

$$\sum_{k=1}^{c} n_{k(r_1 \cdots r_p)} = C_{r_1 \cdots r_p}, \quad \text{for all } (r_1, \ldots, r_p). \tag{5.1.3}$$

Now under the null hypothesis that the c cdf's $F_1, \ldots, F_c$ are all identical, these c (2^p)-contingency tables should be statistically homogeneous and consistent with the basic table. By virtue of (5.1.3), it is thus sufficient to test for the agreement of the c tables (i.e., table k, $k = 1, \ldots, c$). Since the test to be proposed is based on the system of cells demarcated by the pooled sample medians, by analogy with the univariate case (cf. [16]), it is termed the multivariate multisample median test.

Now, under the null hypothesis $H_0 : F_1 = \ldots F_c = F$ (say), and given $\mathbf{\Lambda}_N = \boldsymbol{L}_N$, all possible partitioning of the N rank p-tuplets into the c subsets of sizes $n_1, \ldots, n_c$, respectively, are equally likely, each having the conditional probability $(\Pi_1^c \, n_k \,!/N!)$ which we conventionally put as $\binom{N}{[n_k]}^{-1}$. Let now for the given $\mathbf{\Lambda}_N = \boldsymbol{L}_N$, the realized values of $\{C_{r_1 \cdots r_p}\}$ be $\{c_{r_1 \cdots r_p}\}$. Then by simple logic we derive the conditional probability function of $\left\{n_{k(r_1\ldots r_p)};\ r = 0, 1;\ i = 1, \ldots, p,\ k = 1, \ldots, c\right\}$ conditioned on $\boldsymbol{\Lambda}_N = \boldsymbol{L}_N$ as given by

$$(5.1.4)\qquad \binom{N}{[n_k]}^{-1}\left\{\prod_S \binom{c_{r_1\ldots r_p}}{[n_{k(r_1\ldots r_p)}]}\right\}$$

Thus for any $(N, \boldsymbol{L}_N)$, the expression (5.1.4) is a completely known function. Here also, we shall use a quadraric form in the variables $\left\{n_{k(r_1\ldots r_p)},\ r_i = 0, 1,\ i = 1, \ldots, p;\ k = 1, \ldots, c\right\}$ to formulate our test statistic, and for this we consider first the first and second order moments of these random variables. It follows directly from (5.1.4) that

$$(5.1.5)\qquad E\left\{n_{k(r_1\ldots r_p)}\,\middle|\, \boldsymbol{L}_N, H_0\right\} = \frac{n_k}{N}\, c_{r_1\ldots r_p}, \text{ and}$$

$$(5.1.6)\qquad \operatorname{Cov}\left\{n_{k(r_1\ldots r_p)}, n_{q(r_1', \ldots, r_p')}\,\middle|\, \boldsymbol{L}_N, H_0\right\}$$

$$= n_k(\delta_{kq} N - n_q)c_{r_1\ldots r_p}\left(\delta_{rr'}N - c_{r_1' \ldots r_p'}\right)\Big/ N^2(N-1),$$

where $\delta_{kq} = 1$ if $k = q$, $\delta_{kq} = 0$ if $k \neq q$, and $\delta_{rr'} = 1$ if $(r_1 \ldots r_p) \equiv (r_1' \ldots r_p')$ and $\delta_{rr'}$ is zero otherwise.

5.2. The classification of median tests

The $2^p c$ random variables $\left\{n_{k(r_1\ldots r_p)},\ r_i = 0, 1,\ i = 1, \ldots, p,\ k=1, \ldots, c\right\}$ are subject to 2^p+c-1 constraints, namely (5.1.3) and

$$(5.2.1)\qquad \sum_S n_{k(r_1\ldots r_p)} = n_k, \text{ for } k = 1, \ldots, c;$$

(where the summation S extend over all $(r_1, \ldots, r_p)$,) and where (5.1.3) and (5.2.1) satisfy in turn

$$(5.2.2)\qquad \sum_{k=1}^{c} n_k = N = \sum_S C_{r_1\ldots r_p}.$$

Thus, the effective number of degrees of freedom of this set of random variables is $(2^p-1)(c-1)$. Of course, study of all these d.f. may not be necessary for detecting differences in locations only. For this purpose, we define

(5.2.3) $$n^*_{k(i)} = \sum_{S_{i0}} n_{k(r_1 \cdots r_p)},\ i = 1, \ldots, p,\ k = 1, \ldots, c;$$

where the summation S_{i0} extends over all $(r_1, \ldots, r_p)$ for which $r_{i0} = 0$, i.e., $n^*_{k(i)}$ is the number of observations in the kth sample whose ith variate values do not exceed that of the ath order statistic of pooled sample ith variate values, for $i = 1, \ldots, p$, $k = 1, \ldots, c$. Using (5.1.4) it can be readily shown that

(5.2.4) $$E\{n^*_{k(i)} \mid \boldsymbol{L}_N, H_0\} = an_k/N,\ i = 1, \ldots, p,\ k = 1, \ldots, c.$$

Thus, generalizing the univariate median test [cf. Andrews (1954)], and keeping in mind shifts in locations only, we may base our test on the $p(c-1)$ d.f. carried by the discrepancies $\{N^{-\frac{1}{2}}(n^*_{k(i)} - an_k/N),\ i = 1, \ldots, p,\ k = 1, \ldots, c\}$. Such a median test would be termed a *Type A median test.* It can be shown that the remaining $(2^p-p-1)(c-1)$ d.f. are sensitive not only to shifts in locations but also to any heterogeneity of the association patterns of the different cdf's. Tests based on these $(2^p-p-1)(c-1)$ d.f. may be used to test for the identity of association patterns (assuming the identity of locations) (cf. Chatterjee and Sen (1965), for $p = c = 2$). Such a test would be termed the *Type B median test.* Finally, the test based on all the $(2^p-1)(c-1)$ d.f. is consistent not only against shifts in the median vectors but also against any difference of the association patterns; such a test will be termed the *Type C median test.* Since we are mainly interested in the location problem, we shall consider only type A median test and append very briefly the case of type C median test.

To formulate the test statistic, we write

(5.2.5) $$C^{\alpha, \beta}_{(i,j)} = \sum_{S_{\alpha, \beta, i, j}} C_{r_1 \cdots r_p}$$

where the summation $S_{\alpha, \beta, i, j}$ extends over all possible $r_1, \ldots, r_p$ for which $r_i = \alpha$, $r_j = \beta$; $\alpha, \beta = 0, 1$; and let

(5.2.6) $$d_{ij} = N^{-1} \sum_{\alpha=0}^{1} \sum_{\beta=0}^{1} (-1)^{\alpha+\beta}\, C^{\alpha, \beta}_{(i, j)},\ \text{for } i, j = 1, \ldots, p;$$

(5.2.7) $$\boldsymbol{D} = (d_{ij}).$$

If all $\left\{C_{r_1 \cdots r_p}\right\}$ are positive, it can be shown that $\boldsymbol{D}$ is a positive definite matrix. The positive-definiteness of $\boldsymbol{D}$ may also follow when some of $\left\{C_{r_1 \cdots r_p}\right\}$ are equal to zero. However, it will be seen later on that under some mild restrictions on Ω, $\boldsymbol{D}$ will be positive definite, in probability. If $\boldsymbol{D}$ is not positive definite, as in the rank-sum test, we may work with the highest order positive definite principle minor of it. Thus, on denoting by $\boldsymbol{D}^{-1} = (d^{ij})$ the reciprocal matrix of $\boldsymbol{D}$, we may formulate (following some simple but somewhat lengthy deductions involving the use of (5.1.5), (5.1.6) and (5.2.3), the type A median test-satistic as

$$(5.2.8) \quad M_N = 4 \sum_{k=1}^{c} \sum_{i=1}^{p} \sum_{j=1}^{p} d^{ij} \{n^{*}_{k(i)} - an_k/N\}\{n^{*}_{k(j)} - an_k/N\}/n_k.$$

For small values of $\left(N, \left\{C_{r_1 \cdots r_p}\right\}\right)$, the matrix $\boldsymbol{D}$ can be readily evaluated and the exact null distribution of M_N (given $\boldsymbol{\Lambda}_N = \boldsymbol{L}_N$) may be derived with the aid of (5.1.4). The test function will be essentially similar to (4.1.16), (4.1.17) and (4.1.18). The large sample approach is considered below.

5.3. Large simple distribution of $\boldsymbol{M_N}$ under $\boldsymbol{H_0}$

In this section we make N indefinitely large, subject to (4.2.1). We define $\overline{F}$ as in (4.2.12), and let $\boldsymbol{\eta} = (\eta_1, \ldots, \eta_p)$ follow the cdf $\overline{F}$. Let the medians of $\overline{\eta}_1, \ldots, \overline{\eta}_p$ be denoted by $\bar{\mu}_1, \ldots, \bar{\mu}_p$ and are assumed to be uniquely defined. Also we determine 2^p-cells $\left\{J^0_{r_1 \cdots r_p}\right\}$ by the rule that for any observation belonging to the cell $J^0_{r_1 \cdots r_p}$, r_i is equal to zero (or one) according as $\eta_i \leqslant$ (or $>$) $\bar{\mu}_i$, for $i = 1, \ldots, p$. Let then

$$(5.3.1) \qquad P\left\{\boldsymbol{\eta} \in J^0_{r_1 \cdots r_p}\right\} = \overline{P}^0_{r_1 \cdots r_p}, \text{ for all } (r_1, \ldots, r_p).$$

In the rest of Section 5, we assume the following condition :

Condition B. *Whatever* $\nu_1, \ldots, \nu_c$ $(0 < \nu_k < 1, \sum_1^c \nu_k = 1)$, *for all* $\boldsymbol{F} \epsilon \Omega$,

$$\underset{(r_1 \cdots r_p) \epsilon S}{\text{Inf.}} \quad \overline{P}^0_{r_1 \cdots r_p} > 0.$$

If $\boldsymbol{X}_k = (X_1^{(k)}, \ldots, X_p^{(k)})$ follows the distribution F_k, and

$$P\left\{\boldsymbol{X}_k \epsilon\, J^0_{r_1 \cdots r_p}\right\} = P^{(k)}_{r_1 \cdots r_p}, \tag{5.3.2}$$

then $\overline{P}^0_{r_1 \cdots r_p} = \sum_{k=1}^{c} \nu_k \overline{P}^{(k)}_{r_1 \cdots r_p}$, and hence, the condition B is equivalent to : whatever $\nu_1, \ldots, \nu_c$

$$\underset{(r_1 \cdots r_p) \epsilon S}{\text{Inf}} \quad \underset{k}{\text{Inf}} \left\{P^{(k)}_{r_1 \cdots r_p}\right\} > 0. \tag{5.3.3}$$

If as in Section 2, we write $\mu_1^{(k)}, \ldots, \mu_p^{(k)})$ for the median point of the distributiion F_k, then (5.3.4) will be satisfied, if and only if, none of the 2^p cells formed by taking any arbitrary point within the simplex spanned over the c points $(\mu_1^{(k)}, \ldots, \mu_p^{(k)})$, $k = 1, \ldots, c$ and drawing lines parallel to the axis through that point have zero probability content with respect to all the distributions $F_1, \ldots, F_c$. Then we have the following.

Theorem 5.3.1. *Under* H_0 *in* (2.1) *and subject to the condition* (4.2.1) *and condition* B *in* (5.3.1), *the statistic* M_N, *defined by* (5.2.8), *has asymptotically, in probability, a chi-square distribution with* $p(c-1)$ *d.f.*

Proof. Avoiding the details of proof, we say that under the condition B, as N is increased (subject to (4.2.1), all the $\left\{\left(C_{r_1 \cdots r_p}\right)/N\right\}$ can be made strictly positive, in probability. Consequently, we can apply Stirling's approximations to all the factorials in (5.1.4) if N is taken adequately large. This will lead to the asymptotic multinormality of the set of random variables $\left\{n_k^{-\frac{1}{2}}\left(n_{k(r_1 \cdots r_p)} - (n_k/N)\, c_{r_1 \cdots r_p}\right)\right.$, for all $(r_1, \ldots, r_p)$ and $k = 1, \ldots, c\}$, (under the conditional probability measure, given $\boldsymbol{\Lambda}_N = \boldsymbol{L}_N$.) Since $\{n^*_{k(i)}\}$ are all linear functions of these 2^p-c random variables, by means of linear transformations of variables it can be shown that $[n_k^{-\frac{1}{2}}(n^*_{k(i)} - an_k/N),\ i = 1, \ldots, p,\ k = 1, \ldots, c]$ has (jointly) asymptotically a multinormal distribution with a null mean vector and dispersion matrix $\boldsymbol{D}$, defined by (5.2.6) and (5.2.7). The rest of the proof is simple and is omitted. Hence the theorem.

Corollary 5.3.1. It follows by the same technique that on defining the type C median test statistic as

$$M^*_N = \sum_{k=1}^{c} (N/n_k) \sum_{S} \left\{ n_{k(r_1 \cdots r_p)} - (n_k/N) C_{r_1 \cdots r_p} \right\}^2 \Big/ C_{r_1 \cdots r_p}, \tag{5.3.4}$$

under H_0 in (2.1) and conditioned on $\boldsymbol{\Lambda}_N = \boldsymbol{L}_N$, the statistic M^*_N has asymptotically a chi-square distribution with $(2^p - 1)(c-1)$ d.f.

By virtue of theorem 5.3.1. the large sample type A median test may be formulated as follows. Let $\chi^2_{t,\varepsilon}$ be the upper $100\epsilon\%$ point of a chi-square distribution with t d.f. Then, if

$$M_N \begin{cases} \geqslant \chi^2_{(p-1)(c-1),\,\varepsilon}, & \text{reject } H_0 \text{ in (2.1)} \\ < \chi^2_{(p-1)(c-1),\,\varepsilon}, & \text{accept } H_0. \end{cases} \tag{5.3.5}$$

(For the type C median test, replace M_N by M^*_N and the d.f. $(p-1)(c-1)$ by $(2^p-1)(c-1)$, respectively).

5.4. Consistency of the tests

We define the population medians as in section 2 (cf. (2.3),) and consider the following theorem whose proof is omitted.

Theorem 5.4.1. *The type A median test is consistent against the set of alternatives that the $pc(c-1)$ differences in* (2.3) *are not all zero. The type C median test is consistent against the set of alternatives*

$$\sup_{(k,q)} \left[\sup_{(r_1 \cdots r_p)} \left| P^{(k)}_{r_1 \cdots r_p} - P^{(q)}_{r_1 \cdots r_p} \right| \right] > 0. \tag{5.4.1}$$

and as such, it will be consistent not only to shifts in the median vectors but also to the difference of the association patterns.

By virtue of theorem 5.4.1 for any fixed alternative different from H_0 in (2.1) the power of the tests will be asymptotically equal to unity. Thus, for the study of the asymptotic power properties of the tests we shall again consider the sequence of alternatives in (4.4.1), and this study is made in the following subsection.

5.5. Asymptotic power function of the median tests

As in section 4.4, we shall conceive of a sequence of values of N, and for each N, c subsequences of random variables. The condition (4.2.1) will also be implicit in this section. $F(\boldsymbol{x})$ will be assumed to be absolutely continuous and the corresponding (continuous) density function will be denoted by $f(\boldsymbol{x})$. Further, without any loss of generality, we assume that the median vector of $F(\boldsymbol{x})$ is a null vector. As in the bivariate case (cf. [7]), we require the following conditions :

$$\int_{-\infty}^{\infty} \cdots \int_{-\infty}^{\infty} f\left(x_1, \ldots, x_{p-1}, \frac{1}{N}\right) dx_1, \ldots, dx_{p-1} \tag{5.5.1}$$

$$= \int_{-\infty}^{\infty} \cdots \int_{-\infty}^{\infty} f(x_1, \ldots, x_{p-1}, 0) dx_1 \ldots dx_{p-1} \quad + \; 0(1/N),$$

and similarly for each of the other $p-1$ coordinates in $f(x_1, \ldots, x_p)$. Let then

$$\begin{aligned} \alpha_1 &= \int_{-\infty}^{\infty} \cdots \int_{-\infty}^{\infty} f(0, x_2, \ldots, x_p) dx_2 \ldots dx_p \\ &\cdots \qquad \cdots \qquad \cdots \\ &\cdots \qquad \cdots \qquad \cdots \\ \alpha_p &= \int_{-\infty}^{\infty} \cdots \int_{-\infty}^{\infty} f(x_1, \ldots, x_{p-1}, 0) dx_1 \ldots dx_p; \end{aligned} \tag{5.5.2}$$

We define $P^0_{r_1 \cdots r_p}$ as in (5.3.1) with the only change that $\boldsymbol{\eta}$ has the cdf $F(\boldsymbol{x})$ instead of $\bar{F}(\boldsymbol{x})$. Also we rewrite these as

$$P^0_{0, \cdots 0} = P^0_1; \quad P_{0 \cdots 0, 1} = P^0_2, \; P^0_{0, \cdots 1, 0} = P^0_3,$$

$$P^0_{0, \cdots 1, 1} = P^0_4, \; \ldots, \; P^0_{11 \ldots 1} = P^0_{2^p}.$$

Let us then define a set of p vectors $\boldsymbol{\gamma}_1, \ldots, \boldsymbol{\gamma}_p$ as

(5.5.3) $\gamma_i = \left(e_{i1}\sqrt{P_1^0}, e_{i2}\sqrt{P_2^0}, \ldots e_{i2}\sqrt{P_{2^p}^0}\right); j = 1, \ldots, p_j$

$$\text{where} \quad \left.\begin{aligned} e_{ik} &= \quad 1 \text{ if for } P_k^0, r_i = 0 \\ &= -1 \text{ if for } P_k^0, r_i = 1 \end{aligned}\right\} \text{for} \quad \begin{aligned} i &= 1, \ldots, p \\ k &= 1, \ldots, 2^p. \end{aligned}$$

Then obviously $\gamma_1 \ldots \gamma_p$ are linearly independent of each other and are all of unit length. Also let

$$(5.5.4) \qquad \begin{aligned} \varphi_{ij} &= \gamma_i \cdot \gamma_j, \text{ for } i, j = 1, \ldots, p; \text{ and} \\ \boldsymbol{\Phi} &= (\varphi_{ij}), \boldsymbol{\Phi}^{-1} = (\varphi_{ij})^{-1} = (\varphi^{ij}). \end{aligned}$$

Then it is easily shown that $\boldsymbol{\Phi}$ or $\boldsymbol{\Phi}^{-1}$ are positive definite. Then we have the following.

Theorem 5.5.1. *Under the sequence of alternative in* (4.4.1), *the statistic M_N has asymptotically a non-central χ^2 distribution with $p.(c-1)$ d.f. and with the non-centrality parameter*

$$\Delta_M = 4 \sum_{i=1}^{p} \sum_{j=1}^{p} \phi^{ij}\left(\sum_{k=1}^{c} \nu_k \, \delta_{ik} \, \delta_{jk}\right) \alpha_i \, \alpha_j.$$

Proof. Let $\tilde{X}_1 \ldots \tilde{X}_p$ be the pooled sample medians (a-th largest value; $a = [N/2]$) of the p variates based on the samples (4.4.5) taken from the distributions F_{kN}, $k = 1, \ldots, c$. We define the cells $\left\{J_{r_1 \cdots r_p}, r_i = 0, 1; i = 1, \ldots, p\right\}$ as in the beginning of section 5, i.e., by means of the sample median vector $(\tilde{X}_1, \ldots, \tilde{X}_p)$. Also, let $\boldsymbol{X}_N^{(k)}$ denote a random vector following the cdf F_{kN}, for $k = 0, \ldots, c$, where $F_{0N} = F$. Then we write $p_{r_1 \cdots r_p}^{(k)}$ as the probability content of the cell $J_{r_1 \cdots r_p}$ with respect to the cdf F_{kN}, for $k = 0, \ldots, c$, and the successive derivatives of these probabilities (with respect to $\tilde{X}_i$, $i = 1, \ldots, p$) are denoted by $p_{r_1 \cdots r_p}^{(k); i}$ $p_{r_1 \cdots r_p}^{(k); i, j}, \ldots, p_{r_1 \cdots r_p}^{(k); 1, \cdots, p}$ $(i, j, \ldots = 1, \ldots, p)$, for $k = 0, \ldots, c$; $(r_1 \ldots r_p) \epsilon S$. Then there are 2^p values of $\left\{p_{r_1 \cdots r_p}^{(k)}\right\}$ for each $k = 0, \ldots, c$, $p2^p$ values of $\left\{p_{r_1 \cdots r_p}^{(k); i}\right\}$ for each $k = 0, \ldots, c$, and so on. We now adopt Mood's (1941) technique of finding the joint distribution of the sample medians in a sample from a multivariate population with an extension to the multisample case, and get that the joint probability function of $\left\{n_{k(r_1\cdots r_p)},\ k=1, \ldots, c, (r_1 \ldots r_p)\epsilon S\right\}$ and (the density function of) $\tilde{X}_1, \ldots, \tilde{X}_p$ is the sum of the following terms :

(i) If the pooled sample median vector is determined by p different observations of the pooled sample, then the contribution will be

$$(5.5.5)\quad \left\{\prod_{k=1}^{c}\left[\left(\begin{matrix} n_k \\ \left[n_{k(r_1\cdots r_p)}\right]\end{matrix}\right)\prod_{S}\left(p^{(k)}_{r_1\cdots r_p}\right)^{n_{k(r_1\cdots r_p)}}\right]\right\}.$$

$$\left\{\sum_{S_p}\sum_{S^*}\left[\prod_{j=1}^{p}\left\{n_{k_j(r_1^*\cdots r_p^*)}-\sum_{l=1}^{j-1}\delta_{k_jk_l}\right\}p^{(k_j);j}_{r_1^*\cdots r_p^*}\Big/p^{(k_j)}_{r_1^*\cdots r_p^*}\right]\right\}\prod_{j=1}^{p}d\tilde{X}_j,$$

where $\left\{\delta_{k_jk_l}\right\}$ are Kronecker deltas with $\delta_{k_jk_0}=0$; the product S extends over all the 2^p terms $(r_1 \ldots r_p)\epsilon S$, the sum S_p over the c^p terms : $k_j = 1, \ldots, c$; $j = 1, \ldots, p$; and the sum S^* over all the possible terms like the one within the third bracket succeeding S^*. In the particular case of $c = p = 2$, reference may be made to [7] for some simplification of this procedure.

(ii) If the pooled sample median vector is determined by less than p observations, say $p-h$ observations ($h \geqslant 1$), then proceeding precisely on the same line as in Mood (1941), it can be shown that the corresponding probabilility terms would amount to a term of the order N^{-h}, as compared to (5.5.5). Thus, for $h \geqslant 1$, the contributions of the probability terms may be neglected. Further, it follows from the results on the univariate median test (cf. Mood (1954)] that $N^{\frac{1}{2}}\tilde{X}_j$ is bounded in probability, for $j = 1, \ldots, p$. Hence, it can be shown by simple but somewhat lengthy algebraic manipulations that under the sequence of alternatives in (4.4.1)

(i) $n_{k(r_1\cdots r_p)}\Big/\left[n_k\, p^{(0)}_{r_1\cdots r_p}\right] = 1+O_p(N^{-\frac{1}{2}}),$

(ii) $\left| p^{(k)}_{r_1 \cdots r_p} - p^{(0)}_{r_1 \cdots r_p} \right| = O_p(N^{-\frac{1}{2}}),$

(iii) $\left| p^{(k);\,i}_{r_1 \cdots r_p} - p^{(0);\,i}_{r_1 \cdots r_p} \right| = O_p(N^{-\frac{1}{2}}),$ for $i = 1, \ldots, p$;

(iv) $\left| p^{(0)}_{r_1 \cdots r_p} - p^{(0)}_{1-r_2 \cdots 1-r_p} \right| = O_p(N^{-\frac{1}{2}}),$

for all $(r_1 \ldots r_p)\epsilon S$, $k = 1, \ldots, c$. Hence, it can be shown by some simple but somewhat lengthy algebraic manipulations that

(5.5.6) $$\sum_{S_i} p^{(0);\,i}_{r_1 \cdots r_p} = [\alpha_i + O_p(N^{-\frac{1}{2}})], \quad \text{for } i = 1, \ldots, p;$$

where the summation S_i extends over all possible 2^{p-1} values of $(r_1 \ldots r_p)$; $r_j = 0, 1, j = 1, \ldots, p \ (\neq i)$, for $i = 1, \ldots, p$, and where α_j's are defined by (5.5.2). Hence, from (5.5.5), (5.5.6) and some simplifications, the joint probability function of the set of random variables $\left\{ n_{k(r_1 \cdots r_p)} \,;\, k = 1, \ldots, c, (r_1 \ldots r_p)\epsilon S \right\}$ and $\tilde{X}_1, \ldots, \tilde{X}_p$ asymptotically reduces to

(5.5.7) $$N^p \alpha_1 \ldots \alpha_p \left[\prod_{k=1}^{c} \left\{ \left(\begin{matrix} n_k \\ [n_{k(r_1 \cdots r_p)}] \end{matrix} \right) \prod_{S} \left(p^{k}_{r_1 \cdots r_p} \right)^{n_{1(r_1 \cdots r_p)}} \right\} \right] d\tilde{X}_1 \ldots d\tilde{X}_p + O_p(N^{-1/2}).$$

Again, by using (multivariate) Taylor's expansion it can be shown that

(i) $$N^{\frac{1}{2}} \left| \left(p^{(k)}_{r_1 \cdots r_p} - P^{(k)}_{r_1 \cdots r_p} \right) - \left(p^{(0)}_{r_1 \cdots r_p} - P^{(0)}_{r_1 \cdots r_p} \right) \right| \xrightarrow{P} 0, \ k=1, \ldots, c;$$

(ii) $\left| P^{(0)}_{r_1 \cdots r_p} - P^{(0)}_{r_1 \cdots r_p} \right| = O_p(N^{-\frac{1}{2}}),$

(iii) $p^{(0)}_{r_1 \cdots r_p} = \sum_{k=1}^{c} \nu_k \left[p^{(k)}_{r_1 \cdots r_p} \right] + O_p(N^{-\frac{1}{2}}),$

for all $(r_1 \ldots r_p)\epsilon S$.

Hence, (5.5.7) can be shown to be reducible asymptotically to

$$N^p\alpha_1 \dots \alpha_p \left| |A|/(2\pi)^{2p-1} \right|^{c/2}$$

$$(5.5.8) \qquad \cdot \exp\left\{ -\frac{1}{2} \sum_{k=1}^{c} \sum_{S} \frac{\left[n_{k(r_1\cdots r_p)} - n_k p^{(k)}_{r_1\cdots r_p} \right]^2}{n_k P^{(k)}_{r_1\cdots r_p}} \right\}.$$

$$\prod_{i=1}^{p} d\tilde{X}_i \prod_{k=1}^{c} \prod_{l=1}^{2^p-1} dZ_{k(l)},$$

where A has the principal minors $A_{ii} = 1/P_i^{(0)} + 1/P^{(0)}_{2^p}$, $i = 1, \dots, 2^p-1$, and the off-diagonal minors $A_{ij} = 1/P^{(0)}_{2^p}$, $i \neq j = 1, \dots, 2^p-1$, $Z_{k(j)} = n_k^{-\frac{1}{2}} \, (n_{k(j)} - n_k \, p_j^{(k)})$, $j = 1, \dots, 2^p-1$, $k = 1, \dots, c$; the number i or j $(= 1, \dots, 2^p)$, being attached to $(r_1 \dots r_p)$ in the same convention as in just before (5.5.3). We now write

$$(5.5.9) \qquad d^k_{r_1\cdots r_p} = n_k^{\frac{1}{2}} \left(p^{(k)}_{r_1\cdots r_p} - p^{(0)}_{r_1\cdots r_p} \right), \; k = 1, \dots, c, \;\; (r_1 \dots r_p) \epsilon S,$$

$$(5.5.10) \qquad \xi^k_{(l)} = \left\{ n_{k(l)} - \frac{n_k}{N} \; C_{(l)} \right\} / \sqrt{n_k}$$

for $l = 1, \dots, 2^p-1$; $k = 1, \dots, c$; and

$$W_{(l)} = \{C_{(l)} - N^{(0)}_{p(l)}\} / \sqrt{N} \text{ for } l = 1, \dots, 2^p-1.$$

Then, we have

$$(5.5.11) \qquad Z_{k(l)} = \xi^k_{(l)} - d^k_{(l)} + \sqrt{n_k/N} \; W_{(l)}$$

$$\text{for } l = 1, \dots, 2^p-1; \quad k = 1, \dots, c;$$

$$(5.5.12) \qquad \sum_{k=1}^{c} \sum_{l=1}^{2^p} Z^2_{k(l)} = \sum_{k=1}^{c} \sum_{l=1}^{2^p} [\xi^k_{(l)} - d^k_{(l)}]^2 / P^0_{(l)}$$

$$+ \sum_{l=1}^{2^p} W^2_{(l)} / P^0_{(l)} + o_p(1).$$

By noting that $d^k_{r_1\cdots r_p} = n_k^{\frac{1}{2}} \left(p^{(k)}_{r_1\cdots r_p} - p^{(0)}_{r_1\cdots r_p} \right) + o_p(1)$, for all $k = 1, \dots, c$ and all $(r_1 \dots r_p) \epsilon S$, we get from (5.5.8) through (5.5.12) and on integrating over the range of the variables $\tilde{X}_1, \ldots, \tilde{X}_p$, that the joint distribution of $\xi_{(l)}^{(k)}$, $l = 1, \ldots, 2^p-1$, $k = 1, \ldots, c$ reduces to

$$\text{(5.5.13)} \qquad \left\{ \left| |A| / (2\pi)^{2^{p-1}} \right|^{(c-1)/2} \right.$$

$$\left. \exp \left\{ -\frac{1}{2} \sum_{k=1}^{c} \sum_{S} \frac{\left[\xi^k_{r_1\cdots r_p} - d^k_{r_1\cdots r_p} \right]^2}{P^0_{r_1\cdots r_p}} \right\} \prod_{k=1}^{c} \prod_{i=1}^{2^p-1} d\xi^k_{(i)} \right\}.$$

Hence it follows by some routine algebra that

$$\text{(5.5.14)} \qquad T_N = \sum_{k=1}^{c} \sum_{i=1}^{p} \sum_{j=1}^{p} \phi^{ij} 4\{n^*_{k(i)} - an_k/N\}\{n^*_{k(j)} - an_k/N\}/n_k$$

has asymptotically a noncentral chi-square distribution with $p(c-1)$ d.f. and the noncentrality parameter

$$\text{(5.5.15)} \qquad \Delta_M = 4 \sum_{k=1}^{c} \nu_k \sum_{i=1}^{p} \sum_{j=1}^{p} \varphi^{ij} \alpha_i \alpha_j (\theta_i^{(k)} - \bar{\theta}_i)(\theta_j^{(k)} - \bar{\theta}_j),$$

where α_j's are defined by (5.5.2), $\boldsymbol{\theta}^{(k)}$, $k = i, \ldots, c$ by (4.4.1), and $\boldsymbol{\theta}$ by (4.4.6). Further, it is easy to see that under (4.4.1).

$$\text{(5.5.16)} \qquad \boldsymbol{D} \xrightarrow{P} \boldsymbol{\Phi}, \text{ i.e., } \boldsymbol{D}^{-1} \xrightarrow{P} \boldsymbol{\Phi}^{-1}$$

(as $\boldsymbol{\Phi}$ is assumed to be nonsingular), and hence, from (5.2.8), (5.5.14) and (5.5.16) it follows that under (4.4.1),

$$\text{(5.5.17)} \qquad M_N \overset{P}{\sim} T_N.$$

The rest of the proof follows by some standard procedure and is therefore omitted.

6. ASYMPTOTIC POWER-EFFICIENCY OF THE TWO TESTS

Since, under $\{H_N\}$ in (4.4.1), both the test statistics have noncentral chi-square distributions, differing only in the noncentrality parameters, a comparison of the two noncentrality parameters will reveal their asymptotic relative efficiency (A.R.E.). However, such an A.R.E. depends on the vectors $\boldsymbol{\theta}^{(k)}$, $k = 1, \ldots, c$ as well as on the two matrices $\boldsymbol{\Psi}$ and $\boldsymbol{\Phi}$ entering into the expressions (4.4.6) and (5.5.15). Moreover, interpreted as the *Pitman-efficiency*, it also depends on the level of significance ϵ. Thus, unlike the univariate case, no single measure of efficiency (independent of $\theta^{(k)}$, $k = 1, \ldots, c$ and ϵ) usually exists (unless the p variates in $F(\boldsymbol{x})$ are independent or totally symmetric), and we may have to be satisfied with the assessment of various bounds for the A.R.E. (for any specified $F(\boldsymbol{x})$, which are independent of $\boldsymbol{\theta}^{(k)}$, $k = 1, \ldots, c$. In the particular case of $F(\boldsymbol{x})$ being a p variate normal distribution with a null mean vector, unit variances and a correlation matrix $\boldsymbol{\rho} = (\rho_{ij})$, it follows from well-known results that

$$(6.1) \qquad \psi_{ij} = (6/\pi)\sin^{-1}(_{ij}/2), \; \phi_{ij} = (2/\pi)\sin^{-1}(\rho_{ij}), \quad \text{for } i, j = 1, \ldots, p;$$

and hence, the A.R.E. becomes a function of $\boldsymbol{\theta}^{(k)}$, $k = 1, \ldots, c$ and $\boldsymbol{\rho}$. In the particular case of $p = 2$, it has been shown by the present authors [7] that the A.R.E. of the rank-sum test with respect to the median test is uniformly greater than unity (i.e., $\geqslant 1$, for all $\boldsymbol{\theta}^{(k)}$, $k = 1, 2$ and ϵ). In the multivariate case of $p \geqslant 3$, really the bounds for this A..R.E. depend on the characteristic roots of $\boldsymbol{\Psi\Phi}^{-1}$. Some result of this type has been considered by Bickel (1964) in the multivariate one sample problem, and it appears that the same bounds are also applicable in our case. We therefore omit these results. Finally, comparison of the rank-sum test or the median test with the parametrically optimum test (viz., the likelihood ratio test) for multinormal parent distribution, requires the comparison of the noncentrality parameters in (4.4.6) or (5.5.15) with that of

$$(6.2) \qquad \Delta_H = \sum_{k=1}^{c} \nu_k \sum_{i=1}^{p} \sum_{j=1}^{p} \rho^{ij}(\theta_i^{(k)} - \bar{\theta}_i)(\theta_j^{(k)} - \bar{\theta}_j), \; (\rho^{ij}) = \boldsymbol{\rho}^{-1}.$$

For the case of $p = 2$, the present authors [7] have shown that the A.R.E. of either of the two proposed tests with respect to the Hotelling's T^2 test (equivalent to the likelihood ratio test) is uniformly less than one, (for bivariate normal distributions). Bounds similar to the one given in Bickel's one sample paper, can again be considered for the multisample case when $p > 2$.

7. CONCLUDING REMARKS

In this paper, we have considered the p variate c sample case, for p, $c \geqslant 2$. For the case of p variates and 2 samples, the expressions for the two statistics in (4.1.15) and (5.2.8) can be simplified as

$$(7.1) \quad W_N = \frac{12}{(N+1)n_1 n_2} \sum_{i=1}^{p} \sum_{j=1}^{p} \psi^{ij}(\Lambda_N) \left\{ R_i - \frac{n_1(N+1)}{2} \right\} \left\{ R_j - \frac{n_1(N+1)}{2} \right\}$$

$$M_N = \frac{4N}{n_1 n_2} \sum_{i=1}^{p} \sum_{j=1}^{p} d^{ij} \{n^*_{1(i)} - \tfrac{1}{2}n_1\}\{n^*_{1(j)} - \tfrac{1}{2}n_1\},$$

where R_i is the sum of ranks (with respect to the ith variate) of the first sample observations, for $i = 1, \ldots, p$. Further, if $p = 2$, these expressions can be more simplified as in [7].

The two nonparametric tests developed in this paper have an important role in multivariate analysis. Not only the assumption of multinormality of the parent distribution is waved here, but also the scope of the inference procedures is increased to data where the observations may be available on an ordinal scale, (thus creating much difficulties to the applicability of usual methods); in the field of Psychometry there are numerous instances of this type of data. The authors believe that the proposed tests are also more robust than the usual parametric tests, though the detailed study of this aspect is appreciably computer-dependent and is left to such interested readers.

References

[1] Anderson, T. W. (1958). *An Introduction to Multivariate Analysis*, John Wiley, New York.

[2] Andrews, F. C. (1954). "Asymptotic Behaviour of Some Rank Tests for Analysis of Variance," *Ann. Math. Statist.*, **25**, 724–736.

[3] Bennett, B. M. (1962). "On Multivariate Sign Tests," *J. Roy. Statist. Soc., Ser. B.*, **24**, 159–161.

[4] Bennett, B. M. (1964). "A Bivariate Signed-Rank Test," *J. Roy. Statist. Soc. Ser. B.*, **26**, 457–461.

[5] Bickel, P. J. (1964). "On Some Alternative Estimates for Shift in the p-Variate One Sample Problem," *Ann. Math. Statist.*, **35**, 1079–1090.

[6] Chatterjee, S. K. (1966). "A Bivariate Sign Test for Location," *Ann. Math. Statist.*, **37**, 1771–1782.

[7] Chatterjee, S. K. and Sen, P. K. (1964). "Nonparametric Tests for the Bivariate Two Sample Location Problem,' *Calcutta Statist. Assoc. Bull.*, **13**, 18–58.

[8] Chatterjee, S. K. and Sen, P. K. (1965). "Some Nonparametric Tests for the Two Sample Bivariate Association Problem," *Calcutta Statist. Assoc. Bull.*, **14**, 14–35.

[9] Hodges, J. L. (1955). "A Bivariate Sign Test," *Ann. Math. Statist.*, **26**, 102–106.

[10] Hoeffding, W. (1948). "A Class of Statistics with Asymptotically Normal Distribution," *Ann. Math. Statist.*, **19**, 293–325.

[11] Hoeffding, W. (1951). "A Combinatorial Central Limit Theorem," *Ann. Math. Statist.*, **22**, 558–566.

[12] Hoeffding, W. (1952). "The Large Sample Power of Tests Based on Permutations of Observations," *Ann. Math. Statist.*, **23**, 169-192.

[13] Kruskal, W. H. and Wallis, W. A. (1952). "Use of Ranks in One Criterion Analysis of Variance," *J. Amer. Statist. Assoc.*, **47**, 583–621.

[14] Mood, A. M. (1941). "On the Joint Distribution of the Medians in Samples From a Multivariate Population," *Ann. Math. Statist.*, **12**, 268–278.

[15] Mood, A. M. (1950). *Introduction to the Theory of Statistics*, McGraw-Hill, New York.

[16] Mood, A. M. (1954). "On the Asymptotic Efficiency of Certain Nonparametric Two Sample Tests," *Ann. Math. Statist.*, **25**, 514–522.

[17] Roy, S. N. (1942). "Analysis of Variance for Multivariate Normal Populations ; The Sampling Distribution of the Requisite p-Statistic on the Null and Non-Null Hypotheses," *Sankhyā*, **6**, 35–50.

[18] Wald, A. and Wolfowitz, J. (1944). "Statistical Tests Based on Permutations of Observations," *Ann. Math. Statist.*, **15**, 358–372.

(*Received Jan. 1, 1966.*)

CHAPTER 11

Step-down Multiple Decision Rules

SOMESH DAS GUPTA*, *Indian Statistical Institute***

1. INTRODUCTION

The step-down technique introduced by Roy [11] and Roy and Bargmann [12] for some multivariate testing problems, can be described essentially as follows. A hypothesis is tested by means of testing sequentially (or, in stages) a set of component hypotheses, the inter-section of which is the original hypothesis. The ith stage is considered and the ith hypothesis is tested if, and only if, all the preceding hypotheses are accepted, and the acceptance of all the component hypotheses leads to the acceptance of the original hypothesis. While testing the ith hypothesis the fact that it is tested in a conditional situation, namely, the acceptance of the preceding hypotheses, was not considered by Roy and Roy-Bargmann. Possibly, they ignored the sequential nature of the procedure and their primary objective was to get

*Research sponsored in part by the Office of Naval Research under Contract Number Nonr-266(33), Project Number NR 042-034, Department of Mathematical Statistics, Columbia University, New York. Reproduction in whole or in part is permitted for any purpose of the United States Government.

**Now at the University of Minnesota.

a critical region for the test which happened to be the union of the critical regions corresponding to the component hypotheses. Moreover, in a p-variate situation, the ith hypothesis was formulated in terms of the conditional distribution of the ith variate given the preceding variates. In this conditional situation the test considered for the ith hypothesis might be a good one but nothing is known about the overall performance of the procedure. The parameter space under the alternative hypothesis can be partitioned into subsets such that the ith subset corresponds to the intersection of the ith component alternative hypothesis and the preceding component hypotheses. When the original hypothesis is rejected at the ith stage, the step-down technique *suggests* that the parameter lies in the ith subset of the above-mentioned partition. This good feature of the technique may be properly utilized if the problem is formulated in the following multiple decision setup. The parameter space is partitioned into the subsets of the above-mentioned partition of the set corresponding to the alternative hypothesis and the subset corresponding to the original hypothesis. Then the problem is to decide which one of these subsets include the true value of the parameter. This problem, in a more general setup, is considered in this paper and "good" tests of the component hypotheses are utilized to construct a "good" multiple decision rule. However, a good test of the ith hypotheses is sought under the assumption that all the preceding hypotheses are true. The multiple decision rules considered here are not sequential in nature and the component hypotheses are considered (apparently, in stages) only to construct a good solution for the multiple decision problem.

A similar situation occurs in the analysis of variance problem. For instance, it is customary to pool an interaction sum of squares with the error sum of squares when that interaction is found to be insignificant. When another "effect" is tested against this pooled error sum of squares the corresponding F-test is not valid. However, if we follow the multiple decision approach, the above procedure turns out to be a "good" one. This is discussed in Section 4 along with many other examples.

The above approach of formulating a step-down problem in a multiple decision setup is due to Anderson [1], [2]. He illustrated the theory by the following two examples : (1) The optimum choice of the degree of polynomial regression when the maximum possible degree is specified, and (2) the determination of the order of dependence of a nearly stationary Gaussian stochastic process. The essential inherent structure of a problem for which Anderson's technique would be applicable is briefly discussed in [1]. This is presented systematically with slight modification and generalization in Section 2 of this paper. Anderson's main results along with their proofs are reformulated in Section 3 with some modifications in order to cover the general setup. This is done with the view to focus the basic results in order to discuss many other problems in this light and examine the roles of the different assumptions (*see* the discussion in Section 2). Anderson's method as employed in his two examples has a straightforward extension to the case where the distribution is of exponential type (*see* Notes 3 and 4). This is further illustrated by some concrete cases given in Examples 1–4. (Examples 3 is discussed in [1]; Example 4 is a special case of the problem mentioned in Note 3 and it also follows from the problem treated essentially in [2].) These examples are focussed in order to illustrate the method by some simple and concrete cases and introduce the idea of weak unbiasedness. Examples 5 and 6 deal with the step-down problems mentioned earlier. It is interesting to note that for each of the decision theoretic formulations of the step-down problems discussed in [11] and [12] the component hypotheses are considered in the reverse order; no useful result can be obtained otherwise. These examples also illustrate an extension of Anderson's method to the step-down approach with respect to the variates in a multivariate problem. The notion of unbiased multiple decision rule is discussed in multiparameter situation and a new concept termed as 'weakly unbiased rule' is introduced and its relative advantages are illustrated.

Lehmann [7], [8] considered similar problems from "risk" consideration. The procedure suggested by him [8] utilizes good tests of the component hypotheses and the ith hypothesis was considered if certain prescribed decisions had been obtained for the preceding tests. Most of his results [8] are quite complicated, as one should expect, and the procedures are sequential in nature. Nothing is known about the property of the conditional decision rule [8] suggested by Lehmann.

Hogg [6], for certain testing problems, exploited the same resolution of the parameter space and subsequent reduction by sufficient statistics.

2. THE PROBLEM AND THE ASSUMPTIONS

(In the sequel, the relevant measurability assumptions will not be mentioned although they will be implicitly taken into consideration.)

Let X be a random variable (may be vector-valued) having the probability distribution P_θ, where the parameter θ (may be vector-valued) lies in a set Ω_0. Consider a $(k+1)$-decision problem with decisions $d_1, \ldots, d_{k+1}$, where d_i is the decision: $\theta \epsilon \Theta_i$ and $\Theta_1, \ldots, \Theta_{k+1}$ is a partition of Ω_0. Let

$$(1) \qquad \Omega_i = \Theta_{i+1} \cup \Theta_{i+2} \cup \ldots \cup \Theta_{k+1}, \quad i = 0, 1, \ldots, k$$

Then

$$(2) \qquad \Omega_0 \supset \Omega_1 \supset \ldots \supset \Omega_k,$$

and

$$(3) \qquad \Theta_i = \Omega_{i-1} \overline{\Omega}_i = \Omega_0 \Omega_1 \Omega_2 \ldots \Omega_{i-1} \overline{\Omega}_i, \quad i = 1, \ldots, k,$$

$$\Theta_{k+1} = \Omega_0 \Omega_1 \ldots \Omega_k.$$

Frequently, the subsets Θ_i occur from the decomposition (3), given the subset Ω_i.

A decision rule is given by a measurable function δ of x, where

$$\delta(x) = (\delta_1(x), \ldots, \delta_{k+1}(x)),$$

$$0 \leqslant \delta_i(x) \leqslant 1, \quad i = 1, \ldots, k+1,$$

$$\sum_{i=1}^{k+1} \delta_i(x) = 1, \qquad \text{for all } x. \tag{4}$$

$\delta_i(x)$ denotes the probability of taking the decision d_i, given the observation x.

We shall consider the class of decision rules which satisfy the following conditions.

Condition A. *$E_\theta[\delta_i(X)] = \alpha_i$, $\theta \epsilon \Omega_i$, $i = 1, \ldots, k$, where α_i's are preassigned numbers satisfying*

$$0 < \alpha_i < 1, i = 1, \ldots, k, \quad \sum_{i=1}^{k} \alpha_i < 1. \tag{5}$$

Condition B. *$E_\theta[\delta_i(X)]$, for $\theta \epsilon \Theta_i$, depends on θ only through the function (may be vector-valued) $h_i(\theta)$; $i = 1, \ldots, k$.*

Sometimes, we may also consider decision rules which satisfy

Condition C. *Given $\alpha_1, \ldots, \alpha_k$ satisfying* (5),

$$\begin{aligned} E_\theta[\delta_i(X)] &\leqslant \alpha_i \text{ for } \theta \epsilon \Omega_i \\ &\geqslant \alpha_i \text{ for } \theta \epsilon \Theta_i, \qquad i = 1, 2, \ldots, k. \end{aligned} \tag{6}$$

We shall make the following assumptions on the class of probability distributions $[P_\theta : \theta \epsilon \Omega_0]$, and the subsets Θ_i.

Assumption 1. (a) *For $\theta \epsilon \Omega_0$ there exists a set $\{T_1, T_2, \ldots, T_{k+1}\}$ of sufficient statistics such that $\{T_{i+1}, \ldots, T_{k+1}\}$ is sufficient for the class of distributions when $\theta \epsilon \Omega_i$ ($i = 0, 1, \ldots, k$).*

(b) *When $\theta \epsilon \Omega_i$, the class of distributions of $\{T_{i+1}, \ldots, T_{k+1}\}$ is boundedly complete; $i = 0, 1, \ldots, k$.* Note that T_i's may be vector-valued.

Assumption 2. *For testing $H_i : \theta \epsilon \Omega_i$ against $K_i : \theta \epsilon \Theta_i$ there exists a test ϕ_i^* which satisfies the following :*

(a) *$E[\phi_i(X)] = \beta_i$, $\theta \epsilon \Omega_i$, where β_i is any pre-assigned number satisfying $0 < \beta_i < 1$.*

(b) *$E[\phi_i(X)]$, for $\theta \epsilon \Theta_i = \Omega_{i-1} \overline{\Omega}_i$, depends on θ only through $h_i(\theta)$.*

(c) $E_\theta[\phi_i^*(X)] \geqslant E_\theta[\phi_i(X)]$, $\theta \epsilon \Theta_i$, *for any test* ϕ_i *satisfying* (a) *and* (b). $(i = 1, \ldots, k)$.

We may also refer to the following assumptions :

Assumption 2′. *For testing* $H_i : \theta \epsilon \Omega_i$ *against* $K_i : \theta \epsilon \Theta_i$ *there exists a test* ϕ_i^* *which is UMP unbiased level* $\beta_i (0 < \beta_i < 1)$ *test* ; $i = 1, \ldots, k$.

Assumption 3. (a) *For every* i, $E_\theta[\delta_i(x)]$ *is a continuous function of* θ *which is assumed to lie in a finite-dimensional Euclidean space.*

(b) *For every* i, *every point of* Ω_i *is a boundary point of* Ω_i *and* Θ_i.

Discussion. 1] Given α_i's satisfying (5), we take

$$\beta_1 = \alpha_1, \beta_i = \frac{\alpha_i}{1 - \alpha_1 - \ldots - \alpha_{i-1}}, \quad i = 2, \ldots, k. \tag{7}$$

2] Since $\{T_i, \ldots, T_{k+1}\}$ is sufficient when $\theta \epsilon \Omega_{i-1}$ we consider ϕ_i^* (as defined in Assumption 2 or in Assumption 2′) to be a function of these sufficient statistics.

3] Condition A describes some sort of "similarity" requirement. It is obvious that the above formulation is not symmetric with respect to the sets Θ_i and the ordering of the sets Θ_i is specified by a priori considerations. This ordering along with the relation (2) leads us to call the decision rules considered as *step-down decision rules.* Different examples in Section 4 will clarify this issue to some extent.

4] The main object to consider Assumptions 2′ and 3 is to use them as substitutes for Assumption 2(b); and, in that case, we shall have Conditions A and B replaced by Condition C. Note that Assumption 3 and Condition C imply Condition A.

5] The function $h_i(\theta)$, introduced in Condition B and Assumption 2(b), arises out of invariance consideration. Suppose there exists a group G_i of transformations which leaves the problem of testing H_i against K invariant. Let $h_i(\theta)$ be maximal invariant under the induced group of transformation on the parameter space, when $\theta \epsilon \Theta_i$. In this situation we shall make use of Theorems 4 and 5 in Chapter 6 in [9] when the assumptions made in these theorems are satisfied; then it can be seen that Condition B is equivalent to

$$E_\theta[\delta_i(X) \mid T_i, \ldots, T_{k+1}] = S_i(T_i, \ldots, T_{k+1}) \tag{8}$$

when $\theta \epsilon \Omega_{i-1}$ and where S_i is maximal invariant in the space of sufficient statistics $T_i, \ldots, T_{k+1}$ under the group G_i. This can be seen from Theorem 4 [p. 227–9] subject to the existence of Haar measure on the group G_i.

3. OPTIMAL RULE

We shall call a rule δ^* optimal in a certain class of rules, if

$$E_\theta[\delta_i^*(X)] \geqslant E_\theta[\delta_i(X)], \ \theta \epsilon \Theta_i, \tag{9}$$

$(i = 1, \ldots, k)$ for any rule δ in that class. In this sense, an optimal rule maximizes the probabilities of correct decisions simultaneously.

The proofs of the following results follow essentially the proof given by Anderson [1].

Lemma 1. *For any function* $\delta_j(X)$ *for which*

$$E_\theta[\delta_j(X)] = \alpha_j, \text{ for all } \theta \epsilon \Omega_j,$$

the following relation is satisfied under Assumption 1.

$$E_\theta[\delta_j(X)\phi_i^*(T_i, T_{i+1}, \ldots, T_{k+1})]$$
$$= \alpha_j E_\theta[\phi_i^*(T_i, \ldots, T_{k+1})], \text{ for all } \theta \epsilon \Omega_j,$$

where $j < i$ *and* ϕ_i^* *is a function of* $T_i, \ldots, T_{k+1}$.

Proof. We assume $\theta \epsilon \Omega_j$. Using the bounded completeness of $T_{j+1}, \ldots, T_{k+1}$, we have

$$E[\delta_j(X) \mid t_{j+1}, \ldots, t_{k+1}] = \alpha_j, \text{ a.e.}$$

Hence

$$E_\theta[\delta_j(X)\phi_i^*(T_i, T_{i+1}, \ldots, T_{k+1})]$$
$$= E_\theta[\phi_i^*(T_i, \ldots, T_{k+1}) E(\delta_j(X) \mid T_{j+1}, \ldots, T_{k+1})]$$
$$= \alpha_j E_\theta[\phi_i^*(T_i, \ldots, T_{k+1})].$$

Define

$$\delta^*(X) = (\delta_1^*(X), \ldots, \delta_{k+1}^*(X)),$$

where

$$\delta_1^*(X) = \phi_1^*(X),$$

(10) $$\delta_i^*(X) = \prod_{j=1}^{i-1} [1-\phi_j^*(X)]\, \phi_i^*(X), \quad i = 2, \ldots, k,$$

$$\delta_{k+1}^*(X) = \prod_{j=1}^{k} [1-\phi_j^*(X)].$$

The functions ϕ_i^* used in (10) either satisfy Assumption 2 or Assumption 2′. We shall show that the rule δ^* is optimal in the sense (9). The above construction is suggested by (3).

Theorem 1. *If Assumptions* 1 *and* 2 *hold, then the rule* δ^* *given by* (10), *where* ϕ_i^**'s satisfy Assumption* 2, *satisfies Conditions A and B.*

Proof. The function δ_1^* clearly satisfies the conditions since Assumption 2 holds. Suppose that these conditions are satisfied by $\delta_1^*, \ldots, \delta_{i-1}^*$; then we shall show that they are also satisfied by δ_i^*. For any rule δ, define

(11) $$\psi_i(X : \delta) = \sum_{j=1}^{i} \delta_j(X), \; i = 1, \ldots, k+1.$$

It follows from Lemma 1 that, for $\theta \epsilon \Omega_{i-1}$

(12) $$E_\theta[\psi_{i-1}(X : \delta^*)\phi_i^*(X)] = E_\theta[\phi_i^*(X)] \sum_{j=1}^{i-1} \alpha_j.$$

Note that

(13) $$\delta_i^*(X) = [1-\psi_{i-1}(X : \delta^*)]\phi_i^*(X).$$

Hence, for $\theta \epsilon \Omega_{i-1}$

(14) $$E_\theta[\delta_i^*(X)] = [1- \sum_{j=1}^{i-1} \alpha_j] E_\theta[\phi_i^*(X)].$$

It follows from Assumption 2(b) that δ_i^* satisfies Condition *B*. In particular, for $\theta \epsilon \Omega_i$, we have

(15) $$E_\theta[\phi_i^*(X)] = \beta_i = \frac{\alpha_i}{1- \sum_{j=1}^{i-1} \alpha_j}$$

and

(16) $$E_\theta[\delta_i^*(X)] = [1- \sum_{j=1}^{i-1} \alpha_j]\beta_i = \alpha_i.$$

Thus δ_i^* satisfies Condition A.

In the above proof we considered the statements given in Conditions A and B separately for each i. When we said that δ_i^* satisfied Conditions A and B we meant that those conditions were satisfied for that particular i.

Theorem 2. *If Assumptions* 1, 2′, *and* 3 *hold, then the rule* δ^* *defined by* (10), *where* ϕ_i^**'s satisfy Assumption* 2′, *satisfies Condition* C.

Proof. It is easily seen that Assumptions 2′ and 3 imply

$$E_\theta[\phi_i^*(X)] = \beta_i \quad \text{for } \theta \epsilon \Omega_i, \; i = 1, \ldots, k.$$

It is clear that $E_\theta[\delta_1^*(X)] = \alpha_1$ for $\theta\epsilon\Omega_1$. Suppose $E_\theta[\delta_j^*(X)] = \alpha_j$ for $\theta\epsilon\Omega_j$, $j = 1, 2, \ldots, i-1$; we shall show that $E_\theta[\delta_i^*(X)] = \alpha_i$ for $\theta\epsilon\Omega_i$. Applying Lemma 1 we get

$$E_\theta[\delta_i^*(X)] = \left[1 - \sum_{j=1}^{i-1} \alpha_j\right] E_\theta[\phi_i^*(X)],$$

for $\theta\epsilon\Omega_{i-1}$. In particular, for $\theta\epsilon\Omega_i$

$$E_\theta[\delta_i^*(X)] = \left[1 - \sum_{j=1}^{i-1} \alpha_j\right] \beta_i = \alpha_i.$$

Moreover, for $\theta\epsilon\Theta_i$.

$$E_\theta[\delta_i^*(X)] \geqslant \left[1 - \sum_{j=1}^{i-1} \alpha_j\right] \beta_i = \alpha_i, \tag{17}$$

since

$$E_\theta[\phi_i^*(X)] \geqslant \beta_i.$$

Theorem 3. *If Assumptions* 1 *and* 2 *hold, then the rule* δ^* *defined in* (10) *and in the statement of Theorem* 1 *is optimal in the sense* (9) *among the rules satisfying Conditions* A *and* B.

Proof. For any rule δ satisfying Conditions A and B, define

$$\phi_i(X) = \delta_i(X) + \psi_{i-1}(X\,;\delta)\phi_i^*(X), \tag{18}$$

where ψ_i is given by (11). By Lemma 1, for $\theta\epsilon\Omega_{i-1}$

$$E_\theta[\phi_i(X)] = E_\theta[\delta_i(X)] + E_\theta[\phi_i^*(X)] \sum_{j=1}^{i-1} \alpha_j. \tag{19}$$

Thus ϕ_i satisfies Assumption 2(b). [It can be verified that ϕ_i is a test function.] In particular, for $\theta\epsilon\Omega_i$

$$E_\theta[\phi_i(X)] = \alpha_i+\beta_i \sum_{j=1}^{i-1} \alpha_j = \beta_i \tag{20}$$

Hence ϕ_i satisfies Assumption 2(a). By Assumption 2(c),

$$E_\theta[\phi_i^*(X)] \geqslant E_\theta[\phi_i(X)],\ \theta\epsilon\Theta_i.$$

Thus, for $\theta\epsilon\Theta_i$

$$\begin{aligned} E_\theta[\delta_i^*(X)] &= E_\theta[\{1-\psi_{i-1}(X\ ;\delta^*)\}\,\phi_i^*(X)] \\ &= \left[1-\sum_{j=1}^{i-1}\alpha_j\right] E_\theta[\phi_i^*(X)] \\ &\geqslant E_\theta[\phi_i(X)]-\sum_{j=1}^{i-1}\alpha_j E_\theta[\phi_i^*(X)] \\ &= E_\theta[\delta_i(X)]. \end{aligned}$$

Corollary 1. *Under the assumptions used in Theorem* 3,

$$E_\theta[\delta_i^*(X)] \geqslant \alpha_i,\ \theta\epsilon\Theta_i\ ;\ i = 1, \ldots, k,$$

where δ^* *is defined as in Theorem* 3.

Proof. The above result is obtained by comparing the optimal rule δ^* with the following rule δ defined by

$$\delta_i(x) = \alpha_i, \quad \text{for all } x\ ;\ i = 1, \ldots, k.$$

Theorem 4. *If Assumptions* 1. 2′ *and* 3 *hold, then the rule* δ^* *considered in Theorem* 2 *is optimal in the sense* (9) *among all rules satisfying Condition* C.

Proof. First, note that Assumptions 1 and 3, and Condition C imply Condition A.

Consider a rule δ satisfying Condition C. Following the proof of Theorem 3, we get (19), i.e., for $\theta\epsilon\Omega_{i-1}$

$$E_\theta[\phi_i(X)] = E_\theta[\delta_i(X)]+\sum_{j=1}^{i-1} \alpha_j\, E_\theta[\phi_i^*(X)].$$

Since,

$$\begin{aligned} E_\theta[\phi_i^*(X)] &\geqslant \beta_i,\ \theta\epsilon\Theta_i, \\ &\leqslant \beta_i,\ \theta\epsilon\Omega_i \end{aligned}$$

we have

$$E_\theta[\phi_i(X)] \geqslant \beta_i,\ \theta\epsilon\Theta_i$$
$$\leqslant \beta_i,\ \theta\epsilon\Omega_i$$

Hence, by Assumption 2′

$$E_\theta[\phi_i^*(X)] \geqslant E_\theta[\phi_i(X)],\quad \theta\epsilon\Theta_i.$$

The rest of the proof is similar to that of Theorem 3.

4. EXAMPLES

In the following examples, we shall not go through the explicit verification of the assumptions used. If any assumption is used it will be implied that it has been checked for the situation considered.

Let us first consider the following simple example.

Example 1. Let $X = (X_1, X_2)$, where X_1 and X_2 are independently distributed according to $\mathcal{N}(\theta_1, 1)$ and $\mathcal{N}(\theta_2, 1)$, respectively. The parameter space Ω_0 is taken to be

$$\Omega_0 = [\theta = (\theta_1, \theta_2) \mid \theta_1 \geqslant 0,\ \theta_2 \geqslant 0],$$

and the subsets Θ_i's are given by

$$\Theta_1 = [\theta \mid \theta_1 > 0, \theta_2 \geqslant 0], \Theta_2 = [\theta \mid \theta_1 = 0, \theta_2 > 0], \Theta_3 = [(0,0)].$$

The UMP similar level β_1 test for testing $H_1 : \theta_1 = 0,\ \theta_2 \geqslant 0$ against $K_1 : \theta_1 > 0,\ \theta_2 \geqslant 0$ has the critical region : $X_1 > c_1$, where

$$\beta_1 = \Phi(-c_1),\quad \Phi(z) = \frac{1}{\sqrt{2\pi}} \int_{-\infty}^{z} e^{-\frac{t^2}{2}}\, dt.$$

The UMP level β_2 test for testing $H_2 : \theta_1 = \theta_2 = 0$ against $K_2 : \theta_1 = 0,\ \theta_2 > 0$ has the critical region : $X_2 > c_2$, where $\beta_2 = \Phi(-c_2)$.

It follows that, for the corresponding multiple decision problem among all rules satisfying Condition A, the following rule maximizes the probabilities of correct decisions :

$$\delta_1^*(x) = 1,\quad \text{if } x_1 > c_1,$$
$$\delta_2^*(x) = 1,\quad \text{if } x_1 \leqslant c_1,\ x_2 > c_2,$$
$$\delta_3^*(x) = 1,\quad \text{if } x_1 \leqslant c_1,\ x_2 \leqslant c_2.$$

For the above problem, we required a rule to satisfy Condition A and used Assumptions 1, 2(a), and 2(c) (Assumption 2(c) is altered so as not to include Assumption 2(b) in its statement.)

Note 1. Unbiased multiple decision rule. In the general setup suppose we define an unbiased rule δ by the following : For every $\theta^* \epsilon \Theta_i$ and every $\theta^{**} \epsilon \Theta_j$

$$E_{\theta^*}[\delta_i(X)] \geqslant E_{\theta^{**}}[\delta_i(X)], \tag{21}$$

$i \neq j$; $i, j = 1, 2, \ldots, k+1$. This definition conforms to Neyman-Pearson definition of an unbiased rule in a two-decision problem. It follows from Corollary 1 that (21) holds for $j > i$ and $\delta = \delta^*$; but nothing can be said when $j < i$ even for $\delta = \delta^*$. In fact, it can be easily checked that (21) is not satisfied for the optimum rule δ^* in Example 1. In multiparameter problems the relation (21) is rarely satisfied, with strict inequality in at least one case.

We shall give a weaker definition of unbiased rule, specially suited for multiparameter problems. Let $\theta = (\theta_1, \theta_2, \ldots, \theta_m)$. We call the ith coordinate of θ *free* in a subset Ω of the parameter space if its value is not completely specified in Ω. To meet the requirement (21), we consider $\theta^* \epsilon \Theta_i$ and $\theta^{**} \epsilon \Theta_j$, $(j \neq i)$ such that if any coordinate of θ is free in both Θ_i and Θ_j then it takes the same value in both θ^* and θ^{**}; the comparison (21) is then made with respect to those coordinates of θ which are either non-free in Θ_i but free in Θ_j or free in Θ_i but non-free in Θ_j. If (21) holds for any rule δ with this choice of θ^* and θ^{**} then we call that rule a *weakly unbiased* rule. It turns out that this requirement is satisfied by optimal rules in many problems including Example 1 above, and the strict inequality in (21) holds in many cases for some values of the parameter.

Note 2. Suppose, in Example 1 we take Ω_0 to be

$$\Omega_0 = [\theta \mid -\infty < \theta_1 < \infty, \ -\infty < \theta_2 < \infty],$$

and the subsets Θ_i as

$$\Theta_1 = [\theta \mid \theta_1 > 0], \quad \Theta_2 = [\theta \mid \theta_1 \leqslant 0, \ \theta_2 > 0],$$

$$\Theta_3 = [\theta \mid \theta_1 \leqslant 0, \quad \theta_2 \leqslant 0].$$

In this case,

$$\Omega_1 = [\theta \mid \theta_1 \leqslant 0], \quad \Omega_2 = \Theta_3.$$

It is known [9] that the UMP unbiased test exists for each of the following hypothesis testing problems : (i) $H_1 : \theta \epsilon \Omega_1$; $K_1 : \theta \epsilon \Theta_1$; (ii) $H_2 : \theta \epsilon \Omega_2$; $K_2 : \theta \epsilon \Theta_2$. But the multiple decision rule δ^* defined by (10), where ϕ_i^*'s aɹe the above two UMP unbiased tests (of suitable levels), does not satisfy Condition A or Condition C. Note that Assumption 1 is also violated in the present situation.

Note 3. Example 1 can be extended to the case where X has the following density function with respect to some σ-finite measure :

$$p(x;\theta) = c(\theta)\exp[\sum_{j=1}^{k} \theta_j T_j(x) + \sum_{i=1}^{m} v_i V_i(x)]G(x),$$

where $\theta = (\theta_1, \ldots, \theta_k, v_1, \ldots, v_m)$. We take

$$\Omega_0 = [\theta \mid \theta_i \geqslant 0, \quad i = 1, \ldots, k],$$

$$\Omega_i = [\theta \mid \theta_1 = \ldots = \theta_i = 0\ ;\ \theta_j \geqslant 0,\ j > i], i = 1, \ldots, k.$$

It follows from [9] that, for every i ($i = 1, \ldots, k$), UMP similar level β_i test ϕ_i^* exists for testing $H_i : \theta \epsilon \Omega_i$ against $K_i : \theta \epsilon \Theta_i = \Omega_{i-1}\Omega_i$. Using these test ϕ_i^* if we define δ^* by (10), then (9) holds for any δ satisfying Condition A. It can be seen [9] that the above optimum rule δ^* is "weakly unbiased."

A special case of the above is the normal univariate analysis of variance problem where the corresponding multiple decision setup is for one degree of freedom analysis. One may take θ_1 to be the least important effect and proceed to more important effects.

Note 4. In the above problem we take Ω_0 to the set of all θ for which θ_i's are reals (not necessarily non-negative). Moreover,

$$\Omega_i = [\theta \mid \theta_1 = \ldots = \theta_i = 0], \quad i = 1, \ldots, k.$$

It is known that, for every i, UMP unbiased level β_i test ϕ_i^* exists for testing $H_i : \theta \epsilon \Omega_i$ against $K_i : \theta \epsilon \Omega_{i-1}\Omega_i$. Using these tests

ϕ_i^* we define δ^* by (10). Then δ^* is optimal in the sense (9) among all rules satisfying Condition C.

Example 2. Let $X_1, \ldots, X_n$ be independently distributed random variables having distributions $N(\theta_1, \sigma^2), \ldots, N(\theta_n, \sigma^2)$, $\theta = (\theta_1, \ldots, \theta_n, \sigma^2)$, and

$$\Omega_0 = [\theta \mid \theta_{m+1} = \ldots = \theta_n = 0, \sigma^2 > 0],$$

where $k \leqslant m < n$. Moreover,

$$\Omega_i = [\theta \mid \theta_1 = \ldots = \theta_i = 0,\ \theta_{m+1} = \ldots = \theta_n = 0,\ \sigma^2 > 0],$$

$i = 1, \ldots, k$. Let $X = (X_1, \ldots, X_n)$, and

$$T_i(X) = X_i, \quad i = 1, \ldots, k,$$

$$T_{k+1}(X) = [X_{k+1}, \ldots, X_m, \sum_{i=1}^{n} X_i^2].$$

It is seen that Assumption 1 is satisfied. Note that this problem is a special case of the problem discussed in Note 4. For testing $H_i : \theta \epsilon \Omega_i$ against $K_i : \theta \epsilon \Omega_{i-1} \Omega_i$ the UMP unbiased level β_i test ϕ_i^* is given by

$$\phi_i^*(x) = 1, \text{ if } \frac{|X_i|(n-m+i-1)^{\frac{1}{2}}}{\sqrt{\sum_{j=1}^{i-1} X_j^2 + \sum_{j=m+1}^{n} X_j^2}} > c_i$$
$$= 0, \text{ otherwise},$$

where c is the upper $\beta_i/2$ quantile point of the t-distribution with $(n-m+i-1)$ degrees of freedom. This test is also UMP among all level β_i tests whose power function depend only on $\frac{|\theta_i|}{\sigma}$ when $\theta \epsilon \Theta_i$. It follows from Theorem 3 that the rule δ^* defined by (10) constructed from the above tests ϕ_i^* is optimal, if we set $h_i(\theta) = |\theta_i/\sigma|$, $i = 1, \ldots, k$.

This problem in a different context is discussed in [1].

Example 3. This example is related to the univariate analysis of variance problem. We consider the canonical form of the analysis of variance problem.

Let $X_1, \ldots, X_{k+2}$ be independently distributed random vectors. The distribution of $X_i(r_i \times 1)$ is the r_i-variate normal distribution

$\mathcal{N}(\theta_i, \sigma^2 I_{r_1})$; $i = 1, \ldots, k+2$. Let $\theta' = (\theta'_1, \ldots, \theta'_{k+2}, \sigma^2)$; θ is assumed to be unknown. Let

$$\Omega_0 = [\theta \,|\, \theta_{k+2} = 0, \quad \sigma^2 > 0],$$

$$\Omega_i = [\theta \,|\, \theta_1 = 0, \ldots, \theta_1 = 0, \theta_{k+2} = 0; \sigma^2 > 0],$$

$i = 1, \ldots, k$. To see that Assumption 1 holds, we take,

$$T_i(X) = X_i, \quad i = 1, \ldots, k$$

$$T_{k+1}(X) = (X_{k+1}, \sum_{j=1}^{k+2} X'_j X_j),$$

where $X' = (X'_i, \ldots, X'_{k+2})$. For testing $H_i : \theta \varepsilon \Omega_i$ against K_i : $\theta \varepsilon \Theta_i = \Omega_{i-1} \Omega_i$ the UMP test among all level β_i tests whose power functions depend only on $\theta'_i \theta_i / \sigma^2$ for $\theta \varepsilon \Theta_i$ is given by

$$\phi^*_i(X) = 1, \text{ if } \frac{X'_i X_i}{\sum_{j=1}^{i-1} X'_j X_j + X'_{k+2} X_{k+2}} > c_i \frac{r_i}{\sum_{j=1}^{i-1} r_j + r_{k+2}}$$

$$= \quad 0, \text{ otherise,}$$

where c_i is the upper β_i quantile point of the F-distribution with r_i and $\left(\sum_{j=1}^{i-1} r_j + r_{k+2} \right)$ degrees of freedom. Let δ^* be the rule defined by (10), where ϕ^*_i are given as above. It follows from Theorem 3 and δ^* is optimal in the class of rules satisfying Conditions A and B, if we set $h_i(\theta) = \theta'_i \theta_i / \sigma^2$ $(i = 1, \ldots, k)$.

Example 4. This example is related to the analysis of variance (Model II) problem. We consider the canonical form [9] of the problem.

Let $X_0, X_1, \ldots, X_{k+1}$ be independently distributed random variables. The distribution of X_0 is $\mathcal{N}(\mu, \sigma^2_1)$ and the distribution of X_i / σ^2_i is χ^2 with r_i degrees of freedom $(i = 1, \ldots, k+1)$; the parameters $\mu, \sigma^2_1, \ldots, \sigma^2_{k+1}$ are unknown but it is assumed that $\sigma^2_1 \geqslant \sigma^2_2 \geqslant \ldots \geqslant \sigma^2_{k+1} > 0$. Let

$$\frac{\mu}{\sigma^2_1} = \tau, \quad \frac{1}{2\sigma^2_i} = \sum_{j=0}^{i-1} \tau_j, \quad i = 1, \ldots, k+1.$$

Thus

$$\tau_0 > 0, \quad \tau_j \geqslant 0, \quad j = 1, \dots, k,$$

$$\sigma_1^2 = \dots = \sigma_i^2 \Longleftrightarrow \tau_1 = \dots = \tau_{i-1} = 0.$$

Let

$$\theta = (\tau, \tau_0, \dots, \tau_k),$$

and

$$\Omega_0 = [\theta \,|\, \tau_0 > 0;\ \tau_j \geqslant 0, \quad j = 1, \dots, k]$$

$$\Omega_i = [\theta \,|\, \tau_0 > 0;\ \tau_1 = \dots = \tau_i = 0;\ \tau_j \geqslant 0, j > i],$$

$i = 1, 2, \dots, k$. To see that Assumption 1 holds, we take

$$T_i = X_{i+1} + \dots + X_{k+1}, \quad i = 1, \dots, k$$
$$T_k = (X_0, X_0^2 + X_1 + \dots + X_k + X_{k+1}).$$

This problem is a particular case of the problem considered in Note 3. The UMP similar level β_i test for testing $H_i : \theta \varepsilon \Omega_i\ \Omega_{i-1}$ is given by

$$\phi_i^*(X) = 1, \text{ if } \frac{X_{i+1}}{\sum_{j=1}^{i} X_j} > c_i \frac{r_{i+1}}{\sum_{j=1}^{i} r_j}$$

$$= \quad 0, \text{ otherwise,}$$

where c_i is the upper β_i quantile point of the F-distribution with r_{i+1} and $\sum_{j=1}^{i} r_j$ degrees of freedom. The rule δ^*, defined by (10) using the above tests ϕ_i^*, is optimal in the class of rules satisfying Condition A.

Example 5. This example is related to the Hotelling-T^2 problem and the problem of discrimination.

Let $\boldsymbol{X}_1, \dots, \boldsymbol{X}_n$ be a random sample from the p-variate normal population $\mathcal{N}_p(\boldsymbol{\mu}, \boldsymbol{\Sigma})$, where $\boldsymbol{\mu}$ and $\boldsymbol{\Sigma}$ are unknown and $\boldsymbol{\Sigma}$ is positive definite. Let

$$\boldsymbol{\gamma}' = (\gamma_1, \dots, \gamma_p) = \boldsymbol{\mu}'\boldsymbol{\Sigma}^{-1},$$

$$\boldsymbol{\mu}' = (\mu_1, \dots, \mu_p), \quad \boldsymbol{\mu}'_{[i]} = (\mu_1, \dots, \mu_i),$$

$$\boldsymbol{\Sigma}_{[i]} = \begin{bmatrix} \sigma_{11} & \cdots & \sigma_{1i} \\ \cdots & \cdots & \cdots \\ \sigma_{i1} & \cdots & \sigma_{ii} \end{bmatrix}, \ \boldsymbol{\Sigma} = [\sigma_{ij}].$$

Define

$$\Delta_i^2 = \boldsymbol{\mu}'_{[i]}\, \boldsymbol{\Sigma}_{[i]}^{-1}\, \boldsymbol{\mu}_{[i]}, \qquad i = 1, \ldots, p,$$

$$\lambda_i = \Delta_i^2 - \Delta_{i-1}^2, \qquad i = 2, \ldots, p$$

$$\lambda_1 = \Delta_1^2.$$

It can be seen that

$$\gamma_{i+1} = \ldots = \gamma_p = 0 \Longleftrightarrow \Delta_p^2 = \Delta_i^2$$
$$\Longleftrightarrow \lambda_{i+1} = \ldots = \lambda_p = 0.$$

Let θ stand for $(\boldsymbol{\mu}, \boldsymbol{\Sigma})$, and

$$\Omega_0 = [\theta \,|\, \lambda_1 \geqslant 0, \ldots, \lambda_p \geqslant 0],$$

$$\Omega_i = [\theta \,|\, \lambda_p = \lambda_{p-1} = \ldots = \lambda_{p-i+1} = 0; \lambda_j \geqslant 0,$$
$$j < p-i+1],$$

$i = 1, \ldots, p$. The density of $\boldsymbol{X} = (\boldsymbol{X}_1, \ldots, \boldsymbol{X}_n)$ is

$$\text{Const. } C(\theta) \exp[-\tfrac{1}{2} \operatorname{tr} \boldsymbol{\Sigma}^{-1} \boldsymbol{U} + n\boldsymbol{\gamma}'\bar{\boldsymbol{X}}],$$

where

$$\boldsymbol{U} = \sum_{\alpha=1}^{n} \boldsymbol{X}_\alpha \boldsymbol{X}'_\alpha, \quad \bar{\boldsymbol{X}} = \frac{1}{n} \sum_{\alpha=1}^{n} \boldsymbol{X}_\alpha.$$

In the general formulation of the multiple decision problem we replace k by p in Sections 2 and 3. To see that Assumption 1 holds, we take

$$T_i = \bar{X}_{p-i+1}, \quad i = 1, \ldots, p$$

$$T_{p+1} = \boldsymbol{U} = [u_{ij}],$$

where $\bar{\boldsymbol{X}}' = (\bar{X}_1, \ldots, \bar{X}_p)$. For testing $H_i : \theta \epsilon \Omega_i$ against $K_i : \theta \epsilon \Omega_{i-1}\, \Omega_i$, the UMP unbiased level β_i test is given by [9]

$$\phi_i^*(T_i, \ldots, T_{p+1}) = i, \text{ if } \quad T_i < C_{i1}(T_{i+1}, \ldots, T_{p+1})$$
$$\text{or} \quad T_i > C_{i2}(T_{i+1}, \ldots, T_{p+1})$$
$$= 0, \text{ otherwise},$$

where

$$E[\phi_i^*(T_i, \ldots, T_{p+1}) \,|\, T_{i+1}, \ldots, T_{p+1}] = \beta_i, \ \theta \epsilon \Omega_i$$

$$E[T_i \phi_i^*(T_i, \ldots, T_{p+1}) \,|\, T_{i+1}, \ldots, T_{p+1}]$$
$$= \beta_i\, E[T_i \,|\, T_{i+1}, \ldots, T_{p+1}], \ \theta \epsilon \Omega_i.$$

The above conditional expectations do not depend on θ, since $T_{i+1}, \ldots, T_p$ is sufficient when $\theta \epsilon \Omega_i$. Let

$$\boldsymbol{S} = \boldsymbol{U} - n\overline{\boldsymbol{X}}\overline{\boldsymbol{X}}' = [s_{ij}],$$

$$\mathbf{S}_{[i]} = \begin{bmatrix} s_{11} & \cdots & s_{1i} \\ \cdots & \cdots & \cdots \\ s_{i1} & \cdots & s_{ii} \end{bmatrix}$$

$$\mathbf{s}'_{[i]} = (s_{i1}, \ldots s_{i,\,i-1}), \quad \overline{\boldsymbol{X}}'_{[i]} = (\overline{X}_1, \ldots, \overline{X}_i)$$

Define

$$Z_i = \frac{\overline{\boldsymbol{X}}_{p-i+1} - \mathbf{s}'_{[p-i+1]}\, \mathbf{S}^{-1}_{[p-i]}\, \overline{\boldsymbol{X}}_{[p-i]}}{[1/n + \overline{\boldsymbol{X}}'_{[p-i]}\mathbf{S}^{-1}_{[p-i]}\, \overline{\boldsymbol{X}}_{[p-i]}]^{\frac{1}{2}}.[\mathbf{s}_{p-i+1,\,p-i+1} - \mathbf{s}'_{[p+1]}\, \mathbf{S}^{-1}_{[p-i]}\mathbf{s}_{[p-i+1]}]^{\frac{1}{2}}}$$

$$w_i = \frac{\overline{\boldsymbol{X}}_{p-i+1} - \mathbf{s}'_{[p-i+1]}\mathbf{S}^{-1}_{[p-i]}\, \overline{\boldsymbol{X}}_{[i]}}{[1/n + \overline{\boldsymbol{X}}'_{[p-i]}\mathbf{S}^{-1}_{[p-i]}\overline{\boldsymbol{X}}_{[p-i]}]^{\frac{1}{2}}[u_{p-i+1,p-i+1} - \boldsymbol{u}'_{[p-i+1]}][\boldsymbol{U}^{-1}_{[p-i]}\boldsymbol{u}_{[p-i+1]}]^{\frac{1}{2}}}$$

where $\boldsymbol{U}_{[i]}$, $\boldsymbol{u}_{[i]}$ are defined in the same way as $\mathbf{S}_{[i]}$ and $\mathbf{s}_{[i]}$ are defined. Let

$$d_i^2 = n\overline{\boldsymbol{X}}_{[i]'}\, \mathbf{S}^{-1}_{[i]}\, \overline{\boldsymbol{X}}_{[i]}, \; i = 1, \ldots, p.$$

Then [10],

$$Z_i^2 = \frac{d^2_{p-i+1} - d^2_{p-i}}{1 + d^2_{p-i}}, \quad i = 1, \ldots, p,$$

$$w_i = \frac{Z_i}{\sqrt{1+Z_i^2}}$$
$$= a_i(T_{i+1}, \ldots, T_{p+1})\overline{\boldsymbol{X}}_{p-i+1} + b_i(T_{i+1}, \ldots, T_{p+1}),$$

where $a_i(T_{i+1}, \ldots, T_{p+1}) > 0$, a_i and b_i being functions of $T_{i+1}, \ldots, T_{p+1}$. When $\theta \epsilon \Omega_i$, the distribution of $Z_i\sqrt{n-p+i-1}$ is Student's t-distribution [10] with $n-p+i-1$ degrees of freedom, and thus Z_i is independent of $T_{i+1}, \ldots, T_{p+1}$ [9]. Hence w_i is independent of $T_{i+1}, \ldots, T_{p+1}$, when $\theta \epsilon \Omega_i$. Applying Theorem 1 of Lehmann [9, p. 361], we find that the UMP unbiased level β_i test for testing $H_i : \theta \epsilon \Omega_i$ against $K_i : \theta \epsilon \Omega_{i-1}\Omega_i$ is given by

$$\phi_i^*(X) = 1, \quad \text{if} \quad |Z_i\sqrt{n-p+i-1}| > c_i$$
$$= 0, \text{ otherwise}$$

where c_i is the upper $\beta_i/2$ quantile point of the t-distribution with $n-p+i-1$ degrees of freedom. (We assume, of course, $n > p$). Let δ^* be the rule defined by (10), where ϕ_i^* are defined as above. It follows from Theorem 4 that δ^* is optimal in the class of rules satisfying Condition C. It is also known [4, 3] that the above test ϕ_i^* is UMP similar level β_i test for testing H_i against K_i among all similar level β_i tests whose power functions depend only on λ_{p-i+1}, Δ_{p-i}^2, when $\theta \epsilon \Theta_i$. If we set $h_i(\theta) = (\lambda_{p-i+1}, \Delta_{p-i}^2)$, then it follows from Theorem 3 that δ^* is optimal in the class of rules satisfying Conditions A and B.

Note 5. The step-down procedure [11] suggested by Roy for testing $\boldsymbol{\mu} = \mathbf{0}$ against $\boldsymbol{\mu} \neq \mathbf{0}$ can be described as follows: Let $\boldsymbol{\Sigma} = \boldsymbol{T}\boldsymbol{T}'$, where $\boldsymbol{T}$ is a lower triangular matrix, and let

$$\boldsymbol{\nu}' = \boldsymbol{\mu}'(\boldsymbol{T}')^{-1} = (\nu_1, \ldots, \nu_p).$$

For any given i $(i = 1, \ldots, p)$ the hypothesis $\nu_i = 0$ is tested against $\nu_i \neq 0$ neglecting the variates following the ith and considering the conditional distribution of the ith variate given the preceding variates; the test is the usual t-test. The hypothesis $\boldsymbol{\mu} = \mathbf{0}$ is accepted if, and only if, all the component hypotheses are accepted. Nothing (except some trivial properties) is known about the test procedures for the component hypothesis, and, in particular, about the over-all performance of the technique. Note that, when $\nu_i = 0$ we do not get any further reduction by sufficient statistics.

Note 6. The multiple decision rule δ^* in Example 5 is weakly unbiased. It is known [3] that $E_\theta[\delta^*(X)]$ involves θ only through $\lambda_1, \lambda_2, \ldots, \lambda_p$. Thus the condition for weak unbiasedness reduces to the following. For very i $(i = 1, \ldots, p)$

$$E_\lambda[\partial_i^*(X)] \leqslant E_{\lambda^*}[\partial_i^*(X)]$$

for every λ and λ^* such that

$$\boldsymbol{\lambda}' = (\lambda_1, \ldots, \lambda_{p-i}, \lambda_{p-i+1}, \ldots, \lambda_p),$$

$$\boldsymbol{\lambda}^{*\prime} = (\lambda_1, \ldots, \lambda_{p-i+1}, 0, \ldots, 0).$$

Recall that

$$\delta_i^*(X) = \phi_j^*(X) \prod_{j=1}^{i-1} [1-\phi_i^*(X)].$$

The desired result follows from the fact that $E_\theta[\phi_j^*(X) \mid \boldsymbol{U}_{[p-j]}, \bar{\boldsymbol{X}}_{[p-j]}]$ increases as λ_{p-j+1} increases, and $\phi_i^*(X)$ involves X only through $\boldsymbol{U}_{[p-i+1]}, \bar{\boldsymbol{X}}_{[p-i+1]}$.

Example 6. Let $\boldsymbol{X}_1, \ldots, \boldsymbol{X}_n$ be a random sample from the p-variate normal population $\mathcal{N}_p(\boldsymbol{\mu}, \boldsymbol{\Sigma})$, where $\boldsymbol{\mu}$ and $\boldsymbol{\Sigma}$ are unknown and $\boldsymbol{\Sigma}$ is known to be positive definite. Let $\boldsymbol{X} = (\boldsymbol{X}_1, \ldots, \boldsymbol{X}_n)$ and let $\boldsymbol{\theta}$ stand for $(\boldsymbol{\mu}, \boldsymbol{\Sigma})$. The density of $\boldsymbol{X}$ is

$$C(\theta)\exp[-\tfrac{1}{2}\mathrm{tr}\, \boldsymbol{\Sigma}^{-1}\boldsymbol{U}+n\boldsymbol{\gamma}'\bar{\boldsymbol{X}}],$$

where $\boldsymbol{U}$, $\bar{\boldsymbol{X}}$ and $\boldsymbol{\gamma}$ are defined in the same way as in Example 5. Let ρ_i^2 be the square of the multiple correlation coefficient between the ith variate and its preceding variates, and let R_i^2 be the corresponding sample multiple correlation coefficient ($i = 2, 3, \ldots, p$). Note that

$$\rho_i^2 = 0, \quad i = p, \ p-1, \ldots, p-m+1$$

$$\Longrightarrow \sigma_{ij} = 0, j \neq i; j = 1, \ldots, p; i = p, p-1, \ldots, p-m+1.$$

$$\Longrightarrow \sigma^{ij} = 0, j \neq i; j = 1, \ldots, p; i = p, p-1, \ldots, p-m+1,$$

where

$\boldsymbol{\Sigma}^{-1} = [\sigma^{ij}]$. Let

$$\Omega_0 = [\theta \mid \rho_i^2 \geqslant 0, i = 2, \ldots, p],$$

$$\Omega_i = [\theta \mid \rho_p^2 = \rho_{p-1}^2 = \ldots = \rho_{p-i+1}^2 = 0; \quad \rho_j^2 \geqslant 0, j < p-i+1].$$

$i = 1, \ldots, p-1$. To apply the results in Sections 2 and 3 we replace k by $p-1$. To see that Assumption 1 holds, we take

$$T_i(X) = (u_{p-i+1,1}, \ldots, u_{p-i+1,p-i}), \ i = 1, \ldots, p-1$$

$$T_p(X) = (u_{11}, \ldots, u_{pp}, \bar{X}_1, \bar{X}_2, \ldots, \bar{X}_p)$$

Consider the problem of testing $H_i : \theta\epsilon\Omega_i$ against $K_i : \theta\epsilon\Omega_{i-1}\Omega_i$ at the level β_i. When $\theta\epsilon\Omega_{i-1}$, a set of sufficient statistics is $\{T_i, T_{i+1}, \ldots, T_p\}$. Consider first the class of transformations given by

$$\boldsymbol{X}_\alpha \rightarrow \boldsymbol{X}_\alpha + \boldsymbol{d}, \quad \alpha = 1, \ldots, n,$$

where $\boldsymbol{d}$ is any real $p\times 1$ vector; secondly, consider the class of transformation given by

$$\boldsymbol{X}_\alpha \rightarrow \left[\begin{array}{c|c} \boldsymbol{g}_{p-i} & \boldsymbol{0} \\ \hline \boldsymbol{0} & \boldsymbol{h}_i \end{array}\right] \boldsymbol{X}_\alpha, \quad \alpha = 1, \ldots, n$$

where g_{p-i} is any nonsingular $(p-i)\times(p-i)$ matrix and h_i is any nonsingular $i\times i$ diagonal matrix. These transformations leave the problem of testing H_i against K_i invariant. When $\theta\epsilon\Omega_{i-1}$, a maximal invariant in the space of sufficient statistics T_i, $T_{i+1}, \ldots, T_p$ is R^2_{p-i+1}. Hence [9], the UMP invariant level β_i test for testing H_i against K_i is given by

$$\begin{aligned} \phi_i^*(X) &= 1, \text{ if } R^2_{p-i+1} > C_i \\ &= 0, \text{ otherwise} \end{aligned}$$

where

$$\frac{C_i}{1-C_i} \; \frac{p-i}{n-p+i-1}$$

is the upper β_i quantile point of the F-distribution with $p-i$ and $n-p+i-1$ degrees of freedom. It can be shown [5, 9] that the above test is the UMP test among all level β_i tests whose power functions depend only on ρ^2_{p-i+1}. We set $h_i(\theta) = \rho^2_{p-i+1}$ in in Condition 2. It follows from Theorem 3 that the multiple decision rule δ^* defined by (10) using the above ϕ_i^* is optimal in the class of all rules satisfying Conditions A and B.

Note 7. The multiple decision rule δ^* in Example 6 is weakly unbiased. This follows from the fact that

$$E_\theta[\phi_i^*(X) \mid \boldsymbol{U}_{[p-i]}, \overline{\boldsymbol{X}}_{[p-i]}] \geqslant \beta_i \quad \text{if} \quad \rho^2_{p-i+1} > 0,$$

and $\phi_j^*(X)$ depends on X only through $\boldsymbol{U}_{[p-j+1]}\, \overline{\boldsymbol{X}}_{[p-j+1]}$.

Note 8. Roy and Bargmann [12] considered the step-down technique for testing $\sigma_{ij} = 0$, $i \neq j$.

Many other examples, e.g., equality of dispersion matrices, MANOVA, simultaneous confidence bounds, etc., can be tackled in the above light and weak unbiasedness of some of these rules may be proved as above.

Acknowledgment.

The author is thankful to Professor T. W. Anderson, Dr. J. Sethuraman, and Professor C. R. Rao for some fruitful discussions.

References

[1] Anderson, T. W. (1962). "The Choice of the Degree of a Polynomial Regression and A Multiple Decision Problem," *Ann. Math. Statist.*, **33**, 255–265.

[2] Anderson, T. W. (1963). "Determination of the Order of Dependence in Normally Distributed Time Series," *Time Series Analysis*, edited by M. Rosenblatt, John Wiley and Sons, New York.

[3] Giri, N. (1964). "On the Likelihood Ratio Test of Normal Multivariate Testing Problem," *Ann. Math. Statist.*, **35**, 181–189.

[4] Giri, N. (1965). "On the Likelihood Ratio Test of a Normal Multivariate Testing Problem, II," *Ann. Math. Statist.*, **36**, 1061–1065.

[5] Giri, N. and Kiefer, J. (1964). "Minimax Character of the R^2-Test in the Simplest Case," *Ann. Math. Statist.*, **35**, 1475–1490.

[6] Hogg, R. V. (1961). "On the Resolution of Statistical Hypotheses," *J. Amer. Statist. Assoc.*, **56**, 978–989.

[7] Lehmann, E. L. (1957). "A Theory of Some Multiple Decision Problems, I," *Ann. Math. Statist.*, **28**, 1–25.

[8] Lehmann, E. L. (1957). "A Theory of Some Multiple Decision Problems, II," *Ann. Math. Statist.*, **28**, 547–572.

[9] Lehmann, E. L. (1959). *Testing Statistical Hypotheses*, John Wiley and Sons, New York.

[10] Rao, C. R. (1952). *Advanced Statistical Methods in Biometric Research*, John Wiley and Sons, New York.

[11] Roy, J. (1958). "Step-Down Procedure in Multivariate Analysis," *Ann. Math. Statist.*, **29**, 1177–1187.

[12] Roy, S. N. and Bargmann, R. E. (1958). "Tests of Multiple Independence and the Associated Confidence Bounds," *Ann. Math. Statist.*, **29**, 491–503.

(Received Jan. 18, 1966.)

CHAPTER 12

On the Relation between Union Intersection and Likelihood Ratio Tests

K. R. GABRIEL,* *Hebrew University, Jerusalem, and The University of North Carolina at Chapel Hill*

1. INTRODUCTION AND SUMMARY

Tests of the intersection of a set of hypotheses can be constructed from tests of the individual hypotheses in a number of ways. The best known principles of such construction are the likelihood ratio (LR for short) and the union-intersection (UI for short) methods [6].

It is shown that if the tests of the individual hypotheses are LR tests, the induced LRUI critical region for the intersection hypothesis is contained in the LR critical region for that hypothesis, provided the same critical value is used in all the tests. A condition for equality of LRUI and LR statistics and tests is given. This condition is shown to hold for univariate ANOVA but not for MANOVA, except for hypotheses of rank one.

* This research was supported in part by funds in research grant R01-GM-12868-01 from the National Institutes of Health.

The relation between the two types of statistics is shown to bear on that between Tukey's and Scheffé's methods of multiple comparisons. It bears in a similar way on multivariate methods based, respectively, on LRUI and LR MANOVA statistics.

An earlier discussion of the relation of LRUI and LR tests by Darroch and Silvey is examined critically. A criterion for *reasonableness* of LRUI statistics, as given by these authors, is shown to be of limited use.

2. THE RELATION BETWEEN TWO TYPES OF TESTS

Consider a random variable Y having the density function $f(y; \theta)$ with respect to a σ-finite measure μ, where the parameter θ is assumed to lie in a set ω. Consider a class Ω of hypotheses ω_i: $\theta \epsilon \omega_i$ (ω_i is used to denote both a set in the parameter space, and the hypothesis restricting the parameters to it) where the index i belongs to an index set I. A class Ω of hypotheses ω_i, $i \epsilon I$, is said to be an *L-class* of hypotheses, if

a) $\bigcap_{i \epsilon I} \omega_i \neq \phi$, the empty set,

and

b) the distribution of the LR statistic

$$Z_i = \sup_{\theta \epsilon \omega} f(y; \theta) / \sup_{\theta \epsilon \omega_i} f(y; \theta) \tag{1}$$

is the same for all $\theta \epsilon \omega_i$ and all $i \epsilon I$. Each ω_i of such a class may be tested against the alternative $\theta \notin \omega_i$ by using a critical region $(Z_i > \zeta)$, where ζ is used simultaneously for all $i \epsilon I$. Such a class of critical regions will be called a *LR class* and will be denoted by $\mathfrak{C}_\zeta$. Clearly all these tests have the same size

$$\alpha_\zeta = P_\theta(Z_i > \zeta),\ \theta \epsilon \omega_i. \tag{2}$$

For the intersection hypothesis

$$\omega_0 = \bigcap_{i \epsilon I} \omega_i$$

the LR statistic is

$$Z_0 = \sup_{\theta\epsilon\omega} f(y;\theta) / \sup_{\theta\epsilon\omega_0} f(y;\theta).$$

The LR test of ω_0 having the critical region $C_{0\zeta} = (Z_0 > \zeta)$ will be referred to as a *simultaneous* LR (SLR) test for ω_0 with respect to the LR class of tests $\mathscr{C}_\xi$.

If C_i is a critical region for testing ω_i then the class $\mathscr{C} = \{C_i \mid i\epsilon I\}$ of all critical regions corresponding to an L-class $\Omega = \{\omega_i \mid i\epsilon I\}$ will induce a critical region

$$C_{\mathscr{C}} = \bigcup_{i\epsilon I} C_i$$

for testing the intersection hypothesis ω_0. This principle of constructing tests for ω_0 is called the UI principle; it was suggested by Roy (extended type I test-Section 4 of [6]). The UI critical region induced by an LR class $\mathscr{C}_\zeta$ is denoted by $C_{\Omega\zeta}$ and is given by

$$C_{\Omega\zeta} = \left(\sup_{i\epsilon I} Z_i > \zeta\right).$$

The statistic

(3) $$Z_\Omega = \sup_{i\epsilon I} Z_i$$

will be called the LRUI statistic corresponding to the hypothesis ω_0.

Lemma 1. *For an L-class of hypotheses* $\Omega = \{\omega_i \mid i\epsilon I\}$, *the LRUI statistic* Z_Ω, *corresponding to the intersection hypothesis* $\omega_0 = \bigcap_{i\epsilon I} \omega_i$ *is not greater than the LR statistic* Z_0 *corresponding to* ω_0, *i.e.,*

(4) $$Z_\Omega \leqslant Z_0,$$

with equality in (4) *holding, i.e.,*

(5) $$Z_\Omega = \max_{i\epsilon I} Z_i = Z_0,$$

if, and only if, for (*almost* (μ)) *every sample point* y

(6) $$\sup_{\theta\epsilon\omega_i} f(y;\theta) = \sup_{\theta\epsilon\omega_0} f(y;\theta)$$

holds for some $i = i_y \epsilon I$.

Proof. Since $\omega_0 \subset \omega_i$ for all $i \epsilon I$,

$$\sup_{\theta \epsilon \omega_i} f(y; \theta) \geqslant \sup_{\theta \epsilon \omega_0} f(y; \theta).$$

Hence, for all $i \epsilon I$

$$Z_i \leqslant Z_0.$$

In particular,

$$Z_\Omega = \sup_{i \epsilon I} Z_i \leqslant Z_0.$$

If, for (almost) every y, (6) holds for some $i = i_y \epsilon I$, then

$$Z_{i_y} \leqslant Z_\Omega \leqslant Z_0 = Z_{i_y}.$$

Conversely if $Z_0 = \max_{i \epsilon I} Z_i$ (a.e.), for (almost) every y there exists some $i_y \epsilon I$ for which

$$Z_{i_y} = Z_\Omega = Z_0.$$

Theorem 1. *Let* Ω *be an L-class of hypotheses* ω_i, $i \epsilon I$. *For the intersection hypothesis* $\omega_0 = \bigcap_{i \epsilon I} \omega_i$, *let* $C_{\Omega\zeta}$ *be the LRUI critical region induced by the LR-class of regions* $\mathfrak{C}_\zeta$ *corresponding to* Ω, *and let* $C_{0\zeta}$ *be the SLR critical region for* ω_0 *with respect to* $\mathfrak{C}_\zeta$. *Then*

$$C_{\Omega\zeta} \subseteq C_{0\zeta}. \tag{7}$$

The two regions are μ*-equivalent when*

$$\mu(Z_\Omega < \zeta \leqslant Z_0) = 0, \tag{8}$$

in particular, when $Z_\Omega = Z_0$ *(a.e.).*

Theorem 1 follows immediately from Lemma 1.

If each of the component hypotheses in Ω is tested at the level α then we call α the *per-comparison* level. The level for a test of the intersection hypotheses ω_0 corresponding to Ω is called the *experiment-wise* level.

Corollary 1. *Let* α_ζ *be the per-comparison level of the LR-class of tests* $\mathfrak{C}_\zeta$ *corresponding to an L-class of hypotheses. Let* $\alpha_{\Omega\zeta}$ *and*

$\alpha_{0\zeta}$ be the experiment-wise levels of the UI region $C_{\Omega\zeta}$ and the SLR critical region $C_{0\zeta}$, respectively. Then

$$\alpha_{\Omega\zeta} \leqslant \alpha_{0\zeta}$$

and the same inequality holds for the powers of the two tests against any alternative.

Theorem 2. *Let $\mathfrak{C}_\zeta$ be the LR-class of tests corresponding to an L-class Ω with per-comparison level α_ξ. Let $\alpha_{\Omega\xi'}$, be the level of the LRUI critical region induced by $\mathfrak{C}_{\zeta'}$ and $\alpha_{0\zeta}$ be the level of SLR critical region with respect to $\mathfrak{C}_\zeta$. Then*

$$\alpha_{\Omega\zeta'} = \alpha_{0\zeta} \Rightarrow \zeta' \leqslant \zeta \tag{9}$$

Proof. By Corollary 1, we have

$$\alpha_{0\zeta} = \alpha_{\Omega\zeta'} \leqslant \alpha_{0\zeta'}$$

which implies $\zeta' \leqslant \zeta$.

Each of the LRUI test and SLR test for the intersection hypothesis ω_0 can be used to make inferences on the component hypotheses ω_i of Ω_0. If the LRUI test with critical region $C_{\Omega\zeta} = (Z_\Omega > \zeta)$ accepts ω_0 then we accept each ω_i and if it rejects ω_0 then we accept ω_i if $Z_i > \zeta$ and reject it otherwise. Similar inferences on ω_i can be based on the SLR test with critical region $(Z_0 > \zeta)$.

Corollary 2. *Let Ω be an L-class of hypotheses ω_i, $i \epsilon I$. Consider two classes of LR-tests for Ω with per-comparison levels α and α', respectively. Further, consider the SLR test for $\omega_0 = \bigcap_{i \epsilon I} w_i$ using the same critical value as the α-level LR tests of ω_i, and the LRUI test of ω_0 induced by the L-class of α'-level tests of Ω. Let these two tests have the same* experiment-wise *level α_0, then*

i) *if any ω_i is rejected by the SLR test then it is also rejected by the LRUI test,*

ii) $\alpha \leqslant \alpha'$,

iii) $\beta_i \leqslant \beta_i'$;

where β_i and β_i' are the powers of the two tests for ω_i as induced by the SLR test and the LRUI test, respectively, against any particular alternative.

Proof. Let α and α' correspond to the critical values ζ and ζ' respectively. Thus $\alpha = \alpha_{\zeta}$ and $\alpha' = \alpha_{\zeta'}$. Also, $\alpha_{\Omega\zeta} = \alpha_0$ and $\alpha_{0\zeta'} = \alpha_0$, giving

$$\alpha_{\Omega\zeta} = \alpha_{0\zeta'} \Rightarrow \zeta' \leqslant \zeta,$$

so that

(10) $$(Z_i > \zeta) \Rightarrow (Z_i > \zeta'),$$

establishing (i). Taking probabilities of the two events in (10), (ii) and (iii) follow.

3. ON THE RELATION BETWEEN THE METHODS OF TUKEY AND SCHEFFÉ

Let $\bar{y}_1, \bar{y}_2, \ldots, \bar{y}_k$ be the means of k independent samples, each of size n_0, from normal distributions with expectations $\mu_1, \mu_2, \ldots, \mu_k$, respectively, and let s^2 be the *within* estimate of the common variance, which has $k(n_0-1)$ degrees of freedom. Consider the class Ω of $\binom{k}{2}$ hypotheses

(11) $$\omega_{jg} : \mu_j = \mu_g, \text{ where } j \neq g = 1, 2, \ldots, k$$

whose intersection is the hypothesis

(12) $$\omega_0 : \mu_1 = \mu_2 = \ldots = \mu_k.$$

The maximum likelihoods under the model ω, the intersection hypothesis ω_0, and any particular hypothesis ω_{jg}, are well known to be, respectively,

$$\sup_{\omega} f(y) = (2\pi e/n)^{-n/2} \, [k(n_0-1)s^2]^{-n/2}$$

$$\sup_{\omega_0} f(y) = (2\pi e/n)^{-n/2} \, [k(n_0-1)s^2 + n_0 \sum_{j=1}^{k} (\bar{y}_j - \sum_{g=1}^{k} \bar{y}_g/k)^2]^{-n/2}$$

and

$$\sup_{\omega_{jg}} f(y) = (2\pi e/n)^{-n/2} [k(n_0-1)s^2 + n_0(\bar{y}_j - \bar{y}_g)^2/2]^{-n/2}.$$

where $n = kn_0$, the total number of observations.

The LR statistic for ω_{jg} becomes

$$Z_{jg} = [1+n_0(\bar{y}_j-\bar{y}_g)^2/2k(n_0-1)s^2]^{n/2}$$

and $[k(n_0-1)(Z_{jg}^{2/n}-1)]^{1/2} = (\bar{y}_j-\bar{y}_g)(n_0/2)^{1/2}/s$ is known to have a t distribution with $k(n_0-1)$ degrees of freedom under ω_{jg}. Hence the LR statistics for all $\omega_{jg} \epsilon \Omega$ are equally distributed, so that Ω is an L-class and each ω_{jg} may be tested at level α by use of the critical region

(13) $$(k(n_0-1)(Z_{jg}^{2/n}-1) > t_{1-\alpha/2}^2)$$

where $t_{1-\alpha/2}$ is the upper $\alpha/2$ point of the t distribution with $k(n_0-1)$ d.f.

The LR statistic for ω_0 is

$$Z_0 = [1+n_0 \sum_{j=1}^{k} [\bar{y}_j - \sum_{g=1}^{k} \bar{y}_g/k]^2/k(n_0-1)s^2]^{n/2}$$

so that the SLR test has critical region

(14) $$(k(n_0-1)(Z_0^{2/n}-1) > t_{1-\alpha/2}^2),$$

where $k(n_0-1)(Z_0^{2/n}-1)$ is the ANOVA ratio which is known to have an F distribution with $k-1$ and $k(n_0-1)$ degrees of freedom under ω_0.

On the other hand, the LRUI statistic for ω_0 is

$$Z_\Omega = \max_{j \neq g} [1+n_0(\bar{y}_j-\bar{y}_g)^2/2k(n_0-1)s^2]^{n/2}$$

so that the LRUI test has critical region

(15) $$(k(n_0-1)(Z_\Omega^{2/n}-1) > t_{1-\alpha/2}^2)$$

where

$$[2k(n_0-1)(Z_\Omega^{2/n}-1)]^{1/2} = \max_{j \neq g} \sqrt{n_0}(\bar{y}_j-\bar{y}_g)/s$$

is the ANOVA range statistic which is known to have the studentized range distribution for k means and $k(n_0-1)$ degrees of freedom under ω_0.

Now, from the expressions for the likelihoods, above, one sees that a sample point y satisfies condition (6) only if, for some j' and g'

$$\sum_{j=1}^{k} \left[\bar{y}_j - \sum_{g=1}^{k} \bar{y}_g/k\right]^2 = (\bar{y}_{j'} - \bar{y}_{g'})^2/2.$$

But this occurs only if all other means $\bar{y}_j$ ($j \neq j'$ and $j \neq g'$) are exactly equal to $(\bar{y}_{j'}+\bar{y}_{g'})/2$, an event of probability zero. Hence (6) is not satisfied and the inequalities of Section 2 are strict.

In addition to testing each ω_{jg} of Ω by means of the LR test (13) one may test ω_0 either by means of the SLR test (14) using the usual F ratio of *between* to *within* sums of squares, or by means of the LRUI test (15) which uses the studentized range statistic. From Theorem 1 and Corollary 1 it is clear that for any given *per-comparison* level for each ω_{jg}, the induced F ratio test will reject ω_0 whenever the induced studentized range test does, and hence the level and power of the former will always be larger than those of the latter.

Multiple comparisons techniques start from the other end. Tukey proposed an α_0 *experiment-wise* level studentized range test of ω_0 and simultaneously the use of tests of each ω_{jg} with the same critical value [8]. From Corollary 2 it follows that a particular ω_{jg} that is rejected by Scheffé's technique must also be rejected by Tukey's, but not *vice versa*. Hence, the *per-comparison* error rate for any ω_{jg} is lower with Scheffé's technique than with Tukey's, and so is the power against any alternative to ω_{jg}. Numerical comparisons showing these relations for error rates have been given before [7], [8] and [3].

4. EQUIVALENCE OF THE STATISTICS IN GENERALIZED ANOVA

A sample vector $\boldsymbol{y}$ is considered, having n independent component variables of common but unknown variance. Under the model ω the expectations are orthogonal to an error space $V(\boldsymbol{E})$ generated by the columns of a matrix $\boldsymbol{E}$ of rank ν, i.e.,

$$\omega: \mathcal{E}\boldsymbol{y} \perp V(\boldsymbol{E}).$$

The variance is estimated as

$$s^2 = \| \hat{\boldsymbol{y}}(\boldsymbol{E}) \|^2/\nu$$

where $\hat{\boldsymbol{y}}(\boldsymbol{E})$ indicates the projection of $\boldsymbol{y}$ on $V(\boldsymbol{E})$.

A linear hypothesis ω_0 requires, in addition, that the expectations be also orthogonal to the columns of a further matrix $\boldsymbol{H}$ of rank q whose columns are orthogonal to those of $\boldsymbol{E}$. The combination of model and hypothesis may be written

(16) $$\omega_0 : \mathcal{E}\boldsymbol{y} \perp V(\boldsymbol{E}, \boldsymbol{H}).$$

This hypothesis ω_0 may be obtained as the intersection of all hypotheses on single combinations of columns of $\boldsymbol{H}$. Thus, $\omega_{\boldsymbol{a}}$ is taken as

(17) $$\omega_{\boldsymbol{a}}: \mathcal{E}\boldsymbol{y} \perp V(\boldsymbol{E}, \boldsymbol{Ha})$$

where $\boldsymbol{a}$ is any vector, and clearly $\omega_0 = \bigcap_{\boldsymbol{a}} \omega_{\boldsymbol{a}}$

The maximum likelihoods are well known to be

$$\sup_{\omega} f(\boldsymbol{y}) = (2\pi e/n)^{-n/2} \,||\hat{\boldsymbol{y}}(\mathrm{E})||^{-n},$$

$$\sup_{\omega_0} f(\boldsymbol{y}) = (2\pi e/n)^{-n/2} \,||\hat{\boldsymbol{y}}(\boldsymbol{E},\boldsymbol{H})||^{-n}$$

and

$$\sup_{\omega_{\boldsymbol{a}}} f(\boldsymbol{y}) = (2\pi e/n)^{-n/2} \,||\hat{\boldsymbol{y}}(\boldsymbol{E}, \boldsymbol{Ha})||^{-n}.$$

Hence, for any $\omega_{\boldsymbol{a}}$ the LR statistic is

$$Z_{\boldsymbol{a}} = (||\hat{\boldsymbol{y}}(\boldsymbol{E}, \boldsymbol{Ha})||^2/||\hat{\boldsymbol{y}}(\boldsymbol{E})^2)||^{n/2}$$

$$= (1+||\hat{\boldsymbol{y}}(\boldsymbol{Ha})||^2/\nu s^2)^{n/2}$$

as $\boldsymbol{Ha}$ is orthogonal to $\boldsymbol{E}$. Now, for any $\boldsymbol{a}$, $||\hat{\boldsymbol{y}}(\boldsymbol{Ha})||/s$ has the t distribution with ν degrees of freedom if $\omega_{\boldsymbol{a}}$ holds. Hence Ω is an L-class and the argument of Section 2 applies.

To see that condition (6) is satisfied, note that the projection of any vector on a space $V(\boldsymbol{H})$ is greater than that on any subspace $V(\boldsymbol{Ha})$ unless $\boldsymbol{Ha}$ is itself proportional to $\hat{\boldsymbol{y}}(\boldsymbol{H})$. Since the class Ω includes $\omega_{\boldsymbol{a}}$'s for all vectors $\boldsymbol{a}$ it includes also such $\boldsymbol{a}$'s proportional to $\hat{\boldsymbol{y}}(\boldsymbol{H})$ for which

$$\hat{\boldsymbol{y}}(\boldsymbol{H}\boldsymbol{a}) = \hat{\boldsymbol{y}}(\boldsymbol{H}).$$

For such a vector $\boldsymbol{a}$,

$$\begin{aligned}\|\hat{\boldsymbol{y}}(\boldsymbol{E}, \boldsymbol{H}\boldsymbol{a})\|^2 &= \|\hat{\boldsymbol{y}}(\boldsymbol{E})\|^2 + \|\hat{\boldsymbol{y}}(\boldsymbol{H}\boldsymbol{a})\|^2 \\ &= \|\hat{\boldsymbol{y}}(\boldsymbol{E})\|^2 + \|\hat{\boldsymbol{y}}(\boldsymbol{H})\|^2 \\ &= \|\hat{\boldsymbol{y}}(\boldsymbol{E}, \boldsymbol{H})\|^2,\end{aligned}$$

so that

$$\sup_{\omega_{\boldsymbol{a}}} f(\boldsymbol{y}) = \sup_{\omega_0} f(\boldsymbol{y})$$

and condition (6) is satisfied.

From Lemma 1 it follows that when the class Ω consists of the hypotheses for all possible linear hypotheses $\omega_{\boldsymbol{a}}$, the LR statistic is equal to the LRUI statistic for ANOVA and the SLR and LRUI tests are equivalent. In particular, for one-way ANOVA with equal sample sizes, the F ratio statistic is equal to the LRUI statistic when ω_0 is considered as the intersection of hypotheses on all contrasts in the μ's. Compare this with the result of the previous section, which showed that when the intersection is taken only over pairwise contrasts the LRUI statistic is the studentized range and is smaller than the corresponding F ratio statistic. This example points out the difference in over-all tests and multiple comparisons procedures that result from different choices of the class Ω whose intersection is a given ω_0. If one is interested in inferences on certain individual ω_i's, as well as on ω_0, it would seem appropriate to use a LRUI test based on the intersection of those ω_i's only. One should not augment the critical region by union with critical regions for further ω_i's which are of no interest. For instance, in a multiple comparisons procedure in one-way ANOVA with equal sample sizes in which one is interested only in pairwise comparisons, Tukey's method would seem more suitable, whereas Scheffé's is to be preferred if inferences on other contrasts are also required.

5. ON THE DIFFERENCE OF MANOVA TESTS

Consider a sample of size n from a p-variate normal distribution with common but unknown variance and a linear model for expectations. Denoting the n observations on p variables $y^{(1)}, y^{(2)}, \ldots, y^{(p)}$ by the $(n \times p)$ matrix $\boldsymbol{Y}$, the linear multivariate model corresponding to the univariate model of Section 4, may be stated as

$$\mathcal{E}\, \boldsymbol{Y}^{(j)} \perp V(\boldsymbol{E}) \quad \text{for all } j = 1, 2, \ldots, p$$

where the superscript (j) indicates the jth column of the matrix. Equivalently, the model may be stated as

$$\omega : \mathcal{E}\, \boldsymbol{Yc} \perp V(\boldsymbol{E}) \quad \text{for all vectors } \boldsymbol{c}.$$

Under this model, a rank q hypothesis, corresponding to that of the univariate hypothesis (16), above, can be written

(18) $$\omega_0 : \mathcal{E}\, \boldsymbol{Yc} \perp V(\boldsymbol{E}, \boldsymbol{H}) \text{ for all } \boldsymbol{c}.$$

Again, consider for any $\boldsymbol{a}$ the hypothesis

$$\omega_{\boldsymbol{a}} : \mathcal{E}\, \boldsymbol{Yc} \perp V(\boldsymbol{E}, \boldsymbol{Ha}) \text{ for all } \boldsymbol{c}.$$

Clearly the intersection of all these hypotheses for all possible $\boldsymbol{a}$'s is again ω_0 i.e.,

$$\omega_0 = \bigcup_{\boldsymbol{a}} \omega_{\boldsymbol{a}}$$

Now consider also the rank q univariate hypotheses

(19) $$\omega_{0(\boldsymbol{c})} : \mathcal{E}\boldsymbol{Yc} \perp V(\boldsymbol{E}, \boldsymbol{H})$$

for a given variable $\boldsymbol{Yc}$. Again, the intersection of these univariate hypotheses over all possible $\boldsymbol{c}$'s is ω_0, i.e.,

$$\omega_0 = \bigcap_{\boldsymbol{c}} \omega_{0(\boldsymbol{c})}.$$

Thus ω_0 can be regarded either as the intersection of all single rank p-variate hypotheses $\omega_{\boldsymbol{a}}$ or as the intersection of all univariate hypotheses $\omega_{0(\boldsymbol{c})}$ of rank q. The latter form can be used to demonstrate the relation between the LR statistic for ω_0—Wilks's product of characteristic roots [9]—and the UI intersection statistic based on the LR statistics for univariate hypotheses $\omega_{0(\boldsymbol{c})}$—Roy's maximum characteristic root [6].

To compute the likelihood for any given $\mathbf{c}$, transform the p variables $\boldsymbol{y}' = (y^{(1)}, y^{(2)}, \ldots, y^{(p)})$, to p independent variables $\boldsymbol{X}^{(g)}$ $g = 1, 2, \ldots, p$ such that

$$\mathbf{c}'\boldsymbol{y} = \boldsymbol{X}^{(1)}.$$

This will not change the likelihood but will allow it to be computed as the product of the likelihoods of p independent variables (as in [1], pp. 180–181). Thus

$$\sup_{\omega} f(\boldsymbol{Y}) = (2\pi e/n)^{-n/2} \|\hat{\boldsymbol{X}}^{(1)}(\boldsymbol{E})\|^{-n} \prod_{g=2}^{p} (2\pi e/n)^{-n/2} \|\hat{\boldsymbol{X}}^{(g)}(\boldsymbol{E})\|^{-n}$$

$$\sup_{\omega_0} f(\boldsymbol{Y}) = (2\pi e/n)^{-n/2} \|\hat{\boldsymbol{X}}^{(1)}(\boldsymbol{E}, \boldsymbol{H})\|^{-n} \prod_{g=2}^{p} (2\pi e/n)^{-n/2} \|\hat{\boldsymbol{X}}^{(g)}(\boldsymbol{E},\boldsymbol{H})\|^{-n}$$

$$\sup_{\omega_{0(\mathbf{c})}} f(\boldsymbol{Y}) = (2\pi e/n)^{-n/2} \|\hat{\boldsymbol{X}}^{(1)}(\boldsymbol{E},\boldsymbol{H})\|^{-n} \prod_{g=2}^{p} (2\pi e/n)^{-n/2} \|\hat{\boldsymbol{X}}^{(g)}(\boldsymbol{E})\|^{-n}.$$

Hence, corresponding to hypothesis $\omega_{0(\mathbf{c})}$ the LR statistic is

$$\begin{aligned} Z_{0(\mathbf{c})} &= (\|\hat{\boldsymbol{X}}^{(1)}(\boldsymbol{E},\boldsymbol{H})\|^2 / \|\hat{\boldsymbol{X}}^{(1)}(\boldsymbol{E})\|^2)^{n/2} \\ &= (1 + \|\hat{\boldsymbol{X}}^{(1)}(\boldsymbol{H})\|^2 / \|\hat{\boldsymbol{X}}^{(1)}(\boldsymbol{E})\|^2)^{n/2}. \end{aligned}$$

For any single $\mathbf{c}$ the statistic $\|\hat{\boldsymbol{X}}^{(1)}(\boldsymbol{H})\|^2 / \|\hat{\boldsymbol{X}}^{(1)}(\boldsymbol{E})\|^2$ has q/ν times the F distribution with q and ν degrees of freedom where ν is the rank of $\boldsymbol{E}$. Hence all LR statistics $Z_{0(\mathbf{c})}$ are equally distributed and Ω is an L-class.

Next, to check condition (6), compare $\sup_{\omega_0} f(\boldsymbol{Y})$ with $\sup_{\omega_{0(\mathbf{c})}} f(\boldsymbol{Y})$. Their ratio is

$$\begin{aligned} \sup_{\omega_0} f(\boldsymbol{Y}) / \sup_{\omega_{0(\mathbf{c})}} f(\boldsymbol{Y}) &= \prod_{g=2}^{p} (\|\hat{\boldsymbol{X}}^{(g)}(\boldsymbol{E},\boldsymbol{H})\|^2 / \|\hat{\boldsymbol{X}}^{(g)}(\boldsymbol{E})\|^2)^{n/2} \\ &= \prod_{g=2}^{p} (1 + \|\hat{\boldsymbol{X}}^{(g)}(\boldsymbol{H})\|^2 / \|\hat{\boldsymbol{X}}^{(g)}(\boldsymbol{E})\|^2)^{n/2} \end{aligned}$$

This ratio is greater than one unless $\|\hat{\boldsymbol{X}}^{(g)}(\boldsymbol{H})\|^2 = 0$ for $g = 2, 3, \ldots, p$, i.e., $\boldsymbol{X}^{(2)}, \boldsymbol{X}^{(3)}, \ldots, \boldsymbol{X}^{(p)}$ are all orthogonal to $V(\boldsymbol{H})$.

If $q = 1$, $\boldsymbol{c}$ may be chosen in such a manner that all variables except $\boldsymbol{X}^{(1)}$ are orthogonal to $V(\boldsymbol{H})$. If $q > 1$, the $p-1$ variables $\boldsymbol{X}^{(2)}, \ldots, \boldsymbol{X}^{(p)}$ cannot be confined with probability one to the rank $n-q$ space orthogonal to $V(\boldsymbol{H})$. Hence condition (6) holds only if $q = 1$.

It follows from Lemma 1 that with a MANOVA hypothesis of rank 1 the LRUI and LR statistics coincide—and, indeed, both are known to be equivalent to Hotelling's T^2 statistic—whereas, with a hypothesis of rank greater than one, the two methods induce different statistics. By (4), the LRUI statistic, Roy's maximum characteristic root, will be less than Wilks's likelihood ratio statistic, i.e.,

$$\sup_{\boldsymbol{c}} [1+\|\hat{\boldsymbol{X}}^{(1)}(\boldsymbol{H})\|^2/\|\hat{\boldsymbol{X}}^{(1)}(\boldsymbol{E})\|^2]^{n/2} \leqslant \prod_{g=1}^{p} (1+\|\hat{\boldsymbol{X}}^{(g)}(\boldsymbol{H})\|^2/\|\hat{\boldsymbol{X}}^{(g)}(\boldsymbol{E})\|^2)^{n/2}.$$

This comparison of the two types of MANOVA statistics bears on multivariate tests and multiple comparisons methods, as the comparison of the studentized range and F ratio ANOVA statistics was seen to bear on univariate procedures in Section 3, above. Simultaneous test procedures (STP's [4]) for MANOVA may be constructed by means of either the SLR statistics or the LRUI statistics based on univariate F ratios. These procedures permit simultaneous testing of the over-all MANOVA hypothesis ω_0 (18), each univariate ANOVA hypothesis $\omega_{0(\boldsymbol{c})}$ (19), and the MANOVA hypothesis on any subset of the variables. As in Section 4, the STP further permits simultaneous testing of univariate and multivariate hypotheses of lower rank, down to rank one (as $\omega_{\boldsymbol{a}}$ (17) of Section 4). The appropriate statistics are all compared with the same fixed critical value.

Arguing as in Section 3, we note that if both the simultaneous SLR test procedure and the simultaneous LRUI test procedure have the same *per-variable* significance level for univariate hypotheses $\omega_{0(\boldsymbol{c})}$, then the MANOVA error rates must be higher for the SLR procedure than for the LRUI procedure. Conversely, if the *experiment-wise* error rates for the multivariate ω_0

are fixed to be the same in both procedures, the univariate levels, and powers, will be larger with the LRUI procedure than with the SLR procedure. For a given over-all α_0 the LRUI procedure should be more powerful as regards alternatives to the univariate hypotheses $\omega_{0(c)}$. As in the previous comparison of ANOVA LRUI and SLR statistics for all pairwise contrasts (Section 3), we might conclude here too that the tests themselves have these power properties. Thus the maximum characteristic root statistic would be more powerful than the likelihood ratio statistic against alternatives of deviations from the hypothesis on a single variable. This agrees with what has often been surmised about the relative properties of these two statistics and tends to be confirmed by Monte Carlo studies [5].

6. A NOTE ON A CRITERION PROPOSED BY DARROCH AND SILVEY

Another approach to the comparison of UI and LR tests is that of Darroch and Silvey [2]. These authors consider only finite classes Ω of hypotheses, for which they are concerned whether "the induced (i.e., LRUI) test compares reasonably well with this direct likelihood ratio test". ([2], p. 555). They consider the induced test to be poor if it is possible, in some situations, to have a high probability that all Z_i's are near unity, and hence non-significant, while Z_0 is large, and hence significant.

Darroch and Silvey suggest

$$Z_0 = \prod_{i \in I} Z_i, \tag{20}$$

as "just the kind of relationship we are seeking" ([2], p. 559) to preclude the possibility of the LRUI test's being "poor" in the above sense. Indeed, if (20) holds, and if all Z_i's are near unity, Z_0 cannot be very large for finite Ω.

By the same argument, however, the criterion

$$Z_0 \leqslant \prod_{i \in I} Z_i \tag{21}$$

should be at least as good as (20). In fact, a strict inequality would make the Z_0 even closer to the Z_i's and would make it even less likely that ω_0 be rejected by the LR test when it is not rejected by the LRUI test. Indeed, the case when no contradiction is possible at all, i.e., when the two statistics are equal,

$$Z_0 = Z_\Omega = \max_{\Omega} Z_i,$$

is excluded by criterion (20), except in the unlikely event that all but one of the Z_i's are equal unity. Hence Darroch and Silvey's criterion excludes cases in which the LRUI test would be less "poor" than when (20) holds, and even the ideal case when the two tests were equal, i.e., when the LRUI test is equivalent to the SLR test. It would seem that criterion (20) is of limited usefulness.

As an example, consider again the one way ANOVA case discussed in Section 3, above, with regard to all $\binom{k}{2}$ pairwise comparisons of means. The F ratio LR statistic of (14) for ω_0—equality of all expectations—is not the product of the t statistics of (13) for pairwise comparisons, but can readily be seen to be less than that product (for $k > 2$). Hence criterion (20) does not hold, but a strict inequality holds in (21). And indeed if all pairwise comparison t statistics are small, the SLR F statistic cannot be large either. This shows that the studentized range statistic is not "poor" in Darroch and Silvey's sense, even though their criterion is not satisfied.

Acknowledgments

I am grateful to my colleagues E. Peritz and Ester Samuel for their critical reading of an earlier draft of this paper. I am also obliged to J. N. Darroch and S. D. Silvey for clarifying certain points concerning their joint paper.

References

[1] Anderson, T. W. (1958). *An Introduction to Multivariate Statistical Analysis*, John Wiley and Sons, New York.

[2] Darroch, J. N. and Silvey, S. D. (1963). "On Testing More Than One Hypothesis," *Ann. Math. Statist.*, **34**, 555–567.

[3] Gabriel, K. R. (1964). "A Procedure for Testing the Homogeneity of All Sets of Means in Analysis of Variance," *Biometrics*, **20**, 459–477.

[4] Gabriel, K. R. "Multiple Comparisons in Multivariate Analysis of Variance : Location of Culprit Variables and Deviant Populations," In preparation.

[5] Genizi, A. (1965). "On the power of MANOVA tests," To be published.

[6] Roy, S. N. (1957). *Some Aspects of Multivariate Analysis*, John Wiley and Sons, New York.

[7] Scheffé, Henry. (1953). "A Method for Judging All Contracts in the Analysis of Variance," *Biometrika*, **40**, 87–104.

[8] Scheffé, Henry. (1959). *The Analysis of Variance*, John Wiley and Sons, New York.

[9] Wilks, S. S. (1932). "Certain Generalizations in the Analysis of Variance," *Biometrika*, **24**, 471–494.

(*Received Jan. 1, 1966.*)

CHAPTER 13

Linear Models in Multivariate Analysis

LEON JAY GLESER[1], *Johns Hopkins University, and*
INGRAM OLKIN[2], *Stanford University*

1. INTRODUCTION

Historically, two classes of multivariate linear models have received wide attention—the classical multivariate linear regression model and the "growth curves" model. In the classical multivariate linear regression model we observe a random $n \times p$ matrix $\boldsymbol{Y}$ whose rows are independently distributed, each having a p-variate normal distribution with unknown covariance matrix $\boldsymbol{\psi}$, and having means $\mathcal{E}(\boldsymbol{Y}) = \boldsymbol{X\beta}$, where $\boldsymbol{X}$ is a known $n \times m$ matrix of rank $m \leqslant n$, and $\boldsymbol{\beta}$ is an unknown $m \times p$ matrix of regression parameters. In addition to the problem of estimating $\boldsymbol{\beta}$ and $\boldsymbol{\psi}$, various hypothesis testing problems have been of interest in connection with this model: (i) to test $\boldsymbol{\beta} = \boldsymbol{0}$, (ii) to test $\boldsymbol{\beta}_1 = \boldsymbol{0}$, where $\boldsymbol{\beta} = (\boldsymbol{\beta}_1, \boldsymbol{\beta}_2)$, (iii) to test $\boldsymbol{\beta}_1 = \boldsymbol{0}$ *given* $\boldsymbol{\beta}_2 = \boldsymbol{0}$, or (iv) to test the General Linear Hypothesis $H : \boldsymbol{C\beta D} = \boldsymbol{\xi}_0$, where $\boldsymbol{C}$, $\boldsymbol{D}$, and $\boldsymbol{\xi}_0$ are known matrices.

[1] Supported in part by the Air Force Office of Scientific Research under Contract No. AF 49(638)-1302.

[2] Supported in part by the National Science Foundation Grant GP-3837.

In the "growth curves" model (so named because its principal application has been in the analysis of growth curves), we observe n independent replications of the random $1 \times p$ vector $\boldsymbol{y}$ having a multivariate normal distribution with unknown covariance matrix $\boldsymbol{\psi}$ and unknown mean $\mathcal{E}(\boldsymbol{y}) = \beta \boldsymbol{X}^*$, where $\boldsymbol{X}^*$ is a known $q \times p$ matrix of rank $q \leqslant p$ and $\boldsymbol{\beta}$ is an unknown $1 \times q$ vector of regression parameters. Problems of estimating $\boldsymbol{\beta}$ and $\boldsymbol{\psi}$, and of testing hypotheses of the form $H : \beta \boldsymbol{D} = \boldsymbol{\xi}_0$ have been considered by Cochran and Bliss (1948), Rao (1946, 1959), Olkin and Shrikhande (1954), Giri (1964), and Gleser and Olkin (1964). A k-sample version of this model has been discussed by Rao (1961), and Gleser and Olkin (1966).

In both classes of multivariate linear models, it is useful to transform the random variables (and the parameters) to an equivalent canonical form with a simple structure, from which problems of finding optimal estimators or tests, questions of invariance, etc., can be more easily handled. For example, the canonical form which simplifies the classical multivariate linear regression model (Hsu (1941), Anderson (1958)) has the structure : (1) we observe a random matrix $\boldsymbol{U} : m \times p$ whose rows are independently distributed, each with a p-variate normal distribution having unknown covariance matrix $\boldsymbol{\Sigma}$, and having means

$$\mathcal{E}(\boldsymbol{U}) = \mathcal{E}\begin{pmatrix} \boldsymbol{U}_{11} & \boldsymbol{U}_{12} \\ \boldsymbol{U}_{21} & \boldsymbol{U}_{22} \end{pmatrix} = \begin{pmatrix} \boldsymbol{\Theta}_{11} & \boldsymbol{\Theta}_{12} \\ \boldsymbol{\Theta}_{21} & \boldsymbol{\Theta}_{22} \end{pmatrix},$$

(2) there is an independent estimate $\boldsymbol{V}/n$ of $\boldsymbol{\Sigma}$, where $\boldsymbol{V}$ has the Wishart distribution[1] $\mathcal{W}(\boldsymbol{\Sigma}; p, n-m)$, (3) the hypothesis (iv) is $H : \boldsymbol{\Theta}_{12} = \boldsymbol{0}$.

A canonical form similar in nature to the canonical form for the classical multivariate linear regression model has been utilized in the one-sample "growth curves" model by Olkin and Shrikhande (1954) to obtain the non-null distribution of the likelihood

---

[1] We write $\boldsymbol{V} \sim \mathcal{W}(\boldsymbol{\Sigma}; p, n)$ to mean that the density function of $\boldsymbol{V}$ is

$$p(\boldsymbol{V}) = C(p, n) |\boldsymbol{\Sigma}|^{-n/2} |\boldsymbol{V}|^{(n-p-1)/2} \exp -\tfrac{1}{2} \operatorname{tr} \boldsymbol{\Sigma}^{-1} \boldsymbol{V}, \quad \text{with}$$

$$C(p, n) = \left[2^{pn/2} \pi^{p(p-1)/4} \prod_{1}^{p} \Gamma\left(\frac{n-i+1}{2}\right) \right]^{-1}$$

ratio test, and by Giri (1964) in order to show that the likelihood ratio test for the hypothesis $H : \beta\boldsymbol{\psi}^{-1}\boldsymbol{D} = \boldsymbol{\xi}_0$ is UMP invariant similar. This canonical form is also used by Gleser and Olkin (1964) to find maximum likelihood estimators of β and $\boldsymbol{\psi}$. In a later paper, Gleser and Olkin (1966) use a generalization of this canonical form to investigate the k-sample "growth curves" model, obtaining simultaneous maximum likelihood estimators for the regression parameters $\boldsymbol{\beta}^{(1)}, \ldots, \boldsymbol{\beta}^{(k)}$, of the k samples and for the common covariance matrix $\boldsymbol{\psi}$, the distributions of these estimators, the likelihood ratio test statistic for testing the equality of the regression parameters $\boldsymbol{\beta}^{(1)}, \beta^{(2)}, \ldots, \boldsymbol{\beta}^{(k)}$, the distribution (exact null and asymptotic non-null) of this likelihood ratio test statistic, and the relevant invariance theory for the testing problem.

Recently a more general form of the classical multivariate linear regression model (which subsumes the one and k-sample "growth curves" model) has been considered by Potthoff and Roy (1964) and subsequently by Rao (1965). The Potthoff-Roy model is the following : we observe an $n \times p$ random matrix $\boldsymbol{Y}$ whose n rows are independently distributed, each with a p-variate normal distribution having unknown covariance matrix $\boldsymbol{\psi}$. The means of the elements of $\boldsymbol{Y}$ are assumed to be of the form

$$(1.1) \qquad \mathcal{E}(\boldsymbol{Y}) = \boldsymbol{X}_1\boldsymbol{\beta}\boldsymbol{X}_2,$$

where $\boldsymbol{X}_1$ is a known $n \times m$ matrix of rank $m \leqslant n$, $\boldsymbol{X}_2$ is a known $q \times p$ matrix of rank $q \leqslant p$, and $\boldsymbol{\beta}$ is an unknown $m \times q$ matrix of parameters taking on values in mq-dimensional space.

It is apparent that the classical multivariate linear regression model is a special case of this model with $\boldsymbol{X}_2 = \boldsymbol{I}_p$, $q = p$. The one-sample "growth curves" model is a special case with $m = 1$ and $\boldsymbol{X}_1 = (1, 1, \ldots, 1)' : n \times 1$, while the k-sample model has $m = k$ and $\boldsymbol{X}_1$ block diagonal, $\boldsymbol{X}_1 = \text{diag}\,(\boldsymbol{X}_{11}, \ldots, \boldsymbol{X}_{1k})$, where the

$$\boldsymbol{X}_{1i} = (1, 1, \ldots, 1)' : n_i \times 1, \quad \sum_{i=1}^{k} n_i = n.$$

The problem of estimating $\boldsymbol{\beta}$ and $\boldsymbol{\psi}$ in connection with the model (1.1) has been discussed by Potthoff and Roy (1964) and Rao (1965), but these authors do not explicitly derive the maximum likelihood estimators. Of course in the special cases of the model (1.1) (i.e., the classical and growth curve models), the maximum likelihood estimators and their distributions are known.

The problem of testing the null hypothesis

$$H_0 : \boldsymbol{X}_3\boldsymbol{\beta}\boldsymbol{X}_4 = \boldsymbol{\xi}_0, \tag{1.2}$$

where $\boldsymbol{X}_3$ is a known $u \times m$ matrix of rank $u \leqslant m$, where $\boldsymbol{X}_4$ is a known $q \times v$ matrix of rank $v \leqslant q$, and where $\boldsymbol{\xi}_0$ is a given $u \times v$ matrix of constants, has been considered by Potthoff and Roy (1964) and Rao (1965). Potthoff and Roy discuss *ad hoc* procedures for testing H_0 while Rao indicates how to obtain a conditional likelihood ratio test. The unconditional likelihood ratio test is not explicitly obtained by either author, although it can be shown that Rao's conditional likelihood ratio test is also the unconditional likelihood ratio test.

The purpose of this paper is to obtain a canonical form (Section 2) for the Potthoff-Roy model (1.1) from which the maximum likelihood estimators $\hat{\boldsymbol{\beta}}$ and $\hat{\boldsymbol{\psi}}$ of $\boldsymbol{\beta}$ and $\boldsymbol{\psi}$ (Section 5) and the likelihood ratio test of H_0 (Section 4) can easily be obtained. In the process, we also consider transformations which leave the testing problems invariant (Section 3), obtain the exact null and asymptotic (as $n \rightarrow \infty$) non-null distribution of the likelihood ratio statistic for testing H_0 (Section 4), and determine the distributions of $\hat{\boldsymbol{\beta}}$ and $\hat{\boldsymbol{\psi}}$ (Section 5). Finally, the canonical model which is obtained in Section 2 suggests other related models and testing problems which are briefly discussed in Section 6.

2. A CANONICAL FORM

We repeatedly use the following well-known lemma (e.g., see MacDuffee, (1946), p. 77).

Lemma 2.1. If A is a $k \times l$ matrix of rank r, with $r \leqslant k \leqslant l$, then there exists a non-singular $k \times k$ matrix $\boldsymbol{M}$ and an orthogonal

$l \times l$ matrix $\boldsymbol{\Gamma}$ such that

$$A = M \begin{pmatrix} I_r & 0 \\ 0 & 0 \end{pmatrix} \boldsymbol{\Gamma}.$$

(N.B., the above representation holds with $\begin{pmatrix} I_r & 0 \\ 0 & 0 \end{pmatrix}$ replaced by $\left(P \begin{matrix} I_r & 0 \\ 0 & 0 \end{matrix}\right) Q$, P and Q permutation matrices.)

An application of Lemma 2.1 to X_1 and X_2 yields the representations

$$X_1 = \boldsymbol{\Gamma}_1 \begin{pmatrix} I_m \\ 0 \end{pmatrix} T_1, \qquad X_2 = T_2(I_q, 0)\boldsymbol{\Gamma}_2,$$

so that model (1.1) becomes

$$\mathcal{E}(Y) = \boldsymbol{\Gamma}_1 \begin{pmatrix} I_m \\ 0 \end{pmatrix} T_1 \boldsymbol{\beta} T_2(I_q, 0)\boldsymbol{\Gamma}_2,$$

where $T_1 : m \times m$ and $T_2 : q \times q$ are non-singular matrices, $\boldsymbol{\Gamma}_1 : n \times n$ and $\boldsymbol{\Gamma}_2 : p \times p$ are orthogonal matrices.

Since $\boldsymbol{\Gamma}_1$ and $\boldsymbol{\Gamma}_2$ are known, the matrix

$$Y^* = \boldsymbol{\Gamma}_1' Y \boldsymbol{\Gamma}_2'$$

is observable, and knowledge of the values of Y^* is equivalent to knowledge of the values of Y. The new observation matrix Y^* is a random matrix whose rows are independently distributed, each with a p-variate normal distribution having covariance matrix

$$\boldsymbol{\psi}^* \equiv \boldsymbol{\Gamma}_2 \boldsymbol{\psi} \boldsymbol{\Gamma}_2',$$

and means

$$\mathcal{E}(Y^*) = \begin{pmatrix} I_m \\ 0 \end{pmatrix} T_1 \boldsymbol{\beta} T_2(I_q, 0) \equiv \begin{pmatrix} I_m \\ 0 \end{pmatrix} \boldsymbol{\beta}^*(I_q, 0) = \begin{pmatrix} \boldsymbol{\beta}^* & 0 \\ 0 & 0 \end{pmatrix}. \tag{2.1}$$

Since T_1 and T_2 are known and non-singular, the parameterizations $\boldsymbol{\beta}$ and $\boldsymbol{\beta}^*$ are equivalent. Partition Y^* into

$$Y^* = \begin{pmatrix} Y_1^* \\ Y_2^* \end{pmatrix}; \; Y_1^* : m \times p, \; Y_2^* : (n-m) \times p.$$

Since $\boldsymbol{Y}_1^*$ and $\boldsymbol{Y}_2^*$ are independently distributed, with $\mathcal{E}(\boldsymbol{Y}_1^*)=(\boldsymbol{\beta}^*\boldsymbol{0})$, $\mathcal{E}(\boldsymbol{Y}_2^*) = \boldsymbol{0}$, the pair

$$(\boldsymbol{Y}_1^*, \boldsymbol{Y}_2^{*\prime}\boldsymbol{Y}_2^*) \equiv (\boldsymbol{Y}_1^*, \boldsymbol{S})$$

is a sufficient statistic for $(\boldsymbol{\beta}^*, \boldsymbol{\psi}^*)$ or for $(\boldsymbol{\beta}, \boldsymbol{\psi})$. The $p\times p$ matrix $\boldsymbol{S}$ has the Wishart distribution, $\mathcal{W}(\boldsymbol{\psi}^*;\ p, n-m)$, with parameters $n-m$, p, and $\mathcal{E}\boldsymbol{V} = (n-m)\boldsymbol{\psi}^*$. This distribution is non-singular when $n-m \geqslant p$; we make the assumption that this is the case.

The hypothesis H_0 becomes $H_0 : \boldsymbol{X}_3\boldsymbol{T}_1^{-1}\boldsymbol{\beta}^*\boldsymbol{T}_2^{-1}\boldsymbol{X}_4 = \boldsymbol{\xi}_0$. An application of Lemma 2.1 to $\boldsymbol{X}_3\boldsymbol{T}_1^{-1}$ and $\boldsymbol{T}_2^{-1}\boldsymbol{X}_4$ yields

$$\boldsymbol{X}_3\boldsymbol{T}_1^{-1} = \boldsymbol{T}_3(\boldsymbol{I}_u, \boldsymbol{0})\boldsymbol{\Gamma}_3, \quad \boldsymbol{T}_2^{-1}\boldsymbol{X}_4 = \boldsymbol{\Gamma}_4\begin{pmatrix}\boldsymbol{0}\\ \boldsymbol{I}_v\end{pmatrix}\boldsymbol{T}_4,$$

where $\boldsymbol{T}_3 : u\times u$ and $\boldsymbol{T}_4 : v\times v$ are non-singular matrices, $\boldsymbol{\Gamma}_3 : m\times m$ and $\boldsymbol{\Gamma}_4 : q\times q$ are orthogonal matrices. Thus, H_0 becomes

$$H_0 : \boldsymbol{T}_3(\boldsymbol{I}_u, \boldsymbol{0})\,\boldsymbol{\Gamma}_3\,\boldsymbol{\beta}^*\boldsymbol{\Gamma}_4\begin{pmatrix}\boldsymbol{0}\\ \boldsymbol{I}_v\end{pmatrix}\boldsymbol{T}_4 = \boldsymbol{\xi}_0,$$

or equivalently,

$$H_0 : (\boldsymbol{I}_u, \boldsymbol{0})\boldsymbol{\Gamma}_3\,\boldsymbol{\beta}^*\boldsymbol{\Gamma}_4\begin{pmatrix}\boldsymbol{0}\\ \boldsymbol{I}_v\end{pmatrix} = \boldsymbol{\xi}_0^* \equiv \boldsymbol{T}_3^{-1}\,\boldsymbol{\xi}_0\,\boldsymbol{T}_4^{-1}.$$

Since $\boldsymbol{\Gamma}_3$ and $\boldsymbol{\Gamma}_4$ are non-singular and known, the random matrices

$$\text{(2.2)} \qquad \boldsymbol{Z} = \boldsymbol{\Gamma}_3\boldsymbol{Y}_1^*\begin{pmatrix}\boldsymbol{\Gamma}_4 & \boldsymbol{0}\\ \boldsymbol{0} & \boldsymbol{I}_{p-q}\end{pmatrix}, \; \boldsymbol{V} = \begin{pmatrix}\boldsymbol{\Gamma}_4' & \boldsymbol{0}\\ \boldsymbol{0} & \boldsymbol{I}_{p-q}\end{pmatrix}\boldsymbol{S}\begin{pmatrix}\boldsymbol{\Gamma}_4 & \boldsymbol{0}\\ \boldsymbol{0} & \boldsymbol{I}_{p-q}\end{pmatrix},$$

are observable and equivalent to observing $\boldsymbol{Y}_1^*$ and $\boldsymbol{S}$, respectively. The pair $(\boldsymbol{Z}, \boldsymbol{V})$ is a one-to-one onto transformation of $(\boldsymbol{Y}_1^*, \boldsymbol{S})$ and is thus a sufficient statistic for $(\boldsymbol{\beta}^*, \boldsymbol{\psi}^*)$ and consequently for $(\boldsymbol{\beta}, \boldsymbol{\psi})$. If we define

$$\boldsymbol{\Sigma} = \begin{pmatrix}\boldsymbol{\Gamma}_4' & \boldsymbol{0}\\ \boldsymbol{0} & \boldsymbol{I}_{p-q}\end{pmatrix}\boldsymbol{\psi}^*\begin{pmatrix}\boldsymbol{\Gamma}_4 & \boldsymbol{0}\\ \boldsymbol{0} & \boldsymbol{I}_{p-q}\end{pmatrix},$$

then :

(i) $\boldsymbol{Z} : m\times p$ has rows which are independently distributed, each with a p-variate normal distribution having covariance matrix $\boldsymbol{\Sigma}$ and means

$$\mathcal{E}(Z) = (\Theta, 0); \quad \Theta = \Gamma_3\, \beta^* \Gamma_4 : m \times q,$$

(ii) V has a Wishart distribution with parameters $n-m$, p, and $\mathcal{E}(V) = (n-m)\Sigma$,

(iii) V and Z are independently distributed.

Since Γ_3 and Γ_4 are known and non-singular, the parameterizations (β^*, ψ^*) and (Θ, Σ) are equivalent. In terms of the new model, which we call the canonical model, the null hypothesis becomes

$$H_0 : (I_u, 0) \begin{pmatrix} \Theta_{11} & \Theta_{12} \\ \Theta_{21} & \Theta_{22} \end{pmatrix} \begin{pmatrix} 0 \\ I_v \end{pmatrix} = \Theta_{12} = \xi_0^*, \tag{2.3}$$

where Θ is partitioned as indicated with $\theta_{11} : u \times q-v$. Without loss of generality (since ξ^* is a known matrix of constants), we can assume that $\xi_0^* = 0$.

Thus, the problem of estimating β and Ψ is equivalent to the problem of estimating Θ and Σ, and the problem of testing the hypothesis $H_0 : X_3 \beta X_4 = \xi_0$ reduces to one of testing the null hypothesis $\Theta_{12} = 0$, where inference (for either estimation or testing) is based on the sufficient statistic (Z, V).

Remarks on Rank and Identifiability.

We have assumed in this paper that the matrices X_1, X_2, X_3, X_4 have full rank. Suppose now that X_1 has rank $r_1 < m$, X_2 has rank $r_2 < q$, X_3 has rank $r_3 < u$, and X_4 has rank $r_4 < v$. Using Lemma 2.1, we can repeat the series of transformations discussed above. After the first transformation, the new variables Y^* have the same distributional form and covariances as above, but the means given in (2.1) become instead

$$\mathcal{E}(Y^*) = \begin{pmatrix} I_{r_1} & 0 \\ 0 & 0 \end{pmatrix} \quad \beta^* \quad \begin{pmatrix} I_{r_2} & 0 \\ 0 & 0 \end{pmatrix}.$$

Note that certain of the parameters in the matrix β^* do not appear in the expected value of Y^*. Since β and β^* are equivalent parameterizations, this means that in the model (1.1) some of the elements of β are non-identifiable. This causes no difficulty for hypothesis testing if the identifiable parameters are the only elements of $\boldsymbol{\beta}$ appearing in the null hypothesis. Even then, however, there is an unnecessary over-parameterization, and the matrix $\boldsymbol{\beta}$ may be reduced to a matrix of those parameters which are identifiable (with corresponding changes in the $\boldsymbol{X}_1$ and $\boldsymbol{X}_2$ matrices). The new $\boldsymbol{X}_1$ and $\boldsymbol{X}_2$ matrices are then of full rank. For the problem of estimating $\boldsymbol{\beta}$ and $\boldsymbol{\Psi}$, the non-identifiability of some of the elements of $\boldsymbol{\beta}$ may be resolved by imposing additional linear restrictions on $\boldsymbol{\beta}$ so that the number of free parameters equals $r_1 r_2$, or by reparameterizing to a set of identifiable parameters. It is then possible to redefine $\boldsymbol{X}_1$ and $\boldsymbol{X}_2$ so that these matrices have full rank. A discussion of such techniques for the classical regression model with $\boldsymbol{\Psi} = \sigma^2 \boldsymbol{I}$ appears in Scheffé (1959).

Suppose that $\boldsymbol{X}_1$ and $\boldsymbol{X}_2$ are redefined so as to be of full rank, then $\boldsymbol{Z}$ and $\boldsymbol{V}$ have the same form as in (2.2), but the null hypothesis (2.3) becomes

$$H_0 : \begin{pmatrix} \boldsymbol{I}_{r_3} & 0 \\ 0 & 0 \end{pmatrix} \begin{pmatrix} \boldsymbol{\Theta}_{11} & \boldsymbol{\Theta}_{12} \\ \boldsymbol{\Theta}_{21} & \boldsymbol{\Theta}_{22} \end{pmatrix} \begin{bmatrix} 0 & 0 \\ 0 & \boldsymbol{I}_{r_4} \end{bmatrix} = \begin{bmatrix} 0 & \boldsymbol{\Theta}_{12} \\ 0 & 0 \end{bmatrix} = \boldsymbol{\xi}_0^*,$$

where $\boldsymbol{\Theta}_{11} : r_1 \times (q - r_4)$, $\boldsymbol{\Theta}_{22} : (m-u) \times r_4$. Thus, either there is an inconsistency in the equations defined by H_0, or the null hypothesis simply becomes $\boldsymbol{\Theta}_{12} =$ a constant. It follows that the form of the hypothesis could have been simplified so as to make $r_3 = u$ and $r_4 = v$. Of course, the ranks of $\boldsymbol{X}_3$ and $\boldsymbol{X}_4$ are irrelevant for the problem of estimating $\boldsymbol{\beta}$ and $\boldsymbol{\psi}$, since no matter what the ranks of $\boldsymbol{X}_3$ and $\boldsymbol{X}_4$, the parameterizations $(\boldsymbol{\beta}, \boldsymbol{\psi})$ and $(\boldsymbol{\Theta}, \boldsymbol{\Sigma})$ are equivalent.

Concerning the assumptions $n \geqslant m \geqslant u$, $p \geqslant q \geqslant v$, if $m > n$, then the number of free parameters exceeds the number of observations, so that reasonable estimates of *both* $\boldsymbol{\beta}$ and $\boldsymbol{\psi}$ are unattainable. If $q > p$, then some of the elements of $\boldsymbol{\beta}$ are non-identifiable. If $m < u$ or $q < v$, we can reduce to a canonical model $(\boldsymbol{Z}, \boldsymbol{V})$, but the hypothesis takes the form

$$H_0 : \begin{pmatrix} 0 & \boldsymbol{\Theta}_{12} \\ 0 & 0 \end{pmatrix} = \boldsymbol{\xi}_0^*,$$

which implies that the original hypothesis is either inconsistent or redundant. In the latter case, a non-redundant statement of the hypothesis has $\boldsymbol{X}_3$ and $\boldsymbol{X}_4$ with dimensions $n \leqslant m$, $v \leqslant q$.

3. INVARIANCE THEORY FOR THE HYPOTHESIS TESTING PROBLEM

The elements of the canonical model developed in Section 2 are :

$$\boldsymbol{Z} = \begin{pmatrix} \boldsymbol{Z}_{11} & \boldsymbol{Z}_{12} & \boldsymbol{Z}_{13} \\ \boldsymbol{Z}_{21} & \boldsymbol{Z}_{22} & \boldsymbol{Z}_{23} \end{pmatrix}, \qquad \boldsymbol{V} = \begin{pmatrix} \boldsymbol{V}_{11} & \boldsymbol{V}_{12} & \boldsymbol{V}_{13} \\ \boldsymbol{V}_{21} & \boldsymbol{V}_{22} & \boldsymbol{V}_{23} \\ \boldsymbol{V}_{31} & \boldsymbol{V}_{32} & \boldsymbol{V}_{33} \end{pmatrix},$$

where $\boldsymbol{Z}_{11} : m_1 \times p_1$, $\boldsymbol{Z}_{22} : m_2 \times p_2$, $\boldsymbol{V}_{11} : p_1 \times p_1$, $\boldsymbol{V}_{22} : p_2 \times p_2$;

$$\mathscr{E}(\boldsymbol{Z}) = (\boldsymbol{\Theta}, \boldsymbol{0}) = \begin{pmatrix} \boldsymbol{\Theta}_{11} & \boldsymbol{\Theta}_{12} & \boldsymbol{0} \\ \boldsymbol{\Theta}_{21} & \boldsymbol{\Theta}_{22} & \boldsymbol{0} \end{pmatrix}, \quad \frac{\mathscr{E}(\boldsymbol{V})}{n-m} = \boldsymbol{\Sigma} = \begin{pmatrix} \boldsymbol{\Sigma}_{11} & \boldsymbol{\Sigma}_{12} & \boldsymbol{\Sigma}_{13} \\ \boldsymbol{\Sigma}_{21} & \boldsymbol{\Sigma}_{22} & \boldsymbol{\Sigma}_{23} \\ \boldsymbol{\Sigma}_{31} & \boldsymbol{\Sigma}_{32} & \boldsymbol{\Sigma}_{33} \end{pmatrix},$$

where $(\boldsymbol{\Theta}, \boldsymbol{0})$ and $\boldsymbol{\Sigma}$ are partitioned in the manner of $\boldsymbol{Z}$ and $\boldsymbol{V}$, respectively. For simplicity of notation, we let $m_1 = u$, $m_2 = m-u$, $p_1 = q-v$, $p_2 = v$, and $p_3 = p-q$. Recall that in this model the random matrices $\boldsymbol{Z}$ and $\boldsymbol{V}$ are independently distributed, $\boldsymbol{V}$ has a Wishart distribution $\mathscr{W}(\boldsymbol{\Sigma}; p, n-m)$, and the rows of $\boldsymbol{Z}$ are independently distributed, each with a p-variate normal distribution and covariance matrix $\boldsymbol{\Sigma}$. We wish to test $H_0 : \boldsymbol{\Theta}_{12} = \boldsymbol{0}$ versus alternatives $\boldsymbol{\Theta}_{12} \neq \boldsymbol{0}$. Let

$$\boldsymbol{A} = \begin{pmatrix} \boldsymbol{A}_{11} & \boldsymbol{0} & \boldsymbol{0} \\ \boldsymbol{A}_{21} & \boldsymbol{A}_{22} & \boldsymbol{0} \\ \boldsymbol{A}_{31} & \boldsymbol{A}_{32} & \boldsymbol{A}_{33} \end{pmatrix}, \quad \boldsymbol{F} = \begin{pmatrix} \boldsymbol{F}_{11} & \boldsymbol{0} & \boldsymbol{0} \\ \boldsymbol{F}_{21} & \boldsymbol{F}_{22} & \boldsymbol{0} \end{pmatrix},$$

where the partitioning of $\boldsymbol{A}$ is in the manner of $\boldsymbol{V}$ and $\boldsymbol{\Sigma}$, and the partitioning of $\boldsymbol{F}$ is in the manner of $\boldsymbol{Z}$. Let $\mathscr{A}$ be the group of non-singular matrices of form $\boldsymbol{A}$ and $\mathscr{F}$ the group of matrices of form $\boldsymbol{F}$. Then the transformation $(\boldsymbol{Z}, \boldsymbol{V}) \rightarrow (\boldsymbol{Z}\boldsymbol{A}+\boldsymbol{F}, \boldsymbol{A}'\boldsymbol{V}\boldsymbol{A})$ in the sample space induces the transformation $\boldsymbol{\Sigma} \rightarrow \boldsymbol{A}'\boldsymbol{\Sigma}\boldsymbol{A}$ and

$$\begin{pmatrix} \mathbf{\Theta}_{11} & \mathbf{\Theta}_{12} & \mathbf{0} \\ \mathbf{\Theta}_{21} & \mathbf{\Theta}_{22} & \mathbf{0} \end{pmatrix} \to \begin{pmatrix} \mathbf{\Theta}_{11} & \mathbf{\Theta}_{12} & \mathbf{0} \\ \mathbf{\Theta}_{21} & \mathbf{\Theta}_{22} & \mathbf{0} \end{pmatrix} \mathbf{A}+\mathbf{F} = \begin{pmatrix} \boldsymbol{\eta}_{11} & \boldsymbol{\eta}_{12} & \mathbf{0} \\ \boldsymbol{\eta}_{21} & \boldsymbol{\eta}_{22} & \mathbf{0} \end{pmatrix},$$

in the parameter space. Since $\boldsymbol{\eta}_{12} = \mathbf{\Theta}_{12}\mathbf{A}_{22}$ is $\mathbf{0}$ if and only if $\mathbf{\Theta}_{12} = \mathbf{0}$, the group $\mathcal{A} \times \mathcal{F}$ leaves the problem of testing $\mathbf{\Theta}_{12} = \mathbf{0}$ invariant.

Lemma 3.1. *If $u \leqslant v$, $m \leqslant p-q$, then the maximal invariant in the sample space under the group $\mathcal{A} \times \mathcal{F}$ is the pair*

$$(3.1) \qquad g_1(\mathbf{Z}, \mathbf{V}) \equiv (\mathbf{Z}_{12}, \mathbf{Z}_{13}) \begin{pmatrix} \mathbf{V}_{22} & \mathbf{V}_{23} \\ \mathbf{V}_{32} & \mathbf{V}_{33} \end{pmatrix}^{-1} (\mathbf{Z}_{12}, \mathbf{Z}_{13})',$$

$$g_2(\mathbf{Z}, \mathbf{V}) \equiv \begin{pmatrix} \mathbf{Z}_{13} \\ \mathbf{Z}_{23} \end{pmatrix} \mathbf{V}_{33}^{-1} \begin{pmatrix} \mathbf{Z}_{13} \\ \mathbf{Z}_{23} \end{pmatrix}'.$$

Proof. Consider any invariant function $g(\mathbf{Z}, \mathbf{V})$ (i.e., any g such that $g(\mathbf{Z}, \mathbf{V}) = g(\mathbf{Z}\mathbf{A}+\mathbf{F}, \mathbf{A}'\mathbf{V}\mathbf{A})$ all $\mathbf{A} \epsilon \mathcal{A}$, $\mathbf{F} \epsilon \mathcal{F}$. There exists a matrix $\mathbf{T} \epsilon \mathcal{A}$ satisfying $\mathbf{V}^{-1} = \mathbf{T}\mathbf{T}'$. Let

$$\mathbf{\Gamma} = \begin{pmatrix} \mathbf{\Gamma}_{11} & \mathbf{0} & \mathbf{0} \\ \mathbf{0} & \mathbf{\Gamma}_{22} & \mathbf{0} \\ \mathbf{0} & \mathbf{0} & \mathbf{\Gamma}_{33} \end{pmatrix}$$

be an orthogonal matrix, and let $\mathbf{A}_\Gamma = \mathbf{T}\mathbf{\Gamma}$. Then $\mathbf{A}_\Gamma \epsilon \mathcal{A}$, and for all $\mathbf{F} \epsilon \mathcal{F}$, all $\mathbf{\Gamma}$ of the above form,

$$g(\mathbf{Z}\mathbf{A}_\Gamma+\mathbf{F}, \mathbf{I}) = g(\mathbf{Z}\mathbf{A}_\Gamma+\mathbf{F}, (\mathbf{A}_\Gamma)'\mathbf{V}(\mathbf{A}_\Gamma)) = g(\mathbf{Z}, \mathbf{V}).$$

Thus $g(\mathbf{Z}, \mathbf{V})$ depends upon $(\mathbf{Z}, \mathbf{V})$ only through $\mathbf{Z}\mathbf{A}_\Gamma+\mathbf{F}$. A judicious choice of $\mathbf{F}$ shows that $g(\mathbf{Z}, \mathbf{V})$ is a function only of the quantities

$$(\mathbf{Z}_{12}, \mathbf{Z}_{13}) \begin{pmatrix} \mathbf{T}_{22} \\ \mathbf{T}_{23} \end{pmatrix} \mathbf{\Gamma}_{22}, \quad \begin{pmatrix} \mathbf{Z}_{13} \\ \mathbf{Z}_{23} \end{pmatrix} \mathbf{T}_{33}\, \mathbf{\Gamma}_{33}.$$

Since $u \leqslant v$, $m \leqslant p-q$, the number of rows in each of the above quantities is exceeded by the number of columns. It follows that we can choose $\mathbf{\Gamma}_{22}$ and $\mathbf{\Gamma}_{33}$ in such a way that

$$(\mathbf{Z}_{12}, \mathbf{Z}_{13}) \begin{pmatrix} \mathbf{T}_{22} \\ \mathbf{T}_{23} \end{pmatrix} \mathbf{\Gamma}_{22} = (g_1(\mathbf{Z}, \mathbf{V}), \mathbf{0}), \quad \begin{pmatrix} \mathbf{Z}_{13} \\ \mathbf{Z}_{23} \end{pmatrix} \mathbf{T}_{33}\mathbf{\Gamma}_{33} = (g_2(\mathbf{Z}, \mathbf{V}), \mathbf{0}).$$

Thus, any invariant function $g(\boldsymbol{Z}, \boldsymbol{V})$ depends upon $(\boldsymbol{Z}, \boldsymbol{V})$ only through $g_1(\boldsymbol{Z}, \boldsymbol{V})$ and $g_2(\boldsymbol{Z}, \boldsymbol{V})$. Since $g_1(\boldsymbol{Z}, \boldsymbol{V})$ and $g_2(\boldsymbol{Z}, \boldsymbol{V})$ are invariant under $\mathscr{A} \times \mathscr{F}$, it follows that the pair (g_1, g_2) is the maximal invariant.

A similar proof shows that the maximal invariant in the parameter space is

$$(\boldsymbol{\Theta}_{12}, \boldsymbol{0}) \begin{pmatrix} \boldsymbol{\Sigma}_{22} & \boldsymbol{\Sigma}_{23} \\ \boldsymbol{\Sigma}_{32} & \boldsymbol{\Sigma}_{33} \end{pmatrix}^{-1} \begin{pmatrix} \boldsymbol{\Theta}_{12}' \\ \boldsymbol{0} \end{pmatrix} = \boldsymbol{\Theta}_{12}(\boldsymbol{\Sigma}_{22} - \boldsymbol{\Sigma}_{23}\boldsymbol{\Sigma}_{33}^{-1}\boldsymbol{\Sigma}_{32})\, \boldsymbol{\Theta}_{12}'.$$

When $u > v$ or $m > p-q$, an approach similar to that of Lemma 3.1 allows us to calculate the maximal invariants in the sample space and parameter space. However, the last step in the proof of the lemma does not yield $g_1(\boldsymbol{Z}, \boldsymbol{V})$ and $g_2(\boldsymbol{Z}, \boldsymbol{V})$, but instead yields more complicated quantities.

It should be remarked that the problem of testing $H_0 : \boldsymbol{\Theta}_{12} = \boldsymbol{0}$ is also invariant under the larger group $\mathscr{U} \times \mathscr{A} \times \mathscr{F}$, where $\mathscr{U}$ consists of all $m \times m$ orthogonal matrices of form $\boldsymbol{U} = \begin{pmatrix} \boldsymbol{U}_{11} & \boldsymbol{0} \\ \boldsymbol{0} & \boldsymbol{U}_{22} \end{pmatrix}$, $\boldsymbol{U}_{11}$: $u \times u$, and where transformation takes $(\boldsymbol{Z}, \boldsymbol{V})$ into $(\boldsymbol{UZA}+\boldsymbol{F}, \boldsymbol{A}'\boldsymbol{VA})$, $\boldsymbol{U} \in \mathscr{U}$, $\boldsymbol{A} \in \mathscr{A}$, $\boldsymbol{F} \in \mathscr{F}$. Except in special cases, the form of maximal invariant in the sample space is difficult to characterize explicitly. The approach given in Lemma 3.1, however, indicates that the maximal invariant is a function of $(\boldsymbol{Z}, \boldsymbol{V})$ only through $(\boldsymbol{Z}_{12}, \boldsymbol{Z}_{13}) \begin{pmatrix} \boldsymbol{T}_{22} & \boldsymbol{0} \\ \boldsymbol{T}_{23} & \boldsymbol{T}_{33} \end{pmatrix}$ and $\begin{pmatrix} \boldsymbol{Z}_{13} \\ \boldsymbol{Z}_{23} \end{pmatrix} \boldsymbol{T}_{33}$, where the $\boldsymbol{T}_{ij}$ are defined as in Lemma 3.1.

4. THE LIKELIHOOD RATIO TEST OF H_0

Our first step in determining the likelihood ratio statistic λ for testing $H_0 : \boldsymbol{\Theta}_{12} = \boldsymbol{0}$ is to note that this statistic is equal to the likelihood ratio statistic (LRS) formed by considering only the joint density $p(\boldsymbol{Z}_{12}, \boldsymbol{Z}_{13}, \boldsymbol{Z}_{22}, \boldsymbol{Z}_{23}, \boldsymbol{W})$ of $\boldsymbol{Z}_{12}, \boldsymbol{Z}_{13}, \boldsymbol{Z}_{22}, \boldsymbol{Z}_{23}$, and $\boldsymbol{W} \equiv \begin{pmatrix} \boldsymbol{V}_{22} & \boldsymbol{V}_{23} \\ \boldsymbol{V}_{32} & \boldsymbol{V}_{33} \end{pmatrix} : (p_2+p_3) \times (p_2+p_3)$. The proof of this assertion is reserved for the Appendix. Further, using the arguments leading to Equation (A.4) of the Appendix, λ can be found by considering

$$\sup_{\mathbf{\Theta}_{22}} p(\mathbf{Z}_{12}, \mathbf{Z}_{13}, \mathbf{Z}_{22}, \mathbf{Z}_{23}, \mathbf{W}) \equiv p^*(\mathbf{Z}_{12}, \mathbf{Z}_{13}, \mathbf{Z}_{23}, \mathbf{W}) \tag{4.1}$$

$$\propto \left|\begin{pmatrix}\mathbf{\Sigma}_{22} & \mathbf{\Sigma}_{23}\\ \mathbf{\Sigma}_{32} & \mathbf{\Sigma}_{33}\end{pmatrix}\right|^{-n/2} \exp -\frac{1}{2}\left[\operatorname{tr}\begin{pmatrix}\mathbf{\Sigma}_{22} & \mathbf{\Sigma}_{23}\\ \mathbf{\Sigma}_{32} & \mathbf{\Sigma}_{33}\end{pmatrix}^{-1}\begin{pmatrix}\mathbf{V}_{22} & \mathbf{V}_{23}\\ \mathbf{V}_{32} & \mathbf{V}_{33}\end{pmatrix}\right.$$

$$\left.+\operatorname{tr}\begin{pmatrix}\mathbf{\Sigma}_{22} & \mathbf{\Sigma}_{23}\\ \mathbf{\Sigma}_{32} & \mathbf{\Sigma}_{33}\end{pmatrix}^{-1}(\mathbf{Z}_{12}-\mathbf{\Theta}_{12},\mathbf{Z}_{13})'(\mathbf{Z}_{12}-\mathbf{\Theta}_{12},\mathbf{Z}_{13})+\operatorname{tr}\mathbf{\Sigma}_{33}^{-1}\mathbf{Z}_{23}'\mathbf{Z}_{23}\right].$$

(The symbol "$\propto$" means "proportional to.") Note that $p^*(\mathbf{Z}_{12}, \mathbf{Z}_{13}, \mathbf{Z}_{23}, \mathbf{W})$ is *not* the joint density of $\mathbf{Z}_{12}, \mathbf{Z}_{13}, \mathbf{Z}_{23}$, and $\mathbf{W}$.

The maximum of (4.1) when $\mathbf{\Theta}_{12}$ is unrestricted occurs at

$$\hat{\mathbf{\Theta}}_{12} = \mathbf{Z}_{12}-\mathbf{Z}_{13}\,\mathbf{\Sigma}_{33}^{-1}\mathbf{\Sigma}_{32}, \tag{4.2}$$

(*see* Gleser and Olkin (1966)), in which case we obtain

$$\max_{\mathbf{\Theta}_{12}} p^*(\mathbf{Z}_{12}, \mathbf{Z}_{13}, \mathbf{Z}_{23}, \mathbf{W}) \propto \left|\begin{pmatrix}\mathbf{\Sigma}_{22} & \mathbf{\Sigma}_{23}\\ \mathbf{\Sigma}_{32} & \mathbf{\Sigma}_{33}\end{pmatrix}\right|^{-n/2} \tag{4.3}$$

$$\exp -\frac{1}{2}\left[\operatorname{tr}\begin{pmatrix}\mathbf{\Sigma}_{22} & \mathbf{\Sigma}_{23}\\ \mathbf{\Sigma}_{32} & \mathbf{\Sigma}_{33}\end{pmatrix}^{-1}\mathbf{W}+\operatorname{tr}\mathbf{\Sigma}_{33}^{-1}(\mathbf{Z}_{23}'\mathbf{Z}_{23}+\mathbf{Z}_{13}'\mathbf{Z}_{13})\right].$$

On the other hand,

$$\max_{\mathbf{\Theta}_{12}=\mathbf{0}} p^*(\mathbf{Z}_{12}, \mathbf{Z}_{13}, \mathbf{Z}_{23}, \mathbf{W}) \propto \left|\begin{pmatrix}\mathbf{\Sigma}_{22} & \mathbf{\Sigma}_{23}\\ \mathbf{\Sigma}_{32} & \mathbf{\Sigma}_{33}\end{pmatrix}\right|^{-n/2} \tag{4.4}$$

$$\exp -\frac{1}{2}\left\{\operatorname{tr}\left[\begin{pmatrix}\mathbf{\Sigma}_{22} & \mathbf{\Sigma}_{23}\\ \mathbf{\Sigma}_{32} & \mathbf{\Sigma}_{33}\end{pmatrix}^{-1}[\mathbf{W}+(\mathbf{Z}_{12}, \mathbf{Z}_{13})'(\mathbf{Z}_{12}, \mathbf{Z}_{13})]\right]+\operatorname{tr}\mathbf{\Sigma}_{33}^{-1}\mathbf{Z}_{23}'\mathbf{Z}_{23}\right\}.$$

For either case we need the maximum over $\mathbf{\Sigma}_{22}, \mathbf{\Sigma}_{23}, \mathbf{\Sigma}_{33}$ subject only to the positive definiteness of $\mathbf{\Sigma}$. This maximum is obtained from the following :

Lemma 4.1. *Let*

$$f(\mathbf{\Delta}) = |\mathbf{\Delta}|^{n/2} \exp -\frac{1}{2}[\operatorname{tr}\mathbf{\Delta}^{-1}\mathbf{U}+\operatorname{tr}\mathbf{\Delta}_{22}^{-2}\,\mathbf{Q}], \tag{4.5}$$

where

$$\mathbf{\Delta} = \begin{bmatrix} \mathbf{\Delta}_{11} & \mathbf{\Delta}_{12} \\ \mathbf{\Delta}_{21} & \mathbf{\Delta}_{22} \end{bmatrix}, \qquad \boldsymbol{U} = \begin{bmatrix} \boldsymbol{U}_{11} & \boldsymbol{U}_{12} \\ \boldsymbol{U}_{21} & \boldsymbol{U}_{22} \end{bmatrix},$$

$\mathbf{\Delta}$ and $\boldsymbol{V}$ are $t \times t$, $\mathbf{\Delta}_{11}$ and $\boldsymbol{U}_{11}$ are $l \times l$, $l \leqslant t$, and $\boldsymbol{Q} \geqslant \mathbf{0}$ is $(t-l)\times(t-l)$. The maximum of $f(\mathbf{\Delta})$ over $\mathbf{\Delta} > \mathbf{0}$ is achieved at

$$n\hat{\mathbf{\Delta}}_{22} = \boldsymbol{U}_{22}+\boldsymbol{Q},\ n(\hat{\mathbf{\Delta}}_{11}-\hat{\mathbf{\Delta}}_{12}\hat{\mathbf{\Delta}}_{22}^{-1}\hat{\mathbf{\Delta}}_{21}) = \boldsymbol{U}_{11}-\boldsymbol{U}_{12}\boldsymbol{U}_{22}^{-1}\boldsymbol{U}_{21},$$
$$\hat{\mathbf{\Delta}}_{22}^{-1}\hat{\mathbf{\Delta}}_{21} = \boldsymbol{U}_{22}^{-1}\boldsymbol{U}_{21},$$

or equivalently at

$$n\hat{\mathbf{\Delta}} = \boldsymbol{U}+\begin{pmatrix} \boldsymbol{U}_{12}\boldsymbol{U}_{22}^{-1} \\ \boldsymbol{I} \end{pmatrix}\boldsymbol{Q}(\boldsymbol{U}_{22}^{-1}\boldsymbol{U}_{21}, \boldsymbol{I}),$$

and is equal to

$$\text{(4.6)}\quad \max_{\mathbf{\Delta}>\mathbf{0}} f(\mathbf{\Delta}) = n^{\frac{tn}{2}}\,|\boldsymbol{U}_{22}+\boldsymbol{Q}|^{-\frac{n}{2}}\,|\boldsymbol{U}|^{-\frac{n}{2}}\,|\boldsymbol{U}_{22}|^{\frac{n}{2}}\exp -\tfrac{1}{2}\,nt.$$

Proof. Reparameterize $\mathbf{\Delta}$ by a one-to-one onto transformation to new parameters

$$\mathbf{\Omega}_{22} \equiv \mathbf{\Delta}_{22},\ \mathbf{\Omega}_{11} \equiv (\mathbf{\Delta}^{-1})_{11}^{-1} = \mathbf{\Delta}_{11}-\mathbf{\Delta}_{12}\mathbf{\Delta}_{22}^{-1}\mathbf{\Delta}_{21},\ \mathbf{\Omega}_{21} \equiv \mathbf{\Delta}_{22}^{-1}\mathbf{\Delta}_{21} = \mathbf{\Omega}_{12}'.$$

Then

$$f(\mathbf{\Omega}) = |\mathbf{\Omega}_{11}|^{-\frac{n}{2}}\,|\mathbf{\Omega}_{22}|^{-\frac{n}{2}}\exp -\tfrac{1}{2}[\operatorname{tr}\mathbf{\Omega}_{22}^{-1}(\boldsymbol{Q}+\boldsymbol{U}_{22})$$
$$+\operatorname{tr}\mathbf{\Omega}_{11}^{-1}\boldsymbol{U}_{11}-2\operatorname{tr}\boldsymbol{U}_{12}\mathbf{\Omega}_{21}\mathbf{\Omega}_{11}^{-1}+\operatorname{tr}\mathbf{\Omega}_{12}\boldsymbol{U}_{22}\mathbf{\Omega}_{21}\mathbf{\Omega}_{11}^{-1}].$$

Maximizing with respect to $\mathbf{\Omega}_{11}$ and $\mathbf{\Omega}_{22}$ yields

$$\text{(4.7)}\quad \max_{\mathbf{\Omega}_{11},\,\mathbf{\Omega}_{22}} f(\mathbf{\Omega}) = n^{\frac{tn}{2}}\,|\boldsymbol{U}_{11}-\boldsymbol{U}_{12}\mathbf{\Omega}_{21}-\mathbf{\Omega}_{12}\boldsymbol{U}_{21}+\mathbf{\Omega}_{12}\boldsymbol{U}_{22}\mathbf{\Omega}_{21}|^{-\frac{n}{2}}$$
$$|\boldsymbol{Q}+\boldsymbol{U}_{22}|^{-\frac{n}{2}}\exp -\tfrac{1}{2}\,nt,$$

achieved for $n^{-1}\hat{\mathbf{\Omega}}_{22} = \boldsymbol{Q}+\boldsymbol{U}_{22}$, $n^{-1}\hat{\mathbf{\Omega}}_{11} = \boldsymbol{U}_{11}-\boldsymbol{U}_{12}\mathbf{\Omega}_{21}-\mathbf{\Omega}_{12}\boldsymbol{U}_{21}+\mathbf{\Omega}_{12}\boldsymbol{U}_{22}\mathbf{\Omega}_{21}$. But

$$\boldsymbol{U}_{11}-\boldsymbol{U}_{12}\mathbf{\Omega}_{21}-\mathbf{\Omega}_{12}\boldsymbol{U}_{21}+\mathbf{\Omega}_{12}\boldsymbol{U}_{22}\mathbf{\Omega}_{21}$$
$$= (\boldsymbol{U}_{11}-\boldsymbol{U}_{12}\boldsymbol{U}_{22}^{-1}\boldsymbol{U}_{21})+(\boldsymbol{U}_{12}\boldsymbol{U}_{22}^{-1}-\mathbf{\Omega}_{12})\boldsymbol{U}_{22}(\boldsymbol{U}_{12}\boldsymbol{U}_{22}^{-1}-\mathbf{\Omega}_{12})',$$

whereupon we see that (4.7) is minimized when $\hat{\Omega}_{12} = U_{12}U_{22}^{-1}$. Therefore,

$$\max_{\Omega} f(\Omega) = n^{\frac{tn}{2}} \, |U_{11}-U_{12}U_{22}^{-1}U_{21}|^{-\frac{n}{2}} \, |U_{22}+Q|^{-\frac{n}{2}} \exp -\tfrac{1}{2} nt.$$

The proof is completed by noting that

$$|U| = |U_{22}|\,|U_{11}-U_{12}U_{22}^{-1}U_{21}|.$$

Applying Lemma 4.1 to (4.3) and (4.4) we obtain

$$\lambda^{\frac{2}{n}} = \frac{\max\limits_{\Theta_{12}=0,\,\Sigma} p^*(Z_{12}, Z_{13}, Z_{23}, W)}{\max\limits_{\Theta_{12},\,\Sigma} p^*(Z_{12}, Z_{13}, Z_{23}, W)} = \frac{|W|\,|W_{22}+Z_{13}'Z_{13}|}{|W_{22}|\,|W+(Z_{12}, Z_{13})'(Z_{12}, Z_{13})|},$$

$$(4.8) \qquad = \frac{|I+Z_{13}V_{33}^{-1}Z_{13}'|}{\left|I+(Z_{12}, Z_{13})\begin{pmatrix} V_{22} & V_{23} \\ V_{32} & V_{33}\end{pmatrix}^{-1}(Z_{12}, Z_{13})'\right|}.$$

It is well known that the LRS is invariant under any group which leaves the problem invariant. Thus, the LRS depends upon the observations only through the maximal invariant in the sample space under that group. Therefore, for testing H_0 in the case $u \leqslant v$, $m \leqslant p-q$, the LRS is a function of (Z, V) only through the quantities $g_1(Z, V)$, $g_2(Z, V)$ given in (3.1). Note, however, that the LRS λ does not involve Z_{23}, whereas the maximal invariant (3.1) does. Usually, the LRS involves all of the maximal invariant statistic, so that this result may be a little surprising (*n.b.*, similar remarks hold even when we consider the maximal invariant under the group $\mathscr{U}\times\mathscr{A}\times\mathscr{F}$).

The following argument may serve as a partial explanation. It can be shown that $g_1(Z, V)$ and $g_2(Z, V)$ are functions only of $Z_{12}-Z_{13}V_{33}^{-1}V_{32}$, $V_{22}-V_{23}V_{33}^{-1}V_{32}$, $Z_{13}V_{33}^{-1}Z_{13}'$, $Z_{13}V_{33}^{-1}Z_{23}'$, $Z_{23}V_{33}^{-1}Z_{23}'$. Given Z_{13} and V_{33}, the conditional distribution of $Z_{12}-Z_{13}V_{33}^{-1}V_{32}$ is that of a random matrix whose elements are jointly normally distributed with mean Θ_{12} and covariance matrix $(I+Z_{13}V_{33}^{-1}Z_{13}') \otimes \Lambda_{22}^{-1}$, where $\otimes$ denotes the Kronecker product and $\Lambda_{22}^{-1} = \Sigma_{22}-\Sigma_{23}\Sigma_{33}^{-1}\Sigma_{32}$. The distribution of $V_{22}-V_{23}V_{33}^{-1}V_{32}$ is $\mathscr{W}(\Lambda_{22}^{-1};\ v,$

$n-m-p+q$) conditionally or unconditionally on $\boldsymbol{Z}$, $\boldsymbol{V}_{33}$, and $\boldsymbol{V}_{33}^{-1}\boldsymbol{V}_{32}$. Further, the joint distribution of $\boldsymbol{Z}_{13}\boldsymbol{V}_{33}^{-1}\boldsymbol{Z}_{23}'$ and $\boldsymbol{Z}_{23}\boldsymbol{V}_{33}^{-1}\boldsymbol{Z}_{23}'$ given $\boldsymbol{Z}_{13}$ and $\boldsymbol{V}_{33}$ does not involve either $\boldsymbol{\Theta}_{12}$ or $\boldsymbol{\Lambda}_{22}$. It follows that the only information concerning $\boldsymbol{\Theta}_{12}$ contained in the maximal invariant (g_1, g_2) lies in the statistics $\boldsymbol{Z}_{12}-\boldsymbol{Z}_{13}\boldsymbol{V}_{33}^{-1}\boldsymbol{V}_{32}$, $\boldsymbol{V}_{22}-\boldsymbol{V}_{23}\boldsymbol{V}_{33}^{-1}\boldsymbol{V}_{32}$, and $\boldsymbol{Z}_{13}\boldsymbol{V}_{33}^{-1}\boldsymbol{Z}_{13}'$, and not in that part of the maximal invariant involving $\boldsymbol{Z}_{23}$.

The null distribution of the statistic (4.8) is the same as the distribution of $\prod_{i=1}^{a_1} b_i$, where b_i has the Beta distribution with parameters $\frac{1}{2}(n-m-p+q-v+a_1-i+1)$ and $\frac{1}{2}a_2$, the b_i are independently distributed, $a_1 = \min(v, u)$, and $a_2 = \max(v, u)$ (*see* Gleser and Olkin (1966), Section 3.1). An approximation to this distribution can be obtained by use of Theorem 8.6.2. in Anderson (1958), namely,

$$P\left\{-\frac{2h}{n}\log\lambda \leqslant x\right\} = c_1 P\{\chi^2_{uv} \leqslant x\}+c_2 P\{\chi^2_{uv+4} \leqslant x\}$$
$$+c_3 P\{\chi^2_{uv+8} \leqslant x\}+O(n^{-6}),$$

where $h = n-m-p+q+\frac{1}{2}(u-v+1)$,

$$c_1 = \left(1-\frac{\delta_1}{h^2}\right)^2+\frac{\delta_1}{h^2}-\frac{\delta_2}{h^4}, \quad c_2 = \frac{\delta_1}{h^2}\left(1-\frac{\delta_1}{h^2}\right), \quad c_3 = \frac{1}{2}\frac{\delta_1^2}{h^4}+\frac{\delta_2}{h^4},$$

$$\delta_1 = \frac{uv(v^2+u^2-5)}{48}, \delta_2=\frac{uv}{1920}[3(v^4+u^4)+10u^2v^2-50(v^2+u^2)+159].$$

To find the asymptotic (as $n\to\infty$) non-null distribution of $\lambda^{2/n}$, first note that in the canonical model the covariance matrix $\boldsymbol{\Sigma}$ does not depend upon the sample size n (in fact, $\boldsymbol{\Sigma}$ is a function only of $\boldsymbol{\psi}$, $\boldsymbol{X}_2$, $\boldsymbol{X}_4$, none of which depend upon n). On the other hand, the mean $\boldsymbol{\Theta}$ depends upon n through $\boldsymbol{T}_1$, the non-singular matrix obtained by applying Lemma 2.1 to $\boldsymbol{X}_1'$. We assume that

$$\text{(4.9)} \qquad \lim_{n\to\infty} \boldsymbol{\Theta}/\sqrt{n} = \boldsymbol{\Phi} = \begin{pmatrix} \boldsymbol{\Phi}_{11} & \boldsymbol{\Phi}_{12} \\ \boldsymbol{\Phi}_{21} & \boldsymbol{\Phi}_{22} \end{pmatrix},$$

where $\boldsymbol{\Phi}$ is partitioned in the manner of $\boldsymbol{\Theta}$ and $\boldsymbol{\Phi}_{12} \neq \mathbf{0}$. The dimensions m, p, q, u, and v are assumed fixed.

We may write $\lambda^{2/n}$ in the form

$$\lambda^{-2/n} = |\boldsymbol{I}+\boldsymbol{Q}^{-\frac{1}{2}}(\boldsymbol{Z}_{12}-\boldsymbol{Z}_{13}\boldsymbol{V}_{33}^{-1}\boldsymbol{V}_{32})[\boldsymbol{V}_{22}-\boldsymbol{V}_{23}\boldsymbol{V}_{33}^{-1}\boldsymbol{V}_{32}]^{-1} (\boldsymbol{Z}_{12}-\boldsymbol{Z}_{13}\boldsymbol{V}_{33}^{-1}\boldsymbol{V}_{32})'\boldsymbol{Q}^{-\frac{1}{2}}|$$

where $\boldsymbol{Q} = \boldsymbol{I}+\boldsymbol{Z}_{13}\boldsymbol{V}_{33}^{-1}\boldsymbol{Z}_{13}'$.

Lemma 4.2. *Assume that the random matrices $\boldsymbol{Q}$, $\boldsymbol{R}$, $\boldsymbol{T}$ converge jointly in law to the random matrices $\boldsymbol{Q}^*$, $\boldsymbol{R}^*$, $\boldsymbol{T}^*$, respectively. If $f(q, r, t)$ is any function over the sample space of these random matrices whose discontinuities have probability* 0 *under the joint distribution of $\boldsymbol{Q}^*$, $\boldsymbol{R}^*$, $\boldsymbol{T}^*$, then $f(\boldsymbol{Q}, \boldsymbol{R}, \boldsymbol{T})$ converges in law to $f(\boldsymbol{Q}^*, \boldsymbol{R}^*, \boldsymbol{T}^*)$.*

The result follows by a straightforward application of Chernoff (1956, Theorem 2, p. 5). From Lemma 4.2, noting that plim $\boldsymbol{V}/n = \boldsymbol{\Sigma}$, plim $\boldsymbol{Z}_{13}/\sqrt{n} = \mathbf{0}$, and thus that plim $\boldsymbol{Q} = \boldsymbol{I}$, plim $\boldsymbol{V}_{33}^{-1}\boldsymbol{V}_{32} = \boldsymbol{\Sigma}_{33}^{-1}\boldsymbol{\Sigma}_{32}$, and plim $(\boldsymbol{V}_{22}-\boldsymbol{V}_{23}\boldsymbol{V}_{33}^{-1}\boldsymbol{V}_{32})/n = \boldsymbol{\Sigma}_{22}-\boldsymbol{\Sigma}_{23}\boldsymbol{\Sigma}_{33}^{-1}\boldsymbol{\Sigma}_{32} \equiv \boldsymbol{\Sigma}_{22.3}^{-1}$, we conclude that the limiting distribution of $\log \lambda^{2/n}$ is the same as that of $t = \log|\boldsymbol{I}+\boldsymbol{G}\boldsymbol{G}'|$ where the u rows of $\boldsymbol{G}=n^{-\frac{1}{2}}(\boldsymbol{Z}_{12}-\boldsymbol{Z}_{13}\boldsymbol{\Sigma}_{33}^{-1}\boldsymbol{\Sigma}_{32})\boldsymbol{\Sigma}_{22\cdot3}^{\frac{1}{2}}$ are independent, each with a v-variate normal distribution having covariance matrix $\boldsymbol{I}$. The expected value of $\boldsymbol{G}$ is $\mathcal{E}(\boldsymbol{G}) = n^{-1/2}(\boldsymbol{\Phi}_{12}\Sigma_{22.3}^{-1/2}, \mathbf{0})$. Using the result in Gleser and Olkin (1966, Equation (3.12)) with C_λ in that paper equal to $\boldsymbol{I}$, we have

$$\lim_{n\to\infty} \mathcal{L}(\log \lambda^{2/n}-\log|\boldsymbol{I}+\boldsymbol{\Phi}_{12}\boldsymbol{\Sigma}_{22\cdot3}\boldsymbol{\Phi}_{12}'|) = \mathcal{N}(0, \delta),$$

where

$$\delta = 4\,\mathrm{tr}[(\boldsymbol{I}+\boldsymbol{\Phi}_{12}\boldsymbol{\Sigma}_{22\cdot3}\boldsymbol{\Phi}_{12}')^{-2}\boldsymbol{\Phi}_{12}\boldsymbol{\Sigma}_{22\cdot3}\boldsymbol{\Phi}_{12}'].$$

For $u = 1$, a possible computation formula for the exact non-null distribution of $\lambda^{2/n}$ in terms of a double infinite series has been obtained (Olkin and Shrikhande (1954)). It appears that for $u > 1$, the corresponding formulas for the distribution of $\lambda^{2/n}$ will be more complicated.

5. ESTIMATION OF β AND ψ.

Since (β, ψ) and (Θ, Σ) are equivalent parameterizations (i.e., are related by known non-singular transformations), maximum likelihood estimators of Θ and Σ yield maximum likelihood estimators of β and ψ (*see* Anderson (1958), Lemma 3.2.3). These estimators can be obtained from the results of Section 4, namely,

$$\hat{\Theta} = \begin{pmatrix} Z_{11} & Z_{12} \\ Z_{21} & Z_{22} \end{pmatrix} - \begin{pmatrix} Z_{13} \\ Z_{23} \end{pmatrix} V_{33}^{-1}(V_{31}, V_{32}), \tag{5.1}$$

$$n\hat{\Sigma} = V + \begin{pmatrix} V_{13}V_{33}^{-1} \\ V_{23}V_{33}^{-1} \\ I_{p-q} \end{pmatrix} (Z'_{13}, Z'_{23}) \begin{pmatrix} Z_{13} \\ Z_{23} \end{pmatrix} (V_{33}^{-1}V_{31}, V_{33}^{-1}V_{32}, I_{p-q}). \tag{5.2}$$

The distribution of $\hat{\Theta}$ can be derived as in Section 4.1 of Gleser and Olkin (1966). Let $A \epsilon \mathscr{A}$ be the $p \times p$ lower triangular matrix such that $A'\Sigma A = I$. Letting A and Σ be partitioned as

$$A = \begin{pmatrix} A_{11} & 0 \\ A_{21} & A_{22} \end{pmatrix}, \quad \Sigma = \begin{pmatrix} \Sigma_{11} & \Sigma_{12} \\ \Sigma_{21} & \Sigma_{22} \end{pmatrix},$$

where A_{11} and Σ_{11} are $q \times q$, A_{22} and Σ_{22} are $(p-q)\times(p-q)$, we see that $A_{11}A'_{11} = \Sigma_{11\cdot 2} \equiv (\Sigma_{11} - \Sigma_{12}\Sigma_{22}^{-1}\Sigma_{21})^{-1}$. Now let $Z = (Z_1, Z_2)$,

$$Z_1 \equiv \begin{pmatrix} Z_{11} & Z_{12} \\ Z_{21} & Z_{22} \end{pmatrix} : m \times q, \quad Z_2 \equiv \begin{pmatrix} Z_{13} \\ Z_{23} \end{pmatrix} : m \times (p-q),$$

and make the transformations

$$F = (F_1, F_2) = (Z_1 - \Theta, Z_2)A, \quad W = \begin{pmatrix} W_{11} & W_{12} \\ W_{21} & W_{22} \end{pmatrix} = A'VA,$$

where F and W are partitioned in the manner of Z and Σ, respectively. Then F and W are independent, W has the Wishart distribution $\mathscr{W}(I; p, n-m)$, the rows of F are independently distributed as p-variate normal distributions $\mathscr{N}(0, I)$. Since

$$\tilde{\Theta} \equiv (\hat{\Theta} - \Theta)A_{11} = F_1 - F_2W_{22}^{-1}W_{21},$$

the distribution of $\tilde{\mathbf{\Theta}}$ given $\boldsymbol{W}$ depends upon $\boldsymbol{W}$ only through the (conditional) covariance matrix $\boldsymbol{I}+\boldsymbol{W}_{12}\boldsymbol{W}_{22}^{-2}\boldsymbol{W}_{21}$. If we let $\boldsymbol{L} = \boldsymbol{W}_{12}\ \boldsymbol{W}_{22}^{-1} : q\times(p-q)$, then $p(\tilde{\mathbf{\Theta}}\,|\,\boldsymbol{W}) = p(\tilde{\mathbf{\Theta}}\,|\,\boldsymbol{L})$, and $p(\tilde{\mathbf{\Theta}}) = \int p(\tilde{\mathbf{\Theta}}\,|\,\boldsymbol{L})\, p(\boldsymbol{L})d\boldsymbol{L}$. Consequently

$$(5.3)\qquad p(\tilde{\mathbf{\Theta}})\propto \int_{-\infty\leqslant l_{ij}\leqslant\infty} |\boldsymbol{I}+\boldsymbol{L}\boldsymbol{L}'|^{-\left(\frac{n+q}{2}\right)} \exp-\frac{1}{2}\ \mathrm{tr}\,[\tilde{\mathbf{\Theta}}'\tilde{\mathbf{\Theta}}(\boldsymbol{I}+\boldsymbol{L}\boldsymbol{L}')^{-1}]d\boldsymbol{L}.$$

If $q \leqslant p-q$, then by Hsu's Theorem (Anderson (1958), p. 319), for $\boldsymbol{H} = (\boldsymbol{I}+\boldsymbol{L}\boldsymbol{L}')^{-1}$, (5,3) becomes

$$(5.4)\quad p(\tilde{\mathbf{\Theta}})\propto \int_{\boldsymbol{O}<\boldsymbol{H}<\boldsymbol{I}} |\boldsymbol{H}|^{\frac{(n-p+q-1)}{2}} |\boldsymbol{I}-\boldsymbol{H}|^{\frac{(p-2q-1)}{2}} \exp -\tfrac{1}{2}[\mathrm{tr}\ \tilde{\mathbf{\Theta}}'\tilde{\mathbf{\Theta}}\boldsymbol{H}]d\boldsymbol{H}.$$

For $m = 1$, (5.4) simplifies to

$$p(\tilde{\mathbf{\Theta}}) = C_1 \int_0^1 h^{\frac{n+2q-p}{2}-1}(1-h)^{\frac{p-q}{2}-1} e^{-\frac{1}{2}h\tilde{\mathbf{\Theta}}\tilde{\mathbf{\Theta}}'}dh$$

where $C_1^{-1} = (2\pi)^{\frac{q}{2}}\beta\left(\frac{1}{2}(n+q-p),\ \frac{1}{2}(p-q)\right)$. We note that the distribution of $\tilde{\mathbf{\Theta}}$ depends only on the sum of the roots of the matrix $\tilde{\mathbf{\Theta}}\tilde{\mathbf{\Theta}}'$. In terms of the original model

$$(5.5)\qquad \hat{\boldsymbol{\beta}} = (\boldsymbol{X}_1'\boldsymbol{X}_1)^{-1}\boldsymbol{X}_1'\boldsymbol{Y}\boldsymbol{A}^{-1}\boldsymbol{X}_2'(\boldsymbol{X}_2\boldsymbol{A}^{-1}\boldsymbol{X}_2')^{-1}$$

where $\boldsymbol{A} = \boldsymbol{Y}'(\boldsymbol{I}-\boldsymbol{X}_1(\boldsymbol{X}_1'\boldsymbol{X}_1)^{-1}\boldsymbol{X}_1')\boldsymbol{Y}$. The distribution of $\vec{\boldsymbol{\beta}} = (\boldsymbol{X}_1'\boldsymbol{X}_1)^{\frac{1}{2}}(\hat{\boldsymbol{\beta}}-\boldsymbol{\beta})(\boldsymbol{X}_2\boldsymbol{\psi}\boldsymbol{X}_2')^{-\frac{1}{2}}$ is

$$(5.6)\qquad p(\vec{\boldsymbol{\beta}})\propto \int |\boldsymbol{I}+\boldsymbol{L}\boldsymbol{L}'|^{-\frac{(n+q)}{2}} \exp-\frac{1}{2}\,\mathrm{tr}[\vec{\boldsymbol{\beta}}'\vec{\boldsymbol{\beta}}(\boldsymbol{I}+\boldsymbol{L}\boldsymbol{L}')^{-1}]d\boldsymbol{L}.$$

Turning to the distribution of $\hat{\boldsymbol{\Sigma}}$, we adapt the results of Gleser and Olkin (1966) to obtain

$$(5.7)\ p(\hat{\boldsymbol{\Sigma}}) = \Big[C(p, n-m)\, |\boldsymbol{\Sigma}|^{-\frac{n-m}{2}}\, |\hat{\boldsymbol{\Sigma}}|^{(n-m-p-1)/2} \exp \frac{1}{2} \operatorname{tr} \boldsymbol{\Sigma}^{-1}\hat{\boldsymbol{\Sigma}} \Big]$$

$$\times \frac{|\hat{\boldsymbol{\Sigma}}_{22}|^{m/2}}{|\boldsymbol{\Sigma}_{22}|^{m/2}(2\pi)^{qm/2}} \int\limits_{\mathbf{0} \leqslant \boldsymbol{R}'\boldsymbol{R} \leqslant \boldsymbol{I}} |\boldsymbol{I}+\boldsymbol{R}'\boldsymbol{R}|^{\frac{n-m+2q-p-1}{2}} e^{-\frac{1}{2} tr\, \boldsymbol{D}_\nu \boldsymbol{R}'\boldsymbol{R}} d\boldsymbol{R},$$

where $\hat{\boldsymbol{\Sigma}} > \mathbf{0}$ is partitioned in the manner of $\boldsymbol{\Sigma}$, $\boldsymbol{R} : m \times (p-q)$, $\boldsymbol{D}_\nu = \operatorname{diag}(\nu_1, \ldots, \nu_{p-q})$, the ν's are the characteristic roots of $\Xi\,(\boldsymbol{\Sigma}^{-\frac{1}{2}}\hat{\boldsymbol{\Sigma}}\boldsymbol{\Sigma}^{-\frac{1}{2}})$, and where for a positive definite matrix $\boldsymbol{H} = \begin{pmatrix} \boldsymbol{H}_{11} & \boldsymbol{H}_{12} \\ \boldsymbol{H}_{21} & \boldsymbol{H}_{22} \end{pmatrix}$ partitioned in the manner of $\boldsymbol{\Sigma}$, we have $\Xi\,(\boldsymbol{H}) = \boldsymbol{H}_{22}^{-\frac{1}{2}}\, \boldsymbol{H}_{21}\boldsymbol{H}_{12}\boldsymbol{H}_{22}^{-\frac{1}{2}}$.

The distribution of $\hat{\boldsymbol{\psi}}$ has the form (5.7) with $\hat{\boldsymbol{\Sigma}}$ replaced by $\hat{\boldsymbol{\psi}}$ and $\boldsymbol{\Sigma}$ replaced by $\boldsymbol{\psi}$. In terms of the original model $\hat{\boldsymbol{\psi}}$ is

$$(5.8) \qquad \hat{\boldsymbol{\psi}} = (\boldsymbol{Y}-\boldsymbol{X}_1\hat{\boldsymbol{\beta}}\boldsymbol{X}_2)'\,(\boldsymbol{Y}-\boldsymbol{X}_1\hat{\boldsymbol{\beta}}\boldsymbol{X}_2),$$

where $\hat{\boldsymbol{\beta}}$ is given by (5.5).

6. RELATED MODELS

It is of some interest to note that the techniques used to reduce the model (1.1) and hypothesis testing problem (1.2) to a canonical form when $\boldsymbol{\psi}$ is any unknown positive-definite covariance matrix can also be used in the case that we have knowledge that $\boldsymbol{\psi} = \sigma^2\boldsymbol{I}$, $\sigma^2 > 0$ an unknown constant. Since $\boldsymbol{Z}$ and $\boldsymbol{V}$ are reached from $\boldsymbol{Y}$ only by orthogonal transformations, (Section 2), so that independence in the covariance structure is preserved, it follows that :

(1) The elements of $\boldsymbol{Z}$ are independently distributed, each having a univariate normal distribution with variance σ^2, and means

$$\mathscr{E}\begin{pmatrix} \boldsymbol{Z}_{11} & \boldsymbol{Z}_{12} & \boldsymbol{Z}_{13} \\ \boldsymbol{Z}_{21} & \boldsymbol{Z}_{22} & \boldsymbol{Z}_{23} \end{pmatrix} = \begin{pmatrix} \boldsymbol{\Theta}_{11} & \boldsymbol{\Theta}_{12} & 0 \\ \boldsymbol{\Theta}_{21} & \boldsymbol{\Theta}_{22} & 0 \end{pmatrix}.$$

(2) $\boldsymbol{V}$ has a Wishart distribution $\mathcal{W}(\sigma^2\boldsymbol{I}; p, n-m)$ distributed independently of $\boldsymbol{Z}$.

Since the elements of $\boldsymbol{Z}$ are independent, it follows that

$$\hat{\boldsymbol{\Theta}} = \begin{pmatrix} \boldsymbol{Z}_{11} & \boldsymbol{Z}_{12} \\ \boldsymbol{Z}_{21} & \boldsymbol{Z}_{22} \end{pmatrix}, \; np\hat{\sigma}^2 = \operatorname{tr} \boldsymbol{V} + \operatorname{tr} \boldsymbol{Z}_{13}\boldsymbol{Z}'_{13} + \operatorname{tr} \boldsymbol{Z}_{23}\boldsymbol{Z}'_{23}, \tag{6.1}$$

and these statistics are sufficient for $(\boldsymbol{\Theta}, \sigma^2)$ or $(\boldsymbol{\beta}, \sigma^2)$. Furthermore, to test (1.2) or equivalently $H_0 : \boldsymbol{\Theta}_{12} = \boldsymbol{\xi}_0^*$, the likelihood ratio statistic λ is given by

$$\lambda^{2/np} = \frac{np\hat{\sigma}^2}{np\hat{\sigma}^2 + \operatorname{tr} (\boldsymbol{Z}_{12} - \boldsymbol{\xi}_0^*)(\boldsymbol{Z}_{12} - \boldsymbol{\xi}_0^*)'} \tag{6.2}$$

Now $np\hat{\sigma}^2/\sigma^2$ has the distribution χ^2_{np-mq} and $\operatorname{tr}(\boldsymbol{Z}_{12} - \boldsymbol{\xi}_0^*)(\boldsymbol{Z}_{12} - \boldsymbol{\xi}_0^*)'/\sigma^2$ has the distribution χ^2_{uv} under the null hypothesis H_0. Thus, under H_0, $\lambda^{2/np}$ has the Beta distribution $\beta\left(\frac{np-mq}{2}, \frac{uv}{2}\right)$. The non-null distribution of $\lambda^{2/np}$ is a non central Beta distribution with parameters $\frac{np-mq}{2}, \frac{uv}{2}$, and non-centrality parameter

$$\begin{aligned} \tau &= \sigma^{-2} \operatorname{tr}(\boldsymbol{\Theta}_{12} - \boldsymbol{\xi}_0^*)(\boldsymbol{\Theta}_{12} - \boldsymbol{\xi}_0^*)' \\ &= \sigma^{-2} \operatorname{tr} \boldsymbol{F}(\boldsymbol{X}, \boldsymbol{\beta}) \end{aligned}$$

where

$$\begin{aligned} \boldsymbol{F}(\boldsymbol{X}, \boldsymbol{\beta}) = {} & [\boldsymbol{X}_3(\boldsymbol{X}'_1\boldsymbol{X}_1)^{-1}\boldsymbol{X}'_3]^{-1}(\boldsymbol{X}_3\boldsymbol{\beta}\boldsymbol{X}_4 - \boldsymbol{\xi}_0) \\ & [\boldsymbol{X}'_4(\boldsymbol{X}_2\boldsymbol{X}'_2)^{-1}\boldsymbol{X}_4]^{-1}(\boldsymbol{X}_3\boldsymbol{\beta}\boldsymbol{X}_4 - \boldsymbol{\xi}_0)'. \end{aligned}$$

In terms of the original model (6.1) and (6.2) become

$$\hat{\boldsymbol{\beta}} = (\boldsymbol{X}'_1\boldsymbol{X}_1)^{-1}\boldsymbol{X}'_1\boldsymbol{Y}\boldsymbol{X}'_2(\boldsymbol{X}_2\boldsymbol{X}'_2)^{-1}, \tag{6.3}$$

$$np\hat{\sigma}^2 = \operatorname{tr} \boldsymbol{Y}'[\boldsymbol{I} - \boldsymbol{X}_1(\boldsymbol{X}'_1\boldsymbol{X}_1)^{-1}\boldsymbol{X}'_1]\boldsymbol{Y} + \operatorname{tr} \boldsymbol{Y}[\boldsymbol{I} - \boldsymbol{X}'_2(\boldsymbol{X}_2\boldsymbol{X}'_2)^{-1}\boldsymbol{X}_2]\boldsymbol{Y}',$$

and

$$\lambda^{-2/np} = 1 + (np\,\hat{\sigma}^2)^{-1} \operatorname{tr} \boldsymbol{F}(\boldsymbol{X}, \hat{\boldsymbol{\beta}}). \tag{6.4}$$

These results summarize both the classical results in the univariate analysis of variance (for example, *see* Scheffé (1959)), and more recent work in this area (e.g., Chow (1960)).

The canonical model derived in Section 2 suggests extensions to models involving more complex partitioning of $\mathbf{\Theta}$ and $\mathbf{\Sigma}$. For example, suppose that $\mathbf{Z}$ and $\mathbf{\Theta}$ are partitioned as

$$(6.4)\quad \mathbf{Z}=\begin{pmatrix} \mathbf{Z}_{11} & \mathbf{Z}_{12} & \mathbf{Z}_{13} & \mathbf{Z}_{14} \\ \mathbf{Z}_{21} & \mathbf{Z}_{22} & \mathbf{Z}_{23} & \mathbf{Z}_{24} \\ \mathbf{Z}_{31} & \mathbf{Z}_{32} & \mathbf{Z}_{33} & \mathbf{Z}_{34} \end{pmatrix},\quad \mathbf{\Theta}=\begin{pmatrix} \mathbf{\Theta}_{11} & \mathbf{\Theta}_{12} & \mathbf{\Theta}_{13} & \mathbf{0} \\ \mathbf{\Theta}_{21} & \mathbf{\Theta}_{22} & \mathbf{\Theta}_{23} & \mathbf{0} \\ \mathbf{\Theta}_{31} & \mathbf{\Theta}_{32} & \mathbf{\Theta}_{33} & \mathbf{0} \end{pmatrix},$$

$\mathbf{Z}_{ij}$ and $\mathbf{\Theta}_{ij} : m_i \times p_j$, $i=1, 2, 3$, $j=1, 2, 3, 4$, $\sum_{i=1}^{3} m_i = m$, $\sum_{j=1}^{4} p_j = p$. Let $\mathbf{V}$ and $\mathbf{\Sigma}$ be partitioned in a conformable manner, i.e., $\mathbf{V}=(\mathbf{V}_{ij})$, $\mathbf{\Sigma}=(\mathbf{\Sigma}_{ij})$, $\mathbf{V}_{ij}$ and $\mathbf{\Sigma}_{ij}$: $p_i \times p_j$. As in the canonical model of Section 2, we assume that :

(1) Each of the m rows of $\mathbf{Z}$ has a p-variate normal distribution with unknown covariance $\mathbf{\Sigma}$, unknown means $\mathcal{E}(\mathbf{Z})=\mathbf{\Theta}$.

(2) $\mathbf{V}$ has the Wishart distribution $\mathcal{W}(\mathbf{\Sigma}; p, n-m)$.

(3) The rows of $\mathbf{Z}$ are independently distributed and distributed independently of $\mathbf{V}$.

For such a model the problems of (i) estimating $\mathbf{\Theta}$ and $\mathbf{\Sigma}$, and (ii) testing any of the hypotheses :

(a) $H_1 : \mathbf{\Theta}_{13}=\mathbf{0}$,

(b) $H_2 : \mathbf{\Theta}_{13}=\mathbf{0},\ \mathbf{\Theta}_{23}=\mathbf{0}$,

(c) $H_3 : \mathbf{\Theta}_{12}=\mathbf{0},\ \mathbf{\Theta}_{13}=\mathbf{0},\ \mathbf{\Theta}_{22}=\mathbf{0},\ \mathbf{\Theta}_{23}=\mathbf{0}$,

are in form identical with the corresponding estimation and testing problems of Section 2. On the other hand, the hypothesis

$$H_4 : \mathbf{\Theta}_{12}=\mathbf{0},\quad \mathbf{\Theta}_{13}=\mathbf{0},\quad \mathbf{\Theta}_{23}=\mathbf{0},$$

has a form different from any of the hypotheses H_0, H_1, H_2, H_3. Nevertheless, the development of invariance theory for testing H_4 or the derivation of the likelihood ratio test statistic follows in a direct manner from the methods developed in Section 4. For example, the likelihood ratio statistic for testing H_4 against

$\boldsymbol{\Theta}$ restricted only to the model (6.4) is

$$\lambda_{H_4}^{2/n} = \frac{\left| \boldsymbol{I} + \begin{pmatrix} \boldsymbol{Z}_{14} \\ \boldsymbol{Z}_{24} \end{pmatrix} \boldsymbol{V}_{44}^{-1} (\boldsymbol{Z}'_{14}, \boldsymbol{Z}'_{24}) \right|}{\left| \boldsymbol{I} + \begin{pmatrix} \boldsymbol{Z}_{13} & \boldsymbol{Z}_{14} \\ \boldsymbol{Z}_{23} & \boldsymbol{Z}_{24} \end{pmatrix} \begin{pmatrix} \boldsymbol{V}_{33} & \boldsymbol{V}_{34} \\ \boldsymbol{V}_{43} & \boldsymbol{V}_{44} \end{pmatrix}^{-1} \begin{pmatrix} \boldsymbol{Z}'_{13} & \boldsymbol{Z}'_{23} \\ \boldsymbol{Z}'_{14} & \boldsymbol{Z}'_{24} \end{pmatrix} \right|}$$

$$\times \frac{\left| \boldsymbol{I} + (\boldsymbol{Z}_{13}, \boldsymbol{Z}_{14}) \begin{pmatrix} \boldsymbol{V}_{33} & \boldsymbol{V}_{34} \\ \boldsymbol{V}_{43} & \boldsymbol{V}_{44} \end{pmatrix}^{-1} \begin{pmatrix} \boldsymbol{Z}'_{13} \\ \boldsymbol{Z}'_{14} \end{pmatrix} \right|}{\left| \boldsymbol{I} + (\boldsymbol{Z}_{12}, \boldsymbol{Z}_{13}, \boldsymbol{Z}_{14}) \begin{pmatrix} \boldsymbol{V}_{22} & \boldsymbol{V}_{23} & \boldsymbol{V}_{24} \\ \boldsymbol{V}_{32} & \boldsymbol{V}_{33} & \boldsymbol{V}_{34} \\ \boldsymbol{V}_{42} & \boldsymbol{V}_{43} & \boldsymbol{V}_{44} \end{pmatrix}^{-1} \begin{pmatrix} \boldsymbol{Z}'_{12} \\ \boldsymbol{Z}'_{13} \\ \boldsymbol{Z}'_{14} \end{pmatrix} \right|}.$$

The likelihood ratio statistic $\lambda_{H_4 \mid H_1}$ for testing H_4 against H_1 is given by

$$\lambda_{H_4 \mid H_1}^{2/n} = \frac{\left| \boldsymbol{I} + \begin{pmatrix} \boldsymbol{Z}_{14} \\ \boldsymbol{Z}_{24} \end{pmatrix} \boldsymbol{V}_{44}^{-1} (\boldsymbol{Z}'_{14}, \boldsymbol{Z}'_{24}) \right|}{\left| \boldsymbol{I} + (\boldsymbol{Z}_{12}, \boldsymbol{Z}_{13}, \boldsymbol{Z}_{14}) \begin{pmatrix} \boldsymbol{V}_{22} & \boldsymbol{V}_{23} & \boldsymbol{V}_{24} \\ \boldsymbol{V}_{32} & \boldsymbol{V}_{33} & \boldsymbol{V}_{34} \\ \boldsymbol{V}_{42} & \boldsymbol{V}_{43} & \boldsymbol{V}_{44} \end{pmatrix}^{-1} \begin{pmatrix} \boldsymbol{Z}'_{12} \\ \boldsymbol{Z}'_{13} \\ \boldsymbol{Z}'_{14} \end{pmatrix} \right|}$$

$$\times \frac{\left| \boldsymbol{I} + (\boldsymbol{Z}_{13}, \boldsymbol{Z}_{14}) \begin{pmatrix} \boldsymbol{V}_{33} & \boldsymbol{V}_{34} \\ \boldsymbol{V}_{43} & \boldsymbol{V}_{44} \end{pmatrix}^{-1} \begin{pmatrix} \boldsymbol{Z}'_{13} \\ \boldsymbol{Z}'_{14} \end{pmatrix} \right|^2}{\left| \boldsymbol{I} + \begin{pmatrix} \boldsymbol{Z}_{13} & \boldsymbol{Z}_{14} \\ \boldsymbol{Z}_{23} & \boldsymbol{Z}_{24} \end{pmatrix} \begin{pmatrix} \boldsymbol{V}_{22} & \boldsymbol{V}_{23} \\ \boldsymbol{V}_{32} & \boldsymbol{V}_{33} \end{pmatrix}^{-1} \begin{pmatrix} \boldsymbol{Z}'_{13} & \boldsymbol{Z}'_{23} \\ \boldsymbol{Z}'_{14} & \boldsymbol{Z}'_{24} \end{pmatrix} \right| \left| \boldsymbol{I} + \boldsymbol{Z}_{14} \boldsymbol{V}_{44}^{-1} \boldsymbol{Z}'_{14} \right|},$$

and the likelihood ratio statistic $\lambda_{H_4 \mid H_2}$ for testing H_4 against H_2 is given by

$$\lambda_{H_4 \mid H_2}^{2/n} = \frac{\left| \boldsymbol{I} + (\boldsymbol{Z}_{13}, \boldsymbol{Z}_{14}) \begin{pmatrix} \boldsymbol{V}_{33} & \boldsymbol{V}_{34} \\ \boldsymbol{V}_{43} & \boldsymbol{V}_{44} \end{pmatrix}^{-1} \begin{pmatrix} \boldsymbol{Z}'_{13} \\ \boldsymbol{Z}'_{14} \end{pmatrix} \right|}{\left| \boldsymbol{I} + (\boldsymbol{Z}_{12}, \boldsymbol{Z}_{13}, \boldsymbol{Z}_{14}) \begin{pmatrix} \boldsymbol{V}_{22} & \boldsymbol{V}_{23} & \boldsymbol{V}_{24} \\ \boldsymbol{V}_{32} & \boldsymbol{V}_{33} & \boldsymbol{V}_{34} \\ \boldsymbol{V}_{42} & \boldsymbol{V}_{43} & \boldsymbol{V}_{44} \end{pmatrix}^{-1} \begin{pmatrix} \boldsymbol{Z}'_{12} \\ \boldsymbol{Z}'_{13} \\ \boldsymbol{Z}'_{14} \end{pmatrix} \right|}.$$

Clearly, other variants of the above can be obtained in a similar fashion. For each of the test statistics $\lambda_{H_4}, \lambda_{H_4 \mid H_1}, \lambda_{H_4 \mid H_2}$, etc., the null and non-null distributions are of some interest and provide interesting problems in distribution theory. Consideration of these distributions, however, is beyond the scope of the present exposition.

Appendix

Let $\boldsymbol{X}$ be a $k \times p$ random matrix, each row of which has a multivariate normal distribution with means $\mathcal{E}\boldsymbol{X} = \boldsymbol{\mu}$, and common covariance matrix $\boldsymbol{\Sigma}$, and let $\boldsymbol{S}$ be an independent estimate of $\boldsymbol{\Sigma}$, i.e., $\boldsymbol{S} \sim \mathcal{W}(\boldsymbol{\Sigma}; p, n)$. Suppose that $\boldsymbol{X} = (\dot{\boldsymbol{X}}, \ddot{\boldsymbol{X}})$, $\boldsymbol{\mu} = (\dot{\boldsymbol{\mu}}, \ddot{\boldsymbol{\mu}})$, $\boldsymbol{\Sigma} = \begin{pmatrix} \boldsymbol{\Sigma}_{11} & \boldsymbol{\Sigma}_{12} \\ \boldsymbol{\Sigma}_{21} & \boldsymbol{\Sigma}_{22} \end{pmatrix}$, $\dot{\boldsymbol{X}}, \dot{\boldsymbol{\mu}} : k \times p_1$, $\Sigma_{11} : p_1 \times p_1$, and we wish to test

$$H : \dot{\boldsymbol{\mu}} \in \omega, -\infty < \ddot{\mu}_{ij} < \infty, \boldsymbol{\Sigma} > \boldsymbol{0},$$

versus

$$A : \dot{\boldsymbol{\mu}} \in \Omega, -\infty < \ddot{\mu}_{ij} < \infty, \ \boldsymbol{\Sigma} > \boldsymbol{0}.$$

Note that $\ddot{\boldsymbol{\mu}}$ is unrestricted under H and under A, so that the test involves only the means contained in $\dot{\boldsymbol{\mu}}$.

From the joint distribution of $\boldsymbol{X}$ and $\boldsymbol{S}$, namely,

$$\text{(A.1)} \qquad p(\boldsymbol{X}, \boldsymbol{S}) = c\,|\boldsymbol{S}|^{\frac{n-p-1}{2}}\,|\boldsymbol{\Sigma}|^{-\frac{N}{2}} \exp\{-\tfrac{1}{2}\operatorname{tr} \boldsymbol{\Lambda}[\boldsymbol{S}+(\boldsymbol{X}-\boldsymbol{\mu})'(\boldsymbol{X}-\boldsymbol{\mu})]\},$$

where $\boldsymbol{\Lambda} \equiv \boldsymbol{\Sigma}^{-1}$, we can find the LRS in the usual way as the ratio

$$\text{(A.2)} \qquad \frac{\sup\limits_{\dot{\boldsymbol{\mu}} \in \omega,\, -\infty < \ddot{\mu}_{ij} < \infty,\, \boldsymbol{\Sigma} > \boldsymbol{0}} p(\boldsymbol{X}, \boldsymbol{S})}{\sup\limits_{\dot{\boldsymbol{\mu}} \in \Omega,\, -\infty < \ddot{\mu}_{ij} < \infty,\, \Sigma > \boldsymbol{0}} p(\boldsymbol{X}, \boldsymbol{S})}.$$

We now show that the test based on the LRS for testing H versus A is equivalent to the test based on the LRS for the reduced problem in which we are given $\dot{\boldsymbol{X}}$ and $\boldsymbol{S}_{11}$ and wish to test $H^* : \dot{\boldsymbol{\mu}} \in \omega$, $\boldsymbol{\Sigma}_{11} > \boldsymbol{0}$ versus $A^* : \dot{\boldsymbol{\mu}} \in \Omega$, $\boldsymbol{\Sigma}_{11} > \boldsymbol{0}$. The LRS for the reduced problem is given by

$$\text{(A.3)} \qquad \frac{\sup\limits_{\dot{\boldsymbol{\mu}} \in \omega,\, \Sigma_{11} > \boldsymbol{0}} p(\dot{\boldsymbol{X}}, \boldsymbol{S}_{11})}{\sup\limits_{\dot{\boldsymbol{\mu}} \in \Omega,\, \Sigma_{11} > \boldsymbol{0}} p(\dot{\boldsymbol{X}}, \boldsymbol{S}_{11})}.$$

In (A.1) we first maximize with respect to $\ddot{\boldsymbol{\mu}}$, and obtain in a straightforward manner that

$$\hat{\ddot{\mu}} = \ddot{X} + (\dot{X} - \dot{\mu})\Lambda_{12}\Lambda_{22}^{-1} = \ddot{X} - (\dot{X} - \dot{\mu})\Sigma_{11}^{-1}\Sigma_{12},$$

from which $(\dot{X} - \dot{\mu}, \ddot{X} - \hat{\ddot{\mu}})\, \Lambda (\dot{X} - \dot{\mu}, \ddot{X} - \hat{\ddot{\mu}})'$

$$= (\dot{X} - \dot{\mu})\Sigma_{11}^{-1}(\dot{X} - \dot{\mu})'.$$

Consequently,

$$\text{(A.4)} \qquad \sup_{\ddot{\mu}} p(X, S) = c\,|S|^{\frac{n-p-1}{2}}\, |\Sigma|^{-\frac{N}{2}}$$

$$\exp\left\{-\frac{1}{2}\operatorname{tr}\Lambda S - \frac{1}{2}\operatorname{tr}\Sigma_{11}^{-1}(\dot{X} - \dot{\mu})'(\dot{X} - \dot{\mu})\right\}.$$

We now maximize with respect to Σ_{22} and Σ_{12} over the domain $\Sigma_{22} > \Sigma_{21}\Sigma_{11}^{-1}\Sigma_{12}$ and Σ_{12} unrestricted to obtain the following lemma.

Lemma

$$\sup_{\substack{\Sigma_{22} > \Sigma_{21}\Sigma_{11}^{-1}\Sigma_{12} \\ \Sigma_{12} \text{ unrestricted}}} |\Sigma|^{-\frac{N}{2}} \exp\left\{-\frac{1}{2}\operatorname{tr}\Lambda S\right\}$$

$$= \left(|\Sigma_{11}|^{-\frac{N}{2}} e^{-\frac{1}{2}\operatorname{tr}\Sigma_{11}^{-1}S_{11}}\right)\left(|S_{22} - S_{21}S_{11}^{-1}S_{12}|^{-\frac{N}{2}}\right.$$

$$\left.\exp\left\{-\frac{1}{2}p_1 N\right\}\right).$$

Proof First note that

$$|\Lambda| = |\Lambda_{11} - \Lambda_{12}\Lambda_{22}^{-1}\Lambda_{21}|\,|\Lambda_{22}| = |\Sigma_{11}^{-1}|\,|\Lambda_{22}|,$$

$$\operatorname{tr}\Lambda S = \operatorname{tr}\Sigma_{11}^{-1}S_{11} + (\operatorname{tr}\Lambda S - \operatorname{tr}\Sigma_{11}^{-1}S_{11})$$

$$= \operatorname{tr}\Sigma_{11}^{-1}S_{11} + (2\operatorname{tr}\Lambda_{12}S_{21} + \operatorname{tr}\Lambda_{12}\Lambda_{22}^{-1}\Lambda_{21}S_{11} + \operatorname{tr}\Lambda_{22}S_{22})$$

$$= \operatorname{tr}\Sigma_{11}^{-1}S_{11} + \operatorname{tr}\Lambda_{22}[S_{22} + 2\Lambda_{22}^{-1}S_{21}\Lambda_{12} + \Lambda_{22}^{-1}\Lambda_{21}S_{11}\Lambda_{12}\Lambda_{22}^{-1}]$$

$$= \operatorname{tr}\Sigma_{11}^{-1}S_{11} + \operatorname{tr}\Lambda_{22}[S_{22} + (\Psi + L)S_{11}(\Psi + L)' - LS_{11}L'],$$

where $\Psi = \Lambda_{22}^{-1}\Lambda_{21}$, $L = S_{21}S_{11}^{-1}$. Hence, we need to find

$$\sup_{\substack{\Psi \text{ unrestricted} \\ \Lambda_{22} > 0}} |\Sigma_{11}^{-1}|^{-\frac{N}{2}}\, |\Lambda_{22}|^{-\frac{N}{2}} \exp\left\{-\frac{1}{2}\operatorname{tr}\Sigma_{11}^{-1}S_{11}\right\}$$

$$\times \exp\left\{-\frac{1}{2}\operatorname{tr}\Lambda_{22}[S_{22} + (\Psi + L)S_{11}(\Psi + L)' - LS_{11}L']\right\},$$

where $\boldsymbol{\Lambda}_{22} = (\boldsymbol{\Sigma}_{22} - \boldsymbol{\Sigma}_{21}\boldsymbol{\Sigma}_{11}^{-1}\boldsymbol{\Sigma}_{12})^{-1}$. But now the maximization can be obtained directly as $\hat{\boldsymbol{\Psi}} = -\boldsymbol{L}$ and $\hat{\boldsymbol{\Lambda}}_{22} = \boldsymbol{S}_{22} - \boldsymbol{L}\boldsymbol{S}_{11}\boldsymbol{L}' = \boldsymbol{S}_{22} - \boldsymbol{S}_{21}\boldsymbol{S}_{11}^{-1}\boldsymbol{S}_{12}$.

The equivalence of (A.2) and (A.3) follows by using (A.4) and the Lemma in the computation of (A.2).

References

[1] Anderson T. W. (1958). *Introduction to Multivariate Statistical Analysis,* John Wiley and Sons, New York.

[2] Chernoff, H. (1956). "Large-Sample Theory : Parametric Case," *Ann. Math. Statist.,* **27,** 1–22.

[3] Chow, G. C. (1960). "Tests of Equality Between Sets of Coefficients in Two Linear Regressions," *Econometrica,* **28,** 591–605.

[4] Cochran, W. G. and Bliss, C. I. (1948). "Discriminant Functions With Covariance," *Ann. Math. Statist.,* **19,** 151–176.

[5] Giri, N. C. (1964). "On the Likelihood Ratio Test of a Normal Multivariate Testing Problem," *Ann. Math. Statist.,* **35,** 181–189.

[6] Gleser, L. J. and Olkin, I. (1964). "Estimation for a Regression Model With Covariance," Technical Report No. 15 (NSF Grant GP-214), Stanford University.

[7] Gleser, L. J. and Olkin, I. (1966). "A k-Sample Regression Model With Covariance," *Multivariate Analysis* (P. R. Krishnaiah, Editor). Academic Press, New York.

[8] Hsu, P. L. (1941). "Canonical Reduction of the General Regression Problem," *Ann. Eugenics,* **11,** 42–46.

[9] MacDuffee, C. C. (1946). *The Theory of Matrices,* Second ed., Chelsea Publishing Co., New York.

[10] Olkin, I. and Shrikhande, S. S. (1954). "On a Modified T^2 Problem," *Ann. Math. Statist.,* **25,** 808.

[11] Potthoff, R. F. and Roy, S. N. (1964). "A Generalized Multivariate Analysis of Variance Model Useful Especially for Growth Curve Problems," *Biometrika,* **51,** 313–26.

[12] Rao C. Radhakrishna (1946). "Tests with Discriminant Functions in Multivariate Analysis," *Sankhyā,* **7,** 407–414.

[13] Rao, C. Radhakrishna (1959). "Some Problems Involving Linear Hypotheses in Multivariate Analysis," *Biometrika,* **46,** 49–58.

[14] Rao, C. Radhakrishna (1961). "Some Observations on Multivariate Statistical Methods in Anthropological Research," *Bull. Inst. Internat. Statist.*, **38,** 99–111.

[15] Rao, C. Radhakrishna (1965). "The Theory of Least Squares When the Parameters are Stochastic and Its Application to the Analysis of Growth Curves," *Biometrika,* **52,** 447–458.

[16] Scheffé, H. (1959). *The Analysis of Variance*, John Wiley and Sons, New York.

(Received Feb. 5, 1966.)

CHAPTER 14

S. N. Roy's Interests in and Contributions to the Analysis and Design of Certain Quantitative Multiresponse Experiments

R. GNANADESIKAN[1], *Bell Telephone Laboratories*

1. INTRODUCTION

It is with a keen sense of shared loss and privileged duty that I associate myself with the purposes of this memorial tribute to the late Professor S. N. Roy. First as a student and then as a collaborator, it was a cherished opportunity of mine to get to know Professor Roy not only in the realms of professional activity but also in the informal relationship of person-to-person. His rare and genuine modesty was as exemplary as his quiet and firm intellectual timber was exciting. As the Bard of Avon so aptly said it[2],

His life was gentle, and the elements

[1] Talk presented at the S. N. Roy Memorial Session of the Annual Joint Statistical Meetings at Chicago in December 1964.

[2] *Julius Caesar*, Act V. Sc. 5.

So mix'd in him that Nature might stand up
And say to all the world, "This was a man !"

The late Professor Roy was a person of broad culture and varied interests both in his general outlook and in his specific professional participations and technical activities. It would require an abler person than me to encompass adequately the entire area of Professor Roy's interests and contributions even if these were confined to the recent time period of the past decade. I have decided to limit my remarks to an area of Professor Roy's interests with which I have been directly associated as a collaborator along with several other students and colleagues of his. Specifically, I shall, for lack of both commitment and competence, omit a large body of work, of a nonparametric character, that Professor Roy and some of his co-workers have been developing over the past decade. I shall, in my attempted expository survey of Professor Roy's interests and contributions over the past decade, confine myself to developments relevant to quantitative multiresponse experiments within the familiar framework of normal distribution theory. The material, providing the basis of the present paper, is the content, in part, of an as yet unpublished monograph [10] on multiresponse experiments coauthored by the late Professor Roy, J. N. Srivastava and the present author. Several colleagues, apart from the authors of the monograph, have contributed to various facets of the research endeavour, and among them mention must be made of R. C. Bose, P. R. Krishnaiah, W. F. Mikhail, G. S. Mudholkar, J. Roy, and M. B. Wilk.

Table 1 shows a possible way of structuring the area of concern, namely quantitative multiresponse experiments.

There is, firstly, the broad categorization of the area into questions relating to construction of designs and those relating to development of methods of analysis. Under design questions, one could have the usual designs of univariate statistical theory, such as the familiar factorial experiment, with the statistically important difference that each experimental unit, after applica-

TABLE 1

(*Multiresponse Experiments Normal Theory*) +

DESIGN		ANALYSIS	
USUAL DESIGNS	SPECIAL DESIGNS	FORMAL METHODS	INFORMAL METHODS
E.g. Factorial expt. with multiple response on each experimental unit.	E.g. Deliberate planning for "missing observations" with certain responses.	(i) Estimation (ii) Tests (iii) Multiple comparisons.	E.g. Probability plotting procedures for internal comparisons.

+ The label, *normal theory*, is used merely to distinguish the present concern from the non-parametric work mentioned earlier, and is not meant to suggest that all the statistical methods involved are equally crucially dependent on the assumptions of normality.

tion of a treatment combination, is observed on several response characteristics such as yield, viscosity, etc., instead of on just one response variable. On the other hand, for multiresponse problems, one may also need special designs, as, for example, illustrated by a situation in which the experimenter has background knowledge and information that certain of the factors in his experiment do not significantly affect certain of the responses, and wishes to incorporate such knowledge in the experimental design. Apart from some interesting work of Cuthbert Daniel [2] in this area, the only major systematic effort in this direction appears to have been initiated by the late Professor Roy and is being vigorously carried forward by J. N. Srivastava. In these special designs, which are characterized by a deliberate planning for missing observations with respect to specified responses, the methods of analysis to be used after the data have been collected play an intimate, important and integral role even at the design stage, in that the designs are developed in the light of the method of analysis to be used later.

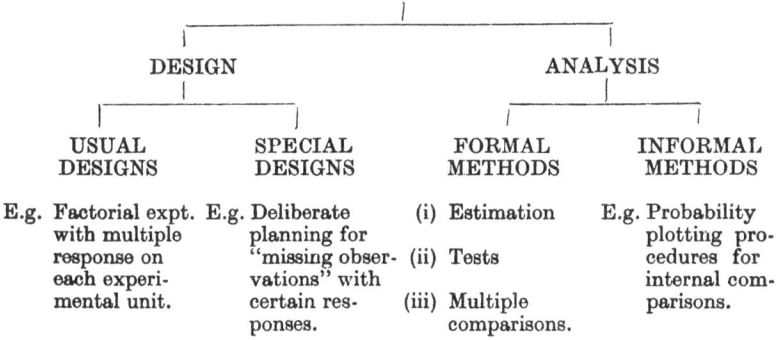

To return to the categorization of the area of concern, with regard to methods of analysis, there is a further subdivision possible into *formal methods* and *informal methods*. The problems of formal statistical methods include those of point estimation, tests of hypotheses within the limited scope of the Neyman-Pearson inferential philosophy, and various multiple comparisons procedures. The informal methods, which are useful for gaining an appreciation of and insight into the structure of multiresponse data, include certain probability plotting procedures proposed as internal comparisons methods by M. B. Wilk and the present author [14], [15], [16], [17], [18].

Although the monograph mentioned heretofore contains a description and discussion of statistical theory and methods pertaining to each of the areas in the above categorization, yet, in the present paper, the emphasis will be on the methods of analysis. One reason for this is the inescapable personal bias of the present author in favour of issues relevant to methods of analysis. On the value and relevance of most of the recent work on construction of designs from a univariate point of view, the late Professor Roy and the present author would be entirely in agreement with the assessment of Professor Kempthorne who, in his invited talk on the current status of the design and analysis of experiments at the 1964 Annual Meetings of the Institute of Mathematical Statistics, stated that (and I paraphrase), these contributions should be judged on their intellectual merit as contributions to pure mathematics rather than on grounds of relevance to statistical practice. This is not to deny the existence of design construction problems of relevance to statistical practice, when one desire to take a multivariate point of view.

The present paper is, in its limited objectives, an attempt to survey briefly certain aspects of both formal and informal methods of analysis for quantitative multiresponse experiments, when the familiar designs of univariate theory are used but multiple responses are observed on each experimental unit.

2. STATEMENT OF MODEL

The usual fixed-effects model may be stated as shown in Table 2.

TABLE 2

Fixed-effects models

(1) Mutually independent p-dimensional observations,

$$\boldsymbol{y}_i' = (y_{i1}, \ldots, y_{ip}), \quad i = 1, 2, \ldots, n,$$

with common unknown $p \times p$ nonsingular covariance matrix, $\boldsymbol{\Sigma}$, and means specified by the linear model,

$$E(\boldsymbol{Y}) = E\begin{pmatrix} \boldsymbol{y}_1' \\ \vdots \\ \boldsymbol{y}_n' \end{pmatrix} = \boldsymbol{A\Theta}.$$

(a) $\boldsymbol{A}$ is $n \times m$ *design matrix* with known elements.

(b) $\boldsymbol{\Theta}$ is $m \times p$ matrix of m (unknown) p-dimensional fixed effects.

(c) $m < n$; $p \leqslant n - \text{rank}(\boldsymbol{A})$.

(2) $\boldsymbol{y}_i'$'s have a p-dimensional normal distribution.

There are n mutually independent p-dimensional observations $\boldsymbol{y}_i'$'s ($i = 1, 2, \ldots, n$), each with the same unknown $p \times p$ nonsingular covariance matrix, $\boldsymbol{\Sigma}$, and means specified by the linear model as shown under item (1) of Table 2. It is assumed that the number of unknowns, m is less than n, the number of observations, and also that the dimensionality of the observations, p, does not exceed the number of observations minus the rank of the design matrix. This last restriction ensures a nonsingular estimate of $\boldsymbol{\Sigma}$ and avoids the so-called high-dimensional problem considered by Dempster [4]. The assumption of a p-dimensional normal distribution for the observations is necessary for some but not for all of the methods to be developed.

It should be pointed out that while the p-dimensional fixed effects (i.e., rows of $\boldsymbol{\Theta}$) are mathematically simple analogues of unidimensional effects, yet there are important conceptual and interpretational difficulties with multidimensional effects. As Tukey [13] has said, "...multiple-response differences are not simple, are usually not easy to think about, are usually not easy to describe."

The assumptions, stated formally in Table 2 as constituting a model, should be viewed as having primary value in generating statistical methods which provide insight into a body of data. Some of them, such as the assumption of a common error covariance matrix for the observations, are more critical for the formal methods than for the informal ones which have the analytically desirable property that they provide insights into the validity of the assumptions, in addition to focusing attention on other interesting facets underlying a specific body of data to which they are applied.

Before we go on to a discussion of methods of analysis, it may be profitable to look at a *rolled-out* version of the model, shown in Table 3, since this version is a useful mathematical representation for developing certain of the methods.

TABLE 3

Rolled-out version

(1) $\boldsymbol{Y}(n\times p) = [\boldsymbol{Y}_1, \boldsymbol{Y}_2, \ldots, \boldsymbol{Y}_p]$; $\boldsymbol{\Theta}(m\times p) = [\boldsymbol{\theta}_1, \boldsymbol{\theta}_2, \ldots, \boldsymbol{\theta}_p]$

$$\boldsymbol{y}^*(pn\times 1) = \begin{bmatrix} \boldsymbol{Y}_1 \\ \boldsymbol{Y}_2 \\ \vdots \\ \boldsymbol{Y}_p \end{bmatrix}; \qquad \boldsymbol{\theta}^*(pm\times 1) = \begin{bmatrix} \boldsymbol{\theta}_1 \\ \boldsymbol{\theta}_2 \\ \vdots \\ \boldsymbol{\theta}_p \end{bmatrix}$$

(2) $E(\boldsymbol{y}^*) = \boldsymbol{A}^*\boldsymbol{\theta}^*$; $\boldsymbol{A}^* = \boldsymbol{A}\otimes\boldsymbol{I}(p)$.

Cov. matrix of $\boldsymbol{y}^* = \boldsymbol{\Sigma}^* = \boldsymbol{I}(n)\otimes\boldsymbol{\Sigma}$.

What is shown here is a rolling out of the $n\times p$ observation matrix by columns (i.e., the n observations on the first response followed by the n observations on the second response and so on until we get to the n observations on the last response) and a similar rolling out by columns of the matrix of effects, $\boldsymbol{\Theta}$. A linear model for the pn-dimensional column vector of observations, $\boldsymbol{y}^*$, in terms of the pm-dimensional column vector of effects, $\boldsymbol{\theta}^*$, may then be written as shown in item (2) of Table 3.

3. FORMAL METHODS

Next formulations of problems are considered to which are addressed the formal statistical methods developed under the aforementioned model.

3.1 Estimation

First we consider the area of statistical point estimation of the effects and of linear functions of them that may be of interest.

First-order constraints and considerations, such as unbiasedness and estimability, which are familiar in univariate theory, generalize directly and simply to the multivariate situation. However, second-order considerations, which are related to assessing the estimates in terms of *dispersion* characteristics, lead to complexities in the multivariate case. The latter issues arise in situations which are characterized not by the response being multidimensional but by an interest of the experimenter in simultaneous estimation of different functions of the effects. Even when the response is unidimensional, multivariate considerations could arise when simultaneous estimation is the concern. In general, simultaneous estimation would necessitate contending both with the different variances of the estimates as well as with the covariance structure among the estimates.

Table 4 shows a two-way breakdown of the problems of point estimation according as the response is unidimensional or multidimensional and according as the estimation interest of the experimenter in the treatment structure is separate estimation of single linear functions or simultaneous estimation of several linear functions.

TABLE 4

Point estimation

Treatment \ Response	Uniresponse	Multiresponse
Single	(i) Classical linear estimation ($\eta = (\mathbf{c}'\boldsymbol{\theta})$ (Least squares; Gauss-Markov).	(iii) Simple extension ($\eta = \mathbf{c}^{*\prime}\boldsymbol{\theta}^{*}$) (Do what is "best" for each response separately).
Simultaneous	(ii) Multivariate criteria ($\boldsymbol{\eta} = \boldsymbol{B\theta}$) (single unknown variance).	(iv) Multivariate criteria ($\boldsymbol{\eta} = \boldsymbol{B}^{*}\boldsymbol{\theta}^{*}$) (unknown covariance matrix).

In Cells (i) and (ii), i.e., uniresponse, in the notation of the model presented earlier, there is a vector, $\boldsymbol{\theta}$, of m univariate effects. In Cell (i) the interest is in estimating a single linearly estimable linear function, $\boldsymbol{\eta}$, of the m effects. In Cell (ii), the interest is in simultaneous estimation of the elements of the vector, $\boldsymbol{\eta}$, which is a set of linearly estimable linear functions of the m effects. Such an interest has provided the motivation for some of the work on optimal designs. Criteria pertaining to the assessment of simultaneous estimates have been used as bases for evaluating *optimality* of a design and for constructing *optimal* designs.

When the response is multidimensional, i.e., Cells (iii) and (iv), using the rolled-out version of the model, one has a pm-dimensional vector, $\boldsymbol{\theta}^*$, of effects, and the interest in Cell (iii) is separate estimation of a single linearly estimable linear function, η, of the elements of $\boldsymbol{\theta}^*$; while in Cell (iv) it is in simultaneous estimation of the elements of a vector, η, which consists of a set of linearly estimable linear functions of the elements of $\boldsymbol{\theta}^*$.

As for the availability of methods and results, they appear to decrease in number from Cells (i) to (iv). Cell (i), namely uniresponse single estimation, is of course the classical case of linear estimation theory with the method of least squares and the Gauss-Markov result concerning the least squares estimates. For Cell (iii), the criteria of assessing an estimate are univariate in nature, and there exist fairly simple extensions of the theoretical results of Cell (i) to this case. In particular, for this multiresponse case, the simple rule is to use the separately *best* estimates for each response in combining them to obtain the *best* estimate of the single linear function, η, of the elements of $\boldsymbol{\theta}^*$.

For Cells (ii) and (iv), which deal with simultaneous estimation, the *dispersion* summary is, in general, provided by a covariance matrix and alternate reductions of this matrix summary into single dimensional criteria exist for assessing the *dispersion* characteristics of the collection of estimates. Among such one-dimensional reductions are : (1) the generalized variance which is the determinant of the covariance matrix; (2) the largest latent root of the covariance matrix; and (3) the trace, or sum of the variances. There are meaningful statistical and physical interpretations for each of these criteria. A probably interesting implication of estimates that are optimal with respect to the third criterion, i.e., those that minimize the trace of the covariance matrix, is that they are also optimal with respect to the other two criteria. Furthermore, as for methods of constructing estimates that satisfy each of the criteria, the usual minimum variance theory and techniques are sufficient for meeting the trace criterion, whereas new statistical methods may be needed for deriving estimates that satisfy the other two criteria without necessarily being optimal with respect to the trace criterion. Simultaneous estimates of interest in practice are more likely to be not ones with sharp optimality with regard to any single narrow criterion, but rather, ones with reasonably good properties with regard to several alternate meaningful criteria.

The covariance structure in Cell (ii) is somewhat simpler than that in Cell (iv), since, in the former, the covariance matrix involves a single unknown variance, σ^2, of the univariate response, while, in the latter, the unknown covariance matrix, $\mathbf{\Sigma}$, of the multiple responses would be involved.

Before leaving the area of point estimation, it may be mentioned that, in the development of the theory and methods here, no explicit use is made of normal theory although, implicitly, there is perhaps some influence in the conceptualization of the criteria for assessing the estimates.

3.2. Tests of Hypotheses

The next category of formal methods is tests of hypotheses, wherein the normal distribution assumption plays a role in developing the methods. Knowledge is sparce concerning the robustness of formal multiresponse statistical procedures which are developed within the framework of normal theory.

With the notation introduced earlier, the general linear hypothesis, for the multiresponse case under the multiresponse fixed-effects model, may be stated as shown in item (1) of Table 5.

TABLE 5

General linear hypothesis

(1) $H_0 : C\Theta = 0$, $H : C\Theta \neq 0$.

(a) C is an $s \times m$ matrix of specified constants.

(b) rank $(C) \leqslant$ rank (A).

(2) *Example* : v treatments with p-dimensional effects, $\tau_1, \tau_2, \ldots, \tau_v$.

$$H_0 : \text{No treatment effects} \Longleftrightarrow \begin{bmatrix} 1 & 0 & 0 & \ldots & 0 & -1 \\ 0 & 1 & 0 & \ldots & 0 & -1 \\ \ldots & & \ldots & & \ldots & \\ 0 & 0 & 0 & \ldots & 1 & -1 \end{bmatrix} \begin{bmatrix} \tau_1' \\ \tau_2' \\ \ldots \\ \tau_v' \end{bmatrix} = 0.$$

The null statement is that a set of linear combinations (subject to being *estimable*) of the p-dimensional effects are zero, i.e., an $s \times m$ matrix of specified constants, C, times the $m \times p$ matrix of effects, Θ, is null. The completely general alternative to this null hypothesis is, of course, the hypothesis that $C\Theta$ is not null. An illustrative example is shown in item (2) of Table 5. If one has v treatments, with their corresponding p-dimensional effects being denoted as column vectors, τ_1, $\tau_2, \ldots, \tau_v$, then the null hypothesis of no treatment effects on any of the response scales could be stated by using the $(v-1) \times v$ matrix, C, shown in item (2) of Table 5, along with a matrix, Θ, whose rows are the v p-dimensional treatment effects.

A general heuristic method of constructing tests was proposed by Roy [9], and this method has since been used extensively for deriving tests of various hypotheses that may be of interest in multiresponse experiments. A fundamental feature of the heuristic principle is that it proposes viewing an overall hypothesis, (which generally tends to be quite complex even if oversimplified in certain other ways), as the *intersection* (in a logical sense) of more elementary hypotheses. The test procedures are then developed as aids to pin-pointing which, if any, of the elementary hypotheses are rejected when the overall integrated hypothesis is rejected. The sensibleness of these procedures that lead to an insight into the structure of the appropriate underlying alternative hypothesis, lies in the fact that generally the null hypothesis is not of prime interest and only serves as a backdrop for studying the deviations from it.

Thus, as shown in Table 6, the proposed viewpoint is to consider an overall null hypothesis, H_0, as the intersection of certain elementary null hypotheses, H_{0i}, and similarly the alternative, H, to H_0 as either the intersection or union of certain other elementary hypotheses, H_i, which are natural alternatives to the H_{0i}. The index set, Γ, to which i belongs and over which the intersections and/or unions are considered may be finite or infinite.

TABLE 6

Decomposition of hypotheses

(1) $H_0 = \bigcap_{i \in \Gamma} H_{0i}$ and $H = \bigcup_{i \in \Gamma} H_i$, (or, $\bigcap_{i \in \Gamma} H_i$).

Γ is an index set.

H_{0i}, H_i are *partials* or *components* of H_0 and H.

Procedure : Accept H_0 *iff* accept every H_{0i}.

(2) *Example* :

(i) $H_0 : \boldsymbol{\tau}_1 = \boldsymbol{\tau}_2 = \ldots = \boldsymbol{\tau}_v$.

(ii) $H_0 = \bigcap_{\boldsymbol{a}} H_{0\boldsymbol{a}} = \bigcap_{\boldsymbol{a}} [\boldsymbol{a}'\boldsymbol{\tau}_1 = \boldsymbol{a}'\boldsymbol{\tau}_2 = \ldots = \boldsymbol{a}'\boldsymbol{\tau}_v]$

$H_{0\boldsymbol{a}}$ is univariate hypothesis and intersection is over all non-null p-dimensional vectors $\boldsymbol{a}$.

(iii) $H_{0\boldsymbol{a}}$ itself is "intersection" over all treatment contrasts.

Let,

$$\boldsymbol{\tau} = [\boldsymbol{\tau}_1, \boldsymbol{\tau}_2, \ldots, \boldsymbol{\tau}_v].$$

Then,

$$H_0 = \bigcap_{\boldsymbol{a}} H_{0\boldsymbol{a}} = \bigcap_{\boldsymbol{a}} \bigcap_{\boldsymbol{b}} H_{0\boldsymbol{ab}} = \bigcap_{\boldsymbol{a}} \bigcap_{\boldsymbol{b}} [\boldsymbol{a}'\boldsymbol{\tau}\boldsymbol{b} = \boldsymbol{0}],$$

where the intersection is over all non-null p-dimensional vectors $\boldsymbol{a}$ and over all v-dimensional vectors $\boldsymbol{b}$ which are such that the sum of their elements is zero (contrasts).

With such a decomposition of H_0 and H into their respective *component* hypotheses, H_{0i} and H_i, the heuristic principle for constructing a test for the overall H_0 is to not reject it if and only if no component H_{0i} of it is rejected. Thus, by setting up a test procedure for the component null hypothesis, H_{0i}, against the component alternative hypothesis, H_i, in the familiar two-decision Neyman-Pearson set-up, one obtains an acceptance region for the overall H_0 by taking the intersection over i of the acceptance regions for the components H_{0i}.

An example may help clarify the approach. Item (2) of Table 6 deals with the example mentioned earlier wherein the null hypothesis specifies equality of v p-dimensional treatment effects. Consider a linear combination of the p unidimensional effects that are elements of each vector of effects, $\boldsymbol{\tau}_1, \ldots, \boldsymbol{\tau}_v$; that is, consider $\boldsymbol{a}'\boldsymbol{\tau}_1, \boldsymbol{a}'\boldsymbol{\tau}_2, \ldots, \boldsymbol{a}'\boldsymbol{\tau}_v$, where $\boldsymbol{a}'$ denotes a p-dimensional row vector whose elements are the coefficients in the linear combination. These would be v univariate treatment effects on the scale of a single response defined as the linear combination, $\boldsymbol{a}'\boldsymbol{y}$, of the p responses. For a specified vector $\boldsymbol{a}$, therefore, one could consider a univariate hypothesis, $H_{0\boldsymbol{a}}$, of no treatment effects on the scale of this derived response which is a linear combination of the p original responses. This is shown as (ii) under item (2) of Table 6. Also, H_0 is true if and only if $H_{0\boldsymbol{a}}$ is true for all non-null choices of the vector, $\boldsymbol{a}$, of coefficients for the linear combination. Thus, H_0 is the *intersection* of $H_{0\boldsymbol{a}}$ over the non-denumerably infinite choices of non-null $\boldsymbol{a}$'s, each of which reduces the p-dimensional response problem to a single-dimensional one.

Going one step further, however, as shown in (iii) of Table 6, $H_{0\boldsymbol{a}}$ itself may be regarded as an *intersection* over all possible contrasts among the v unidimensional treatment effects, $\boldsymbol{a}'\boldsymbol{\tau}_1, \boldsymbol{a}'\boldsymbol{\tau}_2, \ldots, \boldsymbol{a}'\boldsymbol{\tau}_v$. That is, defining a v-dimensional vector, $\boldsymbol{b}$, whose elements add to zero so that they can be used to define a contrast, the hypothesis, $H_{0\boldsymbol{a}}$, is true if and only if the hypothesis, $H_{0\boldsymbol{ab}}$, shown in Table 6, is true for all $\boldsymbol{b}$. Thus H_0 can be written as an intersection over both non-denumerably infinite sets of choices of p-dimensional non-null vectors, $\boldsymbol{a}$, and of v-dimensional non-null vectors, $\boldsymbol{b}$, that define contrasts. Such a decomposition is called a *response-wise* (through $\boldsymbol{a}$) and *contrast-wise* (through $\boldsymbol{b}$) *infinite*, or *doubly infinite* decomposition.

For the univariate null hypothesis, $H_{0\boldsymbol{a}}$, there is the well-known F-test (whose acceptance region is itself the intersection over all t-test acceptance regions for single-degree-of-freedom contrasts). Application of the heuristic principle, mentioned earlier for obtaining a test for the overall H_0, leads to an acceptance region, for H_0, which is the intersection of an infinite number of F-test acceptance regions. This acceptance region is entirely equivalent to the acceptance region of the largest latent root test proposed by Roy, and the above view in terms of a decomposition of hypotheses is merely a convenient derivation of that well-known test.

Other possible decompositions of hypotheses of interest in multiresponse experiments are shown in Table 7.

TABLE 7

Types of decomposition and associated tests

(1) RESPONSE-WISE AND CONTRAST-WISE INFINITE

(2) RESPONSE-WISE INFINITE AND CONTRAST-WISE FINITE

Example : $H_0 = \bigcap_{\boldsymbol{a}} \bigcap_{i \neq j=1}^{v} [\boldsymbol{a}'(\boldsymbol{\tau}_i - \boldsymbol{\tau}_j) = 0]$

(3) RESPONSE-WISE FINITE AND CONTRAST-WISE INFINITE

Example : $H_0 = \bigcap_{i=1}^{p}$ [Step-down univariate linear hypothesis]

(4) RESPONSE-WISE FINITE AND CONTRAST-WISE FINITE

Example : $H_0 = \bigcap_{i=1}^{p}$ [All pair-wise differences of univariate effects are null].

Firstly, there is the doubly infinite decomposition, an example of which was just presented. Secondly, there is the response-wise infinite but contrast-wise finite decomposition. An example of this would be where, in terms of the v unidimensional effects on the derived response scale, namely $\boldsymbol{a}'\boldsymbol{\tau}_1, \ldots, \boldsymbol{a}'\boldsymbol{\tau}_v$, one is interested not in all possible contrasts but merely pair-wise differences. H_0 may then be written as shown under item (2) of Table 7. The heuristic principle leads, in this example, to a test for H_0 whose acceptance region is the intersection of $v(v-1)/2$ Hotelling's T^2 acceptance regions.

Thirdly, there is the response-wise finite but contrast-wise infinite decomposition. An example of this is the so-called *step-wise* or *step-down* approach for multiresponse problems [5], [8]. This involves combined consideration of the marginal behaviour of one response and a hierarchical sequence of conditional behaviours of the other responses. In particular, the overall multiresponse general linear hypothesis, is, in this case, viewed as the intersection of p uniresponse general linear hypotheses, for each of which there is an F-test. Thus, here for the overall hypothesis, the acceptance region is the intersection of p F-test acceptance regions. Under the null hypothesis, the p F-statistics may be shown to be mutually independently distributed.

Finally, there is the doubly finite, or response-wise and contrast-wise finite decomposition of an overall hypothesis. An example of a procedure associated with such a decomposition in multiresponse experiments would be the use of an acceptance region which is the intersection of p step-down univariate acceptance regions, each of which is itself a finite intersection of $v(v-1)/2$ t-test acceptance regions corresponding to pair-wise differences of v treatment effects.

The present state of knowledge concerning the operating characteristics of the alternate test procedures proposed is not sufficiently detailed to be of use in choosing between the procedures in a specific application in the light of its background information. Various authors have established certain global properties, such as admissibility, of some of the tests. Also, several authors, including Das Gupta, Anderson and Mudholkar [3] in addition to some of the earlier mentioned students and colleagues of Roy [11], [12], have shown that a wide class of tests, used in multiresponse analysis of variance, have the monotone power property that the power increases as each noncentral parameter separately increases from its null value of zero.

Unlike in the univariate situation, for multiresponse problems, the statistics that are used for purposes of tests of hypotheses, such as various functions of certain latent roots, are complex derivatives from the data and do not generally have significant value as easily understood summaries of the structure underlying the multiresponse data. It would, therefore, be useful to generate data with certain kinds of meaningful underlying structure and to study the usefulness of the different multiresponse test statistics for uncovering such structures. Such a Monte Carlo approach is currently being pursued by the present author in collaboration with Lauh, Snyder and Yao. [6].

3.3. Multiple Comparison Procedures

For the third and last group of formal methods, namely multiple comparison procedures, there exist various simultaneous confidence estimation procedures which provide the basis for the multiple comparisons. In the multiresponse situation, in order to obtain intervals rather than complicated regions, certain compromises seem to be necessary. Firstly, the intervals are on certain not-always-easy-to-interpret functions of the parameters which, nevertheless, may be considered as distance functions which assume the value zero under null conditions. Secondly, the simultaneous confidence statements have an associated conservative confidence probability ($\geqslant (1-\alpha)$) rather than an exact one. Other than these features, the relationship of the results on simultaneous confidence estimation to their use for purposes of multiple comparisons is quite similar in the multiresponse and uniresponse cases. Procedures have been proposed appropriate to each type of decomposition of hypotheses considered earlier.

4. INFORMAL METHODS

The late Professor Roy, coming out of a background of formal training in physics with its strong empirical traditions and heritage, accepted, in a natural manner, the pragmatic principle of usefulness in practice as the touchstone of work that is claimed to be of relevance to statistical practice. He was explicit in recognizing that the objectives of statistical analysis are neither so narrow nor so formal as described and implied by some statistical theories of estimation and testing hypotheses. There is a significant need for, and much value in, developing informal statistical methods for handling multiresponse experimental data in such a way that the statistical analysis and resultant summary, (i) takes some account of the multivariate structure, and (ii) is comprehensible in the sense of stimulating insight into the experimental situation (as distinct from carrying out often artificial and occasionally even pointless tests of hypotheses).

The late Professor Roy shared this concern for developing such informal methods and enthusiastically supported the work of M. B. Wilk and the present author in their attempts to develop various graphical internal comparisons methods for supplementing the analysis of variance both when the response is univariate and when it is multivariate. Statistical techniques for simultaneous comparisons among comparable quantities with the aid of a statistical standard that is, at least in part, generated internal to the data on hand have been called *internal comparisons* methods.

A convenient categorization of orthogonal analysis of variance situations is shown in Table 8.

TABLE 8

Internal comparisons methods

Response Structure / Decomp. of treatment structure	Uniresponse	Multiresponse
All 1 d.f. contrasts	(i) Half-normal $\sim\chi^2(1)$ $\sim$Gamma ($\eta = 1/2$)	(iv) Gamma (estimated η)
All $\nu(>1)$ d.f. groupings	(ii) $\chi^2(\nu)\sim$Gamma ($\eta = \nu/2$)	(v) In progress
Mixed d.f. groupings	(iii) Unequal components probability plots.	(vi) ?

The breakdown by columns is according as the response structure is uniresponse or multiresponse, while the breakdown by rows is for various decompositions of the treatment structure : all single-degree-of-freedom contrasts, all $\nu(>1)$ degrees-of-freedom groupings and, lastly, mixed-degrees-of-freedom groupings. In the body of the table are shown various graphical internal comparisons methods that have been proposed to date for the different categories. Thus for Cell I (uniresponse with all single-degree-of-freedom contrasts), one has the half-normal procedure advocated by Daniel [1]. For Cells II, III and IV probability plotting procedures have been suggested by M. B. Wilk and the present author [14], [15], [16], [17], [18]. The work for Cell V is now in progress [7] and Cell VI remains for future investigation.

5. CONCLUDING REMARKS

The objective of the present paper has been to convey, through a brief survey of certain research interests and technical contributions, a memorial portrait of, and timely tribute to, the late Professor S. N. Roy. To lend authenticity to this endeavour, direct personal experience and shared professional efforts provided the basis for the paper. Yet, I am forcefully reminded of Benjamin Disraeli's words :

> "Experience is the child of Thought, and Thought is the child of Action. We can not learn men from books."[3]

References

[1] Daniel, C. (1959). "Use of Half-Normal Plots in Interpreting Factorial Two-Level Experiments," *Technometrics*, **1**, 311–41.

[2] Daniel, C. (1960). "Parallel Fractional Replicates," *Technometrics*, **2**, 263–8.

[3] Das Gupta, S., Anderson, T. W. and Mudholkar, G. S. (1964). "Monotonicity of the Power Functions of Some Tests of the Multivariate Linear Hypothesis," *Ann. Math. Statist.*, **35**, 200–5.

[4] Dempster, A. P. (1958). "A High Dimensional Two Sample Significance Test," *Ann. Math. Statist.*, **29**, 995–1010.

[5] Dempster, A. P. (1963). "Multivariate Theory for General Stepwise Methods," *Ann. Math. Statist.*, **34**, 873–83.

---

[3] *Vivian Grey*, Book V, Chapter I.

[6] Gnanadesikan, R., Lauh, E., Snyder, M. and Yao, Y. (1964). "Efficiency Comparisons of Certain Multivariate Analysis of Variance Test Procedures," Unpublished manuscript. (Abstract No. 28, *Ann. Math. Statist.*, **36**, 356.)

[7] Rappeport, M. A. (1965). Ph.D. Thesis, Courant Inst. of Math. Sciences of New York University.

[8] Roy, J. (1958). "Step-Down Procedure in Multivariate Analysis," *Ann. Math. Statist.*, **29**, 1177–87.

[9] Roy, S. N. (1953). "On a Heuristic Method of Test Construction and its Use in Multivariate Analysis," *Ann. Math. Statist.*, **24**, 220–38.

[10] Roy, S. N., Gnanadesikan, R. and Srivastava, J. N. (1964). *Analysis and Design of Certain Quantitative Multiresponse Experiments*. Unpublished monograph.

[11] Roy, S. N. and Mikhail, W. F. (1961). "On the Monotonic Character of the Power Functions of Two Multivariate Tests," *Ann. Math. Statist.*, **32**, 1145–51.

[12] Srivastava, J. N. (1964). "On the Monotonicity Property of the Three Main Tests for Multivariate Analysis of Variance," *J. Roy. Statist. Soc. Ser.* B, **26**, 77-81.

[13] Tukey, J. W. (1962). The Future of Data Analysis. *Ann. Math. Statist.*, **33**, 1-67.

[14] Wilk, M. B. and Gnanadesikan, R. (1961). "Graphical Analysis of Multiresponse Experimental Data Using Ordered Distances," *Proc. Nat. Acad. Sci. USA*, **47**, 1209–12.

[15] Wilk, M. B. and Gnanadesikan, R. (1964*a*). "Graphical Methods for Internal Comparisons in Multiresponse Experiments," *Ann. Math. Statist.*, **35**, 613–31.

[16) Wilk, M. B. and Gnanadesikan, R. (1964*b*). "A Probability Plotting Procedure for Internal Comparisons in a General Analysis of Variance," Invited paper presented at Royal Statistical Society Meetings in Cardiff, Wales.

[17] Wilk, M. B. and Gnanadesikan, R. (1964*c*). "Internal Comparison Methods in the Analysis of Variance," Invited paper presented at the joint statistical meetings in Chicago.

[18] Wilk, M. B., Gnanadesikan, R. and Huyett, M. J. (1962). "Probability Plots for the Gamma Distribution," *Technometrics*, **4**, 1–20.

(*Received Jan. 1, 1966.*)

CHAPTER 15

An Example of the Analysis of a Series of Response Curves and an Application of Multivariate Multiple Comparisons

JAMES E. GRIZZLE
The University of North Carolina at Chapel Hill

1. INTRODUCTION

It is not uncommon in biometry to encounter data which consist of a long series of observations made over a period of time or a series of observations made over a succession of doses on each of several subjects. These observations are not expected to be independent and, therefore, some type of multivariate analysis would be deemed appropriate. If the conventional multivariate analysis of variance was made, the number of variates would be so large as to make the analysis unwieldy, and perhaps great sacrifices would be made in the power of the tests due to inclusion of variables which contain only a small amount of information about the differences among treatments.

Furthermore, in addition to their purely statistical relationships, the observations made on each individual might have a strictly functional relationship to the stimulus which gave rise to each observation. When this functional relationship exists it can be utilized in the analysis with great benefit as follows : the parameters in the same functional relationship are estimated for each individual in the study. Thus we summarize the data for each individual into a small number of estimated parameters which describe succinctly the relationships among the original observations. Now the problem becomes how are the estimated parameters related to the treatments ? Usually the estimated parameters will not be independent. Therefore, we analyze the estimated parameters by multivariate analysis of variance. This analysis should be asymptotically valid for estimated parameters that occur in a large variety of types of functions including those of the non-linear variety, and if we wish, data other than the estimated parameters can be included in the analysis.

There is a good chance that at least two benefits can be realized by using the method of analysis outlined. The information contained in the original observations can be summarized succinctly by a much smaller number of variables which may have a physical interpretation, and by reducing a large number of original variables to a smaller number of high information content we can increase the power of the tests made. One of the main services statisticians render to investigators in the experimental sciences is reduction of voluminous data to a smaller number of summary statistics that can be comprehended and manipulated easily. Thus the reduction of the many variates collected on each individual to a smaller number which still retain physical meaning is worthwhile even if the second mentioned benefit is not realized. While not set forth in exactly this light, these ideas came from reading the paper by Potthoff and Roy [5].

In this paper we wish to give an example in which ten to forty observations made on each subject were reduced to four new variables which were then analyzed by multivariate analysis of variance, and some multivariate multiple comparisons were made.

2. THE DATA

The purpose of the experiment was to investigate the potentialities of Beta-amino propionitrile as a preventive of joint stiffness. The right knee joints of 250 gram female Sprague Dawley rats were immobilized in 40° flexion with an internal extra-articular splint. The data* on the following five treatment groups were analyzed.

Group 1 : Knees immobilized three weeks—no drug given.

Group 2 : Knees immobilized three weeks—drug given all three weeks.

Group 3 : Knees immobilized three weeks—drug given the first week.

Group 4 : Knees immobilized three weeks—drug given the third week.

Group 5 : Knees immobilized two weeks—no drug given.

At the end of the period of treatment, the rats were sacrificed, and all the soft tissue was removed from the leg except the joint capsule and collateral ligaments. The femur was mounted on the movable arm of a goniometer, and the tibia was maintained in a horizontal position so that the force produced by adding weights to the distal end of the tibia would always be perpendicular to the horizontal plane. Half gram weights were added every ten seconds, and the angle between the tibia and the femur was recorded after the addition of each weight. In most cases a point was reached at which the joint continued to extend without the addition of more weight. After the voluntary extension had stopped, addition of weight was resumed until an angle of 135° was reached. A plot of the results for a typical group is shown in Figure 1. The angles formed between the femur and the tibia after the addition of each one half gram weight are the original variables in this example.

---

*The author is indebted to Leonard T. Furlow, Jr., M.D. and Erle E. Peacock, Jr., M.D. of the Department of Surgery, University of North Carolina School of Medicine, Chapel Hill, North Carolina for the use of these data.

TABLE 1

Summary of data obtained from the regression lines

Group	a_1	b_1	a_2	b_2	w_1	w_2
	0.3113	0.0332	0.7777	0.0092	5.5	17.5
	0.3202	0.0287	0.7438	0.0140	9.5	13.0
	0.3786	0.0219	0.7675	0.0094	11.5	17.0
1	0.3921	0.0270	0.5520	0.0236	13.5	16.0
	0.3836	0.0402	0.7800	0.0196	6.5	8.0
	0.3869	0.0383	0.7092	0.0174	5.5	13.5
	0.4524	0.1372	0.7487	0.0255	2.5	7.5
	0.3634	0.1013	0.8189	0.0162	1.5	7.0
2	0.3734	0.0569	0.7619	0.0165	3.5	10.5
	0.3756	0.0495	0.7598	0.0136	4.5	13.0
	0.4361	0.0568	0.8174	0.0192	1.5	6.0
	0.3596	0.0604	0.8521	0.0103	5.5	7.0
	0.4001	0.0468	0.8139	0.0106	8.0	10.5
3	0.3861	0.0374	0.6014	0.0229	8.0	15.0
	0.3968	0.0412	0.7283	0.0226	6.0	9.0
	0.3963	0.0286	0.7662	0.0110	8.5	15.0
	0.3603	0.0470	0.6917	0.0437	3.5	5.5
	0.3452	0.0381	0.7473	0.0172	5.5	10.5
4	0.3962	0.0407	0.7848	0.0165	6.5	8.5
	0.3763	0.0434	0.8062	0.0184	3.5	6.5
	0.3910	0.0406	0.7334	0.0293	3.5	6.5
	0.5005	0.0725	0.7767	0.0355	1.5	4.5
	0.5699	0.0699	0.7391	0.0269	2.5	7.5
5	0.6114	0.0670	0.8009	0.0176	3.5	7.0
	0.3803	0.0417	0.7540	0.0270	3.5	6.5
	0.3951	0.0765	0.8518	0.0137	2.0	5.0

3. THE ANALYSIS

Figure 1 shows that the first portion of the response curve for each animal was approximately linear while the second portion was often definitely non-linear. One of the forms of Hooke's law is that the extension of a spring is proportional to the weight attached to it. The joint and capsule do not behave like a spring

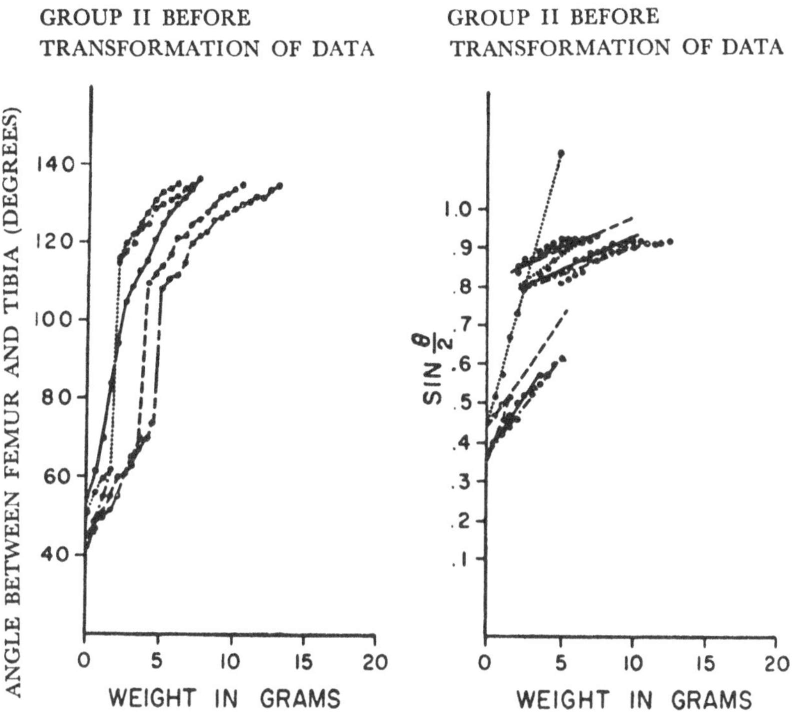

Figure 1

since the femur does not resume its original position after the weight is removed, but plastic substances often obey Hooke's law with the exception in behaviour noted. The quantity $y = \sin\frac{\theta}{2}$ is proportional to the distance from the point at which the leg is suspended to the point at which the weight is applied, and thus provides the transformation for changing the angular measurement to a linear measurement. Since the points of suspension and the points of addition of the one half gram weights were constant among animals apart from experimental error in setting up the apparatus, the fact that we do not know the constant of proportionality is no handicap. Plotting the new variable $y = \sin\frac{\theta}{2}$ against weight resulted in two lines, approximately straight, for each animal, the first being the response line before the spontaneous movement without addition of extra weight began and the second being the response after the spontaneous movement had taken place and the addition of weights had been resumed again. The discontinuity between the two lines is the point at which the spontaneous movement took place.

Two straight lines were fitted to the response curves for each rat. These curves fit exceedingly well; the per cent of the total sum of squares accounted for by regression was always over 95% and usually exceeded 98%. Fortunately the estimated regression coefficients have a physical interpretation. They are the rates of increase of the distance between the point of attachment of the leg to the apparatus and the point of addition of weight per half gram of weight.

The parameters of the regression lines do not provide a complete summary of the data in this experiment. The weight at which the discontinuity occurred and the weight at which an angle of 135° was reached are of interest and perhaps reflect differences in effects of the treatments. Therefore, for an adequate summary of the original data we shall propose the vector

$$\mathbf{Z}' = (\log b_1, \log b_2, w_1, w_2),$$

where b_1 is the slope of the first component of the response line, b_2 of the second, w_1 is the weight at which the discontinuity occurred and w_2 is the weight at which an angle of 135° extension was reached. We shall explain why $\log b$ rather than b was used in the analysis in succeeding paragraphs. In a few cases the point at which the spontaneous extension occurred, and thus w_1 was observed, was not clear from the data. In these cases lines were fitted corresponding to each of the possible choices of assignment of the point to the first line or to the second. The point of division between the first and second lines was chosen as the one which yielded the smallest *total residual sum of squares*.

The regression coefficients and w_1 and w_2 are tabulated in Table 1. The intercepts of the two lines, a_1 and a_2, are tabulated also, but they were not included in any of the analyses. The means and variances of b_1, b_2, w_1, and w_2 are tabulated in Table 2.

TABLE 2

Means and variances of b_1, b_2, w_1 *and* w_2

Group	b_1		b_2		w_1		w_2	
	Mean	Variance ($\times 10^2$)	Mean	Variance ($\times 10^3$)	Mean	Variance	Mean	Variance
1	0.0316	0.004901	0.0155	0.03304	8.7	11.37	14.2	12.47
2	0.0803	0.1431	0.0182	0.02058	2.7	1.70	8.8	8.25
3	0.0429	0.01140	0.0155	0.04412	7.2	7.30	11.3	12.95
4	0.0420	0.001145	0.0250	0.1362	4.5	2.00	7.5	4.00
5	0.0655	0.01900	0.0242	0.07416	2.6	0.80	6.1	1.68
$\frac{s^2 \text{ max.}}{s^2 \text{ min.}}$	124.97		6.62		14.2		7.71	

Bartlett's test of homogeneity of the variances was not made. Instead the statistics s^2_{max}/s^2_{min} shown on the line of Table 2 were used [4]. The critical value for this test is 25.2 at the 0.05 significance level. The means and variances of the logarithms of b_1 and b_2 are shown in Table 3. The variances of the logarithm of b_1 are heterogeneous even after the transformation.

TABLE 3

Mean and variance of $\log_{10}(b_1 \times 10^2)$ *and* $\log_{10}(b_2 \times 10^2)$

Group	$\log_{10}(b_1 \times 10^2)$		$\log_{10}(b_2 \times 10^2)$	
	Mean	Variance	Mean	Variance
1	0.4897	0.009906	0.1646	0.02856
2	0.8694	0.03680	0.2501	0.01047
3	0.6191	0.014361	0.1587	0.03287
4	0.6217	0.001222	0.3650	0.03370
5	0.8069	0.01135	0.3587	0.02728
$\frac{s^2 \text{ max.}}{s^2 \text{ min.}}$	30.11		3.22	

A test of normality of the deviations of log b_1 from their group means was made by examining the third and fourth cumulants as outlined in [2]. In spite of the heterogeneity of variances there was no significant departure from normality. The test statistics are respectively 0.55 and 0.20. Before the transformation they were 2.67 and 4.23. Plots of b_2 and log b_2 on normal probability paper showed that log b_2 produced a straighter line than b_2. Therefore, it was decided to continue the analysis using the vector ($\log b_1, \log b_2, w_1, w_2$).

Another issue could be raised about the analysis. Given values of w_1 and w_2 the denominator in the usual variance formula of a regression co-efficient can be calculated. Therefore, if w_1 and w_2 are random variables, so are the sums of squares of weights which appear in the variances of the regression co-efficients. If we adopt this point of view, grave problems arise in trying to develop an analysis of the b's. However, it seems equally valid to regard the b's simply as samples from some distribution. That is, the b's are simply some observations that happen to be at our disposal. Then their variances are not conditional on w_1 and w_2. This is the way we shall regard them in the analysis when making probability statements, but we shall regard them as rates of change when a physical interpretation of the results is made.

The remainder of the analysis is exceedingly simple. It is merely a multivariate analysis of the between-within type, although anything we shall do in the analysis could be extended to other designs by using the general model

$$\mathcal{E}(\boldsymbol{Z}) = \boldsymbol{A}\boldsymbol{\xi}.$$

TABLE 4

Mean products among and within treatments

	among $= \boldsymbol{S}_H$			
	$\log b_1$	$\log b_2$	w_1	w_2
$\log b_1$	.12795	.04045	−2.0926	−2.0189
$\log b_2$		.052470	−1.1294	−1.5626
w_1			39.871	43.642
w_2				56.373
	within $= \boldsymbol{S}_E$			
	$\log b_1$	$\log b_2$	w_1	w_2
$\log b_1$	.014495	.0031338	−.12931	−.20332
$\log b_2$		.026696	−.028403	−.14041
w_1			3.9111	3.1040
w_2				8.1016

The matrices S_H, and S_E, denoting between and within mean squares respectively, are shown in Table 4. The characteristic roots of $S_H S_E^{-1}$ are 12·196, 5·0429, 0·48319 and 0·012449. If we divide each diagonal element of S_H by its corresponding element in S_E, we obtain the 4 F-statistics that are appropriate for testing homogeneity of group means of each variable individually. The F-statistics for the variables in the order of their presentation in Table 4 are 8·83, 1·97, 10·19 and 6·96; each has 4 and 21 degrees of freedom. Each F, with the exception of the second, is significant at far beyond the ·01 level. The test statistic for largest root criterion is $\boldsymbol{\theta} = \frac{4}{21}(12{\cdot}196)/\left(1+\frac{4}{21}\,12{\cdot}196\right) = 0{\cdot}6991$ which, as expected, is significant at beyond the ·01 level [3]. If, instead, one wants to use the likelihood ratio test criterion, the statistic

$$\Lambda = \left(25 - \frac{4+4+1}{2}\right) \sum_{j=1}^{4} \ln\left(1+\frac{4}{21}\,\lambda_j\right),$$

where λ_j is the jth characteristic root of $S_H S_E^{-1}$, has approximately a χ^2-distribution with 16 degrees of freedom. The observed value, 20·5 (1·923) = 39·42, is significant at far beyond the ·01 level. The more exact test suggested by Bartlett [6] confirms the χ^2 approximation. These tests lead to the conclusion that the five treatment groups are not homogeneous and that the slope of the second portion of the curve does not contribute to the difference among treatments.

A repeat analysis of the data omitting log b_2 yields characteristic roots of $S_H S_E^{-1}$ of 11·931, 4·6778, and 0·31609. Notice that both the largest root and the trace of $S_H S_E^{-1}$ are only slightly changed.

This raises an important point that is, in the author's opinion, often not appreciated by practitioners of multivariates analysis of variance : the chance of finding significant differences by a multivariate test is lessened when dependent variables which do not contain information about differences among treatments are included in the analysis. The way this comes about is that the deletion of a non-significant variable does not change the largest root much. Hence the test criteria will be about the same with or without the non-significant variable, and degrees of freedom are lost. The degrees of freedom associated with the largest root criterion are

$$\gamma_1 = (|\text{d.f. for hypothesis}-p|-1)/2,$$

$$\gamma_2 = \frac{1}{2}(\text{d.f. for error}-p-1),$$

$$k = \min(\text{d.f. for hypothesis}, p).$$

In this example when all 4 variables are included

$$\gamma_1 = (|4-4|-1)/2 = -\frac{1}{2},$$

$$\gamma_2 = \frac{1}{2}(21-4-1) = 8,$$

$$k = \min(4, 4) = 4,$$

and when only three variables are included

$$\gamma_1 = (|4-3|-1)/2 = 0,$$

$$\gamma_2 = \frac{1}{2}(21-3-1) = 8{\cdot}5,$$

$$k = \min(4, 3) = 3.$$

The critical values of the test of the largest root of $S_H S_E^{-1}$ at the ·05 level are ·575 for $p = 4$ and ·520 for $p = 3$. The observed values of the test criteria are 0·6991 and 0·6944. Notice that the critical value for $p = 3$ is 90% of that for $p = 4$ while the observed test statistic when $p = 3$ is 99% of that for $p = 4$. At the ·01 level the critical values are ·660 and ·610. Similar remarks apply for the likelihood ratio criterion. Thus it would seem to be a good policy to list the variables to be subjected to analysis in the design of an experiment, and to include in this list only those variables that have a reasonable expectation of yielding information about differences among treatments.

It is possible to make the general statement that if adherence to customary significance levels is rigid, multivariate procedures are poor for exploratory work. Some experimenters who have a nodding acquaintance with multivariate procedures seem to feel that if one is not quite sure of exactly which variables contain information about differences among groups, then one should measure as many variables as possible and analyze the whole body by multivariate analysis. They would be better advised to stick to univariate procedures in these cases or to use multivariate analysis with the realization that unless the interpretation of the tests of significance is tempered with good judgment, real differences in some of the variables in the analysis can be overlooked.

4. MULTIPLE COMPARISONS

Having found that the group means are not homogeneous, we wish to go further and make some comparisons among the groups.

The same problems encountered in univariate multiple comparisons occur in multivariate multiple comparisons, and the generalization from one variate to p variates raises new issues that must be faced. When the treatment effects have been declared non-homogeneous by a multivariate test of significance, should the multiple comparison procedure be applied to each of the p variates separately or should some overall measure be used in making the comparison ? Sometimes the physical nature of the problem will provide the answer, but frequently it will not. Some may say that we should not want to control the frequency of Type I errors for all p variates, but should instead control it for each of the p variates individually. It is the author's opinion that this issue should be decided by the investigator who performed the experiment. However, experience teaches that it is often difficult to explain to the investigator the issues involved with enough clarity for him to realize the implications of his choice. If the choice has been to control the error rate for all p variables simultaneously, there remains the problem of the dearth of multivariate multiple comparison procedures. Fortunately any univariate multiple comparison procedure can be converted into a conservative multivariate analog by the use of Bonferroni inequalities by simply using $\alpha = \alpha/p$ rather than α as the size of the critical region. This technique often gives shorter bounds than the omnibus multivariate procedures that are presently available. A considerable amount of work remains to be done in this area.

The most interesting comparison to make on these data is to compare each treatment with the control, i.e., compare groups 2, 3, 4 and 5 with 1. We shall assume that we desire to control the error for these 4 comparisons on three variables simultaneously. This amounts to the multivariate extension of Dunnett's test [1].

Siotani [7] has tabulated approximate percentage points of $(\bar{\boldsymbol{Y}}_i-\bar{\boldsymbol{Y}}_k)'\boldsymbol{S}_e^{-1}(\bar{\boldsymbol{Y}}_i-\bar{\boldsymbol{Y}}_k) = D_i^2$, $i = 1, \ldots, k-1$, for $p = 2$ and $k-1 = 16$, and he and others have indicated how percentage points can be obtained still more approximately for any p by using Bonferroni inequalities. The percentage points computed from the first term of Bonferroni inequalities can be used for bounds on $E[\boldsymbol{a}'(\bar{\boldsymbol{Y}}_i-\bar{\boldsymbol{Y}}_k)]$ for $k-1$ comparisons and all $p\times 1$ vectors $\boldsymbol{a}$. One can compute bounds on all linear functions of the p variates by using the technique just presented. Siotani assumes the sample sizes are equal in each group in his table, but it is not necessary to make this assumption when using the inequality. The approximate percentage points are obtained by

$$\left(\frac{1}{n_i}+\frac{1}{n_k}\right)n_e\cdot\frac{p}{n_e-p+1}\,F\left(\frac{\alpha}{k-1}, p, n_e-p+1\right),$$

where n_e is the number of degrees of freedom for error. This gives only the first term of the Bonferroni inequality and is therefore conservative. Comparison among the D^2 can be made directly, or approximate confidence intervals for $E[\boldsymbol{a}'(\bar{\boldsymbol{Y}}_i-\bar{\boldsymbol{Y}}_k)]$ can be computed by

$$\boldsymbol{a}'(\bar{\boldsymbol{Y}}_i-\bar{\boldsymbol{Y}}_k) \pm \left[\left(\frac{1}{n_i}+\frac{1}{n_k}\right)\frac{n_e p}{n_e-p+1}\cdot F\left(\frac{\alpha}{k-1}, p, n_e-p+1\right)\boldsymbol{a}'\boldsymbol{S}_e\boldsymbol{a}\right]^{\frac{1}{2}}$$

The four D^2 are shown in Table 5, and the differences in the individual means are shown in Table 6. The critical values of F were determined by graphical interpolation and are only approximate.

TABLE 5

D^2 for comparing Groups 2—5 with Group 1

Comparison	D^2	$F = \frac{n_e-p+1}{n_e p} D^2 \frac{n_i \; n_k}{n_i+n_k}$
Group 5—Group 1	11.69	9.62
Group 4—Group 1	6.96	5.72
Group 3—Group 1	1.37	0.11
Group 2—Group 1	12.65	10.40

$F_{(\frac{.05}{4}, 3, 19)} = 4.70$, $F_{(\frac{.01}{4}, 3, 19)} = 6.80$

TABLE 6

Comparison of individual means to control value

Comparison	$\log b_1$		w_1		w_2	
	Mean Difference	Test Statistic	Mean Difference	Test Statistic	Mean Difference	Test Statistic
Group 5—Group 1	0.3172	4.35	−6.1	5.10	−8.1	4.70
Group 4—Group 1	0.1321	1.81	−4.2	3.51	−6.7	3.89
Group 3—Group 1	0.1294	1.77	−1.5	1.25	−2.9	1.68
Group 2—Group 1	0.3797	5.20	−5.0	4.18	−5.4	3.13

$\left(\frac{n_e p}{n_e-p+1} F_{(\frac{.05}{4}, 3, 19)}\right)^{\frac{1}{2}} = 3.95$, $\left(\frac{n_e p}{n_e-p+1} F_{(\frac{.01}{4}, 3, 19)}\right)^{\frac{1}{2}} = 4.75$

$t_{D(\frac{.05}{3}, 21)} = 3.20$ $\qquad$ $t_{D(\frac{.01}{3}, 21)} = 4.00$

determined by extrapolation.

Two sets of critical values are given for Table 6. The first is obtained from the above bounds, and the second is obtained by applying the Bonferroni inequality to Dunnett's test. To get this value we use the critical value for Dunnett's test, denoted by t_D, at significance level α/p. The bounds obtained from t_D are shorter than the bounds for $\mathcal{E}[\boldsymbol{a}'(\bar{\boldsymbol{Y}}_i-\bar{\boldsymbol{Y}}_k)]$ for all $\boldsymbol{a}$. In this case we are interested in only one particular set of $\boldsymbol{a}'$, viz.,

$$\boldsymbol{a}_1' = (1, 0, 0),$$

$$\boldsymbol{a}_2' = (0, 1, 0),$$

$$\boldsymbol{a}_3' = (0, 0, 1).$$

This is the only set of $\boldsymbol{a}$ covered by t_D, and therefore t_D gives the shorter bounds.

It should be noted in passing that the comparison of the procedures to ascertain which yields the shortest interval or the smallest critical values can be made without looking at the data. If it is not already obvious from the nature of the functions which are covered by the bounds, the computation of constants in the bounds desired can be made and the procedure producing the shortest bounds or the smallest critical value selected.

5. INTERPRETATION

The five treatment groups are not homogeneous. They are different in regard to the slope of the first position of the line, the weight at which the discontinuity occurred and the weight, at which an angle of 135° was reached. The slopes of the second portions of the line were homogeneous.

From the nature of the data one expects that log b_1 will be inversely related to w_1 and w_2. This turns out to be the case. The correlations are, respectively, $-0{\cdot}54$ and $-0{\cdot}59$. Both are significant at beyond the ·01 level; w_1 and w_2 have a correlation of 0·55. These results imply that the more mobile the joint the sooner the points w_1 and w_2 are reached.

We can use the D^2 as a measure of similarity of groups 2–5 to group 1. This shows that the animals who had their knee joints immobilized three weeks and who were given drug the first week are the most like the controls. None of the means are different from the control values, and D^2 does not approach significance. The rats given drug the third week only rank next in similarity to group 1. They do not differ in regard to the slope, but w_1 and w_2 are different from group 1. The group of rats whose knees were immobilized for two weeks is quite different from group 1, and so is the group that received treatment during all three weeks of immobilization.

6. SUMMARY

We have illustrated the following three main points in this paper.

(1) When there is a functional relationship among the elements of a vector in a multivariate observation, the parameters in the functional relationship should be estimated if possible and the estimated parameters analyzed further by multivariate analysis of variance.

(2) Variates which do not have a reasonable expectation of containing information about differences among treatments should not be included in the analysis in order to prevent loss of power. Multivariate procedures may be poor for exploratory work unless a considerable amount of care is exercised.

(3) When in doubt, the procedure yielding the shortest confidence bounds for a function of the parameters among several that may be available can be selected by resorting to a few simple computations and without looking at data. This statement applies to approximate as well as exact bounds.

Acknowledgments

The author is indebted to Dr. Richard F. Potthoff for many helpful conversations during the course of preparation of this paper. This work was supported by National Institutes of Health, Institute of General Medical Sciences, Grant No. GM-12868-01.

References

[1] Dunnett, C. W. (1955). "A Multiple Comparison Procedure for Comparing Several Treatments to a Control," *J. Amer. Statist. Assoc.*, **50**, 1096–1121.

[2] Fisher, R. A. (1948). *Statistical Methods for Research Workers.* Oliver and Boyd, Edinburgh.

[3] Heck, D. L. (1960). "Charts of Some Upper Percentage Points of the Distribution of the Largest Characteristic Root," *Ann. Math. Statist.*, **31**, 625–642.

[4] Pearson, E. S. and Hartley, H. O. (1958). *Biometrika Tables for Statisticians*, Cambridge University Press, Table 31, p. 179.

[5] Potthoff, Richard F. and Roy, S. N. (1964). "A Generalized Multivariate Analysis of Variance Model Useful Especially for Growth Curve Problems," *Biometrika*, **51**, 313–326.

[6] Rao, C. R. (1952). *Advanced Statistical Methods in Biometric Research.* John Wiley and Sons, New York.

[7] Siotani, Minoru (1960). "Notes on Multivariate Confidence Bounds," *Ann. Inst. Statist. Math.*, **11**, 167–182.

(Received Jan. 1, 1966.)

CHAPTER 16

On Some Selection and Ranking Procedures with Applications to Multivariate Populations*

SHANTI S. GUPTA AND WILLIAM J. STUDDEN,
Purdue University

1. INTRODUCTION AND SUMMARY

This paper is concerned with ranking and selection of k multivariate normal populations. The selection and ranking problem is formulated in terms of suitably defined scalarfunctions. For k multivariate normal populations with mean vectors μ_i, $i = 1, 2, \ldots, k$, each of which has p components, a function that arises naturally, is the scalar quantity $\lambda_i = \mu_i' \Sigma_i^{-1} \mu_i$ where Σ_i is the covariance matrix of the ith population. With suitably defined statistics the ranking of multivariate normal population in terms of λ_i can be reduced to the ranking of non-centrality parameters of non-central chi-square or non-central F distributions.

* Research supported by Contract NONR-1100(26) with the Office of Naval Research and by Contract AF 33(657)11737 with the Aerospace Research Laboratories. Reproduction in whole or in part permitted for any purpose of the United States Government.

We are interested in selecting the populations with large (small) values of the parameters λ_i. The procedures to be defined select a non-empty subset which is small and yet large enough to guarantee a certain basic probability requirement. This requirement is that the population with the largest (smallest) value of the parameter is included in the selected subset with probability at least equal to a given number $P^*(1/k < P^* < 1)$. This type of problem has been studied in a number of recent papers. For a rather complete bibliography, reference should be made to Gupta (1966).

In Section 2, a formal statement of the problem is given and procedures are defined for selecting populations with the largest and smallest parameters. Probability of a correct selection and its partial infimum are evaluated.

Section 3 deals with a general result concerning the infimum of the probability of a correct selection. In Section 4 applications to multivariate populations are given.

2. FORMAL STATEMENT OF THE PROBLEM

Suppose each of the k populations, $\pi_1, \pi_2, \ldots, \pi_k$ has an observable random variable Y_i, $i = 1, 2, \ldots, k$, whose density function is $f_{\lambda_i}(y)$, $y \geqslant 0$, $\lambda_i \geqslant 0$. We assume that the density function $f_\lambda(y)$ has a monotone likelihood ratio. This implies that the expected value of Y is a monotone increasing function of λ. In all specific cases to be considered the mean value will be a linear increasing function of λ.

Let the ranked λ's be denoted by

$$\lambda_{[1]} \leqslant \lambda_{[2]} \leqslant \cdots \leqslant \lambda_{[k]}. \tag{2.1}$$

It is assumed that there is no a priori information available about the correct pairing of the ordered $\lambda_{[i]}$ values and the k given populations. Any population associated with $\lambda_{[k]}$ ($\lambda_{[1]}$) will be called a best population. A correct selection is defined as the selection of any subset of the k populations which includes a best population. Our problem is to define a selection procedure which selects a small, non-empty subset of the k populations and guarantees that a best population has been included with probability at least $P^*(1/k < P^* < 1)$. If CS stands for a correct selection then our goal is to define a decision rule R such that

$$\inf_{\Omega} P\{CS \mid R\} \geqslant P^* \tag{2.2}$$

where $\Omega = \{(\lambda_1, \lambda_2, \ldots, \lambda_k) : \lambda_i \geqslant 0, \text{ all } i\}$.

Selection Procedures

Let y_i be an observation on Y_i, $i = 1, 2, \ldots, k$. Then the procedures for selecting the population with the largest value $\lambda_{[k]}$ is

R : Select π_i iff

$$cy_i \geqslant y_{\max}, \quad c > 1 \tag{2.3}$$

where $c = c(k, P^*)$ is the minimal value for which (2.2) is satisfied. Similarly, the procedure R' for selecting a subset containing the population with the smallest value $\lambda_{[1]}$ is defined to be

R' : Select π_i iff

$$y_i \leqslant b\, y_{\min}, \quad b > 1 \tag{2.4}$$

where $b = b(k, P^*)$ is again the minimal value for which (2.2) is satisfied.

Probability of a Correct Selection and Its Infimum

We will now derive an expression for the probability of a correct selection and its infimum. Let $y_{(i)}$, $i = 1, 2, \ldots, k$, be the observation which has come from the population $\pi_{(i)}$ with parameter $\lambda_{[i]}$. It should be noted that $y_{(i)}$ is one of the numbers y_i, $i = 1, 2, \ldots, k$, though it is not known to us. For selecting the population associated with $\lambda_{[k]}$, we then have

$$\begin{aligned} P\{\text{Selecting } \pi_{(k)} \mid R\} &= P\{cy_{(k)} \geqslant y_{\max}\} \\ &= P\{cy_{(k)} \geqslant y_{(j)}, j = 1, 2, \ldots, k-1\} \\ &= \int_0^\infty \Big[\prod_{j=1}^{k-1} F_{\lambda_{[j]}}(cy) \Big] f_{\lambda_{[k]}}(y)dy. \end{aligned} \tag{2.5}$$

Since $f_\lambda(y)$ is assumed to have a monotone likelihood ratio, it follows that $F_\lambda(y) \leqslant F_{\lambda'}(y)$ for all $\lambda > \lambda'$ and each y. In this case

$$(2.6) \qquad P\{\text{Selecting } \pi_{(k)}\} \geqslant \int_0^\infty \left[F_{\lambda_{[k]}}(cy) \right]^{k-1} f_{\lambda_{[k]}}(y)dy.$$

Since $P\{CS \mid R\} \geqslant P\{\text{Selecting } \pi_{(k)} \mid R\}$, we conclude that

$$(2.7) \qquad \inf_{\Omega} P\{CS \mid R\} \geqslant \inf_{\lambda} \int_0^\infty F_\lambda^{k-1}(cy) f_\lambda(y)dy.$$

For the problem of selecting the population with the smallest value $\lambda_{[1]}$, a similar argument shows that

$$(2.8) \qquad \inf_{\Omega} P\{CS \mid R'\} \geqslant \inf_{\lambda} \int_0^\infty \left[1 - F_\lambda\left(\frac{y}{b}\right) \right]^{k-1} f_\lambda(y)dy.$$

In the next section we discuss a general theorem dealing with the infima of the expressions on the right sides of (2.7) and (2.8).

3. A RESULT CONCERNING THE INFIMA OF PROBABILITY OF A CORRECT SELECTION

Let $g_j(x)$, $j = 0, 1, 2, \ldots$ be a sequence of density functions on the interval $[0, \infty)$ and define

$$(3.1) \qquad f_\lambda(x) = \sum_{j=0}^{\infty} \frac{\lambda^j e^{-\lambda}}{j\,!} g_j(x), \quad x \geqslant 0.$$

For a fixed integer $k \geqslant 2$ and $c > 1$ let

$$(3.2) \qquad I(\lambda) = \int_0^\infty [F_\lambda(cx)]^{k-1} f_\lambda(x)dx.$$

and

$$(3.3) \qquad J(\lambda) = \int_0^\infty \left[1 - F_\lambda\left(\frac{x}{c}\right) \right]^{k-1} f_\lambda(x)dx.$$

The purpose of this section is to provide sufficient conditions on the sequence $g_j(x)$, $j = 0, 1, \ldots$ which guarantee that the functions $I(\lambda)$ and $J(\lambda)$ attain their minimum value on $[0, \infty)$ at the point $\lambda = 0$.

Theorem 3.1

(*i*) *If for each integer* $l \geqslant 0$

$$\sum_{i=0}^{l} \frac{1}{i\,!\,(l-i)\,!} \{(G_{i+1}(cx)-G_i(cx))g_{l-i}(x) \tag{3.4}$$
$$-cg_i(cx)(G_{l-i+1}(x)-G_{l-i}(x))\} \geqslant 0$$

then the functions $I(\lambda)$ *and* $J(\lambda)$ *defined in* (3.2) *and* (3.3) *are non-decreasing in* λ.

(*ii*) *If strict inequality holds in* (3.4) *for some integer* l *then* $I(\lambda)$ *and* $J(\lambda)$ *are strictly increasing.*

Corollary 3.1.a.

Let

$$g_j(x) = \frac{x^{\mu+j-1}e^{-x}}{\Gamma(\mu+j)}, \quad j = 0, 1, \ldots \tag{3.5}$$

where $\mu > 0$. *Then the functions* $I(\lambda)$ *and* $J(\lambda)$ *defined in* (3.2) *and* (3.3) *are strictly increasing.*

Proof. For $g_j(x)$ defined by (3.5), integrating by parts we see that

$$G_i(x)-G_{i+1}(x) = g_{i+1}(x), \quad i = 0, 1, \ldots . \tag{3.6}$$

For $l \geqslant 1$ we insert the above expression in (3.4) and combine the terms i and $l-i$. It may easily be shown that (3.4) reduces to

$$\sum_{i=0}^{\left[\frac{l-1}{2}\right]} \frac{e^{-x(c+1)}\,x^{2\mu+l-1}\,(l-2i)\,c^{i+\mu}(c^{l-2i}-1)}{\Gamma(\mu+i+1)\,\Gamma(\mu+l-i+1)\,i\,!\,(l-i)\,!}.$$

For $l \geqslant 1$ the above expression is strictly positive. For $l = 0$ equation (3.4) reduces to zero. The corollary thus follows from Theorem 3.1.

Corollary 3.1.b. *Let*

$$g_j(x) = \frac{\Gamma(\mu+\nu+j)}{\Gamma(\nu)\,\Gamma(\mu+j)}\,\frac{x^{\mu+j-1}}{(1+x)^{\mu+\nu+j}}, \quad x \geqslant 0, \tag{3.7}$$

where $\mu > 0$, $\nu > 0$ *and* $j = 0, 1, \ldots$. *Then the functions* $I(\lambda)$ *and* $J(\lambda)$ *defined in* (3.2) *and* (3.3) *are strictly increasing in* λ.

Proof. The proof in this case proceeds as in Corollary 3.1.a. The expression corresponding to (3.6) is

$$G_j(x)-G_{j+1}(x) = \frac{\Gamma(\mu+\nu+j)}{\Gamma(\nu)\,\Gamma(\mu+j+1)}\,\frac{x^{\mu+j}}{(1+x)^{\mu+\nu+j}}. \tag{3.8}$$

Combining terms as in Corollary 3.1.a, equation (3.4) can be reduced to

$$\sum_{i=0}^{\left[\frac{l-1}{2}\right]} \frac{\Gamma(\mu+\nu+l-i)\,(l-2i)}{i\,!\,(l-i)\,!\,\Gamma(\mu+l-i+1)} \frac{x^{2\mu-2i-1}\,c^{\mu+i}}{[(1+x)(1+cx)]^{\mu+\nu+i}}$$

$$\left[\left(\frac{cx}{1+cx}\right)^{l-2i}-\left(\frac{x}{1+x}\right)^{l-2i}\right].$$

For $l \geqslant 1$ the above expression is positive since the function $[x/(1+x)]$ is strictly increasing in x. For $l = 0$, (3.4) can be checked separately.

In order to prove Theorem 3.1 we first consider a number of elementary lemmas. For each integer $\alpha \geqslant 0$, we define $A(\alpha)$ as the set of k-tuples $(\alpha_1, \alpha_2, \ldots, \alpha_k)$ where $\alpha_i\, (i = 1, \ldots, k)$ are non-negative integers and $\sum_{i=1}^{k} \alpha_i = \alpha$. The multinomial coefficient $\frac{\alpha\,!}{\alpha_1\,!\,\alpha_2\,!\ldots\alpha_k\,!}$ will be denoted by $\binom{\alpha}{\alpha_1, \alpha_2 \ldots, \alpha_k}$ as usual.

Lemma 3.1. *The functions $I(\lambda)$ and $J(\lambda)$ defined in (3.2) and (3.3) can be expressed as*

$$I(\lambda) = e^{-\lambda k} \sum_{\alpha=0}^{\infty} a_\alpha \lambda^\alpha \tag{3.9}$$

and

$$J(\lambda) = e^{-\lambda k} \sum_{\alpha=0}^{\infty} b_\alpha \lambda^\alpha \tag{3.10}$$

where

$$\alpha\,!\,a_\alpha = \sum_{A(\alpha)} \binom{\alpha}{\alpha_1, \ldots, \alpha_k} \int_0^\infty \left\{\prod_{i=1}^{k-1} G_{\alpha_i}(cx)\right\} g_{\alpha_k}(x)dx \tag{3.11}$$

and

$$\alpha\,!\,b_\alpha = \sum_{A(\alpha)}^{\infty} \binom{\alpha}{\alpha_1, \ldots, \alpha_k} \int_0^\infty \left\{\prod_{i=1}^{k-1}\left[1-G_{\alpha_i}\left(\frac{x}{c}\right)\right]\right\} g_{\alpha_k}(x)dx. \tag{3.12}$$

Proof. Equation (3.9) follows easily by inserting the expression for $f_\lambda(x)$ from (3.1) in (3.2). Equation (3.10) follows in the same manner after observing that

$$1-F_\lambda(x) = \sum_{j=0}^{\infty} \frac{\lambda^j e^{-\lambda}}{j!} (1-G_j(x)).$$

Lemma 3.2. *The functions $I(\lambda)$ is nondecreasing provided*

(3.13) $$(\alpha+1)a_{\alpha+1}-ka_\alpha \geqslant 0, \quad \alpha = 0, 1, \ldots .$$

If strict inequality holds for some α then $I(\lambda)$ is strictly increasing. Similar statements hold for $J(\lambda)$ if a_α is replaced by b_α.

Proof. The above statements follow readily by differentiating the expressions (3.9) and (3.10).

Lemma 3.3.

(*i*) *For each set of integers $\alpha_1, \alpha_2, \ldots, \alpha_k$ we have*

(3.14) $$\int_0^\infty \prod_{i=1}^{k-1} G_{\alpha_i}(cx) g_{\alpha_k+1}(x)dx = \int_0^\infty \prod_{i=1}^{k-1} G_{\alpha_i}(cx) g_{\alpha_k}(x)dx$$

$$-\int_0^\infty \frac{d}{dx}\left[\prod_{i=1}^{k-1} G_{\alpha_i}(cx)\right](G_{\alpha_k+1}(x)-G_{\alpha_k}(x))dx$$

(*ii*) *Equation* (3.14) *remains true if the $k-1$ functions G_{α_i}, $i = 1, \ldots, k-1$ are replaced by $1-G_{\alpha_i}$, $i = 1, \ldots, k-1$.*

Proof. To prove part (i) we first integrate the left side of (3.14) by parts to obtain.

(3.15) $$\int_0^\infty \prod_{i=1}^{k-1} G_{\alpha_i}(cx) g_{\alpha_k+1}(x)dx = 1-\int_0^\infty \frac{d}{dx}\left[\prod_{i=1}^{k-1} G_{\alpha_i}(cx)\right] G_{\alpha_k+1}(x)dx.$$

The right side of (3.15) is then written as

(3.16) $$1-\int_0^\infty \frac{d}{dx}\left[\prod_{i=1}^{k-1} G_{\alpha_i}(cx)\right] G_{\alpha_k}(x)dx$$

$$-\int_0^\infty \frac{d}{dx}\left[\prod_{i=1}^{k-1} G_{\alpha_i}(cx)\right](G_{\alpha_k+1}(x)-G_{\alpha_k}(x))dx.$$

Now applying (3.15), with α_k+1 replaced by α_k, to the first two terms of (3.16), the desired result (3.14) follows. Part (ii) is obtained by a similar argument.

We now proceed with the Proof of Theorem 3.1. We first show that if (3.4) holds then $I(\lambda)$ is nondecreasing. From (3.11) we have

$$(3.17)\quad (\alpha+1)\,!\,a_{\alpha+1}=\sum_{A(\alpha+1)}\binom{\alpha+1}{\alpha_1,\ldots,\alpha_k}\int_0^\infty\left\{\prod_{i=1}^{k-1}G_{\alpha_i}(cx)\right\}g_{\alpha_k}(x)dx.$$

Since $\binom{\alpha+1}{\alpha_1,\ldots,\alpha_k}=\binom{\alpha}{\alpha_1-1,\alpha_2,\ldots,\alpha_k}+\binom{\alpha}{\alpha_1,\alpha_2-1,\alpha_2,\ldots,\alpha_k}$

$$+\cdots+\binom{\alpha}{\alpha_1,\ldots,\alpha_{k-1},\alpha_k-1}$$

we rewrite (3.17), after a simple change of variables in the sums, as

$$(\alpha+1)\,!\,a_{\alpha+1}=\sum_{A(\alpha)}\binom{\alpha}{\alpha_1,\ldots,\alpha_k}\left[\sum_{j=1}^{k-1}\int_0^\infty G_{\alpha_j+1}(cx)\left\{\prod_{\substack{i=1\\i\neq j}}^{k-1}G_{\alpha_i}(cx)\right\}\right.$$

$$\left.g_{\alpha_k}(x)dx+\int_0^\infty\left\{\prod_{i=1}^{k-1}G_{\alpha_i}(cx)\right\}g_{\alpha_k+1}(x)dx\right].$$

Then

$$(\alpha+1)\,!\,a_{\alpha+1}=(k-1)\alpha\,!\,a_\alpha$$

$$+\sum_{A(\alpha)}\binom{\alpha}{\alpha_1,\ldots,\alpha_k}\left[\sum_{j=1}^{k-1}\int_0^\infty(G_{\alpha_j+1}(cx)-G_{\alpha_j}(cx))\left\{\prod_{\substack{i=1\\i\neq j}}^{k-1}G_{\alpha_i}(cx)\right\}g_{\alpha_k}(x)dx\right.$$

$$\left.+\int_0^\infty\left\{\prod_{i=1}^{k-1}G_{\alpha_i}(cx)\right\}g_{\alpha_k+1}(x)dx\right].$$

For the last integral in the above expression we insert its value from Lemma 3.3 to obtain

$$(\alpha+1)\,!\,a_{\alpha+1}=k\,\alpha\,!\,a_\alpha$$

$$+\sum_{A(\alpha)}\binom{\alpha}{\alpha_1,\ldots,\alpha_k}\left[\sum_{j=1}^{k-1}\int_0^\infty\Big(G_{\alpha_j+1}(cx)-G_{\alpha_j}(cx)\Big)\left\{\prod_{\substack{i=1\\i\neq j}}^{k-1}G_{\alpha_i}(cx)\right\}g_{\alpha_k}(x)dx\right.$$

$$\left.-\int_0^\infty\frac{d}{dx}\left\{\prod_{i=1}^{k-1}G_{\alpha_i}(cx)\right\}\Big(G_{\alpha_k+1}(x)-G_{\alpha_k}(x)\Big)dx\right]$$

$$=k\alpha\,!\,a_\alpha+\sum_{A(\alpha)}\binom{\alpha}{\alpha_1,\ldots,\alpha_k}\left[\sum_{j=1}^{k-1}\int_0^\infty\left\{\prod_{\substack{i=1\\i\neq j}}^{k-1}G_{\alpha_j}(cx)\right\}\right.$$

$$\left[\left(G_{\alpha_j+1}(cx)-G_{\alpha_j}(cx)\right) g_{\alpha_k}(x)-cg_{\alpha_j}(cx)\left(G_{\alpha_k+1}(x)-G_{\alpha_k}(x)\right)\right]dx\Big].$$

We now interchange the summations and fix α_i ($i = 1, \ldots, k-1$, $i \neq j$).

Summing over α_j and α_k with $\alpha_j+\alpha_k = l = \alpha-\sum_{\substack{i=1 \\ i\neq j}}^{k-1} \alpha_i$ we find that

$$\alpha!\,[(\alpha+1)a_{\alpha+1}-ka_{\alpha}] \geqslant 0 \tag{3.18}$$

provided (3.4) holds. Therefore the function $I(\lambda)$ is nondecreasing whenever (3.4) is satisfied for all integers $l (l \geqslant 0)$. Moreover if strict inequality holds for some l in (3.4) then strict inequality holds for some α in (3.18) and hence $I(\lambda)$ is strictly increasing.

The proof of the statements concerning the function $J(\lambda)$ are analogous and will be omitted.

4. SELECTION AND RANKING OF MULTIVARIATE NORMAL POPULATIONS IN TERMS OF $\lambda_i = \mu_i' \Sigma_i^{-1} \mu_i$

Let $\pi_i : N(\mu_i, \Sigma_i)$, $i = 1, 2, \ldots, k$ be p-variate normal populations with mean vectors μ_i and covariance matrix Σ_i, respectively. Let $\lambda_i = \mu_i' \Sigma_i^{-1} \mu_i$.

Case 1. Σ_i known, $i = 1, 2, \ldots, k$.

We take a sample of n independent observations from each of the k populations. Let x_{ij} denote the jth observation of the p-dimensional random vector on the ith population; then for each j, $j = 1, 2, \ldots, n$, we compute

$$y_{ij} = x_{ij}' \Sigma_i^{-1} x_{ij}, \quad i = 1, 2, \ldots, k; \ \ j = 1, 2, \ldots, n. \tag{4.1}$$

Since, y_{ij}, $j = 1, 2, \ldots, n$ correspond to the n independent observations on a non-central χ^2 with non-centrality parameter λ_i and degrees of freedom p, then

$$y_i = \sum_{j=1}^{n} y_{ij}$$

is distributed as a non-central χ^2 with non-centrality parameter $\lambda_i' = n\lambda_i = n\mu_i' \Sigma_i^{-1} \mu_i$ and degrees of freedom $p' = np$. The proposed selection rule for the population with the largest value of λ_i is :

R : Select π_i iff

$$c \sum_{j=1}^{n} y_{ij} \geqslant \max_{i}\left\{ \sum_{j=1}^{n} y_{ij};\, i = 1, 2, \ldots, k\right\}$$

where the constant $c = c(k, np, P^*)$ $(c > 1)$, is determined to satisfy

$$\inf_{\lambda'} \int_0^\infty F_{\lambda'}^{k-1}(cy) f_{\lambda}'(y)dy = P^* \tag{4.2}$$

where, now, $F_{\lambda}(\cdot)$ and $f_{\lambda}'(\cdot)$ are the *cdf* and the density function of a non-central χ^2 with np d.f. Since the infimum of the above integral takes place when $\lambda' = 0$, by Corollary 3.1.a, we have, the equation determining c

$$\int_0^\infty H_{p'}^{k-1}(cy) h_{p'}(y)dy = P^*, \quad p' = np \tag{4.3}$$

where

$$H_{2p'}(x) = \int_0^x \frac{e^{-y}}{\Gamma(p')}\, y^{p'-1}dy \text{ and } \frac{d}{dx}H_{p'}(x) = h_{p'}(x).$$

The values of $c' = 1/c$ satisfying (4.3) are given by Gupta (1963) for selected values of p' and P^* (*see* Table 1, $p' = \nu/2$). Approximate c' values (obtained by using Wilson-Hilferty cube root transformation) are given by Gupta (1966) where the result concerning the infimum of $P\{CS \mid R\}$ is proved for the case $k = 2$. Armitage and Krishnaiah (1964) have extensive tables for c'.

The rule for selecting the population with the minimum value of λ, is defined by

R' : Select π_i iff

$$\sum_{j=1}^{n} y_{ij} \leqslant b \min_{i}\left\{ \sum_{j=1}^{n} y_{ij}; \quad i = 1, 2, \ldots, k \right\}.$$

It follows from the Corollary 3.1.a that the constant $b = b\,(k, np, P^*)$ is given by

$$\int_0^\infty \left[1 - H_{p'}\left(\frac{y}{b}\right)\right]^{k-1} h_{p'}(y)dy = P^*, \quad p' = np. \tag{4.4}$$

The values $b' = 1/b$ satisfying (4.4) are tabulated in Gupta and Sobel (1962) for selected values of p' and P^* and more extensively by Krishnaiah and Armitage (1964).

Case 2. $\mathbf{\Sigma}_i$ unknown, $i = 1, 2, \ldots, k$.

If $\mathbf{\Sigma}_i$'s are not known, we modify the rules R and R' as follows. Let $z_i = \bar{\boldsymbol{x}}_i' \mathbf{S}_i^{-1} \bar{\boldsymbol{x}}_i$ where $\bar{\boldsymbol{x}}_i$ is the sample mean vector of the ith population and where $\mathbf{S}_i$ is the usual sample covariance matrix with $(n-1)$ as the divisor.

R : Select π_i iff

$$c_1 z_i \geqslant z_{\max}$$

where $z_{\max} = \max(z_1, z_2, \ldots, z_k)$ and where $c_1 = c_1(k, p, n, P^*)$ is a constant (greater than unity) which satisfies

$$\int_0^\infty F_{p,\, n-p}^{k-1}(c_1 x)\, f_{p,\, n-p}(x) dx = P^* \tag{4.5}$$

where $f_{p,n-p}$ is given by (3.7) with $j = 0$, $\mu = p/2$ and $\nu = (n-p)/2$ i.e., it is the density of a random variable which is $\frac{p}{n-p}$ times the central F random variable. $F_{p,n-p}(\cdot)$ is the corresponding cdf. The modified procedure R' is

R' : Select π_i iff

$$z_i \leqslant b_1 x_{\min}$$

where $b_1 = b_1(k, p, n, P^*)$ is a constant (greater than unity) determined by

$$\int_0^\infty [1 - F_{p,n-p}(x/b_1)]^{k-1} f_{p,n-p}(x) dx = P^*. \tag{4.6}$$

It should be pointed out that (4.5) and (4.6) are consequences of Corollary 3.1.b and the fact that each Kz_i $(i = 1, \ldots, k)$, $(K = \text{Constant})$ has the density (non-central F) given by (3.1) in conjunction with (3.7).

Remark 1. It should be pointed out that the procedures R and R' discussed under case 1 are not "strictly analogous" to those given for case 2. If we use procedures based on $\bar{\boldsymbol{x}}_i' \mathbf{\Sigma}_i^{-1} \bar{\boldsymbol{x}}_i$ in case 1, the corresponding constants c and b turn out to be independent of the number of observations which is undesirable.

Remark 2. The efficiency of these procedures in terms of expected size or related criteria has not been investigated here. Also the "indifference zone" approach, a different type of formulation, due to Bechhofer (1954) has not been discussed here.

References

[1] Armitage, J. V. and Krishnaiah, P. R. (1964). "Tables for the Studentized Largest Chi-Square Distribution and Their Applications," ARL 64-188, Aerospace Research Laboratories, Wright-Patterson Air Force Base, Ohio.

[2] Bechhofer, R. E. (1954). "A Single-Sample Multiple Decision Procedure for Ranking Means of Normal Populations with Known Variances," *Ann. Math. Statist.*, **25**, 16–29.

[3] Gupta, S. S. (1963). "On a Selection and Ranking Procedure for Gamma Populations," *Ann. Inst. Statist. Math. Tokyo*, **14**, 199–216.

[4] Gupta, S. S. and Sobel, M. (1962). "On the Smallest of Several Correlated *F*-Statistics," *Biometrika*, **49**, 509–523.

[5] Gupta, S. S. (1966). "On Some Selection and Ranking Procedures For Multivariate Normal Populations Using Distance Functions," *Multivariate Analysis*, Ed. P. R. Krishnaiah, Academic Press, N. Y., 457-475.

[6] Krishnaiah, P. R. and Armitage, J. V. (1964). "Distribution of the Studentized Smallest Chi-Square With Tables and Applications," ARL 64-218, Aerospace Research Laboratories, Wright-Patterson Air Force Base, Ohio.

(Received Jan. 1, 1966.)

CHAPTER 17

On Characterizing Dependence in Joint Distributions

W. J. HALL,[1]

The University of North Carolina at Chapel Hill

SUMMARY.

Ways of characterizing the dependence of one random variable on another (or several others) are investigated. In particular, an *index of dependence* of X on Y is introduced which (i) always exists, (ii) lies between zero and unity inclusive, (iii) is zero if and only if X and Y are independent, (iv) is unity if X is a function of Y (and only if whenever X has finite variance), (v) may assume every value between zero and unity inclusive by varying the joint distribution but holding the marginal distributions fixed (assuming Y continuously distributed), (vi) is invariant under linear transformation of X and one-to-one transformation of Y, and (vii) equals k/m whenever X and Y are sums of (non-degenerate) independent and identically distributed random variables $Z_1, Z_2, \ldots, X$ being the sum of the first mZ's and Y the sum of the first kZ's ($m \geqslant k$). When the correlation ratio exists, its square cannot exceed the dependence index, and when (X, Y) is either bivariate normal or trinomial in distribution then the index equals the square of the correlation coefficient. The index is derived by first introducing and investigating a *dependence characteristic*, defined as the correlation ratio of $\exp(itX)$ on Y as a function of t. A *correlation characteristic* and *index* are also introduced. A brief survey of correlation and regression theory for complex-valued random variables is included. (No statistical aspects of dependence are considered).

[1] Research supported in part by the National Institutes of Health under Grant GM-10397.

1. INTRODUCTION

We shall be concerned with a pair of real-valued (except in Section 7) random variables (X, Y) defined on a fixed probability space. How can the dependence of X on Y, or the relationship between X and Y, be simply characterized ? A number of indices of relationship or dependence have been considered in the literature, the most popular being the *correlation coefficient* ρ and the *correlation ratio* η. A historical survey of these and other measures of association is included in papers by Kruskal (1958) and Goodman and Kruskal (1959).

A. Renyi (1959) explored the properties of these and several other indices, especially the *maximal correlation coefficient* $\bar{\rho}$, introduced by Gebelein in 1941[2]. Renyi proposed seven axioms (A–G) that an index of relationship might be required to satisfy, and showed that none of the indices except the maximal correlation coefficient has all seven of the axiomatic properties.

The maximal correlation is defined as the least upper bound, over choices of measurable functions f and g for which $f(X)$ and $g(Y)$ have finite positive variances, on the correlation between $f(X)$ and $g(Y)$. Sarmanov (1958) and Renyi (1959) have independently considered its evaluation (see remark at the end of this section—page 345). However, their methods are only applicable under certain regularity conditions; otherwise, nothing is known about the evaluation of $\bar{\rho}$, and it need not even be attained —that is, it may exceed the correlation between every pair of random variables $f(X)$, $g(Y)$.

Another drawback of this index $\bar{\rho}$ is that it too easily equals unity. It is sufficient that *some* (non-constant) function $f(X)$ equal *some* function $g(Y)$ a.s. (almost surely). For example, if (X, Y) is distributed arbitrarily over the square with lower left corner at the origin and upper right corner at (N, N) with probability $1-\epsilon$, and arbitrarily over the upper right quadrant

[2] Maximal correlation has also been considered in a series of papers by O. V. Sarmanov (1958–1960) (brought to my attention by Professor V. V. Petrov), who was apparently unaware of Gebelein work.

with corner (N, N) with probability ϵ, then $\bar{\rho} = 1$; this is seen to be so by letting $f(x) = 1$ whenever $x > N$ and $= 0$ otherwise, for then $f(X) = f(Y)$ a.s.. Here $\bar{\rho} = 1$ even though X and Y may be conditionally independent within the square and within the upper right quadrant. Thus the maximal correlation may be unity while a "minimal correlation"—e.g., that of (X, Y) truncated to the square—is zero. Other examples with $\bar{\rho} = 1$ appear as Examples 1, 3 and 7 in Section 6.

Before further discussion, we shall list Renyi's seven axioms for an index $\rho = \rho(X, Y)$ of relationship between X and Y. Renyi calls it a "measure of dependence" but we prefer to reserve the word "dependence" for directional relationships—of X on Y, say—rather than for indices which treat X and Y symmetrically.

Axiom A : ρ is defined for any non-trivial X and Y (neither X nor Y being constant a.s.).

Axiom B : $\rho(X, Y) = \rho(Y, X)$.

Axiom C : $0 \leqslant \rho \leqslant 1$.

Axiom D : $\rho = 0$ iff (if and only if) X and Y are independent.

Axiom E : $\rho = 1$ if either $X = g(Y)$ a.s. or $Y = f(X)$ a.s. for some measurable functions f and g.

Axiom F[3] : If f and g are measurable functions then $\rho(f(X), g(Y)) \leqslant \rho(X, Y)$, with equality holding if both f and g are 1–1.

Axiom G : If (X, Y) are bivariate normal in distribution, then $\rho(X, Y)$ is the absolute value of the correlation coefficient of (X, Y).

We shall seek an *index of dependence* of X on Y—rather than an *index of relationship* between X and Y. Such dependence indices are more relevant in the context of predicting X from Y, for example. That there is a distinction is apparent from considering a r.v. Y, assuming both positive and negative values, and letting $X = Y^2$. Then X is surely completely dependent

[3] We have taken the liberty of adding the harmless but useful inequality statement to Renyi's Axiom *F*.

on Y but not conversely, at least in the sense that perfect prediction of X from Y, but not of Y from X, is possible.

We shall seek an index $\delta = \delta(X \mid Y)$ of dependence of X on Y which satisfies the following modifications of Renyi's axioms, with the correspondence $\delta = \rho^2$. (We follow Renyi's pragmatic approach of formulating axioms which we are later successful in satisfying.)

We slightly strengthen his Axiom A to :

Axiom A′ : δ is defined for any (X, Y).

We discard Axiom B, the symmetry axiom, and retain unaltered his Axioms C and D for δ. We only require part of Axiom E :

Axiom E′ : $\delta = 1$ if there exists a r.v. $g(Y)$ for which $X = g(Y)$ a.s.

(As already noted, the maximal correlation equals unity if $f(X) = g(Y)$ a.s., and thus satisfies a stronger version of Axiom E than seems desirable.) We have not been successful in defining a satisfactory index which would equal unity *only* if X is a function of Y a.s. (except when X has finite variance), so Axiom E′ is not an "iff" requirement. Nor is it clear that it should be an "iff" requirement, for when knowledge of Y reduces the variance from infinity to a finite quantity that may be considered as complete a state of dependence as when knowledge of Y reduces the variance from a finite positive quantity to zero. (See Section 5.)

We prefer not to permit distortion of the X-scale, as does the maximal correlation $\bar{\rho}$.[4] For such distortions permit inflation of an index of dependence by permitting an exaggeration of that part of the x-axis where X is highly dependent on Y. We thus replace Axiom F with :

[4] All direct measurements are, in practice, typically lattice-valued; it is only for mathematical convenience that more general random variables are introduced, but non-linear transformation will destroy the actual lattice structure; thus, restriction to linear transformation will guarantee preservation of the lattice property of real data.

Axiom F′ : (a) If g is a measurable function then

$$\delta(X \mid g(Y)) \leqslant \delta(X \mid Y), \text{ with equality holding if } g \text{ is 1–1.}$$

(b) $\delta(aX+b \mid Y) = \delta(X \mid Y)$ for arbitrary real a and b ($a \neq 0$). Thus, δ should be invariant under 1–1 linear transformation of X and 1–1 measurable transformation of Y.

Axiom G serves as a basis for scaling the range of the index. We shall replace it with another scaling axiom which also provides a basis for interpreting intermediate values of the index; it presumably implies Axiom G for $\delta = \rho^2$ (this is easily proved if we confine attention to bivariate normal distributions with rational-valued ρ^2).

Axiom G′ : If $Z_1, Z_2, \ldots,$ are independent and identically distributed non-trivial r.v.'s, and $Y = Z_1+\ldots+Z_k$ and $X = Z_1+\ldots+Z_{k+l}$ for arbitrary non-negative intergers k and l ($k+l > 0$), then $\delta(X \mid Y) = k/(k+l)$.

We shall introduce an index δ which has all of the axiomatic properties A′, C, D, E′, F′ and G′ (and G). We should like to add a *continuity axiom,* but are not yet prepared to do so. That is, we should like to require that if the joint and conditional distributions of (X_n, Y_n) and X_n given Y_n converge (in some appropriate sense) to those of (X, Y) and X given Y, then $\delta(X_n \mid Y_n) \to \delta(X \mid Y)$. The maximal correlation certainly does not satisfy such an axiom (*see* Example 7 in Section 6); it also appears doubtful whether the dependence index introduced herein satisfies such an axiom completely generally although it typically does.

Other properties of the index will also be derived; for example, it will be shown that, when Y has a continuous distribution, all values between zero and unity are possible values for δ, whatever the fixed marginal distributions.

The correlation ratio η of X on Y satisfies many of these axioms, namely C, E′, F′ and G′, but not A′ or D—it only exists when X has finite variance and it equals zero too easily. But we may consider ways of extending and modifying it to overcome these shortcomings. One way to assure its existence is to truncate X to X_n, vanishing outside the interval $(-n\ n)$, and consider some limiting characteristic of the sequence of squared correlation ratios of X_n on Y $(n = 1, 2, \dots)$. More generally, we may consider some family of bounded functions of X, say $f(t, X)$ with parameter t, such that $f(t, X)$ approximates X as t approaches a limit, and consider the squared correlation ratio, $\eta^2(t)$, of $f(t, X)$ on Y. (The truncation method is a special case.) We find it convenient to consider the complex-valued function $f(t, X) = e^{itX}$, t real and positive, which is bounded but may behave like X for t near the origin. We do find that $\eta^2(0+)$ is a generalization of η^2 which always exists (satisfies Axiom A′). A modification is successfully found which satisfies Axiom D as well, and hence serves as an *index of dependence.*

The function $\eta^2(t)$, called the *dependence characteristic,* is studied in Section 4. Various indices derived from it are considered in Section 5. Illustrative examples appear in Sections 6 and 8; in the latter section it is shown that all of the various indices considered, including $\bar{\rho}^2$, coincide with the squared linear correlation coefficient when (X, Y) is trinomially distributed—just as in the case of bivariate normal (X, Y). In Section 9, we consider an analogous correlation characteristic and correlation index-symmetric in X and Y. Besides characterizing the linear relationship between X and Y, they provide lower bounds on the dependence characteristic and index.

In Section 7, we note that virtually everything said about the dependence of X on Y remains true if Y is vector-valued, so that the dependence of one r.v. on several r.v.'s is thus also characterized, as in *multiple correlation* and *multiple regression* theory. Extensions to vector-valued X, as in *canonical correlation* theory, are not considered here (but see Sarmanov and Zaharov (1960)). We conclude with miscellaneous remarks and interpretations in Section 10.

We begin (Sections 2 and 3) by reviewing the basic notions of correlation and regression, stated here for complex-valued r.v.'s, for they will be needed in the sequel.

A final introductory comment concerns the possible relevance of all this. One context of relevance is the prediction of future r.v.'s from knowledge of current observations—the context of classical regression theory. Another context is that of Bayesian inference where X is a parameter in the distribution of the data Y, or X is a future r.v. connected with Y through the intermediary of a common parameter with a prior distribution. A third context of possible relevance is that of approximation, a dependence index providing a measure of how well Y approximates X, or rather how well a convenient Y will do in lieu of the more desirable X. See also papers by Sarmanov (1960), Lehmann (1966), Kruskal (1958) and Goodman and Kruskal (1959). The latter three papers emphasize statistical interpretations and applications, which are not considered here.

Remark on maximal correlation : Let $H(x, y)$ denote the joint distribution function of (X, Y) and $H_0(X, Y)$ the product of the marginal distribution functions of X and Y. Suppose H is absolutely continuous w.r.t. H_0 with density (Radon-Nikodym derivative) denoted $k(x, y) = dH(x, y)/dH_0(x, y)$, and suppose k is integrable (dH). We then say the *mean square contingency exists*: it is defined as the positive square root of $\int(k-1)dH = \int(k-1)^2 dH_0 = Ek(X, Y)-1$ (*see* Renyi (1959)). Renyi showed that these conditions imply the existence of functions $f(X)$ and $g(Y)$ with correlation $\bar{\rho}$.

Slightly earlier (but apparently not known by Renyi), Sarmanov (1958) considered maximal correlation under these assumptions—that is, when the mean square contingency exists. He noted that $k(x, y)$ has the bilinear decomposition

$$(0) \qquad k(x, y) = dH(x, y)/dH_0(x, Y) \sim 1+\Sigma\rho_i f_i(x)g_i(y)$$

(with eigenfunctions (1,1) and $\{f_i(x), g_i(y)\}$ and corresponding eigenvalues 1 and $\lambda_i^2 = \rho_i^{-2}$, $1 < \lambda_1^2 \leqslant \lambda_2^2 \leqslant \ldots$), where the r.v.'s $\{f_1(X), g_1(Y), f_2(X), g_2(Y), \ldots\}$ each have mean zero and variance unity, and every pair is uncorrelated except $E f_i(X)g_i(Y) = \rho_i$ for $i = 1, 2, \ldots$. Moreover, $\bar{\rho} = |\rho_1|$.

Sarmanov also showed how ρ can be calculated iteratively. Let $g^{(0)}(Y)$ be an arbitrary function with mean zero and finite positive variance, and define $f^{(2k+1)}(x) = E[g^{(2k)}(Y)\,|\,x]$, $g^{(2k+2)}(y) = E[f^{(2k+1)}(X)\,|\,y]$, $k = 0, 1, 2, \ldots$ — successively regressing on X and Y alternatively. Then $f_1(X) = a \cdot \lim_{k\to\infty} f^{(2k+1)}(x)\rho_1^{-2k}$ and $g_1(y) = b \cdot \lim_{k\to\infty} g^{(2k)}(y)\rho_1^{-2k}$ for some a and b. Hence, if k is large enough, $\bar{\rho}^2 = \rho_1^2 \approx f^{(2k+1)}(x)/f^{(2k-1)}(x) \approx g^{(2k)}(y)/g^{(2k-2)}(y)$, all x and y. (Incidentally, combining results of Saramanov and Renyi, it may be noted that, when defined, the mean square contingency $\int(k-1)^2 dH_0$ is, by (0), simply the square root of $\Sigma\rho_i^2$, and hence it certainly exceeds $\bar{\rho} = |\rho_1|$.)

Sarmanov also proves that if the regressions of X on Y, and of Y on X are linear (still assuming the mean square contingency exists), and variances are finite and positive, then $\bar{\rho}$ equals, in magnitude, the linear correlation coefficient ρ of (X, Y).

2. PRELIMINARIES ON CORRELATION THEORY FOR COMPLEX-VALUED RANDOM VARIABLES

In this section we briefly review the basic results of correlation theory for a pair of random variables (X, Y) with a joint probability distribution defined. We shall present the results for complex-valued r.v.'s but shall use only standard elementary notions of probability in doing so.

Suppose S, T, U and V are real-valued r.v.'s defined on a probability space, and define the complex-valued r.v.'s

$$X = S + iT, \qquad Y = U + iV.$$

(Equivalently, X and Y are complex-valued measurable functions on the basic probability space.) We temporarily assume that all of these r.v.'s have finite second moments and that neither X nor Y is a constant a.s.. The *mean*, *variance*, and *standard deviation* of X (and likewise for Y), and the *covariance* and *correlation* of (X, Y) are defined by (a horizontal bar denotes complex conjugate):

$$EX = ES+iET; \quad X' = X-EX, \quad Y' = Y-EY;$$

$$\text{var } X = EX' \,\bar{X}' = E|X'|^2 = \text{var } S+\text{var } T; \sigma_X = \sqrt{\text{var } X}$$

$$\text{cov}(X, Y) = EX' \,\bar{Y}' = \text{cov}(S, U)+\text{cov}(T, V)$$
$$+i \text{ cov}(T, U)-i \text{ cov}(S, V);$$

$$\text{corr }(X, Y) = \text{cov}(X, Y)/\sigma_X\sigma_Y.$$

The following elementary properties are readily verified; here $a, b, c,$ and d denote complex numbers.

$E(aX+b) = aEX+b$, var $X = E|X|^2-|EX|^2$,

$$\text{var}(aX+b) = |a|^2 \text{ var } X,$$

var $X \geqslant 0$ with equality iff X is a constant a.s.,

$\text{cov}(aX+b,\ cY+d) = a\bar{c} \text{ cov}(X, Y)$,

$\text{cov}(Y,X) = \overline{\text{cov}(X,Y)}$, $\text{corr}(aX+b,\ cY+d) = (a\bar{c}/|a|\,|c|)$ $\text{corr }(X,Y)$,

$\text{corr}(Y,X) = \overline{\text{corr}(X,Y)}$, $\text{var}(aX+bY) = |a|^2 \text{ var } X+|b|^2$. var $Y+a\bar{b} \text{ cov}(X,Y)+\bar{a}b \text{ cov}(Y,X)$, $\text{cov}(X,Y) = \text{corr}(X, Y) = 0$ whenever X and Y are independent (i.e., the vectors (S,T) and (U,V) are independent) and $|\text{corr}(X,Y)| \leqslant 1$ with equality iff $X = aY+b$.

The last property (essentially Schwarz's inequality) will be proved analogously to the elementary proof for real r.v.'s appearing, for example, in Feller (1957, pg. 222) : Let $X'' = (X-EX)/\sigma_X$ and $Y'' = (Y-EY)/\sigma_Y$, and denote $\text{corr}(X,Y) = \rho e^{i\theta} = \text{corr }(X'',Y'')$ where $\rho = |\text{corr}(X,Y)|$. Then the variance of $X''-e^{i\theta}Y''$ is found to be $2(1-\rho)$ which is zero ($\rho = 1$) iff $X'' = e^{i\theta}Y''$, i.e., $X = (\sigma_X/\sigma_Y)e^{i\theta}Y+EX-(\sigma_X/\sigma_Y)e^{i\theta}EY$.

All of the above properties may be viewed as elementary results in Hilbert space theory, the elements of the space being the complex-valued r.v.'s (defined up to a.s. equivalence), the *inner product* being the covariance, and the *norm* being the standard deviation. We shall, however, continue to use only the most elementary notions of probability theory for real r.v.'s together with the arithmetic of complex numbers.

3. PRELIMINARIES ON REGRESSION THEORY FOR COMPLEX-VALUED RANDOM VARIABLES

We now briefly review basic results in regression theory for two complex-valued random variables X and Y. (Actually, the range space of Y is completely arbitrary in this section except in Assertion 5b). Now only X, and not necessarily Y, will be assumed to have a finite and positive variance. $E(h(X, Y)\,|\,Y)$ denotes conditional expectation and $\text{var}(X\,|\,Y)$ the variance of the (regular) conditional probability distribution of X given Y.

Assertion 1 : $EX = EE(X\,|\,Y)$; in fact, $Eh(X, Y) = EE(h(X,Y)\,|\,Y)$ if the LHS exists.

Assertion 2 : $\text{var } X = \text{var } E(X\,|\,Y)+E \text{ var}(X\,|\,Y)$.

This is proved by writing $\text{var } X = E|X'|^2 = EE(|X'|^2|\,Y)$ and substituting $(X'-E(X'\,|\,Y))+E(X'\,|\,Y)$ for X', as in the usual proof for the case of real r.v.'s.

Definition : The *regression function* of X on Y, denoted $r(Y)$, is that measurable complex-valued function $g(Y)$ for which $E|X-g(Y)|^2$ is a minimum.

Assertion 3 : $r(Y) = E(X\,|\,Y)$a.s.

This is proved by considering $E|X-g(Y)|^2$, substracting and adding $E(X\,|\,Y)$ to X, to obtain $E|X-E(X\,|\,Y)|^2+E|E(X\,|\,Y)-g(Y)|^2$ (the cross-product terms vanishing upon iterating the expectations as in Assertion 1). This is minimized by the choice $g(Y) = E(X\,|\,Y)$.

Assertion 4 : $\text{cov}(X, r(Y)) = \text{var } r(Y)$.

Iterating the expectation (Assertion 1) in the LHS quickly yields the RHS.

Definition : The *squared correlation ratio* of X on Y, denoted η^2 or $\eta^2(X\,|\,Y)$, is defined as $\eta^2 = |\text{corr}(X, r(Y))|^2$.

Assertion 4a :
$$\eta^2 = \frac{\text{var } r(Y)}{\text{var } X} = 1-\frac{E \text{ var}(X\,|\,Y)}{\text{var } X}$$
$$= 1-\frac{\text{var}(X-r(Y))}{\text{var } X}.$$

This follows from Assertion 4, using Assertions 2 and 3.

Assertion 4b : $0 \leqslant \eta^2 \leqslant 1$; $\eta^2 = 0$ iff $r(Y)$ is constant a.s. ($= EX$); in particular, $\eta^2 = 0$ if (but not only if) X and Y are independent; $\eta^2 = 1$ iff X is a function of Y a.s. ($X = r(Y)$).

This follows from Assertion 2, 3 and 4a, and the consequence of a variance being zero.

Assertion 5 : Among all complex-valued r.v.'s $g(Y)$ with finite variances, the regression function uniquely (up to linear functions and a.s. equivalence) maximizes $|\text{corr}(X, g(Y))|$.

Hence, $|\text{corr}(X, g(Y))| < |\text{corr}(X, r(Y))|$ unless $g(Y) = ar(Y)+b$ a.s. A proof, following Renyi (1959), is as follows : Letting X have mean zero and variance unity and $g(Y)$ have mean μ and variance σ^2, we have

$$\begin{aligned}\text{corr}(X, g(Y)) &= |E(X\overline{(g(Y)-\mu)})|/\sigma \\ &= |E(r(Y)\overline{(g(Y)-\mu)})|/\sigma \text{ upon iterating} \\ &= |\text{corr}(r(Y), g(Y))|\sqrt{\text{var } r(Y)} \leqslant \sqrt{\text{var } r(Y)}\end{aligned}$$

with equality holding iff $g(Y) = ar(Y)+b$ a.s.

Assertion 5a : $\eta^2 = \max|\text{corr}(X, g(Y))|^2$ where "max" indicates least upper bound over all g for which the correlation is defined.

Assertion 5b : If Y (complex-valued) also has finite and positive variance, then $\eta^2 \geqslant |\text{corr}(X,Y)|^2$ with equality iff $r(Y)$ is linear a.s.

This assertion remains valid if Y is vector-valued and corr(X, Y) denotes the multiple correlation coefficient.

We finally note that all such results on regression are special cases of theorems on projection operators on Hilbert spaces; for a conditional expectation operation is completely equivalent to a projection operation which is non-negative (monotonic) and constant preserving (Bahadur, 1955). However, as noted before, we confine attention to elementary notions and methods of probability theory here.

4. THE DEPENDENCE CHARACTERISTIC

We now assume that X and Y are real-valued r.v.'s on a probability space. No assumptions about existing means and variances are imposed; we shall assume X is non-trivial (not a constant a.s.), taking care of the trivial case separately. We shall apply results of regression theory to the complex-valued r.v. e^{itX} (playing the role of X in Section 3) for arbitrary real t.

We first consider only the marginal distribution of X. Let $\phi(t)$ denote the c.f. (*characteristic function*) of X, the expectation of e^{itX}, and let $\gamma(t)$ denote the variance of e^{itX} ; thus

$$(1) \qquad \gamma(t) = \operatorname{var} e^{itX} = 1 - |\phi(t)|^2.$$

Now $|\phi(t)|^2$ is also a c.f., namely that of the distribution of the difference between two independent versions of X. Hence, the behaviour of $\gamma(t)$ is immediately deducible from well-known properties of c.f.'s (see Lukacs (1960), e.g.). Thus, γ is uniformly continuous, real and symmetric about $t = 0$ where it vanishes, and $0 \leqslant \gamma(t) \leqslant 1$. Also, γ vanishes nowhere but at the origin unless X has a lattice distribution. In this case $\gamma(t)$ is a periodic function symmetric about $t = 0$, and hence symmetric within each period and vanishing on a t-lattice, say at $t = mt_0$ for $m = 0, \pm 1, \pm 2, \ldots$. Thus, in the lattice case, $\gamma(t)$ is completely defined by its values for $0 \leqslant t \leqslant t_0/2$, vanishing at $t = 0$ and strictly positive for $0 < t \leqslant t_0/2$. (If $X = a + Zb$ where Z is integer-valued a.s., there being at least two successive integers with positive probability, then $t_0 = 2\pi/b$.) On the other hand, if X is absolutely continuous, $\gamma(t)$ tends to unity as t approaches $+\infty$.

The behaviour of $\gamma(t)$ near the origin determines all existing central moments of even order. In particular, if X has finite variance σ^2, then

$$(2) \qquad \gamma(t) = \sigma^2 t^2 + o(t^2) \text{ as } t \to 0.$$

$\gamma(t)$ cannot approach zero any more rapidly than a multiple of t^2 unless it is identically zero, which occurs iff X is a constant a.s.

But if the variance of X is infinite, $\gamma(t)$ approaches zero more slowly than any multiple of t^2. For example, for a Cauchy distribution, $\gamma(t)$ approaches zero linearly on either side of the origin. (The behaviour of $\gamma(t)$ at the origin will be related to work by Pitman (1961) on the behaviour of the real and imaginary parts of the c.f. at the origin in Section 10.) It will be convenient to introduce the following terminology : Let X and X' be two r.v.'s with corresponding γ and γ'; then X is said to have a *higher order of variability* than X' if $\gamma'(t)/\gamma(t) \to 0$ as $t \to 0$. Furthermore, we say that every trivial r.v. (constant a.s.) has *zero order variability* and every r.v. with finite positive variance has *first order variability*; we will not classify any higher orders of variability corresponding to r.v.'s with infinite variances.

We now return to consideration of the joint behavior of X and Y. Denote the conditional c.f. of X given Y by $\phi(t \mid Y)$, the regression function of e^{itX} on Y; then

$$(3) \qquad \operatorname{var} \phi(t \mid Y) = E\,|\phi(t \mid Y)|^2 - |\phi(t)|^2 = \gamma(t) - E\,\gamma(t \mid Y)$$

where $\gamma(t \mid Y) = \operatorname{var}(e^{itX} \mid Y) = 1 - |\phi(t \mid Y)|^2$ as in (1). We shall be considering the squared correlation ratio, $\eta^2(t)$, of e^{itX} on Y; let us first note that, when $\sigma^2 = \operatorname{var} X$ is finite, $\eta^2(t) = \eta^2 + o(1)$ as $t \to 0$ where η^2 is the squared correlation ratio of X on Y. For, denoting the conditional variance by $\sigma^2(Y)$, $\gamma(t \mid Y) = \sigma^2(Y)t^2 + o(t^2)$ as in (2), and therefore, from (3) and Assertion 2 of Section 3, $\operatorname{var} \phi(t \mid Y) = \operatorname{var} r(Y).t^2 + o(t^2)$. Hence $\operatorname{var} \phi(t \mid Y)/\operatorname{var} e^{itX} = \eta^2 + o(1)$, but (Assertion 4a) the LHS is $\eta^2(t)$. Thus, the function $\eta^2(t)$ is indeed an extension of the concept of η^2. Now we shall consider $\eta^2(t)$ more generally.

First note that, for each y-value, $|\phi(t \mid y)|^2$ is a c.f., namely that of the conditional distribution of $X_1^* - X_2^*$, where X_1^* and X_2^* are conditionally (given $Y = y$) independent each having as distribution the conditional distribution of X. Hence $E\,|\phi(t \mid Y)|^2$ is also a c.f.—of the (symmetric) marginal distribution of $X_1^* - X_2^*$. This expectation is therefore real, symmetric and uniformly continuous in t. Thus, $\eta^2(t)$, the ratio of (3) to (1), is real, symmetric and continuous, being defined for all t for which $\gamma(t)$ does not vanish. For any positive $t\,(\neq mt_0$ if X has a lattice distribution), $\eta^2(t)$ is thus defined. Since it is continuous and bounded (between 0 and 1 inclusive), it may also be defined at $t = 0$ (and also at $t = mt_0$ in the lattice case) so as to be continuous there. Because of the symmetry around $t = 0$ (and around $t = mt_0$), we need not specify whether limits are from the right or left. We thus are able to define :

Definition : The *dependence characteristic* of X on Y, denoted $\eta^2(t)$, is the squared correlation ratio of e^{itX} on Y, as a function of t, being defined by continuity at t-values where var e^{itX} vanishes; it is defined for all $t \geqslant 0$. If X is a constant a.s., $\eta^2(t)$ is identically zero.

Thus

$$\eta^2(t) = \frac{E\,|\phi(t\,|\,Y)\,|^2 - |\phi(t)\,|^2}{1 - |\phi(t)\,|^2} = \frac{\gamma(t) - E\gamma(t\,|\,Y)}{\gamma(t)}. \tag{4}$$

Since $\eta^2(t) = \eta^2(-t)$, we confine attention to the domain $t \geqslant 0$. We now list some of the properties of the dependence characteristic, immediately derivable from properties of c.f.'s and of correlation ratios.

Property 1 : $\eta^2(t)$ is defined for arbitrary (X, Y); $\eta^2(\cdot)$ maps, continuously, the non-negative reals into the closed unit interval. If X has a lattice distribution, $\eta^2(t)$ is periodic and symmetric within each period; if (X, Y) is absolutely continuous,[5] $\eta^2(t) \rightarrow 0$ as $t \rightarrow \infty$.

Property 2 : If var $X < \infty$, then $\eta^2(t) = \eta^2 + o(1)$ as $t \rightarrow 0$, where η^2 is the squared correlation ratio of X on Y.

Property 3 : $\eta^2(t) = 0$ for all t iff X and Y are independent. For, if $\eta^2(t) = 0$ at a fixed t-value, then $\phi(t\,|\,Y)$ is constant a.s. (Assertion 4b) and hence equals $\phi(t)$ a.s.; and if this is so for every positive t-value (except possibly on a lattice) then it is so for all

[5] It is not sufficient that X be absolutely continuous; *see* Example 3 in Section 6.

real t and the conditional distribution therefore coincides (a.s.) with the unconditional one. But this is a necessary and sufficient condition for the independence of X and Y. The converse is likewise valid.

It will be seen by example (Examples 4a and 4b, Section 6) that $\eta^2(t)$ may vanish throughout a finite interval even though X and Y are not independent; it is thus necessary in Property 3 to state that $\eta^2(t)$ vanishes for all t. This is associated with the fact that c.f.'s of differing distributions may coincide throughout a finite interval.

Property 4 : $\eta^2(t) = 1$ for all t iff X is a function of Y a.s. In fact, $\eta^2(t) = 1$ at two incommensurable t-values implies that X is a function of Y a.s.

For, as in Assertion 4b of Section 3, $\eta^2(t) = 1$ iff var $(e^{itX} \mid Y) = 0$ a.s., i.e., iff $e^{itX} = \phi(t \mid Y)$ so that $|\phi(t \mid Y)| = 1$ a.s. But the modulus of a c.f. equals unity at two incommensurable t-values iff the distribution is degenerate (Lukacs, 1960, pg. 25). Hence, the conditional distribution is degenerate a.s.—that is, X is a function of Y a.s.

However, $\eta^2(0) = 1$ does not imply that X is a function of Y except when var $X < \infty$. It does in this case since $\eta^2 = 1$ iff X is a function of Y a.s. The implications of $\eta^2(0) = 1$ will be pursued in the next section. Examples 5a and 3 (Section 6) are relevant. In Example 1, $\eta^2(t) = 1$ at every odd multiple of π but is less than unity at the origin and elsewhere.

Property 5 : The dependence characteristic of $aX+b$ is $\eta^2(|a|t)$.

Property 6 : If $\eta_g^2(t)$ denotes the dependence characteristic of X on $g(Y)$ for some measurable g, then $\eta^2(t) \geqslant \eta_g^2(t)$ for all t, with equality holding for all t iff either g is 1–1 a.s. or X and Y are independent.

This is a consequence of Assertion 5 of Section 3. For, $\eta^2(t) = |\text{corr}(e^{itx}, \phi(t \mid Y))|^2$ which is greater than $|\text{corr}(e^{itX}, h(Y))|^2$

for any other complex-valued $h(Y)$, including $h(Y) = E(e^{itX} \mid g(Y))$. This extremal property of $\eta^2(t)$ makes possible the determination of lower bounds on it, and also on indices derived therefrom.

We also have a limited continuity property, derivable from the convergence of the characteristic functions in (4) :

Property 7 : If the joint distribution of (X_n, Y_n) converges weakly to the distribution of (X, Y), if the conditional distribution of X_n given $Y_n = y$ converges weakly to the conditional distribution of X given $Y = y$ a.s., and if t_0 is such that var $e^{itX} > 0$, then the dependence characteristic of X_n on Y_n at t_0 converges to the dependence characteristic of X on Y at t_0.

But whether such a property holds at $t = 0$ is not known.

We next attempt to gain some understanding of the scaling of the range of $\eta^2(t)$. To this end, let H be the joint distribution function of (X, Y), with marginal distribution functions F and G of X and Y, respectively, and let $H_0(x, y) = F(x)\, G(y)$. Let $H_\pi = \pi H + (1-\pi) H_0$ for $0 \leqslant \pi \leqslant 1$, and denote by (X_π, Y_π) r.v.'s with joint distribution function H_π; X_π and Y_π have marginal distribution functions F and G respectively.

Property 8 : The dependence characteristic of X_π on Y_π equals π^2 times the dependence characteristic of X on Y.

To prove Property 8, we first evaluate

$$\text{var}\, \phi_\pi(t \mid Y_\pi) = E \mid \phi_\pi(t \mid Y_\pi) \mid^2 + \mid \phi_\pi(t) \mid^2.$$

Now $\phi_\pi(t \mid y) = \pi \phi(t \mid y) + (1-\pi)\phi(t)$ and $\phi_\pi(t) = \phi(t)$; after simplification, we find

$$\text{var}\, \phi_\pi(t \mid Y_\pi) = \pi^2\, \text{var}\, \phi(t \mid Y),$$

from which Property 8 immediately follows.

The joint distribution H_0 has the same marginals as does H but with dependence characteristic zero. We shall now prove :

Property 9 : If Y has a continuous distribution function, there exists (X^0, Y^0) with the same marginal distributions as (X, Y) but with dependence characteristic identically unity.

Either of the extremal distributions introduced by Hoeffding (1940), and again by Frechèt (1951), will serve : $H^0(x, y) = \min [F(x), G(y)]$ or $H^{00}(x, y) = \max [0, F(x)+G(y)-1]$. (These may be shown to be bivariate distribution functions; they obviously have the correct marginals F and G and are such that $H^{00}(x, y) \leqslant H(x, y) \leqslant H^0(x, y)$ for *any* joint distribution function H with marginals F and G.) The points of increase of the joint distribution function H^0 may be shown to be contained within a set C with the following properties : C is a curve moving northeasterly across the (x, y)-plane, possibly with some vertical and horizontal segments (occurring only at discontinuity points of $F(x)$ and $G(y)$, respectively), and for any (x, y) in C both $G(y-) \leqslant F(x)$ and $F(x-) \leqslant G(y)$ hold. (If both F and G are continuous and increasing, the curve C is the locus of points where $F(x) = G(y)$; the general case is somewhat tedious and the proofs are omitted here[6]). If G is continuous, then $F(X-) \leqslant G(Y) \leqslant F(X)$ a.s. according to H^0; hence, X is completely determined by Y a.s., and $\eta^2(t)$ is therefore identically unity by Property 4. If we consider H^{00} instead, we only need replace F by $1-F$. That X need not be a function of Y a.s. when G is not continuous is seen in Example 2. It also seems intuitively clear that it may not be possible to associate with each discrete mass point of the Y distribution a unique X-value—it is certainly not possible when Y has only a finite number of possible values and X has a continuous distribution.

As a consequence of Properties 8 and 9, we conclude that, given F and G, G continuous, and a particular t-value t_0, $\eta^2(t_0)$ can assume any value between zero and unity inclusive, depending on the specific joint distribution H with the specified marginals. In this sense, it may be said that the dependence characteristic reflects the dependence of X on Y, but is little affected by the individual distributions of X and of Y.

[6] A not completely valid proof of this appears in Frechèt (1951). Valid proofs have been derived by J. Th. Runnenburg and myself (unpublished). I wish to acknowledge conversations with Professor Runnenburg regarding these extremal distributions, which he originally brought to my attention.

Now let us suppose the mean square contingency exists (*see* remarks on page 7). Then the conditional expectation of any $u(X)$ given Y may be obtained by taking the unconditional expectation of $u(X)k(X, y)$, as seen from (0). Hence $\phi(t|y) = \phi(t)+\Sigma\rho_i g_i(y)\phi_i(t)$ where $\phi_i(t) = E(e^{itX} f_i(X))$. Note incidentally that $|\phi_i(t)|^2 \leqslant E^2|f_i(X)| \leqslant Ef_i^2 = 1$. We then find var $\phi(t|Y) = \Sigma\rho_i^2|\phi_i(t)|^2$, using the orthogonality properties, so that $\eta^2(t)$ has the representation given by:

Property 10 : If the mean square contingency exists, then

$$\eta^2(t) = \frac{\sum_{i=1}^{\infty} \rho_i^2|\phi_i(t)|^2}{1-|\phi(t)|^2}$$

where the ρ_i^{-2}'s are the eigenvalues corresponding to the eigenfunctions $f_i(x)$, and $\phi_i(t) = E(e^{itX} f_i(X))$.

Since the eigenfunctions $g_i(y)$ of the expansion (0) play no explicit role here, it may be preferable to obtain the f_i's and ρ_i's (as in Sarmanov (1958*b*)) by expanding $E(k(x, Y)\, k(z, Y)) \sim 1+\Sigma\rho_i^2 f_i(x) f_i(z)$.

In the next section we shall consider various indices, or coefficients, derived from the function $\eta^2(t)$ which will serve as numerical indices of the dependence of X on Y; the dependence characteristic, however, characterizes this dependence more fully.

5. INDICES OF DEPENDENCE

We shall first consider $\eta_0^2 = \eta^2(0)$, where $\eta^2(t)$ is the dependence characteristic of X on Y, as a possible index of dependence. Thus, $\eta_0^2 = \lim_{t\to 0} \eta^2(t)$ and $\eta^2(t)$ is the squared correlation ratio of e^{itX} on Y (4); if X is a constant a.s., then $\eta_0^2 = 0$ by definition. The following six properties of η_0^2 are readily verified from the corresponding six properties of $\eta^2(t)$ in Section 4 :

(1) η_0^2 is defined for any pair of r.v.'s (X, Y).

(2) $\eta_0^2 = \eta^2$ if η^2 is defined (i.e., if var $X < \infty$).

(3) $\eta_0^2 = 0$ if (but not only if—see Examples 1, 4a, 4b and 5a) X and Y are independent.

(4) $\eta_0^2 = 1$ if (but not only if — see Examples 5a and 3—unless var $X < \infty$) X is a function of Y a.s.

(5) η_0^2 is invariant under 1–1 linear transformation of X.

(6) η_0^2 is not increased if Y is replaced by $g(Y)$ for some measurable Y; it is unchanged if (but not only if) g is 1–1 a.s.

We thus see that Axioms A′, C, E′, and F′ of Section 1 (and G because of property (2)) are satisfied by η_0^2, but Axiom D is not satisfied. We shall now verify that Axiom G′ is also satisfied by η_0^2.

Let $\psi(t)$ denote the squared modulus of the c.f. of Z (in the notation of Axiom G′). We find after a simple calculation that

$$\eta^2(t) = \psi(t)^l \,\frac{1-\psi(t)^k}{1-\psi(t)^{k+l}}$$

and the RHS $= k/(k+l)+o(1)$ as $t \to 0$ since $\psi(t)$ tends to unity from below as $t \to 0$. Incidentally, a stronger version of Axiom G′ may be proved, namely, that if $Y = Z_1+\ldots+Z_k+Z_{k+l+1}+\cdots+Z_{k+l+m}$ and $X = Z_1+\ldots+Z_{k+l}$ then $\eta_0^2 \geqslant k^2/(k+l)(k+m)$ with equality holding if var $X < \infty$, or if the Z_i's are symmetrically distributed, or if $m = 0$ (proof omitted).

We now introduce another possible index of dependence, namely

$$\bar{\eta}^2 = \max \eta^2(t)$$

where "max" indicates least upper bound over all $t \geqslant 0$. It is readily verified that $\bar{\eta}^2$ satisfies Axiom A′, Axiom C, Axiom D (from Property 3 of $\eta^2(t)$), and Axiom G (see Example 6). That $\bar{\eta}^2$ also satisfies Axiom G′ will now be proved by showing that, for X and Y as defined in Axiom G′, $\bar{\eta}^2 = \eta_0^2$ and it has already been shown that $\eta_0^2 = k/(k+l)$. It has been seen above that $\eta^2(t) = h(\psi(t))$ where $h(v) = v^l \,\dfrac{1-v^k}{1-v^{k+l}}$. Now $\psi(t)$ achieves its maximum, namely unity, at $t = 0$. It is therefore sufficient to show that $h(\cdot)$ is an increasing function on the unit interval, for then $\eta^2(t)$ achieves its maximum at $t = 0$. (We confine attention to the non-trivial case $k > 0$.)

Let $u = v^k$, $r = l/k$, and $g(u) = h(u^{1/k}) = u^r(1-u)/(1-u^{r+1})$. It will be sufficient to show that g is increasing on the unit interval. It may be shown that $dg/du = u^{r-1}G(u)(1-u^{r+1})^{-2}$ where $G(u) = r(1-u^{r+1})-(r+1)(u-u^{r+1})$. Since $dG/du = -(r+1)(1-u^r) < 0$, $G(u)$ is decreasing and hence is greater than $G(1) = 0$ for u inside the unit interval. Hence, dg/du is positive inside the unit interval, as was to be proved.

We have thus shown that $\bar{\eta}^2$ has all the axiomatic properties we originally considered in Section 1 for an index of dependence. However, it does not share with η_0^2 the desirable property that, when var $X < \infty$, it equals unity *only* if X is a function of Y a.s. (*see* Example 1). Since $0 \leqslant \eta_0^2 \leqslant (\eta_0^2+\bar{\eta}^2)/2 \leqslant \bar{\eta}^2 \leqslant 1$, we are led to define :

Definition : $\delta = \delta(X \mid Y) = \frac{1}{2}\eta_0^2+\frac{1}{2}\bar{\eta}^2$ is the *index of dependence* of X on Y.[7]

Since $\delta = 0$ iff $\bar{\eta}^2 = 0$, δ shares axiomatic property D with $\bar{\eta}^2$. And since $\delta = 1$ iff $\eta_0^2 = 1$, it shares with η_0^2 any implication of a value unity. Since both η_0^2 and $\bar{\eta}^2$ satisfy Axioms A′, C, E′, F′, G and G′, so does δ.

That δ does not satisfy an "iff"version of Axiom E′ is demonstrated by Examples 3 and 5b. In these examples, $X = Y+Z$ where Y has infinite variance and Z is independent of Y with finite variance; all that matters is that Y have a higher order of variability than Z (*see* Section 4). For, in the notation and terminology of Section 4, if $\eta_0^2 = 1$ (and hence $\delta = 1$), then $E[\gamma(t \mid Y)/\gamma(t)] \to 0$, and hence, on the average (w.r.t. the distribution of Y), the conditional distribution of X given Y has a lower order of variability than does the unconditional distribution of X. Hence, an "iff" version of Axiom E′ is : $\delta = 1$ iff the order of variability of the conditional distribution has, on the average, been reduced.

---

[7] A quotation from a 1924 paper by Harald Cramer (appearing in Page 140 of Goodman and Kruskal (1959)) is appropriate: : "Every attempt to measure a conception like this by a single number must necessarily contain a certain amount of arbitrariness and suffer from certain inconveniences."

The implications of Properties 8 and 9 of the dependence characteristic on these indices are as follows: In the notation of Property 8, each of the indices η_0^2, $\bar{\eta}^2$, and δ is reduced by a factor π^2 when (X, Y) is replaced by (X_π, Y_π). That the ordinary linear correlation coefficient is likewise reduced (by a factor π, its square being reduced by π^2) may be directly and simply verified. Also, given F and G (G continuous), and given an arbitrary number, say π^2, in the closed unit interval, there exists a joint distribution function H with marginals F and G and with $\delta = \pi^2$; the same could be said for η_0^2 or $\bar{\eta}^2$. This is seen to be so by taking $H = \pi H + (1-\pi)H_0$; then the dependence characteristic is identically π^2. Thus, the range of the index of dependence is indeed the full unit interval, whatever the marginal distributions so long as the Y-distribution is continuous. If the Y-distribution is not continuous, the range will still be an interval but the maximum value may possibly be less than unity (Example 2). It is not known in general what this maximum value is, nor whether it is attained by one of the extremal distributions. (That ρ and $-\rho$ attain their maxima for the extremal distributions is obvious from a formula of Hoeffding (1940) for the covariance of X and Y; see Lemma 2 in Lehmann (1966).[8])

We now raise the question: are there other indices derivable from $\eta^2(t)$ which would be unity iff X is a function of Y a.s.? Letting $\underline{\eta}^2$ denote the greatest lower bound on $\eta^2(t)$, the index $\frac{1}{2}\underline{\eta}^2 + \frac{1}{2}\bar{\eta}^2$ is indeed unity (zero, resp.) iff $\eta^2(t)$ is identically unity (zero, resp.), and hence is unity iff X is a function of Y a.s. and zero iff X and Y are independent. But it is not very satisfactory otherwise for, for absolutely continuous (X, Y), its range is confined to the interval $[0, 1/2]$ since $\underline{\eta}^2 = 0$. We might consider instead averaging $\eta^2(t)$ w.r.t. some absolutely continuous probability distribution on $[0, \infty)$, but we should then have to give up scale invariance because of Property 5 of Section 4.

[8] These facts were originally brought to my attention by Professor J. Th. Runnenburg.

Another possibility is to consider a limit of $T^{-1}\int_0^T \eta^2(t)dt$ as $T\to\infty$. This way an "iff" version of Axiom E' could be satisfied, but at the expense of Axiom D (e.g., see Example 4b), as well as some of the other axioms.

The relationship of δ and $\bar{\eta}^2$ to $\bar{\rho}$ (the maximal correlation) is not clearcut. If, in defining $\bar{\rho}$, we maximized over complex-valued functions f and g instead of only real-valued f and g, then $\bar{\rho}^2$ could not be exceeded by $\bar{\eta}^2$ or δ.

6. EXAMPLES

Example 1 : (X, Y) has a 3- or 4-point distribution, with probabilities $p, q, r/2$, and $r/2$ at the points in the (x, y)-plane with coordinates $(0, 1)$, $(0, -1)$, $(1, 0)$ and $(-1, 0)$, respectively, where $p+q+r = 1$ and either p or q may be zero. The r.v. X has a symmetric distribution, both conditionally and unconditionally, so that $\eta^2 = 0$ (and hence $\rho = 0$). The maximum correlation coefficient $\bar{\rho} = 1$ since $|X|$ and $1-|Y|$ are perfectly correlated (coinciding with certainty). We now evaluate the dependence characteristic. We find

$$\phi(t) = 1-r(1-\cos t),$$

$$\phi(t\,|\,y) = 1 \text{ if } y = \pm 1 \text{ and } = \cos t \text{ if } y = 0,$$

$$E\,|\phi(t\,|\,Y)\,|^2 = 1-r(1-\cos^2 t),\ \eta^2(t) = \frac{(1-r)(1-\cos t)}{2-r(1-\cos t)}$$

(after some reduction).

Hence, $\eta^2(t)$ oscillates continuously between 0 (at $t = 0$ and all even multiples of π) and 1 (at all odd multiples of π). Consequently, $\eta_0^2 = 0$, $\bar{\eta}^2 = 1$ and $\delta = 1/2$. Whether or not $1/2$ is a reasonable index of dependence of X on Y, it seems clear that neither 0 nor 1 are very satisfactory indices of either dependence or relationship.

Example 2 : X and Y each have 2-point distributions : $X = 1$ with probability p and $= 0$ otherwise while $Y = 1$ with probability q and $= 0$ otherwise ($0 < p < 1$, $0 < q < 1$). The joint distribution is then completely determined by specifying one of the four probabilities, as in a 2×2 table; let r be the probability that $X = Y = 1$ ($r \geqslant 0$). The correlation coefficient ρ is readily evaluated and found to be $(r-pq)/\sqrt{pq(1-p)(1-q)}$, and $\eta^2 = \rho^2$. Since both X and Y have 2-point marginal distributions, the only functions $f(X)$ and $g(Y)$ with positive variances are 1–1 functions (a.s.), and these are equivalent to linear functions; hence, $\bar{\rho}^2 = \rho^2$. To evaluate the dependence characteristic, we require :

$$\phi(t) = 1-p+pe^{it},\ \phi(t\,|\,0) = \frac{1-p-q+r}{1-q} + \frac{p-r}{1-q}\,e^{it},$$

$$\phi(t\,|\,1) = \frac{q-r}{q} + \frac{r}{q}\,e^{it}.$$

and, after simplification, one finds $\eta^2(t) = \rho^2$, all t. Hence $\eta_0^2 = \bar{\eta}^2 = \delta = \rho^2$.

For fixed marginals (fixed p and q), we now examine the effect of differing values of r. If $p = q$ or $= 1-q$, there is a 2-point joint distribution ($r = p = q$ or $r = 0$) consistent with the fixed marginals; then X is a (linear) function of Y a.s. If $p \neq q$, and $\neq 1-q$, the maximal common value of ρ^2, η^2 and δ is attained at one of the 3-point joint distributions (verification omitted), and is thus less than unity. These 3-point joint distributions are the extremal distributions.

Example 3 : The (X, Y)-distribution is confined to two equi-probable diagonal lines; specifically, $X = Y+1$ and $Y-1$, respectively, with probabilities 1/2 and 1/2. The conditional distribution of X given y is a 2-point distribution, symmetric around y and with unit variance. We further assume that Y has infinite variance.

We first show that $\bar{\rho}$ is unity by considering the correlation ratio η_n of $f_n(X)$, a symmetric truncation of X at $\pm n$, on Y, and showing that $\eta_n^2 \rightarrow 1$ as $n \rightarrow \infty$. Since $f_n(X)$ is known within ± 1 given y, its conditional variance is at most unity. On the other hand, the unconditional variance of $f_n(X)$ is arbitrarily large for n sufficiently large. Hence, η_n^2 is arbitrarily close to unity.

We now evaluate the dependence characteristic. Let $\psi(t)$ be the c.f. of Y. We find

$$\phi(t\,|\,y) = e^{ity}\cos t, \quad \phi(t) = \psi(t)\cos t, \quad \text{and}$$

$$\eta^2(t) = \frac{\cos^2 t - \cos^2 t\,|\,\psi(t)\,|^2}{1-\cos^2 t\,|\,\psi(t)\,|^2}$$

Since Y has infinite variance, $|\,\psi(t)\,|^2$ approaches unity (as $t \to 0$) slower than does $1-ct^2$ for any c, and hence slower than does $\cos^2 t$. Consequently, the trigonometric term can be ignored in the denominator of $\eta^2(t)$ as $t \to 0$ and it factors out of the numerator so that $\eta^2(t) \to 1$. Hence $\eta_0^2 = 1$ and therefore $\bar{\eta}^2 = \delta = 1$. In fact, $\eta^2(k\pi) = 1$ for $k = 0, 1, 2, \ldots.$ Also, $\eta^2[(k+\frac{1}{2})\pi] = 0$ for $k = 0, 1, 2, \ldots$. Hence, $\eta^2(t)$ oscillates between 0 and 1, but not necessarily periodically.

In this example, η^2 is not defined, but $\bar{\rho}^2 = \eta_0^2 = \bar{\eta}^2 = \delta = 1$ although X is not a function of Y a.s.

Example 4a : Suppose $Y = 0$ and 1, each with probability $\frac{1}{2}$, and the conditional distributions of X given y have the Polya-type characteristic functions (Lukacs, 1960) :

$$\phi(t\,|\,0) = \begin{cases} 1-|t|, & 0 \leqslant |t| \leqslant 1/2 \\ 1/4|t|, & t \geqslant 1/2 \end{cases}$$

$$\phi(t\,|\,1) = \begin{cases} 1-|t|, & |t| \leqslant 1 \\ 0, & |t| \geqslant 1. \end{cases}$$

Hence, for $t \geqslant 0$,

$$\phi(t) = \begin{cases} 1-t, & 0 \leqslant t \leqslant 1/2 \\ (1-t)/2+1/8t, & 1/2 \leqslant t \leqslant 1 \\ 1/8t, & t \geqslant 1 \end{cases}$$

and

$$E\,|\,\phi(t\,|\,Y)\,|^2 = \begin{cases} (1-t)^2, & 0 \leqslant t \leqslant 1/2 \\ (1-t)^2/2+1/32t^2, & 1/2 \leqslant t \leqslant 1 \\ 1/32t^2, & t \geqslant 1. \end{cases}$$

Hence, $\eta^2(t)$ is seen to vanish throughout the interval $0 \leqslant t \leqslant 1/2$, but is positive for all $t > 1/2$. It may be shown that a unique maximum occurs at $t = 1$, namely $\bar{\eta}^2 = 1/63$; since $\eta_0^2 = 0$, $\delta = 1/126$; η^2 is not defined. The value of $\bar{\rho}^2$ is unknown, but it is at least 1/33 since the squared correlation ratio of $\cos tX$ on Y may be shown to be (utilizing the symmetry) $(\text{var } e^{itX}/\text{var} \cos tX)\eta^2(t)$ which may be shown to have a maximum value of 1/33 (at $t = 1$).

Example 4b : This example is similar to 4a, with the same Y, the same $\phi(t \mid 0)$, but with $\phi(t \mid 1) = \phi(t \mid 0)^2$. Then, $\phi(t) = \frac{1}{2}(1-t)(2-t)$ for $0 \leqslant t \leqslant 1$ and $= 0$ for $t \geqslant 1$, and $E\,|\phi(t \mid Y)|^2 = \frac{1}{2}(1-t)^2+\frac{1}{2}(1-t)^4$ for $0 \leqslant t \leqslant 1$ and $= 0$ for $t \geqslant 1$. Hence, $\eta^2(t)$ is found, after algebraic reduction, to be $= t(1-t)^2/(12-13t+6t^2-t^3)$ for $0 \leqslant t \leqslant 1$ and $= 0$ for $t \geqslant 1$. Thus, $\eta^2(t)$ vanishes for $t = 0$ ($\eta_0^2 = 0$) and all $t \geqslant 1$ but is positive for $0 < t < 1$. It has a unique maximum, occurring approximately at $t = 0{\cdot}425$ and $\bar{\eta}^2$ is approximately 0·019 so that δ is approximately 0·009. As in Example 4a, $\bar{\rho}^2$ can be shown to exceed $\bar{\eta}^2$ (and hence δ) since the squared correlation ratio of $\cos tX$ on Y exceeds $\eta^2(t)$ for $0 < t < 1$; but the exact value of $\bar{\rho}^2$ is not known.

Example 5a : Y and Z are independent standard normal and Cauchy r.v.'s, respectively, with c.f.'s $\psi(t) = e^{-t^2/2}$ and $\chi(t) = e^{-|t|}$, and $X = Y+Z$. Then

$$\phi(t) = \exp(-\tfrac{1}{2}\,t^2-|t|), \quad \phi(t \mid y) = \exp(ity-|t|),$$

$$|\phi(t)|^2 = \exp(-t^2-2\,|t|), \quad E\,|\phi(t \mid Y)|^2 = \exp(-2\,|t|),$$

and

$$\eta^2(t) = e^{-2t}\,\frac{1-e^{-t^2}}{1-e^{-t^2-2t}} = \tfrac{1}{2}\,t+o(t) \;\text{ as }\; t \to 0.$$

Thus, $\eta_0^2 = 0$. The absolute maximum occurs at approximately 0·581; $\bar{\eta}^2$ is approximately 0·115 and δ approximately 0·058; η^2 is not defined; $\bar{\rho}^2 > 0$ since, e.g., the squared correlation ratio of $\cos X$ on Y is positive, but its value is unknown.

Example 5b : This is the same as 5a with the roles of Y and Z interchanged. Then

$$\phi(t\,|\,y) = \exp(ity-\tfrac{1}{2}t^2), \quad E\,|\,\phi(t\,|\,Y)\,|^2 = \exp(-t^2),$$

and

$$\eta^2(t) = e^{-t^2}\,\frac{1-e^{-2t}}{1-e^{-t^2-2t}} = 1-t+o(t) \text{ as } t\to 0.$$

Thus, $\eta_0^2 = \bar{\eta}^2 = \delta = 1 = \rho^2$, the latter holding since the squared correlation ratio of $\cos tX$ on Y tends to unity as t tends to zero (details omitted).

Example 6 : $(X,\ Y)$ is bivariate normal with correlation coefficient ρ (and zero means and unit variances for convenience). Then

$$\phi(t\,|\,y) = \exp(i\rho yt-\tfrac{1}{2}(1-\rho^2)t^2), \quad \phi(t) = \exp(-\tfrac{1}{2}t^2),$$

and

$$\eta^2(t) = (e^{\rho^2t^2}-1)/(e^{t^2}-1)$$

which decreases monotonically from ρ^2 at $t = 0$ to 0 at $t = \infty$; hence $\bar{\eta}_0^2 = \bar{\eta}^2 = \delta = \eta^2 = \rho^2$. That $\bar{\rho}^2 = \rho^2$ has been shown by Gebelein (1941); it also follows from Sarmanov's (1958b) result, stated above in the last paragraph of Section 1.

Example 7 : X and Y are independent with a common absolutely continuous distribution with density everywhere positive—and symmetric about the origin for convenience (for example, normal, Cauchy, or Laplace distributions). For each $n = 1, 2, \ldots$, let

$$(X_n,\ Y_n) = \begin{cases} (X,\ Y) \text{ if } (X,\ Y)\,\epsilon A_n = \{x, y\,|\,|x| \leqslant n, |y| \leqslant n\} \\ (-2n,\ -2n) \text{ if } (X,\ Y) \notin A_n \text{ and } X+Y < 0 \\ (2n,\ 2n) \text{ if } (X,\ Y) \notin A_n \text{ and } X+Y \geqslant 0. \end{cases}$$

Note that $(X_n,\ Y_n)\to(X,\ Y)$ in distribution (in fact, *strongly*—that is, for every Borel set B in the plane, $\Pr[(X_n,\ Y_n)\epsilon B]\to \Pr[(X,\ Y)\epsilon B]$).

Setting $f_n(x) = x$ if $|x| < n$ and $= 0$ otherwise, we have $f_n(X_n) = f_n(Y_n)$ a.s. so that $(X_n,\ Y_n)$ has maximal correlation unity for every n, while $(X,\ Y)$ has maximal correlation zero.

Let F be the marginal distribution function of X and denote

$$p_n = \int_{-n}^{n} dF(x), \quad \chi_n(t) = \int_{-n}^{n} e^{itx}\, dF(x).$$

Then

$$\phi_n(t\ y) = \begin{cases} e^{ity} & \text{if } \ y = \pm\, 2n \\ \chi_n(t)/p_n & \text{if } \ |y| \leqslant n \end{cases}$$

and $\phi_n(t) = \frac{1}{2}(1-p_n^2)\,(e^{-2int}+e^{2int})+p_n\chi_n(t)$ since the marginal distribution of X_n (and also of Y_n) assigns probability $\frac{1}{2}(1-p_n^2)$ at $\pm 2n$ and has density $p_n dF(x)/dx$ over the interval $(-n, n)$. It is thus readily verified that the dependence characteristic of X_n on Y, denoted $\eta_n^2(t)$, tends to zero as $n \to \infty$ for every $t > 0$.

Also, a tedious but straightforward calculation yields

$$\eta_n^2(t) = \left(1+ \frac{p_n^2}{1-p_n^2} \cdot \frac{\sigma_n^2}{4n^2}\right)^{-1} + O(t^2) \ \text{ as } \ t \to 0$$

where $\sigma_n^2 = \text{var}(X_n|y)$ for $|y| \leqslant n$. Since $\sigma_n^2 < 4n^2$ and $p_n^2 \to 1$, we have $\eta_n^2(0) \to 0$ as $n \to \infty$. Hence, the squared correlation ratio of X_n on Y tends to zero.

That $\bar{\eta}_n^2$ also tends to zero may be seen as follows. For large t, $\eta_n^2(t) \sim c_n \dfrac{1-c_n \cos^2 2nt}{1-c_n^2 \cos^2 2nt} \leqslant c_n$ where $c_n = 1-p_n^2$ so that, given ϵ, $\eta_n^2(t) < c_n+\epsilon$ for $t > t_\epsilon$ (say). Since $c_n \to 0$ as $n \to \infty$, $\max\limits_{t > t_\epsilon} \eta_n^2(t) < 2\epsilon$ for n sufficiently large. Since $\eta_n^2(t) \to 0$ as $n \to \infty$ for each $t \epsilon [0, t_\epsilon]$, its maximum over this interval also tends to zero. We thus conclude that $\bar{\eta}_n^2 \to 0$ and also δ_n, the dependence index of X_n on Y, tends to zero, the dependence index of X on Y.

Quite generally, $\eta_n^2(t) \to \eta^2(t)$ for each $t > 0\,(t \neq mt_0)$ if the joint and conditional distributions of (X_n, Y_n) converge to those of (X, Y); but a general proof that $\eta_n^2(0)$ and $\max\limits_t \eta_n^2(t)$ converge is not available. Thus, a continuity axiom is satisfied in this example, but general results are not available.

A multinomial example is given in Section 8.

7. CHARACTERIZING MULTIPLE DEPENDENCE

Let (X, Y) have a joint distribution where now X is real-valued and Y is vector-valued. The assertions about correlation ratios of Section 3 all remain valid. The *dependence characteristic* of X on Y is defined precisely as before, as the correlation ratio of e^{itX} on Y (when existent and by continuity at other t-values), and all of its properties of Section 4 remain valid.

The *dependence index* of X on Y is also defined in an identical way and all of its properties carry over except that Axiom G' must be modified slightly (see below). Axiom G also remains valid : δ coincides with the squared multiple correlation coefficient when (X, Y) are multinormal. An immediate consequence of Axiom F′ is that the dependence index of X on $(Y_1, \ldots, Y_k)$ is no greater than the dependence index of X on $(Y_1, \ldots, Y_l)$ for $k < l$.

Axiom G' may be replaced by:

Axion G″ : If $Z_1, Z_2, \ldots,$ are independent and identically distributed non-trivial random vaiables, and if $X = Z_1+\ldots+Z_{k+l}$ and $Y = (Z_1, \ldots, Z_k)$, then $\delta(X \mid Y) = k/(k+l)$.

That δ satisfies this Axiom is immediate since the dependence characteristic of X on $(Z_1, \ldots, Z_k)$ is seen to be identical with that of X on $Z_1+\ldots+Z_k$ (which is displayed in Section 5); in fact, the conditional distribution of X is the same whether $(Z_1, \ldots, Z_k)$ or $Z_1+\ldots+Z_k$ is given.

8. MULTINOMIAL EXAMPLE

Suppose (X, Y) has a trinomial distribution :

$$\Pr(X = x,\ Y = y) = \binom{n}{x,\, y,\, n-x-y} p^x q^y (1-p-q)^{n-x-y}$$

for x, y non-negative integers and $x+y \leqslant n$, where n is a positive integer and p, q are positive numbers for which $p+q \leqslant 1$; X and Y may be thought of as the frequencies of two of the categories in a trichotomy or in any multichotomy. Marginally, X and Y are binomially distributed with parameters (n, p) and

(n, q), respectively, and X given $Y = y$ is conditionally binomial with parameters $(n-y,\ p/(1-q))$. The correlation coefficient for (X, Y) is $\rho = -\sqrt{pq/(1-p)(1-q)}$.

The maximal correlation coefficient $\bar{\rho}$ is $|\rho|$—that is, there is no pair of r.v.'s $f(X)$, $g(Y)$ with correlation coefficient existing and exceeding $|\rho|$. This is so since both regression functions are linear (and the mean square contingency exists—see end of Section 1).

We now proceed to calculate the dependence characteristic. We find $|\phi(t)|^2 = [1-2p(1-p)s]^n$ where $s = 1-\cos t$ and likewise for $|\phi(t\,|\,y)|^2$ with p replaced by $p/(1-q)$ and n by $n-y$. Since $Eu^Y = (1-q+qu)^n$, we find $E|\phi(t\,|\,Y)|^2 = \left[1-2p\dfrac{1-p-q}{1-q}s\right]^n$ after reduction. Therefore

$$\eta^2(t) = (A^n-B^n)/(1-B^n)$$

where

$$A = 1-2p\,\frac{1-p-q}{1-q}\,s, \quad B = 1-2p(1-p)s.$$

Recalling that $s = 1-\cos t$, we see that $\eta^2(t)$ is periodic with period 2π and symmetric within each period. Displaying the dependence on n, we write

$$\eta^2(t) = g_n(s(t)).$$

The function $g_n(s)$ on $[0, 2]$ completely determines the dependence characteristic $\eta^2(t)$.

We readily find the following properties of $g_n(s)$:

(i) $g_1(s) = pq/(1-p)(1-q) = \rho^2$, all s.

(ii) $g_n(0+) = \rho^2$, all n.

(iii) For $n > 1$, $g_n(s)$ is decreasing in s for $0 < s < 2$.

We sketch the proof of (iii). The sign of $dg_n(s)/ds$ is found to be the same as that of $H(A, B) = B^{n-1}(1-B)-A^{n-1}(1-A) +A^{n-1}B^{n-1}(B-A)$. It is readily verified that $0 < B < A < 1$. Now $H(B, B) = H(1, B) = 0$ while $H(A, B)$ as a function of A for fixed B has a unique stationary point, a minimum; consequently, $H(A, B) < 0$ for $B < A < 1$, so that $g_n(s)$ is decreasing, as claimed.

From these properties we conclude that $\eta^2(t)$ is U-shaped for $0 \leqslant t \leqslant 2\pi$ (for $n < 1$ and constant for $n = 1$) and hence $\eta_0^2 = \bar{\eta}^2 = \delta = \rho^2$.

This example permits us to introduce anothers caling axiom, analogous to Renyi's Axiom G and satisfied by $\bar{\rho}^2$, η_0^2, $\bar{\eta}^2$, and δ :

> **Axiom H :** If (X, Y) is trinomially distributed with parameters (n, p, q), $p > 0$, $q > 0$, $p+q \leqslant 1$, then $\delta(X \mid Y)$ (or $\rho^2(X, Y)$) coincides with the squared correlation coefficient.

It may also be verified that $\eta^2(t)$ is decreasing in n for each fixed t ($\neq 2k\pi$, k integral.) We proceed by induction, having already established that $g_2(s) < g_1(s)$ for $0 < s < 2$ (by (i) and (iii)). Assuming $g_n(s) > g_{n+1}(s)$, or equivalently (upon clearing fractions)

$$A^n B^n (A-B) > B^n(1-B) - A^n(1-A),$$

we wish to prove that $A^{n+1}B^{n+1}(A-B) - B^{n+1}(1-B) + A^{n+1}(1-A)$ is positive. Using the previous inequality, this last expression exceeds

$$AB[B^n(1-B) - A^n(1-A)] - B^{n+1}(1-B) + A^{n+1}(1-A)$$
$$= (1-A)(1-B)(A^{n+1} - B^{n+1}) \geqslant 0,$$

completing the proof.

For any t other than a multiple of 2π, $\eta^2(t)$ tends to zero monotonically with n, but at multiples of 2π, $\eta^2(t)$ remains equal to ρ^2.

Finally, we note that if $(X_1, X_2, \ldots, X_m)$ is multinomially distributed with $X_1 + X_2 + \ldots + X_m = n$ and parameters $p_1, p_2, \ldots, p_m$ (summing to unity), then the conditional distribution of $X = X_1 + \ldots + X_k$ given $Y = (X_{k+1}, X_{k+2}, \ldots, X_{k+l})$ where $k+l < m$ is identical with the conditional distribution given $Z = X_{k+1} + \ldots + X_{k+l}$—namely, binomial $(n-z,\ (p_1 + \ldots + p_k)/(1 - p_{k+1} - \ldots - p_{k+l}))$—and hence the dependence characteristic of X on Y, defined in Section 7, is identical with that of X on Z.

9. CORRELATION CHARACTERISTIC AND INDEX

Recall that $\eta^2(t) = \max_g |\text{corr}(e^{iXt}, g(Y))|^2$. Hence $\eta^2(t) \geqslant |\text{corr}(e^{iXt}, g(Y))|^2$ for *any* complex-valued $g(Y)$ with finite positive variance. In particular, it might be useful to consider $g(Y) = e^{iYu}$ for some u for which var $e^{iYu} > 0$. This would indeed provide a lower bound on $\eta^2(t)$, and from it lower bounds on the various dependence indices could also be derived.

In this section, we shall examine the correlation between e^{itX} and e^{iuY}, as a means of characterizing the linear relationship between X and Y. We shall not pursue it too deeply, however, nor extract explicit bounds on the dependence indices.

Let $\rho(t, u)$ be the correlation coefficient for the random variables e^{itX} and e^{iuY}. Let ϕ_1 and ϕ_2 be the c.f.'s of X and Y, respectively, and ϕ the joint c.f. Letting t_0 be the smallest positive t-value (or$+\infty$) for which $\phi_1(t) = 1$, and likewise for u_0 and $\phi_2(u)$, we see that $\rho(t, u)$ is defined for (t, u) such that $0 < |t| < t_0$, $0 < |u| < u_0$. (If either X or Y is degenerate, we can let $\rho(t, u) = 0$ identically). We call this complex-valued function $\rho(t, u)$ the *correlation characteristic* of (X, Y); it may be calculated from the formula

$$\rho(t, u) = \frac{\phi(t, -u) - \phi_1(t)\phi_2(-u)}{(1-|\phi_1(t)|^2)^{1/2}(1-|\phi_2(u)|^2)^{1/2}}.$$

Some properties of $\rho(t, u)$ are listed below; all but the ninth are immediate consequences of properties of c.f.'s and the ninth is proved below :

(i) $\rho(t, u)$ exists for any (X, Y).

(ii) $|\rho(t, u)| \leqslant 1$.

(iii) $\rho(t, u) = \overline{\rho(-t, -u)}$.

(iv) $\rho(t, u)$ is continuous.

(v) $\rho(t, u)$ is identically zero iff X and Y are independent.

(vi) $\rho(t, u)$, together with specification of the marginal distributions of X and Y, completely determine the joint distribution of (X, Y).

(vii) If X and Y have finite positive variances, then

$$\rho(t, u) = \rho \cdot tu/|tu| \text{ for } (t, u) \text{ near } (0, 0)$$

where $\rho = \text{corr}(X, Y)$ (upon expanding the c.f.'s).

(viii) If $X' = aX+b$ and $Y' = AY+B$, $a \neq 0$, $A \neq 0$, then the correlation characteristic of (X', Y') is

$$\rho'(t, u) = e^{itb-iuB}\, \rho(at, Au).$$

(ix) $|\rho(t, ct)| = 1$ for all t for which it is defined and some $c \neq 0$ iff $X = cY+b$ a.s. for some b.

(x) If the mean square contingency exists, then the numerator in $\rho(t, u)$ may be represented

$$\sum_{i=1}^{\infty} \rho_i E(e^{iXt} f_i(X)) \cdot E(e^{-iYu} g_i(Y))$$

where the ρ_i's, f_i's and g_i's are defined in (0) (page 7).

To prove (ix), we first note that if $X = cY+b$ a.s., then e^{itX} is linear in e^{iuY} for $u = ct$ so that the correlation between e^{itX} and e^{ictY}, namely $\rho(t, ct)$, has modulus unity (Section 2). Conversely, if $\rho(t, ct)$ has modulus unity, then $e^{itX} = k(t)e^{ictY}+l(t)$ a.s. (Section 2). If $l(t')$ is not identically zero, let t' be such that $l(t') \neq 0$. Then $e^{it'X}$ and $k(t')e^{ict'Y}+l(t')$ are each complex valued r.v.'s confined to non-concentric circles. Since they are equal a.s. each has a two-point distribution. Since this must likewise hold for an interval of t values, it can be concluded (details omitted) that X and Y have two-point distributions and X is a function of Y—and hence a linear function a.s. On the other hand, if $l(t) = 0$, all t, then $e^{itX} = e^{ictY+i\theta(t)}$ (say) and therefore $tX - ctY - \theta(t)$ is a multiple of 2π a.s. Since this is true for all $t > 0$, $X - cY - \theta(t)/t = 0$, all t, and hence $X = cY+b$ (say) a.s.

Thus, the correlation characteristic is an extension of the notion of ρ, but it is a function of two variables rather than a single number. We shall make the following definitions, to reduce the scope of $\rho(t, u)$:

$$\rho_0(c) = \lim_{t \to 0+} |\rho(t, ct)| \quad (c \neq 0),$$

$$\rho_0 = \max_c \rho_0(c).$$

The index ρ_0 will be shown to serve as a generalization of ρ; we call it the *linear correlation index*, although other indices might surely be considered iustead. Although the existence of a linear correlation coefficient formally depends on the finiteness of both variances, it could be defined more generally by some kind of truncation. The approach here appears to be more convenient. This introduction of c, and then its elimination, will lead to invariance under linear transformation. We have elected to consider only an unsigned index, but a sign could possibly be retained with slight modification. This index is somewhat analogous to η_0 of Section 5, rather than δ.

We are now prepared to list properties of ρ_0; the last two will be proved below.

(a) ρ_0 exists.

(b) $0 \leqslant \rho_0 \leqslant 1$.

(c) ρ_0 is symmetric in the roles of X and Y.

(d) $\rho_0 = |\rho|$ if X and Y have finite positive variances and correlation coefficient ρ.

(e) $\rho_0 = 0$ if (but not only if) X and Y are independent.

(f) $\rho_0 = 1$ if $X = aY + b$ a.s. and X and Y are non-degenerate (but not only if—see Example 5b—unless X and Y have finite variances).

(g) ρ_0 is invariant under 1–1 linear transformation of X and of Y.

(h) If X and Y are each sums of non-degenerate independent and identically distributed r.v.'s $Z_1, Z_2, \ldots, X$ having l summands, Y having m summands, and k of the summands being common to X and Y, then

$$\rho_0 = k/\sqrt{lm}(k \geqslant 0,\ l > 0,\ m > 0).$$

To prove (*g*), we use the notation of property (viii) and note that

$$|\rho'(t, ct)| = |\rho(at, Act)|$$

which, by (iii), equals $|\rho(-at, -Act)|$ and hence equals $\left|\rho\left(t', \frac{Ac}{a}t'\right)\right|$ where $t' = |a|t$. Therefore $\rho_0'(c) = \lim_t |\rho'(t, ct)| = \rho_0(Ac/a)$, and $\rho_0' = \rho_0$.

Property (h) is a generalization of the property described in Axiom G′ of Section 1, and will now be proved. We assume $k > 0$, the case $k = 0$ being trivial. Let ψ be the common c.f. of the Z_j's. We find

$$|\rho(t, ct)|^2 = \frac{|\psi(t-ct)^k\psi(t)^{l-k}\psi(-ct)^{m-k}-\psi(t)^l\psi(-ct)^m|^2}{(1-|\psi(t)|^{2l})(1-|\psi(-ct)|^{2m})}$$

Let $h(t) = 1-\psi(t)$ and note that, for t near 0 and s a positive integer, $\psi(t)^s \sim 1-sh(t)$ and hence $|\psi(t)|^{2s} \sim 1-2sH(t)$ where $H(t) = \mathcal{R}h(t)$. Hence, near $t = 0$, we find (after reduction)

$$|\rho(t, ct)|^2 \sim \frac{k^2}{lm}\,\frac{|h(t)+h(-ct)-h(t-ct)|^2}{4H(t)H(-ct)} = \frac{k^2}{lm}\cdot K(c, t), \text{ say.}$$

Note that $K(c, t)$ is free of k, l and m. Hence, this formula is also valid when $k = l = m$ in which case the corresponding $|\rho(t, ct)|^2 \sim K(c, t)$. Hence $\lim_{t\to 0} K(c, t) \leqslant 1$. But, from its definition, $K(1, t) = |h(t)+h(-t)|^2/4H(t)\cdot H(-t)$. Since $\psi(-t) = \bar{\psi}(t)$, we have $h(-t) = \bar{h}(t)$ and $H(-t) = H(t)$, and therefore $K(1, t) = 1$. We thus conclude that $\rho_0^2(c) \leqslant k^2/lm$, all c, whereas $\rho_0^2(1) = k^2/lm$. Hence, $\rho_0^2 = k^2/lm$ as was to be proved.

Properties (f), (g) and (h) give some justification for calling ρ_0 an index of *linear* relationship. It is not very satisfactory as an index of general relationship since it equals zero too easliy and is invariant only under linear transformation. An alternative index is $\bar{\rho}_0 = \max_{t,u}|\rho(t, u)|$, or some kind of a mixture of ρ_0 and $\bar{\rho}_0$. Such indices vanish only if (X, Y) are independent, but other properties will not be pursued here.

As an example, if (X, Y) has a bivariate normal distribution with zero means and unit variances and correlation coefficient ρ, we find

$$\rho(t, u) = e^{-\frac{1}{2}(t^2+u^2)}(e^{\rho tu}-1)(1-e^{-t^2})^{-\frac{1}{2}}(1-e^{-u^2})^{-\frac{1}{2}}$$

Hence, $\rho_0(c) = |\rho|$ for all c and $\rho_0 = |\rho|$, as it must by (d).

10. MISCELLANEOUS COMMENTS

This section will consist of a few miscellaneous comments—firstly, on the interpretation of η_0 and ρ_0, secondly on prediction theory, thirdly on work of Pitman on the behavior of characteristic functions near the origin, and finally, on the use of moment generating kernels instead of characteristic kernels.

The variability of the distribution of a r.v. is characterized by the behavior of its c.f. near the origin—the more variable the distribution the more rapid the approach to unity. This has been utilized in the two indices η_0 and ρ_0. The first, a generalization of the correlation ratio, is an index of the average variability in the conditional distribution—high dependence of X on Y being consistent with low variability in the conditional distribution relative to the unconditional, and a large value of η_0. When variances are finite, we only need recall that η_0^2 is the difference between the unconditional variance and the average (expected) conditional variance, normalized so as to lie between zero and unity. Similarly the index ρ_0 is small when the approach to unity of the c.f. of any linear combination of X and Y (since $\phi(t, -ct)$ is the c.f. of $X-cY$) is substantially slower relative to the corresponding approach if X and Y were independent. In fact, the magnitude of ρ, of which ρ_0 is a generalization, may be described as the magnitude of the difference between the variance of a linear combination of X and Y when independent and the variance of the same linear combination based on the joint distribution—normalized so as to lie between zero and unity; specifically,

$$|\rho| = \left| \frac{\mathrm{var}(aX_0+bY_0+c)-\mathrm{var}(aX+bY+c)}{2(\mathrm{var}\, aX\cdot \mathrm{var}\, bY)^{1/2}} \right|$$

where X_0 and Y_0 are independent with the same marginal distributions as X and Y, respectively. Each of these, η^2 and ρ^2, were generalized by replacing certain r.v.'s by characteristic kernels (replacing a r.v. Z by e^{itZ}) and observing the behavior near $t = 0$.

An alternative way of looking at these indices is in the context of prediction. Recall (Assertion 4a in Section 4) that $1-\eta^2$ is the normalized variance of the difference between X and its "best predictor" from Y, namely $E(X \mid Y) = r(Y) : 1-\eta^2 = \text{var}(X-r(Y))/\text{var } X$; "best" is defined in the sense of minimizing the mean square error of prediction $E|X-r(Y)|^2$. Similarly, ρ may be defined in the context of *linear* prediction of X from Y. For $1-\rho^2$ is the normalized variance of the difference between X and its "best linear predictor" from Y, namely $l(Y) = EX + \rho(\text{var } X/\text{var } Y)^{1/2}(Y-EY)$, and $1-\rho^2 = \text{var } (X-l(Y))/\text{var } X$. The indices η_0^2 and ρ_0^2 are analogous except we are predicting e^{itX} from Y, or linearly predicting e^{itX} from e^{ictY} respectively, for t near the origin.

The dependence characteristic $\eta^2(t)$ was seen to be a function of certain characteristic functions. Hence η_2^0 may be evaluated by examining the behavior of the characteristic functions near the origin. Such behavior when the variance of X is infinite was studied by Pitman (1961), and we shall briefly consider it here.

Now $\eta^2(t)$ can be defined in terms of the c.f. of X and the conditional c.f. of X given Y or else in terms of the c.f.'s of certain derived symmetric distributions. We consider the latter first.

Let $Z = X_1-X_2$ and $Z^* = X_1^*-X_2^*$, defined in Section 4 (before (4)), with (real) c.f.'s χ and χ^*. In this notation we have

$$\eta^2(t) = \frac{\chi^*(t)-\chi(t)}{1-\chi(t)} \; .$$

Pitman gave asymptotic formulas for the real part (his $U(t)$) of a c.f.; one of the forms, for example, was $1-\chi(t) \sim cH(1/t)$, where H is (here) the *tail sum* of the intermediate distribution function of Z—specifically, $H(z) = 2\,\Pr(X_1-X_2 > z)+\Pr(X_1-X_2 = z)$—and c depends on the rate at which H decreases to zero. Defining H^* analogously, we obtain $\eta^2(t) \sim 1-c^*H^*(1/t)/cH(1/t)$ as $t \to 0$, under appropriate conditions on H and H^*.

The other possibility is to work directly with the c.f. $\phi(t)$ of X and the conditional c.f. $\phi(t\,|\,y)$ of X given $Y=y$. Now $\eta^2(t)$ depends only on the magnitude of these c.f.'s (see (4)). Writing $\phi(t) = U(t)+iV(t) = \lambda(t)e^{i\theta(t)}$ where U and V are the real and imaginary parts, $\lambda(t)$ is the modulus and $\theta(t)$ the argument, it is apparent that $\theta(t) \sim V(t)$ and $1-\lambda(t) \sim 1-U(t)-\frac{1}{2}V(t)^2$ near the origin (since $U(t) \to 1$ and $V(t) \to 0$). Hence, $\gamma(t) = 1-\lambda(t)^2 \sim 2(1-U(t))-V(t)^2$, and likewise for $\gamma(t\,|\,Y)$. After reduction, we thus have (formally)

$$\eta^2(t) \sim \frac{\text{var } V(t\,|\,Y)}{2-2U(t)-V(t)^2} \text{ as } t \to 0.$$

(In particular, η_0^2 vanishes if the conditional distributions are all symmetric around zero (or any other constant), since then $V(t\,|\,y)$ is identically zero.) Pitman's asymptotic formulas for $U(t)$, $V(t)$ and $V(t\,|\,Y)$ may therefore be useful in evaluating η_0^2.

Likewise, ρ_0 is defined solely in terms of certain univariate c.f.'s at the origin, namely those of X, of Y, and of $X-cY$. Again, Pitman's results may be useful.

Other dependence and correlation characteristics and indices could be defined by considering moment generating kernels instead of characteristic kernels, that is, comparing var e^{tX} with E var$(e^{tX}\,|\,Y)$ or correlating e^{tX} and e^{uY}. The advantage is that the corresponding $\eta^2(t)$ and $\rho(t, u)$ are completely determined by their values near the origin; a further convenience can be derived from the fact that e^{tX} is one-to-one in X for each t. The disadvantage is, of course, that moment generating functions need not exist.

References

[1] Bahadur, R. R. (1955). "Measurable Sub-Spaces and Sub-Algebras," *Proc. Amer. Math. Soc.*, **6**, 565–570.

[2] Feller, William (1957). *An Introduction to Probability Theory and Its Applications*, I. 2nd ed., John Wiley and Sons, New York.

[3] Fréchet, Maurice (1951). "Sur les Tableaux de Corrélation dont les Marges sont Données," *Annales de L'Université de Lyon, Section A*, **14**, 53–77.

[4] Gebelein, H. (1941). "Das Statistische Problem der Korrelation als Variationsund Eigenwert-problem und sein Zusammenhang mit der Ausgleichungsrechnung," *Zeitschrift für Angew. Math. und Mech.*, **21**, 364–379.

[5] Goodman, Leo A. and Kruskal, William H. (1959). "Measures of Association for Cross Classifications. II : Further Discussion and References," *J. Amer. Statist. Assoc.*, **54**, 123–163.

[6] Hoeffding, Wassily (1940). "Masstabinvariante Korrelationstheorie," *Schriften des Mathematischen Instituts und des Instituts für Angewandte Mathematik der Universitat Berlin*, **5**, 179–233.

[7] Kruskal, William H. (1958). "Ordinal Measures of Association," *J. Amer. Statist. Assoc.*, **53**, 814–861.

[8] Lehmann, E. L. (1966). "Some Concepts of Dependence," *Ann. Math. Statist.*, **37**, 1137–1153.

[9] Lukacs, Eugene (1960). *Characteristic Functions*, Charles Griffin and Co., London.

[10] Pitman, E. J. G. (1961). "Some Theorems on Characteristic Functions of Probability Distributions," *Proceedings of the Fourth Berkeley Symposium on Mathematical Statistics and Probability*, II, 393–402. University of California Press, Berkeley.

[11] Renyi, A. (1959). "On Measures of Dependence," *Acta Math. Acad. Sci. Hung.*, **10**, 441–451.

[12] Sarmanov, O. V. (1958*a*). "Maximum Correlation Coefficient (Symmetric Case)," *Dokl. Akad. Nauk SSSR*, **120**, 715–718. English translation appears in *Selected Translations in Mathematical Statistics and Probability*, **4**, (1963), 271–275, Amer. Math. Soc., Providence.

[13] Sarmanov, O. V. (1958*b*). "Maximum Correlation Coefficient (Nonsymmetric Case)," *Dokl. Akad. Nauk SSSR*, **121**, 52–55. English translation appears in *Selected Translations in Mathematical Statistics and Probability*, **2**, 207–210, Amer. Math. Soc., Providence.

[14] Sarmanov, O. V. (1960). "Characteristic Correlation Functions and Their Applications in the Theory of Stationary Markov Processes," *Dokl. Akad. Nauk SSSR*, **132**, 769–772. English translation appears in *Soviet Math.*, **1**, 651–654.

[15] Sarmanov, O. V. and Zaharov, V. K. (1960). "Maximum Coefficients of Multiple Correlation," *Dokl. Akad. Nauk SSSR*, **130**, 269–271. English translation appears in *Soviet Math.*, **1**, 51–53.

(*Received Jan. 1, 1966.*)

CHAPTER 18

A General Purpose Test of Censoring of Extreme Sample Values

N. L. JOHNSON,
The University of North Carolina at Chapel Hill

1. INTRODUCTION

The problem considered in this paper is one of a kind which arises when there is some suspicion that sample values may have been *censored* by exclusion of a certain number of the greatest and/or least of the observed values. We will consider a relatively simple case, in which it is supposed that the population distribution of random variables representing observed values is known, and where the original (uncensored) sample is randomly chosen.

2. STATEMENT OF PROBLEM

The problem may be stated formally in the following way. There are available r observations, which may be regarded as values of continuous random variables. These are, possibly, a subset $(x_1', x_2', \ldots, x_r')$ from n independent and identically distributed continuous random variables $x_1', x_2', \ldots, x_n'$, corresponding to a complete random sample of size $n = r+s_0+s_r$, from which the s_0 smallest and s_r largest values have been removed ("censored"). Denoting this last statement by the hypothesis

H_{s_0, s_r}, it is desired to test the hypothesis $H_{0,0}$—corresponding to no censoring—against the set of alternatives $\{H_{s_0, s_r}\}$, where s_0, s_r may have any non-negative integer values.

In [1], it was shown that the most powerful test of $H_{0,0}$ against a *specified* alternative H_{s_0, s_r} has a critical region of form

$$(1) \qquad y_1^{\theta}(1-y_r) \geqslant K_{\alpha}(\theta)$$

where $\theta = s_0/s_r$; $y_j = \int_{-\infty}^{x_j} f(x')dx'$; $x_1 = \min(x_1', x_2', \ldots x_r')$; $x_r = \max(x_1', x_2', \ldots, x_r')$, $f(x')$ is the probability density function of each x_i' and $K_{\alpha}(\theta)$ satisfies the equation

$$\Pr [y_1^{\theta}(1-y_r) \geqslant K_{\alpha}(\theta) \,|\, H_{0,0}] = \alpha,$$

α being the significance level of the test.

It follows from (1) that, for any given value of $\theta \geqslant 0$, there is a uniformly most powerful test of $H_{0,0}$ with respect to *all* alternatives H_{s_0, s_r} for which $s_0 = \theta s_r$. (If $s_r = 0$, we take $\theta = \infty$, and the critical region is of form "$y_1 \geqslant$ constant".)

It is clear however, that there is not a uniformly most powerful test of $H_{0,0}$ with respect to the complete set of alternatives $\{H_{s_0, s_r}\}$. The construction of a *general purpose* test, with (hopefully) reasonable power against such general alternative hypotheses, is the main purpose of this paper. In order to get a test to consider, we will make use of a heuristic principle stated formally and studied by Roy [2], and used incidentally by Neyman and Pearson [3]. It will be found that, although quite drastic approximations are used in applying the principle, a simple, and intuitively appealing, test is obtained. The value of the principle here lies in the fact that it does direct attention to a particular test, out of a number of intuitively appealing tests, as being worthy of further study.

3. DERIVATION OF THE TEST

We will use a particular form of the *union-intersection* principle of Roy [2]. We seek to find a critical region which is the union (over all $\theta \geqslant 0$) of critical regions (1) of most powerful tests with respect to $\{H_{\theta s_r, s_r}\}$, with common significance level α'. This will not, of course, be the same as the significance level, α, of the general purpose test. It is clear, in fact, that $\alpha > \alpha'$.

In [1] it is shown that $K_\alpha(\theta)$ satisfies the equation

$$\alpha' = r \int (1-K_{\alpha'}(\theta)y^{-\theta}-y)^{r-1}\, dy, \tag{2}$$

the interval of integration corresponding to $1-K_{\alpha'}(\theta)y^{-\theta}-y \geqslant 0$. Points (y_1, y_r) on the boundary of the critical region must satisfy the equations

$$y_1^\theta(1-y_r) = K_{\alpha'}(\theta) \tag{3}$$

$$\frac{\partial}{\partial\theta}[y_1^\theta(1-y_r)] = \frac{\partial K_{\alpha'}(\theta)}{\partial\theta} \tag{4}$$

whence

$$\log y_1 = \frac{\partial}{\partial\theta}(\log K_{\alpha'}(\theta)). \tag{5}$$

Equations (3) and (5) are parametric equations of the boundary. In principle, for a given value of θ, y_1 can be determined from (5), and then y_r can be calculated from (3). However, the evaluation of $\partial(\log K_{\alpha'}(\theta))/\partial\theta$ is a matter of some technical difficulty. Here we will obtain an approximate solution, remembering that we are using a heuristic principle and that the test finally obtained will be judged on its own merits.

We first write down some easily derived properties of $K_{\alpha'}(\theta)$. These are

$$K_{\alpha'}(0) = 1-\alpha'^{1/r} \tag{6}$$

$$K_{\alpha'}(\theta)^{-1} = [K_{\alpha'}(\theta)]^{1/\theta} \text{ (from symmetry).} \tag{7}$$

Also (see [1])

$$K_{\alpha'}(1) \doteqdot \tfrac{1}{4}(1-\alpha'^{1/r})^2. \tag{8}$$

(We note that from (6) and (7)

$$K_{\alpha'}(\theta) \sim (1-\alpha'^{1/r})^\theta \quad \text{for large } \theta).$$

From (2), noting that the integrand is equal to zero at each end-point of the interval of integration, we find, on differentiating with respect to θ,

$$(9) \qquad r(r-1)\int(1-Ky^{-\theta}-y)^{r-2}(Ky^{-\theta}\log y-\frac{\partial K}{\partial\theta}y^{-\theta})dy=0.$$

(For convenience $K_{\alpha'}(\theta)$ is abbreviated to K.)
From (9),

$$(10) \qquad \frac{\partial \log K}{\partial\theta}=\frac{\int y^{-\theta}(1-Ky^{-\theta}-y)^{r-2}\log y\,dy}{\int y^{-\theta}(1-Ky^{-\theta}-y)^{r-2}dy}$$

We notice that the right-hand side of (10) is equal to the expected value, $E(z)$, of a random variable $z(=\log y)$ with probability density function proportional to

$$(11) \qquad e^{-(\theta-1)z}(1-Ke^{-\theta z}-e^z)^{r-2}$$

over the interval $1-Ke^{-\theta z}-e^z\geqslant 0$, and zero elsewhere.

We now introduce our first approximation. In place of $E(z)$ we use the modal value of z, z_0 say, which maximizes (11). This satisfies the equation

$$(r-2)(K\theta e^{-\theta z_0}-e^{z_0})=(\theta-1)(1-Ke^{-\theta z_0}-e^{z_0})$$

or

$$(12) \qquad [(r-1)\theta-1]Ke^{-\theta z_0}-(r-\theta-1)e^{z_0}=\theta-1$$

Using z_0 as an approximation to $E(z)$, we obtain from (5),

$$(13) \qquad [(r-1)\theta-1]Ky_1^{-\theta}-(r-\theta-1)\,y_1\doteqdot\theta-1.$$

Using (3), this can be rewritten

$$(14) \qquad [(r-1)\theta-1]y_r+(r-\theta-1)y_1\doteqdot(r-2)\theta.$$

We now make a further approximation, supposing that r is large enough for (14) to imply

$$(15) \qquad \theta y_r+y_1\doteqdot\theta,$$

that is $\quad y_1/\theta\doteqdot 1-y_r.$

Then from (3)

$$K_{\alpha'}(\theta)\doteqdot y_1^{\theta+1}/\theta$$

and so

$$\frac{\partial}{\partial\theta}(\log K_{\alpha'}(\theta))\doteqdot\log y_1-\theta^{-1}+(\theta+1)\frac{\partial\log y_1}{\partial\theta}.$$

Again using (5) we have

$$\frac{\partial \log y_1}{\partial \theta} \doteqdot \frac{1}{\theta(\theta+1)}$$

whence

(16) $$y_1 \doteqdot C_{\alpha'}\, \theta/(\theta+1)$$

where $C_{\alpha'}$ depends on α but not on θ.

From (15) and (16) we obtain the simple critical region

(17) $$y_1+(1-y_r) \geqslant C_{\alpha'}$$

which our analysis suggests to be worthy of consideration.

It is interesting to note that (16) implies that

$$K_{\alpha'}(\theta) \doteqdot C_{\alpha'}^{\theta+1}\, \theta^{\theta}(\theta+1)^{-(\theta+1)}$$

and that this formula satisfies (7). From (6), it appears that $C_{\alpha'}$ might be taken equal to $(1-\alpha'^{1/r})$. If this be done then

$$K_{\alpha'}(1) \doteqdot \tfrac{1}{4}\,(1-\alpha'^{1/r})^2$$

agreeing with the approximate formula (8). These results may be regarded as some indirect support for the approximation used in this section.

Henceforward, however, we shall be concerned with the test with critical region (17), suggested by the approximate analysis, and not with the approximations themselves.

4. APPLICATION AND PROPERTIES

If $H_{0,0}$ is valid then $v = y_1+(1-y_r)$ is distributed as $\chi_4^2/(\chi_4^2+\chi_{2(r-1)}^2)$, where χ_4^2 and $\chi_{2(r-1)}^2$ are mutually independent. The probability density function of v is

$$p(v) = r(r-1)v(1-v)^{r-2} \quad (0 \leqslant v \leqslant 1)$$

To obtain a significance level α, therefore, $C_{\alpha'}$ must satisfy the equation

$$r(1-C_{\alpha'})^{r-1} - (r-1)(1-C_{\alpha'})^r = 1-\alpha$$

or, in terms of the incomplete beta function ratio

(18) $$I_{C_{\alpha'}}(2, r-1) = 1-\alpha.$$

Table 1 gives some values of $C_{\alpha'}$ (based on Table 16 of Pearson and Hartley [4]), for $\alpha = 0{\cdot}05,\ 0{\cdot}01$.

TABLE 1

Significance limits for $y_1+(1-y_r)$

$r =$	2	3	4	5	6	7	8
5% limit	0.975	0.865	0.751	0.657	0.582	0.521	0.471
1% limit	0.995	0.941	0.859	0.778	0.706	0.643	0.590

$r =$	9	10	12	15	20	30	60
5% limit	0.429	0.394	0.339	0.279	0.216	0.148	0.077
1% limit	0.544	0.504	0.440	0.368	0.288	0.201	0.105

The test may also be applied by calculating the observed value of

$$g = \frac{(r-1)(y_1+1-y_r)}{2(y_r-y_1)} \tag{19}$$

and comparing with the F-distribution with $2, 4(r-1)$ degrees of freedom.

The power of the test is easily calculated. If H_{s_0, s_r} is valid, g is distributed as $\{\frac{1}{2}(s_0+s_r)+1\}\times[F$ with $2(s_0+s_r+2)$, $2(r-1)$ degrees of freedom]. The power with respect to H_{s_0, s_r} is

$$\beta_{s_0, s_r} = 1-I_{C_{\alpha'}}(s_0+s_r+2, r-1) = I_{1-C_{\alpha'}}(r-1, s_0+s_r+2) \tag{20}$$

As r increases the distribution of g, when H_{s_0, s_r} is valid (s_0, s_r remaining fixed) tends to that of $\frac{1}{2}(\chi^2$ with $2(s_0+s_r+2)$ degrees of freedom). We note the interesting result that as $r\to\infty$, s_0 and s_r remaining fixed, the power tends to the limiting value

$$\lim_{r\to\infty} \beta_{s_0, s_r} = \Pr\left[\chi^2_{2(s_0+s_r+2)} > \chi^2_{4,\alpha}\right]$$

where $\chi^2_{4,\alpha}$ is the upper $100\alpha\%$ point of the χ^2 distribution with four degrees of freedom. $\chi^2_{4,\alpha}$ satisfies the equation

$$(1+\tfrac{1}{2}\chi^2_{4,\alpha})\exp(-\tfrac{1}{2}\chi^2_{4,\alpha}) = \alpha$$

and the limiting power can be expressed as

$$\alpha + \left\{ \sum_{j=2}^{s_0+s_r+1} (\tfrac{1}{2}\chi^2_{4,\alpha})^{j'}/j\,! \right\} \exp(-\tfrac{1}{2}\chi^2_{4,\alpha}). \tag{21}$$

Table 2 gives some values of β_{s_0, s_r} for $r = 4$, 30 and ∞, $\alpha = 0{\cdot}05$ and various values of (s_0+s_r). The power depends on s_0 and s_r only through the function (s_0+s_r).

TABLE 2

Power β_{s_0, s_r} of the general purpose test $(\alpha = 0.05)$

$s_0+s_r =$	2	4	6	8	10	12	14	16	18	20
$r = 4$	0.167	0.318	0.470	0.605	0.716	0.799	0.862	0.906	0.938	0.959
$r = 30$	0.281	0.608	0.845	0.953	0.989	—	—	—	—	—
$r = \infty$	0.303	0.661	0.892	0.976	0.996	—	—	—	—	—

For $r > 30$ it appears that the $r = \infty$ values give a useful approximation. It may also be noted that β_{s_0, s_r} gives a lower bound for the power of the most powerful test (1) when the appropriate hypothesis is valid. Table 3 gives comparable values of $\bar{\beta}_{s_0, 0}$ for the most powerful test when $\theta = \infty$ $(s_r = 0)$.

TABLE 3

Power $\bar{\beta}_{s_0, 0}$ of most powerful test when $s_r = 0$ $(\alpha = 0.05)$

$s_0 =$	2	4	6	8	10
$r = 4$	0.294	0.576	0.780	0.897	0.955
$r = 30$	0.403	0.781	0.949	0.991	0.9989
$r = \infty$	0.423	0.815	0.967	0.996	0.9997

5. ALTERNATIVE DERIVATION

An alternative (*post factum*) derivation of test (17) may be of some interest as it demonstrates another *general purpose* feature of the test.

Suppose we seek a test of $H_{0,0}$, with specified significance level α which will maximize the sum (or average) of all β_{s_0, s_r} for which $s_0+s_r = s$, whatever be the value of s. The critical region, w, must satisfy the conditions

$$r(r-1) \int\int_w (y_r-y_1)^{r-2}dy_1dy_r = \alpha \tag{22}$$

and

$$\sum_{j=0}^{s} \beta_{j,s-j} \text{ maximum, where} \tag{23}$$

$$\beta_{j\ s-j} = \frac{(r+s)!}{(r-2)\,!\,j\,!\,(s-j)\,!} \int\int_w y_1^j(1-y_r)^{s-j}(y_r-y_1)^{r-2}dy_1dy_r.$$

Now $\displaystyle\sum_{j=0}^{s} \beta_{j,s-j} = \frac{(r+s)\,!}{(r-2)\,!\,s\,!} \int\int_w (y_1+1-y_r)^s(y_r-y_1)^{r-2}dy_1dy_r$

and, using the Neyman-Pearson lemma, this is maximized, subject to (22) if w is defined by the inequality

$$\frac{(r+s)\,!}{(r-2)\,!\,s\,!} (y_1+1-y_r)^s(y_r-y_1)^{r-2} \geqslant C'r(r-1)(y_r-y_1)^{r-2}$$

where C' is a constant.

This inequality is equivalent to (17), so our general purpose test maximizes the average power.

References

[1] Johnson, N. L. (1966). "Tests of Sample Censoring," *Trans. 20th Conf. Amer. Soc. Qual. Control*, 699-703.

[2] Roy, S. N. (1957). *Some Aspects of Multivariate Analysis* (Chapter 2), John Wiley and Sons, New York.

[3] Neyman, J. and Pearson, E. S. (1928). "On the Use and Interpretation of Certain Test Criteria for Purposes of Statistical Inference," *Biometrika*, **20A**, 174–240.

[4] Pearson, E. S. and Hartley, H. O. (1958). *Biometrika Tables for Statisticians*, Vol. 1, Cambridge University Press.

(*Received Jan. 1, 1966.*)

CHAPTER 19

On the Connection between Multiplicity Theory and O. Hanner's Time Domain Analysis of Weakly Stationary Stochastic Processes

G. KALLIANPUR AND V. MANDREKAR,
University of Minnesota

1. INTRODUCTION

In an early paper full of original ideas, O. Hanner [5] obtained a decomposition for a mean-continuous, purely non-deterministic, weakly stationary process depending on a continuous time parameter. Such a representation had already been derived by K. Karhunen in his 1947 paper [8] and is now generally known by his name. The interest of Hanner's work arises from the fact that his method is based on a time-domain analysis of the process itself and is entirely free of spectral considerations. In recent years, in the light of the extensive development of multivariate stationary processes, it has appeared desirable to separate the time-domain analysis from spectral studies, and the interest in the former has revived. As an example, we mention the paper of P. Masani and J. Robertson [9] whose approach considered the discretized process corresponding to the given continuous one and makes essential use of the Cayley transform associated with the unitary group of the process. The extension of this method to (finite-dimensional) multivariate stationary processes has been carried out by J. Robertson in his thesis [11].

We deduce in Section 2 as an easy corollary of the Karhunen representation theorem the fact that a purely non-deterministic stationary process has multiplicity one of Lebesgue type. A proof of this result does not seem to have been given in the literature. Our main concern, in this paper, however, is to show how a modification of Hanner's original proof directly leads to this result. We show how this is done in Theorem 1 (Section 3). The Karhunen representation is then derived as a consequence.

The purpose of studying this multiplicity question directly is to bring out in the analysis the separation of two issues (1) the form of the spectral type, (2) the multiplicity. The fact that the process has Lebesgue type is intimately connected with the stationarity and hence with the construction of the element $Z(I_a^b)$ of Hanner [5], (p. 166). On the other hand, the dimension of the process governs the determination of multiplicity. The modification of Hanner's technique which gives us both the above properties lies in recasting his arguments in the form of multiplicity theory. In so doing we can bring out the generality of the reasoning by which he solves the first problem. It will be shown that the univariate character of the process enters into the arguments only to establish the fact that the multiplicity is one. It is interesting to note that the reasoning of Hanner used in determining the multiplicity assumes the fact that every non-trivial process has multiplicity at least one. Hence the problem is to exploit the univariate character of the process to show that the multiplicity does not exceed one. Thus an extension of Hanner's method for multivariate stationary processes lies in the study of the spectral types and multiplicity. We have undertaken such a study in another paper in which using these ideas we are able to study the corresponding time domain problems for a large class of infinite-dimensional stationary processes [7]. It is now also obvious that the recent work of H. Cramér [1], [2], [3] and T. Hida [6] on the multiplicity theory of stochastic processes is an extension of Hanner's ideas to non-stationary processes.

2. THE MULTIPLICITY OF A WEAKLY STATIONARY STOCHASTIC PROCESS

Let $x_t(-\infty < t < \infty)$ be a one dimensional (complex valued) weakly stationary (w.s.) process with mean assumed zero and satisfying the following assumptions (A) :

A— (i). x_t is purely non deterministic, and

A—(ii). x_t is continuous in quadratic mean.

Following [5], let $L_2(x)$ be the Hilbert space of the process $\{x_t\}$, and let $L_2(x; a) = \mathcal{S}\{x_t, t \leqslant a\}$ where $\mathcal{S}\{\ldots\}$ denote the closed linear subspace of $L_2(x)$ spanned by the set of random variables in $\{\ldots\}$. We shall also write

$L_2(x; a, b) = L_2(x; b) \ominus L_2(x; a)(a \leqslant b)$. As we proceed further, whenever necessary, we shall draw on Hanner's notation and also on some of the notation and terminology of [6].

The following expresentation was obtained in [8] for a w.s. process $\{x_t\}$ satisfying (A).

$$x_t = \int_{-\infty}^{t} G(u-t)d\xi(u), \tag{2.1}$$

where ξ is a homogeneous random set function, i.e., a random set function with the property

$$\mathcal{E}[\xi(\Delta_1)\overline{\xi(\Delta_2)}] = \mu(\Delta_1 \cap \Delta_2),$$

μ being Lebesgue measure and Δ_1, Δ_2 any two sets of finite μ measure;

$L_2(x; t) = L_2(\xi; t) = \mathcal{S}\{\xi(\Delta), \Delta \subseteq (-\infty, t)\}$ for every t and $\int_{-\infty}^{0} |G(u)|^2 d\mu(u)$ is finite.

It has been shown in Hida in [6] that every second order process $\{x_t\}$ satisfying (A) and not necessarily stationary has the representation

$$x_t = \sum_{n=1}^{N} \int_{-\infty}^{t} F_n(t, u) dE(u) f^{(n)} \tag{2.2}$$

where (i) $\{E(u)\}$ $(-\infty < u < \infty)$ is the resolution of the identity determined by the projection operator with range $L_2(x; u)$, (ii) N is the multiplicity of x_t (N may be finite or ∞), (iii) $f^{(n)}$ $(n = 1, \ldots, N)$ are elements of $L_2(x)$ with the following properties: (a) the processes $Z_n(\Delta) = E(\Delta) f^{(n)}$, where $\Delta = (a, b]$ and $E(\Delta) = E(b) - E(a)$ have orthogonal increments and are mutually orthogonal, (b) the variance function of $Z_n(\Delta)$ is given by $\rho_{f^{(n)}}(\Delta)$ where $\rho_{f^{(n)}}(\Delta) = \| E(\Delta) f^{(n)} \|^2$, (c) $\rho_{f(1)} \gg \rho_{f(2)} \gg \cdots \rho_{f(N)}$, (d) $F_n(t, \cdot)$ is square integrable with respect to $\rho_{f(n)}$, and

(e) $\sum_{h=1}^{N} \int_{-\infty}^{+\infty} |F_n(t, u)|^2 d\rho_{f(n)}(u) < \infty$. The representation (2.2) can be chosen so as to satisfy

$$L_2(x; t) = \sum_{n=1}^{N} \oplus L_2(z_n; t) \quad \text{for each } t. \tag{2.3}$$

It is convenient at this point to recall some of the terminology of multiplicity theory in a separable Hilbert space H. We shall introduce just those ideas that will be used in this paper. Let A be any self-adjoint operator with spectral measure function $E(\cdot)$. For any element f of H let ρ_f be the finite measure on the Borel sets of the line given by $\rho_f(\Delta) = \| E(\Delta) f \|^2$. The family of all finite measures on the line is divided into equivalence classes by the relation of equivalence between measures (equivalence here means mutual absolute continuity). If ρ is used to denote the equivalence class to which the measure ρ_f belongs, ρ will be called the spectral type of f with respect to A. ρ is also referred to as the spectral type belonging to A. If elements f and g are such that $\rho_f \equiv \rho_g$, they obviously have the same spectral type ρ. If for every measure ν belonging to ρ, $\nu \equiv \mu$, the Lebesgue measure then ρ is called the Lebesgue type. We shall say that the spectral type ρ dominates the spectral type $\sigma(\rho > \sigma, \sigma < \rho)$ if any (and thus every) measure belonging to σ is absolutely continuous with respect to any measure belonging to ρ. ρ and σ are said to be independent spectral types if for any spectral type ν such that $\nu < \rho$ and $\nu < \sigma$ we have $\nu = 0$. An element f is said to be of maximal spectral type ρ (with respect to A) if for every g in H, $\rho_g \ll \rho_f$. The subspace $\mathfrak{S}\{E(\Delta)f, \Delta$ over all finite intervals$\}$ is called the cyclic subspace with respect to A generated by f. If this space coincides with H, A is said to be cyclic and f is called a cyclic or generating element of A. A has multiplicity one in this case. Also, if f is a generating element of A then it is of maximal spectral type with respect to A. The spectral type of a generating element of A is referred to as the spectral type of A. The reader is referred to the article by A. I. Plessner and V. A. Rohlin [10] for further details.

Assuming the Karhunen representation (2.1) we shall show that the multiplicity of $\{x_t\}$ is one of Lebesgue type (Proposition, see also [3]). It might seem obvious that $N = 1$ on comparing (2.1) and (2.2). Although this is true it should be noted that in (2.1) the function ξ has for its variance function the Lebesgue measure μ, so that it is not possible to write $\xi(u) = E(u)f$ (for all u) for some f in $L_2(x)$.

Proposition. *The multiplicity of a weakly stationary process is one of Lebesgue type.*

Lemma 1 : *If x_t is a w.s. process, satisfying assumption A—(ii), then for any element $f \epsilon L_2(x)$ $\rho_f \ll \mu$.*

Proof : Let $\{T_h\}$ $(-\infty < h < +\infty)$, be the strongly continuous unitary group of the x_t-process ([2] (p. 55)). We recall from [5] that for every h, and any t real, we have

$$T_h E(t-h) = E(t) T_h. \tag{2.4}$$

Now $\rho_f(\Delta - h) = \|E(\Delta - h)f\|^2$ where $\Delta - h = \{u - h \mid u \epsilon \Delta\}$ and Δ is a Borel measurable set. Therefore, by (2.4) $\rho_f(\Delta - h) = \|E(\Delta)T_h f\|^2$. By strong continuity of the group,

$$\rho_f(\Delta - h) \to \rho_f(\Delta) \text{ as } h \to 0.$$

The assertion of the lemma is now an immediate consequence of a theorem due to N. Weiner and R. C. Young (see [12], p. 91).

Proof of the Proposition : By the property of the Karhunen representation and (2.3), we have

$$L_2(x; t) = L_2(\xi; t) = \sum_1^N \oplus L_2(z_n; t) \text{ for all } t.$$

Hence (see Doob [4], pp. 425–428), $z_n(t) = \int_{-\infty}^{t} h_n(t, u) d\xi(u)$. Since $\{z_n(t), -\infty < t < +\infty\}$ are mutually orthogonal processes with orthogonal increments, holding t fixed, we have for any finite interval $\Delta \subseteq (-\infty, t)$, $z_n(\Delta) = \int_\Delta h_n(t, u) d\xi(u)$ by (2.3), and hence $(n \neq 1)$, $\mathcal{E}[z_1(\Delta)\overline{z_n(\Delta)}] = \int_\Delta h_1(t, u)\overline{h_n(t, u)} d\mu(v) = 0$. This implies that

$$\mu\{v \epsilon (-\infty, t] \,|\, h_1(t, v)\overline{h_n(t, v)} \neq 0\} = 0. \tag{2.5}$$

Define for $u \epsilon (-\infty, t]$, $S_n^{(t)} = \{u \,|\, h_n(t, u) \neq 0\}$ for $n = 1, 2, \ldots, N$. By (2.5), $\mu\{S_1^{(t)} \cap S_n^{(t)}\} = 0$, and therefore by Lemma 2.1, $\rho_{f(n)}(S_1^{(t)} \cap S_n^{(t)}) = 0$ for all $n \neq 1$. Clearly, $\rho_{f(1)}([S_1^{(t)}]^c) = 0$ and $\rho_{f(n)}([S_n^{(t)}]^c) = 0$ and hence by the maximality of $f^{(1)}$, $\rho_{f(n)}([S_1^{(t)}]^c) \cup [S_n^{(t)}]^c) = 0$ for $n \neq 1$ where $(\quad)^c$ denotes complementation with respect to $(-\infty, t]$. Thus $\rho_{f(n)}\{(-\infty, t]\} = 0$ for every t, so that $\rho_{f(n)} = 0$ $(n \neq 1)$. This gives $L_2(\xi; t) = L_2(z_1; t)$ i.e., $\xi(t) = \int_{-\infty}^{t} \nu_t(u) dz_1(u)$. From this and the proper canonical property, for each finite interval $\Delta \subseteq (-\infty, t]$, $\mu(\Delta) = \mathcal{E}|\xi(\Delta)|^2 = \int_\Delta |\nu_t(u)|^2 d\rho_{f(1)}(u)$. For each t, $\mu \ll \rho_{f(1)}$ on $(-\infty, t)$ and hence $\mu \ll \rho_{f(1)}$. This along with Lemma 2.1 proves the proposition.

3. A NEW PROOF OF THE MULTIPLICITY RESULT AND THE KARHUNEN REPRESENTATION

A new proof of the Karhunen representation using multiplicity arguments is presented in this section. The following main theorem is obtained essentially using the modified arguments of Hanner.

Theorem 1. *The self-adjoint operator A of a weakly stationary continuous parameter, purely non-deterministic univariate process continuous in q.m. is cyclic and has Lebesgue spectral type.*

Proof : Since $L_2(x)$ is separable there is an element $f^{(1)}$ of maximal spectral type with respect to A. Given $u > 0$ and $a < b$, let Δ_0 be a subinterval of $(0, u]$, $A < a-u$, $B > b$ and define

$$(3.1) \qquad g_a^b = E(a, b] \int_A^B T_h E(\Delta_0) f^{(1)} dh.$$

Then g_a^b depsnds on u and Δ_0 but not on A and B. As a matter of fact let A' $(< A)$ then $\int_{A}^{B} T_h E(\Delta_0) f^{(1)} dh \in E(-\infty, a] L_2(x)$, since for each $(A' \leqslant h \leqslant A)$, $T_h E(\Delta_0) f^{(1)} = E(\Delta_0 + h) T_h f^{(1)} \in E(u+A)$ $L_2(x)$. Thus $E(a, b) \int_{A'}^{A} T_h E(\Delta_0) f^{(1)} dh = 0$. A similar argument holds if $(A' > A)$. Let $B' > B$, then $\int_{B}^{B'} T_h E(\Delta_0) f^{(1)} \in E(b, \infty)$ $L_2(x)$ by a similar consideration. As Hanner showed for $a < b < c$

$$(3.2) \qquad \begin{cases} g_a^b + g_b^c = g_a^c,\ g_a^b \perp g_b^c \text{ and for any real } t \\ T_t g_a^b = g_{a+t}^{b+t}. \end{cases}$$

From these properties it follows that $\|g_a^b\|^2 = \tau(b-a)$, where τ is a non-negative number which does not depend on a and b. In fact, there is a subinterval, say Δ_0 of $(0, u)$ such that for all a, b $(a < b)$, $\|g_a^b\|^2 = (b-a)$. It can be shown that there exists a u such that $\|g_0^u\|^2 \neq 0$. The proof of this fact is on the same lines as that of Hanner's Proposition C and hence is omitted.

In the definition of g_a^b in (3.1), we shall choose this Δ_0. Now if for $0 < b < u$, we consider $g_0^b = E(0, b] \int_{A'}^{B'} T_h E(\Delta_0) f^{(1)} dh$ where $A' < -u$, $B' > b$, then by definition of g_0^b, $g_0^b = E(0, b] g_0^u$. Also $g_0^u = g_0^b + g_b^u$ with $g_b^u \perp g_0^b$. Hence $\|g_0^u\|^2 = \|g_0^b\|^2 + \|g_b^u\|^2$. If $g_0^b = 0$, we get $\tau(u-b) = \tau u$. The τ in the left hand side is positive by the choice of Δ_0. This however leads to a contradiction since $b < u$. For $b > u$ $\|g_0^b\|^2 \geqslant \|g_0^u\|^2 > 0$. If $b' < 0$, let $\beta = -b'$ and consider $T_\beta g^0_{b'} = g^\beta_0$. But $\|g^\beta_0\|^2 \neq 0$ by previous arguments and hence $\|g^0_{b'}\|^2 \neq 0$. It is easy to deduce that $g^d_c \neq 0$ for all intervals $(c, d]$, c, d real. Thus by this choice of Δ_0, we get $\|g^b_a\|^2 = \tau(b-a)$ where τ is independent of a, b and $\tau > 0$. Without loss of generality we assume $\|g^b_a\|^2 = (b-a)$.

Our definition of g^b_a differs from Hanner's definition of $Z(I^b_a)$ in the following respect. Hanner defines $Z(I^b_a) = E(a, b] \int_A^B T_h z dh$, where z is any non-zero element of $L_2(x; 0, u)$. We choose a particular z of the form $E(\Delta_0)f^{(1)}$, $f^{(1)}$ being an element with maximal spectral type. The advantage of this consists in the fact that by bringing $f^{(1)}$ explicitly in the evidence in the construction of g^b_a we are able to prove that the maximal spectral type is Lebesgue.

Lemma 3.1. *The maximal spectral type of A is equivalent to μ, the Lebesgue measure.*

Proof: It suffices to show that $\rho_{f^{(1)}} \equiv \mu$ where $f^{(1)}$ is the element introduced above. Now for $I = (a, b]$, $\rho_{g^b_a}$ the spectral measure of g^b_a is given by $\rho_{g^b_a}(\Delta) = \mu^I(\Delta)$ for every finite interval Δ, where the measure μ^I is defined as $\mu^I(S) = \mu(I \cap S)$ for every measurable set S on the real line. Hence $\rho_{g^b_a} = \mu^I$. Since $f^{(1)}$ is an element of maximal spectral type, $\rho_{f^{(1)}} \gg \rho_{g^b_a} = \mu^I$ for every I. Hence $\mu \ll \rho_{f^{(1)}}$. An application of Lemma 1 (Section 2) now comyletes the proof.

Let us define $H_1(t) = \mathfrak{S}\{g^b_a, -\infty < a \leqslant b \leqslant t\}$ for each t, and $H_1 = H_1(\infty)$. With P_{H_1} denoting the projection onto H_1, let $x^{(1)}_t = P_{H_1} x_t$; then $x^{(1)}_t$ is weakly stationary, continuous in q.m. and purely non-deterministic. Further defining $y_t = x_t - x^{(1)}_t$ we have

$$L_2(x; t) = L_2(x^{(1)}; t) \oplus L_2(y; t) \quad \text{for every } t \text{ and} \tag{3.3}$$

$$L_2(x^{(1)}; +\infty) \perp L_2(y; +\infty)$$

Lemma 3.2. (i) $L_2(x^{(1)}; a, b) = \mathcal{S}\{E(\alpha, \beta]g_a^b, a < \alpha \leqslant \beta \leqslant b\}$
$= H_1(a, b)$ (a, b real)

(ii) *For all a, b, $L_2(x^{(1)}; a, b)$ is the cyclic subspace generated by g_a^b.*

(iii) *A_I has the spectral type μ^I, where for any interval I, A_I is the restriction of A to $L_2(x^{(1)}; I)$.*

Proof : For $a < t \leqslant b$, since $x_t \perp H_1 \ominus H_1(t)$ and by (3.3), $x_t^{(1)} - P_{L_2(x^{(1)}; a)} x_t^{(1)} = P_{H_1(t)} x_t - P_{L_2(x; a)} P_{H_1(t)} x_t$. But

$P_{L_2(x; a)} P_{H_1(t)} = P_{L_2(x; a)} (P_{H_1(a)} + P_{H_1(t) \ominus H_1(a)}) = P_{L_2(x; a)} P_{H_1(a)}$ $= P_{H_1(a)}$. Hence $x_t^{(1)} - P_{L_2(x^{(1)}; a)} x_t^{(1)} = (P_{H_1(t)} - P_{H_1(a)}) x_t$ and $L_2(x^{(1)}; a, b) \subseteq H_1(a, b)$. The other side of the equality is obvious since for all α, β, $E(\alpha, \beta]g_a^b \epsilon\, L_2(x; a, b)$ and $E(\alpha, \beta)g_a^b \perp L_2(y; \infty)$. This proves (i).

For any interval Δ, $E(\Delta)g_a^b = E(\Delta \cap (a, b])g_a^b = 0$ if Δ and $(a, b]$ are disjoint and $E(\Delta)g_a^b = E(\alpha, \beta]g_a^b$ if $\Delta \cap (a, b] = (\alpha, \beta]$. This fact and (i) together imply that $L_2(x^{(1)}; a, b) = \mathcal{S}\{E(\Delta)g_a^b$, Δ any finite interval of the real line], which is assertion (ii) of the lemma. It is now obvious that A is reduced by $L_2(x^{(1)}; a, b)$. Conclusion (iii) follows from this and the fact that $\rho_{g_a^b} = \mu^I$.

Lemma 3.3. *$L_2(x^{(1)})$ reduces A and $A^{(1)}$, the restriction of A to $L_2(x^{(1)})$ is a cyclic operator of Lebesgue type.*

Proof : If $\omega \epsilon DA \cap L_2(x^{(1)})$ (which is non-empty) where D_A denotes the domain of A, then for $n = 0, \pm 1, \ldots$.

$$(3.2) \quad E(n-1, n]A\omega = AE(n-1, n]\omega = AP_{L_2(x^{(1)}; n-1, n)} \omega \epsilon L_2(x^{(1)}: n-1, n)$$

Hence $A\omega \epsilon\, L_2(x^{(1)})$ i.e., A is reduced by $L_2(x^{(1)})$. Now if I_j ($j = 1, 2, \ldots$) are disjoint finite intervals whose union is the real line we have $L_2(x^{(1)}) = \sum_j \oplus L_2(x^{(1)}; I_j)$. Hence

$$A^{(1)} = \sum_{j=1}^{\infty} A_{I_j}^{(1)} P_{L_2(x^{(1)}, I_j)}.$$

Each $A^{(1)}_{I_j}$ is cyclic and the corresponding spectral types μ_2 are independent (see definition given in Section 2). Hence ([10], p. 152) it follows that $A^{(1)}$ is cyclic with spectral type equivalent to μ. The following lemma gives multiplicity conditions under which a stationary process has Karhunen representation.

Lemma 3.4. *Let $\{u_t, -\infty < t < +\infty\}$ be a weakly stationary purely non-deterministic process continuous in q.m., and B be the self-adjoint operator of the process given by the resolution of the identity $\{\beta(t), -\infty < t < +\infty\}$ where $\beta(t)$ denotes the projection of $L_2(u)$ onto $L_2(u; t)$. If B is cyclic and has Lebesgue spectral type then $u_t = \int_{-\infty}^{t} G(u-t)d\xi(u)$ where $\{\xi(u), -\infty < u < +\infty\}$ is a process with stationary orthogonal increments such that $L_2(u; t) = \mathcal{S}\{\xi(\Delta), \Delta$ a sub-interval of $(-\infty, t]\} = L_2(\xi; t)$.*

Proof : Suppose that B is cyclic and let f be a generating element of A. Let $\rho_f^{(h)}(S) = \| \beta(S)T_h f \|^2$ for every measurable S in $(-\infty, t]$. Clearly by the generating property of f

$$T_h f = \int_{-\infty}^{+\infty} r(h, u)d\beta(u)f \text{ giving } \beta(\Delta)T_h f = \int_\Delta r(h, u)d\beta(u)f$$

for every finite interval Δ. Hence

$$d\beta(u)T_h f = r(h, u)d\beta(u)f. \tag{3.4}$$

This implies that $\rho_f^{(h)} \ll \rho_f$ for all h and $r(h, u) = \left[\dfrac{d\rho_{f(u)}^{(h)}}{d\rho_f}\right]^{\frac{1}{2}}$

where $\dfrac{d\rho_f^{(h)}}{d\rho_f}$ denotes the Radon-Nikodym derivative of $\rho_f^{(h)}$ with respect to ρ_f. Again, from the fact that f is a generating element and $u_0 \epsilon L_2(u)$ one has

$$u_0 = \int_{-\infty}^{0} F(0, u)d\beta(u)f \text{ with } \int_{-\infty}^{0} |F(0, u)|^2 d\rho_f(u) < \infty.$$

and therefore

$$u_t = T_t u_0 = \int_{-\infty}^{t} F(0, u-t)d\beta(u)T_t f \tag{3.5}$$

Now since $\rho_f \equiv \mu$, we can define an orthogonal random set function ξ by

$$\xi(\Delta) = \int_\Delta \left[\frac{d\rho_f}{d\mu}(u)\right]^{-\frac{1}{2}} d\beta(u)f \tag{3.6}$$

having Lebesgue measure as its variance function. Inverting (3.6) we get

$$d\beta(u)f = \left[\frac{d\rho_f}{d\mu}(u)\right]^{\frac{1}{2}} d\xi(u) \tag{3.7}$$

From (3.4) and (3.7), we can write (3.5) as

$$u_t = \int_{-\infty}^{t} F(0, u-t)\left[\frac{d\rho_f^{(t)}}{d\mu}\right]^{\frac{1}{2}} d\xi(u) \tag{3.8}$$

However, in view of (2.4)

$$\rho_f^{(t)}(\Delta) = \rho_f(\Delta - t) \text{ and thus } \frac{d\rho_f^{(t)}}{d\mu}(u) = \frac{d\rho_f}{d\mu}(u-t).$$

If we define $G(u-t) = F(0, u-t)\left[\frac{d\rho_f}{d\mu}(u-t)\right]^{\frac{1}{2}}$, (3.8) has the form (2.1), since by (2.3), (3.6) and (3.7) it follows that $L_2(u; t) = L_2(z; t) = L_2(\xi; t)$. Now in the above Lemma consider $x_t^{(1)}$ as u_t and $A^{(1)}$ as B then we obtain

$$x_t^{(1)} = \int_{-\infty}^{t} G_1(u-t)\, d\xi_1(u),$$

where $\xi_1(t)$-process has stationary orthogonal increments with $L_2(x^{(1)}; t) = L_2(\xi_1; t)$. Consider now the y_t-process. Our object is to show that $y_t = 0$ for each t[i.e., $L_2(y; t) = \{0\}$ for every t]. If this is not the case, since y_t is a weakly stationary, purely non-deterministic process continuous in q.m. we get by repeating previous arguments [for x_t] that $y_t = y_t^{(1)} + z_t$ where $L_2(z) \perp L_2(y^{(1)})$ and $y_t^{(1)} = \int_{-\infty}^{t} G_2(u-t)\, d\xi_2(u)$ with $L_2(y^{(1)}; t) = L_2(\xi_2; t)$.

Thus we have

$$\left\{\begin{aligned} x_t &= \int_{-\infty}^{t} G_1(u-t)d\xi_1(u) + \int_{-\infty}^{t} G_2(u-t)d\xi_2(u) + z_t, \text{ where} \\ L_2(x; t) &= L_2(x^{(1)}; t) \oplus L_2(y^{(1)}; t) \oplus L_2(z; t) \\ &= L_2(\xi_1; t) \oplus L_2(\xi_2; t) \oplus L_2(z; t) \end{aligned}\right. \tag{3.9}$$

and

$$L_2(x; 0) = \mathfrak{S}\{x_t, t \leqslant 0\}. \tag{3.10}$$

These relations enable us to obtain a contradiction to $y_t \neq 0$, essentially using the last arguments of Hanner's Proposition D (pp. 172–173), by construction an element in $L_2(x; 0)$ of the form,

$$W_s = \int_s^0 \overline{G_2(u-s)}d\xi_1(u) - \int_s^0 \overline{G_1(u-s)}d\xi_2(u).$$

Thus $y_t = 0$, $A = A^{(1)}$ and the theorem is proved.

As a corollary to the above result we can obtain the Karhunen representation.

Corollary. A weakly stationary continuous in q.m. continuous parameter purely non-deterministric process has the Karhunen representation.

The proof follows immediately from Theorem 1 and Lemma 3.4.

References

[1] Cramér, H. (1961). "On Some Classes of Non-Stationary Processes," *Proc. Fourth Berkeley Symp. Math. Statist. Prob.*, **2**, 57–77.

[2] Cramér, H. (1961). "On the Structure of Purely Non-Deterministic Processes," *Arkiv For Math.*, **4**, 249–266.

[3] Cramér, H. (1964). "Stochastic Processes as Curves in Hilbert Space," *Theoriya Veryatnosteii i ee Promeneniya IX*, **2**, 1964.

[4] Doob, J. L. (1953). *Stochastic Processes*, John Wiley and Sons, New York.

[5] Hanner, O. (1950). "Deterministic and Non-Deterministic Processes," *Arkiv For Math.*, **1**, 161–177.

[6] Hida, T. (1960). "Canonical Representations of Gaussian Processes and Their Applications," *Mem. Coll. Sci. Kyoto, Ser. A.*, **33**, 109–155.

[7] Kallianpur, G. and Mandrekar, V. (1965). "Multiplicity and Representation Theory of Purely Non-Deterministic Stochastic Processes," *Theoriya Veryatnosteii i ee Promeneniya*, 1965.

[8] Karhunen, K. (1947). "Uber lineare Methoden in der Wahrscheinlichkeitsrechnung," *Ann. Ac. Sci. Fennicae, Ser. A*, I, No. 37.

[9] Masani, P. and Robertson, J. (1962). "The Time-Domain Analysis of a Continuous Parameter Weakly Stationary Stochastic Process," *Pacific J. Math.*, **12**, 1361–1378.

[10] Plessner, A. I. and Rohlin, V. A. (1946). "Spectral Theory of Linear Operators II," (Russian) *Uspehi Mat. Nauk I*, No. 1, 71–191.

[11] Robertson, J. (1963). "Multivariate Continuous Parameter Weakly Stationary Stochastic Processes," Thesis, Indiana University, June.

[12] Saks, S. (1938). *Theory of the Integral* (translation by L. C. Young), New York.

(*Received Nov. 12, 1965.*)

CHAPTER 20

Extensions of Fractile Graphical Analysis to Higher Dimensional Data

P. C. MAHALANOBIS, *Indian Statistical Institute*

1. INTRODUCTION

The technique of fractile graphical analysis (F.G.A.), in the form of bivariate graphs, was developed and applied in the analysis of economic data in a series of papers (Mahalanobis, 1958a, 1958b, 1960, 1962). The wide applicability of this method was further demonstrated in recent papers by Linder (1963), Rhea Das (1960a, b, 1964) and others. Some conjectures made by the author were studied in a number of theoretical papers by Kawada (1961), Kitagawa (1960), Mitrofanova (1961), Sethuraman (1961, 1963) and Takeuchi (1961). The computational aspects through the use of unit record machines were examined by Roy and Kalyanasundaram (1963). The object of the present paper is to provide further theoretical foundations of F.G.A. and extend it to higher dimensional data.

Genesis of the problem. Let us suppose that we wish to study the differences in the distribution of per capita consumption of cereals between two different regions or over time in a single region. It is a common practice in such cases to compare the mean values, standard deviations and other measures characterizing the distributions. Such overall comparisons may not be meaningful or completely informative especially when differences in the consumption of cereals are not the same for all 'comparable sub-sets' of the populations under comparison (in two regions or at different points of time). Thus it may be pertinent to ask whether there is any *differential* increase in the consumption of cereals between the 'poor' and 'rich' sections of the populations. Such a problem leads us to define comparable subsets of populations such as poor, rich and so on. For this purpose we use a suitable concomitant variable such as per capita income of an individual as an indication of economic status. We note that values of income in real terms, that is, making adjustments for changes in prices, at different times or in different regions are not, however, comparable and therefore comparable subsets cannot be defined as groups of individuals having the same per capita income. But groups of individuals having the same relative economic status in the two populations, as defined by ranks with respect to income within a population, may be amenable to meaningful comparisons. In other situations there may be other ways of defining comparable subsets of the populations. After choosing a number of comparable subsets on the basis of a concomitant variable, we examine the difference in the distributions of a main variable for every pair of comparable subsets. Fractile graphical analysis is a convenient technique by which the desired comparisons can be made through appropriate graphs drawn on the basis of sample data. The actual computations involved when there are one main variable and one concomitant variable are briefly explained in Section 2, and certain generalizations to cases with more than one concomitant variable are discussed in other sections.

2. BIVARIATE FRACTILE GRAPH

A fractile graph in the case of two dimensional data is defined as follows. Let (Y, X) denote two variables and (Y_1, X_1), ..., (Y_N, X_N), be N independent observations. One particular variate, say X, is selected for ranking the observations in ascending order. Replacing the X values by ranks and arranging the observations in ascending order of X, we obtain

$$(Y_{(1)}, 1), (Y_{(2)}, 2), \ldots, (Y_{(N)}, N) \tag{2.1}$$

where $Y_{(i)}$ is the Y value associated with the X value of rank i. Now divide the observations (2.1) into a chosen number, g, of groups such that each group consists of $h = N/g$ consecutive observations. These are called fractile groups. The ith fractile group represented by $[i]$ consists of the observations

$$(Y_{(ih-h+1)}, ih-h+1), (Y_{(ih-h+2)}, ih-h+2), \ldots, (Y_{(ih)}, ih) \tag{2.2}$$

which are replaced by the pair

$$(Y_{[i]}, i) \tag{2.3}$$

where i represents the ith fractile group and $Y_{[i]}$ is any statistic (such as the mean, median, maximum, etc.) based on the Y values of the observations in $[i]$,

$$Y_{(ih-h+1)}, \ldots, Y_{(ih)}. \tag{2.4}$$

The g pairs

$$(Y_{[1]}, 1), \ldots, (Y_{[g]}, g) \tag{2.5}$$

provide the fractile graph, by plotting $Y_{[i]}$ against i, $i = 1, \ldots, g$ and joining the successive points by straight lines.

The graph so obtained is represented by G_g. The fractile graph for the entire population, using the same procedure for all members of the population, may be designated Γ_g. As the sample size increases, G_g will provide a consistent estimator of Γ_g.

Separation between graphs. To compare two fractile graphs based on independent samples from two different populations it is necessary to consider the difference between graphs of parallel samples from the same population. For this purpose we divide each sample into two independent halves. The graphs of the two half samples from the first population are denoted by G_{g1} and G_{g2} and that of the entire sample by G_g. Similarly we have the corresponding graphs G'_{g1}, G'_{g2} and G'_g for a sample from the second population. We then choose a measure of separation $\|A-B\|$ between any two graphs A and B. To test the significance of the observed difference between G_g and G'_g we may use a statistic of the type

$$(2.6) \qquad M = 4\frac{\sqrt{\frac{n\,n'}{n+n'}}\,\|G_g-G'_g\|}{(\sqrt{n}\|G_{g1}-G_{g2}\|+\sqrt{n'}\|G'_{g1}-G'_{g2}\|)}$$

The overall measure of separation, proposed in the earlier papers, is the area between graphs. The exact distribution of M is unknown.

The purpose of drawing the fractile graphs is not, however, only to make an overall comparison over the whole range as provided by a statistic of the type (2.6). It also seems possible to compare the graphs at each fractile point or at sets of consecutive fractile points and draw inferences. If necessary, a test of the type (2.6) may be used for particular sections of the graphs to examine the significance of the observed differences.

In practice a significance test is hardly necessary when there is a clear separation of the graphs, over the whole range or over any portion of the graphs, indicated by the fractile points for the two halves of one sample being completely above or below those for the other sample.

The F.G.A. can be used in any situation in which parallel or interpenetrating network of samples (I.P.N.S.) can be drawn, for example, in the study of consumption of cereals in India (Mahalanobis, 1962) or in testing the normality of frequency distributions by Linder (1963). The F.G.A. can also be used to test the significance of differences in concentration curves (Mahalanobis, 1960)[1].

[1] It may be noted that the F.G.A. is based on concepts which have no connection with concentration curves. The innovation in F.G.A. is the introduction of the concept of a graphical error. It is therefore possible to carry out tests of significance with F.G.A. which is not possible with a concentration curve. This crucial point of using the graphical error for the tests of significance was missed by Swamy (1963).

Surmises. The exact distribution of the separation in FGA is not known. The author made some surmises (Mahalanobis 1958b) which have been later approximately verified by model sampling experiments.

The error area e between two fractile graphs (with sample sizes N_1 and N_2, and g the fixed number of fractile groups for both graphs) would tend to decrease as $\sqrt{(N_1+N_2)/N_1N_2}$, and to increase proportionately to g. Also, as each fractile group consists of $h = N/g$ observations, it follows that, when sample sizes are kept constant, then the error area e would tend to vary approximately as $g^{3/2}$. Also, if g is changed to gk, then changes in the error area would vary proportionately to $k^{3/2}$, which is a most useful property in testing the significance of the separation by changing values of g.

An Example from the National Sample Survey of India. An example of F.G.A., based on data for per capita expenditure on foodgrains, estimated from household surveys in rural areas in the 7th Round (October 1953-March 1954) and the 9th Round (May–November 1955), is shown in the Chart on the next page. The design in both Rounds was multi-stage, interpenetrating (IPNS), probability sub-samples, with replacement, using households as last stage units.

In each sub-sample, the households were ranked, in ascending order of expenditure in rupees per person per 30 days, and were divided into twenty groups, each consisting of an equal number of estimated[2] households, to supply the 'five per cent fractile groups'. The sub-samples were then combined and ranked; and twenty 'five per cent fractile' groups were formed for the combined sample, separately, for each of both 7th and 9th Rounds. Average values of per capita expenditure on foodgrains were plotted for each five-percentile group, and adjoining points were joined by straight lines, to obtain "fractile graphs", for the two

---

[2] The size of the samples must, however, be taken as the number of households for which data are collected for each Round.

sub-samples and for the combined sample respectively, separately for the 7th and the 9th Rounds.

The error area is shown in the chart by the shaded portion between the two sub-sample graphs for each Round. The 'separation', which is given by the area between the two combined graphs for the 7th and the 9th Round respectively, is clearly significant, because the fractile graphs for the 9th

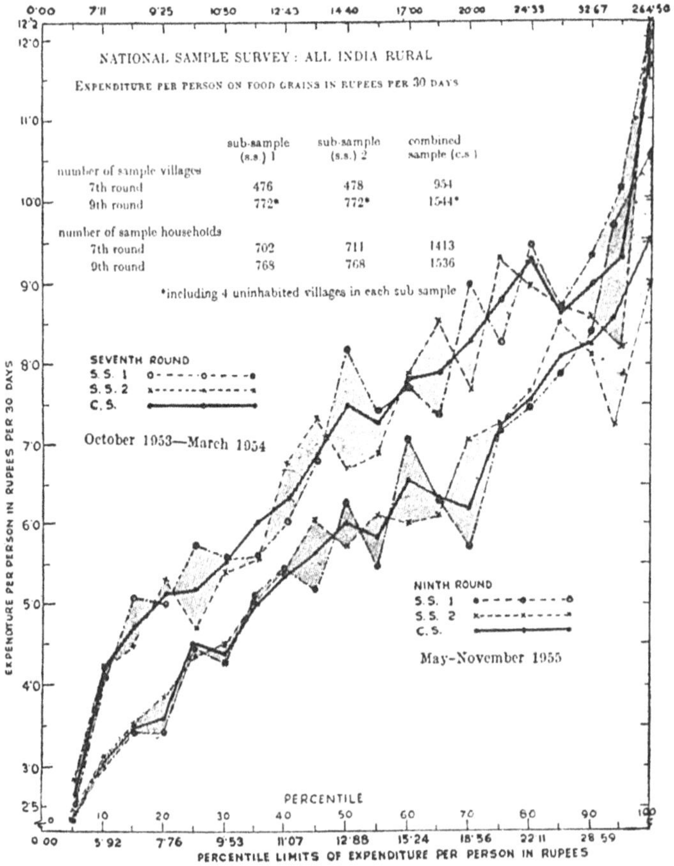

Round all stand well above the corresponding fractile graphs for the 7th Round, indicating that per capita expenditure on foodgrains was significantly higher in the 9th Round, for practically the whole rural population except for rich households at the top.

3. EXTENSION TO MULTIVARIATE CASE

Let us consider N observations from a $(k+1)$ variate distribution and denote the variates by $(Y, X_1, \ldots, X_k)$, where $X_1, \ldots, X_k$ are in the nature of k concomitant variables. There are several natural ways of defining comparable subsets. We shall consider some of them.

Hierarchical group ranking of observations. An observation is said to belong to the group ranking $[i_1, \ldots, i_k]$ if it is in the i_kth fractile when ranked on X_k considering only the observations belonging to the group ranking $[i_1, \ldots, i_{k-1}]$, which is similarly defined with respect to $[i_1, \ldots, i_{k-2}]$, and so on. The group ranking $[i_1]$ simply defines the i_1th fractile when ranked on X_1 only. If the number of fractiles considered for X_j is g_j, then the number of observations in the group $[i_1, \ldots, i_m]$ is

$$h_m = \frac{N}{g_1 \cdots g_m}, \quad m = 1, \ldots, k.$$

Since i_m can take the g_m values, $1, \ldots, g_m$, there are $g_1, \ldots, g_k$ possible group rankings and the vectors defining them are lattice points in a k dimensional Euclidean space. Every observation can be uniquely identified by a group ranking. Let $Y_{[i_1 \cdots i_k]}$ be a statistic, such as the mean, median, etc., of the Y values of the h_k observations belonging to the group ranking $[i_1, \ldots, i_k]$. We thus have a $(k+1)$ dimensional vector

$$(Y'_{[i_1 \cdots i_k]}, i_1, \ldots, i_k) \tag{3.1}$$

as a generalization of (2.3). There are $N/h_k = g_1 \ldots g_k = g$ vectors of fractile points (3.1) which can be plotted in a $(k+1)$ dimensional space.

As in the bivariate case we need to define a measure of separation between two sets of fractile points corresponding to two different samples (either parallel or from different sources). Let us represent the fractile points of another sample by

$$(Y'_{[i_1 \cdots i_k]}, i_1, \ldots, i_k).$$

We suggest two ways of defining a graphical measure. It may be seen that by fixing the coordinates $i_1, \ldots, i_{k-1}$, we have the conditional bivariate graphs generated by the pairs $(Y_{[i_1 \ldots i_{k1}]}, i_k)$, $i_k = 1, \ldots, g_k$. The area between such bivariate graphs for fixed $i_1, \ldots, i_{k-1}$ can be measured, which is represented by

$$a_{i_1 \ldots i_{k-1}}$$

The separation between the two sets of fractile points in the $(k+1)$ dimensional space may be defined by

$$\|G_g - G'_g\| = \Sigma a_{i_1 \ldots i_{k-1}} \tag{3.2}$$

where the summation in (3.2) is over all the combinations of $(i_1, \ldots, i_{k-1})$.

As in the bivariate case let $\|G_{g1} - G_{g2}\|$ be the separation between two parallel halves of one sample of size n and $\|G'_{g1} - G'_{g2}\|$, that between two parallel halves of another sample of size n' and $\|G_g - G'_g\|$, that between two samples. Then to test the hypothesis that the samples belong to two different populations we may use the statistic

$$M = \frac{\sqrt{\dfrac{4n\,n'}{n+n'}}\,\|G_g - G'_g\|}{\sqrt{n}\|G_{g1} - G_{g2}\| + \sqrt{n'}\|G'_{g1} - G'_{g2}\|} \tag{3.3}$$

analogous to (2.6) in the bivariate case.

Another approach is to consider a *simplex* consisting of 2^k points

$$(i_1 + j_1, i_2 + j_2, \ldots, i_k + j_k)$$

with $(i_1, \ldots, i_k)$ as one corner, where each j_m can take two possible values 0 or 1 and graduate the corresponding Y points by a ruled surface, i.e., fit an equation of the type

$$Y = a + \Sigma b_i x_i + \Sigma\, \Sigma c_{ij} x_i x_j + \ldots + m x_1 \ldots x_k$$

The local surfaces fitted to all the simplexes define a surface over the $g_1 \ldots g_k$ lattice points, which may be called a *fractile surface*. For two samples, there are two fractile surfaces and the *volume* between them, which is a natural extension of the area in the bivariate case, may be defined as the distance $\|G_g - G'_g\|$ between two fractile graphs, G_g and G'_g. We can define the statistic M as in (3.3) with the new definition of separation as the volume between the fitted ruled surfaces.

In the method of hierarchical ranking and construction of fractile surface, the results depend on the order in which the concomitant variates are considered. In multi-variate sampling the order in which the different variates would be taken into consideration is a question which, in principle, cannot be decided on *a priori* grounds. In the case of two correlated variates, either variate may be treated as the independent variate; the choice has to be made to suit the purpose in view. In the multivariate case it is also necessary to choose a serial order of the variates which is likely to serve the purpose in view.

Reduction to a single concomitant variable. Another approach is to reduce the k dimensional concomitant variable to one dimension by considering a suitable function, in which case the theory of bivariate graphs applies. The reduction to one variable can be achieved by constructing a discriminant function, based on the concomitant variables only, between samples to be compared. The discriminant function provides the *best separation* with respect to the concomitant variables and ranking based on the *discriminant function* removes the differences due to the concomitant variables.

Applications of higher dimensional fractile graphical analysis are being considered and it is hoped to discuss them elsewhere.

In conclusion I wish to thank Dr. C. R. Rao for his useful suggestions and the discussions I had with him on the subject matter and for his help in the preparation of this paper.

References

Das, Rhea S. (1960*a*). "Applications of Fractile Graphical Analysis to Psychometry : I—Item Analysis," *Psychological Studies*, **5**, 11–18.

Das, Rhea S. and Sharma, K. N. (1960*b*). "Applications of Fractile Graphical Analysis to Psychometry : II—Reliability," *Psychological Studies*, **5**, 71–77.

Das, Rhea S. (1964). "Item Analysis by Probit and Fractile Graphical Methods," *Brit. J. Statist. Psych.*, **17**, 51–64.

Iyengar, N. S. and Bhattacharya, N. (1965). "Some Observations on Fractile Graphical Analysis," *Econometrika*, **33**, 644–645.

Kawada, Y. (1961). "Some Remarks Concerning the Expectation of the Error Area in Fractile Analysis," *Sankhyā, Ser. A*, **23**, 155–160.

Kitagawa, T. (1960). "Sampling Distributions of Statistics Associated with a Fractile Graphic Method," *Bull. Math. Statist.*, Fukuoka, Japan, **9**, 10–42.

Linder, A. (1963). Convocation address. Indian Statistical Institute.

Mahalanobis, P. C. (1958a). "A Method for Fractile Graphical Analysis with Some Surmises of Results," *Transactions of the Bose Research Institute*, Calcutta, **22**, 223–230.

Mahalanobis, P. C. (1958b). Lectures in Japan : Fractile Graphical Analysis, Indian Statistical Institute.

Mahalanobis, P. C. (1960). "A Method of Fractile Graphical Analysis," *Econometrika*, **28**, 325–351.

Mahalanobis, P. C. (1962). "A Preliminary Note on the Consumption of Cereals in India," *Bull. Inst. Internat. Statist.*, **39**, 53–76.

Mahalanobis, P. C. and Lahiri, D. B. (1961). "Analysis of Errors in Censuses and Surveys with Special Reference to Experience in India," *Sankhyā, Ser. A*, **23**, 325–358.

Mitrofanova, N. M. (1961). "On Some Problems of Fractile Graphical Analysis," *Sankhyā, Ser. A*, **23**, 145–154.

Parthasarathy, K. R. and Bhattacharya, P. K. (1961). "Some Limit Theorems in Regression Theory," *Sankhyā, Ser. A*, **23**, 91–102.

Rao, C. R., Mitra, S. K. and Mattai, A. (1966). *Formulae and Tables for Statistical Work*, Statistical Publishing Society, Calcutta.

Roy, J. and Kalyanasundaram, G. (1963). "Punched Card Processing of Sample Survey Data for Fractile Graphical Analysis," *Contributions to Statistics*, (Volume presented to Professor P. C. Mahalanobis on the occasion of his 70th birthday), Statistical Publishing Society, Calcutta, 411–418.

Sethuraman, J. (1961). "Some Limit Distributions Connected with Fractile Graphical Analysis," *Sankhyā, Ser. A*, **23**, 79–90.

Sethuraman, J. (1963). "Fixed Interval Analysis and Fractile Analysis," *Contributions to Statistics*, (Volume presented to Professor P. C. Mahalanobis on the occasion of his 70th birthday) Statistical Publishing Society, Calcutta, 449–470.

Swamy, S. (1963). "Notes on Fractile Graphical Analysis," *Econometrika*, **31**, 551–554.

Takeuchi, K. (1961). "On Some Properties of Error Area in the Fractile Graph Method," *Sankhyā, Ser. A*, **23**, 65–78.

(Received April 2, 1969.)

CHAPTER 21

A Review of Contingency Tables

MARVIN A. KASTENBAUM*, *Mathematics Division, Oak Ridge National Laboratory*

INTRODUCTION AND SUMMARY

This report is a modification of an invited address presented at the Gordon Research Conference on Statistics in Chemistry and Chemical Engineering in New Hampton, New Hampshire on July 14, 1965, Kastenbaum (1965). It is primarily a review of the elementary principles and recent literature on contingency tables. The appended bibliography is a chronological listing of papers on the subject of contingency tables that have appeared in the literature primarily during the past ten or eleven years. This list is by no means complete. It results from a partial literature search of the journals in the library of the Oak Ridge National Laboratory, and includes only a few papers from statistics journals in the social sciences.

The report begins with some discussion of the sampling distributions which underlie the construction and formation of

* Research sponsored by the U.S. Atomic Energy Commission under contract with the Union Carbide Corporation.

contingency tables. This is followed by a brief review of the history and theory of the χ^2 test criterion which is applied when testing hypotheses with contingency tables. Then the report covers such topics as partitioning of X^2, contingency tables of more than two dimensions, measure of association, missing and mixed-up values in contingency tables, contingency tables which come about as a result of consumer preference testing, Markov chains in contingency tables, the Bayesian approach to contingency tables, and alternative analyses to the χ^2 in contingency tables. Illustrative examples appear throughout the paper.

1. STRUCTURE OF 2×2 CONTINGENCY TABLES

The basic structure of contingency tables, especially those which are most frequently encountered, is described with admirable lucidity by G. A. Barnard (1947) and E. S. Pearson (1947). Although these authors concentrate primarily on the 2×2 contingency table, their logical approach to the whole problem is stated in such elementary terms that both papers should be read by all statistical practitioners. A systematic elaboration and application of the ideas of Barnard and Pearson appears in a series of papers by Roy, Mitra, and Kastenbaum in 1955 and 1956.

As Barnard points out, the theory of statistical significance tests deals with abstraction of experimental results. The fact that the figures in question may happen to be results of a public opinion survey, counts of radio-active particles, on-the-job accidents, or number of defectives is ignored in carrying out the test. For the purpose of the statistical theory, the experiment in question could just as well be represented by an experiment involving drawing balls from an urn. Once the abstract picture has been formed, the analysis of it is largely a matter of pure mathematics. As Barnard points out, "What distinguishes the statistician from the pure mathematician in this connection should be the statistician's ability to form valid abstract pictures of concrete cases and his clear recognition of the limits of validity of his abstract pictures."

Consider the simplest type of contingency table, namely the 2×2, or four-fold table, as Table 1a is sometimes referred to.

TABLE 1*a*

The 2×2 *contingency table—cell frequencies*

	C	S	Total
G	n_{11}	n_{12}	$n_{1\cdot}$
B	n_{21}	n_{22}	$n_{2\cdot}$
Total	$n_{\cdot 1}$	$n_{\cdot 2}$	N

TABLE 1*b*

The 2×2 *contingency table—cell probabilities*

	C	S	Total
G	p_{11}	p_{12}	$p_{1\cdot}$
B	p_{21}	p_{22}	$p_{2\cdot}$
Total	$p_{\cdot 1}$	$p_{\cdot 2}$	

This table displays a sample of total size N divided into four mutually exclusive and jointly exhaustive cells, such that $n_{11}+n_{12}+n_{21}+n_{22} = N$. Such a table might have come about in any one of the following ways :

1.1 One multinomial, or two variates (responses)

The results of a testing procedure among grade school children show (C), the presence or (S) the absence of a certain characteristic. G and B represent girls and boys respectively. All identification and test result information is punched into data cards and stored in no apparent order. On the basis of a sample of size N we wish to determine if the percentage of boys with a particular characteristic is equal to the percentage of girls with the same characteristic. The sample of size N is taken at random from the card files in which all the data cards are stored, and it is then stored. If the true proportion of cards in category CG is p_{11} while p_{12}, p_{21} and p_{22} represents the proportions of the other data cards, then the probability associated with the table is given by the general term of the multinomial expansion $(p_{11}+p_{12}+p_{21}+p_{22})^N$, i.e.,

$$P = \frac{N\,!}{n_{11}\,!\,n_{12}\,!\,n_{21}\,!\,n_{22}\,!}\,p_{11}^{n_{11}}\,p_{12}^{n_{12}}\,p_{21}^{n_{21}}\,p_{22}^{n_{22}}. \tag{1.1.1}$$

1.2 Two binomial samples, or one variate (response) and one way of classification (factor)

Perhaps a more realistic way in which the table may be constructed under a similar set of circumstances would be if $n_{1\cdot}$ cards on females and $n_{2\cdot}$ cards on males were examined for the proportions with the desired characteristics. In this case, of course, all the data cards would have been sorted first by sex. Then two random samples of sizes $n_{1\cdot}$ and $n_{2\cdot}$ respectively would have been selected.

If the probability of selecting a girl with the desired characteristic is p_{11}, and the probability of selecting a boy with the same characteristic is p_{21}, then the probability of the arrangement in the table is equal to the probability of finding n_{11} positives in a sample of $n_{1\cdot}$ girls and n_{21} positives in a sample of $n_{2\cdot}$ boys.

This probability may be written as the product of two binomials, namely

$$P = \frac{n_{1\cdot}!}{n_{11}\,!\,n_{12}\,!}\,p_{11}^{n_{11}}\,p_{12}^{n_{12}}\,\frac{n_{2\cdot}\,!}{n_{21}\,!\,n_{22}\,!}\,p_{21}^{n_{21}}\,p_{22}^{n_{22}}, \tag{1.2.1}$$

where $p_{12} = 1-p_{11}$, $p_{22} = 1-p_{21}$, $n_{1\cdot} = n_{11}+n_{12}$, and $n_{2\cdot} = n_{21}+n_{22}$.

1.3 One hypergeometric, or two ways of classification (factors)

A third way in which the data in our table might occur is perhaps the most artificial of the three structures which we will discuss. Its artificiality comes not from the fact that such a structure does not occur in real life, but rather that such a structure occurs less frequently than the two structures already discussed. The classical example of such a structure occurs in the "lady tasting tea" experiment proposed by Sir Ronald Fisher, but this structure also arises in testing for extra-sensory perception. In this case all the values in the marginal totals as well as the total sample size are fixed.

Here the experimenter has two types of objects of known frequencies, $n_{1.}$ and $n_{2.}$. He tells his subject that these are among N objects, $n_{1.}$ of type G and $n_{2.}$ of type B, and asks him to identify these objects accordingly. The probability that the subject will place n_{11} observations in the first cell given that the marginal totals are fixed in this way is given by the underlying hypergeometric distribution as follows :

$$(1.3.1) \qquad P = \frac{n_{1.}\,!\, n_{2.}\,!\, n_{.1}\,!\, n_{.2}\,!}{N!\, n_{11}\,!\, n_{12}\,!\, n_{21}\,!\, n_{22}\,!} = \frac{\binom{n_{1.}}{n_{11}}\binom{n_{2.}}{n_{21}}}{\binom{N}{n_{.1}}}$$

where $n_{.1} = n_{1.}$, and $n_{.2} = n_{2.}$.

The three underlying distributions represented by equations (1.1.1), (1.2.1), and (1.3.1) account for almost all contingency tables which are encountered in experimental situations.

2. DEGREES OF FREEDOM

The number of degrees of freedom is equal to the total number of cells in the table, minus the number of independent linear constraints on the observations, minus the number of independent parameters to be estimated from the data. In the case of a 2×2 table consider the situation for the three structures which are discussed above.

2.1 One Multinomial

The total number of cells is four, the number of independent parameters to be estimated from the data are two, and the number of linear constraints on the observations is one. This comes about as follows : $p_{11}+p_{12}+p_{21}+p_{22} = 1$. However, these four values are not independent because they are subject to the following two constraints : $p_{1.}+p_{2.} = 1$, and $p_{.1}+p_{.2} = 1$. Therefore, only two independent parameters $p_{1.}$ (or $p_{2.}$) and

$p_{\cdot 1}$ (or $p_{\cdot 2}$) need to be estimated. The one linear constraint on the observations is that $n_{11}+n_{12}+n_{21}+n_{22} = N$.

2.2 Two Binomials

In the case of two binomials there is one independent parameter to be estimated from the data, and there are two linear contraints on the observations. The independent parameter comes from the fact that although one set of marginal totals is fixed, the other set is variable and must be estimated subject to the constraints $p_{\cdot 1}+p_{\cdot 2} = 1$. The two linear constraints are $n_{11}+n_{12} = n_{1\cdot}$ and $n_{21}+n_{22} = n_{2\cdot}$.

2.3 One Hypergeometric

In this case there are no independent parameters to be estimated from the data. Instead, there are four constraints on the observations, three of which are independent. The values in the two rows and the two columns are constrained by their respective marginal totals, but the fact that both sets of marginal totals are also constrained to add to N reduces the number of independent constraints to three.

In all three cases there are four cells and a combination of three linear constraints or independent parameters to be estimated from the data. Thus the number of degrees of freedom is equal to one in all three cases. This is one very interesting result in the two-way table. Another pertinent and interesting fact is that in the two-way table all three structures result in the same large sample χ^2 test of the null hypothesis. These two results, which are perhaps a bit of mathematical good fortune, may be the cause of some of the confusion associated with the interpretation of data in two-way tables.

3. STRUCTURE OF $r \times s$ CONTINGENCY TABLES

Let $i = 1, 2, \ldots, r$ rows and $j = 1, 2, \ldots, s$ columns.

3.1 One Multinomial

$$\sum_{i=j}^{r} \sum_{j=1}^{s} n_{ij} = N; \sum_{i=1}^{r} \sum_{j=1}^{s} p_{ij} = \sum_{i=1}^{r} p_{i.} = \sum_{j=1}^{s} p_{.j} = 1.$$

(3.1.1) $$P = \frac{N!}{\prod_{i=1}^{r} \prod_{j=1}^{s} n_{ij}!} \prod_{i=1}^{r} \prod_{j=1}^{s} p_{ij}^{n_{ij}}.$$

Number of degrees of freedom : $rs-[(r-1)+(s-1)]-1 = (r-1)(s-1)$.

3.2 *r* Multinomial Samples

$$\sum_{j=1}^{s} n_{ij} = n_{i.} \text{ for all } i\, ;\ \sum_{i=1}^{r} n_{i.} = N;\ \sum_{j=1}^{s} p_{.j} = 1$$

(3.2.1) $$P = \prod_{i=1}^{r} \left[n_{i.}! \prod_{j=1}^{s} p_{ij}^{n_{ij}} / n_{ij}! \right].$$

Number of degrees of freedom : $rs-(s-1)-r = (r-1)(s-1)$.

3.3 One Hypergeometric

$$\sum_{j=1}^{s} n_{ij} = n_{i.},\ \sum_{i=1}^{r} n_{ij} = n_{.j},\ \sum_{i=1}^{r} n_{i.} = \sum_{j=1}^{s} n_{.j} = N$$

(3.3.1) $$P = \prod_{i=1}^{r} n_{i.}! \prod_{j=1}^{s} n_{.j}! \Big/ N! \prod_{i=1}^{r} \prod_{j=1}^{s} n_{ij}!$$

Number of degrees of freedom : $rs-(r+s-1) = (r-1)(s-1)$.

4. THE NULL HYPOTHESIS IN THE TWO-WAY TABLE

4.1 One Multinomial

The composite hypothesis of interest is that the two variates are *independent*, i.e., $H_0{:}\, p_{ij}=p_{i.}p_{.j}$, against the alternative $H \neq H_0$, where $p_{i.}$ and $p_{.j}$ (for $i=1, 2, \ldots, r$ and $j=1, 2, \ldots, s$) are arbitrary positive parameters subject to the constraints $\sum_i p_{i.} = \sum_j p_{.j} = 1$. This is analogous to the hypothesis of no correlation ($\rho_{xy} = 0$) in a bivariate normal population. Under H_0 the likelihood function is given by

(4.1.1) $$P_0 = \frac{N!}{\prod_{i=1}^{r} \prod_{j=1}^{s} n_{ij}!} \prod_{i=1}^{r} p_{i.}^{n_{i.}} \prod_{j=1}^{s} p_{.j}^{n_{.j}}$$

4.2 r Multinomials

In this case, there are r independent samples of fixed sizes $n_{1\cdot}, n_{2\cdot}, \ldots, n_{r\cdot}$, with p_{ij}, the probability of an observation in cell (ij), such that $\sum_j p_{ij} = p_{i.} = 1$.

The composite hypothesis of interest is that p_{ij}, for any j, is independent of i; i.e., the probability of an observation being in the jth position of row i is the same for all i. This may be specified as

$$H_0 : p_{ij} = p_{\cdot j} \text{ against } H \neq H_0,$$

where $p_{.j}$ $(j = 1, \ldots, s)$ is an arbitrary positive parameter subject to the constraint $\sum_j p_{.j} = 1$. This is referred to as the hypothesis of "homogeneity." It is analogous to and a generalization of the hypothesis of equality of means for r homoscedastic univariate normal populations. For random samples from normal populations, $N(\mu_i, \sigma^2)$, $i = 1, 2, \ldots, r$, the usual hypothesis of interest is:

$$H_0 : \mu_1 = \mu_2 = \ldots = \mu_r = \mu,$$

$$H : \mu_i \neq \mu.$$

The standard test of this hypothesis is the F test of equality of means.

The likelihood function under the hypothesis of homogeneity of r multinomial samples is

$$P_0 = \prod_{i=1}^{r} n_{i.}\,! \prod_{j=1}^{s} p_{.j}^{n_{.j}} \Big/ \prod_{i=1}^{r} \prod_{j=1}^{s} n_{ij}\,! \tag{4.2.1}$$

4.3 One Hypergeometric

In this case, the likelihood function given by (3.3.1) is unchanged, and the null hypothesis is one of "independence" or "inability to discriminate." The question posed under this structure is specific, as in the case of the "lady tasting tea" experiment.

Note that if we were to start with (3.1.1) and let $H_0 : p_{ij} = p_{i.}\,p_{.j}$, and then find the conditional probability under H_0 subject to $n_{i.}$ and $n_{.j}$ being fixed, this probability would be the hypergeometric probability given by (3.3.1).

5. TESTS OF HYPOTHESES IN THE TWO-WAY TABLE

Karl Pearson (1900) proposed as a criterion for testing hypotheses, such as those mentioned above, the statistic

$$X^2 = \sum_{i=1}^{r} \sum_{j=1}^{s} [(n_{ij}-Np_{ij})^2/Np_{ij}]. \tag{5.1}$$

Pearson suggested that in the limit, as N becomes large, this statistic has the χ^2 distribution with $rs-1$ degrees of freedom. He further suggested that the number of degrees of freedom remains unchanged when p_{ij} are estimated from the data. This we now know is wrong. Fisher (1922, 1924) pointed out Pearson's error, and went on to give proof of the limiting distribution of X^2 which avoids most of the mathematical complexities in Pearson's proof. A fully rigorous proof of the limiting distribution of X^2 is given by Cramer (1946). Cramer assumes, among other things, that the p_{ij} are estimated by the method of maximum likelihood.

The maximum likelihood estimates of p_{ij} under the null hypothesis are found to be $p_{ij} = n_{i.}n_{.j}/N^2$, so that the usual form of this statistic for contingency tables is

$$X^2 = \sum_{i=1}^{r} \sum_{j=1}^{s} \left(n_{ij} - \frac{n_{i.}n_{.j}}{N}\right)^2 \Big/ \frac{n_{i.}n_{.j}}{N}. \tag{5.2}$$

This statistic is distributed in the limit, as N becomes large, as chi-square with $(r-1)(s-1)$ degrees of freedom.

The most pertinent question at this point is, "Is the same χ^2 test to be used for all three structures?" The answer is, "Yes, when N is large." However, this is not so for small samples. When the sample size, N, is small, Fisher (1934) recommends that the exact probabilities obtained from the hypergeometric distribution (3.3.1) be used. The greatest objection of Fisher's recommendation is that a loss of power may result if the hypergeometric probabilities are computed when the data actually arise from a single or from several multinomial samples.

K. D. Tocher (1950) has proposed a modification of Fisher's exact test that is most powerful, in the Neyman–Pearson sense, for one-tailed tests with data from any of the three structures. The modification, which is a randomized test procedure, is best illustrated by an example.

Consider the hypothetical example (Table 2) involving the "Yes-No" responses of two groups to an item on an attitude inventory.

TABLE 2

Hypothetical example

	Yes	No	Total
Group 1	9	2	11
Group 2	7	6	13
Total	16	8	24

The null hypothesis is that an individual's response ("yes" or "no") is not influenced by the group from which he comes. Given these data, we wish to make a one-tailed test at the 5% level. The two possible sets of data which deviate more from the null hypothesis are

$$\begin{array}{ccc} 10 & 1 & 11 \\ 6 & 7 & 13 \\ \hline 16 & 8 & 24 \end{array} \quad \text{and} \quad \begin{array}{ccc} 11 & 0 & 11 \\ 5 & 8 & 13 \\ \hline 16 & 8 & 24 \end{array}$$

In Fisher's exact test, the probabilities of the three tables as computed from the hypergeometric are summed. This gives

$$\frac{8!\ 11!\ 13!\ 16!}{24!}\left[\frac{1}{2!\ 6!\ 7!\ 9!}+\frac{1}{6!\ 7!\ 10!}+\frac{1}{5!\ 8!\ 11!}\right]$$

$$= 0{\cdot}12833+0{\cdot}02567+0{\cdot}00175$$

$$= 0{\cdot}15575.$$

This value is regarded as the significance probability.

In Tocher's modification the total probability of all more extreme cases is computed; that is,

$$0{\cdot}02567+0{\cdot}00175 = 0{\cdot}02742.$$

If the numbers, 0·15575 and 0·02742, are both *below* a pre-assigned significance level (0·05, say) reject the hypothesis. If they are both *above* 0·05, accept it. If one is above and one is below, as in this example, calculate

$$\frac{0{\cdot}05-0{\cdot}02742}{0{\cdot}12833} = 0{\cdot}17595.$$

Now, draw a random number between 0 and 1. If this number is less than 0·17595, reject; if greater, accept.

Rationale : If H_0 is rejected only when the two most extreme cases occur, the significance level is 0·02742. The third most extreme case, represented by the data, occurs with prob. 0·12833. Tocher's modification declares as "significant" a fraction 0·17595 of the cases in which the observed data are encountered.

The Pearson chi-square statistic for these data is $X^2 = 2{\cdot}098$. Inasmuch as the square root of a chi-square variable with one degree of freedom is distributed as a Normal with zero mean and unit variance, the probability is $P[\chi^2 \geqslant 2{\cdot}098] = 2P[\chi \geqslant 1{\cdot}448] = 0{\cdot}148$. For comparison with the exact procedure we must consider only one-tail of the normal distribution so that

$$P[\chi \geqslant 1{\cdot}448] = 0{\cdot}074.$$

5.1 Yates' Correction for Continuity

The distribution of the Pearson X^2 statistic is discontinuous. When all expectations are small the chi-square approximation may be poor. The correction proposed by Yates (1934) amounts to reading the chi-square table, not at X_0^2 but at a point between X_0^2 and the value of X^2 immediately below X_0^2 in the discrete series of values. One form of the formula for the 2×2 table is

$$X_c^2 = \frac{N\left[\,|n_{11}\,n_{22}-n_{12}\,n_{21}|-\frac{N}{2}\right]^2}{n_{.1}\,n_{.2}\,n_{1.}\,n_{2.}}. \tag{5.1.1}$$

If equation (5.1.1) is applied to the data in Table 2, the resulting statistic is $X_c^2 = 1{\cdot}028$, and the associated probability is

$P[\chi^2 \geqslant 1\cdot028] = 2P[\chi > 1\cdot014] = 0\cdot310$. The corresponding probability for the one-tailed test is $P[\chi > 1\cdot014] = 0\cdot155$. The following table gives comparable probabilities for a one-tailed test of the data in Table 2 :

Test	*One-tailed probability*
Fisher exact	0·156
X^2 uncorrected	0·074
X_c^2 corrected	0·155.

Caution : If a number of X^2 values, each with a d.f. are added to form a total X^2, the individual values should *not* be corrected for continuity. The total X^2 may be corrected, after it has been obtained, by a procedure given by Cochran (1952).

5.2 Cochran's Recommendations Concerning Analyses of Two-way Tables

a. The 2×2 table : For $N < 20$, or $20 < N < 40$ and the smallest expectation less than 5, use Fisher's exact test. For $N > 40$, use X^2, corrected for continuity if the smallest expectation is less than 5.

b. Tables with degrees of freedom between 2 and 60 and *all* expectations less than 5 : For small N, use Fisher's exact test. Otherwise use X^2, considering whether the continuity correction is needed.

c. Tables with degrees of freedom greater than 60 and *all* expectations less than 5 : Try to obtain the exact mean and variance of X^2, and use the normal approximation to the exact distribution. See Haldane (1937).

d. Tables with more than 1 degree of freedom and *some* expectations greater than 5 : Use X^2 uncorrected for continuity.

e. Continuous data : To test goodness of fit group the data using enough cells to keep expectations down to 12 per cell for $N = 200$, 20 per cell for $N = 400$, 30 per cell for $N = 1000$. Pool at the tails so that the minimum expectation is no smaller than 1.

6. ALTERNATIVE ANALYSES TO THE TRADITIONAL CHI-SQUARE

In recent years many authors have discussed analyses of contingency tables other than the traditional Pearson chi-square analysis. These alternatives fall into two major categories; namely

a. the logit transformation in conjunction with standard least squares, and

b. the likelihood ratio test.

6.1 The Logit Transformation

This method was proposed by Woolf (1955), and elaborated on by Plackett (1962), Gart (1962), Goodman (1963, 1964), Lindley (1964) and others. It may be demonstrated for the $2 \times s$ table.

Let n_{ij} be the observation in the ith row and j^{th} column of a $2 \times s$ contingency tables, where $i = 1, 2$, and $j = 1, 2, \ldots, s$ Let

$$x_j = n_{1j}/n_{2j},\ z_j = ln\ x_j, \text{ and } \frac{1}{u_j} = \frac{1}{n_{1j}} + \frac{1}{n_{2j}},$$

Then

(6.1.1) $$X^2 = \sum_{j=1}^{s} u_j(z_j - \bar{z})^2$$

is distributed asymptotically as χ^2 with $(s-1)$ degrees of freedom, where

(6.1.2) $$\bar{z} = \sum_{j=1}^{s} u_j z_j \Big/ \sum_{j=1}^{s} u_j.$$

Equation (6.1.1) may also be written as

(6.1.3) $$X^2 = \sum_{j=1}^{s} u_j z_j^2 - \left(\sum_{j=1}^{s} u_j z_j\right)^2 \Big/ \sum_{j=1}^{s} u_j.$$

Example. The data in Table 3 are the hypothetical responses of a sample of state legislators to the proposition that, "People should not be allowed to vote unless they can pass a literacy test." Is there evidence from these responses that the probability of agreeing with the proposition is independent of the region in which the legislator resides ?

TABLE 3

Response to the proposition by geographic region

	Region			
	East	Midwest	South	Total
Disagree	17	17	12	46
Agree	36	24	60	120
Total	53	41	72	166

Arithmetic Necessary for Calculating Logit Test Statistics :

$$x_1 = \frac{17}{36} ; x_2 = \frac{17}{24} ; x_3 = \frac{12}{60}$$

$$z_1 = -0{\cdot}75031; \quad z_2 = -0{\cdot}34494; \quad z_3 = -1{\cdot}60943$$

$$\frac{1}{u_1} = \frac{1}{17}+\frac{1}{36} = 0{\cdot}08660; \frac{1}{u_2} = \frac{1}{17}+\frac{1}{24} = 0{\cdot}10049;$$

$$\frac{1}{u_3} = \frac{1}{12}+\frac{1}{60} = 0{\cdot}1$$

$$u_1 = 11{\cdot}54736; \quad u_2 = 9{\cdot}95124; \quad u_3 = 10{\cdot}0.$$

$$\sum_{j=1}^{3} u_j = 31{\cdot}499; \sum_{j=1}^{3} u_j z_j = -28{\cdot}191; \sum_{j=1}^{3} u_j z_j^2 = 33{\cdot}587.$$

Logit $X^2 = 8{\cdot}357$; Pearson $X^2 = 8{\cdot}760$.

6.2. The Likelihood-Ratio Test

This test, proposed by Wilks (1935), has been investigated by Woolf (1957), Chakravarti and Rao (1959), and Kullback, Kupperman and Ku (1962). In their 1962 papers, Kullback, Kupperman and Ku resort to an information theory approach and define a minimum discrimination information statistic (M.D.I.S.) with the following properties :

a. Distributed asymptotically as chi-square under the null hypothesis, and as noncentral chi-square under the alternative hypothesis, with appropriate degrees of freedom and noncentrality parameter, *b*. Additivity,

c. Convexity.

This statistic is $2\hat{I} = -2 \ln \lambda$, where $\ln \lambda = \log_e \lambda$, and where λ is the likelihood ratio. For the $r \times s$ contingency table,

$$(6.2.1)\quad 2\hat{I} = \sum_{i=1}^{r} \sum_{i=j}^{r} 2n_{ij} \ln n_{ij} + 2N \ln N$$

$$- \sum_{i=1}^{r} 2n_{i\cdot} \ln n_{i\cdot} - \sum_{j=1}^{s} 2n_{\cdot j} \ln n_{\cdot j}\,.$$

If this appears to be a formidable alternative to the Pearson X^2 statistic, the authors give us reassurance by presenting tables of $2X \ln X$ for values of X from 1 to 10,000.

Applying the likelihood ratio approach to the data in Table 3 yields

$$\sum_{i=1}^{2} \sum_{j=1}^{3} 2n_{ij} \ln n_{ij} = 1154{\cdot}177; \quad 2N \ln N = 1697{\cdot}180$$

$$\sum_{i=1}^{2} 2n_{i\cdot} \ln n_{i\cdot} = 1501{\cdot}233; \sum_{j=1}^{3} 2n_{\cdot j} \ln n_{\cdot j} = 1341{\cdot}204.$$

The three comparable test statistics are

$$2\hat{I} = 8{\cdot}920,$$

$$\text{Logit } X^2 = 8{\cdot}357,$$

$$\text{Pearson } X^2 = 8{\cdot}760.$$

In each case the probability of observing a value of chi-square greater than X^2, $P[\chi^2 \geqslant X^2] < 0{\cdot}02$. If the level of significance for the test is 5%, then the hypothesis of independence between geographic region and opinion on this proposition would be rejected. Thus all these procedures lead to the conclusion that opinion on this proposition differs among the three regions under consideration.

7. PARTITIONING OF CONTINGENCY TABLES

Let us assume there was reason to believe, prior to the survey, that the opinion of legislators is the same in the East and Midwest, but different in the South. Can these hypotheses be tested using the same set of data ?

This question often arises in the analysis of variance, where tools are available for partitioning the sum of squares associated with the test of hypothesis concerning the equality of s means into at most $(s-1)$ orthogonal sum of squares. With each of the $(s-1)$ sums of squares so derived a hypothesis may be tested. Moreover, the total sum of squares with $(s-1)$ degrees of freedom is equal to the sum of all its orthogonal parts.

Applications of an analogous technique to contingency tables are given by Lancaster (1949), Irwin (1949), Kimball (1954), and Kastenbaum (1960). The test statistic proposed by these authors is the standard Pearson X^2. Kullback, Kupperman, and Ku (1962) show how the same additive partitions may be tested with the minimum discrimination information statistic.

Example. Partition of Data in Table 3

	Regions		
	East	Midwest	Total
Disagree	17	17	34
Agree	36	24	60
Total	53	41	94

	Regions		
	East and Midwest	South	Total
Disagree	34	12	46
Agree	60	60	120
Total	94	72	166

Person X^2

$$X_1^2 = \frac{(166)^2}{(46)(120)}\left[\frac{(17)^2}{53}+\frac{(17)^2}{41}-\frac{(34)^2}{94}\right] = 1{\cdot}017 \quad (p > 0{\cdot}3)$$

$$X_2^2 = \frac{(166)^2}{(46)(120)}\left[\frac{(34)^2}{94}+\frac{(12)^2}{72}-\frac{(46)^2}{166}\right] = 7{\cdot}743 \quad (p < 0{\cdot}01)$$

$1{\cdot}017+7{\cdot}743 = 8{\cdot}760 =$ Pearson's X^2.

M.D.I.S.

$$2\hat{I}_1 = 0{\cdot}879 \ (p > 0{\cdot}3) \qquad 2\hat{I}_2 = 8{\cdot}041 \ (p < 0{\cdot}01)$$

$$0{\cdot}879+8{\cdot}041 = 8{\cdot}920 = 2\hat{I}$$

In each case the conclusion would be that the opinion of the legislators is the same in the East and Midwest, but different in the South.

Alternatively, the contingency table may be partitioned into the same component parts, with the standard Pearson X^2 analysis on each part. This procedure will not result in additive X^2. However this approximate partition is adequate for most tests of significance. Moreover, it has *not* been shown that the additive partition is really preferable to the approximate partition in small samples. This property of additive partitions, in fact, may be the principal motivation for much of the work on contingency tables which has appeared in the literature in recent years. Consider only that satisticians have known for a long time about a random variable whose distribution may be specified by the chi-square probability density function, and that the sum of such random variables is also distributed as chi-square. Add to this the knowledge that a test statistic exists which is, at least asymptotically distributed as chi-square. The result is a desire to add things up or to separate things into their component parts.

8. CONTINGENCY TABLES OF MORE THAN TWO DIMENSIONS

Except for some brief references in two or three statistical texts, the subject of multidimensional contingency tables was all but ignored until fifteen years ago. Indeed, in his text which was published in 1954, O. L. Davies raises and summarily dismisses the three-dimensional case as follows : "Such examples may be treated by an extension of the methods already explained which the reader will have no difficulty in making if he has understood the principles." Unfortunately, things are not as simple as Davies indicates they might be.

The transition from two dimensions to three dimensions necessitates a full understanding of the underlying structure of the data, a clear and concise idea of what null hypotheses are to be tested, and knowledge of the appropriate estimators associated with these null hypotheses in order to calculate the corresponding test statistic. Contrary to Davies' brief, new conceptual problems are posed in going from two-way to three-way tables. On the other hand, the extension from three to higher dimensional contingency tables does not pose any new problems. The theory of multidimensional contingency tables is presented in the many references cited in the bibliography. Therefore what follows is only a brief description of some of the highlights.

Let n_{ijk} denote the observed frequency and p_{ijk}, the probability of having an observation in cell (ijk) of a three-way table, where $i = 1, 2, \ldots, r$ rows, $j = 1, 2, \ldots, s$ columns, and $k = 1, 2, \ldots, t$ layers. Also, let the marginal frequencies be denoted by

$$n_{\cdot jk} = \sum_{i=1}^{r} n_{ijk} \; ; \; n_{\cdot\cdot k} = \sum_{i}^{r} \sum_{j}^{s} n_{ijk} = \sum_{j}^{s} n_{\cdot jk} = \sum_{i}^{r} n_{i\cdot k}$$

$$n_{i\cdot k} = \sum_{j=1}^{s} n_{ijk} \; ; \; n_{\cdot j\cdot} = \sum_{i}^{r} \sum_{k}^{t} n_{ijk} = \sum_{i}^{r} n_{ij\cdot} = \sum_{k}^{t} n_{\cdot jk}$$

$$n_{ij\cdot} = \sum_{k=1}^{t} n_{ijk} \; ; \; n_{i\cdot\cdot} = \sum_{j}^{s} \sum_{k}^{t} n_{ijk} = \sum_{j}^{s} n_{ij\cdot} = \sum_{k}^{t} n_{i\cdot k}$$

$$N = \sum_{i=1}^{r} \sum_{j=1}^{s} \sum_{k=1}^{t} n_{ijk}$$

For a single multinomial sample of size N, corresponding summations over the p_{ijk} are similarly denoted, and

$$\sum_{i=1}^{r} \sum_{j=1}^{s} \sum_{k=1}^{t} p_{ijk} = 1.$$

8.1 Hypotheses Concerning the Two-Way Marginals of a Three Way Table

Pairwise Independence

(8.1.1) $H_0 : p_{ij\cdot} = p_{i\cdot\cdot}\, p_{\cdot j\cdot}$. Independence of I and J,

(8.1.2) $H_0 : p_{i\cdot k} = p_{i\cdot\cdot}\, p_{\cdot\cdot k}$. Independence of I and K,

(8.1.3) $H_0 : p_{\cdot jk} = p_{\cdot j\cdot}\, p_{\cdot\cdot k}$. Independence of J and K.

8.2 Hypotheses Which Have No Analogue in a Two-Way Table

Conditional Independence (Partial Independence)

(8.2.1) $$H_0 : \frac{p_{ijk}}{p_{\cdot\cdot k}} = \frac{p_{i\cdot k}}{p_{\cdot\cdot k}} \cdot \frac{p_{\cdot jk}}{p_{\cdot\cdot k}}, \text{ or } p_{ijk} = \frac{p_{i\cdot k}\, p_{\cdot jk}}{p_{\cdot\cdot k}} .$$

Equation (8.2.1) is the hypothesis of conditional independence of I and J given K. This condition is analogous to the hypothesis of zero partial correlation between I and J in a three-variate Normal population. It does not imply the independence of I and K or of J and K. However, if I is independent of K, and J is independent of K, and if (8.2.1) also holds, we have the condition of mutual independence :

Mutual Independence

(8.2.2) $$H_0 : p_{ijk} = p_{i\cdot\cdot}\, p_{\cdot j\cdot}\, p_{\cdot\cdot k}.$$

Multiple Independence

(8.2.3) $H_0 : p_{ijk} = p_{ij\cdot}\, p_{\cdot\cdot k}$. Indenpedence of K and (IJ).

This condition is analogous to the hypothesis of zero multiple correlation in a three-variate Normal population. It implies independence between I and K and between J and K. The converse, however, is not true. That is to say (8.2.3) implies (8.1.2) and (8.1.3), but (8.1.2) and (8.1.3) do not imply (8.2.3).

It was at this point, in late 1954, that Roy and Kastenbaum found a need for additional theoretical concepts. For, in the special case of the multivariate *normal* population, not only does zero multiple correlation imply zero correlation between all pairs of variables, but also conversely. Obviously this did not hold for a three-variate contingency table. Therefore, the primary objective of their research was to find a set of conditions which, when superimposed on the conditions of independence between pairs of variables, would jointly yield the condition of multiple independence. Their investigations led them to the condition of

"No three-factor interaction"

$$H_0 : p_{rst}p_{ijt}p_{isk}p_{rjk} = p_{ist}p_{rjt}p_{rsk}p_{ijk}. \tag{8.2.4}$$

Equation (8.2.4) can be shown to be a generalization of the condition proposed by Bartlett (1935) for the $2\times2\times2$ and the $2\times2\times3$ tables.

There has been considerable discussion in the literature of the past ten years concerning this hypothesis. For a summary of the theory and philosophy concerning some of the hypotheses which may be tested in a three-way table the reader is referred to the paper by B. N. Lewis (1962). This paper also gives techniques for analysis of data in multidimensional contingency tables. In addition the recent work of Bhapkar and Koch (1961, 1965, 1966) suggests that other hypotheses might be more relevant and appropriate for certain types of contingency tables. For detailed reading on the subject of multiway contingency tables see Bartlett (1935), Norton (1945), Simpson (1951), Lancaster (1951), Roy and Kastenbaum (1956), Roy and Mitra (1956), Kastenbaum and Lamphiear (1959), Lancaster (1960), Darroch (1962), Kullback, Kupperman and Ku (1962), Plackett (1962), Birch (1963), Goodman (1963), and Goodman (1946, *Ann. Math. Stat.*; 1964, *JASA*). For a numerical demonstration of the techniques (other than higher-order interactions) of analyzing data in multidimensional contingency tables, see the paper by P. N. Ries and Harry Smith (1963).

Some of the difficulties which O. L. Davies chose to dismiss in his statement which was quoted earlier may be pointed up in remarks such as this one, extracted from Goodman (1964, *JASA*). "In 1951 Lancaster suggested, on heuristic grounds, a rather simple procedure for testing the hypothesis of zero three-factor interaction. Unfortunately, the distribution of the test statistic suggested by Lancaster is not necessarily distributed as supposed (i.e., as chi-square asymptotically).... Lancaster's statement which has been quoted by Kendall and Stuart [Vol. 2, 1961, p. 584] and by Snedecor (1958), that this test and Bartlett's are "asymptotically equal" is in error".

9. SPECIAL TOPICS IN CONTINGENCY TABLES

9.1. Missing and "Mixed-Up" Values

Bross (1954) considers the problem of misclassification in 2×2 contingency tables, and Watson (1956) considers the problem of missing and "mixed-up" values in contingency tables. In his paper, Watson presents an iterative procedure similar to the "missing-plot technique" in analysis of variance for estimating the missing values in the contingency table before carrying out the standard Pearson X^2 test. Kastenbaum (1958) demonstrates that in some $r \times s$ tables, explicit algebraic formulas can be found for the missing values in contingency tables. The most recent work on these subjects have been done by Asano (1965) and by Mote and Anderson (1965).

9.2 Consumer Preference

An interesting structure for contingency tables is discussed by Anderson (1959) in a problem involving consumer preference studies. One lot of each of three varieties (v_1, v_2, v_3) of snap beans was displayed in retail stores, and each of 123 consumers was asked to rank the beans according to first, second, and third choices. The actual data are given in Table 4.

TABLE 4

Consumer rankings of three varieties of snap beans

Variety	Rank			Total
	1	2	3	
v_1	42	64	17	123
v_2	31	16	76	123
v_3	50	43	30	123
Total	123	123	123	369

The question is, "Does the usual X^2 test of independence of ranks and varieties with four degrees of freedom apply?" That is, does each variety have the same chance (1/3) of receiving a given rank, regardless of rank ? This is not the usual problem of a contingency table with fixed border totals, because repeated sampling is not a random rearrangement of (3×123) items subject to border restrictions. For $i = 1, 2, \ldots, r$ varieties and $j = 1, 2, \ldots, r$ ranks, Anderson shows that the appropriate test statistic is

$$\frac{(r-1)}{r} Q^2 = \frac{(r-1)}{n} \sum_{i=1}^{r} \sum_{j=1}^{r} \left[n_{ij} - \frac{n}{r} \right]^2 \tag{9.2.1}$$

which is distributed asymptotically as chi-square with $(r-1)^2$ degrees of freedom.

9.3 Markov Chains

The structures for contingency tables which have been discussed involve some assumptions of independence of successive sample observations. There frequently arise practical situations in which these assumptions are not valid. One class of such situations involves dependent observations resulting from realizations of states of a simple stationary Markov chain, (Billingsley, 1961)]. In this situation the matrix of transition probabilities is given by a stochastic matrix which is square and whose row totals add to 1. The corresponding frequencies form a square contingency table. Analyses of such data, including tests of hypotheses of a specified matrix of transition probabilities, Markovity, and homogeneity of several realizations of Markov chains are given by Kullback, Kupperman, and Ku (1962). Tests auxiliary to chi-square in Markov chains are given by Gold (1963).

9.4 Measures of Association

The two areas of study covered by the broad title of "Statistical Inference" are hypothesis tests and estimation. All of this report has been devoted to the former and none to the latter. In contingency tables the lack of independence should give some indication of the degree of association. Measures of association in contingency tables have been proposed for at least as long as tests of hypotheses. Indeed they have been used and abused rather widely in the social sciences. The series of papers by Kruskal and Goodman (*JASA* 1954, 1959, 1963) on measures of association for cross classification and the recent papers by Goodman (1963) on interactions in multidimensional contingency tables provide an excellent history and summary of the subject.

9.5 Bayesian Analysis

Finally, as if not to be outdone, contingency tables have most recently been given consideration by the Bayesian statisticians. Lindley (1964), and Good (1965) describe how data from a multinomial distribution, and in particular, data in the form of a contingency table, may be studied using a prior distribution of parameters and expressing the results in the form of a posterior distribution of the parameters. And most recently John J. Gart in a paper submitted to the *JRSS* (*B*) in April 1965 shows how a Bayesian argument may be used for choosing a critical region for the exact test.

References

1966

Bhapkar, V. P., "A Note on the Equivalence of Two Test Criteria for Hypotheses in Categorical Data, *J. Amer. Statistist. Assn.*, **16**, 228–235.

1965

Asano, C., "On Estimating Multinomial Probabilities by Pooling Incomplete Samples," *Ann. Inst. Statist. Math.*, **17**, 1.

Bhapkar, V. P. and Koch, Gary G., "On the Hypothesis of 'No-Interaction' in Three-Dimensional Contingency Tables," *Inst. of Statistics*, Univ. of North Carolina, Mimeo Series No. 440.

Bhapkar, V. P. and Koch, Gary, G., "Hypothesis of No Interaction in Four-Dimensional Contingency Tables," *Inst. of Statistics*, Univ. of North Carolina, Mimeo Series No. 449.

Good, I. J., "*The Estimation of Probabilities; An Essay on Modern Bayesian Methods*," MIT Research Monograph No. 30.

Kastenbaum, Marvin A., "Contingency Tables; A Review," Technical Summary Report No. 596, Mathematics Research Center, Univ. of Wisconsin.

Katti, S. K. and Sastry, A. N., "Biological Examples of Small Expected Frequencies and the Chi-Square Test," *Biometrics*, **21**, 49–54.

Lewontin, R. C. and Felsenstein, J., "The Robustness of Homogeneity Tests in $2 \times n$ Tables," *Biometrics*, **21**, 19–33.

Mote, V. L. and Anderson, R. L., "An Investigation of the Effect of Misclassification on the Properties of χ^2-Tests in the Analysis of Categorical Data," *Biometrika*, **52**, 95–109.

Radhakrishna, S., "Combination of Results from Several 2×2 Contingency Tables," *Biometrics*, **21**, 86–98.

1964

Allison, Harry E., "Computational Forms for Chi-Square," *American Statistician*, **18**, 17–18.

Bennett, B. M. and Nakamura, E., "Tables for Testing Significance in a 2×3 Contingency Table," *Technometrics*, **6**, 439–458.

Bross, I. D. J., "Taking a Covariable into Account," *J. Amer. Statist. Assn.*, **59**, 725–736.

Chew, Victor, "Application of the Negative Binomial Distribution with Probability of Misclassification," *Va. J. Sci.*, **15**, 34–40.

Goodman, Leo A., "Simultaneous Confidence Limits for Cross-Product Ratios in Contingency Tables," *J. Roy. Statist. Soc., Ser. B*, **26**, 86–102.

Goodman, Leo A., "Simple Methods for Analyzing Three-Factor Interaction in Contingency Tables," *J. Amer. Statist. Assn.*, **59**, 319–352.

Goodman, Leo A., "Interactions in Multidimensional Contingency Tables," *Ann. Math. Statist.*, **35**, 632–646.

Goodman, Leo A., "Stimultaneous Confidence Intervals for Contrasts Among Multinomial Populations," *Ann. Math. Statist.*, **35**, 716–725.

Harkness, W. L. and Katz, L., "Comparison of the Power Functions for the Test of Independence in 2×2 Contingency Tables," *Ann. Math. Statist.*, **35**, 1115–1127.

Lindley, D. V., "The Bayesian Analysis of Contingency Tables," *Ann. Math. Statist.*, **35**, 1622–1643.

Plackett, R. L., "The Continuity Correction in 2×2 Tables," *Biometrika*, **21**, 327–338.

Putter, Joseph, "The χ^2 Goodness-of-Fit Test for a Class of Cases of Dependent Observations," *Biometrika*, **15**, 250–252.

Somers, R. H., "Simple Measures of Association for the Triple Dichotomy," *J. Roy. Statist. Soc., Ser. A.*, **127**, 409–415.

Tallis, G. M., "The Use of Models in the Analysis of Some Classes of Contingency Tables," *Biometrics*, **24**, 832–839.

1963

Bennett, B. M. and Nakamura, E., "Tables for Testing Significance in a 2×3 Contingency Tables," *Technometrics*, **5**, 501–511.

Birch, M. W., "Maximum Likelihood in Three-Way Contingency Tables," *J. Roy. Statist., Soc., Ser. B*, **25**, 220–233.

Darroch, J. N. and Silvey, S. D., "On Testing More than One Hypothesis," *Ann. Math. Statist.*, **34**, 555–567.

Diamond, Earl L., "The Limiting Power of Categorical Data Chi-Square Tests Analogous to Normal Analysis of Variance," *Ann. Math. Statist.*, **34**, 1432–1441.

Edwards, A. W. F., "The Measure of Association in a 2×2 Table," *J. Roy. Statist. Soc., Ser. A*, **126**, 109–114.

Feldman, S. E. and Klinger, E., "Short Cut Calculation of the Fisher-Yates Exact Test," *Psychometrika*, **28**, 289–291.

Gold, Ruth A., "Tests Auxiliary to χ^2 Tests in a Markov Chain," *Ann. Math. Statist.*, **34**, 56–74.

Good, I. J., "Maximum Entropy for Hypothesis Formulation, Especially for Multidimensional Contingency Tables," *Ann. Math. Statist.*, **34**, 911–934.

Goodman, Leo A., "On Methods for Comparing Contingency Tables," *J. Roy. Statist. Soc., Ser. A*, **126**, 94–108.

Goodman, Leo A., "On Plackett's Test for Contingency Table Interactions," *J. Roy. Statist. Soc., Ser. B*, **25**, 179–188.

Goodman, Leo A., and Kruskal, W. H., "Measures of Association for Cross Classification III : Approximate Sampling Theory," *J. Amer. Statist. Assn.*, **58**, 310–364.

Ku, H. H., "A Note on Contingency Tables Involving Zero Frequencies and the $2\hat{I}$ Test," *Technometrics*, **5**, 398–400.

Mantel, Nathan, "Chi-Square Tests with One Degree of Freedom : Extensions of the Mantel-Haenszel Procedure," *J. Amer. Statist. Assn.*, **58**, 690–700.

Newell, D. J., "Misclassification in 2×2 Tables," *Biometrics*, **19**, 187–188.

Okamato, Masashi, "Chi-Square Statistic Based on the Pooled Frequencies of Several Observations," *Biometrika*, **50**, 524–528.

Ries, P. N., and Smith, Harry, "The Use of Chi-Square for Preference Testing in Multidimensional Problems," *Chem. Eng. Prog. Symposium Series*, **59**, 39–43.

Walsh, John E., "Loss in Test Efficiency due to Misclassification for 2×2 Tables," *Biometrics*, **19**, 158–162.

1962

Daly, C., "A Simple Test for Trends in a Contingency Tables," *Biometrics*, **18**, 114–119.

Darroch, J. N., "Interactions in Multi-Factor Contingency Tables," *J. Roy. Statist. Soc., Ser. B*, **24**, 251–263.

Fisher, Sir Ronald A., "Confidence Limits for a Cross-Product Ratio," *Austral. J. Statist.*, **4**, 41.

Gart, J. J., "Approximate Confidence Limits for Relative Risks," *J. Roy. Statist. Soc., Ser. B*, **24**, 454–463.

Gart, J. J., "On the Combination of Relative Risks," *Biometrics*, **18**, 601–610.

Kincaid, W. M., "The Combination of $2 \times m$ Contingency Tables," *Biometrics*, **18**, 224–228.

Kullback, S., Kupperman, M. and Ku, H. H., "An Application of Information Theory to the Analysis of Contingency Tables with a Table of $2N \ln N$, $N = 1(1)10,000$," *Jour. Res. Nat. Bur. Stds. B*, **66**, 217–243.

Kullback, S., Kupperman, M., and Ku, H. H., "Tests for Contingency Tables and Markov Chains," *Technometrics*, **4**, 573–608.

Lewis, B. N., "On the Analysis of Interaction in Multi-Dimensional Contingency Tables," *J. Roy. Statist. Soc., Ser. A*, **125**, 88–117.

Plackett, R. L., "A Note on Interactions in Contingency Tables," *J. Roy. Statist. Soc., Ser. B*, **24**, 162–166.

Tallis, G. M., "The Maximum Likelihood Estimation of Correlation from Contingency Tables," *Biometrics*, **18**, 342–353.

1961

Berger, Agnes, "On Comparing Intensities of Association Between Two Binary Characteristics in Two Different Populations," *J. Amer. Statist. Assn.*, **56**, 889–908.

Bhapkar, V. P., "Some Tests for Categorical Data," *Ann. Math. Statist.*, **32**, 72–83.

Billingsley, P., *Statistical Inference for Markov Processes*, Statistical Research Monographs, Vol. 2, The Univ. of Chicago Press, Chicago.

Claringbold, P. J., "The Use of Orthogonal Polynomials in the Partition of Chi-Square," *Austral. J. Statist.*, **3**, 48–63.

Friedlander, D., "A Technique for Estimating a Contingency Table, Given the Marginal Totals and Some Supplementary Data," *J. Roy. Statist. Soc., Ser. A*, **124**, 412–420.

Gregory, G., "Contingency Tables with a Dependent Classification," *Austral. J. Statist.*, **3**, 42–47.

Grizzle, James, E., "A New Method for Testing Hypotheses and Estimating Parameters for the Logistic Model," *Biometrics*, **17**, 372–385.

Kendall, M. G. and Stuart, A., *The Advanced Theory of Statistics*, Vol. 2, Charles Griffin, London.

Okamato, M. and Ishii, G., "Test of Independence in Intraclass 2×2 Tables," *Biometrika*, **48**, 181–190.

Rogot, E., "A Note on Measurement Errors and Detecting Real Differences," *J. Amer. Statist. Assn.*, **56**, 314–319.

Schull, William J., "Some Problems of Analysis of Multi-Factor Tables," *Bull. Inst. Internat. Statist.*, **28**, 259–270.

Yates, F., "Marginal Percentages in Multiway Tables of Quantal Data with Disproportionate Frequencies," *Biometrics*, **17**, 1–9.

1960

Bennett, B. M. and Hsu, P., "On the Power Function of the Exact Test for the 2×2 Contingency Table," *Biometrika*, **47**, 393–398.

Gridgeman, N. T., "Card-Matching Experiments : A Conspectus of Theory," *J. Roy. Statist. Soc., Ser. A*, **23**, 45–49.

Ishii, G., "Intraclass Contingency Tables," *Ann. Inst. Statist. Math.*, **12**, 161–207; corrections, p. 279.

Kastenbaum, M. A., "A Note on the Additive Partitioning of Chi-Square in Contingency Tables," *Biometrics*, **16**, 416–422.

Kupperman, Morton, "On Comparing Two Observed Frequency Counts," *Appl. Statist.*, **9**, 37–42.

Lancaster, H. O., "On Tests on Independence in Several Dimensiona," *J. Austral. Math. Soc.*, **1**, 241–254.

Robertson, W. H., "Programming Fisher's Exact Method of Comparing Two Percentages," *Technometrics*, **2**, 103–107.

1959

Anderson, R. L., "Use of Contingency Tables in the Analysis of Consumer Perference Studies," *Biometrics*, **15**, 582–590.

Chakravarti, I. M. and Rao, C. R., "Tables for Some Small Sample Tests of Significance for Poisson Distributions and 2×3 Contingency Tables," *Sankhyā*, **21**, 315–326.

Goodman, Leo A. and Kruskal, W. H., "Measures of Association for Cross Classification II : Further Discussion and References," *J. Amer. Statist. Assn.*, **54**, 123–163.

Haldane, J. B. S., "The Analysis of Heterogeneity, I," *Sankhyā*, **21**, 209–216.

Hoyt, C. J., Krishnaiah, P. R. and Torrance, E. P., "Analysis of Complex Contingency Data," *J. Exper. Ed.*, **27**, 187–194.

Kastenbaum, M. A. and Lamphiear, D. E., "Calculation of Chi-Square to Test the No Three-Factor Interaction Hypothesis," *Biometrics*, **15**, 107–115.

Kullback, S., *Information Theory and Statistics*, John Wiley and Sons, New York.

Kupperman, Morton, "A Rapid Significance Test for Contingency Tables," *Biometrics*, **15**, 625–628.

Nass, C. A. G., "The χ^2-Test for Small Expectations in Contingency Tables, with Special Reference to Accidents and Absenteeism," *Biometrika*, **46**, 365–385.

Silvey, S. D., "The Lagrangian Multiplier Test," *Ann. Math. Statist.*, **30**, 389–407.

Somers, Robert H., "The Rank Analogue of Product-Moment Partial Correlation and Regression, with Application to Manifold, Ordered Contingency Tables," *Biometrika*, **47**, 241–246.

Steyn, H. S., "On χ^2-Tests for Contingency Tables of Negative Binomial Type," *Statist. Neerlandica*, **13**, 433–444.

Weiner, Irving B., "A Note of the Use of Mood's Likelihood Ratio Test for Item Analyses Involving 2×2 Tables with Small Samples," *Psychometrika*, **24**, 371–372.

1958

Blalock, H. M. Jr., "Probabilistic Interpretations for the Mean Square Contingency," *J. Amer. Statist. Assn.*, **53**, 102–105.

Kastenbaum, M. A., "Estimation of Relative Frequencies of Four Sperm Types in *Drosophila melanogaster*," *Biometrics*, **14**, 223–228.

Mitra, S. K., "On the Limiting Power Function of the Frequency Chi-Square Test," *Ann. Math. Statist.*, **29**, 1221–1233.

Snedecor, G. W., "Chi-Square of Bartlett, Mood and Lancaster in a 2^3 Contingency Table," *Biometrics*, **14**, 560–562.

1957

Bross, Irwin, D. J. and Kasten, Ethel L., "Rapid Analysis of 2×2 Tables," *J. Amer. Statist. Assn.*, **52**, 18–28.

Corsten, L. C. A., "Partition of Experimental Vectors Connected with Multinomial Distributions," *Biometrics*, **13**, 451–484.

Edwards, J. H., "A Note on the Practical Interpretation of 2×2 Tables," *Brit. J. Prev. Soc. Med.*, **11**, 73–78.

Lancaster, H. O., "Some Properties of the Bivariate Normal Distribution Considered in the Form of a Contingency Table," *Biometrika*, **44**, 289–292.

Mote, V. L., "An Investigation of the Effect of Misclassification of the Chi-Square Tests in the Analysis of Categorial Data," Ph.D. Dissertation, Univ. North Carolina at Raleigh, Institute of Statistics, Mimeo Series No. 182.

Roy, S. N., *Some Aspects of Multivariate Analysis*, John Wiley and Sons, New York.

Sakoda, J. M. and Cohen, B. H., "Exact Probabilities for Contingency Tables Using Binomial Coefficients," *Psychometrika*, **22**, 83–86.

Woolf, Barnet, "The Log Likelihood Ratio Test (The G-Test). Methods and Tables for Tests of Heterogeneity in Contingency Tables," *Ann. Human Genetics*, **21**, 397–409.

1956

Fishman, J. A., "A Note on Jenkins' 'Improved Method for Tetrachoric *r*'," *Psychometrika*, **20**, 305.

Good, I. J., "On the Estimation of Small Frequencies in Contingency Tables," *J. Roy. Statist. Soc., Ser. B*, **13**, 113–124.

Gridgeman, N. T., "A Testing Experiment," *Appl. Statist.*, **15**, 106–112.

Leander, E. K. and Finney, D. J., "An Extension of the Use of the χ^2-Test," *Appl. Statist.*, **5**, 132–136.

Mainland, D., Herrera, L. and Sutcliffe, M. I., "Statistical Tables for Use with Binomial Samples-Contingency Tests, Confidence Limits, and Sample Size Estimates," New York University College of Medicine, New York.

Roy, S. N. and Kastenbaum, Marvin, A., "On the Hypothesis of No 'Interaction' in a Multiway Contingency Table," *Ann. Math. Statist.*, **27**, 749–757.

Roy, S. N., and Mitra, S. K., "An Introduction to Some Nonparametric Generalizations of Analysis of Variance and Multivariate Analysis," *Biometrika*, **43**, 361–376.

Watson, G. S., "Missing and 'Mixed-Up' Frequencies in Contingency Tables," *Biometrics*, **12**, 47–50.

1955

Armitage, P., "Tests for Linear Trends in Proportions and Frequencies," *Biometrics*, **11**, 375–386.

Armsen, P., "Tables for Significance Tests of 2×2 Contingency Tables," *Biometrika*, **42**, 494–505.

Cochran, W. G., "A Test of Linear Function of the Deviations Between Observed and Expected Numbers," *J. Amer. Statist. Assn.*, **50**, 377–397.

Haldane, J. B. S., "Substitutes for χ^2," *Biometrika*, **42**, 265–266.

Haldane, J. B. S., "A Problem in the Significance of Small Numbers." *Biometrika*, **42**, 266–267.

Haldane, J. B. S., "The Rapid Calculation of χ^2 as a Test of Homogeneity from a $2\times n$ Table," *Biometrika*, **42**, 519–520.

Jenkins, W. L., "An Improved Method for Tetrachoric r," *Psychometrika*, **20**, 253–258.

Kastenbaum, Marvin A., "Analysis of Data in Multiway Contingency Tables," Unpublished Doctoral Dissertation, North Carolina State College.

Leslie, P. H., "A Simple Method of Calculating the Exact Probability in 2×2 Contingency Tables with Small Marginal Totals," *Biometrika*, **42**, 522–523.

Mitra, S. K., "Contributions to the Statistical Analysis of Categorical Data," Univ. North Carolina, Institute of Statistics, Mimeo. Series, No. 142.

Roy, S. N. and Kastenbaum, Marvin A., "A Generalization of Analysis of Variance and Multivariate Analysis to Data Based on Frequencies in Qualitative Categorical or Class Intervals," Univ. North Carolina, Institute of Statistics, Mimeo Series, No. 131.

Roy, S. N. and Mitra, S. K., "An Introduction to Some Nonparametric Generalizations of Analysis of Variance and Multivariate Analysis," Univ. North Carolina Institute of Statistics, Mimeo Series, No. 139.

Sekar, C. Chandra, Agarivala, S. P. and Chakraborty, P. N., "On the Power Function of a Test of Significance for the Difference Between Two Proportions," *Sankhyā*, **15**, 381–390.

Stuart, Alan, "A Test of Homogeneity of the Marginal Distributions in a Two-Way Classification," *Biometrika*, **42**, 412–416.

Woolf, Barnet, "On Estimating the Relation Between Blood Group and Disease," *Ann. Human Genetics*, **19**, 251–253.

Yates, F., "A Note on the Application of the Combination of Probabilities Test to a Set of 2×2 Tables," *Biometrika*, **42**, 401–411.

Yates, F., "The Use of Transformations and Maximum Likelihood in the Analysis of Quantal Experiments Involving Two Treatments, "*Biometrika*, **42**, 382–403.

1954

Bross, Irwin, D. J., "Misclassification in 2×2 Tables," *Biometrics*, **10**, 478–486.

Cochran, W. G., "Some Methods for Strengthening the Common Chi-Square Tests," *Biometrics*, **10**, 417–451.

Dawson, R. B., "A Simplified Expression for the Variance of the χ^2 Function on a Contingency Table," *Biometrika*, **41**, 280.

Goodman, Leo A., and Kruskal, W. H., "Measures of Association for Cross Classification," *J. Amer. Statist. Assn.*, **49**, 732–764.

Kimball, A. W., "Short-Cut Formulas for the Exact Partition of Chi-Square in Contingency Tables," *Biometrics*, **10**, 452–458.

McGill, W. J., "Multivariate Information Transmission." *Psychometrika*, **19**, 97–116.

1952

Cochran, William G., "The χ^2-Test of Goodness of Fit," *Ann. Math. Statist.*, **23**, 315–345.

Dyke. G. V. and Patterson, H. D., "Analysis of Factorial Arrangements When the Data are Proportions," *Biometrics*, **8**, 1–12.

1951

Lancaster, H. O., "Complex Contingency Tables Treated by the Partition of Chi-Square," *J. Roy. Statist. Soc., Ser. B*, **13**, 242–249.

Simpson, C. H., "The Interpretation of Interaction in Contingency Tables," *J. Roy. Statist. Soc., Ser. B.*, **13**, 238–241.

1950

Tocher, K. D., "Extension of the Neyman-Pearson Theory of Tests to Discontinuous Variates," *Biometrika*, **37**, 130–144.

1949

Hsu, P. L., "The Limiting Distributions of Functions of Sample Means and Application to Testing Hypotheses," *Proc. Berkeley Symposium on Mathematical Statistics and Probability* (1945, 1946), University of California Press, Berkeley and Los Angles.

Irwin, J. O., "A Note on the Subdivision of Chi-Square into Components," *Biometrika,* **36**, 130–134.

Lancaster, H. O., "The Derivation and Partition of Chi-Square in Certain Discrete Distributions," *Biometrika,* **36**, 117–129.

1948

Yates, F., "The Analysis of Contingency Tables with Groupings Based on Quantitative Characters," *Biometrika,* **35**, 176–181.

1947

Barnard, G. A., "Significance Tests for 2×2 Tables," *Biometrika,* **34**, 123–138.

Pearson, E. S., "The Choice of Statistical Tests Illustrated on the Interpretation of Data Classed in a 2×2 Table," *Biometrika,* **34**, 139–167.

1946

Cramer, H., *Mathematical Methods of Statistics,* Princeton University Press, Princeton.

1945

Norton, H. W., "Calculation of Chi-Square for Complex Contingency Tables," *J. Amer. Statist. Assn.,* **40**, 251–258.

1937

Haldane, J. B. S., "The Exact Value of the Moments of the Distribution of χ^2 Used as a Test of Goodness of Fit, When Expectations Are Small," *Biometrika,* **29**, 133–143.

1935

Bartlett, M. S., "Contingency Table Interactions," *J. Roy. Statist. Soc. Supp.,* **2**, 248–252.

Wilks, S. S., "The Likelihood Test of Independence in Contingency Tables," *Ann. Math. Statist.,* **6**, 190–196.

1934

Fisher, R. A., *Statistical Methods for Research Workers,* 5th. and subsequent editions, Oliver and Boyd, Edinburgh.

Yates, F., "Contingemcy Tables Involving Small Numbers and the χ^2-Test," *J. Roy. Statist. Soc., Supp.,* 1, **1**, 217–235.

1924

Fisher, R. A., "The Conditions Under Which Chi-Square Measures the Discrepancy Between Observation and Hypothesis," *J. Roy. Statist. Soc.*, **87**, 442–450.

1922

Fisher, R. A., "On the Interpretation of Chi-Square from Contingency Tables, and the Calculation of *P*," *J. Roy. Statist. Soc.*, **85**, 87–94.

1900

Pearson, Karl, "On the Criterion that a Given System of Deviations from the Probable in the Case of a Correlated System of Variables Is Such That It Can Be Reasonably Supposed To Have Arisen from Random Sampling," *Philos. Mag.*, Ser. 5, **50**, 157–172.

(*Received Sept. 12, 1965. Revised June 13, 1966.*)

CHAPTER 22

On a Multivariate F Distribution

P. R. KRISHNAIAH AND J. V. ARMITAGE, *Aerospace Research Laboratories, Wright-Patterson Air Force Base, Ohio*

1. INTRODUCTION

The multivariate F distribution discussed here plays an important role in simultaneous tests of hypotheses under the ANOVA and MANOVA models and in ranking and selection procedures. This distribution was considered by Krishnaiah [9–11] earlier. A special case of this distribution is the multivariate t^2 distribution. In this paper, we discuss the evaluation of the probability integrals of the multivariate F distribution. Also, upper 5% and 1% points of the distribution of the maximum of t^2 variates which are jointly distributed as the multivariate t^2 are tabulated. Some applications of these tables are also discussed.

2. MULTIVARIATE F DISTRIBUTION

We will first give the following known (*see* [9, 11]) definition of the multivariate F distribution.

Let $X : n \times p$ be a matrix of n independent random row vectors which are distributed identically as a p-variate normal with a common covariance matrix $\boldsymbol{\Sigma} = (\sigma_{ij})$ and mean vector $\boldsymbol{\mu}$. Also, let $S = (s_{ij}) = X'X$ where X' is the transpose of X. In addition let s^2/σ^2, where $E(s^2) = m\sigma^2$, be a chi-square variate with m degrees of freedom distributed independently of s_{11}, $s_{22}, \ldots, s_{pp}$. Then, the joint distribution of $F_1, F_2, \ldots, F_p$ where $F_i = m\sigma^2 s_{ii}/n\sigma_{ii}s^2$, $(i = 1, 2, \ldots, p)$, is a non-central (central) p-variate F distribution with (n, m) degrees of freedom and with Σ as the covariance matrix of the "accompanying" p-variate normal when $\boldsymbol{\mu} \neq \mathbf{0}$ $(\boldsymbol{\mu} = \mathbf{0})$. Here we note that the joint distribution of $s_{11}/\sigma_{11}, \ldots, s_{pp}/\sigma_{pp}$ is given in [15] when $\boldsymbol{\mu} = \mathbf{0}$.

The p-variate F distribution is known [11] to be a singular or non-singular distribution according as $\boldsymbol{\Sigma}$ is singular or non-singular. Unless otherwise stated, the discussion in the sequel is restricted to the non-singular case.

The frequency function of the central bivariate F distribution is known [10] to be

$$f(F_1, F_2) = \frac{m^{m/2}(1-\rho_{12}^2)^{(m+n)/2}}{\Gamma(m/2)\,\Gamma(n/2)} \tag{2.1}$$

$$\sum_{j=0}^{\infty} \frac{\rho_{12}^{2j}\,\Gamma[n+(m/2)+2j]\,n^{n+2j}(F_1F_2)^{(n/2)+j-1}}{j\,!\,\Gamma(n/2)+j][m(1-\rho_{12}^2)+n(F_1+F_2)]^{n+(m/2)+2j}}$$

where $\rho_{12} = \sigma_{12}/[\sigma_{11}\sigma_{22}]^{\frac{1}{2}}$. The frequency function of the multivariate F distribution is given in [9]. When $n = 1$, the multivariate F distribution reduces to the multivariate t^2 distribution. When the off-diagonal elements of $\boldsymbol{\Sigma}$ are equal to zero, the distribution of the maximum (minimum) of $F_1, \ldots, F_p$ reduces to the distribution of the studentized largest (smallest) chi-square. Tables of the studentized largest and smallest chi-square distributions are given in the literature (see [1, 7, 8, 13]).

3. PROBABILITY INTEGRALS OF THE MULTIVARIATE *F* DISTRIBUTION

In the central bivariate case, it is known [10] that

$$\int_0^c \int_0^c f(F_1, F_2)dF_1 dF_2 = \frac{(1-\rho_{12}^2)^{n/2}}{\Gamma(m/2)\Gamma(n/2)} \sum_{j=0}^{\infty} \frac{\rho_{12}^{2j}\,\Gamma[n+(m/2)+2j]}{j\,!\,\Gamma[(n/2)+j]} L_j$$

where

$$L_j = \int_0^a \int_0^a \frac{(xy)^{(n/2)+j-1}}{(1+x+y)^{n+2j+(m/2)}}\, dx\, dy,\; a = cn/m(1-\rho_{12}^2).$$

Using (2.1) we can similarly show that

$$\int_c^{\infty} \int_c^{\infty} f(F_1, F_2)dF_1\, dF_2 = \frac{(1-\rho_{12}^2)^{n/2}}{\Gamma(n/2)} \sum_{j=0}^{\infty} \frac{\rho_{12}^{2j}\,\Gamma[n+(m/2)+2j]}{\Gamma[(n/2)+j]} L_j^*$$

where

$$L_j^* = \int_a^{\infty} \int_a^{\infty} \frac{(xy)^{(n/2)+j-1}}{(1+x+y)^{n+2j+(m/2)}}\, dxdy.$$

The integrals L_j and L_j^* are respectively similar to the probability integrals associated with the studentized largest and smallest chi-square distributions. They can be evaluated by using the techniques discussed in [1, 7, 8, 13]. When $p > 2$, the following approximation in suggested in [10] :

$$P_0 \cong 1-P_1+P_2 \tag{3.1}$$

where

$$P_0 = P[F_i \leqslant c;\quad i = 1, 2, \ldots, p],$$

$$P_1 = \sum_{i=1}^{p} P[F_i > c],$$

$$P_2 = \sum_{i \neq j=1}^{p} P[F_i > c,\; F_j > c].$$

The above approximation holds even when the multivariate F distribution is non-central and (or) singular if $\rho_{ij} \neq 1$, $(i \neq j = 1, 2, \ldots, p)$, where $\rho_{ij} = \sigma_{ij}/[\sigma_{ii}\sigma_{jj}]^{\frac{1}{2}}$. The right side of (3.1) actually gives an upper bound on P_0. A lower bound for P_0 is given by $1-P_1$. Similar bounds can be given for P_0^* where

$$P_0^* = P[F_i \geqslant c; \quad i = 1, 2, \ldots, p].$$

The following three tables give the exact and approximate upper 5% values of the multivariate F distribution for $n = 1$ and for selected values of m, p and ρ where $\rho = \rho_{ij}$ for $i \neq j = 1, 2, \ldots, p$. The exact values given in Table I are taken from [1] while the exact values in Tables II and III are taken from the tables given at the end of this paper. The approximate values in all three tables are computed using (3.1). On the basis of the empirical evidence, we feel that the approximation (3.1) becomes better as m becomes larger and ρ and p becomes smaller.

We will now consider the exact evaluation of the probability integrals of the non-central multivariate F distribution when $n = 1$ and the covariance matrix of the "accompanying" multivariate normal is of particular structure.

TABLE I

Comparison of the exact and approximate upper 5% Values for $\rho = 0{\cdot}0$

m \ p		3	4	5	6	7	8	9	10	11	12
15	E	7.12	7.87	8.46	8.96	9.39	9.77	10.11	10.42	10.69	10.95
	A	7.12	7.86	8.45	8.94	9.35	9.72	10.05	10.34	10.60	10.84
20	E	6.73	7.41	7.95	8.40	8.78	9.12	9.42	9.70	9.95	10.18
	A	6.73	7.40	7.94	8.38	8.76	9.10	9.39	9.66	9.91	10.13
25	E	6.51	7.14	7.65	8.07	8.44	8.75	9.04	9.29	9.52	9.74
	A	6.51	7.14	7.65	8.07	8.42	8.74	9.02	9.27	9.50	9.71
30	E	6.36	6.98	7.46	7.87	8.21	8.51	8.79	9.03	9.25	9.46
	A	6.36	6.98	7.46	7.87	8.21	8.51	8.78	9.02	9.24	9.44
35	E	6.26	6.86	7.33	7.72	8.06	8.35	8.61	8.85	9.06	9.26
	A	6.26	6.86	7.33	7.72	8.05	8.34	8.60	8.84	9.05	9.25

E denotes exact value
A denotes approximate value

TABLE II

Comparison of the exact and approximate upper 5%

Vaules for $\rho = 0{\cdot}3$

m \ p		3	4	5	6	7
15	E	7.02	7.72	8.27	8.74	9.13
	A	7.00	7.67	8.19	8.60	8.94
20	E	6.63	7.27	7.78	8.20	8.56
	A	6.62	7.24	7.72	8.11	8.43
25	E	6.42	7.02	7.50	7.90	8.24
	A	6.41	7.00	7.46	7.83	8.14
30	E	6.28	6.86	7.32	7.70	8.03
	A	6.27	6.84	7.28	7.65	7.95
35	E	6.18	6.75	7.20	7.57	7.88
	A	6.17	6.73	7.16	7.52	7.81

E denotes exact value

A denotes approximate value

TABLE III

Comparison of the exact and approximate upper 5%

Values for $\rho = 0{\cdot}5$

m \ p		3	4	5	6	7
15	E	6.81	7.44	7.93	8.34	8.68
	A	6.74	7.25	7.57	7.74	7.77
20	E	6.45	7.03	7.48	7.85	8.16
	A	6.40	6.88	7.20	7.40	7.50
25	E	6.25	6.80	7.22	7.57	7.87
	A	6.20	6.67	6.99	7.20	7.33
30	E	6.12	6.65	7.06	7.40	7.68
	A	6.08	6.53	6.85	7.07	7.21
35	E	6.03	6.54	6.94	7.27	7.55
	A	5.99	6.44	6.75	6.98	7.12

E denotes exact value

A denotes approximate value

We will first consider the evaluation of the probablity integrals of the non-central multivariate chi-squrae distribution with one degree of freedom since this is needed in the sequel.

Let $x_1, x_2, \ldots, x_p$ be jointly distributed as a p-variate normal with mean vector $\boldsymbol{\mu}' = (\mu_1, \mu_2, \ldots, \mu_p)$ and covariance matrix $C = (c_{ij})$ where $c_{ii} = 1$, $(i = 1, 2, \ldots, p)$. Also let C be such that $c_{ij} = c_i c_j$ for $i \neq j$, and $0 \leqslant c_i < 1$. Then, it is known [4] that x_i's can be expressed as

$$x_i - \mu_i = d_i y_i - c_i y_0 \qquad i = 1, 2, \ldots, p$$

where $y_1, y_2, \ldots, y_p$ and y_0 are independently distributed normal variates with zero means and unit variances and $d_i = \sqrt{1-c_i^2}$. So

$$P[z_i \leqslant h_i; \quad i = 1, 2, \ldots, p]$$

$$(3.2) \quad = P\left[\frac{-\sqrt{h_i}+c_i y_0 - \mu_i}{d_i} \leqslant y_i \leqslant \frac{\sqrt{h_i}+c_i y_0-\mu_i}{d_i};\ i=1,2,\ldots,p\right]$$

$$= \frac{1}{\sqrt{2\pi}} \int_{-\infty}^{\infty} e^{-y_0^2/2}\, \psi(y_0) dy_0$$

where

$$z_i = x_i^2$$

$$\psi(y_0) = \prod_{i=1}^{p} \int_{\lambda_i}^{\delta_i} \frac{e^{-y^2/2}}{\sqrt{2\pi}} dy,$$

$$\lambda_i = \frac{-\sqrt{h_i}+c_i y_0 - \mu_i}{d_i}, \quad \delta_i = \frac{\sqrt{h_i}+c_i y_0-\mu_i}{d_i}.$$

But $\qquad \psi(y_0) = \prod_{i=1}^{p} [I_{1i}+I_{2i}]$, where

$$I_{1i} = \pm \frac{1}{2\sqrt{\pi}} \int_0^{\delta_i^2/2} e^{-z} z^{-\frac{1}{2}}\, dz \text{ according as } \delta_i \text{ is positive or negative.}$$

and

$$I_{2i} = \pm \frac{1}{2\sqrt{\pi}} \int_0^{\lambda_i^2/2} e^{-z} z^{-\frac{1}{2}}\, dz \text{ according as } \lambda_i \text{ is negative or positive.}$$

A method is suggested in [1] for the evaluation of the incomplete gamma integrals. Using that method, we can compute $\psi(y_0)$ for any given value of y_0. Hence, using Gauss-Hermite quadrature formula, we can evaluate the integral in (3.2). Here we note that if $\mu_i = 0$, $(i = 1, 2, \ldots, p)$, then

$$\frac{1}{\sqrt{2\pi}} \int_{-\infty}^{\infty} e^{-y_0^2/2} \psi(y_0) dy_0 = \frac{\sqrt{2}}{\sqrt{\pi}} \int_{0}^{\infty} e^{-y_0^2/2} \psi(y_0) dy_0$$

since $\psi(y_0) = \psi(-y_0)$ in this special case.

We will now consider the evaluation of the probability integral of the distribution of the maximum of t^2 variates whose joint distribution is multivariate t^2. Let $W = \dfrac{m \max(z_1, \ldots, z_p)}{w^*}$ where $z_1, z_2, \ldots, z_p$ are defined earlier and w^* is a central chi-square variate with m degrees of freedom distributed independently of $z_1, z_2, \ldots, z_p$ and $E(w^*) = m$. Then

$$P[W \leqslant c] = P\left[\frac{mz_i}{w^*} \leqslant c; \quad i = 1, 2, \ldots, p\right]$$

$$= \int_0^{\infty} \frac{\exp(-w) w^{(m/2)-1}}{\Gamma(m/2)} G(w) dw \tag{3.3}$$

where

$$G(w) = \int_0^{2cw/m} \cdots \int_0^{2cw/m} f(z_1, z_2, \ldots, z_p) \prod_{i=1}^{p} dz_i \tag{3.3a}$$

and $f(z_1, z_2, \ldots, z_p)$ is the joint density function of $z_1, z_2, \ldots, z_p$. We can compute $G(w)$ for any given value of w, by using the method discussed earlier. When $p+m$ is even, the whole integral on the right side of (3.3) can be evaluated by using Gauss-Laguerre quadrature formula with the weight function e^{-x}. When $p+m$ is odd, the derivatives of the integrand (excluding e^{-x}) on the right side of (3.3), need not exist at the origin and hence the error due to approximating the integral on the right side of (3.3) by the Gauss-Laguerre quadrature formula with weight function e^{-x} may be unbounded. So, in this case, we use Generalized Gauss-Laguerre quadrature formula with the weight function $e^{-x}x^{-\frac{1}{2}}$ to compute the integral on the right side of (3.3) the zeros and weights associated with this formula are given in [17].

4. CONSTRUCTION OF TABLES

In this paper, we are interested in constructing tables of the values of c for given values of p, m, ρ, and α where

$$\int_0^\infty \frac{e^{-w}\, w^{(m/2)-1}}{\Gamma(m/2)}\, G^*(w)dw = (1-\alpha), \tag{4.1}$$

and $G^*(w)$ is equivalent to $G(w)$ when $\mu_i = 0$ for $i = 1, 2, \ldots, p$. Holding p, m and ρ fixed, we computed the values of α for different values of c by using the method discussed in the previous section. In computing these values, we used the combination of 40 point Gauss-Hermite quadrature formula and 32-point Gauss-Laguerre (or Generalized Gauss-Laguerre) quadrature formula. (Tables for 32 point Gauss-Laguerre quadrature formula are available in the literature (e.g., see [16]) and tables for 40 point Gauss-Hermite quadrature formula are given in [2]. Then, using cubic interpolation, we computed the values for c for the desired values of α. Tables of c are given at the end of this paper for $\alpha = 0{\cdot}05$, $0{\cdot}01$, $m = 5(1)35$, $p = 1(1)10$ and $\rho = 0(0{\cdot}1)0{\cdot}9$ where $\rho_{ij} = c_{ij}/[c_{ii}\, c_{jj}]^{\frac{1}{2}} = \rho$ for $i \neq j = 1, 2, \ldots, p$. (For more extensive tables, the reader is referred to a technical report by the authors [14]). The entries in these tables are correct to two decimals except for a few values which may differ from actual values by at most one unit in the second decimal. Dunnett [6] has constructed the tables of $\sqrt{c}$ for $\alpha = 0{\cdot}05$, $0{\cdot}01$, $\rho = 0{\cdot}5$ and for different values of m and p. He computed exact values for $m = 5$, 10, 20, and ∞ and interpolated the figures for intermediate values of m. His exact values, when squared, agree with the corresponding values given at the end of this paper if we take round off error into account. The values in the first columns of the tables given at the end of this paper agree with the corresponding values in the F tables. The entries in tables for $\rho = 0$ agree with the corresponding values given in [1] and hence are not reported here. Siotani [18] computed the values of $(1-\alpha)$ for $p = 2$, $|\rho| = 0{\cdot}0(0{\cdot}1)0{\cdot}9$, $0{\cdot}95$, $\sqrt{c} = 2{\cdot}0(0{\cdot}5)4{\cdot}5$ and $m = 10(2)50(5)90$, 100, 120, 150, 200, ∞. Recently, Dunn and Massey [3] constructed tables for the values of $\sqrt{c}$ when $1-\alpha = 0{\cdot}50$, $0{\cdot}60$, $0{\cdot}70$, $0{\cdot}80$, $0{\cdot}90$, $0{\cdot}95$, $0{\cdot}975$, $0{\cdot}99$ $p = 2, 6, 10, 20$, $m = 4, 10, 30, \infty$ and $\rho = 0(0{\cdot}1)1{\cdot}0$.

5. APPLICATION OF THE TABLES

Consider the model equation

$$y_t = \alpha+\beta t+\epsilon_t \tag{5.1}$$

where $\{\epsilon_t\}$ is a Gaussian, stationary hth order Markov process with zero mean value. Let $H_1 : \alpha = 0$ and $H_2 : \beta = 0$. Krishnaiah and Murthy [12] proposed a test to test H_1 and H_2 simultaneously. The tables given at the end of this paper are very useful in the application of this test procedure.

Krishnaiah [9–11] proposed test procedures for the multiple comparisons of means and mean vectors under general Analysis of Variance and Multivariate Analysis of Variance models respectively. The tables given in this paper are very useful in the application of these test procedures. We will now discuss how these tables can be used for the application of a special case of the test procedure considered in [10] for the univariate case.

Consider k normal populations with means $a_1, a_2, \ldots, a_k$ and a common variance σ^2. Let

$$H_i : \sum_{t=1}^{k} v_{it}\, a_t = 0 \qquad i = 1, 2, \ldots, q$$

where v_{it}'s are subject to the restrictions $\sum_{t=1}^{k} v_{it}^2 = 1$ for $i = 1, 2, \ldots, q$. Also, let y_{tu} denote the observed value on uth individual in tth group. In addition, let $y_{t.} = \sum_{u=1}^{N_t} y_{tu}$, $\bar{y}_{t\cdot} = y_{t\cdot}/N$ and $N = \sum_{t=1}^{k} N_t$, where N_t denotes the size of tth sample. Then, according to the test procedure in [10], we accept or reject H_i according as

$$F_i \lesseqgtr F_a$$

where

$$F_i = (N-k)s_i^2/s_0^2, \qquad i = 1, 2, \ldots, q$$

$$s_i^2 = \left(\sum_{t=1}^{k} v_{it}\,\bar{y}_{t\cdot}\right)^2 \Big/ \left(\sum_{t=1}^{k} v_{it}^2/N_t\right),$$

$$s_0^2 = \sum_{t=1}^{k} \sum_{u=1}^{N_t} (y_{tu}-\bar{y}_{t\cdot})^2,$$

and F_a is chosen such that

$$P\left[F_i \leqslant F_a\,;\ i = 1, 2, \ldots, q \,\Big|\, \bigcap_{i=1}^{q} H_i\right] = (1-\alpha). \tag{5.1}$$

When $\bigcap_{i=1}^{q} H_i$ is true, the joint distribution of $F_1, F_2, \ldots, F_q$ is a central q-variate F distribution with $(1, N-k)$ degrees of freedom and with $\mathbf{\Omega} = (\omega_{tu})$ as the covariance matrix of the "accompanying" multivariate normal; the distribution is singular or non-singular according as the rank of $\mathbf{\Omega}$ is less than or equal to q. Here, we note that

$$\omega_{ij} = \begin{cases} \sigma^2 \sum_{t=1}^{k} v_{it}^2/N_t & \text{if } i = j \\ \sigma^2 \sum_{t=1}^{k} v_{it}\, v_{jt}/N_t & \text{if } i \neq j. \end{cases}$$

If, in particular, $N_1 = N_2 = \ldots = N_k = N_0$ and $\sum_{t=1}^{k} v_{it}\, v_{jt} = b$ for $i \neq j = 1, 2, \ldots, q$, then we can obtain the upper 5% and

1% critical values F_a in the non-singular case from the tables given at the end of this paper, by letting $p = q$, $m = N-k$, $\rho = b$ and $n = 1$.

Acknowledgment

The authors are grateful to the referee for his helpful suggestions in the preparation of this paper.

References

[1] Armitage, J. V. and Krishnaiah, P. R. (1964). *Tables for the Studentized Largest Chi-Square Distribution and Their Applications.* ARL-64-188, Aerospace Research Laboratories, Wright-Patterson Air Force Base, Ohio.

[2] Baber, L., Krishnaiah, P. R. and Armitage, J. V. (1964). "New Tables for the Zeros and Weights of Hermite Polynomials. (Unpublished Manuscript).

[3] Dunn, O. J. and Massey, Jr., F. J. (1965). "Estimation of Multiple Contrasts Using t-Distributions," *J. Amer. Statist. Assoc.*, **60**, 573–583.

[4] Dunnett, C. W. and Sobel, M. (1955). "Approximations to the Probability Integral and Certain Percentage Points of a Multivariate Analogue of Student's t-Distribution," *Biometrika*, **42**, 258–260.

[5] Dunnett, C. W. (1955). "A Multiple Comparison Procedure for Comparing Several Treatments with a Control," *J. Amer. Statist. Assoc.*, **50**, 1091–1121.

[6] Dunnett, C. W. (1964). "New Tables for Multiple Comparisons with a Control," *Biometrics*, **20**, 482–491.

[7] Gupta, S. S. and Sobel, M. (1962). "On the Smallest of Several Correlated F Statistics," *Biometrika*, **49**, 509–523.

(8) Gupta, S. S. (1962). "On Selection and Ranking Procedure for Gamma Populations," *Ann. Inst. Statist. Math.*, **14**, 199–216.

[9] Krishnaiah, P. R. (1964). *Multiple Comparison Tests in Multivariate Case.* ARL 64-124, Aerospace Research Laboratories, Wright-Patterson Air Force Base, Ohio.

[10] Krishnaiah, P. R. (1965). "On the Simultaneous ANOVA and MANOVA Tests," *Ann. Inst. Statist. Math.*, **17**, 35–53.

[11] Krishnaiah, P. R. (1965). "Multiple Comparison Tests in Multi-Response Experiments." *Sankhyā*, **27**, 65–72.

[12] Krishnaiah, P. R. and Murthy, V. K. (1966). "Simultaneous Tests for Trend and Serial Correlations for Gaussian Markov Residuals," *Econometrica*, **34**, 472–480.

[13] Krishnaiah, P. R. and Armitage, J. V. (1964). *Distribution of the Studentized Smallest Chi-Square, with Tables and Applications*. ARL 64-218, Aerospace Research Laboratories, Wright-Patterson Air Force Base, Ohio.

[14] Krishnaiah, P. R. and Armitage, J. V. (1965). *Probability Integrals of the Multivariate F Distribution, with Tables and Applications*. ARL 65-236, Aerospace Research Laboratories, Wright-Patterson Air Force Base, Ohio.

[15] Krishnamoorthy, A. S. and Parthasarathy, M. (1951). "A Multivariate Gamma Type Distribution," *Ann. Math. Statist.*, **22**, 549–557.

[16] Krylov, V. I. (1962). *Approximate Calculation of Integrals*, (translated by A. H. Strowd). The Macmillan Company, New York.

[17] Shao, T. S., Chen, T. C. and Frank, R. M. (1964). "Tables of Zeros and Gaussian Weights of Certain Associated Laguerre Polynomials and the Related Generalized Hermite Polynomials," *Math. Comp.*, **18**, 598–616.

[18] Siotani, M. (1964). "Interval Estimation for Linear Combination of Means," *J. Amer. Statist. Assoc.*, **59**, 1161–1164.

(*Received Oct. 1, 1965. Revised Jan. 10, 1967.*)

TABLE IV*

Upper 5% points of the distribution of W for $\rho = 0.1$

m/p	1	2	3	4	5	6	7	8	9	10
5	6.61	9.54	11.53	13.05	14.30	15.35	16.27	17.08	17.81	18.46
6	5.99	8.49	10.17	11.45	12.49	13.37	14.13	14.81	15.42	15.97
7	5.59	7.83	9.32	10.44	11.36	12.13	12.80	13.39	13.92	14.40
8	5.32	7.38	8.73	9.75	10.58	11.28	11.88	12.42	12.89	13.33
9	5.12	7.05	8.31	9.26	10.02	10.66	11.22	11.71	12.15	12.55
10	4.96	6.80	7.99	8.88	9.60	10.20	10.72	11.18	11.59	11.97
11	4.84	6.60	7.74	8.58	9.26	9.84	10.33	10.77	11.16	11.51
12	4.75	6.45	7.53	8.35	9.00	9.55	10.02	10.43	10.80	11.14
13	4.67	6.32	7.37	8.15	8.78	9.31	9.76	10.16	10.52	10.84
14	4.60	6.21	7.23	7.99	8.60	9.11	9.54	9.93	10.27	10.59
15	4.54	6.11	7.11	7.85	8.44	8.94	9.36	9.74	10.07	10.37
16	4.49	6.04	7.01	7.73	8.31	8.79	9.21	9.57	9.90	10.19
17	4.45	5.97	6.92	7.63	8.19	8.67	9.07	9.43	9.74	10.03
18	4.41	5.91	6.85	7.54	8.09	8.56	8.95	9.30	9.61	9.89
19	4.38	5.85	6.78	7.46	8.01	8.46	8.85	9.19	9.50	9.77
20	4.35	5.81	6.72	7.39	7.93	8.37	8.76	9.09	9.39	9.66
21	4.32	5.76	6.66	7.33	7.86	8.30	8.67	9.00	9.30	9.56
22	4.30	5.73	6.62	7.27	7.79	8.23	8.60	8.92	9.21	9.48
23	4.28	5.69	6.57	7.22	7.73	8.16	8.53	8.85	9.14	9.40
24	4.26	5.66	6.53	7.17	7.68	8.11	8.47	8.79	9.07	9.33
25	4.24	5.63	6.50	7.13	7.63	8.05	8.41	8.73	9.01	9.26
26	4.22	5.60	6.46	7.09	7.59	8.01	8.36	8.67	8.95	9.20
27	4.21	5.58	6.43	7.06	7.55	7.96	8.31	8.62	8.90	9.14
28	4.19	5.56	6.40	7.02	7.51	7.92	8.27	8.58	8.85	9.09
29	4.18	5.54	6.38	6.99	7.48	7.88	8.23	8.53	8.80	9.05
30	4.17	5.52	6.35	6.96	7.45	7.85	8.19	8.49	8.76	9.00
31	4.16	5.50	6.33	6.94	7.42	7.82	8.16	8.46	8.72	8.96
32	4.15	5.48	6.31	6.91	7.39	7.79	8.12	8.42	8.68	8.92
33	4.14	5.47	6.29	6.89	7.36	7.76	8.09	8.39	8.65	8.89
34	4.13	5.45	6.27	6.87	7.34	7.73	8.07	8.36	8.62	8.85
35	4.12	5.44	6.25	6.85	7.32	7.71	8.04	8.33	8.59	8.82

* $W = m \max (z_1, \ldots, z_p)/w^*$ where $z_0, \ldots, z_p$ are independently distributed central chi-square variates with one degree of freedom and w^* is another central chi-square variate with m degrees of freedom distributed independently of z_i's. Also, $E(z_i) = 1$, for $i = 1, 2, \ldots, p$ and $E(w^*) = m$.

TABLE IV (*Contd.*)

Upper 5% points of the distribution of W for ρ = 0·2

m/p	1	2	3	4	5	6	7	8	9	10
5	6.61	9.50	11.44	12.93	14.14	15.16	16.04	16.82	17.52	18.15
6	5.99	8.46	10.10	11.35	12.36	13.21	13.95	14.60	15.18	15.71
7	5.59	7.80	9.26	10.35	11.24	11.99	12.64	13.21	13.72	14.18
8	5.32	7.35	8.68	9.68	10.48	11.16	11.74	12.26	12.72	13.13
9	5.12	7.02	8.26	9.19	9.93	10.56	11.10	11.57	12.00	12.38
10	4.96	6.78	7.94	8.81	9.51	10.10	10.61	11.05	11.45	11.81
11	4.84	6.58	7.69	8.52	9.19	9.74	10.22	10.64	11.02	11.36
12	4.75	6.42	7.49	8.29	8.93	9.46	9.92	10.32	10.68	11.00
13	4.67	6.30	7.33	8.10	8.71	9.22	9.66	10.05	10.40	10.71
14	4.60	6.19	7.19	7.94	8.53	9.03	9.45	9.83	10.16	10.46
15	4.54	6.10	7.08	7.80	8.38	8.86	9.28	9.64	9.96	10.26
16	4.49	6.02	6.98	7.68	8.25	8.72	9.12	9.48	9.79	10.08
17	4.45	5.95	6.89	7.58	8.14	8.60	8.99	9.34	9.65	9.92
18	4.41	5.89	6.81	7.50	8.04	8.49	8.87	9.21	9.52	9.79
19	4.38	5.84	6.75	7.42	7.95	8.39	8.77	9.11	9.40	9.67
20	4.35	5.79	6.69	7.35	7.87	8.31	8.68	9.01	9.30	9.56
21	4.32	5.75	6.63	7.29	7.80	8.23	8.60	8.92	9.21	9.47
22	4.30	5.71	6.59	7.23	7.74	8.16	8.53	8.85	9.13	9.38
23	4.28	5.67	6.54	7.18	7.68	8.10	8.46	8.77	9.05	9.31
24	4.26	5.64	6.50	7.13	7.63	8.05	8.40	8.71	8.99	9.24
25	4.24	5.62	6.47	7.09	7.59	8.00	8.35	8.65	8.93	9.17
26	4.22	5.59	6.43	7.05	7.54	7.95	8.30	8.60	8.87	9.11
27	4.21	5.57	6.40	7.02	7.50	7.91	8.25	8.55	8.82	9.06
28	4.19	5.54	6.38	6.98	7.47	7.87	8.21	8.51	8.77	9.01
29	4.18	5.52	6.35	6.95	7.43	7.83	8.17	8.46	8.73	8.96
30	4.17	5.50	6.33	6.93	7.40	7.79	8.13	8.42	8.69	8.92
31	4.16	5.49	6.30	6.90	7.37	7.76	8.10	8.39	8.65	8.88
32	4.15	5.47	6.28	6.88	7.34	7.73	8.06	8.35	8.61	8.84
33	4.14	5.45	6.26	6.85	7.32	7.70	8.03	8.32	8.58	8.81
34	4.13	5.44	6.24	6.83	7.29	7.68	8.01	8.29	8.55	8.78
35	4.12	5.43	6.23	6.81	7.27	7.65	7.98	8.27	8.52	8.75

TABLE IV (*Contd.*)

Upper 5% points of the distribution of W for $\rho = 0{\cdot}3$

m/p	1	2	3	4	5	6	7	8	9	10
5	6.61	9.42	11.30	12.73	13.88	14.85	15.68	16.42	17.08	17.67
6	5.99	8.40	9.98	11.18	12.15	12.96	13.66	14.27	14.82	15.32
7	5.59	7.75	9.16	10.21	11.06	11.77	12.39	12.93	13.41	13.84
8	5.32	7.30	8.59	9.55	10.32	10.97	11.52	12.01	12.45	12.84
9	5.12	6.98	8.18	9.07	9.78	10.38	10.89	11.35	11.75	12.11
10	4.96	6.74	7.87	8.71	9.38	9.94	10.42	10.84	11.22	11.56
11	4.84	6.54	7.62	8.42	9.06	9.59	10.05	10.45	10.81	11.13
12	4.75	6.39	7.43	8.19	8.81	9.32	9.75	10.14	10.48	10.79
13	4.67	6.26	7.26	8.01	8.60	9.09	9.51	9.88	10.21	10.51
14	4.60	6.15	7.13	7.85	8.42	8.90	9.31	9.66	9.98	10.27
15	4.54	6.06	7.02	7.72	8.27	8.74	9.13	9.48	9.79	10.07
16	4.49	5.99	6.92	7.60	8.15	8.60	8.99	9.33	9.63	9.90
17	4.45	5.92	6.83	7.51	8.04	8.48	8.86	9.19	9.49	9.75
18	4.41	5.86	6.76	7.42	7.94	8.38	8.75	9.07	9.36	9.62
19	4.38	5.81	6.69	7.34	7.86	8.28	8.65	8.97	9.25	9.51
20	4.35	5.76	6.63	7.27	7.78	8.20	8.56	8.87	9.15	9.40
21	4.32	5.72	6.58	7.21	7.71	8.13	8.48	8.79	9.07	9.31
22	4.30	5.68	6.54	7.16	7.65	8.06	8.41	8.72	8.99	9.23
23	4.28	5.65	6.49	7.11	7.60	8.00	8.35	8.65	8.92	9.16
24	4.26	5.62	6.45	7.07	7.55	7.95	8.29	8.59	8.85	9.09
25	4.24	5.59	6.42	7.02	7.50	7.90	8.24	8.53	8.79	9.03
26	4.22	5.56	6.39	6.99	7.46	7.85	8.19	8.48	8.74	8.97
27	4.21	5.54	6.36	6.95	7.42	7.81	8.14	8.43	8.69	8.92
28	4.19	5.52	6.33	6.92	7.39	7.77	8.10	8.39	8.64	8.87
29	4.18	5.50	6.30	6.89	7.35	7.74	8.06	8.35	8.60	8.83
30	4.17	5.48	6.28	6.86	7.32	7.70	8.03	8.31	8.56	8.79
31	4.16	5.46	6.26	6.84	7.29	7.67	8.00	8.28	8.53	8.75
32	4.15	5.44	6.24	6.81	7.27	7.64	7.96	8.24	8.49	8.71
33	4.14	5.43	6.22	6.79	7.24	7.62	7.94	8.21	8.46	8.68
34	4.13	5.41	6.20	6.77	7.22	7.59	7.91	8.18	8.43	8.65
35	4.12	5.40	6.18	6.75	7.20	7.57	7.88	8.16	8.40	8.62

TABLE IV (*Contd.*)

Upper 5% points of the distribution of W for $\rho = 0{\cdot}4$

m/p	1	2	3	4	5	6	7	8	9	10
5	6.61	9.32	11.11	12.45	13.53	14.43	15.20	15.88	16.50	17.03
6	5.99	8.31	9.82	10.95	11.86	12.61	13.26	13.83	14.34	14.80
7	5.59	7.67	9.01	10.01	10.81	11.48	12.05	12.55	13.00	13.40
8	5.32	7.24	8.46	9.37	10.10	10.70	11.22	11.67	12.08	12.44
9	5.12	6.92	8.06	8.91	9.58	10.14	10.62	11.04	11.42	11.75
10	4.96	6.68	7.76	8.55	9.19	9.72	10.17	10.56	10.91	11.23
11	4.84	6.49	7.52	8.28	8.88	9.38	9.81	10.19	10.52	10.82
12	4.75	6.33	7.33	8.06	8.64	9.12	9.53	9.89	10.21	10.50
13	4.67	6.21	7.17	7.88	8.44	8.90	9.30	9.64	9.95	10.23
14	4.60	6.10	7.04	7.73	8.27	8.72	9.10	9.44	9.74	10.01
15	4.54	6.02	6.93	7.60	8.13	8.56	8.94	9.26	9.55	9.82
16	4.49	5.94	6.83	7.49	8.00	8.43	8.80	9.12	9.40	9.65
17	4.45	5.87	6.75	7.39	7.90	8.32	8.67	8.99	9.26	9.51
18	4.41	5.81	6.68	7.31	7.81	8.22	8.57	8.87	9.15	9.39
19	4.38	5.76	6.61	7.23	7.72	8.13	8.47	8.77	9.04	9.28
20	4.35	5.72	6.56	7.17	7.65	8.05	8.39	8.69	8.95	9.18
21	4.32	5.68	6.51	7.11	7.59	7.98	8.31	8.61	8.87	9.10
22	4.30	5.64	6.46	7.06	7.53	7.92	8.25	8.54	8.79	9.02
23	4.28	5.61	6.42	7.01	7.48	7.86	8.19	8.47	8.72	8.95
24	4.26	5.58	6.38	6.97	7.43	7.81	8.13	8.41	8.66	8.89
25	4.24	5.55	6.35	6.93	7.38	7.76	8.08	8.36	8.61	8.83
26	4.22	5.52	6.32	6.89	7.34	7.72	8.03	8.31	8.56	8.78
27	4.21	5.50	6.29	6.86	7.31	7.68	7.99	8.27	8.51	8.73
28	4.19	5.48	6.26	6.83	7.27	7.64	7.95	8.22	8.47	8.68
29	4.18	5.46	6.23	6.80	7.24	7.60	7.92	8.19	8.42	8.64
30	4.17	5.44	6.21	6.77	7.21	7.57	7.88	8.15	8.39	8.60
31	4.16	5.42	6.19	6.75	7.18	7.54	7.85	8.12	8.35	8.56
32	4.15	5.41	6.17	6.72	7.16	7.52	7.82	8.09	8.32	8.53
33	4.14	5.39	6.15	6.70	7.13	7.49	7.79	8.06	8.29	8.50
34	4.13	5.38	6.13	6.68	7.11	7.47	7.77	8.03	8.26	9.47
35	4.12	5.36	6.12	6.66	7.09	7.44	7.74	8.00	8.23	8.44

TABLE IV (*Contd.*)

Upper 5% points of the distribution of W for $\rho = 0{\cdot}5$

m/p	1	2	3	4	5	6	7	8	9	10
5	6.61	9.18	10.84	12.08	13.07	13.89	14.60	15.21	15.77	16.25
6	5.99	8.20	9.61	10.65	11.48	12.17	12.76	13.28	13.74	14.16
7	5.59	7.57	8.83	9.75	10.49	11.10	11.62	12.07	12.48	12.84
8	5.32	7.14	8.29	9.14	9.81	10.36	10.84	11.25	11.62	11.95
9	5.12	6.83	7.91	8.69	9.31	9.83	10.27	10.65	11.00	11.30
10	4.96	6.60	7.61	8.35	8.94	9.43	9.84	10.20	10.52	10.81
11	4.84	6.41	7.38	8.09	8.65	9.11	9.51	9.85	10.16	10.43
12	4.75	6.26	7.20	7.88	8.42	8.86	9.24	9.57	9.86	10.12
13	4.67	6.14	7.05	7.71	8.23	8.65	9.02	9.34	9.62	9.87
14	4.60	6.04	6.92	7.56	8.07	8.48	8.84	9.15	9.42	9.66
15	4.54	5.95	6.81	7.44	7.93	8.34	8.68	8.98	9.25	9.49
16	4.49	5.87	6.72	7.33	7.81	8.21	8.55	8.84	9.10	9.34
17	4.45	5.81	6.64	7.24	7.71	8.10	8.43	8.72	8.98	9.20
18	4.41	5.75	6.57	7.16	7.63	8.01	8.33	8.61	8.86	9.09
19	4.38	5.70	6.51	7.09	7.55	7.92	8.24	8.52	8.77	8.99
20	4.35	5.66	6.45	7.03	7.48	7.85	8.16	8.44	8.68	8.90
21	4.32	5.62	6.40	6.97	7.42	7.78	8.09	8.36	8.60	8.82
22	4.30	5.58	6.36	6.92	7.36	7.72	8.03	8.30	8.53	8.74
23	4.28	5.55	6.32	6.87	7.31	7.67	7.97	8.24	8.47	8.68
24	4.26	5.52	6.28	6.83	7.26	7.62	7.92	8.18	8.41	8.62
25	4.24	5.49	6.25	6.80	7.22	7.57	7.87	8.13	8.36	8.56
26	4.22	5.47	6.22	6.76	7.18	7.53	7.83	8.08	8.31	8.52
27	4.21	5.44	6.19	6.73	7.15	7.49	7.79	8.04	8.27	8.47
28	4.19	5.42	6.16	6.70	7.12	7.46	7.75	8.00	8.23	8.43
29	4.18	5.40	6.14	6.67	7.09	7.43	7.72	7.97	8.19	8.39
30	4.17	5.39	6.12	6.65	7.06	7.40	7.68	7.93	8.15	8.35
31	4.16	5.37	6.10	6.62	7.03	7.37	7.65	7.90	8.12	8.32
32	4.15	5.35	6.08	6.60	7.01	7.34	7.63	7.87	8.09	8.29
33	4.14	5.34	6.06	6.58	6.99	7.32	7.60	7.85	8.06	8.26
34	4.13	5.32	6.04	6.56	6.96	7.30	7.58	7.82	8.04	8.23
35	4.12	5.31	6.03	6.54	6.94	7.27	7.55	7.80	8.01	8.20

TABLE IV (*Contd.*)

Upper 5% points of the distribution of W for $\rho = 0{\cdot}6$

m/p	1	2	3	4	5	6	7	8	9	10
5	6.61	9.00	10.51	11.62	12.50	13.23	13.86	14.40	14.88	15.31
6	5.99	8.04	9.33	10.27	11.01	11.63	12.15	12.61	13.01	13.37
7	5.59	7.44	8.58	9.42	10.08	10.62	11.08	11.49	11.84	12.16
8	5.32	7.02	8.07	8.84	9.44	9.93	10.36	10.72	11.05	11.34
9	5.12	6.72	7.70	8.42	8.98	9.44	9.83	10.17	10.47	10.74
10	4.96	6.49	7.42	8.10	8.63	9.06	9.43	9.75	10.04	10.29
11	4.84	6.31	7.20	7.85	8.35	8.77	9.12	9.43	9.70	9.94
12	4.75	6.16	7.03	7.65	8.13	8.53	8.87	9.17	9.43	9.66
13	4.67	6.05	6.88	7.48	7.95	8.34	8.67	8.95	9.21	9.43
14	4.60	5.95	6.76	7.35	7.80	8.18	8.50	8.77	9.02	9.24
15	4.54	5.86	6.66	7.23	7.68	8.04	8.35	8.62	8.86	9.08
16	4.49	5.79	6.57	7.13	7.57	7.93	8.23	8.49	8.73	8.94
17	4.45	5.72	6.49	7.04	7.47	7.82	8.12	8.38	8.61	8.81
18	4.41	5.67	6.43	6.97	7.39	7.74	8.03	8.28	8.51	8.71
19	4.38	5.62	6.37	6.90	7.32	7.66	7.95	8.20	8.42	8.61
20	4.35	5.58	6.31	6.84	7.25	7.59	7.87	8.12	8.34	8.53
21	4.32	5.54	6.27	6.79	7.19	7.53	7.81	8.05	8.27	8.46
22	4.30	5.50	6.22	6.74	7.14	7.47	7.75	7.99	8.20	8.39
23	4.28	5.47	6.19	6.70	7.09	7.42	7.69	7.93	8.14	8.33
24	4.26	5.44	6.15	6.66	7.05	7.37	7.64	7.88	8.09	8.28
25	4.24	5.42	6.12	6.62	7.01	7.33	7.60	7.83	8.04	8.23
26	4.22	5.39	6.09	6.59	6.98	7.29	7.56	7.79	8.00	8.18
27	4.21	5.37	6.06	6.56	6.94	7.26	7.52	7.75	7.96	8.14
28	4.19	5.35	6.04	6.53	6.91	7.22	7.49	7.72	7.92	8.10
29	4.18	5.33	6.02	6.50	6.88	7.19	7.46	7.68	7.88	8.06
30	4.17	5.31	5.99	6.48	6.86	7.17	7.43	7.65	7.85	8.03
31	4.16	5.30	5.97	6.46	6.83	7.14	7.40	7.62	7.82	8.00
32	4.15	5.28	5.96	6.44	6.81	7.11	7.37	7.60	7.79	7.97
33	4.14	5.27	5.94	6.42	6.79	7.09	7.35	7.57	7.77	7.94
34	4.13	5.25	5.92	6.40	6.77	7.07	7.33	7.55	7.74	7.92
35	4.12	5.24	5.91	6.38	6.75	7.05	7.30	7.53	7.72	7.89

TABLE IV (*Contd.*)

Upper 5% points of the distribution of W for $\rho = 0{\cdot}7$

m/p	1	2	3	4	5	6	7	8	9	10
5	6.61	8.76	10.08	11.04	11.80	12.42	12.94	13.40	13.80	14.17
6	5.99	7.84	8.97	9.78	10.42	10.95	11.39	11.78	12.12	12.43
7	5.59	7.26	8.27	8.99	9.56	10.03	10.42	10.76	11.07	11.34
8	5.32	6.86	7.79	8.45	8.97	9.40	9.76	10.07	10.35	10.60
9	5.12	6.57	7.44	8.06	8.55	8.95	9.28	9.57	9.83	10.06
10	4.96	6.35	7.17	7.77	8.22	8.60	8.92	9.19	9.44	9.65
11	4.84	6.17	6.97	7.53	7.97	8.33	8.64	8.90	9.13	9.34
12	4.75	6.03	6.80	7.35	7.77	8.12	8.41	8.66	8.88	9.08
13	4.67	5.92	6.66	7.19	7.60	7.94	8.22	8.47	8.68	8.88
14	4.60	5.82	6.55	7.06	7.47	7.79	8.07	8.31	8.52	8.70
15	4.54	5.74	6.45	6.96	7.35	7.67	7.94	8.17	8.37	8.56
16	4.49	5.67	6.37	6.86	7.25	7.56	7.82	8.05	8.25	8.43
17	4.45	5.61	6.30	6.78	7.16	7.47	7.72	7.95	8.15	8.32
18	4.41	5.56	6.23	6.71	7.08	7.38	7.64	7.86	8.05	8.23
19	4.38	5.51	6.18	6.65	7.02	7.31	7.56	7.78	7.97	8.14
20	4.35	5.47	6.13	6.59	6.95	7.25	7.50	7.71	7.90	8.07
21	4.32	5.43	6.08	6.54	6.90	7.19	7.44	7.65	7.83	8.00
22	4.30	5.40	6.04	6.50	6.85	7.14	7.38	7.59	7.77	7.94
23	4.28	5.37	6.01	6.46	6.81	7.09	7.33	7.54	7.72	7.88
24	4.26	5.34	5.97	6.42	6.77	7.05	7.29	7.49	7.67	7.83
25	4.24	5.31	5.94	6.39	6.73	7.01	7.25	7.45	7.63	7.79
26	4.22	5.29	5.92	6.36	6.70	6.98	7.21	7.41	7.59	7.75
27	4.21	5.27	5.89	6.33	6.67	6.94	7.18	7.38	7.55	7.71
28	4.19	5.25	5.87	6.30	6.64	6.91	7.14	7.34	7.52	7.67
29	4.18	5.23	5.84	6.28	6.61	6.89	7.12	7.31	7.49	7.64
30	4.17	5.21	5.82	6.26	6.59	6.86	7.09	7.28	7.46	7.61
31	4.16	5.20	5.81	6.23	6.57	6.84	7.06	7.26	7.43	7.58
32	4.15	5.18	5.79	6.22	6.55	6.81	7.04	7.23	7.41	7.56
33	4.14	5.17	5.77	6.20	6.53	6.79	7.02	7.21	7.38	7.53
34	4.13	5.16	5.76	6.18	6.51	6.77	7.00	7.19	7.36	7.51
35	4.12	5.14	5.74	6.16	6.49	6.75	6.98	7.17	7.34	7.49

TABLE IV (*Contd.*)

Upper 5% points of the distribution of W for $\rho = 0{\cdot}8$

m/p	1	2	3	4	5	6	7	8	9	10
5	6.61	8.43	9.52	10.30	10.90	11.39	11.80	12.16	12.47	12.76
6	5.99	7.56	8.50	9.16	9.67	10.09	10.44	10.74	11.01	11.25
7	5.59	7.01	7.85	8.44	8.90	9.27	9.58	9.86	10.09	10.31
8	5.32	6.63	7.41	7.95	8.37	8.71	9.00	9.25	9.46	9.66
9	5.12	6.36	7.08	7.60	7.99	8.31	8.58	8.81	9.01	9.19
10	4.96	6.15	6.84	7.33	7.70	8.00	8.26	8.48	8.67	8.84
11	4.84	5.99	6.65	7.11	7.47	7.76	8.01	8.22	8.40	8.57
12	4.75	5.85	6.49	6.94	7.29	7.57	7.81	8.01	8.19	8.35
13	4.67	5.74	6.37	6.80	7.14	7.41	7.64	7.84	8.01	8.16
14	4.60	5.65	6.26	6.69	7.02	7.28	7.50	7.70	7.86	8.01
15	4.54	5.58	6.17	6.59	6.91	7.17	7.39	7.57	7.74	7.88
16	4.49	5.51	6.09	6.50	6.82	7.07	7.29	7.47	7.63	7.77
17	4.45	5.45	6.03	6.43	6.74	6.99	7.20	7.38	7.54	7.68
18	4.41	5.40	5.97	6.37	6.67	6.92	7.12	7.30	7.46	7.60
19	4.38	5.36	5.92	6.31	6.61	6.85	7.06	7.23	7.39	7.52
20	4.35	5.32	5.87	6.26	6.56	6.80	7.00	7.17	7.32	7.46
21	4.32	5.28	5.83	6.21	6.51	6.74	6.94	7.12	7.26	7.40
22	4.30	5.25	5.79	6.17	6.46	6.70	6.90	7.07	7.21	7.35
23	4.28	5.22	5.76	6.14	6.42	6.66	6.85	7.02	7.17	7.30
24	4.26	5.19	5.73	6.10	6.39	6.62	6.81	6.98	7.13	7.25
25	4.24	5.17	5.70	6.07	6.35	6.58	6.78	6.94	7.09	7.21
26	4.22	5.15	5.67	6.04	6.32	6.55	6.74	6.91	7.05	7.18
27	4.21	5.13	5.65	6.02	6.30	6.52	6.71	6.88	7.02	7.15
28	4.19	5.11	5.63	5.99	6.27	6.50	6.68	6.85	6.99	7.11
29	4.18	5.09	5.61	5.97	6.25	6.47	6.66	6.82	6.96	7.09
30	4.17	5.07	5.59	5.95	6.23	6.45	6.63	6.79	6.94	7.06
31	4.16	5.06	5.57	5.93	6.20	6.43	6.61	6.77	6.91	7.04
32	4.15	5.04	5.56	5.91	6.19	6.41	6.59	6.75	6.89	7.01
33	4.14	5.03	5.54	5.90	6.17	6.39	6.57	6.73	6.87	6.99
34	4.13	5.02	5.53	5.88	6.15	6.37	6.55	6.71	6.85	6.97
35	4.12	5.01	5.51	5.87	6.14	6.35	6.54	6.69	6.83	6.95

TABLE IV (*Contd.*)

Upper 5% points of the distribution of W for $\rho = 0{\cdot}9$

m/p	1	2	3	4	5	6	7	8	9	10
5	6.61	7.96	8.73	9.28	9.67	10.03	10.28	10.55	10.72	10.95
6	5.99	7.16	7.81	8.27	8.59	8.89	9.09	9.32	9.45	9.64
7	5.59	6.65	7.25	7.67	7.98	8.25	8.45	8.66	8.79	8.98
8	5.32	6.30	6.85	7.24	7.52	7.76	7.94	8.12	8.24	8.39
9	5.12	6.05	6.57	6.93	7.20	7.42	7.60	7.76	7.89	8.02
10	4.96	5.85	6.35	6.69	6.95	7.16	7.33	7.48	7.61	7.73
11	4.84	5.70	6.18	6.51	6.76	6.96	7.12	7.27	7.39	7.50
12	4.75	5.58	6.04	6.36	6.60	6.80	6.96	7.10	7.22	7.32
13	4.67	5.48	5.93	6.24	6.47	6.66	6.82	6.95	7.07	7.17
14	4.60	5.39	5.83	6.14	6.37	6.55	6.70	6.83	6.95	7.05
15	4.54	5.32	5.75	6.05	6.28	6.46	6.61	6.73	6.85	6.95
16	4.49	5.26	5.68	5.98	6.20	6.37	6.52	6.65	6.76	6.86
17	4.45	5.20	5.62	5.91	6.13	6.30	6.45	6.57	6.68	6.78
18	4.41	5.16	5.57	5.86	6.07	6.24	6.39	6.51	6.61	6.71
19	4.38	5.12	5.53	5.81	6.02	6.19	6.33	6.45	6.56	6.65
20	4.35	5.08	5.48	5.76	5.97	6.14	6.28	6.40	6.50	6.59
21	4.32	5.05	5.45	5.72	5.93	6.10	6.23	6.35	6.46	6.55
22	4.30	5.02	5.41	5.69	5.89	6.06	6.19	6.31	6.41	6.50
23	4.28	4.99	5.38	5.65	5.86	6.02	6.16	6.27	6.37	6.46
24	4.26	4.97	5.36	5.62	5.83	5.99	6.12	6.24	6.34	6.43
25	4.24	4.94	5.33	5.60	5.80	5.96	6.09	6.21	6.31	6.40
26	4.22	4.92	5.31	5.57	5.77	5.93	6.06	6.18	6.28	6.37
27	4.21	4.90	5.29	5.55	5.75	5.91	6.04	6.15	6.25	6.34
28	4.19	4.89	5.27	5.53	5.73	5.88	6.02	6.13	6.23	6.31
29	4.18	4.87	5.25	5.51	5.70	5.86	5.99	6.11	6.20	6.29
30	4.17	4.85	5.23	5.49	5.69	5.84	5.97	6.08	6.18	6.27
31	4.16	4.84	5.22	5.47	5.67	5.82	5.95	6.06	6.16	6.25
32	4.15	4.83	5.20	5.46	5.65	5.81	5.94	6.05	6.14	6.23
33	4.14	4.81	5.19	5.44	5.64	5.79	5.92	6.03	6.12	6.21
34	4.13	4.80	5.17	5.43	5.62	5.78	5.90	6.01	6.11	6.19
35	4.12	4.79	5.16	5.42	5.61	5.76	5.89	6.00	6.09	6.18

TABLE V

Upper 1% points of the distribution of W for ρ = 0·1

m/p	1	2	3	4	5	6	7	8	9	10
5	16.26	22.07	26.01	29.09	31.55	33.75	35.54	37.28	38.70	40.15
6	13.75	18.22	21.22	23.52	25.38	26.97	28.33	29.57	30.62	31.65
7	12.25	15.97	18.43	20.30	21.82	23.11	24.22	25.21	26.09	26.91
8	11.26	14.50	16.62	18.22	19.52	20.62	21.57	22.41	23.17	23.85
9	10.56	13.47	15.36	16.78	17.93	18.90	19.74	20.48	21.14	21.74
10	10.04	12.71	14.43	15.72	16.76	17.64	18.40	19.06	19.66	20.21
11	9.65	12.13	13.73	14.92	15.88	16.68	17.38	17.99	18.54	19.04
12	9.33	11.68	13.17	14.28	15.18	15.93	16.58	17.15	17.66	18.12
13	9.07	11.31	12.72	13.77	14.62	15.32	15.93	16.47	16.95	17.37
14	8.86	11.00	12.35	13.35	14.16	14.83	15.41	15.92	16.37	16.78
15	8.68	10.75	12.04	13.00	13.77	14.41	14.97	15.45	15.89	16.28
16	8.53	10.53	11.78	12.71	13.44	14.06	14.59	15.06	15.48	15.85
17	8.40	10.34	11.56	12.45	13.16	13.76	14.27	14.72	15.12	15.49
18	8.28	10.18	11.36	12.23	12.92	13.50	13.99	14.43	14.82	15.17
19	8.18	10.04	11.19	12.03	12.71	13.27	13.75	14.17	14.55	14.89
20	8.10	9.91	11.04	11.86	12.52	13.07	13.54	13.95	14.32	14.65
21	8.02	9.80	10.90	11.71	12.35	12.89	13.35	13.75	14.11	14.43
22	7.95	9.70	10.78	11.57	12.20	12.73	13.18	13.57	13.92	14.24
23	7.88	9.61	10.67	11.45	12.07	12.58	13.02	13.41	13.75	14.06
24	7.82	9.53	10.57	11.34	11.95	12.45	12.88	13.26	13.60	13.91
25	7.77	9.45	10.48	11.24	11.84	12.33	12.76	13.13	13.46	13.76
26	7.72	9.38	10.40	11.15	11.74	12.22	12.64	13.01	13.34	13.63
27	7.68	9.32	10.33	11.06	11.64	12.13	12.54	12.90	13.22	13.51
28	7.64	9.26	10.26	10.98	11.56	12.03	12.44	12.80	13.12	13.40
29	7.60	9.21	10.19	10.91	11.48	11.95	12.35	12.70	13.02	13.30
30	7.56	9.16	10.13	10.85	11.41	11.87	12.27	12.62	12.93	13.21
31	7.53	9.11	10.08	10.78	11.34	11.80	12.19	12.54	12.84	13.12
32	7.50	9.07	10.03	10.73	11.28	11.73	12.12	12.46	12.77	13.04
33	7.47	9.03	9.98	10.67	11.22	11.67	12.06	12.39	12.69	12.96
34	7.44	8.99	9.94	10.62	11.16	11.61	11.99	12.33	12.63	12.89
35	7.42	8.96	9.90	10.58	11.11	11.56	11.94	12.27	12.56	12.83

TABLE V (*Contd.*)

Upper 1% points of the distribution of W for $\rho = 0{\cdot}2$

m/p	1	2	3	4	5	6	7	8	9	10
5	16.26	21.99	25.85	28.86	31.24	33.37	35.09	36.78	38.11	39.53
6	13.75	18.17	21.11	23.35	25.17	26.71	28.02	29.22	30.24	31.23
7	12.25	15.93	18.34	20.17	21.65	22.90	23.98	24.94	25.79	26.58
8	11.26	14.46	16.55	18.12	19.39	20.46	21.38	22.20	22.93	23.59
9	10.56	13.44	15.30	16.69	17.82	18.76	19.57	20.29	20.94	21.52
10	10.04	12.69	14.38	15.65	16.66	17.52	18.26	18.91	19.49	20.01
11	9.65	12.11	13.68	14.85	15.79	16.58	17.25	17.85	18.39	18.87
12	9.33	11.65	13.13	14.22	15.10	15.83	16.47	17.02	17.52	17.97
13	9.07	11.29	12.68	13.72	14.55	15.24	15.83	16.36	16.82	17.24
14	8.86	10.98	12.32	13.30	14.09	14.75	15.31	15.81	16.26	16.66
15	8.68	10.73	12.01	12.96	13.71	14.34	14.88	15.36	15.78	16.16
16	8.53	10.51	11.75	12.66	13.39	13.99	14.51	14.97	15.38	15.74
17	8.40	10.33	11.53	12.41	13.11	13.70	14.20	14.64	15.03	15.39
18	8.28	10.16	11.33	12.19	12.87	13.44	13.92	14.35	14.73	15.08
19	8.18	10.02	11.16	12.00	12.66	13.21	13.68	14.10	14.47	14.80
20	8.10	9.90	11.01	11.83	12.47	13.01	13.47	13.88	14.24	14.56
21	8.02	9.78	10.88	11.68	12.31	12.84	13.29	13.68	14.03	14.35
22	7.95	9.68	10.76	11.54	12.16	12.68	13.12	13.51	13.85	14.16
23	7.88	9.59	10.65	11.42	12.03	12.53	12.97	13.35	13.68	13.99
24	7.82	9.51	10.55	11.31	11.91	12.41	12.83	13.20	13.54	13.83
25	7.77	9.44	10.46	11.21	11.80	12.29	12.71	13.07	13.40	13.69
26	7.72	9.37	10.38	11.12	11.70	12.18	12.59	12.96	13.28	13.57
27	7.68	9.31	10.31	11.03	11.61	12.08	12.49	12.85	13.16	13.45
28	7.64	9.25	10.24	10.96	11.52	11.99	12.40	12.75	13.06	13.34
29	7.60	9.20	10.17	10.89	11.45	11.91	12.31	12.65	12.96	13.24
30	7.56	9.15	10.12	10.82	11.37	11.83	12.23	12.57	12.87	13.15
31	7.53	9.10	10.06	10.76	11.31	11.76	12.15	12.49	12.79	13.06
32	7.50	9.06	10.01	10.70	11.25	11.70	12.08	12.42	12.71	12.98
33	7.47	9.02	9.96	10.65	11.19	11.63	12.02	12.35	12.64	12.91
34	7.44	8.98	9.92	10.60	11.13	11.58	11.95	12.28	12.58	12.84
35	7.42	8.95	9.88	10.55	11.08	11.52	11.90	12.22	12.51	12.78

TABLE V (*Contd.*)

Upper 1% points of the distribution of W for $\rho = 0{\cdot}3$

m/p	1	2	3	4	5	6	7	8	9	10
5	16.26	21.86	25.59	28.48	30.74	32.78	34.38	36.00	37.21	38.58
6	13.75	18.07	20.93	23.08	24.82	26.29	27.54	28.68	29.65	30.59
7	12.25	15.85	18.20	19.96	21.38	22.58	23.61	24.52	25.33	26.06
8	11.26	14.40	16.43	17.95	19.17	20.19	21.07	21.85	22.54	23.17
9	10.56	13.39	15.20	16.55	17.63	18.53	19.31	20.00	20.61	21.17
10	10.04	12.64	14.29	15.52	16.50	17.32	18.03	18.65	19.21	19.71
11	9.65	12.07	13.60	14.74	15.64	16.40	17.05	17.63	18.14	18.60
12	9.33	11.62	13.06	14.12	14.97	15.68	16.29	16.82	17.30	17.73
13	9.07	11.25	12.62	13.62	14.43	15.10	15.67	16.17	16.62	17.02
14	8.86	10.95	12.25	13.22	13.98	14.62	15.16	15.64	16.07	16.45
15	8.68	10.70	11.95	12.87	13.61	14.22	14.74	15.20	15.61	15.97
16	8.53	10.48	11.69	12.58	13.29	13.88	14.38	14.82	15.21	15.57
17	8.40	10.30	11.47	12.34	13.02	13.59	14.07	14.50	14.88	15.22
18	8.28	10.14	11.28	12.12	12.78	13.33	13.81	14.22	14.59	14.92
19	8.18	10.00	11.11	11.93	12.58	13.11	13.57	13.97	14.33	14.65
20	8.10	9.87	10.96	11.76	12.39	12.92	13.37	13.76	14.11	14.42
21	8.02	9.76	10.83	11.61	12.23	12.75	13.18	13.57	13.91	14.21
22	7.95	9.66	10.71	11.48	12.09	12.59	13.02	13.39	13.73	14.03
23	7.88	9.57	10.61	11.36	11.96	12.45	12.87	13.24	13.57	13.86
24	7.82	9.49	10.51	11.25	11.84	12.32	12.74	13.10	13.42	13.71
25	7.77	9.42	10.42	11.15	11.73	12.21	12.62	12.97	13.29	13.58
26	7.72	9.35	10.34	11.06	11.63	12.11	12.51	12.86	13.17	13.45
27	7.68	9.29	10.27	10.98	11.54	12.01	12.41	12.75	13.06	13.34
28	7.64	9.23	10.20	10.91	11.46	11.92	12.31	12.65	12.96	13.23
29	7.60	9.18	10.14	10.84	11.39	11.84	12.23	12.57	12.87	13.13
30	7.56	9.13	10.08	10.77	11.32	11.76	12.15	12.48	12.78	13.05
31	7.53	9.08	10.03	10.71	11.25	11.69	12.07	12.40	12.70	12.96
32	7.50	9.04	9.98	10.66	11.19	11.63	12.01	12.33	12.62	12.88
33	7.47	9.00	9.93	10.60	11.13	11.57	11.94	12.27	12.55	12.81
34	7.44	8.97	9.89	10.55	11.08	11.51	11.88	12.20	12.49	12.74
35	7.42	8.93	9.85	10.51	11.03	11.46	11.82	12.14	12.43	12.68

TABLE V (*Contd.*)

Upper 1% points of the distribution of W for $\rho = 0{\cdot}4$

m/p	1	2	3	4	5	6	7	8	9	10
5	16.26	21.67	25.22	27.95	30.06	31.96	33.44	34.95	36.04	37.33
6	13.75	17.94	20.66	22.70	24.34	25.72	26.89	27.95	28.85	29.71
7	12.25	15.74	17.99	19.66	21.01	22.13	23.09	23.94	24.70	25.38
8	11.26	14.31	16.25	17.70	18.86	19.82	20.65	21.37	22.02	22.61
9	10.56	13.31	15.05	16.38	17.36	18.22	18.95	19.60	20.17	20.69
10	10.04	12.57	14.16	15.33	16.27	17.05	17.71	18.30	18.82	19.29
11	9.65	12.00	13.48	14.57	15.43	16.15	16.77	17.31	17.79	18.22
12	9.33	11.56	12.95	13.97	14.78	15.45	16.03	16.53	16.98	17.39
13	9.07	11.20	12.52	13.48	14.25	14.89	15.43	15.91	16.33	16.71
14	8.86	10.90	12.16	13.09	13.82	14.42	14.94	15.40	15.80	16.16
15	8.68	10.65	11.86	12.75	13.45	14.04	14.53	14.97	15.36	15.70
16	8.53	10.44	11.61	12.47	13.15	13.71	14.19	14.61	14.98	15.31
17	8.40	10.25	11.39	12.23	12.88	13.43	13.89	14.30	14.66	14.98
18	8.28	10.10	11.21	12.02	12.65	13.18	13.63	14.03	14.37	14.69
19	8.18	9.96	11.04	11.83	12.45	12.97	13.41	13.79	14.13	14.44
20	8.10	9.83	10.89	11.67	12.27	12.78	13.21	13.58	13.91	14.21
21	8.02	9.72	10.76	11.52	12.12	12.61	13.03	13.40	13.72	14.01
22	7.95	9.62	10.65	11.39	11.98	12.46	12.87	13.23	13.55	13.83
23	7.88	9.54	10.54	11.27	11.85	12.32	12.73	13.08	13.39	13.67
24	7.82	9.46	10.45	11.17	11.73	12.20	12.60	12.95	13.25	13.53
25	7.77	9.38	10.36	11.07	11.63	12.09	12.48	12.82	13.13	13.40
26	7.72	9.32	10.28	10.98	11.53	11.99	12.37	12.71	13.01	13.28
27	7.68	9.25	10.21	10.90	11.45	11.89	12.28	12.61	12.90	13.17
28	7.64	9.20	10.14	10.83	11.37	11.81	12.19	12.51	12.81	13.07
29	7.60	9.15	10.08	10.76	11.29	11.73	12.10	12.43	12.72	12.97
30	7.56	9.10	10.03	10.70	11.22	11.66	12.03	12.35	12.63	12.89
31	7.53	9.05	9.97	10.64	11.16	11.59	11.95	12.27	12.55	12.81
32	7.50	9.01	9.92	10.58	11.10	11.53	11.89	12.20	12.48	12.73
33	7.47	8.97	9.88	10.53	11.04	11.47	11.83	12.14	12.41	12.66
34	7.44	8.94	9.84	10.48	10.99	11.41	11.77	12.08	12.35	12.60
35	7.42	8.90	9.79	10.44	10 94	11.36	11.71	12.02	12.29	12.54

TABLE V (*Contd.*)

Upper 1% points of the distribution of W for $\rho = 0{\cdot}5$

m/p	1	2	3	4	5	6	7	8	9	10
5	16.26	21.41	24.73	27.25	29.19	30.92	32.26	33.63	34.60	35.77
6	13.75	17.74	20.30	22.20	23.71	24.98	26.05	27.01	27.84	28.61
7	12.25	15.59	17.70	19.27	20.51	21.54	22.43	23.20	23.89	24.51
8	11.26	14.18	16.02	17.37	18.44	19.33	20.10	20.76	21.36	21.89
9	10.56	13.20	14.84	16.05	17.00	17.80	18.48	19.07	19.60	20.07
10	10.04	12.47	13.98	15.08	15.95	16.68	17.29	17.83	18.31	18.74
11	9.65	11.91	13.32	14.34	15.15	15.82	16.39	16.89	17.34	17.73
12	9.33	11.47	12.80	13.76	14.52	15.15	15.69	16.15	16.57	16.94
13	9.07	11.12	12.38	13.29	14.01	14.61	15.12	15.56	15.95	16.30
14	8.86	10.83	12.03	12.90	13.59	14.16	14.65	15.07	15.44	15.78
15	8.68	10.58	11.74	12.58	13.24	13.79	14.25	14.66	15.02	15.34
16	8.53	10.37	11.49	12.31	12.95	13.47	13.92	14.31	14.66	14.97
17	8.40	10.19	11.28	12.07	12.69	13.20	13.64	14.02	14.35	14.65
18	8.28	10.04	11.10	11.87	12.47	12.97	13.39	13.76	14.08	14.38
19	8.18	9.90	10.94	11.69	12.28	12.76	13.17	13.53	13.85	14.14
20	8.10	9.78	10.79	11.53	12.10	12.58	12.98	13.33	13.64	13.92
21	8.02	9.67	10.67	11.39	11.95	12.42	12.81	13.16	13.46	13.73
22	7.95	9.57	10.55	11.26	11.82	12.27	12.66	13.00	13.30	13.56
23	7.88	9.48	10.45	11.15	11.69	12.14	12.52	12.85	13.15	13.41
24	7.82	9.40	10.36	11.04	11.58	12.02	12.40	12.73	13.01	13.27
25	7.77	9.33	10.27	10.95	11.48	11.92	12.29	12.61	12.89	13.15
26	7.72	9.27	10.20	10.87	11.39	11.82	12.18	12.50	12.78	13.03
27	7.68	9.21	10.13	10.79	11.30	11.73	12.09	12.40	12.68	12.93
28	7.64	9.15	10.06	10.71	11.23	11.65	12.00	12.31	12.59	12.83
29	7.60	9.10	10.00	10.65	11.15	11.57	11.92	12.23	12.50	12.74
30	7.56	9.05	9.95	10.59	11.09	11.50	11.85	12.15	12.42	12.66
31	7.53	9.01	9.89	10.53	11.03	11.43	11.78	12.08	12.35	12.58
32	7.50	8.97	9.84	10.48	10.97	11.37	11.72	12.01	12.28	12.51
33	7.47	8.93	9.80	10.43	10.92	11.32	11.66	11.95	12.21	12.45
34	7.44	8.89	9.76	10.38	10.86	11.26	11.60	11.89	12.15	12.38
35	7.42	8.86	9.72	10.34	10.82	11.21	11.55	11.84	12.10	12.33

TABLE V (*Contd.*)

Upper 1% *points of the distribution of W for* $\rho = 0{\cdot}6$

m/p	1	2	3	4	5	6	7	8	9	10
5	16.26	21.06	24.10	26.36	28.10	29.63	30.82	32.01	32.88	33.88
6	13.75	17.49	19.84	21.56	22.92	24.04	25.01	25.84	26.59	27.26
7	12.25	15.38	17.33	18.75	19.88	20.80	21.60	22.28	22.90	23.44
8	11.26	14.01	15.70	16.94	17.91	18.71	19.40	19.99	20.52	20.99
9	10.56	13.04	14.57	15.67	16.54	17.26	17.87	18.40	18.87	19.29
10	10.04	12.33	13.73	14.75	15.54	16.20	16.75	17.24	17.67	18.05
11	9.65	11.79	13.09	14.04	14.78	15.39	15.90	16.35	16.75	17.11
12	9.33	11.36	12.59	13.48	14.18	14.75	15.23	15.66	16.03	16.37
13	9.07	11.01	12.18	13.03	13.69	14.23	14.70	15.10	15.45	15.77
14	8.86	10.72	11.85	12.66	13.29	13.81	14.25	14.63	14.97	15.28
15	8.68	10.48	11.57	12.35	12.96	13.46	13.88	14.25	14.57	14.87
16	8.53	10.28	11.33	12.08	12.67	13.16	13.57	13.92	14.24	14.52
17	8.40	10.10	11.12	11.86	12.43	12.90	13.30	13.64	13.95	14.22
18	8.28	9.95	10.95	11.66	12.22	12.67	13.06	13.40	13.69	13.96
19	8.18	9.81	10.79	11.49	12.03	12.48	12.86	13.18	13.47	13.73
20	8.10	9.69	10.65	11.33	11.87	12.31	12.68	13.00	13.28	13.53
21	8.02	9.59	10.53	11.20	11.72	12.15	12.51	12.83	13.11	13.35
22	7.95	9.49	10.42	11.08	11.59	12.01	12.37	12.68	12.95	13.19
23	7.88	9.41	10.32	10.97	11.47	11.89	12.24	12.54	12.81	13.05
24	7.82	9.33	10.23	10.87	11.37	11.78	12.12	12.42	12.69	12.92
25	7.77	9.26	10.15	10.78	11.27	11.67	12.02	12.31	12.57	12.80
26	7.72	9.19	10.07	10.70	11.18	11.58	11.92	12.21	12.47	12.70
27	7.68	9.14	10.00	10.62	11.10	11.50	11.83	12.12	12.37	12.60
28	7.64	9.08	9.94	10.55	11.03	11.42	11.75	12.03	12.28	12.51
29	7.60	9.03	9.88	10.49	10.96	11.34	11.67	11.95	12.20	12.42
30	7.56	8.98	9.83	10.43	10.89	11.28	11.60	11.88	12.12	12.35
31	7.53	8.94	9.78	10.37	10.84	11.21	11.53	11.81	12.05	12.27
32	7.50	8.90	9.73	10.32	10.78	11.16	11.47	11.75	11.99	12.21
33	7.47	8.86	9.69	10.27	10.73	11.10	11.42	11.69	11.93	12.14
34	7.44	8.83	9.64	10.23	10.68	11.05	11.36	11.63	11.87	12.09
35	7.42	8.79	9.61	10.19	10.64	11.00	11.31	11.58	11.82	12.03

TABLE V (*Contd.*)

Upper 1% points of the distribution of W for $\rho = 0{\cdot}7$

m/p	1	2	3	4	5	6	7	8	9	10
5	16.26	20.61	23.28	25.24	26.74	28.03	29.06	30.03	30.79	31.60
6	13.75	17.15	19.23	20.73	21.91	22.87	23.69	24.40	25.04	25.59
7	12.25	15.10	16.84	18.09	19.06	19.86	20.54	21.13	21.65	22.12
8	11.26	13.77	15.29	16.37	17.22	17.92	18.51	19.02	19.47	19.87
9	10.56	12.84	14.20	15.18	15.94	16.56	17.09	17.55	17.95	18.31
10	10.04	12.14	13.40	14.30	15.00	15.57	16.06	16.47	16.84	17.17
11	9.65	11.62	12.79	13.63	14.28	14.81	15.26	15.65	16.00	16.31
12	9.33	11.20	12.31	13.10	13.72	14.22	14.64	15.01	15.33	15.62
13	9.07	10.86	11.92	12.67	13.26	13.74	14.14	14.49	14.80	15.07
14	8.86	10.58	11.60	12.32	12.88	13.34	13.73	14.06	14.36	14.62
15	8.68	10.34	11.33	12.03	12.57	13.01	13.38	13.70	13.99	14.24
16	8.53	10.15	11.10	11.78	12.30	12.73	13.09	13.40	13.67	13.92
17	8.40	9.97	10.90	11.56	12.07	12.49	12.84	13.14	13.40	13.64
18	8.28	9.82	10.73	11.37	11.87	12.27	12.62	12.91	13.17	13.40
19	8.18	9.69	10.58	11.21	11.69	12.09	12.42	12.71	12.97	13.19
20	8.10	9.58	10.45	11.06	11.54	11.93	12.25	12.54	12.79	13.01
21	8.02	9.47	10.33	10.93	11.40	11.78	12.10	12.38	12.63	12.84
22	7.95	9.38	10.22	10.82	11.28	11.65	11.97	12.24	12.48	12.70
23	7.88	9.30	10.13	10.71	11.17	11.54	11.85	12.12	12.35	12.56
24	7.82	9.22	10.04	10.62	11.07	11.43	11.74	12.00	12.23	12.44
25	7.77	9.15	9.96	10.53	10.98	11.33	11.64	11.90	12.13	12.33
26	7.72	9.09	9.89	10.45	10.89	11.25	11.55	11.80	12.03	12.23
27	7.68	9.03	9.82	10.38	10.81	11.17	11.46	11.72	11.94	12.14
28	7.64	8.98	9.76	10.32	10.74	11.09	11.38	11.64	11.86	12.06
29	7.60	8.93	9.71	10.25	10.68	11.02	11.31	11.56	11.78	11.98
30	7.56	8.88	9.65	10.20	10.62	10.96	11.25	11.50	11.71	11.91
31	7.53	8.84	9.61	10.15	10.56	10.90	11.19	11.43	11.65	11.84
32	7.50	8.80	9.56	10.10	10.51	10.85	11.13	11.37	11.59	11.78
33	7.47	8.76	9.52	10.05	10.46	10.79	11.08	11.32	11.53	11.72
34	7.44	8.73	9.48	10.01	10.42	10.75	11.03	11.27	11.48	11.67
35	7.42	8.70	9.44	9.97	10.37	10.70	10.98	11.22	11.43	11.62

TABLE V (*Contd.*)

Upper 1% points of the distribution of W for $\rho = 0{\cdot}8$

m/p	1	2	3	4	5	6	7	8	9	10
5	16.26	19.98	22.20	23.78	25.00	26.00	26.85	27.57	28.22	28.78
6	13.75	16.67	18.40	19.63	20.58	21.35	22.01	22.57	23.07	23.51
7	12.25	14.72	16.17	17.20	17.99	18.63	19.18	19.65	20.06	20.43
8	11.26	13.44	14.71	15.61	16.30	16.87	17.34	17.75	18.11	18.43
9	10.56	12.54	13.69	14.50	15.13	15.64	16.07	16.43	16.76	17.04
10	10.04	11.87	12.94	13.69	14.26	14.73	15.12	15.46	15.76	16.02
11	9.65	11.36	12.36	13.06	13.60	14.04	14.41	14.73	15.00	15.25
12	9.33	10.96	11.91	12.57	13.08	13.49	13.84	14.14	14.40	14.64
13	9.07	10.64	11.54	12.17	12.66	13.05	13.38	13.67	13.92	14.14
14	8.86	10.37	11.24	11.84	12.31	12.69	13.01	13.28	13.52	13.74
15	8.68	10.14	10.98	11.57	12.02	12.39	12.69	12.96	13.19	13.39
16	8.53	9.95	10.76	11.33	11.77	12.13	12.43	12.68	12.91	13.10
17	8.40	9.78	10.58	11.13	11.56	11.91	12.19	12.44	12.66	12.86
18	8.28	9.64	10.42	10.96	11.38	11.71	12.00	12.24	12.45	12.64
19	8.18	9.51	10.27	10.81	11.21	11.54	11.82	12.06	12.27	12.45
20	8.10	9.40	10.15	10.67	11.07	11.39	11.66	11.90	12.10	12.28
21	8.02	9.30	10.03	10.55	10.94	11.26	11.53	11.76	11.96	12.14
22	7.95	9.21	9.93	10.44	10.83	11.14	11.40	11.63	11.83	12.00
23	7.88	9.13	9.84	10.34	10.72	11.03	11.29	11.51	11.71	11.88
24	7.82	9.06	9.76	10.25	10.63	10.94	11.19	11.41	11.60	11.77
25	7.77	8.99	9.69	10.17	10.55	10.85	11.10	11.32	11.51	11.67
26	7.72	8.93	9.62	10.10	10.47	10.77	11.02	11.23	11.42	11.58
27	7.68	8.87	9.56	10.03	10.40	10.69	10.94	11.15	11.34	11.50
28	7.64	8.82	9.50	9.97	10.33	10.62	10.87	11.08	11.26	11.43
29	7.60	8.77	9.45	9.91	10.27	10.56	10.80	11.01	11.19	11.35
30	7.56	8.73	9.40	9.86	10.22	10.50	10.74	10.95	11.13	11.29
31	7.53	8.69	9.35	9.81	10.16	10.45	10.69	10.89	11.07	11.23
32	7.50	8.65	9.31	9.76	10.11	10.40	10.63	10.84	11.02	11.17
33	7.47	8.61	9.27	9.72	10.07	10.35	10.59	10.79	10.96	11.12
34	7.44	8.58	9.23	9.68	10.03	10.31	10.54	10.74	10.92	11.07
35	7.42	8.55	9.19	9.64	9.99	10.26	10.50	10.70	10.87	11.03

TABLE V (*Contd.*)

Upper 1% points of the distribution of W for $\rho = 0{\cdot}9$

m/p	1	2	3	4	5	6	7	8	9	10
5	16.26	19.04	20.63	21.71	22.59	23.19	23.86	24.21	24.81	24.97
6	13.74	15.95	17.20	18.06	18.71	19.25	19.68	20.07	20.39	20.69
7	12.25	14.11	15.16	15.87	16.43	16.86	17.24	17.55	17.83	18.08
8	11.26	12.91	13.86	14.48	15.01	15.36	15.75	15.97	16.28	16.43
9	10.56	12.06	12.92	13.46	13.95	14.23	14.62	14.76	15.10	15.15
10	10.04	11.44	12.23	12.74	13.18	13.46	13.79	13.94	14.24	14.31
11	9.65	10.96	11.70	12.21	12.60	12.90	13.18	13.38	13.61	13.75
12	9.33	10.58	11.28	11.76	12.11	12.41	12.64	12.85	13.01	13.19
13	9.07	10.27	10.94	11.40	11.75	12.04	12.26	12.47	12.63	12.81
14	8.86	10.02	10.66	11.11	11.44	11.71	11.93	12.13	12.29	12.45
15	8.68	9.81	10.43	10.86	11.18	11.44	11.66	11.84	12.00	12.14
16	8.53	9.62	10.23	10.65	10.96	11.21	11.43	11.60	11.76	11.90
17	8.40	9.47	10.06	10.47	10.77	11.02	11.22	11.40	11.55	11.69
18	8.28	9.33	9.91	10.31	10.61	10.85	11.05	11.22	11.37	11.50
19	8.18	9.21	9.78	10.17	10.46	10.70	10.90	11.06	11.21	11.34
20	8.10	9.10	9.66	10.05	10.34	10.57	10.76	10.93	11.07	11.20
21	8.02	9.01	9.56	9.94	10.22	10.45	10.64	10.80	10.94	11.07
22	7.95	8.93	9.47	9.84	10.12	10.35	10.53	10.69	10.83	10.95
23	7.88	8.85	9.38	9.75	10.03	10.25	10.43	10.59	10.73	10.85
24	7.82	8.78	9.31	9.67	9.95	10.16	10.35	10.50	10.64	10.76
25	7.77	8.72	9.24	9.60	9.87	10.09	10.27	10.42	10.55	10.67
26	7.72	8.66	9.18	9.53	9.80	10.01	10.19	10.34	10.48	10.59
27	7.68	8.61	9.12	9.47	9.74	9.95	10.13	10.28	10.41	10.52
28	7.64	8.56	9.07	9.41	9.68	9.89	10.06	10.21	10.34	10.46
29	7.60	8.51	9.02	9.36	9.62	9.83	10.01	10.15	10.28	10.40
30	7.56	8.47	8.97	9.31	9.57	9.78	9.95	10.10	10.23	10.34
31	7.53	8.43	8.93	9.27	9.53	9.73	9.90	10.05	10.18	10.29
32	7.50	8.40	8.89	9.23	9.48	9.69	9.86	10.00	10.13	10.24
33	7.47	8.36	8.85	9.19	9.44	9.65	9.81	9.96	10.08	10.19
34	7.44	8.33	8.82	9.15	9.41	9.61	9.77	9.92	10.04	10.15
35	7.42	8.30	8.79	9.12	9.37	9.57	9.74	9.88	10.00	10.11

CHAPTER 23

Testing a Table of Random Numbers

ARTHUR LINDER, *University of Geneva and Swiss Federal Institute of Technology.*

1. INTRODUCTORY REMARKS

The results obtained by drawing numbers in a lottery have been placed at our disposal. From these we have constructed a table of random numbers. A first table consisting of 10,000 digits has been published (Linder, 1953) and later reproduced (Linder, 1961). Recently it was possible to compile a further 16,000 digits. The whole table consisting of 26,000 digits will be published in a forthcoming third edition of (Linder, 1953). In this paper we will consider tests for randomness for the 26,000 digits of the complete table.

Our table of random numbers is arranged in 13 pages, each containing 40 rows with 50 digits in every row. The random numbers in the table are given in the same sequence as they occurred in the lottery.

2. TESTS FOR RANDOMNESS

Many tests for randomness have been described in the literature. Programs for several of these tests were prepared to be used on an IBM 1620. In this preliminary report only four tests will be presented.

The *frequency test* was suggested by Kendall and Babington Smith (1938) in their well-known paper. It is designed to test the hypothesis that each of the digits 0, 1, 2, ..., 9 has a probability of 1/10 to occur in the table of random numbers.

The *sum test* has been used by Yule (1938) in order to test if the sums of five consecutive digits are in accordance with the assumption of equal probabilities for all digits.

If *a*, *b*, *c*, *d* and *e* designate different digits, the *poker test* given by Kendall and Babington Smith (1938) tests whether the five consecutive digits of the following types occur with the probabilities given below, which are obtained under the assumption of equal probability for each digit.

Type	Probability × 10,000
a a a a a	1
a a a a b	45
a a a b b	90
a a a b c	720
a a b b c	1,080
a a b c d	5,040
a b c d e	3,024
Total	10,000

As a fourth test we consider the *H test of rank order* given by Kruskal and Wallis (1952). If we consider the digits 0 as forming one sample, the digits 1 as forming another sample and so on to digits 9, and if N_j is the number of digits in the jth sample, R_j the sum of ranks of the jth sample, then

$$H = 12 \sum_j (R_j^2/N_j)/N(N+1) - 3(N-1),$$

where $\sum_j N_j = N$, is distributed as χ^2 with 9 degrees of freedom.

When applying the H test the rank was replaced by the number indicating the position of the digit in the table.

3. HOW TO TEST

Having decided on the tests which shall be applied, the question arises as to the order and grouping of the digits.

If we assume for the moment being that we have chosen the grouping of the digits, the order in which the digits are presented is clearly irrelevant for the frequency test. On the contrary, the three other tests may give different results if applied to the rows or to the columns of the table of random numbers. Tables of random numbers may be used in different ways, sometimes reading off digits along rows, sometimes along columns. In view of this we have chosen to test the table of random numbers both by rows and by columns. In this paper we will give the results of testing by rows only.

The question of grouping requires some comment. Given the table of 26,000 digits, the simplest thing would be to apply the test to the whole set, to assume a level of significance and to judge the result of the test for the table as a whole. This procedure is open to criticism. We could, in fact, well obtain a verdict of non-significance for the whole table, while in some parts of it the test would show significant departure from expectation. This would be a serious indication against randomness especially if those parts which yield significant deviations were all situated, say, at the end of the table.

Many authors have for this reason advocated a study of randomness in different parts of the table of random numbers. The problem is to decide how small or how large the groups should be, to which the test shall be applied. To be more specific, should we, in our case, form

5	groups	of	5,200	digits,
10	,,	,,	2,600	,, ,
20	,,	,,	1,300	,, ,
52	,,	,,	500	,, ,
260	,,	,,	100	,, ?

The greater the number of digits in one group, the better will the test detect deviations from expectation. With only 100 digits in one group, the test will not reveal even considerable differences between observed and expected frequencies. From this point of view it would clearly be preferable to choose 5 groups of 5,200 digits or 10 groups with 2,600 digits.

On the other hand, if we want to study local randomness in reasonable detail, the groups should be rather small. We would then prefer groups with 100 or 500 digits.

It has sometimes been argued that groups within a table which yield small probabilities by any test should be deleted. This policy seems to me questionable. It overlooks the fact that *a truly random arrangement of digits should give as many significant results as are called for by the level of significance which has been chosen.* We can go even further: If the digits are produced in a truly random way, the test results, e.g., values of χ^2, should conform to the χ^2 distribution. As a consequence, we would prefer to form as many groups as possible so as to be able to compare the values of χ^2 obtained with the theoretical distribution.

There are thus two conflicting tendencies for the choice of the number and size of groups. On one hand we would prefer to have large groups so as to increase the sensitivity of the individual test, on the other hand the size of the groups should be small so that we could compare the results of a greater number of tests with expectation. As a compromise we have chosen 52 groups of 500 digits in our table of 26,000 digits. The sensitivity of the individual test is still sufficiently high, while with 52 values of χ^2 we are also in a position to make a comparison between observation and expectation under the assumption of randomness.

4. SOME RESULTS

4.1 Frequency test

Let us recall that each of the 13 pages of our table of random numbers contains 40 rows with 50 digits each. We have formed

4 groups on each page simply by taking the first, second, third and fourth groups of 10 rows. To each of the 52 groups—containing 500 digits—we applied the frequency test. In the first group the number of 0's, 1's, etc., observed and expected was as follows.

Digit	0	1	2	3	4	5	6	7	8	9	Total
Observed frequencies	51	41	50	60	46	48	39	58	62	45	500
Expected frequencies	50	50	50	50	50	50	50	50	50	50	500
Difference, d	1	−9	0	10	−4	−2	−11	8	12	−5	0
d^2	1	81	0	100	16	4	121	64	144	25	556

We thus get $\chi^2 = 11.120$, and this has $n = 9$ degrees of freedom.

In order to compare the 52 χ^2's with the χ^2-distribution, we classified them in 10 groups whose limits are the deciles $\chi^2_{0.1}$, $\chi^2_{0.2}$, etc. The results are shown below :

Deciles of χ^2		4.17	5.38	6.39	7.36	8.34	9.41	10.66	12.24	14.68	Total
Observed frequencies	4	3	4	4	6	6	7	5	7	6	52
Expected frequencies	5.2	5.2	5.2	5.2	5.2	5.2	5.2	5.2	5.2	5.2	52
Difference, d	−1.2	−2.2	−1.2	−1.2	0.8	0.8	1.8	−0.2	1.8	0.8	0.0
d^2	1.44	4.84	1.44	1.44	0.64	0.64	3.24	0.04	3.24	0.64	17.60

The calculated χ^2's conform sufficiently well with expectation; this may be tested by χ^2, for which we obtain $\chi^2 = 3{\cdot}385$ with $n = 9$.

There is thus no indication of nonrandomness by the frequency test. This finding is confirmed if we plot the 52 χ^2's in the order in which the groups occur in the table, which is also the order in which the random numbers were produced by the drawings in the lottery. This plot does not indicate any non-random behaviour.

4.2 Sum test (Yule)

The fifty digits in every row of the table form 10 groups of 5 digits. In the sum test the frequency distribution of the total of the digits in these groups of five digits is compared with expectation. It is easy to see that the probability distribution of the total of five digits is as follows :

Total		Probability (× 100,000)	Total		Probability (× 100,000)	Total		Probability (× 100,000)
0	45	1	8	37	495	16	29	3,795
1	44	5	9	36	715	17	28	4,335
2	43	15	10	35	996	18	27	4,840
3	42	35	11	34	1,340	19	26	5,280
4	41	70	12	33	1,745	20	25	5,631
5	40	126	13	32	2,205	21	24	5,875
6	39	210	14	31	2,710	22	23	6,000
7	38	330	15	30	3,246			

A total of 10 (or of 35) is expected to occur in a sample of 100 groups with a frequency of about 1. When comparing the frequencies of totals in the table with expectation, we have pooled the classes for 0, 1, 2, ..., 9, and for 36, 37, 38, ..., 45.

As an example we give the frequency distribution of totals of 5 digits in the first group of 500 digits.

	Frequency			Frequency	
Total	Observed	Expected	Total	Observed	Expected
0–9	1	2.002	23	15	6.000
10	1	0.996	24	7	5.875
11	2	1.340	25	3	5.631
12	1	1.745	26	7	5.280
13	1	2.205	27	2	4.840
14	–	2.710	28	3	4.335
15	5	3.246	29	4	3.795
16	3	3.795	30	3	3.246
17	4	4.335	31	3	2.710
18	4	4.840	32	–	2.205
19	5	5.280	33	2	1.745
20	6	5.631	34	2	1.340
21	8	5.875	35	1	0.996
22	4	6.000	36–45	3	2.002
			Sum	100	100.000

The resulting $\chi^2 = 28{\cdot}048$ has 27 degrees of freedom.

The 52 χ^2's obtained in this way were classified, using the deciles of χ^2 with $n = 27$.

Deciles of χ^2		18.1	20.7	22.7	24.5	26.3	28.2	30.3	32.9	36.7	Total
Observed frequencies	8	3	6	4	4	5	4	7	5	6	52
Expected frequencies	5.2	5.2	5.2	5.2	5.2	5.2	5.2	5.2	5.2	5.2	52
Difference, d	2.8	−2.2	0.8	−1.2	−1.2	−0.2	−1.2	1.8	−0.2	0.8	0.0
d	7.84	4.84	0.64	1.44	1.44	0.04	1.44	3.24	0.04	0.64	21.60

The comparison between observed and expected frequencies leads to $\chi^2 = 4{\cdot}154$ with $n = 9$.

When we studied the 52 χ^2's in the order in which they occur in the table, we found no indication of non-randomness.

4.3 Poker test

For the first 500 digits of the table we found the following frequencies of the seven different types of arrangements of five digits.

Type	Frequency Observed	Frequency Expected
a a a a a	–	0·01
a a a a b	1	0·45
a a a b b	2	0·90
a a a b c	7	7·20
a a b b c	9	10·80
a a b c d	48	50·40
a b c d e	33	30·24
Total	100	100·00

Instead of using the familiar χ^2 test we preferred to compute the likelihood ratio G, because of the small expectations in the first two classes. The test value G is distributed as χ^2 with 6 degrees of freedom. For the example given above we obtain $G = 2{\cdot}196$.

The distribution of the 52 values of G, compared to χ^2 with 6 degrees of freedom is shown below.

Deciles of χ^2		2.20	3.07	3.83	4.57	5.35	6.21	7.23	8.56	10.64	Total
Observed frequencies	8	5	10	5	6	5	2	6	2	3	52
Expected frequencies	5.2	5.2	5.2	5.2	5.2	5.2	5.2	5.2	5.2	5.2	52
Difference, d	2.8	−0.2	4.8	−0.2	0.8	−0.2	−3.2	0.8	−3.2	−2.2	0.0
d^2	7.84	0.04	23.04	0.04	0.64	0.04	10.24	0.64	10.24	4.84	57.60

Here again the $\chi^2 = 11{\cdot}077$ seems to indicate that the observed frequencies conform reasonably well with expectation. Here also the χ^2's were studied in the sequence in which they occur in the table, without finding any anomaly.

4.4 *H*-test (10 sample test of rank order)

For the first 500 digits we obtained, on calculating the 10 sample test of rank orders, $H = 11{\cdot}781$. This quantity is distributed as χ^2 with 9 degrees of freedom if the arrangement of the digits is random. The distribution of the 52 H-values is as follows.

Deciles of χ^2		4.17	5.38	6.39	7.36	8.34	9.41	10.66	12.24	14.68	Total
Observed frequencies	7	10	3	4	2	2	5	7	8	4	52
Expected frequencies	5.2	5.2	5.2	5.2	5.2	5.2	5.2	5.2	5.2	5.2	52
Difference, d	1.8	4.8	−2.2	−1.2	−3.2	−3.2	−0.2	1.8	2.8	−1.2	0.0
d^2	3.24	23.04	4.84	1.44	10.24	10.24	0.04	3.24	7.84	1.44	65.60

The comparison between observed and expected frequencies gives $\chi^2 = 12{\cdot}615$, a satisfactory value. No significant departure from randomness was found when the values of H were studied in the order given by the table.

5. COMPARISON OF TESTS

Having obtained values for four different tests in each of 52 groups of 500 digits we made a rough, preliminary comparison of the four tests. To this end we used the test values of two tests as rectangular coordinates in a plane, obtaining 52 points. A straight line was drawn parallel to the first axis, so that 26 points were situated on each side of it. Similarly, a straight line parallel to the second axis separated the points in two groups comprising 26 points. If there were independence we should expect 13 points in each quadrant formed by the two straight lines.

The actual numbers obtained when comparing χ^2's for the *frequency test* and the *sum test* are :

		Frequency test	
		Low values	High values
Sum test	Low values	19	7
	High values	7	19

Taking the totals in the diagonals we get 38 and 14, against an expected value of 26. The discrepancy is clearly significant.

If we compare the *frequency test* and the *poker test* by the same procedure, we get :

		Frequency test	
		Low values	High values
Poker test	Low values	17	9
	High values	9	17

Here again we find a dependence between the two tests which is significant.

By this admittedly crude method all other comparisons between two out of the four tests were performed but none was found to be significant.

These results seem to indicate that there is a correlation between the frequency test and the sum test. This may be expected because a disturbance in the frequencies will in general also give a distortion in the sums of five digits.

Similarly, disturbed frequencies of the digits will affect the poker test.

While both the sum test and the poker test are correlated with the frequency test, our results seem to indicate no strong correlation between sum test and poker test.

The H-test (10 sample rank order test) seems to be almost independent of the three other tests.

Acknowledgment

Mr. Heinz Kres has written the test programs and carried out the computations on the IBM 1620. I wish to thank him also for valuable suggestions.

References

Kendall, M. G. and Babington Smith, B. (1938). "Randomness and Random Sampling Numbers," *J. Roy. Statist. Soc.*, **101**, 147–166.

Kruskal, W. H. and Wallis, W. A. (1952). "Use of Ranks in One-Criterion Variance Analysis. *J. Amer. Statist. Assoc.*, **47**, 583–621.

Linder, A. (1953). *Planen und Auswerten von Versuchen.* Birkhauser, Basel.

Linder, A. (1961). *Handliche Sammlung mathematisch-statistischer Tafeln*, Birkhauser, Basel.

Yule, G. U. (1938). "A Test of Tippett's Random Sampling Numbers," *J. Roy. Statist. Soc.*, **101**, 167–172.

(*Received Oct. 30, 1965.*)

CHAPTER 24

On Gammaization of the Variance Ratio

SUJIT KUMAR MITRA, *Indian Statistical Institute and*
BRIJ MOHAN MAHAJAN, *Planning Commission,*
New Delhi

1. INTRODUCTION

Let X and Y be independent Gamma variables with parameters m and n respectively. The c.d.f. of a Gamma variable with parameter m will be denoted by $G_m(x)$. Consider the random variable $U = nX/Y$ and denote its c.d.f. by $F_{mn}(x)$. It is well-known that as $n \to \infty$, $F_{mn}(x) \to G_m(x)$ at each x, suggesting thereby that when n is large it is often a sufficient approximation to assume that the distribution of U is Gamma (m). For small or moderately large values of n however the Gamma approximation could be hardly exact enough, and it may be necessary to transform the variable U suitably so that the Gamma approximation can be applied with more confidence. Such a transformation is not hard to visualize. For example $V = G_m^{-1}F_{mn}(U) = s(U)$ has precisely a Gamma (m) distribution for each n, large or small. The difficulty consists in obtaining $s(U)$ in a neat closed form. It is seen that $s(u)$ defines a monotone increasing, continuous and differentiable function of the real variable u. The object of the present paper is to derive a suitable polynomial approximation for $s(U)$. This may be called the problem of *Gammaization* of U (or equivalently of the variance ratio. Note that if $\nu_1 = 2m$ and $\nu_2 = 2n$ are integers, U/m has the variance ratio distribution with ν_1 d.f. for numerator and ν_2 for the denominator).

2. GAMMAIZATION OF THE VARIANCE RATIO

Let the p.d.f.'s of the Gamma (m) variable and of U be denoted by $g_m(x)$ and $f_{mn}(x)$, respectively. If $s(U)$ has the Gamma (m) distribution, we have, for any real u,

$$f_{mn}(u) = g_m(s(u))s'(u) \tag{2.1}$$

Hence, if

$$s(u) = (1-a_1)u-a_2u^2-a_3u^3-\dots$$

we have

$$f_{mn}(u) = (1-\Sigma ia_i\, u^{i-1})g_m(u-\Sigma a_i\, u^i)$$
$$= g_m(u)B(u)$$

where
$$B(u) = B_0+B_1u+B_2u^2+\dots$$

with coefficients depending upon the a_i's.

Consider now the finite (formal) Laguerre series expansion of $f_{mn}(u)$ given in the Appendix, which by a slight rearrangement of terms can be written as

$$f_{mn}(u) \approx g_m(u)P(u). \tag{2.3}$$

where

$$P(u) = P_0+P_1u+P_2u^2+\dots .$$

The coefficients a_i are obtained by equating B_i with P_i in succession $(i = 0, 1, 2, \dots)$.

For the calculations given in this paper, terms upto and including $O\left(\frac{1}{n^6}\right)$ have been retained in the Laguerre series expansion, leading to the following expansion for $s(U)$:

$$V = s(U) \approx A_0(U) - \frac{A_1(U)}{n-1} + \frac{A_2(U)}{(n-1)_2} - \frac{A_3(U)}{(n-1)_3} + \cdots, \tag{2.4}$$

where

$$A_0(U) = U$$

$$A_1(U) = \frac{U^2-(m-1)U}{2}$$

$$A_2(U) = \frac{8U^3-(7m-19)U^2-(m^2+12m-13)U}{24},$$

$$A_3(U) = [12U^4-(10m-58)U^3-(m^2+42m-91)U^2$$
$$-(m^3+5m^2+47m-53)U]/48,$$

$$A_4(U) = [1152U^5-(938m-9578)U^4-(78m^2+7196m-28394)U^3$$
$$-(63m^3+657m^2+18393m-36393)U^2$$
$$-(73m^4+600m^3+1870m^2+16680m-19223)U]/5760,$$

$$A_5(U) = [1920U^6-(1544m-24584)U^5$$
$$-(114m^2+18748m-119662)U^4$$
$$-(84m^3+1472m^2+83796m-277352)U^3$$
$$-(79m^4+1128m^3+7042m^2+166488m-312977)U^2$$
$$-(99m^5+1241m^4+5970m^3+14870m^2+132171m$$
$$-154351)U]/11520,$$

$$A_6(U) = [414720U^7-(331144m-7588744)U^6$$
$$-(22444m^2+5832856m-55206980)U^5$$
$$-(15724m^3+413576m^2+40115468m-203840768)U^4$$
$$-(13484m^4+294120m^3+2989280m^2+135489240m$$
$$-403930444)U^3-(13799m^5+267433m^4+2186186m^3$$
$$+10605334m^2+229105487m-416360839)U^2$$
$$-(18125m^6+324324m^5+2303301m^4+8429400m^3$$
$$+18381615m^2+165428676m-194885441)U]/2903040.$$

3. THE INVERSE TRANSFORMATION $U = s^{-1}(V)$

The inverse transformation transforming a Gamma (m) variable to U is expressible as $U = F_{mn}^{-1}G_m(V) = s^{-1}(V)$.

From (2.4), by series inversion, we have

$$U = s(V) \approx C_0(V) + \frac{C_1(V)}{n-1} + \frac{C_2(V)}{(n-1)_2} + \dots, \tag{3.1}$$

where

$$C_0(V) = V,$$

$$C_1(V) = \frac{V^2-(m-1)V}{2},$$

$$C_2(V) = \frac{4V^3-(11m+1)V^2+(7m^2-7)V}{24},$$

$$\begin{aligned}C_3(V) = {} & [2V^4-(10m+14)V^3+(17m^2+36m-5)V^2 \\ & -(9m^3+21m^2-9m-21)V]/48,\end{aligned}$$

$$\begin{aligned}C_4(V) = {} & [48V^5-(362m+1078)V^4+(1098m^2+5236m+4226)V^3 \\ & -(1527m^3+8553m^2+10257m-3057)V^2 \\ & +(743m^4+4320m^3+5450m^2-4320m-6193)V]/5760,\end{aligned}$$

$$\begin{aligned}C_5(V) = {} & [16V^6-(164m+796)V^5+(736m^2+5912m+10152)V^4 \\ & -(1804m^3+17536m^2+48204m+28456)V^3 \\ & +(2291m^4+23670m^3+76220m^2+65370m-29311)V^2 \\ & -(1075m^5+11145m^4+36850m^3+31350m^2-37925m \\ & -42495)V]/11520,\end{aligned}$$

$$\begin{aligned}C_6(V) = {} & [576V^7-(7544m+52936)V^6+(45124m^2+537832m \\ & +1473364)V^5-(160676m^3+2383624m^2+ \\ & +10808692m+13863008)V^4+(538904m^4 \\ & +5008536m^3+32572568m^2+63863784m \\ & +30588368)V^3-(438667m^5+7106213m^4 \\ & +41450890m^3+99316310m^2+66061483m \\ & -40191163)V^2+(203723m^6+3250800m^5 \\ & +18936687m^4+46116000m^3+28109697m^2 \\ & -49366800m-47250107)V]/2903040.\end{aligned}$$

4. ASYMPTOTIC EXPANSIONS FOR THE FRACTILES OF THE VARIANCE RATIO

The Fisher-Cornish expansion [1, 2] for the fractiles of $Z = e^{2F}$, from which a similar expansion for F can easily be derived, depends for the rapidity of its convergence on the largeness of both the numerator and denominator degrees of freedom. This is because the expansion is based on the normal approximation to the distribution of Z and corrections provided by the successive terms of an Edgeworth series. The expansion developed in Section 3 is suitable in situations where one of the degrees of freedom is small. Numerical tables giving the coefficients of relevant terms for 1% and 5% levels of significance are given below. The expansion for a fractile of F is obtained by dividing (3.1) throughout by m and substituting for V the corresponding fractile of Gamma (m). The coefficients given below are for the situation where ν_2 ($= 2n$, the denominator degrees of freedom) is large while ν_1 ($= 2m$, the numerator degrees of freedom) is small. A case where ν_1 is large and ν_2 small, could be covered in the usual way noting that $1/F$ has also a variance ratio distribution with degrees of freedom interchanged.

To determine the accuracy of fractile values computed using Table 4.1 a comparison is made in Table 4.2 of values so determined with those given in Merrington Thompson's Table [3]. Nine cases where a discrepancy is noted in the fifth digit are marked by an asterisk. In all these 9 cases an independent calculation of the fractile values was made using fractiles of Beta determined to 10 places of decimal on a IBM 1401 system. These are shown in brackets at appropriate places.

References

[1] Cornish, E. A. and Fisher, R. A. (1937). "Moments and Cumulants in the Specification of Distribution," *Rev. Inst. Int. Statist.*, **4**, 1–14.

[2] Fisher, R. A. and Cornish, E. A. (1960). "The Percentage Points of Distributions Having Known Cumulants," *Technometrics*, **2**, 209–226.

[3] Merrington, Maxine and Thompson, Catherine M. (1943). "Tables on Percentage Points of the Inverted Beta (F) Distribution," *Biometrika*, **33**, 73–88.

TABLE 4.1

Coefficients for calculating fractiles of F

5% upper tail values

Coefficient of	numerator degrees of freedom			
	1	2	3	4
1	3.841460	2.995735	2.604910	2.371932
$\frac{1}{n-1}$	4.649569	4.487214	4.437939	4.440098
$\frac{1}{(n-1)_2}$	—0.476650	—0.00638	0.15642	0.18858
$\frac{1}{(n-1)_3}$	−0.2933	−1.1122	−1.3922	−1.3939
$\frac{1}{(n-1)_4}$	2.092	4.241	5.004	4.897
$\frac{1}{(n-1)_5}$	10.31	−18.00	−20.93	−20.30

1% upper tail values

1	6.634900	4.605170	3.781633	3.319175
$\frac{1}{n-1}$	12.664200	10.603795	9.780155	9.357335
$\frac{1}{(n-1)_2}$	4.757354	5.673631	6.017348	6.166573
$\frac{1}{(n-1)_3}$	−8.25669	−8.88461	−9.16151	−9.26954
$\frac{1}{(n-1)_4}$	19.7977	20.2487	20.7593	21.1131
$\frac{1}{(n-1)_5}$	−63.813	−62.832	−64.715	−67.134
$\frac{1}{(n-1)_6}$	261.00	248.16	259.26	287.03

TABLE 4.1 (*Contd.*)

5% lower tail values

Coefficient of	numerator degrees of freedom			
	1	2	3	4
1	0.00393214	0.5129350	0.1172820	0.17768025
$\frac{1}{n-1}$	0.00098691	0.0013155	−0.0190042	−0.05726986
$\frac{1}{(n-1)_2}$	−0.00086225	−0.001293	0.0283194	0.0987001
$\frac{1}{(n-1)_3}$	0.001569	0.002564	−0.070033	−0.271074
$\frac{1}{(n-1)_4}$	−0.004378	−0.00765	0.24253	1.0193
$\frac{1}{(n-1)_5}$	0.0165		−1.081	−4.8618
$\frac{1}{(n-1)_6}$			5.892	27.871

1% lower tail values

Coefficient of	1	2	3	4
1	0.0001570879	0.01005035	0.03827733	0.07427750
$\frac{1}{n-1}$	0.000039275	0.00005050	−0.00847047	−0.0316216
$\frac{1}{(n-1)_2}$	−0.000034366	−0.00005033	0.0123738	0.056915
$\frac{1}{(n-1)_3}$	0.000062594	0.0001005	−0.0304709	−0.151168
$\frac{1}{(n-1)_4}$	−0.0001747	−0.0003557	0.10546	0.571782
$\frac{1}{(n-1)_5}$	0.0006584		−0.47030	−2.74144
$\frac{1}{(n-1)_6}$	−0.003131			15.910

Sufficient number of places have been retained to ensure accuracy in the fifth significant figure for denominator d.f. $\geqslant 30$ (i.e., $n \geqslant 15$) except for 5% lower tail values corresponding to numerator d.f. 3 and 4 (i.e., $m = 1{\cdot}5$ and 2), where the convergence being relatively slow requires the denominator d,f. to be greater than or equal to 40 for a five digit accuracy.

TABLE 4.2

Fractiles of F (upper tail values)

(A) based on Table 4.1 and
(B) corresponding entries in Merrington Thompson Table

probability level	denominator degrees of freedom		numerator degrees of freedom 1	2	3	4
5%	30	A	4.17087	3.31581	2.92226	2.68960
		B	4.1709	3.3158	2.9223	2.6896
	40	A	4.08476	3.23173	2.83874	2.60597
		B	4.0848	3.2317	2.8387	2.6060
	60	A	4.00119	3.15041	2.75808	2.52522
		B	4.0012	3.1504	2.7581	2.5252
	120	A	3.92013	3.07178	2.68017	2.44724
		B	3.9201	3.0718	2.6802	2.4472
1%	30	A	7.56252	5.39038	4.50980*	4.01793
		B	7.5625	5.3904	4.5097 (4.50974)	4.0179
	40	A	7.31411	5.17851	4.31259	3.82830
		B	7.3141	5.1785	4.3126	3.8283
	60	A	7.07711	4.97743	4.12591	3.64904
		B	7.0771	4.9774	4.1259	3.6491* (3.64905)
	120	A	6.85090	4.78651	3.94911	3.47953
		B	6.8510* (6.85090)	4.7865	3.9493* (3.94910)	3.4796* (3.47953)

probability level	numerator degrees of freedom		denominator degrees of freedom 1	2	3	4
5%	30	A	250.093	19.4623	8.61650*	5.74571*
		B	250.09	19.462	8.6166 (8.61658)	5.7459 (5.74588)
	40	A	251.141	19.4707	8.59443	5.71700
		B	251.14	19.471	8.5944	5.7170
	60	A	252.195+	19.4790	8.57201	5.68776
		B	252.20	19.479	8.5720	5.6878
	120	A	253.252	19.4873	8.54936	5.65812
		B	253.25	19.487	8.5494	5.6581
1%	30	A	6260.74	99.4657	26.5050	13.8370*
		B	6260.7	99.466	26.505	13.838 (13.8377)
	40	A	6286.80	99.4740	26.4108	13.7453
		B	6286.8	99.474	26.411	13.745
	60	A	6313.02	99.4824	26.3163	13.6521
		B	6313.0	99.483* (99.4824)	26.316	13.652
	120	A	6339.37	99.4906	26.2211	13.5581
		B	6339.4	99.491	26.221	13.558

Appendix

A Laguerre series expansion for the distribution of U

Consider formally the expansion

$$f_{mn}(u) \approx g_m(u)\left[1+\sum_{k=1}^{n} c_k\, L_k^{(m)}(u)\right]$$

where c_k's are certain constants and the $L_k^{(m)}(u)$'s are Laguerre polynomials defined by*

$$\left(\frac{d}{du}\right)^k u^{k+m-1}\, e^{-u} = (-1)^k\; k\,!\; L_k^{(m)}(u)\; u^{m-1}\, e^{-u}$$

Multiplying both sides of the expansion by $L_k^{(m)}(u)$, integrating over the range of u and utilizing the well-known orthogonality property of the Laguerre polynomials we have

$$c_k = (-1)^k \sum_{i=1}^{k} (-1)^i \frac{n^i}{(n-1)_i} \binom{k}{i}.$$

We shall now prove the following result concerning the order of c_k.

Result : *c_k is of order* $O\left(\frac{1}{n^d}\right)$

where $d = \left[\frac{k+1}{2}\right]$, *the greatest integer* $\leqslant \frac{k+1}{2}$.

To prove this result, we shall write

$$c_k = (-1)^k \sum_{i=1}^{k} (-1)^i\, n^i\, (n-i-1)_{k-i} \binom{k}{i} / (n-1)_k$$

and substitute,**

$$(n-i-1)_{k-i} = \sum_{j=0}^{k-i} \binom{k-i}{j} n^j\, B_{k-i-j}^{(k-i+1)}(-i)$$

*See e.g., Cramér, H. *Mathematical Methods of Statistics*, Princeton, University Press, Princeton, p. 133, 1946.

** See e.g., Frisch, R. *Sur les semi-invariants et moments employé's dans l'étude des distributions statistiques*. Skrifter af det Norske Videnskaps Academie, II, Hist-Filos, Klasse, No. 3. Oslo, 1926.

where $B_r^{(p)}(x)$ is the Bernoulli polynomial of order p and degree r in x. We have thus

$$c_k = (-1)^k \sum_{s=0}^{k} d_{ks} \frac{n^{k-s}}{(n-1)_k}$$

$$d_{ks} = \binom{k}{s} \sum_{r=0}^{k-s} (-1)^r \binom{k-s}{r} B_s^{(k-r+1)}(-r)$$

Note that the Bernoulli polynomial $B_s^{(k-r+1)}(-r)$ can always be expressed as

$$a_0 + a_1(r)_1 + a_2(r)_2 + \ldots + a_s(r)_s$$

where a_i's are constant not depending on r and that

$$\sum_{r=0}^{t} (-1)^r \binom{t}{r} (r)_s = 0 \qquad \text{if } s < t$$

Hence $d_{ks} = 0$ if $s < k-s$ that is $s < \dfrac{k}{2}$

which proves the above result.

Given below are the values of d_{ks} for computing c_k ($k = 1, 2, \ldots, 12$).

VALUES OF d_{ks}

k	$k-s=$						
	0	1	2	3	4	5	6
1	1						
2	2	1					
3	6	7					
4	24	46	3				
5	120	326	55				
6	720	2556	760	15			
7	5040	22212	9856	525			
8	40320	212976	128492	12460	105		
9	362880	2239344	1731276	257068	5985		
10	3628800	25659360	24403176	5031460	217350	945	
11	39916800	318540960	361732536	97266620	6536530	79695	
12	479001600	4261576320	5649316992	1896760536	179386900	4109490	10395

(*Received Jan. 1, 1966.*)

CHAPTER 25

Some Tchebycheff Type Inequalities for Matrix Valued Random Variables

GOVIND S. MUDHOLKAR, *University of Rochester*

1. SUMMARY

Let $\boldsymbol{x}_1, \boldsymbol{x}_2, \ldots, \boldsymbol{x}_n$ be n jointly distributed random p-vectors with $E\boldsymbol{x}_i = 0$ and $E\boldsymbol{x}_i\boldsymbol{x}_i' = \boldsymbol{\Sigma}_i$ $(p \times p)$, $i = 1, 2, \ldots, n$. Write $\boldsymbol{\Sigma} = \boldsymbol{\Sigma}_1 + \boldsymbol{\Sigma}_2 + \ldots + \boldsymbol{\Sigma}_n$ and $\boldsymbol{x}(p \times n) = (\boldsymbol{x}_1, \boldsymbol{x}_2, \ldots, \boldsymbol{x}_n)$. Suppose that $c_1, c_2, \ldots, c_p$ are the p characteristic roots of $\boldsymbol{XX}'$ and $\sigma_1, \sigma_2, \ldots, \sigma_p$ are the characteristic roots of $\boldsymbol{\Sigma}$. Let $f(c_1, c_2, \ldots, c_p) \geqslant 0$ be a concave, symmetric function of $(c_1, c_2, \ldots, c_p)$. In Section 2 we have established the main result of this note, which is the probability inequality

$$\Pr[f(c_1, c_2, \ldots, c_p) \geqslant K] \; \frac{f(\sigma_1, \sigma_2, \ldots, \sigma_p)}{K}.$$

In Section 3 we have discussed some examples.

2. THE MAIN INEQUALITY

The proof of the main result of this note depends upon the following Lemma 1, which is contained in Theorem 1 of [2].

Lemma 1. *Let $c_1, c_2, \ldots, c_p$ be the characteristic roots of a symmetric positive semidefinite matrix $\boldsymbol{A}$. Let $f(t_1, t_2, \ldots, t_p) \geqslant 0$ be a concave symmetric function of $(t_1, t_2, \ldots, t_p)$ on the positive orthant of R^p. Then we have*

$$\min_{y_i'y_i=\delta_{ij}} f(y_1'Ay_1, y_2'Ay_2, \ldots, y_p'Ay_p') = f(c_1, c_2, \ldots, c_p),$$

that is, $f(c_1, c_2, \ldots, c_p)$ is the minimum of $f(y_1'Ay_1, y_2'Ay_2, \ldots, y_p'Ay_p)$ over all orthonormal sets of p-vectors $y_1, y_2, \ldots, y_p$.

Theorem. *Let $X(p\times n)$ be a random matrix whose columns $x_1, x_2, \ldots, x_n$ are jointly distributed with zero means and covariance matrices $\Sigma_1, \Sigma_2, \ldots, \Sigma_n$ respectively. Let $c_1, c_2, \ldots, c_p$ be the characteristic roots of XX'; and $\sigma_1, \sigma_2, \ldots, \sigma_p$ be the characteristic roots $\Sigma = \Sigma_1+\Sigma_2+\ldots+\Sigma_n$. Then for any non-negative, symmetric concave function $f \geqslant 0$ on the positive orthant of R^p, and $K \geqslant 0$ we have,*

$$\Pr[f(c_1, c_2, \ldots, c_p) \geqslant K] \leqslant \frac{f(\sigma_1, \sigma_2, \ldots, \sigma_p)}{K}.$$

Proof. Let $XX' = A$. Then by the Lemma 1 we have $Pr[f(c_1, c_2, \ldots, c_p) \geqslant K] \leqslant \Pr[f(y_1'Ay_1, y_2'Ay_2, \ldots, y_p'Ay_p) \geqslant K]$, for any orthormal set of p-vectors $y_1, y_2, \ldots, y_p$. Therefore,

$$\Pr[f(c_1, c_2, \ldots, c_p) \geqslant K] \leqslant \frac{Ef(y_1'Ay_1, y_2'Ay_2, \ldots, y_p'Ay_p)}{K},$$

by standard arguments of the proofs of Tchebycheff-type inequalities. But because of the concavity of f (Jensen inequality) we have

$$Ef(y_1'Ay_1, y_2'Ay_2, \ldots, y_p'Ay_p) \leqslant f(Ey_1'Ay_1, Ey_2'Ay_2, \ldots, Ey_p'Ay_p)$$

$$= f(y_1'\Sigma y_1, y_2'\Sigma y_2, \ldots, y_p'\Sigma y_p).$$

Hence

$$\Pr[f(c_1, c_2, \ldots, c_p) \geqslant K] \leqslant \min_{y_i'y_j=\delta_{ij}} \frac{f(y_1'\Sigma y_1, y_2'\Sigma y_2, \ldots, y_p'\Sigma y_p)}{K}$$

$$= \frac{f(\sigma_1, \sigma_2, \ldots, \sigma_p)}{K},$$

by Lemma 1, which completes the proof.

3. EXAMPLES

3.1. Determinants

It is known that $\left(\prod_{i=1}^{p} t_i\right)^{1/p}$ is a concave function of $(t_1, t_2, \ldots, t_x)$ over the positive orthant. Therefore,

$$\Pr[|\boldsymbol{XX}'| \geqslant K] = \Pr\left[\left(\prod_{i=1}^{p} c_i\right)^{1/p} \geqslant K^{1/p}\right] \leqslant \frac{|\boldsymbol{\Sigma}|^{1/p}}{K^{1/p}}.$$

This inequality may, more simply, be proved by using the representation,

$$|\boldsymbol{A}|^{1/p} = \min_{|\boldsymbol{B}|=1} \frac{\operatorname{tr} \boldsymbol{AB}}{p}$$

of a symmetric positive semidefinite $(p \times p)$ matrix $\boldsymbol{A} = \boldsymbol{XX}'$ as follows :

$$\Pr[|\boldsymbol{XX}'| \geqslant K] \leqslant \Pr\left[\frac{\operatorname{tr} \boldsymbol{XX}'\boldsymbol{B}}{p} \geqslant K^{1/p}\right], \quad |\boldsymbol{B}| = 1$$

$$\leqslant \frac{E \operatorname{tr} \boldsymbol{XX}'\boldsymbol{B}}{pK^{1/p}}, \quad |\boldsymbol{B}| = 1$$

$$= \frac{\operatorname{tr} \boldsymbol{\Sigma B}}{pK^{1/p}},$$

for all $\boldsymbol{B}$ such that $|\boldsymbol{B}| = 1$. Hence, by above representation for determinants

$$\Pr[|\boldsymbol{XX}'| \geqslant K] \leqslant \frac{|\Sigma|^{1/p}}{K^{1/p}}.$$

3.2. Sums of powers

It is well known that $\left(\sum_{i=1}^{p} t_i^r/p\right)^{1/r}$ is a concave function of $(t_1, t_2, \ldots, t_p)$ over the positive orthant if $r \leqslant 1$. Thus

$$\Pr\left[\left(\sum_{i=1}^{p} c_i^r/p\right)^{1/r} \geqslant K\right] \leqslant \frac{\left(\sum_{i=1}^{r} \sigma_i^r/p\right)^{1/r}}{K}$$

For $r = 1$ this reduces to

$$\Pr\left[\frac{1}{p} \text{ tr } \boldsymbol{XX}' \geqslant K\right] \leqslant \frac{1/p \text{ tr } \boldsymbol{\Sigma}}{K}$$

Also taking limits as $r \to 0$ and $r \to -\infty$, respectively we get

$$\Pr\left[\, |\boldsymbol{XX}'|^{1/p} \geqslant K\right] \leqslant \frac{|\boldsymbol{XX}'|^{1/p}}{K};$$

and

$$\Pr\left[\text{Ch}_{\min}(\boldsymbol{XX}') = c_p \geqslant K\right] \leqslant \frac{\text{Ch}_{\min}(\boldsymbol{\Sigma})}{K} = \frac{\sigma_p}{K}.$$

3.3. Partial sums of the ordered characteristic roots.

Let $t_{(1)} \leqslant t_{(2)} \leqslant \ldots \leqslant t_{(p)}$ denote ordered values of $(t_1, t_2, \ldots, t_p)$. Then it is easy to verify that $\sum_{i=1}^{q} t_{(i)}$, $1 \leqslant q \leqslant p$, is a concave function of $(t_1, t_2, \ldots, t_p)$. Therefore, for $l_1, \geqslant l_2 \geqslant \ldots \geqslant l_p \geqslant 0$,

$$l_1 t_{(1)} + l_2 t_{(2)} + \ldots + l_p t_{(p)} = \sum_{q=1}^{p-1} (l_q - l_{q+1}) \sum_{i=1}^{q+1} t_{(i)} + l_1 t_{(1)}$$

is also a concave symmetric function of $(t_1, t_2, \ldots, t_p)$. Thus

$$\Pr\left[\sum_{i=1}^{p} l_i c_{(i)} \geqslant K\right] \leqslant \frac{\sum_{i=1}^{p} l_i \sigma_{(i)}}{K}.$$

For $l_1 = l_2 = \ldots = l_p = 1/p$, this reduces to,

$$\Pr\left[1/p \text{ tr } \boldsymbol{XX}' \geqslant K\right] \leqslant \frac{1/p \text{ tr } \boldsymbol{\Sigma}}{K}.$$

For $l_1 = 1$, $l_2 = l_3 = \ldots = l_p = 0$ the inequality reduces to

$$\Pr\left[\text{Ch}_{\min}(\boldsymbol{XX}') \geqslant K\right] \leqslant \frac{\text{Ch}_{\min}(\boldsymbol{\Sigma})}{K}$$

3.4. Elementary symmetric functions

Let $\epsilon_r = \epsilon_r(t_1, t_2, ..., t_p)$ denote the rth elementary symmetric function of $t_1, t_2, ..., t_p$. Then $\epsilon_r^{1/r}$ is a concave function of $(t_1, t_2, ..., t_p)$ on the positive orthant. Thus

$$\Pr[\epsilon_r(c_1, c_2, ..., c_p) \geqslant K)] \leqslant \frac{\epsilon_r^{1/r}(\sigma_1, \sigma_2, ..., \sigma_p)}{K^{1/r}}.$$

When $r = 1$ and $r = p$, this inequality reduces to the inequalities involving $\operatorname{tr}(\boldsymbol{SS}')$ and $|\boldsymbol{XX}'|$ respectively. Furthermore, if $1 \leqslant r \leqslant s \leqslant p$, then $(\epsilon_s/\epsilon_{s-r})^{1/r}$ is also a concave function of $(t_1, t_2, ..., t_p)$ on the non-negative orthant. Therefore,

$$\Pr\left[\frac{\epsilon_s(c_1, c_2, ..., c_p)}{\epsilon_{r-s}(c_1, c_2, ..., c_p)} \geqslant K\right] \leqslant \frac{1}{K^r} \frac{\epsilon_r^{1/r}(\sigma_1, \sigma_2, ..., \sigma_p)}{\epsilon_{r-s}^{1/r}(\sigma_1, \sigma_2, ..., \sigma_p)}.$$

If we take $s = p$, $r = p-1$ we have a probability inequality for $|\boldsymbol{XX}'|/\operatorname{tr}(\boldsymbol{XX}')$.

3.5. Power sums

Let $T_{(r,k)} = T_{(r,k,p)}(t_o, t_2, ..., t_p)$ be defined by

$$T_{(r,k)} = \sum_{i_1+i_2+\cdots+i_p=r} \delta_{i_1}\delta_{i_2} \cdots \delta_{i_p} t_1^{i_1} t_2^{i_2} \cdots t_p^{i_p},$$

where $i_1, i_2, ..., i_p$ and r are nonnegative integers, k is a positive number ($p < k+1$, if k is not an integer), and $\delta_i = \binom{k}{i}$. If $k = 1$ then $T_{(r,k)} = \epsilon_r$. It is known that $T_{(r,k)}^{1/r}$ is a concave function of $(t_1, t_2, ..., t_p)$ on the nonnegative orthant. Thus

$$\Pr[T_{(r,k)}(c_1, c_2, ..., c_p) \geqslant K] \leqslant \frac{T_{(r,k)}^{1/r}(\sigma_1, \sigma_2, ..., \sigma_p)}{K^{1/r}}.$$

Now, if $f_1, f_2, ..., f_n$ are n concave functions of $(t_1, t_2, ..., t_p)$; and $g(u_1, u_2, ..., u_n)$ is a concave nondecreasing function of $(u_1, u_2, ..., u_n)$ then it follows that $g(f_1, f_2, ..., f_n)$ is a concave function of $(t_1, t_2, \ldots, t_p)$. Using this result one may generate more examples.

Remark. I feel that the probability inequalities in this note are *sharp*. However, I have not proved this. For a comprehensive and up to date account of *sharpness* of Tchebycheff type inequalities, see Kemperman [1]. I wish to record my thanks to Kemperman for the manuscript of [1], which has inspired this note.

References

[1] Kemperman, J. H. B. (1965). "On the Sharpness of Tchebycheff Type Inequalities. I, II, III," *Nederl. Akad. Wetensch. Proc. Ser. A.*, 68 = *Indag. Math.*, **27**, 554–571; 572–587; 588–601.

[2] Marcus, M. (1957). "Convex Functions of Quadratic Forms," *Duke Math. J.*, **24**, 321–325.

(Received Sept. 20, 1065.)

m-Associate Cyclical Association Schemes

H. K. NANDI AND B. ADHIKARY, *Calcutta University*

1. INTRODUCTION

The class of partially balanced incomplete block designs (PBIBD) of Bose and Nair (1939), later generalized by Nair and Rao (1942), is very wide and includes many designs suggested from practical considerations from time to time. Bose and Shimamoto (1952) have recognized some distinct sub-classes of two-associate PBIBD's by the distinguishing nature of the association scheme e.g., group divisible, cyclical, triangular, L_2, etc. The group divisible association scheme has been generalized to m-associate classes by Roy (1953, 1955, 1962) to hierarchical group divisible schemes, HGD_m, as also orthogonal group divisible schemes, $OGDL_i$. It is the object of this article to generalize the concept of two-associate cyclical scheme to higher associate classes. The general association scheme will be formulated, and, in particular, the three-associate PBIB designs of this class will be studied in some details. To study the combinatorial properties of these three-associate designs the determinant $|\boldsymbol{NN}'|$ where $\boldsymbol{N}$ is the incidence matrix of the design, will be evaluated. The Hasse-Minkowski p-invariant $C_p(\boldsymbol{NN}')$ has also been determined.

The two-associate cyclical scheme has been defined by Bose and Shimamoto (1952) as follows :—Let the $v-1$ non-zero elements of a module M be divided into two sets containing respectively the elements $(d_1, d_2, \ldots, d_{n_1})$ and $(e_1, e_2, \ldots, e_{n_2})$ where the first set of elements is such that among the $n_1(n_1-1)$ non-zero differences arising out of them, the elements $(d_1, d_2, \ldots, d_{n_1})$ are repeated α times and the elements $(e_1, e_2, \ldots, e_{n_2})$ β times each.

[It may be noted that though it has not been explicit, it is necessary to assume $(d_1, d_2, \ldots, d_{n_1}) \equiv (-d_1, -d_2, \ldots, -d_{n_1})$ if $\alpha = \beta$.] Then the first associates of any treatment i are $i+d_1, i+d_2, \ldots, i+d_{n_1}$ and its second associates $i+e_1, i+e_2, \ldots, i+e_{n_2}$.

For convenience, the group operation will be taken here to be "multiplication" instead of "addition". The "differences" will change to "ratios" and the null element "0" to the unit element "1". In this multiplicative notation, the above two-associate cyclical scheme may be restated as follows :—Consider an abelian group G consisting of v elements. Let it be possible to divide the non-unit elements of G into two disjoint sets A and B with n_1 and n_2 elements respectively, one containing its inverse elements also. Let the elements of the set A be such that among the $n_1(n_1-1)$ non-units ratios arising out of them, the elements of A are repeated α times and those of B β times each. Then the first associates of any element θ are θA and its second associates θB. The parameters of the association scheme are known to be

$$v,\ n_1,\ n_2\ ;$$

$$(1.1) \qquad (p^1_{jk}) = \begin{pmatrix} \alpha & n_1-\alpha-1 \\ n_1-\alpha-1 & n_2-n_1+\alpha+1 \end{pmatrix}$$

$$(p^2_{jk}) = \begin{pmatrix} \beta & n_1-\beta \\ n_1-\beta & n_2-n_1+\beta-1 \end{pmatrix}$$

The only restriction on α and β is

$$(1.2) \qquad n_1\alpha+n_2\beta = n_1(n_1-1).$$

For various values of α, β subject to (1.2) different sub-classes of the above will be obtained. Though all the examples given by Bose and Shimamoto (1952) correspond to $n_1 = n_2$ and $\beta = n_1 - \alpha - 1$, this is not any necessary condition. It may be observed that corresponding to (i) $\beta = 0$ and (ii) $\alpha = n_1 - n_2 - 1$, the cyclical association scheme reduces to a group divisible one. Further, it may be noted that the L_2 association scheme is a particular sub-class of the cyclical one. To see this, consider the abelian group G of n^2 elements formed by the powers of a, b and their products where $a^n = b^n = 1$. Take the set A to be $(a, a^2, \ldots, a^{n-1}, b, b^2, \ldots, b^{n-1})$ and $B = G - 1 - A$. Writing the elements in the form of a square as

$$\begin{array}{ccccc} 1 & a & a^2 & \cdots & a^{n-1} \\ b & ab & a^2b & \cdots & a^{n-1}b \\ \cdots & \cdots & \cdots & \cdots & \cdots \\ b^{n-1} & ab^{n-1} & a^2b^{n-1} & \cdots & a^{n-1}b^{n-1} \end{array}$$

and remembering that the first and second associates of θ are θA and θB respectively, it is easily seen that two elements in a row or column are first associates and those not in a row or column second associates.

2. GENERAL ASSOCIATION SCHEME

We first prove a lemma which will be used later.

Lemma. *In a group consisting of n elements, among the set of $n(n-1)$ ratios that arise by taking the ratios of distinct elements of the group, the non-unit elements of the group are each repeated n times.*

Proof. Let c be any non-unit element of G. This will occur among the ratios as many times as the number of distinct pairs of elements a, b satisfying $ac^{-1} = b$. As a takes the n distinct values, b also takes n distinct values. Hence the result.

Consider an abelian group G consisting of v elements. Let it be possible to decompose G into $(m-l)$ direct factors as

$$G = G_1 \otimes G_2 \otimes \cdots \otimes G_{m-l} \tag{2.1}$$

where G_i consists of m_i elements. Let among these $(m-l)$ groups, a set of l groups, say the i_1th, i_2th, ..., i_lth groups admit the following further decomposition

$$G_{i_j}-1 = A_{i_j} \cup B_{i_j}, \quad j = 1, 2, ..., l \tag{2.2}$$

where $A_{i_j} \cap B_{i_j} = 0$, A_{i_j} consists of m'_{i_j} distinct elements and B_{i_j}, m''_{i_j} distinct elements and further A_{i_j} contains also its inverse elements. Obviously,

$$m'_{i_j} + m''_{i_j} = m_{i_j} - 1 \tag{2.3}$$

Let, further, the elements of A_{i_j} be such that among the non-unit ratios arising out of them, the elements of A_{i_j} appear α_{i_j} times and the elements of B_{i_j} β_{i_j} times each. So

$$m'_{i_j} \alpha_{i_j} + m''_{i_j} \beta_{i_j} = m'_{i_j}(m'_{i_j} - 1). \tag{2.4}$$

Without any loss of generality, it may be assumed that $i_1 < i_2 < ... < i_l$.

Let us define the different associates of any element θ as follows :

1st associates	:	$\theta(G_1 - 1)$
2nd ,,	:	$\theta\, G_1 \otimes (G_2 - 1)$
....		
(i_1-1)th ,,	:	$\theta G_1 \otimes G_2 \otimes \dots \otimes (G_{i_1-1} - 1)$
i_1th ,,	:	$\theta G_1 \otimes G_2 \otimes \dots \otimes G_{i_1-1} \otimes A_{i_1}$
(i_1+1)th ,,	:	$\theta G_1 \otimes G_2 \otimes \dots \otimes G_{i_1-1} \otimes B_{i_1}$
(i_1+2)th ,,	:	$\theta G_1 \otimes G_2 \otimes \dots \otimes G_{i_1} \otimes (G_{i_1+1} - 1)$

(2.5) ....

i_2th ,, : $\theta G_1 \otimes G_2 \otimes \dots \otimes (G_{i_2-1}-1)$

(i_2+1)th ,, : $\theta G_1 \otimes G_2 \otimes \dots \otimes G_{i_2-1} \otimes A_{i_2}$

(i_2+2)th ,, : $\theta G_1 \otimes G_2 \otimes \dots \otimes G_{i_2-1} \otimes B_{i_2}$

(i_2+3)th ,, : $\theta G_1 \otimes G_2 \otimes \dots \otimes (G_{i_2+1}-1)$

....

mth ,, : $\theta G_1 \otimes G_2 \otimes \dots \otimes (G_{m-l}-1)$

Theorem 1. *The association scheme defined by* (2.1) *to* (2.5) *above is a PBIB association scheme with the parameters of the first kind as*

$$v = m_1 m_2 \dots m_{m-l}$$

$$n_1 = m_1 - 1,\ n_2 = m_1(m_2-1),\ \dots,\ n_{i_1-1} = m_1 m_2 \dots m_{i_1-2}(m_{i_1-1}-1),$$

$$n_{i_1} = m_1 m_2 \dots m_{i_1-1} m'_{i_1},\ n_{i_1+1} = m_1 m_2 \dots m_{i_1-1} m''_{i_1},$$

$$n_{i_1+2} = m_1 m_2 \dots m_{i_1}(m_{i_1+1}-1),\ \dots$$

$$n_{i_2} = m_1 m_2 \dots m_{i_2-2}(m_{i_2-1}-1),$$

$$n_{i_2+1} = m_1 m_2 \dots m_{i_2-1} m'_{i_2},$$

$$n_{i_2+2} = m_1 m_2 \dots m_{i_2-1} m''_{i_2},$$

$$n_{i_2+3} = m_1 m_2 \dots m_{i_2}(m_{i_2+1}-1),\ \dots$$

$$n_m = m_1 m_2 \dots m_{m-l-1}(m_{m-l}-1).$$

$$(p^w_{jk}) = \left[\begin{array}{c:c:c} \mathbf{0}_{(w-1)\times(w-1)} & \boldsymbol{x}_{(w-1)} & \mathbf{0}_{(w-1)\times(m-w)} \\ \hdashline \boldsymbol{x}'_{(w-1)} & & \\ \hdashline \mathbf{0}_{(m-w)\times(w-1)} & \multicolumn{2}{c}{\boldsymbol{D}_{(m-w+1)\times(m-w+1)}} \end{array}\right]$$

for $w = 1, 2, \dots, m$ except $i_1, i_1+1, \dots, i_l+(l-1), i_l+l$; where $\mathbf{0}_{i\times j}$ is an $i\times j$ null matrix, $\boldsymbol{x}'_{(i)} = (n_1, n_2, \dots, n_i)$ and $\boldsymbol{D}_{(m-w+1)\times(m-w+1)}$

is a diagonal matrix whose diagonal elements are $m_1 m_2 \ldots m_{w-1}$ (m_w-2), $n_{w+1}, \ldots, n_m$ respectively.

$$(p_{jk}^t) = \begin{pmatrix} \boldsymbol{A}_{(t+1)\times(t+1)} & \boldsymbol{0}_{(t+1)\times(m-t-1)} \\ \boldsymbol{0}_{(m-t-1)\times(t+1)} & \boldsymbol{D}_{(m-t-1)\times(m-t-1)} \end{pmatrix}$$

$$(p_{jk}^{t+1}) = \begin{pmatrix} \boldsymbol{B}_{(t+1)\times(t+1)} & \boldsymbol{0}_{(t+1)\times(m-t-1)} \\ \boldsymbol{0}_{(m-t-1)\times(t+1)} & \boldsymbol{D}_{(m-t-1)\times(m-t-1)} \end{pmatrix}$$

where

$$\boldsymbol{A}_{(t+1)\times(t+1)} = \left[\begin{array}{c|c:c} \boldsymbol{0}_{(t-1)\times(t-1)} & \boldsymbol{x}_{(t-1)} & \boldsymbol{0}_{(t-1)\times 1} \\ \hdashline \boldsymbol{x}'_{(t-1)} & \multicolumn{2}{c}{} \\ \cdots & \multicolumn{2}{c}{\boldsymbol{X}} \\ \boldsymbol{0}_{1\times(t-1)} & \multicolumn{2}{c}{} \end{array}\right]$$

$$\boldsymbol{B}_{(t+1)\times(t+1)} = \left[\begin{array}{c|c:c} \boldsymbol{0}_{(t-1)\times(t-1)} & \boldsymbol{0}_{(t-1)\times 1} & \boldsymbol{x}_{(t-1)} \\ \hdashline \boldsymbol{0}_{1\times(t-1)} & \multicolumn{2}{c}{} \\ \cdots & \multicolumn{2}{c}{\boldsymbol{Y}} \\ \boldsymbol{x}'_{(t-1)} & \multicolumn{2}{c}{} \end{array}\right]$$

$$\boldsymbol{X} = \begin{pmatrix} m_1 m_2 \ldots m_{t-1}\alpha_t & m_1 m_2 \ldots m_{t-1}(m'_t-\alpha_t-1) \\ m_1 m_2 \ldots m_{t-1}(m'_t-\alpha_t-1) & m_1 m_2 \ldots m_{t-1}(m'_t-m'_t+\alpha_t+1) \end{pmatrix}$$

$$\boldsymbol{Y} = \begin{pmatrix} m_1 m_2 \ldots m_{t-1}\beta_t & m_1 m_2 \ldots m_{t-1}(m'_t-\beta_t) \\ m_1 m_2 \ldots m_{t-1}(m'_t-\beta_t) & m_1 m_2 \ldots m_{t-1}(m'_t-m'_t+\beta_t-1) \end{pmatrix}$$

$\boldsymbol{D}_{(m-t-1)\times(m-t-1)}$ is a diagonal matrix with diagonal elements as $n_{t+2}, n_{t+3}, \ldots, n_m$; other matrices have been defined already and

$$(t, \alpha_t) = (i_1, \alpha_{i_1}), (i_2+1, \alpha_{i_2}), \ldots, (i_l+(l-1), \alpha_{i_l})$$

$$(t, \beta_t) = (i_1, \beta_{i_1}), (i_2+1, \beta_{i_2}), \ldots, (i_l+(l-1), \beta_{i_l}).$$

Proof. We prove the different parts of the theorem separately.

(a) We first prove the symmetry of associateship. Let ϕ be any treatment which is ith associate of θ. We are to prove that θ is an ith associate of ϕ. As ϕ is an ith associate of θ, so

$\phi \epsilon \theta G_1 \otimes G_2 \otimes \ldots \otimes (G_i - 1)$, say. Hence ϕ can be expressed as $\phi = \theta g_1 g_2 \ldots g_{i-1} g_i'$ where $g_j \epsilon G_j$, $j = 1, 2, \ldots, i-1$ and g_i is a non-unit element of G_i. Since $G_1, G_2, \ldots, G_i$ are all groups, $g_1, g_2, \ldots, g_{i-1}, g_i'$ have unique inverses in the corresponding groups. So $\theta = \phi(g_1 g_2 \ldots g_{i-1} g_i')^{-1} = \phi(g_i')^{-1} g_{i-1}^{-1} \ldots g_2^{-1} g_1^{-1} \epsilon \phi(G_i - 1) \otimes G_{i-1} \otimes \ldots \otimes G_2 \otimes G_1$. Hence $\theta \epsilon \phi G_1 \otimes G_2 \otimes \ldots \otimes (G_i - 1)$, by definition of direct product, and is an ith associate of ϕ. For other types of ith associates the proof is similar, remembering the sets A_{i_j}, B_{i_j} are such by definition that they contain their respective inverses.

(b) As regards the number of associates of different types the result is evident by counting. In fact we can count the number of ith associates as the number of elements contained in $\theta G_1 \otimes G_2 \otimes \ldots \otimes (G_i - 1)$, which is n_i, given above, independently of the element θ.

(c) Here we shall actually calculate some p^i_{jk} values in course of which it will be automatically proved that p^i_{jk} are constants independently of the elements.

For values of w and t as stated in the theorem

p^w_{ww} = No. of elements common between the wth associates of θ and the wth associates of ϕ when θ and ϕ are wth associates of each other

= No. of elements common between $\theta G_1 \otimes G_2 \otimes \ldots (G_w - 1)$ and $\phi G_1 \otimes G_2 \otimes \ldots \otimes (G_w - 1)$ when $\phi \epsilon \theta G_1 \otimes G_2 \otimes \ldots (G_w - 1)$.

As $\phi \epsilon \theta G_1 \otimes G_2 \otimes \ldots \otimes (G_w - 1)$, ϕ can be expressed as $\phi = \theta g_1 g_2 \ldots g_{w-1} g_w'$ where $g_j \epsilon G_j$, $j = 1, 2, \ldots, w-1$ and g_w' is a non-unit element of G_w.

So

p^w_{ww} = No. of elements common between $\theta G_1 \otimes G_2 \otimes \ldots \otimes (G_w - 1)$ and $\theta g_1 g_2 \ldots g_{w-1} g_w' \, G_1 \otimes G_2 \otimes \ldots \otimes (G_w - 1)$

= No. of elements common between $\theta G_1 \otimes G_2 \otimes \ldots (G_w - 1)$ and $\theta G_1 \otimes G_2 \otimes \ldots \otimes G_{w-1}(G_w - g_w')$

$= m_1 m_2 \ldots m_{w-1}(m_w - 2)$.

For $i \neq w$,

p^w_{iw} = No. of elements common between $\theta G_1 \otimes G_2 \otimes \ldots (G_i - 1)$ and $\phi G_1 \otimes G_2 \otimes \ldots \otimes (G_w - 1)$ when $\phi = \theta g_1 g_2 \ldots g_{w-1} g'_w$

= No. of elements common between $\theta G_1 \otimes G_2 \otimes \ldots \otimes (G_i - 1)$ and $\theta G_1 \otimes G_2 \otimes \ldots \otimes G_{w-1} \otimes (G_w - g'_w)$

$= m_1 m_2 \ldots m_{i-1} = n_i$ for $w > i$

$= 0$ for $i > w$.

p^t_{tt} = No. of elements common between $\theta G_1 \otimes G_2 \otimes \ldots G_{t-1} \otimes A_t$ and $\phi G_1 \otimes G_2 \otimes \ldots \otimes G_{t-1} \otimes A_t$ when $\phi = \theta g_1 g_2 \ldots g_{t-1} a_t$, $g_j \in G_j$, $a_t \in A_t$.

$= m_1 m_2 \ldots m_{t-1}$ [No. of elements common between A_t and $a_t A_t$]

$= m_1 m_2 \ldots m_{t-1} \alpha_t$.

Other p^i_{jk} parameters of the theorem can be verified in the same way, making use of the lemma proved earlier.

Corollary 1. If $l = 0$, then the m-associate cyclical scheme reduces to a m-associate hierarchical group divisible association scheme (HGD_m), and conversely.

Proof. In this case it is found that the parameters can be identified with those of HGD_m by renaming the different associates. As Roy (1955, 1962) has proved the uniqueness of the HGD_m association scheme, the results follow immediately.

Corollary 2. If $\alpha_t = m'_t - 1$, $\beta_t = 0$, $m''_t = m'_t(m'_t + 1)$ for all t, α_t, β_t, m'_t in Theorem 1, then again the association scheme reduces to that of HGD_m with $N_{m-t} = N_{m-t-1}$ (in Roy's notation), and conversely.

Proof. Same as in Corollary 1.

3. THREE ASSOCIATE CYCLICAL SCHEMES

The above generalisation of the cyclical association scheme gives rise to a large number of possibilities. Thus, for $m = \xi$, we may take $l = 0, 1, 2, \ldots, \xi/2$ or $[(\xi-1)/2]$ according as ξ is even or odd. Again there are $\binom{\xi-l}{l}$ distinct types depending on the positions of the l groups to be decomposed further. So, there are $\sum_l \binom{\xi-l}{l}$ distinct types of cyclical schemes for the ξ-associate scheme. Thus for the 2-associate scheme there are only two types of cyclical association schemes, viz., when $l = 0$ and 1. For the 3-associate case also, there are two values of l, viz. $l = 0, 1$; but there are altogether three types of generalizations. For $l = 0$, we get the well-known *HGD* designs of Roy (1953). For $l = 1$, there are two distinct types according as the first or second direct factor group is decomposed. Conventionally, we may suppose that the group G_2 is decomposed in both types of generalization.

Consider an abelian group G consisting of $v = m_1 n$ elements. Let it be possible to decompose G into two direct factors: $G = G_1 \otimes G_2$ where G_1 consists of m_1 elements $1, d_1, d_2, \ldots, d_{m_1-1}$, while the non-unit elements of G_2 can be divided into two disjoint sets A and B of m_2 and m_3 elements respectively, i.e., $A = (e_1, e_2, \ldots, e_{m_2})$ and $B = (f_1, f_2, \ldots, f_{m_3})$. Further let the elements of A be such that all the inverse elements are also in A and that among the $m_2(m_2-1)$ non-unit ratios arising out of them, the elements of A are repeated α times and those of B β times each. Obviously, $n = m_2+m_3+1$, $m_2\alpha+m_3\beta = m_2(m_2-1)$.

First type of generalization.

Let the first associates of any treatment θ be $\theta(G_1-1)$, its second associates $\theta G_1 \otimes A$ and its third associates $\theta G_1 \otimes B$.

The parameters of the association scheme are :

(3.1) $$v = m_1(m_2+m_3+1) = m_1 n,\ n_1 = m_1-1,\ n_2 = m_1 m_2,\ n_3 = m_1 m_3$$

(3.2) $$(p^1_{ij}) = \begin{bmatrix} m_1-2 & 0 & 0 \\ 0 & m_1 m_2 & 0 \\ 0 & 0 & m_1 m_3 \end{bmatrix}$$

$$(3.3) \qquad (p^2_{ij}) = \begin{bmatrix} 0 & m_1-1 & 0 \\ m_1-1 & m_1\alpha & m_1(m_2-\alpha-1) \\ 0 & m_1(m_2-\alpha-1) & m_1(m_3-m_2+\alpha+1) \end{bmatrix}$$

$$(3.4) \qquad (p^3_{ij}) = \begin{bmatrix} 0 & 0 & m_1-1 \\ 0 & m_1\beta & m_1(m_2-\beta) \\ m_1-1 & m_1(m_2-\beta) & m_1(m_3-m_2+\beta-1) \end{bmatrix}$$

It follows from the above that if θ and ϕ are ith associates of each other, then all the first associates of θ (with the exception of ϕ in case $i = 1$) are ith associates of ϕ and conversely; $i = 1, 2, 3$.

Second type of generalization.

Let the first associates of any treatment θ be θA, second associates θB and third associates $\theta G_2 \otimes (G_1-1)$. The parameters of the association scheme are

$$(3.5) \qquad v = m_1(m_2+m_3+1), \quad n_1 = m_2, \quad n_2 = m_3$$
$$n_3 = (m_1-1)(m_2+m_3+1).$$

$$(3.6) \quad (p^1_{jk}) = \begin{bmatrix} \alpha & m_2-\alpha-1 & 0 \\ m_2-\alpha-1 & m_3-m_2+\alpha+1 & 0 \\ 0 & 0 & (m_1-1)(m_2+m_3+1) \end{bmatrix}$$

$$(3.7) \quad (p^2_{jk}) = \begin{bmatrix} \beta & m_2-\beta & 0 \\ m_2-\beta & m_3-m_2+\beta-1 & 0 \\ 0 & 0 & (m_1-1)\,(m_2+m_3+1) \end{bmatrix}$$

$$(3.8) \quad (p^3_{jk}) = \begin{bmatrix} 0 & 0 & m_2 \\ 0 & 0 & m_3 \\ m_2 & m_3 & (m_1-2)(m_2+m_3+1) \end{bmatrix}$$

It immediately follows that if θ and ϕ are two treatments which are third associates, then all the first and second associates of θ

(including θ itself) are third associates of ϕ and conversely. But if θ and ϕ are first or second associates, then all the third associates of θ are third associates of ϕ and conversely.

Parameters of both the generalizations depend on the values of α and β. Three major sub-classes of them will be as follows :—

(I) $\alpha = 0, \quad \beta \neq 0$

(II) $\alpha \neq 0, \quad \beta = 0$

(III) $\alpha \neq 0, \quad \beta \neq 0.$

Of these three sub-classes, sub-class (III) will be discussed hereafter, while sub-classes (I) and (II) will be discussed elsewhere.

Some examples of PBIB designs with this three-associate cyclical association scheme will be given below.

Design 1 (First type). Consider the group G formed by products of a, b, c where $a^3 = b^3 = c^2 = 1$. Take $G_1 = (1, a^2b, ab^2)$; $A = (a^2b^2, ab)$ and $B = (c, abc, a^2b^2c)$. Here $m_1 = 3$, $m_2 = 2$, $m_3 = 3$, $\alpha = 1$, $\beta = 0$. Among the non-unit ratios of the elements of A, only the elements of A appear once each. The p^i_{jk} parameters are as follows :

$$(p^1_{jk}) = \begin{bmatrix} 1 & 0 & 0 \\ 0 & 6 & 0 \\ 0 & 0 & 9 \end{bmatrix}, \qquad (p^2_{jk}) = \begin{bmatrix} 0 & 2 & 0 \\ 2 & 3 & 0 \\ 0 & 0 & 9 \end{bmatrix}$$

$$(p^3_{jk}) = \begin{bmatrix} 0 & 0 & 2 \\ 0 & 0 & 6 \\ 2 & 6 & 0 \end{bmatrix}.$$

Let us consider the initial block $(1, a, b, a^2b^2)$. It may be verified that the block gives the following ratios :

a^2b and ab^2 each repeated thrice

a^2b^2, ab, a, b, a^2 and b^2 each repeated once.

Hence on developing the block we get the solution of the design :

$$v = b = 18, \; r = k = 4, \; \lambda_1 = 3, \; \lambda_2 = 1, \; \lambda_3 = 0$$

$$n_1 = 2, \; n_2 = 6 \text{ and } n_3 = 9.$$

The layout is

$(1, a, b, a^2b^2)$	(c, ac, bc, a^2b^2c)
(a, a^2, ab, b^2)	(ac, a^2c, abc, b^2c)
$(a^2, 1, a^2b, ab^2)$	(a^2c, c, a^2bc, ab^2c)
(b, ab, b^2, a^2)	(bc, abc, b^2c, a^2c)
$(b^2, ab^2, 1, a^2b)$	(b^2c, ab^2c, c, a^2bc)
$(ab, a^2b, ab^2, 1)$	(abc, a^2bc, ab^2c, c)
(a^2b^2, b^2, a^2, ab)	$(a^2b^2c, b^2c, a^2c, abc)$
(ab^2, a^2b^2, a, b)	(ab^2c, a^2b^2c, ac, bc)
(a^2b, b, a^2b^2, a)	(a^2bc, bc, a^2b^2c, ac)

Design 2 (First type). Consider the group formed by a, b, c and their products when $a^3 = b^5 = c^2 = 1$. Let $G_1 = (1, c)$; $A = (a, a^2)$ and $B = (b, b^2, b^3, b^4, ab, ab^2, ab^3, ab^4, a^2b, a^2b^2, a^2b^3, a^2b^4)$. Here $m_1 = 2$, $m_2 = 2$, $m_3 = 12$, $\alpha = 1$, $\beta = 0$. The parameters (p^i_{jk}) are :

$$(p^1_{jk}) = \begin{bmatrix} 0 & 0 & 0 \\ 0 & 4 & 0 \\ 0 & 0 & 24 \end{bmatrix}; \quad (p^2_{jk}) = \begin{bmatrix} 0 & 1 & 0 \\ 1 & 2 & 0 \\ 0 & 0 & 24 \end{bmatrix};$$

$$(p^3_{jk}) = \begin{bmatrix} 0 & 0 & 1 \\ 0 & 0 & 4 \\ 1 & 4 & 18 \end{bmatrix}$$

Consider the initial block $(1, a, ac)$. On developing the block we get the solution of the design :

$$v = b = 30, \quad r = k = 3, \quad \lambda_1 = 2, \quad \lambda_2 = 1, \quad \lambda_3 = 0,$$
$$n_1 = 1, \quad n_2 = 4, \quad n_3 = 24.$$

Design 3 (Second type). Consider the group formed by a, b and their different powers, when $a^6 = b^2 = 1$. Let G_1

$= (1, b)$, $A = (a^2, a^4)$ $B = (a, a^3, a^5)$. So $m_1 = 2$, $m_2 = 2$, $m_3 = 3$, $\alpha = 1$, $\beta = 0$. The (p^i_{jk}) parameters are

$$(p^1_{jk}) = \begin{bmatrix} 1 & 0 & 0 \\ 0 & 3 & 0 \\ 0 & 0 & 6 \end{bmatrix}; \quad (p^2_{jk}) = \begin{bmatrix} 0 & 2 & 0 \\ 2 & 0 & 0 \\ 0 & 0 & 6 \end{bmatrix};$$

$$(p^3_{jk}) = \begin{bmatrix} 0 & 0 & 2 \\ 0 & 0 & 3 \\ 2 & 3 & 0 \end{bmatrix}$$

Let us consider the initial blocks $(1, a^2, a^4, b)$, $(1, a^2, a^4, ab)$. On developing these blocks we get the solution of the design :

$v = 12$, $b = 24$, $r = 8$, $k = 4$, $\lambda_1 = 6$, $\lambda_2 = 0$, $\lambda_3 = 1$,
$n_1 = 2$, $n_2 = 3$, $n_3 = 6$.

Design 4 (Second type). Consider the group formed by a, b and their powers with $a^{15} = b^2 = 1$. Let $G_1 = (1, b)$; $A = (a^5, a^{10})$; $B = (a, a^2, a^3, a^4, a^6, a^7, a^8, a^9, a^{11}, a^{12}, a^{13}, a^{14})$. So, $m_1 = 2$, $m_2 = 2$, $m_3 = 12$, $\alpha = 1$, $\beta = 0$. The parameters (p^i_{jk}) are

$$(p^1_{jk}) = \begin{bmatrix} 1 & 0 & 0 \\ 0 & 12 & 0 \\ 0 & 0 & 15 \end{bmatrix}; \quad (p^2_{jk}) = \begin{bmatrix} 0 & 2 & 0 \\ 2 & 9 & 0 \\ 0 & 0 & 15 \end{bmatrix};$$

$$(p^3_{jk}) = \begin{bmatrix} 0 & 0 & 2 \\ 0 & 0 & 12 \\ 2 & 12 & 0 \end{bmatrix}.$$

Let us consider the set of 5 blocks :

$(1, ab, a^6b, a^{11}b)$, $(1, a^2b, a^7b, a^{12}b)$,
$(1, a^3b, a^8b, a^{13}b)$, $(1, a^4b, a^9b, a^{14}b)$, $(1, a^5, a^{10}, b)$.

On developing these blocks we get the solution of the design :

$v = 30$, $b = 150$, $r = 20$, $k = 4$, $\lambda_1 = 15$, $\lambda_2 = 0$, $\lambda_3 = 2$,
$n_1 = 2$, $n_2 = 12$, $n_3 = 15$.

4. COMBINATORIAL PROPERTIES OF CYCLICAL THREE-ASSOCIATE *PBIB* DESIGNS

As is well-known, many combinatorial properties of PBIB designs follow from the positive semi-definiteness of the determinant $|NN'|$. Further, the Hasse-Mankowski p-invariant $C_p(NN')$ also helps in proving the non-existence of certain classes of designs. Accordingly, we shall evaluate here the determinant $|NN'|$ and $C_p(NN')$ for both the types of generalizations. It may be noted that only sub-class (III) is being considered in this paper.

Evaluation of $|NN'|$. (First type of generalization)—Sub-class (III).

$$|NN'| = \begin{vmatrix} \boldsymbol{A} & \boldsymbol{B}_1 \ldots \boldsymbol{B}_1 & \boldsymbol{B}_2 \ldots \boldsymbol{B}_2 \\ \boldsymbol{B}_1 & \boldsymbol{A} \ldots \ldots \ldots & \ldots \ldots \\ \ldots & \ldots \ldots \ldots \ldots & \ldots \ldots \\ \boldsymbol{B}_1 & \ldots \ldots \ldots \boldsymbol{A} & \ldots \ldots \\ \boldsymbol{B}_2 & \ldots \ldots \ldots \ldots & \boldsymbol{A} \ldots \\ \ldots & \ldots \ldots \ldots \ldots & \ldots \ldots \\ \boldsymbol{B}_2 & \ldots \ldots \ldots \ldots & \ldots \boldsymbol{A} \end{vmatrix} : \begin{array}{l} m_1(m_2+m_3+1) \\ \times m_1(m_2+m_3+1) \end{array} \tag{4.1}$$

where

$$\boldsymbol{A} = \begin{bmatrix} r & \lambda_1 & \ldots & \lambda_1 \\ \lambda_1 & r & \ldots & \lambda_1 \\ \ldots & \ldots & \ldots & \ldots \\ \lambda_1 & \lambda_1 & \ldots & r \end{bmatrix} : \quad m_1 \times m_1$$

$$\boldsymbol{B}_1 = \begin{bmatrix} \lambda_2 & \lambda_2 & \ldots & \lambda_2 \\ \ldots & \ldots & \ldots & \ldots \\ \lambda_2 & \lambda_2 & \ldots & \lambda_2 \end{bmatrix} : \quad m_1 \times m_1$$

$$B_2 = \begin{bmatrix} \lambda_3 & \lambda_3 & \cdots & \lambda_3 \\ \cdots & \cdots & \cdots & \cdots \\ \lambda_3 & \lambda_3 & \cdots & \lambda_3 \end{bmatrix} : \quad m_1 \times m_1$$

and in each row and column of the above determinant, $\boldsymbol{A}$ occurs in the main diagonal position and other (m_2+m_3) positions are occupied by m_2 $\boldsymbol{B}_1$'s and m_3 $\boldsymbol{B}_2$'s. Each column of the determinant is a m_1 $(m_2+m_3+1)\times m_1$ matrix and will be referred to as a sub-matrix. Consider the first sub-matrix

$$\begin{bmatrix} \boldsymbol{A} \\ \boldsymbol{B}_1 \\ \vdots \\ \boldsymbol{B}_1 \\ \boldsymbol{B}_2 \\ \vdots \\ \boldsymbol{B}_2 \end{bmatrix} \tag{4.2}$$

Substract the last column of (4.2) from its preceding (m_1-1) columns. Repeat this operation on the other (m_2+m_3) sub-matrices. These operations do not affect the value of the determinant (4.1). Finally, in the transformed determinant add

1st, 2nd, ..., (m_1-1)th rows to the m_1th row

(m_1+1)th, ..., $(2m_1-1)$th rows to the $(2m_1)$th row

....

$m_1(m_2+m_3)$th, ..., $[m_1(m_2+m_3+1)-1]$th rows to $m_1(m_2+m_3+1)$th row.

Then (4.1) reduces to

$$(r-\lambda)^{(m_1-1)(m_2+m_3+1)} |\boldsymbol{D}|, \tag{4.3}$$

where

(4.4)

$$
D = \begin{bmatrix} r+\lambda_1(m_1-1) & m_1\lambda_2 & \cdots & m_1\lambda_2 & \cdots & m_1\lambda_3) \\ m_1\lambda_2 & r+\lambda_1(m_1-1) & \cdots & \cdots & \cdots \\ \cdots\cdots & \cdots\cdots & \cdots\cdots & \cdots\cdots & \cdots\cdots & \cdots\cdots \\ m_1\lambda_3 & \cdots\cdots\cdots & & & & r+\lambda_1(m_1-1) \end{bmatrix}
$$

D is a $(m_2+m_3+1)\times(m_2+m_3+1)$ matrix and can be easily identified as the NN' of a two-associate PBIBD of the following form

$$
\begin{bmatrix} R & \Lambda_1 \cdots & \Lambda_1 & \Lambda_2 \cdots & \Lambda_2 \\ \Lambda_1 & R & \cdots & \cdots & \cdots \\ \cdots & \cdots & \cdots & \cdots & \cdots \\ \Lambda_1 & \cdots & R & \cdots & \cdots \\ \cdots & \cdots & \cdots & \cdots & \cdots \\ \Lambda_2 & \cdots & \cdots & \cdots & R \end{bmatrix}
$$

where

$$R = r+\lambda_1(m_1-1),\ \Lambda_{i-1} = \lambda_i m_1,\quad i = 2, 3.$$

Thus

$$|D| = (R-Z_1)^{\alpha_1}(R-Z_2)^{\alpha_2}[R+m_2\Lambda_1+m_3\Lambda_2] \tag{4.5}$$

where the process of evaluating Z_i's and α_i's has been given by Connor and Clatworthy (1954). Following these authors, we obtain $\alpha_1 = \dfrac{m_3}{1+m_2}$ and $\alpha_2 = m_2+m_3-\dfrac{m_3}{1+m_2}$. As α_1 and α_2 are integers, put $\dfrac{m_3}{1+m_2} = p$ (an integer). Then we have

$$
\begin{aligned}
|NN'| &= rk(r-\lambda_1)^{(m_1-1)(m_2+m_3+1)}[r+\lambda_1(m_1-1)-m_1\lambda_2]^p. \\
&\qquad \times(rk-v\lambda_3)^{m_2+m_3-p}, \quad \text{if } \lambda_2 < \lambda_3 \\
&= rk(r-\lambda_1)^{(m_1-1)(m_2+m_3+1)}[r+\lambda_1(m_1-1)-m_1\lambda_2]^{m_2+m_3-p} \\
&\qquad \times(rk-v\lambda_3)^p, \quad \text{if } \lambda_2 > \lambda_3.
\end{aligned}
$$

Second type of generalization-sub-class (III)

Here $|\boldsymbol{NN'}|$ can be written as

$$(4.6)\qquad |\boldsymbol{NN'}| = \begin{vmatrix} \boldsymbol{A} & \boldsymbol{B} & \boldsymbol{B} & \dots & \boldsymbol{B} \\ \boldsymbol{B} & \boldsymbol{A} & \boldsymbol{B} & \dots & \boldsymbol{B} \\ \dots & \dots & \dots & \dots & \dots \\ \boldsymbol{B} & \boldsymbol{B} & \boldsymbol{B} & \dots & \boldsymbol{A} \end{vmatrix} \quad : m_1(m_2+m_3+1) \times m_1(m_2+m_3+1)$$

where

$$\boldsymbol{A} = \begin{bmatrix} r & \lambda_1 & \dots & \lambda_1 & \lambda_2 & \dots & \lambda_2 \\ \lambda_1 & \dots & \dots & \dots & & \dots & \dots \\ \dots & \dots & \dots & \dots & & \dots & \dots \\ \lambda_1 & \dots & \dots & \dots & & \dots & \dots \\ \lambda_2 & \dots & \dots & \dots & & \dots & \dots \\ \dots & \dots & \dots & \dots & & \dots & \dots \\ \lambda_2 & \dots & \dots & \dots & & \dots & r \end{bmatrix} : (m_2+m_3+1) \times (m_2+m_3+1)$$

$$\boldsymbol{B} = \begin{bmatrix} \lambda_3 & \lambda_3 & \dots & \dots & \dots & \lambda_3 \\ \lambda_3 & \lambda_3 & \dots & \dots & \dots & \lambda_3 \\ \dots & \dots & \dots & \dots & \dots & \dots \\ \lambda_3 & \lambda_3 & \dots & \dots & \dots & \lambda_3 \end{bmatrix} : (m_2+m_3+1) \times (m_2+m_3+1)$$

Hence

$$|\boldsymbol{NN'}| = |\boldsymbol{A}-\boldsymbol{B}|^{m_1-1} |\boldsymbol{A}+(m_1-1)\boldsymbol{B}|$$

Putting

$$R = r-\lambda_3,\ \Lambda_1 = \lambda_1-\lambda_3,\ \Lambda_2 = \lambda_2-\lambda_3,$$

and

$$R' = r+(m_1-1)\lambda_3,\ \Lambda_1' = \lambda_1+(m_1-1)\lambda_3,\ \Lambda_2' = \lambda_2+(m_1-1)\lambda_3$$

it is found that both $|\boldsymbol{A}-\boldsymbol{B}|$ and $|\boldsymbol{A}+(m_1-1)\ \boldsymbol{B}|$ are of the same form as $|\boldsymbol{A}|$ when $(r, \lambda_1, \lambda_2)$ is replaced by $(R, \Lambda_1, \Lambda_2)$ or

$(R', \Lambda_1', \Lambda_2')$. But $|A|$ is evaluated easily as it is the determinant $|NN'|$ of a 2-associate PBIBD. Thus we obtain (4.6) as

$$|NN'| = rk\,[r+m_2\lambda_1+m_3\lambda_2-(1+m_2+m_3)\lambda_3]^{m_1-1}[r-\lambda_1]^{\frac{m_1 m_3}{1+m_2}}$$
$$\times[r+m_2\lambda_1-(1+m_2)\lambda_2]^{m_1\left(m_2+m_3-\frac{m_3}{1+m_2}\right)}, \text{ if } \lambda_1 < \lambda_2$$
$$= rk\,[r+m_2\lambda_1+m_3\lambda_2-(1+m_2+m_3)\lambda_3]^{m_1-1}$$
$$\times[r-\lambda_1]^{m_1\left(m_2+m_3-\frac{m_3}{1+m_2}\right)}$$
$$\times[r+m_2\lambda_1-(1+m_2)\lambda_2]^{\frac{m_1 m_3}{1+m_2}}, \text{ if } \lambda_1 > \lambda_2.$$

Evaluation of $C_p\,(NN')$. We shall derive the Hasse-Minkowski p-invariant of NN' for sub-class (III) of both the generalizations, but it can be easily shown that the same expression holds true for the sub-classes (I) and (II) as well.

We shall make use of the results of Ogawa (1959), Shrikhande and Jain (1962) and Singh and Shukla (1963) without further reference.

Let $M = NN'$ be a positive definite matrix of order v. Let $\rho_0, \rho_1, \rho_2, \rho_3$ be the 4 distinct and rational eigenvalues of M with multiplicities $\alpha_0, \alpha_1, \alpha_2, \alpha_3$ respectively. Let X_i be a matrix : $(v\times\alpha_i)$ such that the columns X_{ij} generate the eigen space of M corresponding to ρ_i, $i = 0, 1, 2, 3$.

Consider the following set of $1+(m_1-1)(m_2+m_3+1)+\alpha_2$ v-vectors.

It is easy to verify that $X_0 = (X_{01})$ is the eigen vector corresponding to $\rho_0 = rk$, $X_1 = (X_{11}, X_{12}, \ldots, X_{1\{(m_1-1)(m_2+m_3+1)\}})$ is the eigen space corresponding to the eigen value $\rho_1 = r-\lambda_1$, $X_2 = (X_{21}, X_{22}, \ldots, X_{2\alpha_2})$ the eigen space corresponding to $\rho_2 = r+(m_1-1)\lambda_1-m_1\lambda_2$, for the first type of generalization.

Let

$$S = (X_0, X_{11}, \ldots, X_{1\{(m_1-1)(m_2+m_3+1)\}}; X_{21}, \ldots, X_{2\alpha_2}; X_{31}, \ldots, X_{3\alpha_3})$$

$$
\begin{array}{lllll}
 & \overbrace{}^{m_1} & \overbrace{}^{m_1} & & \overbrace{}^{m_1} \\
X'_{01} = & (1, 1, \ldots, 1, & 1, 1, \ldots, 1, & \ldots \ldots & 1, 1, \ldots, 1) \\
X'_{11} = & (1, -1, 0, \ldots, 0, & 0, 0, \ldots, 0, & \ldots \ldots & 0, 0, \ldots, 0) \\
X'_{12} = & (1, 1, -2, 0, \ldots, 0, & 0, 0, \ldots, 0, & \ldots \ldots & 0, 0, \ldots, 0) \\
\vdots & & & & \\
X'_{1m_1-1} = & (1, 1, \ldots, 1, -(m-1), & 0, 0, \ldots, 0, & \ldots \ldots & 0, 0, \ldots, 0) \\
X'_{1m_1} = & (0, 0, \ldots, 0, & 1, -1, 0, \ldots, 0, & \ldots \ldots & 0, 0, \ldots, 0) \\
X'_{1(m_1+1)} = & (0, 0, \ldots, 0, & 1, 1, -2, 0, \ldots, 0 & \ldots \ldots & 0, 0, \ldots, 0) \\
\vdots & & & & \\
X'_{1(2m_1-2)} = & (0, 0, \ldots, 0, & 1, 1, \ldots, 1, -(m_1-1), & \ldots \ldots & 0, 0, \ldots, 0) \\
\vdots & & & & \\
X'_{1\{(m_1-1)(m_2+m_{3+2})\}} = & (0, 0, \ldots, 0, & 0, 0, \ldots, 0, & \ldots \ldots & 1, -1, 0, \ldots, 0) \\
\vdots & & & & \\
X'_{1\{(m_1-1)(m_2+m_3+1\}} = & (0, 0, \ldots, 0, & 0, 0, \ldots, 0, & \ldots \ldots & 1, \ldots, 1, -(m_1-1) \\
X'_{21} = & (1, 1, \ldots, 1, & -1, \ldots, -1, 0, \ldots, 0, & \ldots \ldots & 0, 0, \ldots, 0) \\
X'_{22} = & (1, 1, \ldots, 1, & 1, \ldots, 1, -2, -2, \ldots, -2, 0, \ldots, 0, & & 0, \ldots, 0) \\
\vdots & & & & \\
X'_{2\alpha_2} = & (1, 1, \ldots, 1, & 1, \ldots, 1, 1, \ldots, 1, \ldots, -\alpha_2, \ldots, -\alpha_2, & & 0, \ldots, 0)
\end{array}
$$

where $X_{31}, \ldots, X_{3\alpha_3}$ constitute a set of α_3 vectors orthogonal to the vector space generated by $X_0, X_{11}, \ldots, X_{1\{(m_1-1)(m_2+m_3+1)\}}$, $X_{21}, \ldots, X_{2\alpha_2}$. Then S is a non-singular $v \times v$ matrix with rational elements.

Putting

$$Q_i = X_i'X_i, \quad i = 0, 1, 2, 3$$

we have

$$|Q_0| = v$$

$$|Q_1| = [1.2.2.3. \ldots (m_1-1)m_1]^{m_2+m_3+1}$$

$$= [m_1\{(m_1-1)\,!\,\}^2]^{m_2+m_3+1}$$

$$\sim m_1^{m_2+m_3+1}$$

$$|Q_2| = [1.2.m_1.\ 2.3.m_1 \ldots \alpha_2(\alpha_2+1)m_1]^{\alpha_2}$$

$$= \left(m_1^{\alpha_2}\right)^{\alpha_2} [(\alpha_2+1)!\ \alpha_2!\ ]^{\alpha_2}$$

$$\sim (\alpha_2+1)^{\alpha_2}$$

Since

$$S'MS = \begin{bmatrix} \rho_0 v & 0 & 0 & 0 \\ 0 & \rho_1 Q_1 & 0 & 0 \\ 0 & 0 & \rho_2 Q_2 & 0 \\ 0 & 0 & 0 & \rho_3 Q_3 \end{bmatrix}$$

or,

$$M \sim \begin{bmatrix} \rho_0 v & 0 & 0 & 0 \\ 0 & \rho_1 Q_1 & 0 & 0 \\ 0 & 0 & \rho_2 Q_2 & 0 \\ 0 & 0 & 0 & \rho_3 Q_3 \end{bmatrix}$$

we find

$$C_p(M) = (-1, -1)_p \left(\rho_0, -v\rho_1^{\alpha_1}\rho_2^{\alpha_2}\rho_3^{\alpha_2}\right)_p (\rho_1, \rho_2)_p^{\alpha_1\alpha_2}$$

(4.7)
$$(\rho_1, \rho_3)_p^{\alpha_1\alpha_3} (\rho_2, \rho_3)_p^{\alpha_2\alpha_3} (\rho_1, -1)_p^{\frac{\alpha_1(\alpha_1+1)}{2}} (\rho_2, -1)_p^{\frac{\alpha_2(\alpha_2+1)}{2}}$$

$$(\rho_3, -1)_p^{\frac{\alpha_3(\alpha_3+1)}{2}} (\rho_1, |Q_1|)_p (\rho_2, |Q_2|)_p$$

$$(\rho_3, v|Q_1|\,|Q_2|)_p$$

As all the terms on the righthand side of (4.7) are known, $C_p(NN')$ can be evaluated.

For the second type of generalization the same expression (4.7) will be obtained but with different Q_1, Q_2, Q_3, as we find that the following set of $1+(m_1-1)+m_1(m_2+m_3-\alpha_2')$ v-vectors form the eigen space corresponding to $\rho_0 = rk$, $\rho_1 = r+m_2\lambda_1+m_3\lambda_2 -(1+m_2+m_3)\lambda_3$ and $\rho_2 = r-\lambda_1$ where $\alpha_2' = m_3/(1+m_2)$.

$$
\begin{aligned}
X'_{01} &= (\overbrace{1, 1, \ldots, 1,}^{m_2+m_3+1} \quad \overbrace{1, 1, \ldots, 1,}^{m_2+m_3+1} \quad \ldots \ldots \quad \overbrace{1, 1, \ldots, 1}^{m_2+m_3+1}) \\
X'_{11} &= (1, 1, \ldots, 1, \quad -1, -1, \ldots, -1, \quad \ldots \ldots \quad 0, 0, \ldots, 0) \\
&\vdots \\
X'_{1(m_1-1)} &= (1, 1, \ldots, 1, \quad 1, \ldots, 1, \quad \ldots \ldots -(m_1-1), \ldots, -(m_1-1)) \\
X'_{21} &= (1, -1, 0, \ldots, \quad 0, \ldots, 0, \quad \ldots \ldots \quad 0, 0, \ldots, 0) \\
&\vdots \\
X'_{2(m_2+m_3-\alpha'_2)} &= (1, 1, \ldots, 1, \quad 1 \ldots, 1, -(m_2+m_3-\alpha'_2), \quad 0, \ldots, 0) \\
&\vdots \\
X'_{2\{(m_1-1)(m_2+m_3-\alpha'_2+1)\}} &= \\
&(0, \ldots, 0, \quad 0, \ldots, 0, \ldots, 1, -1, \quad \ldots \ldots \quad 0, \ldots, 0) \\
&\vdots \\
X'_{2\{m_1(m_2+m_3-\alpha'_2)\}} &= \\
&(0, \ldots, 0, \quad 0, \ldots, 0, \ldots, 1, 1, \ldots, 1, -(m_2+m_3-\alpha'_2) \quad 0, \ldots, 0)
\end{aligned}
$$

Here

$$|Q_0| = v$$

$$|Q_1| \sim [m_1(m_2+m_3+1)]^{m_1-1}$$

$$|Q_2| \sim (m_2+m_3+1-\alpha'_2)^{m_1}$$

References

Bose, R. C. and Nair, K. R. (1939). "Partially Balanced Incomplete Block Designs," *Sankhyā*, **4**, 337–372.

Bose, R. C. and Shimamoto, T. (1952). "Classification and Analysis of Partially Balanced Designs with Two Associate Classes," *J. Amer. Statist. Assoc.*, **47**, 151–184.

Connor, W. S. and Clatworthy, W. H. (1954). "Some Theorems for Partially Balanced Incomplete Block Designs," *Ann. Math. Statist.*, **25**, 100–112.

Nair, K. R. and Rao, C. R. (1942). "A Note on Partially Balanced Incomplete Block Designs," *Science and Culture*, **7**, 568–569.

Ogawa, J. (1959). "A Necessary Condition for Existence of Regular and Symmetrical Experimental Designs of Triangular Type, With Partially Balanced Incomplete Blocks," *Ann. Math. Statist.*, **30**, 1063–1071.

Roy, P. M. (1953). "Hierarchical Group Divisible Incomplete Block Designs with *m*-Associate Classes," *Science and Culture*, **19**, 210–211.

Roy, P. M. (1955). "On Some Combinatorial Problems in the Design of Experiments," D.Phil. thesis of the Calcutta University (unpublished).

Roy, P. M. (1962). "On the Properties and Construction of H.G.D. Designs with m-Associate Classes," *Cal. Statist. Assoc. Bull.*, **11**, 10–38.

Shrikhande, S. S. and Jain, N. C. (1962). "The Non-Existence of Some Partially Balanced Incomplete Block Designs with Latin Square Type Association Schemes," *Sankhyā*, *Ser. A*, **24**, 259–268.

Singh, N. K. and Shukla, G. C. (1963). "The Non-Existence of Some Partially Balanced Incomplete Block Designs with Three Associate Classes," *J. Ind. Statist. Assoc.*, **1**, 71–77.

(Received Jan. 1, 1966.)

CHAPTER 27

On the Null-Distribution of the F-Statistic in a Randomized Partially Balanced Incomplete Block Design with Two Associate Classes under the Neyman Model

JUNJIRO OGAWA, SADAO IKEDA* AND MOTOYASU OGASAWARA
Nihon University, Tokyo

SUMMARY AND INTRODUCTION

One of the authors, J. Ogawa, showed that for a complete block design [3], a Latin-square design [4], and for a balanced incomplete block design [5] the null-distribution of the F-statistic (i.e., the variance ratio) in their analysis of variance can be approximated by the familiar central F-distribution even under the Neyman model, i.e., an intra-block analysis model with both the unit errors and technical errors, if they are randomized. In the present article, the authors consider the null-distribution of the F-statistic in a randomized partially balanced incomplete block design

* Sadao Ikeda is now with Waseda University, Tokyo.

with two associate classes under the Neyman model and they reached the same conclusion as mentioned above, though the calculations are much heavier than those for previous cases.

N. Giri [2] treated the same problem under the Fisher model for a special class of randomized partially balanced imcomplete block designs with two associate classes when $\lambda_1 = 0$ and $\lambda_2 \neq 0$.

Thus the result presented in the present paper seems to be the most general one among those thus far published along the lines mentioned above.

The extension of the reasoning presented in this paper to a randomized partially balanced incomplete block design with m associate classes is not difficult and will be presented in a forthcoming paper [8].

If one looks at those results obtained from the point of view of the usual normal theory, this may be regarded as the so-called "robustness" of the usual regression model with normal errors ignoring the unit errors completely.

In Section 1 the spectral decomposition of the matrix $\boldsymbol{NN}'$, $\boldsymbol{N}$ being the incidence matrix of the design under consideration, is given and this is useful for discussions in later sections. The null-distribution of the F-statistic before the randomization under the Neyman model is presented in Section 2 and this turns out to be a non-central F-distribution whose non-centrality parameter depends upon the quantity θ which is a quadratic form of the unit errors. In Section 3 the exact mean and variance of the quantity θ with respect to the permutation distribution due to the randomization are calculated. The calculations in this part are quite complicated. Then, in Section 4, if the number of blocks is sufficiently large and certain uniformity conditions on the within-block moments of the 2nd and 4th order of the unit errors are satisfied, the permutation distribution of θ is shown to be approximated by a suitable Beta-distribution. Finally, in Section 5, it is shown that the null-distribution of the F-statistic after the randomization can be approximated by a familiar central F-distribution provided the two conditions mentioned above are satisfied.

1. SPECTRAL DECOMPOSITION OF THE MATRIX NN'

As for the definition of a partially balanced incomplete block design and notations being used in the present section, references should be made to [1] and [6].

Let the association matrices be $\boldsymbol{A}_0 = \boldsymbol{I}_v$, $\boldsymbol{A}_1$, $\boldsymbol{A}_2$, and let their regular representations be

$$\mathcal{P}_0 = \boldsymbol{I}_3,\ \mathcal{P}_1 = \begin{Vmatrix} 0 & 1 & 0 \\ n_1 & p^1_{11} & p^2_{11} \\ 0 & p^1_{21} & p^2_{21} \end{Vmatrix},\ \mathcal{P}_2 = \begin{Vmatrix} 0 & 0 & 1 \\ 0 & p^1_{12} & p^2_{12} \\ n_2 & p^1_{22} & p^2_{22} \end{Vmatrix}$$

respectively, then it is known [6] that there exists a non-singular matrix

$$C = \begin{Vmatrix} 1 & 1 & 1 \\ c_{10} & c_{11} & c_{12} \\ c_{20} & c_{21} & c_{22} \end{Vmatrix}$$

such that

$$C\mathcal{P}_uC^{-1} = \begin{Vmatrix} n_u & 0 & 0 \\ 0 & z_{1u} & 0 \\ 0 & 0 & z_{2u} \end{Vmatrix},\ u = 0, 1, 2 \tag{1.1}$$

simultaneously, where

$$\begin{aligned} z_{11} &= (a+d)/2, & z_{21} &= (a-d)/2, \\ z_{12} &= -(a+2+d)/2, & z_{22} &= -(a+2-d)/2, \end{aligned} \tag{1.2}$$

with

$$(1.3)\quad \begin{aligned} a &= p_{21}^2 - p_{12}^1 - 1, \\ d &= (p_{12}^2 - p_{12}^1 - 1)^2 + 4p_{21}^2 = (p_{12}^1 - p_{21}^2 - 1)^2 + 4p_{12}^1. \end{aligned}$$

One may put as follows :

$$(1.4)\quad \begin{aligned} c_{10} &= 1, \quad c_{11} = z_{11}/n_1, \quad c_{12} = z_{12}/n_2, \\ c_{20} &= 1, \quad c_{21} = z_{21}/n_1, \quad c_{22} = z_{22}/n_2. \end{aligned}$$

Three orthogonal idempotents of the association algebra are given by

$$(1.5)\quad \begin{cases} \boldsymbol{A}_0^{\#} = (1/v)\boldsymbol{G}_v, \\ \boldsymbol{A}_1^{\#} = [1+(z_{11}^2/n_1)+(z_{12}^2/n_2)]^{-1}[\boldsymbol{A}_0+(z_{11}/n_1)\boldsymbol{A}_1+(z_{12}/n_2)\boldsymbol{A}_2], \\ \boldsymbol{A}_2^{\#} = [1+(z_{21}^2/n_1)+(z_{22}^2/n_2)]^{-1}[\boldsymbol{A}_0+(z_{21}/n_1)\boldsymbol{A}_1+(z_{22}/n_2)\boldsymbol{A}_2] \end{cases}$$

with respective ranks $\alpha_0 = 1$, α_1 and α_2.

Let the incidence matrix of the design be $\boldsymbol{N}$, then it is known that

$$(1.6)\quad \boldsymbol{NN}'(= r\boldsymbol{A}_0+\lambda_1\boldsymbol{A}_1+\lambda_2\boldsymbol{A}_2) = \rho_0\boldsymbol{A}_0^{\#}+\rho_1\boldsymbol{A}_1^{\#}+\rho_2\boldsymbol{A}_2^{\#},$$

where

$$(1.7)\quad \rho_0 = rk,\ \rho_1 = r+z_{11}\lambda_1+z_{12}\lambda_2,\ \rho_2 = r+z_{21}\lambda_1+z_{22}\lambda_2$$

and

$$(1.8)\quad \boldsymbol{A}_0^{\#}+\boldsymbol{A}_1^{\#}+\boldsymbol{A}_2^{\#} = \boldsymbol{I}_v.$$

Thus (1.6) gives the spectral decomposition of the symmetric matrix $\boldsymbol{NN}'$ and ρ_0, ρ_1, ρ_2 are the characteristic roots of the matrix with multiplicities α_0, α_1, α_2 respectively. Column vectors of $\boldsymbol{A}_u^{\#}$ are characteristic vectors corresponding to the characteristic root ρ_u of $\boldsymbol{NN}'$.

2. THE NULL-DISTRIBUTION OF F-STATISTIC BEFORE THE RANDOMIZATION UNDER THE NEYMAN MODEL

We shall be dealing with a partially balanced incomplete block design with two associate classes which has v treatments having such an association as explained in the previous section, b blocks of size k each, r replications of each treatment, and the numbers of incidence of any pair of treatments λ_1 or λ_2 according as they are 1st associates or 2nd associates. As for the notations being used in this section, references should be made to [1] and [6].

Let the incidence matrices of treatment and blocks be $\boldsymbol{\Phi}$ and $\boldsymbol{\Psi}$ respectively, then the Neyman model assuming no interaction between treatments and units is given by

(2.1) $$\boldsymbol{x} = \gamma\mathbf{1}+\boldsymbol{\Phi\tau}+\boldsymbol{\Psi\beta}+\pi+\boldsymbol{e},$$

where $\boldsymbol{x}' = (x_1, \ldots, x_n)$ is the observation vector, $\boldsymbol{\tau}' = (\tau_1, \ldots, \tau_v)$ and $\boldsymbol{\beta}' = (\beta_1, \ldots, \beta_b)$ are treatment-effects and block-effects subjected to the restrictions

$$\tau_1+\ldots+\tau_v = 0 \text{ and } \beta_1+\ldots+\beta_b = 0$$

respectively, and $\boldsymbol{\pi}' = (\pi_1, \ldots, \pi_n)$ stands for the unit errors. Finally, $\boldsymbol{e}' = (e_1, \ldots, e_n)$ is the technical error vector being distributed as $N(\mathbf{0}'\sigma^2, \boldsymbol{I}_n)$.

Now we are interested in testing the null-hypothesis

(2.2) $$H_0 : \boldsymbol{\tau} = \mathbf{0}.$$

Likewise one may consider the testing the partial null-hypotheses

$$H_0^{(u)} : \boldsymbol{A}_u^{\#}\,\boldsymbol{\tau} = \mathbf{0}, \quad u = 1, 2.$$

Although the arguments for these hypotheses are similar to the one presented in this paper for H_0, we confine ourselves to the null-hypothesis H_0 for the moment and testing of $H_0^{(u)}$ will be dealt with in the forthcoming paper [8].

Sums of squares due to treatments adjusted and errors are given by

(2.3) $$S_t^2 = \boldsymbol{x}'(\boldsymbol{V}_1^{\#}+\boldsymbol{V}_2^{\#})\boldsymbol{x},$$

$$S_e^2 = \boldsymbol{x}'\left(\boldsymbol{I}-\frac{1}{k}\boldsymbol{B}-\boldsymbol{V}_1^{\#}-\boldsymbol{V}_2^{\#}\right)\boldsymbol{x}$$

respectively, where

$$\boldsymbol{V}_u^{\#} = \frac{k}{r(rk-\rho_u)}\left(\boldsymbol{T}_u^{\#}-\frac{1}{k}\boldsymbol{B}\boldsymbol{T}_u^{\#}\right)\left(\boldsymbol{T}_u^{\#}-\frac{1}{k}\boldsymbol{T}_u^{\#}\boldsymbol{B}\right)$$

$$= \frac{k}{rk-\rho_u}\left(\boldsymbol{I}-\frac{1}{k}\boldsymbol{B}\right)\boldsymbol{T}_u^{\#}\left(\boldsymbol{I}-\frac{1}{k}\boldsymbol{B}\right),\quad u=1,2$$

with $\boldsymbol{T}_u^{\#} = \boldsymbol{\Phi}\boldsymbol{A}_u^{\#}\boldsymbol{\Phi}'$ and $\boldsymbol{B} = \boldsymbol{\Psi}\boldsymbol{\Psi}'$.

Under the null-hypothesis H_0, they can be expressed as follows:

(2.5) $$S_t^2 = \boldsymbol{\pi}'\left(\boldsymbol{I}-\frac{1}{k}\boldsymbol{B}\right)\boldsymbol{\Phi}(c_1\boldsymbol{A}_1^{\#}+c_2\boldsymbol{A}_2^{\#})\boldsymbol{\Phi}'\left(\boldsymbol{I}-\frac{1}{k}\boldsymbol{B}\right)\boldsymbol{\pi}$$

$$+2\boldsymbol{\pi}'\left(\boldsymbol{I}-\frac{1}{k}\boldsymbol{B}\right)\boldsymbol{\Phi}(c_1\boldsymbol{A}_1^{\#}+c_2\boldsymbol{A}_2^{\#})\boldsymbol{\Phi}'\left(\boldsymbol{I}-\frac{1}{k}\boldsymbol{B}\right)\boldsymbol{e}$$

$$+\boldsymbol{e}'\left(\boldsymbol{I}-\frac{1}{k}\boldsymbol{B}\right)\boldsymbol{\Phi}(c_1\boldsymbol{A}_1^{\#}+c_2\boldsymbol{A}_2^{\#})\boldsymbol{\Phi}'\left(\boldsymbol{I}-\frac{1}{k}\boldsymbol{B}\right)\boldsymbol{e},$$

$$S_e^2 = \boldsymbol{\pi}'\left(\boldsymbol{I}-\frac{1}{k}\boldsymbol{B}\right)[\boldsymbol{I}-\boldsymbol{\Phi}(c_1\boldsymbol{A}_1^{\#}+c_2\boldsymbol{A}_2^{\#})\boldsymbol{\Phi}']\left(\boldsymbol{I}-\frac{1}{k}\boldsymbol{B}\right)\boldsymbol{\pi}$$

$$+2\boldsymbol{\pi}'\left(\boldsymbol{I}-\frac{1}{k}\boldsymbol{B}\right)[\boldsymbol{I}-\boldsymbol{\Phi}(c_1\boldsymbol{A}_1^{\#}+c_2\boldsymbol{A}_2^{\#})\boldsymbol{\Phi}']\left(\boldsymbol{I}-\frac{1}{k}\boldsymbol{B}\right)\boldsymbol{e}$$

$$+\boldsymbol{e}'\left(\boldsymbol{I}-\frac{1}{k}\boldsymbol{B}\right)[\boldsymbol{I}-\boldsymbol{\Phi}(c_1\boldsymbol{A}_1^{\#}+c_2\boldsymbol{A}_2^{\#})\boldsymbol{\Phi}']\left(\boldsymbol{I}-\frac{1}{k}\boldsymbol{B}\right)\boldsymbol{e},$$

where

(2.6) $$c_1 = k/(rk-\rho_1),\quad c_2 = k/(rk-\rho_2).$$

The null-distribution of the variate

$$\chi_1^2 = S_t^2/\sigma^2$$

before the randomization is the non-central chi-square distribution of degrees of freedom $v-1$ and with the non-centrality parameter

(2.7) $$\kappa_1 = \boldsymbol{\pi}'\boldsymbol{\Phi}(c_1\boldsymbol{A}_1^{\#}+c_2\mathrm{A}_2^{\#})\boldsymbol{\Phi}'\boldsymbol{\pi}/\sigma^2.$$

Hence its probability element is

$$(2.8)\qquad \exp\left(-\frac{\kappa_1}{2}\right)\sum_{\mu=0}^{\infty}\frac{\left(\frac{\kappa_1}{2}\right)^{\mu}}{\mu!}\frac{\left(\frac{\chi_1^2}{2}\right)^{\frac{v-1}{2}+\mu-1}}{\Gamma\left(\frac{v-1}{2}+\mu\right)}\exp\left(-\frac{\chi_1^2}{2}\right)d\left(\frac{\chi_1^2}{2}\right).$$

The null-distribution of the variate

$$\chi_2^2 = S_e^2/\sigma^2$$

before the randomization is the non-central chi-square distribution of degrees of freedom $n-v-b+1$ and with the non-centrality parameter

$$(2.9)\qquad \kappa_2 = \boldsymbol{\pi}'\left(\boldsymbol{I}-\frac{1}{k}\boldsymbol{B}\right)[\boldsymbol{I}-\boldsymbol{\Phi}(c_1\boldsymbol{A}_1^{\#}+c_2\boldsymbol{A}_2^{\#})\boldsymbol{\Phi}']\left(\boldsymbol{I}-\frac{1}{k}\boldsymbol{B}\right)\boldsymbol{\pi}/\sigma^2$$

$$= \Delta/\sigma^2-\kappa_1,\quad \Delta = \boldsymbol{\pi}'\boldsymbol{\pi}.$$

Hence its probability element is given by

$$(2.10)\qquad \exp\left(\frac{\kappa_2}{2}\right)\sum_{\nu=0}^{\infty}\frac{\left(\frac{\kappa_2}{2}\right)^{\nu}}{\nu!}\frac{\left(\frac{\chi_2^2}{2}\right)^{\frac{n-v-b+1}{2}+\nu-1}}{\Gamma\left(\frac{n-v-b+1}{2}+\nu\right)}\exp\left(-\frac{\chi_2^2}{2}\right)d\left(\frac{\chi_2^2}{2}\right).$$

Since χ_1^2 and χ_2^2 are stochastically independent of each other, the null-distribution of the F-statistic

$$(2.11)\qquad F = \frac{n-v-b+1}{v-1}\frac{S_t^2}{S_e^2}$$

before the randomization is, after a little algebra, given by

$$\frac{\Gamma\left(\frac{n-b}{2}\right)}{\Gamma\left(\frac{v-1}{2}\right)\Gamma\left(\frac{n-v-b+1}{2}\right)}\left(\frac{v-1}{n-v-b+1}F\right)^{\frac{v-1}{2}-1}$$

$$\left(1+\frac{v-1}{n-v-b+1}F\right)^{-\frac{n-b}{2}}d\left(\frac{v-1}{n-v-b+1}F\right)$$

(2.12)

$$\times \exp\left(-\frac{\Delta}{2\sigma^2}\right)\sum_{l=0}^{\infty}\frac{\left(\frac{\Delta}{2\sigma^2}\right)^l}{l!}\left(1+\frac{v-1}{n-v-n+1}F\right)^{-l}$$

$$\sum_{\mu+\nu=l}\frac{l!}{\mu!\,\nu!}\theta^{\mu}(1-\theta)^{\nu}\left(\frac{v-1}{n-v-b+1}F\right)^{\mu}$$

$$\times\frac{\Gamma\left(\frac{v-1}{2}\right)\Gamma\left(\frac{n-v-b+1}{2}\right)\Gamma\left(\frac{n-b}{2}+l\right)}{\Gamma\left(\frac{n-b}{2}\right)\Gamma\left(\frac{v-1}{2}+\mu\right)\Gamma\left(\frac{n-v-b+1}{2}+\nu\right)},$$

where

$$\theta=\Delta^{-1}\boldsymbol{\pi}'\boldsymbol{\Phi}(c_1\boldsymbol{A}_1^{\#}+c_2\boldsymbol{A}_2^{\#})\boldsymbol{\Phi}'\boldsymbol{\pi}. \tag{2.13}$$

The null-distribution of the F after the randomization should be obtained as

$$\frac{\Gamma\left(\frac{n-b}{2}\right)}{\Gamma\left(\frac{v-1}{2}\right)\Gamma\left(\frac{n-v-b+1}{2}\right)}\left(\frac{v-1}{n-v-b+1}\right)^{\frac{v-1}{2}-1} \tag{2.14}$$

$$\left(1+\frac{v-1}{n-v-b+1}F\right)^{-\frac{n-b}{2}}d\left(\frac{v-1}{n-v-b+1}F\right)$$

$$\exp\left(-\frac{\Delta}{2\sigma^2}\right)\sum_{l=0}^{\infty}\frac{\left(\frac{\Delta}{2\sigma^2}\right)^l}{l\,!}\left(1+\frac{v-1}{n-v-b+1}F\right)^{-l}$$

$$\sum_{\mu+\nu=l}\frac{l!}{\mu!\,\nu!}\mathscr{E}[\theta^{\mu}(1-\theta)^{\nu}]\left(\frac{v-1}{n-v-b+1}F\right)^{\mu}$$

$$\times\frac{\Gamma\left(\frac{v-1}{2}\right)\Gamma\left(\frac{n-v-b+1}{2}\right)\Gamma\left(\frac{n-b}{2}+l\right)}{\Gamma\left(\frac{n-b}{2}\right)\Gamma\left(\frac{v-1}{2}+\mu\right)\Gamma\left(\frac{n-v-b+1}{2}+\nu\right)},$$

where $\mathscr{E}[\theta^{\mu}(1-\theta)^{\nu}]$ stands for the expectation with respect to the permutation distribution of θ due to the randomization.

3. THE CALCULATION OF THE MEAN AND VARIANCE OF THE QUANTITY θ WITH RESPECT TO THE PERMUTATION DISTRIBUTION DUE TO RANDOMIZATION

Since

$$c_1\boldsymbol{A}_1^{\#}+c_2\boldsymbol{A}_2^{\#} = \mu_0\boldsymbol{A}_0+\mu_1\boldsymbol{A}_1+\mu_2\boldsymbol{A}_2,$$

with

$$(3.1)\quad \begin{cases} \mu_0 = \dfrac{k}{(rk-\rho_1)\left(1+\dfrac{z_{11}^2}{n_1}+\dfrac{z_{12}^2}{n_2}\right)}+\dfrac{k}{(rk-\rho_2)\left(1+\dfrac{z_{21}^2}{n_1}+\dfrac{z_{22}^2}{n_2}\right)}, \\ \mu_1 = \dfrac{kz_{11}}{(rk-\rho_1)\left(1+\dfrac{z_{11}^2}{n_1}+\dfrac{z_{12}^2}{n_2}\right)}+\dfrac{kz_{21}}{(rk-\rho_2)\left(1+\dfrac{z_{21}^2}{n_1}+\dfrac{z_{22}^2}{n_2}\right)}, \\ \mu_2 = \dfrac{kz_{12}}{(rk-\rho_1)\left(1+\dfrac{z_{11}^2}{n_1}+\dfrac{z_{12}^2}{n_2}\right)}+\dfrac{kz_{22}}{(rk-\rho_2)\left(1+\dfrac{z_{21}^2}{n_1}+\dfrac{z_{22}^2}{n_2}\right)}, \end{cases}$$

let us put

$$(3.2)\qquad \boldsymbol{T}^{\#} = \boldsymbol{\Phi}(c_1\boldsymbol{A}_1^{\#}+c_2\boldsymbol{A}_2^{\#})\,\boldsymbol{\Phi}' = \mu_0\boldsymbol{T}_0+\mu_1\boldsymbol{T}_1+\mu_2\boldsymbol{T}_2.$$

3.1. Necessary notations

Now we use the numbering of the whole units such that the ith unit in the pth block bears the number $f = (p-1)k+i$.

Let

$$(3.3)\qquad \boldsymbol{T}_u = \| T_{pq}^{(u)} \|,\quad \boldsymbol{T}_{pq}^{(u)} = \| t_{ij}^{(u)pq} \|,\qquad p, q = 1, \ldots, b \qquad i, j = 1, \ldots, k,$$

where

$$t_{ij}^{(u)pq} = \begin{cases} 1, & \text{if the } i\text{th unit in the } p\text{th block and the } j\text{th unit in the } q\text{th block receive treatments which are } u\text{th associates,} \\ 0, & \text{otherwise,} \end{cases}$$

and further let

$$(3.4)\qquad \boldsymbol{T}^{\#} = \| \mathrm{T}_{pq}^{\#} \|,\quad \boldsymbol{T}_{pq}^{\#} = \| t_{ij}^{\#pq} \|,$$

where

$$(3.5)\qquad t_{ij}^{\#pq} = \mu_0 t_{ij}^{(0)pq}+\mu_1 t_{ij}^{(1)pq}+\mu_2 t_{ij}^{(2)pq}.$$

Notations necessary for the calculations in this section are introduced here for the cases $k \geqslant 4$. The cases in which $k=2$ or 3 shall be examined later.

(i)

Number of ordered pairs of treatments being of the uth associates which occur in the pth block is denoted by

$$\lambda_{pp}^{(1)uu} = \sum_{i \neq j} t_{ij}^{(u)pp}, \quad u = 1, 2.$$

One can see that

$$\lambda_{pp}^{(1)11} + \lambda_{pp}^{(1)22} = k(k-1).$$

(ii)

Number of ordered triplets of treatments in the pth block, in which the 1st and the second are uth associates and the 1st and 3rd are vth associates is denoted by

$$\lambda_{pp}^{(2)uv} = \sum_{i \neq j \neq l} t_{ij}^{(u)pp} t_{il}^{(v)pp}, \quad u, v = 1, 2.$$

One gets the following relations immediately :

$$\lambda_{pp}^{(2)uv} = \lambda_{pp}^{(2)vu},$$

$$\begin{cases} \lambda_{pp}^{(2)11} + \lambda_{pp}^{(2)12} = (k-2)\lambda_{pp}^{(1)11}, \\ \lambda_{pp}^{(2)12} + \lambda_{pp}^{(2)22} = (k-2)\lambda_{pp}^{(1)22}. \end{cases}$$

(iii)

Number of two pairs of treatments in the pth block, of which two treatments of the one pair are uth associates and those of the other pair are vth associates is denoted by

$$\lambda_{pp}^{(3)uv} = \sum_{i \neq j \neq l \neq m} t_{ij}^{(u)pp} t_{lm}^{(v)pp}, \quad u, v = 1, 2.$$

Trivial relations are

$$\lambda_{pp}^{(3)uv} = \lambda_{pp}^{(3)vu},$$

$$\begin{cases} \lambda_{pp}^{(3)11} + \lambda_{pp}^{(3)12} = (k-2)(k-3)\lambda_{pp}^{(1)11}, \\ \lambda_{pp}^{(3)12} + \lambda_{pp}^{(3)22} = (k-2)(k-3)\lambda_{pp}^{(1)22}. \end{cases}$$

Other similar 12 notations with their trivial relations are listed below.

(iv)

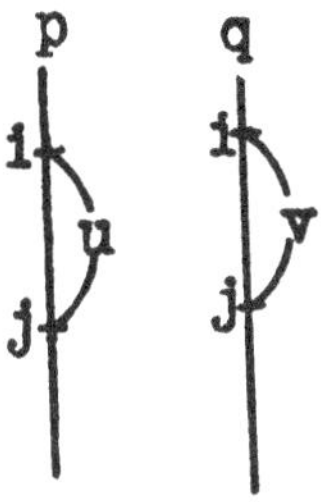

$$\lambda_{pq}^{(4)uv} = \sum_{i \neq j} t_{ij}^{(u)pp}\, t_{ij}^{(v)qq}, \quad u, v = 1, 2.$$

$$\begin{cases} \lambda_{pq}^{(4)11} + \lambda_{pq}^{(4)12} = \lambda_{pp}^{(1)11}, \\ \lambda_{pq}^{(4)21} + \lambda_{pq}^{(4)22} = \lambda_{pp}^{(1)22}, \end{cases}$$

$$\begin{cases} \lambda_{pq}^{(4)11} + \lambda_{pq}^{(4)21} = \lambda_{qq}^{(1)11}, \\ \lambda_{pq}^{(4)12} + \lambda_{pq}^{(4)22} = \lambda_{qq}^{(1)22}. \end{cases}$$

(v)

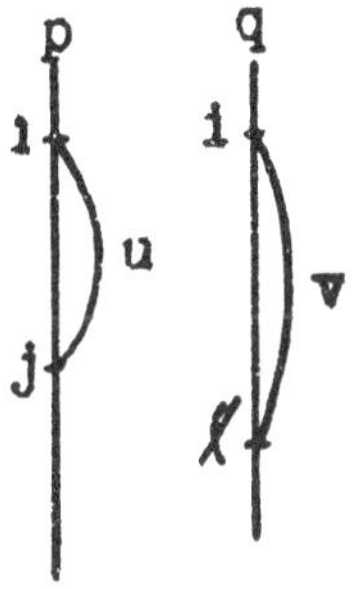

$$\lambda_{pq}^{(5)uv} = \sum_{i \neq j \neq l} t_{ij}^{(u)pp}\, t_{il}^{(v)qq}, \quad u, v = 1, 2.$$

$$\begin{cases} \lambda_{pq}^{(5)11} + \lambda_{pq}^{(5)12} = (k-2)\lambda_{pp}^{(1)11}, \\ \lambda_{pp}^{(5)21} + \lambda_{pq}^{(5)22} = (k-2)\lambda_{pp}^{(1)11}, \end{cases}$$

$$\begin{cases} \lambda_{pq}^{(5)11} + \lambda_{pq}^{(5)21} = (k-2)\lambda_{qq}^{(1)22}, \\ \lambda_{pq}^{(5)12} + \lambda_{pq}^{(5)22} = (k-2)\lambda_{qq}^{(1)22}. \end{cases}$$

(vi)

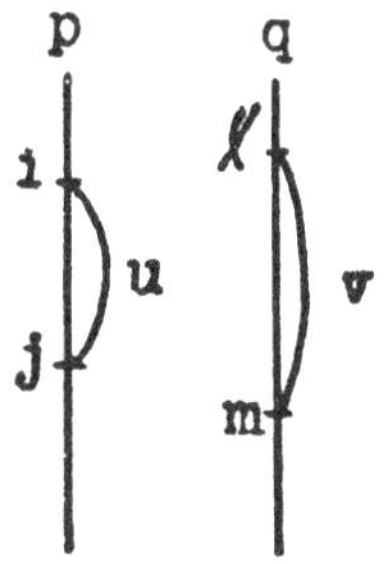

$$\lambda_{pq}^{(6)uv} = \sum_{i \neq j \neq l \neq m} t_{ij}^{(u)pp}\, t_{lm}^{(v)qq}, \quad u, v = 1, 2.$$

$$\begin{cases} \lambda_{pq}^{(6)11} + \lambda_{pq}^{(6)12} = (k-2)(k-3)\lambda_{pp}^{(1)11}, \\ \lambda_{pq}^{(6)21} + \lambda_{pq}^{(6)22} = (k-2)(k-3)\lambda_{pp}^{(1)22}, \end{cases}$$

$$\begin{cases} \lambda_{pq}^{(6)11} + \lambda_{pq}^{(6)21} = (k-2)(k-3)\lambda_{qq}^{(1)11}, \\ \lambda_{pq}^{(6)12} + \lambda_{pq}^{(6)22} = (k-2)(k-3)\lambda_{qq}^{(1)22}. \end{cases}$$

(vii)

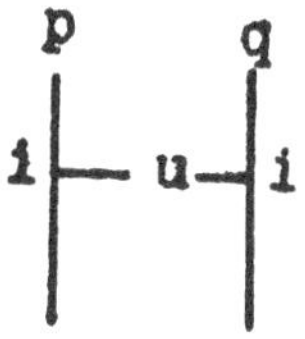

$$\lambda_{p\jmath}^{(7)uu} = \sum_{i=1}^{k} t_{ii}^{(u)pq}, \quad u = 0, 1, 2.$$

$$\sum_{u=0}^{2} \lambda_{pq}^{(7)uu} = k.$$

(viii)

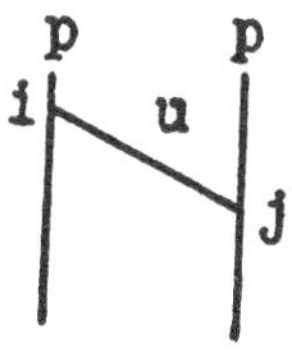

$$\lambda_{pq}^{(8)uu} = \sum_{i \neq j} t_{ij}^{(u)pq}, \quad u = 0, 1, 2.$$

$$\sum_{u=0}^{2} \lambda_{pq}^{(8)uu} = k(k-1).$$

(ix)

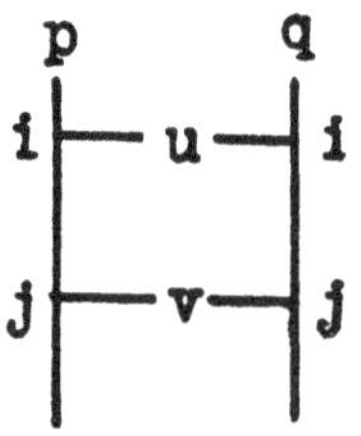

$$\lambda_{pq}^{(9)uv} = \sum_{i \neq j} t_{ii}^{(u)pq} t_{jj}^{(v)pq}, \quad u, v = 0, 1, 2.$$

$$\lambda_{pq}^{(9)uv} = \lambda_{pq}^{(9)vu},$$

$$\sum_{u=0}^{2} \lambda_{pq}^{(9)uv} = (k-1)\lambda_{pq}^{(7)uu}, \quad u = 0, 1, 2.$$

(x)

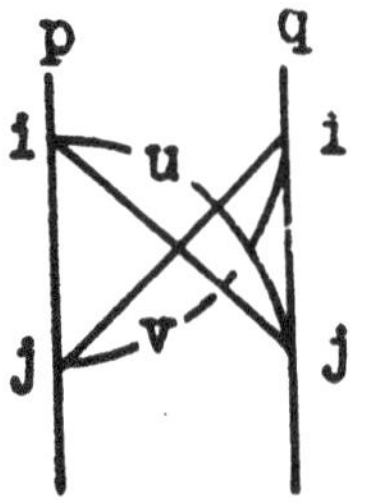

$$\lambda_{pq}^{(10)uv} = \sum_{i \neq j} t_{ij}^{(u)pq} t_{ji}^{(v)pq}, \quad u, v = 0, 1, 2.$$

$$\lambda_{pq}^{(10)uv} = \lambda_{pq}^{(10)vu},$$

$$\sum_{v=0}^{2} \lambda_{pq}^{(10)uv} = \lambda_{pq}^{(8)uu}, \quad u = 0, 1, 2.$$

(xi)

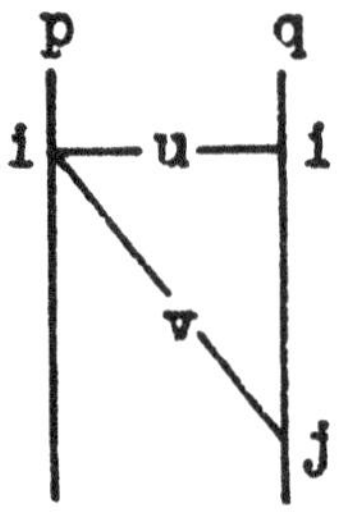

$$\lambda_{pq}^{(11)uv} = \sum_{i \neq j} t_{ii}^{(u)pq} t_{ij}^{(v)pq}, \quad u, v = 0, 1, 2.$$

$$\lambda_{pp}^{(11)00} = 0,$$

$$\begin{cases} \sum_{v=0}^{2} \lambda_{pq}^{(11)uv} = (k-1)\lambda_{pq}^{(7)uu}, \quad u = 0, 1, 2. \\ \sum_{u=0}^{2} \lambda_{pq}^{(11)uv} = \lambda_{pq}^{(8)vv}, \quad v = 0, 1, 2. \end{cases}$$

(xii)

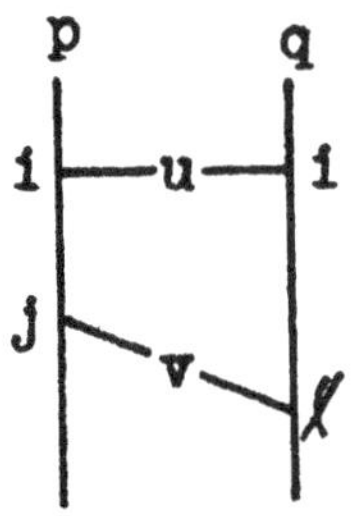

$$\lambda_{pq}^{(12)uv} = \sum_{i \neq j \neq l} t_{ii}^{(u)pq} t_{jl}^{(v)pq}, \quad u, v = 0, 1, 2.$$

$$\begin{cases} \sum_{v=0}^{2} \lambda_{pq}^{(12)uv} = (k-1)(k-2)\lambda_{pq}^{(7)uu}, \quad u = 0, 1, 2, \\ \sum_{u=0}^{2} \lambda_{pq}^{(12)uv} = (k-1)\lambda_{pq}^{(8)vv}, \quad v = 0, 1, 2. \end{cases}$$

(xiii)

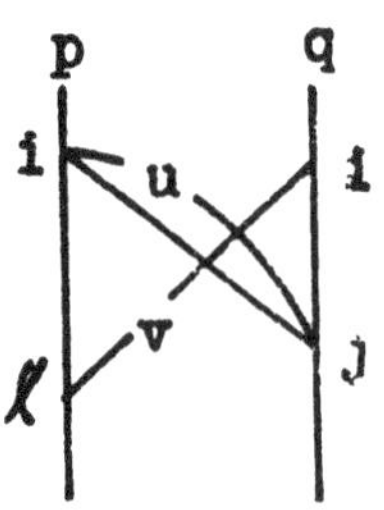

$$\lambda_{pq}^{(13)uv} = \sum_{i \neq j \neq l} t_{ij}^{(u)pq} t_{il}^{(v)pq}, \quad u, v = 0, 1, 2.$$

$$\lambda_{pq}^{(13)uv} = \lambda_{pq}^{(13)vu},$$

$$\sum_{v=0}^{2} \lambda_{pq}^{(13)uv} = (k-2)\lambda_{pq}^{(8)uu}, \quad u = 0, 1, 2.$$

(xiv)

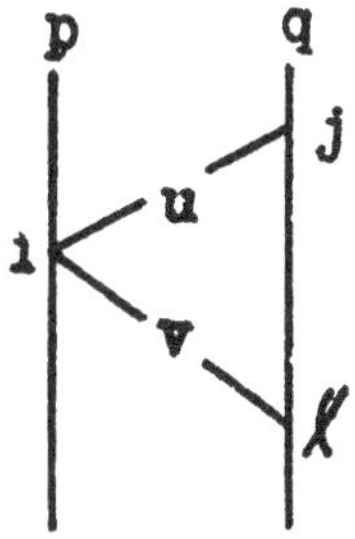

$$\lambda_{pq}^{(14)uv} = \sum_{i \neq k \neq l} t_{ij}^{(u)pq} \, t_{il}^{(v)pq}, \quad u, v = 0, 1, 2.$$

$$\lambda_{pq}^{(14)00} = 0,$$

$$\sum_{v=0}^{2} \lambda_{pq}^{(14)uv} = (k-2)\lambda_{pq}^{(8)uu}, \quad u = 0, 1, 2.$$

(xv)

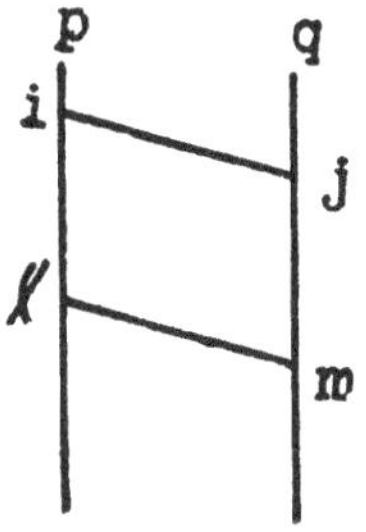

$$\lambda_{pq}^{(15)uv} = \sum_{i \neq j \neq l \neq m} t_{ij}^{(u)pq} \, t_{lm}^{(v)pq}, \quad u, v = 0, 1, 2.$$

$$\lambda_{pq}^{(15)uv} = \lambda_{pq}^{(15)vu},$$

$$\sum_{v=0}^{2} \lambda_{pq}^{(15)uv} = (k-2)(k-3)\lambda_{pq}^{(8)uu}, \quad u = 0, 1, 2.$$

We list the relationships holding among those parameters of 15 classes.

$$\begin{cases} (\lambda_{pp}^{(1)11})^2 = 2\lambda_{pp}^{(1)11} + 4\lambda_{pp}^{(2)11} + \lambda_{pp}^{(3)11}, \\ \lambda_{pp}^{(1)22} = k(k-1) - \lambda_{pp}^{(1)11}, \end{cases}$$

$$\begin{cases} \lambda_{pp}^{(2)12} = (k-2)\lambda_{pp}^{(1)11} - \lambda_{pp}^{(2)11}, \\ \lambda_{pp}^{(2)22} = k(k-1)(k-2) - 2(k-2)\lambda_{pp}^{(1)11} + \lambda_{pp}^{(2)11}, \end{cases}$$

$$\begin{cases} \lambda_{pp}^{(3)12} = (k-2)(k-3)\lambda_{pp}^{(1)11} - \lambda_{pp}^{(3)11}, \\ \lambda_{pp}^{(3)22} = k(k-1)(k-2)(k-3) - 2(k-2)(k-3)\lambda_{pp}^{(1)11} + \lambda_{pp}^{(3)11}, \end{cases}$$

$$\begin{cases} \lambda_{pq}^{(4)22} = k(k-1) - \lambda_{pp}^{(1)11} - \lambda_{qq}^{(1)11} + \lambda_{pq}^{(4)11}, \\ \lambda_{pq}^{(5)22} = k(k-1)(k-2) - (k-2)\lambda_{pp}^{(1)11} - (k-2)\lambda_{qq}^{(1)11} + \lambda_{pq}^{(5)11}, \end{cases}$$

$$\lambda_{pp}^{(1)11} \lambda_{qq}^{(1)11} = 2\lambda_{pq}^{(4)11} + 4\lambda_{pq}^{(5)11} + \lambda_{pq}^{(6)11},$$

$$\begin{cases} \lambda_{pq}^{(6)22} = k(k-1)(k-2)(k-3) - (k-2)(k-3)\lambda_{pp}^{(1)11} \\ \qquad\qquad - (k-2)(k-3)\lambda_{qq}^{(1)11} + \lambda_{pq}^{(6)11}, \\ \lambda_{pq}^{(7)22} = k - \lambda_{pq}^{(7)00} - \lambda_{pq}^{(7)11}, \\ \lambda_{pq}^{(8)22} = k(k-1) - \lambda_{pq}^{(8)00} - \lambda_{pq}^{(8)11}, \\ \lambda_{pq}^{(9)22} = k(k-1) - 2(k-1)\lambda_{pq}^{(7)00} - 2(k-1)\lambda_{pq}^{(7)11} + \lambda_{pq}^{(9)00} \\ \qquad\qquad + 2\lambda_{pq}^{(9)01} + \lambda_{pq}^{(9)11}, \\ \lambda_{pq}^{(10)22} = k(k-1) - 2\lambda_{pq}^{(8)00} - 2\lambda_{pq}^{(8)11} + \lambda_{pq}^{(10)00} + 2\lambda_{pq}^{(10)01} + \lambda_{pq}^{(10)11} \end{cases}$$

$$\lambda_{pq}^{(11)20} = \lambda_{pq}^{(8)00} - \lambda_{pq}^{(11)10},$$

$$\begin{cases} \lambda_{pq}^{(11)12} = k(k-1)-(k-1)\lambda_{pq}^{(7)00}-(k-1)\lambda_{pq}^{(7)11}-\lambda_{pq}^{(8)00}-\lambda_{pq}^{(8)11} \\ \qquad +\lambda_{pq}^{(11)01}+\lambda_{pq}^{(11)10}+\lambda_{pq}^{(11)11}, \\ \lambda_{pq}^{(12)22} = k(k-1)(k-2)-(k-1)(k-2)\lambda_{pq}^{(7)00}-(k-1)(k-2)\lambda_{pq}^{(7)11} \\ \qquad -(k-2)\lambda_{pq}^{(8)00}-(k-2)\lambda_{pq}^{(8)11}+\lambda_{pq}^{(12)00}+\lambda_{pq}^{(12)01}+\lambda_{pq}^{(12)10}+\lambda_{pq}^{(12)11}, \\ \lambda_{pq}^{(13)22} = k(k-1)(k-2)-2(k-2)\lambda_{pq}^{(8)00}-2(k-2)\lambda_{pq}^{(8)11} \\ \qquad +\lambda_{pq}^{(13)00}+2\lambda_{pq}^{(13)01}+\lambda_{pq}^{(13)11}, \\ \lambda_{pq}^{(14)22} = k(k-1)(k-2)-2(k-2)\lambda_{pq}^{(8)00}-2(k-2)\lambda_{pq}^{(8)11} \\ \qquad +2\lambda_{pq}^{(14)01}+\lambda_{pq}^{(14)11}, \\ \lambda_{pq}^{(15)22} = k(k-1)(k-2)(k-3)-2(k-2)(k-3)\lambda_{pq}^{(8)00}-2(k-2) \\ \qquad (k-3)\lambda_{pq}^{(8)11}+\lambda_{pq}^{(15)00}+2\lambda_{pq}^{(15)01}+\lambda_{pq}^{(15)11}, \end{cases}$$

$$(\lambda_{pq}^{(7)00})^2 = \lambda_{pq}^{(7)00}+\lambda_{pq}^{(9)00},$$

$$\lambda_{pq}^{(7)00}\lambda_{pq}^{(7)11} = \lambda_{pq}^{(9)01},$$

$$(\lambda_{pq}^{(7)11})^2 = \lambda_{pq}^{(7)11}+\lambda_{pq}^{(9)11},$$

$$\lambda_{pq}^{(7)00}\lambda_{pq}^{(8)11} = \lambda_{pq}^{(12)00},$$

$$\lambda_{pq}^{(7)00}\lambda_{pq}^{(8)11} = \lambda_{pq}^{(11)01}+\lambda_{qp}^{(11)01}+\lambda_{pq}^{(12)01},$$

$$\lambda_{pq}^{(7)11}\lambda_{pq}^{(8)00} = \lambda_{pq}^{(11)10}+\lambda_{qp}^{(11)10}+\lambda_{pq}^{(12)10},$$

$$\lambda_{pq}^{(7)11}\lambda_{pq}^{(8)11} = \lambda_{pq}^{(11)11}+\lambda_{qp}^{(11)11}+\lambda_{qp}^{(12)11},$$

$$(\lambda_{pq}^{(8)00})^2 = \lambda_{pq}^{(8)00}+\lambda_{pq}^{(10)00}+\lambda_{pq}^{(14)00}+\lambda_{qp}^{(14)00}+2\lambda_{pq}^{(13)00}+\lambda_{pq}^{(15)00},$$

$$(\lambda_{pq}^{(8)11})^2 = \lambda_{pq}^{(8)11}+\lambda_{pq}^{(10)11}+\lambda_{pq}^{(14)11}+\lambda_{qp}^{(14)11}+2\lambda_{pq}^{(13)11}+\lambda_{pq}^{(15)11},$$

$$\lambda_{pq}^{(8)00}\lambda_{pq}^{(8)11} = \lambda_{pq}^{(10)01}+\lambda_{pq}^{(14)01}+\lambda_{qp}^{(14)01}+2\lambda_{pq}^{(13)01}+\lambda_{pq}^{(15)01},$$

$$(\lambda_{pq}^{(7)00}+\lambda_{pq}^{(8)00})^2 = \lambda_{pq}^{(7)00}+\lambda_{pq}^{(8)00}+\lambda_{pq}^{(9)00}+\lambda_{pq}^{(10)00}+2\lambda_{pq}^{(12)00} \\ +2\lambda_{pq}^{(13)00}+\lambda_{pq}^{(15)00},$$

$$(\lambda_{pq}^{(7)00}+\lambda_{pq}^{(8)00})(\lambda_{pq}^{(7)11}+\lambda_{pq}^{(8)11}) = \lambda_{pq}^{(9)01}+\lambda_{pq}^{(10)01}+2\lambda_{pq}^{(11)01} \\ +\lambda_{pq}^{(11)10}+\lambda_{pq}^{(12)01}+\lambda_{pq}^{(12)10}+2\lambda_{pq}^{(13)01}+2\lambda_{pq}^{(14)01}+\lambda_{pq}^{(15)01},$$

$$(\lambda_{pq}^{(7)11}+\lambda_{pq}^{(8)11})^2 = \lambda_{pq}^{(7)11}+\lambda_{pq}^{(8)11}+\lambda_{pq}^{(9)11}+\lambda_{pq}^{(10)11}+2\lambda_{pq}^{(11)11} \\ +2\lambda_{pq}^{(12)11}+2\lambda_{pq}^{(13)11}+2\lambda_{pq}^{(14)11}+\lambda_{pq}^{(15)11}.$$

3.2. The mean value of θ

Let us write

$$\theta = \Theta/\Delta,$$

where

(3.2.1) $$\Theta = \boldsymbol{\pi}'\boldsymbol{T}^{\#}\boldsymbol{\pi} = \boldsymbol{\pi}'(\mu_0\boldsymbol{T}_0+\mu_1\boldsymbol{T}_1+\mu_2\boldsymbol{T}_2)\boldsymbol{\pi}.$$

Following Ogawa [5], we calculate $\mathcal{E}(\Theta)$ as follows :

(3.2.2) $$\mathcal{E}(\Theta) = \frac{1}{(k!)^b} \sum_{\sigma_1,\dots,\sigma_b} \boldsymbol{\pi}'\boldsymbol{U}_{\sigma}'\boldsymbol{T}^{\#}\boldsymbol{U}_{\sigma}\boldsymbol{\pi}$$

$$= \frac{1}{(k!)^b} \sum_{\sigma_1,\dots,\sigma_b} \Big[\sum_{p=1}^{b} \boldsymbol{\pi}^{(p)\prime}\boldsymbol{S}_{\sigma_p}'\boldsymbol{T}_{pp}^{\#}\boldsymbol{S}_{\sigma_p}\boldsymbol{\pi}^{(p)}$$

$$+ \sum_{p\neq q} \boldsymbol{\pi}^{(p)\prime}\boldsymbol{S}_{\sigma_p}'\boldsymbol{T}_{pq}^{\#}\boldsymbol{S}_{\sigma_q}\boldsymbol{\pi}^{(q)} \Big].$$

Now, since

$$\boldsymbol{\pi}^{(p)\prime}\boldsymbol{S}_{\sigma_p}'\boldsymbol{T}_{pp}^{\#}\boldsymbol{S}_{\sigma_p}\boldsymbol{\pi}^{(p)} = \sum_{i,j=1}^{k} t_{ij}^{\#pp}\,\boldsymbol{\pi}_{\sigma(i)}^{(p)}\,\boldsymbol{\pi}_{\sigma(j)}^{(p)}$$

$$= \sum_{i=1}^{k} t_{ii}^{\#pp}\,\pi_{\sigma(i)}^{(p)^2} + \sum_{i\neq j} t_{ij}^{\#pp}\,\pi_{\sigma(i)}^{(p)}\,\pi_{\sigma(j)}^{(p)},$$

and

$$\boldsymbol{\pi}^{(p)\prime}\boldsymbol{S}_{\sigma_p}'\boldsymbol{T}_{pp}^{\#}\boldsymbol{S}_{\sigma_q}\boldsymbol{\pi}^{(q)} = \sum_{i=1}^{k} t_{ii}^{\#pq}\,\pi_{\sigma(i)}^{(q)}\,\pi_{(i)}^{(q)} + \sum_{i\neq j} t_{ij}^{\#pq}\,\pi_{\sigma(i)}^{(p)}\,\pi_{(j)}^{(q)},$$

where we have put $\sigma_p = \sigma$ and $\sigma_q = \tau$, and

$$\mathcal{E}(\pi_{\sigma(i)}^{(p)^2}) = \frac{1}{k}\,\Delta_p,$$

$$\mathcal{E}(\pi_{\sigma(i)}^{(p)}\pi_{\sigma(j)}^{(p)}) = \frac{-1}{k(k-1)}\,\Delta_p,$$

$$\mathcal{E}(\pi_{\sigma(i)}^{(p)}\pi_{\tau(i)}^{(q)}) = \mathcal{E}(\pi_{\sigma(i)}^{(p)}\,\pi_{\tau(j)}^{(q)}) = 0,$$

we get

$$(3.2.3)\quad \mathcal{E}(\Theta)=\frac{1}{k}\sum_{p=1}^{b}\Delta_p\sum_{i=1}^{k}t_{ii}^{\#pp}-\frac{1}{k(k-1)}\sum_{p=1}^{b}\Delta_p\Big(\sum_{i\neq j}t_{ij}^{\#pp}\Big),$$

where $\Delta_p=\boldsymbol{\pi}^{(p)\prime}\boldsymbol{\pi}^{(p)}$.

Now it can be seen that

$$\sum_{i=1}^{k}t_{ii}^{\#pp}=\sum_{i=1}^{k}(\mu_0 t_{ii}^{(0)pp}+\mu_1 t_{ii}^{(1)pp}+\mu_2 t_{ii}^{(2)pp})$$

$$=\mu_0\sum_{i=1}^{k}t_{ii}^{(0)pp}=k\mu_0,$$

and

$$\sum_{i\neq j}t_{ij}^{\#pp}=\sum_{i\neq j}(\mu_0 t_{ij}^{(0)pp}+\mu_1 t_{ij}^{(1)pp}+\mu_2 t_{ij}^{(2)pp})$$

$$=\mu_1\sum_{i\neq j}t_{ij}^{(1)pp}+\mu_2\sum_{i\neq j}t_{ij}^{(2)pp}$$

$$=\mu_1\lambda_{pp}^{(1)11}+\mu_2\lambda_{pp}^{(1)22}$$

$$=\mu_1\lambda_{pp}^{(1)11}+\mu_2[k(k-1)-\lambda_{pp}^{(1)11}]$$

$$=k(k-1)\mu_2+(\mu_1-\mu_2)\lambda_{pp}^{(1)11}.$$

Hence we have

$$(3.2.4)\qquad \mathcal{E}(\Theta)=(\mu_0-\mu_2)\Delta-\frac{\mu_1-\mu_2}{k(k-1)}\sum_{p=1}^{b}\Delta_p\lambda_{pp}^{(1)11},$$

and consequently

$$(3.2.5)\qquad \mathcal{E}(\theta)=\mu_0-\mu_2-\frac{\mu_1-\mu_2}{k(k-1)}\frac{1}{\Delta}\sum_{p=1}^{b}\Delta_p\lambda_{pp}^{(1)11}.$$

As a special case, we shall examine the case of BIBD. In this case, since the spectral decomposition of the matrix $\boldsymbol{NN}'$ is

$$\boldsymbol{NN}'=rk\frac{1}{v}\boldsymbol{G}_v+(r-\lambda)\Big(\boldsymbol{I}_v-\frac{1}{v}\boldsymbol{G}_v\Big),$$

and therefore

$$\rho_0=rk,\quad \rho_1=\rho_2=r-\lambda.$$

Whence it follows that

$$\mu_0 = k(v-1)/(v^2\lambda), \quad \mu_1 = \mu_2 = -k/(v^2\lambda).$$

Thus for a BIBD, (3.2.5) reduces

$$\mathcal{E}(\theta) = k/(v\lambda),$$

which confirms the earlier result given by Ogawa [5].

3.3. The variance of θ

$$\Theta^2 = (\boldsymbol{\pi}'\boldsymbol{U}_\sigma'\boldsymbol{T}^{\#}\boldsymbol{U}_\sigma\boldsymbol{\pi})^2$$

$$= \left(\sum_{p=1}^{b} \boldsymbol{\pi}^{(p)'}\boldsymbol{S}_{\sigma_p}'\boldsymbol{T}_{pp}^{\#}\boldsymbol{S}_{\sigma_p}\boldsymbol{\pi}^{(p)} + \sum_{p\neq q} \boldsymbol{\pi}^{(p)'}\boldsymbol{S}_{\sigma_p}'\boldsymbol{T}_{pq}^{\#}\boldsymbol{S}_{\sigma_q}\boldsymbol{\pi}^{(q)}\right)^2$$

$$= \left(\sum_{p=1}^{b} \boldsymbol{\pi}^{(p)'}\boldsymbol{S}_{\sigma_p}'\boldsymbol{T}_{pp}^{\#}\boldsymbol{S}_{\sigma_p}\boldsymbol{\pi}^{(p)}\right)^2$$

$$+2\left(\sum_{p=1}^{b} \boldsymbol{\pi}^{(p)'}\boldsymbol{S}_{\sigma_p}'\boldsymbol{T}_{pp}^{\#}\boldsymbol{S}_{\sigma_p}\boldsymbol{\pi}^{(p)}\right)\left(\sum_{r\neq s} \boldsymbol{\pi}^{(r)'}\boldsymbol{S}_{\sigma_r}'\boldsymbol{T}_{rs}^{\#}\boldsymbol{S}_{\sigma_s}\boldsymbol{\pi}^{(s)}\right)$$

$$+\left(\sum_{r\neq s} \boldsymbol{\pi}^{(r)'}\boldsymbol{S}_{\sigma_r}'\boldsymbol{T}_{rs}^{\#}\boldsymbol{S}_{\sigma_s}\boldsymbol{\pi}^{(s)}\right)^2.$$

This can be expanded as follows:

$$\Theta^2 = \sum_{p} \boldsymbol{\pi}^{(p)'}\boldsymbol{S}_{\sigma_p}'\boldsymbol{T}_{pp}^{\#}\boldsymbol{S}_{\sigma_p}\boldsymbol{\pi}^{(p)}\boldsymbol{\pi}^{(p)'}\boldsymbol{S}_{\sigma_p}'\boldsymbol{T}_{pp}^{\#}\boldsymbol{S}_{\sigma_p}\boldsymbol{\pi}^{(p)}$$

$$+ \sum_{p\neq q} \boldsymbol{\pi}^{(p)'}\boldsymbol{S}_{\sigma_p}'\boldsymbol{T}_{pp}^{\#}\boldsymbol{S}_{\sigma_p}\boldsymbol{\pi}^{(p)}\boldsymbol{\pi}^{(q)'}\boldsymbol{S}_{\sigma_q}'\boldsymbol{T}_{qq}^{\#}\boldsymbol{S}_{\sigma_q}\boldsymbol{\pi}^{(q)}$$

$$+2\Big[\sum_{p\neq q} \boldsymbol{\pi}^{(p)'}\boldsymbol{S}_{\sigma_p}'\boldsymbol{T}_{pp}^{\#}\boldsymbol{S}_{\sigma_p}\boldsymbol{\pi}^{(p)}\Big(\boldsymbol{\pi}^{(p)'}\boldsymbol{S}_{\sigma_p}'\boldsymbol{T}_{pq}^{\#}\boldsymbol{S}_{\sigma_q}\boldsymbol{\pi}^{(q)}$$

$$+\boldsymbol{\pi}^{(q)'}\boldsymbol{S}_{\sigma_q}'\boldsymbol{T}_{qp}^{\#}\boldsymbol{S}_{\sigma_p}\boldsymbol{\pi}^{(r)}\Big)$$

$$+\sum_{p\neq q\neq r} \boldsymbol{\pi}^{(p)'}\boldsymbol{S}_{\sigma_p}'\boldsymbol{T}_{pp}^{\#}\boldsymbol{S}_{\sigma_p}\boldsymbol{\pi}^{(p)}\boldsymbol{\pi}^{(q)'}\boldsymbol{S}_{\sigma_q}\boldsymbol{T}_{qr}^{\#}\boldsymbol{S}_{\sigma_r}\boldsymbol{\pi}^{(p)}\Big]$$

$$+\sum_{p\neq q} \boldsymbol{\pi}^{(p)'}\boldsymbol{S}_{\sigma_p}'\boldsymbol{T}_{pq}^{\#}\boldsymbol{S}_{\sigma_q}\boldsymbol{\pi}^{(q)}\Big[\boldsymbol{\pi}^{(p)'}\boldsymbol{S}_{\sigma_p}'\boldsymbol{T}_{pq}^{\#}\boldsymbol{S}_{\sigma_q}\boldsymbol{\pi}^{(q)}$$

$$+\boldsymbol{\pi}^{(q)'}\boldsymbol{S}_{\sigma_q}'\boldsymbol{T}_{pq}^{\#}\boldsymbol{S}_{\sigma_p}\boldsymbol{\pi}^{(p)}\Big]$$

$$+\sum_{p\neq q\neq r} \pi^{(p)'}S'_{\sigma_p}T^{\#}_{pq}S_{\sigma_q}\pi^{(q)}\Big[\pi^{(p)'}S'_{\sigma_p}T^{\#}_{pr}S_{\sigma_r}\pi^{(r)} + \pi^{(r)'}S'_{\sigma_r}T^{\#}_{rp}S_{\sigma_p}\pi^{(p)}\Big]^*$$

$$+\sum_{p\neq q\neq r} \pi^{(p)'}S'_{\sigma_p}T^{\#}_{pq}S_{\sigma_q}\pi^{(q)}\Big[\pi^{(p)'}S'_{\sigma_p}T^{\#}_{pr}S_{\sigma_r}\pi^{(r)} + \pi^{(r)'}S'_{\sigma_r}T^{\#}_{rq}S_{\sigma_q}\pi^{(q)}\Big]^*$$

$$+\sum_{p\neq r\neq s\neq q} \pi^{(p)'}S'_{\sigma_p}T^{\#}_{pq}S_{\sigma_q}\pi^{(q)}\pi^{(r)'}S'_{\sigma_r}T^{\#}_{rs}S_{\sigma_s}\pi^{(s)*}.$$

Since the terms which are linear with respect to some S_σ such as those with* in the above vanish by taking their expectations, we have

$$\mathscr{E}(\Theta^2) = \mathscr{E}(A)+\mathscr{E}(B)+\mathscr{E}(C)+\mathscr{E}(C'),$$

where

$$(3.3.1)\quad \begin{cases} A = \sum_{p} \pi^{(p)'}S'_{\sigma_p}T^{\#}_{pp}S_{\sigma_p}\pi^{(p)}\pi^{(p)'}S'_{\sigma_p}T^{\#}_{pp}S_{\sigma_p}\pi^{(p)}, \\ B = \sum_{p\neq q} \pi^{(p)'}S'_{\sigma_p}T^{\#}_{pp}S_{\sigma_p}\pi^{(p)}\pi^{(q)'}S'_{\sigma_q}T^{\#}_{qq}S_{\sigma_q}\pi^{(q)}, \\ C = \sum_{p\neq q} \pi^{(p)'}S'_{\sigma_p}T^{\#}_{pq}S_{\sigma_q}\pi^{(q)}\pi^{(p)'}S'_{\sigma_p}T^{\#}_{pq}S_{\sigma_q}\pi^{(q)}, \\ C' = \sum_{p\neq q} \pi^{(p)'}S'_{\sigma_p}T^{\#}_{pq}S_{\sigma_q}\pi^{(q)}\pi^{(q)'}S'_{\sigma_q}T^{\#}_{qp}S_{\sigma_p}\pi^{(p)}. \end{cases}$$

Since $C = C'$, it follows that

$$(3.3.2)\qquad \mathscr{E}(\Theta^2) = \mathscr{E}(A)+\mathscr{E}(B)+2\mathscr{E}(C).$$

Now we shall calculate $\mathscr{E}(A)$, $\mathscr{E}(B)$ and $\mathscr{E}(C)$ in this order. We present the process of the calculation of $\mathscr{E}(A)$ in detail.

Since

$$(\pi^{(p)'}S'_{\sigma_p}T^{\#}_{pp}S_{\sigma_p}\pi^{(p)})^2$$

$$= \Big(\sum_{i=1}^{k} t^{\#pp}_{ii}\,\pi^{(p)^2}_{\sigma(i)} + \sum_{i\neq j} t^{\#pp}_{ij}\,\pi^{(p)}_{\sigma(i)}\,\pi^{(p)}_{\sigma(j)}\Big)^2$$

$$= \sum_i t_{ii}^{\#pp^2} \pi_{\sigma(i)}^{(p)^4} + \sum_{i \neq j} t_{ii}^{\#pp} t_{jj}^{\#pp} \pi_{\sigma(i)}^{(p)^2} \pi_{\sigma(j)}^{(p)^2} + 4 \sum_{i \neq j} t_{ii}^{\#pp} t_{ii}^{\#pp} \pi_{\sigma(i)}^{(p)^3} \pi_{\sigma(j)}^{(p)}$$

$$+2 \sum_{i \neq j \neq l} t_{ii}^{\#pp} t_{jl}^{\#pp} \pi_{\sigma(i)}^{(p)^2} \pi_{\sigma(j)}^{(p)} \pi_{\sigma(l)}^{(p)} + 2 \sum_{i \neq j} t_{ij}^{\#pp} t_{ij}^{\#pp} \pi_{\sigma(i)}^{(p)^2} \pi_{\sigma(j)}^{(p)^2}$$

$$+4 \sum_{i \neq j \neq l} t_{ij}^{\#pp} t_{ij}^{\#pp} \pi_{\sigma(i)}^{(p)^2} \pi_{\sigma(j)}^{(p)} \pi_{\sigma(l)}^{(p)}$$

$$+ \sum_{i \neq j \neq l \neq m} t_{ij}^{\#pp} t_{lm}^{\#pp} \pi_{\sigma(i)}^{(p)} \pi_{\sigma(j)}^{(p)} \pi_{\sigma(l)}^{(p)} \pi_{\sigma(m)}^{(p)},$$

and

$$(A_1) \qquad \mathcal{E}(\pi_{\sigma(i)}^{(p)^4}) = \Gamma_p/k \text{ where } \Gamma_p = \sum_{i=1}^{k} \pi_i^{(p)^4},$$

$$\sum_{i=1}^{k} t_{ii}^{\#pp^2} = k\mu_0^2,$$

$$(A_2) \qquad \mathcal{E}(\pi_{\sigma(i)}^{(p)^2} \pi_{\sigma(j)}^{(p)^2}) = (\Delta_p^2 - \Gamma_p)/k(k-1),$$

$$\sum_{i \neq j} t_{ii}^{\#pp} t_{jj}^{\#pp} = \sum_{i \neq j} \left(\sum_{v=0}^{2} \mu_v t_{ii}^{(v)pp} \right) \left(\sum_{v=0}^{2} \mu_v t_{jj}^{(v)pp} \right)$$

$$= \sum_{i \neq j} \mu_0^2 = k(k-1)\mu_0^2,$$

$$(A_3) \qquad \mathcal{E}(\pi_{\sigma(i)}^{(p)^3} \pi_{\sigma(j)}^{(p)}) = -\Gamma_p/k(k-1)$$

$$\sum_{i \neq j} t_{ii}^{\#pp} t_{ij}^{\#pp} = \sum_{i \neq j} \left(\sum_{v=0}^{2} \mu_v t_{ii}^{(v)pp} \right) \left(\sum_{v=0}^{2} \mu_v t_{ij}^{(v)pp} \right)$$

$$= \mu_0 \mu_1 \sum_{i \neq j} t_{ij}^{(1)pp} + \mu_0 \mu_2 \sum_{i \neq j} t_{ij}^{(2)pp}$$

$$= \mu_0 \mu_1 \lambda_{pp}^{(1)11} + \mu_0 \mu_2 \lambda_{pp}^{(1)22} = \mu_0 \sum_{s=1}^{s} \mu_s \lambda_{pp}^{(1)ss},$$

$$(A_4) \quad \mathcal{E}(\pi_{\sigma(i)}^{(p)^2} \pi_{\sigma(j)}^{(p)} \pi_{\sigma(l)}^{(p)}) = \frac{-1}{k(k-1)(k-2)} (\Delta_p^2 - 2\Gamma_p),$$

$$\sum_{i\neq j\neq l} t_{ii}^{\#pp}\, t_{jl}^{\#pp} = \sum_{i\neq j\neq l}\left(\sum_{v=0}^{2}\mu_v\, t_{ii}^{(v)pp}\right)\left(\sum_{v=0}^{2}\mu_v\, t_{jl}^{(v)pp}\right)$$

$$= (k-2)\mu_0 \sum_{j\neq l}\left(\sum_{v=0}^{2}\mu_v\, t_{jl}^{(v)pp}\right)$$

$$= (k-2)\mu_0 \sum_{s=1}^{2}\mu_s\, \lambda_{pp}^{(1)ss},$$

$$(A_5)\quad \mathcal{E}(\pi_{\sigma(i)}^{(p)^2}\, \pi_{\sigma(j)}^{(p)^2}) = \frac{1}{k(k-1)}(\Delta_p^2-\Gamma_p),$$

$$\sum_{i\neq j} t_{ij}^{\#pp^2} = \sum_{i\neq j}\left(\sum_{v=0}^{2}\mu_v\, t_{ij}^{(v)pp}\right)^2 = \sum_{i\neq j}\left(\sum_{v=1}^{2}\mu_v^2\, t_{ij}^{(v)pp}\right) = \sum_{s=1}^{2}\mu_s^2\, \lambda_{pp}^{(1)ss},$$

$$(A_6)\quad \mathcal{E}(\pi_{\sigma(i)}^{(p)^2}\, \pi_{\sigma(j)}^{(p)}\, \pi_{\sigma(l)}^{(p)}) = \frac{-1}{k(k-1)(k-2)}(\Delta_p^2-2\Gamma_p),$$

$$\sum_{i\neq j\neq l} t_{ij}^{\#pp}\, t_{il}^{\#pp} = \sum_{i\neq j\neq l}\left(\sum_{v=0}^{2}\mu_v\, t_{ij}^{(v)pp}\right)\left(\sum_{v=0}^{2}\mu_v\, t_{il}^{(v)pp}\right)$$

$$= \sum_{s,t=0}^{2}\mu_s\,\mu_t \sum_{i\neq j\neq l} t_{ij}^{(s)pp}\, t_{il}^{(t)pp} = \sum_{s,t=1}^{2}\mu_s\,\mu_t\, \lambda_{pp}^{(2)st},$$

$$(A_7)\quad \mathcal{E}(\pi_{\sigma(i)}^{(p)}\, \pi_{\sigma(j)}^{(p)}\, \pi_{\sigma(l)}^{(p)}\, \pi_{\sigma(m)}^{(p)}) = \frac{1}{k(k-1)(k-2)(k-3)}(\Delta_p^2-2\Gamma_p),$$

$$\sum_{j\neq j\neq l\neq m} t_{ij}^{\#pp}\, t_{lm}^{\#pp} = \sum_{i\neq j\neq l\neq m}\left(\sum_{v=0}^{2}\mu_v\, t_{ij}^{(v)pp}\right)\left(\sum_{v=0}^{2}\mu_v\, t_{lm}^{(v)pp}\right)$$

$$= \sum_{s,t=0}^{2}\mu_s\,\mu_t \sum_{i\neq j\neq l\neq m} t_{ij}^{(s)pp}\, t_{lm}^{(t)pp}$$

$$= \sum_{s,t=1}^{2}\mu_s\,\mu_t\, \lambda_{pp}^{(3)st}.$$

Hence we obtain

$$\mathcal{E}(A) = \sum_{p=1}^{b}\Big[\mu_0^2\Gamma_p+\mu_0^2(\Delta_p^2-\Gamma_p)-\frac{4}{k(k-1)}\mu_0\Gamma_p\sum_{s=1}^{2}\mu_s\,\lambda_{pp}^{(1)ss}$$

(3.3.3) $$-\frac{2}{k(k-1)}\mu_0(\Delta_p^2-2\Gamma_p)\sum_{s=1}^{2}\mu_s\lambda_{pp}^{(1)ss}$$

$$+\frac{2}{k(k-1)}(\Delta_p^2-\Gamma_p)\sum_{s=1}^{2}\mu_s^2\lambda_{pp}^{(1)ss}$$

$$-\frac{4}{k(k-1)(k-2)}(\Delta_p^2-2\Gamma_p)\sum_{s,t\geq 1}\mu_s\mu_t\lambda_{pp}^{(2)st}$$

$$+\frac{3}{k(k-1)(k-2)(k-3)}(\Delta_p^2-2\Gamma_p)\sum_{s,t\geq 1}\mu_s\mu_t\lambda_{pp}^{(3)st}\Bigg]$$

$$=\sum_{p=1}^{b}\Bigg[\frac{1}{k(k-1)}\Delta_p^2\sum_{s=1}^{2}(\mu_0-\mu_s)^2\lambda_{pp}^{(1)ss}$$

$$+(\Delta_p^2-2\Gamma_p)\frac{1}{k(k-1)}\sum_{s=1}^{2}\mu_p^2\lambda_{pp}^{(1)ss}$$

$$-\frac{4}{k(k-1)(k-2)}\sum_{s,t\geq 1}\mu_s\mu_t\lambda_{pp}^{(2)st}$$

$$+\frac{3}{k(k-1)(k-2)(k-3)}\sum_{s,t\geq 1}\mu_s\mu_t\lambda_{pp}^{(3)st}\Bigg].$$

In similar manner one can obtain

(3.3.4) $$\mathcal{E}(B)=\sum_{p\neq q}\Delta_p\Delta_q\Bigg[\frac{1}{k(k-1)}\sum_{s=1}^{2}(\mu_0-\mu_s)^2\lambda_{pp}^{(1)ss}$$

$$-\frac{1}{k(k-1)}\sum_{s=1}^{2}\mu_s^2\lambda_{pp}^{(1)ss}$$

$$+\frac{2}{k^2(k-2)^2}\sum_{s,t\geq 1}\mu_s\mu_t\lambda_{pq}^{(4)st}$$

$$+\frac{4}{k^2(k-1)^2}\sum_{s,t\geq 1}\mu_s\mu_t\lambda_{pq}^{(5)st}$$

$$+\frac{1}{k^2(k-1)^2}\sum_{s,t\geq 1}\mu_s\mu_t\lambda_{pq}^{(6)st}\Bigg]$$

$$= \mathcal{E}^2(\Theta) - \sum_p \Delta_p^2 \left[\frac{1}{k(k-1)} \sum_{s=1}^{2} (\mu_0 - \mu_s)^2 \lambda_{pp}^{(1)ss} \right.$$

$$- \frac{1}{k(k-1)} \sum_{s=1}^{2} \mu_s^2 \lambda_{pp}^{(1)ss}$$

$$+ \frac{2}{k^2(k-1)^2} \sum_{s=1}^{2} \mu_s^2 \lambda_{pp}^{(1)ss}$$

$$+ \frac{4}{k^2(k-1)^2} \sum_{s,t \geq 1} \mu_s \mu_t \lambda_{pp}^{(2)st}$$

$$\left. + \frac{1}{k^2(k-1)^2} \sum_{s,t \geq 1} \mu_s \mu_t \lambda_{pp}^{(3)st} \right]$$

and

$$(3.3.5) \quad \mathcal{E}(C) = \sum_{p \neq q} \Delta_p \Delta_q \left[\frac{1}{k^2} \sum_{s=1}^{2} \mu_s^2 (\lambda_{pq}^{(7)ss} + \lambda_{pq}^{(8)ss}) \right.$$

$$+ \frac{1}{k^2(k-1)^2} \sum_{s,t \geq 1} \mu_s \mu_t (\lambda_{pq}^{(9)st} + \lambda_{pq}^{(10)st} + 2\lambda_{pq}^{(12)st} + 2\lambda_{pq}^{(13)st} + \lambda_{pq}^{(15)st}$$

$$\left. - \frac{2}{k^2(k-1)} \sum_{s,t \geq 1} \mu_s \mu_t (2\lambda_{pq}^{(11)st} + \lambda_{pq}^{(14)st}) \right].$$

Therefore one obtains the variance of θ as

$$\mathrm{Var}\,\theta = \Delta^{-2}\,\mathrm{Var}\,(\Theta) = [\mathcal{E}(A) + \mathcal{E}(B) + 2\mathcal{E}(C) - \mathcal{E}^2(\Theta)]/\Delta^2$$

$$= \sum_{p=1}^{b} \frac{\Delta_p^2 - 2\Gamma_p}{\Delta^2} \left[\frac{1}{k(k-1)} \sum_{s=1}^{2} \mu_s^2 \lambda_{pp}^{(1)ss} - \frac{4}{k(k-1)(k-2)} \sum_{s,t \geq 1} \mu_s \mu_t \lambda_{pp}^{(2)st} \right.$$

$$\left. + \frac{3}{k(k-1)(k-2)(k-3)} \sum_{s,t \geq 1} \mu_s \mu_t \lambda_{pp}^{(3)st} \right]$$

$$+ \sum_{p=1}^{b} \frac{\Delta_p^2}{\Delta^2} \left[\frac{1}{k(k-1)} \sum_{s=1}^{2} \mu_s^2 \lambda_{pp}^{(1)ss} - \frac{2}{k^2(k-1)^2} \sum_{s=1}^{2} \mu_s^2 \lambda_{pp}^{(1)ss} \right.$$

$$-\frac{4}{k^2(k-1)^2}\sum_{s,\,t\geqslant 1}\mu_s\,\mu_t\,\lambda_{pp}^{(2)st}-\frac{1}{k^2(k-1)^2}\sum_{s,\,t\geqslant 1}\mu_s\,\mu_t\,\lambda_{pp}^{(3)st}\Big]$$

$$+\sum_{p\neq q}\Big[\frac{\Delta_p\Delta_q}{\Delta^2}\,\frac{2}{k^2}\sum_{s=1}^{2}\mu_s^2\,(\lambda_{pq}^{(7)ss}+\lambda_{pq}^{(8)st})$$

$$+\frac{2}{k^2(k-1)^2}\sum_{s,\,t\geqslant 1}\mu_s\mu_t(\lambda_{pq}^{(9)st}+\lambda_{pq}^{(10)st}+2\lambda_{pq}^{(12)st}+2\lambda_{pq}^{(13)st}+\lambda_{pq}^{(15)st})$$

$$-\frac{4}{k^2(k-1)}\sum_{s,\,t\geqslant 1}\mu_s\,\mu_t(2\lambda_{pq}^{(11)st}+\lambda_{pq}^{(14)st})\Big].$$

For the case of BIBD, one may put as

$$\lambda_{pp}^{(s)uv}=0,\ s=1,2,3,\quad\text{and}\quad\lambda_{pq}^{(s)uv}=0,\quad 4\leqslant s\leqslant 15,$$

if $u=2$ or $v=2$, and hence the following relationships are obtained from those in Section 3.1 :

$$\lambda_{pp}^{(1)11}=k(k-1),$$

$$\lambda_{pp}^{(2)11}=k(k-1)(k-2),$$

$$\lambda_{pp}^{(3)11}=k(k-1)(k-2)(k-3),$$

$$\lambda_{pq}^{(9)00}+\lambda_{pq}^{(10)00}+2\lambda_{pq}^{(12)00}+2\lambda_{pq}^{(13)00}+\lambda_{pq}^{(15)00}=\lambda_p^2-\lambda_{pq},$$

$$\lambda_{pq}^{(9)10}+\lambda_{pq}^{(10)10}+2\lambda_{pq}^{(12)10}+2\lambda_{pq}^{(13)10}+\lambda_{pq}^{(15)10}+\lambda_{pq}^{(9)01}+\lambda_{pq}^{(10)01}$$
$$+2\lambda_{pq}^{(12)01}+2\lambda_{pq}^{(13)01}+\lambda_{pq}^{(15)01}$$
$$=2\lambda_{pq}(k^2-\lambda_{pq})-4(\lambda_{pq}^{(11)01}+\lambda_{pq}^{(11)10}+\lambda_{pq}^{(14)01}),$$

$$\lambda_{pq}^{(9)11}+\lambda_{pq}^{(10)11}+2\lambda_{pq}^{(12)11}+2\lambda_{pq}^{(13)11}+\lambda_{pq}^{(15)11}$$
$$=(k^2-\lambda_{pq})^2-(k^2-\lambda_{pq})-2(2\lambda_{pq}^{(11)11}+\lambda_{pq}^{(14)11}),$$

$$\lambda_{pq}^{(11)01}+\lambda_{pq}^{(11)10}+\lambda_{pq}^{(14)01}=(k-1)\lambda_{pq},$$

$$2\lambda_{pq}^{(11)11}+\lambda_{pq}^{(14)11}=(k-1)(k^2-2\lambda_{pq}),$$

where we have put

$$\lambda_{pq}=\lambda_{pq}^{(7)00}+\lambda_{pq}^{(8)00}=\text{number of treatments common to the } p\text{th and } q\text{th blocks},$$

By these equalities, (3.3.6) reduces

$$\mathrm{Var}\ \theta = 2(v\lambda\Delta)^{-2} \sum_{p \neq q} \Delta_p \Delta_q \left[\lambda_{pq} + \frac{1}{(k-1)^2} (\lambda_{pq}^2 - \lambda_{pq}) \right],$$

which confirms the Ogawa result for a BIBD once again [5].

4. AN APPROXIMATION TO THE PERMUTATION DISTRIBUTION OF θ BY A CERTAIN BETA-DISTRIBUTION WHEN THE NUMBER b OF BLOCKS IS SUFFICIENTLY LARGE AND CERTAIN UNIFORMITY CONDTIONS ARE IMPOSED ON THE UNIT ERRORS

In this section, we assume that the following uniformity conditions

$$\Delta_p = \Delta_0 \text{ and } \Gamma_p = \Gamma_0 \quad \text{for } p = 1, 2, \ldots, b \tag{4.1}$$

are imposed on the unit errors. Under such conditions the mean of θ and the variance of θ, Var θ, can be presented as follows :

$$\mathcal{E}(\theta) = \frac{1}{bk(k-1)} \sum_{s=1}^{2} (\mu_0 - \mu_s) \sum_p \lambda_{pp}^{(1)ss}, \tag{4.2}$$

$$\begin{aligned} \mathrm{Var}\ \theta = \left(1 - \frac{2\Gamma_0}{\Delta_0^2}\right) & \left[\frac{1}{b^2 k(k-1)} \sum_{s=1}^{2} \mu_s^2 \sum_p \lambda_{pp}^{(1)ss} \right. \\ & - \frac{4}{b^2 k(k-1)(k-2)} \sum_{s,t \geqq 1} \mu_s \mu_t \sum_p \lambda_{pp}^{(2)st} \\ & \left. + \frac{3}{b^2 k(k-1)(k-2)k-3)} \sum_{s,t \geqq 1} \mu_s \mu_t \sum_p \lambda_{pp}^{(3)st} \right] \\ & + \left[\frac{1}{b^2 k(k-1)} \sum_{s=1}^{2} \mu_s^2 \sum_p \lambda_{pp}^{(1)ss} - \frac{2}{b^2 k^2 (k-1)^2} \sum_{s=1}^{2} \mu_s^2 \sum_p \lambda_{pp}^{(1)ss} \right. \\ & - \frac{4}{b^2 k^2 (k-1)^2} \sum_{s,t \geqq 1} \mu_s \mu_t \sum_p \lambda_{pp}^{(2)st} \\ & \left. - \frac{1}{b^2 k^2 (k-1)^2} \sum_{s,t \geqq 1} \mu_s \mu_t \sum_p \lambda_{pp}^{(3)st} \right] \end{aligned} \tag{4.3}$$

$$+\frac{2}{b^2k^2}\sum_{A=1}^{2}\mu_s^2(\lambda_{pq}^{(7)ss}+\lambda_{pq}^{(8)ss})$$

$$+\frac{2}{b^2k^2(k-1)^2}\sum_{s,t\geqslant 1}\mu_s\mu_t\sum_{p\neq q}(\lambda_{pq}^{(9)st}+\lambda_{pq}^{(10)st}+2\lambda_{pq}^{(12)st}+2\lambda_{pq}^{(13)st}$$

$$+\lambda_{pq}^{(15)st})$$

$$-\frac{4}{b^2k^2(k-1)}\sum_{s,t\geqslant 1}\mu_s\mu_t\sum_{p\neq q}(2\lambda_{pq}^{(11)st}+\lambda_{pq}^{(14)st}).$$

In order to express Var θ in terms of the parameters of the design under consideration, we show that

(i) $\sum_{p}\lambda_{pp}^{(1)ss}=\lambda_s n_s v, \quad s=1,2,$

(ii) $\sum_{p\neq q}(\lambda_{pq}^{(7)ss}+\lambda_{pq}^{(8)ss})=(r^2-\lambda_s)n_s v, \quad s=0,1,2,$

(4.4) (iii) $\sum_{p\neq q}(\lambda_{pq}^{(9)st}+\lambda_{pq}^{(10)st}+2\lambda_{pq}^{(12)st}+2\lambda_{pq}^{(13)st}+\lambda_{pq}^{(15)st})$

$$=\sum_{\alpha,\beta=0}\lambda_\alpha\lambda_\beta\sum_{\gamma=0}^{2}p_{\alpha\beta}^{\gamma}p_{st}^{\gamma}n_\gamma v-\delta_{st}\lambda_s n_s v$$

$$-\delta_{s0}(1-\delta_{t0})(k-2)\lambda_t n_t v-\delta_{t0}(k-2)\lambda_s n_s v-2\lambda_{st}^{(1)}-\lambda_{st}^{(2)},$$
$$s,t=0,1,2,$$

(iv) $\sum_{p\neq q}(2\lambda_{p}^{(11)st}+\lambda_{pq}^{(14)st})=r\sum_{\alpha=0}^{2}\lambda_\alpha p_{st}^{\alpha}n_\alpha v$

$$-\delta_{s0}(1-\delta_{t0})\lambda_t n_t v-\delta_{t0}(1-\delta_{s0})\lambda_s n_s v-\lambda_{st}^{(1)}, \quad s=0,1,2,$$

where we have put

(4.5) $$\lambda_{st}^{(1)}=\sum_{p}\lambda_{pp}^{(2)st} \quad\text{and}\quad \lambda_{st}^{(2)}=\sum_{p}\lambda_{pp}^{(3)st},$$

and δ_{st} stands for the Kronecker delta.

Indeed,

(4.4) (i) : $\sum_{p}\lambda_{pp}^{(1)ss}=\sum_{p}\sum_{i\neq j}t_{ij}^{(s)pp}=\sum_{p}\sum_{\alpha,\beta}n_{\alpha p}a_{\alpha s}^{\alpha}n_{\beta p}=\operatorname{tr}(\boldsymbol{N}'\boldsymbol{A}_s\boldsymbol{N})$

$$=\operatorname{tr}(\boldsymbol{N}\boldsymbol{N}'\boldsymbol{A}_s)=\operatorname{tr}\left(\sum_{\alpha}\lambda_\alpha\sum_{\beta}p_{\alpha s}^{\beta}\boldsymbol{A}_s\right)=\lambda_s n_s v.$$

(4.4) (ii) : $\sum_{p\neq q}(\lambda_{pq}^{(7)ss}+\lambda_{pq}^{(8)ss}) = \sum_{p\neq q}\left(\sum_{i,j} t_{ij}^{(s)pq}\right)$

$$= \sum_{p,q}\sum_{\alpha,\beta} n_{\alpha p}a_{\alpha s}^{\beta}n_{\beta q} - \sum_{p}\sum_{\alpha,\beta} n_{\alpha p}a_{\alpha s}^{\beta}n_{\beta q}$$

$$= \mathbf{1}'\boldsymbol{N}'\boldsymbol{A}_s\boldsymbol{N}\mathbf{1} - \mathrm{tr}(\boldsymbol{N}'\boldsymbol{A}_s\boldsymbol{N}) = r^2 n_s v - \lambda_s n_s v = (r^2-\lambda_s)n_s v.$$

(4.4) (iii) : It was shown that

$$\sum_{p\neq q}(\lambda_{pq}^{(7)00}+\lambda_{pq}^{(8)00})^2 = \sum_{p\neq q}(\lambda_{pq}^{(7)00}+\lambda_{pq}^{(8)00})$$

$$+\sum_{p\neq q}(\lambda_{pq}^{(9)00}+\lambda_{pq}^{(10)00}+2\lambda_{pq}^{(12)00}+2\lambda_{pq}^{(13)00}+\lambda_{pq}^{(15)00}).$$

Now, let us consider the following identity relation :

$$\sum_{p,q}\left(\sum_{\alpha,\beta} n_{\alpha\beta}a_{\alpha 0}^{\beta}n_{\beta q}\right)^2 = \mathrm{tr}(\boldsymbol{N}'\boldsymbol{A}_0\boldsymbol{N})^2.$$

Since

$$\sum_{\alpha,\beta} n_{\alpha p}a_{\alpha 0}^{\beta}n_{\beta q} = \begin{cases} k, & \text{if } p=q, \\ \lambda_{pq}^{(7)00}+\lambda_{pq}^{(8)00}, & \text{if } p\neq q, \end{cases}$$

the left-hand side of the identity is shown to be

$$\sum_{p,q}\left(\sum_{\alpha,\beta} n_{\alpha\beta}a_{\alpha 0}^{\beta}n_{\beta q}\right)^2 = bk^2+\sum_{p\neq q}(\lambda_{p}^{(7)00}+\lambda_{pq}^{(8)00}).$$

The right-hand side of the identity is

$$\mathrm{tr}(\boldsymbol{N}'\boldsymbol{A}_0\boldsymbol{N})^2 = (\boldsymbol{N}'\boldsymbol{N})^2 = \mathrm{tr}\left(\sum_{\alpha,\beta,\gamma}\lambda_\alpha\lambda_\beta p_{\alpha\beta}^{\gamma}A_\gamma\right) = \sum_{\alpha}\lambda_\alpha^2 n_\alpha v.$$

Hence we obtain the relation

$$\sum_{p\neq q}(\lambda_{pq}^{(9)00}+\lambda_{pq}^{(10)00}+2\lambda_{pq}^{(12)00}+2\lambda_{pq}^{(13)00}+\lambda_{pq}^{(15)00})$$

$$= \sum_{\alpha}\lambda_\alpha^2 n_\alpha v - r(k-1)v.$$

It was also shown that

$$\sum_{p\neq q}(\lambda_{pq}^{(7)00}+\lambda_{pq}^{(8)00})(\lambda_{pq}^{(7)ss}+\lambda_{pq}^{(8)ss}) = 2\sum_{p\neq q}(2\lambda_{pq}^{(11)0s}+\lambda_{pq}^{(14)0s})$$

$$+\sum_{p\neq q}(\lambda_{pq}^{(9)0s}+\lambda_{pq}^{(10)0s}+2\lambda_{pq}^{(12)0s}+2\lambda_{pq}^{(13)0s}+\lambda_{pq}^{(15)0s}), \quad s=1, 2.$$

Now,

$$\sum_{p,q}\sum_{\alpha,\beta,\gamma} n_{\alpha p}a^{\beta}_{\alpha 0}a^{\gamma}_{\alpha s}n_{\beta q}n_{\gamma q}=\sum_{p}\sum_{\alpha,\gamma} n_{\alpha p}a^{\gamma}_{\alpha s}n_{\gamma q}+\sum_{p\neq q}\sum_{\alpha,\gamma} n_{\alpha p}a^{\gamma}_{\alpha s}n_{\alpha q}n_{\gamma q}$$

$$=\sum_{p}\lambda^{(1)ss}_{pp}+\sum_{p\neq q}(2\lambda^{(11)0s}_{pq}+\lambda^{(14)0s}_{pq}),$$

and, on the other hand,

$$\sum_{p,q}\sum_{\alpha,\beta,\gamma} n_{\alpha p}a^{\beta}_{\alpha 0}a^{\gamma}_{\alpha s}n_{\beta q}n_{\gamma q}=\sum_{q}\sum_{\alpha,\beta,\gamma} a^{\beta}_{\alpha 0}a^{\gamma}_{\alpha s}n_{\beta q}n_{\gamma q}\sum_{p} n_{\alpha p}$$

$$=r\sum_{q}\sum_{\alpha,\gamma} n_{\alpha q}a^{\gamma}_{\alpha s}n_{\gamma q}=r\operatorname{tr}(\boldsymbol{N}'\boldsymbol{A}_s\boldsymbol{N})=r\lambda_s n_s v.$$

Thus one gets the relation

$$\sum_{p\neq q}(2\lambda^{(11)0s}_{pq}+\lambda^{(14)0s}_{pq})=(r-1)\lambda_s n_s v.$$

Since

$$\sum_{p,q}\left(\sum_{\alpha,\beta} n_{\alpha p}a^{\beta}_{\alpha 0}n_{\beta q}\right)\left(\sum_{\gamma,\delta} n_{\gamma p}a^{\delta}_{\gamma s}n_{\delta q}\right)$$

$$=\sum_{p}(\sum_{\alpha} n_{\alpha p})\left(\sum_{\gamma,\delta} n_{\gamma p}a^{\delta}_{\gamma s}n_{\delta q}\right)+\sum_{p\neq q}\left(\sum_{\alpha,\beta} n_{\alpha p}a^{\beta}_{\alpha 0}a_{\beta q}\right)\left(\sum_{\gamma,\delta} n_{\gamma p}a^{\delta}_{\gamma s}n_{\delta q}\right)$$

$$=k\sum_{p}\lambda^{(1)ss}_{pp}+\sum_{p\neq q}(\lambda^{(7)00}_{pq}+\lambda^{(8)00}_{pq})(\lambda^{(7)ss}_{pq}+\lambda^{(8)ss}_{pq}),$$

and

$$\sum_{p,q}\left(\sum_{\alpha,\beta} n_{\alpha p}a^{\beta}_{\alpha 0}n_{\beta q}\right)\left(\sum_{\gamma,\delta} n_{\gamma p}a^{\delta}_{\gamma s}n_{\delta q}\right)=\operatorname{tr}(\boldsymbol{N}'\boldsymbol{A}_0\boldsymbol{N}\boldsymbol{N}'\boldsymbol{A}_s\boldsymbol{N})$$

$$=\operatorname{tr}(\boldsymbol{N}\boldsymbol{N}'\boldsymbol{N}\boldsymbol{N}'\boldsymbol{A}_s)=\operatorname{tr}\left(\sum_{\alpha,\beta}\lambda_\alpha\lambda_\beta\sum_{\gamma} p^{\gamma}_{\alpha\beta}\boldsymbol{A}_\gamma\boldsymbol{A}_s\right)$$

$$=\operatorname{tr}\left(\sum_{\alpha,\beta}\sum_{\gamma,\delta}\lambda_\delta\lambda_\beta p^{\gamma}_{\alpha\beta}p^{\delta}_{\gamma s}\boldsymbol{A}_\delta\right)=\sum_{\alpha,\beta,\gamma}\lambda_\alpha\lambda_\beta p^{\gamma}_{\alpha\beta}p^{0}_{\gamma s}v=\sum_{\alpha,\beta}\lambda_\alpha\lambda_\beta p^{s}_{\alpha\beta}n_s v$$

$$=\sum_{\alpha,\beta=1}^{2}\lambda_\alpha\lambda_\beta p^{s}_{\alpha\beta}n_s v+2\lambda_s n_s v,$$

one obtains the relation

$$\sum_{p\neq q} (\lambda_{pq}^{(9)0s}+\lambda_{pq}^{(10)0s}+2\lambda_{pq}^{(12)0s}+2\lambda_{pq}^{(13)0s}+\lambda_{pq}^{(15)0s})$$

$$=\sum_{\alpha,\beta\geq 1} \lambda_\alpha\lambda_\beta p_{\alpha\beta}^s n_s v-(k-2)\lambda_s n_s v.$$

Similarly, the two way calculations of

$$\sum_{p,q}\sum_{\alpha,\beta,\gamma,\delta} n_{\alpha p}n_{\beta p}a_{\delta s}^{\gamma}a_{\beta t}^{\delta}n_{\gamma q}n_{\delta q}$$

yield the relation

$$\sum_{p\neq q} (2\lambda_{pq}^{(11)st}+\lambda_{pq}^{(14)st}) = r\sum_{\alpha}\lambda_\alpha p_{st}^\alpha n_\alpha v-\sum_p \lambda_{pp}^{(2)st},$$

and consequently

$$\sum_{p\neq q} (\lambda_{pq}^{(9)st}+\lambda_{pq}^{(10)st}+2\lambda_{pq}^{(12)st}+2\lambda_{pq}^{(13)st}+\lambda_{pq}^{(15)st})$$

$$=\sum_{\alpha,\beta\geq 1}\sum_{\gamma=0}^{2}\lambda_\alpha\lambda_\beta p_{\alpha\beta}^{\gamma}p_{st}^{\gamma}n_\gamma v-\delta_{st}\lambda_s n_s v-2\sum_p \lambda_{pp}^{(2)st}-\sum_p \lambda_{pp}^{(3)st}.$$

Thus one obtains

$$\begin{aligned}
\operatorname{Var}\theta = \left(\frac{2(2k-3)}{k(k-1)}\ \frac{2\Gamma_0}{\Delta_0^2}\right)\Bigg[&\frac{v}{b^2k(k-1)}\sum_{s\geq 1}\mu_s^2\lambda_s n_s \\
&-\frac{4}{b^2k(k-1)(k-2)}\sum_{s,t\geq 1}\mu_s\mu_t\lambda_{st}^{(1)} \\
&+\frac{3}{b^2k(k-1)(k-2)(k-3)}\sum_{s,t\geq 1}\mu_s\mu_t\lambda_{st}^{(2)}\Bigg] \\
&-\frac{2v}{b^2k(k-1)^2}\sum_{s\geq 1}(\mu_0-\mu_s)^2\lambda_s n_s \\
&+\frac{2v}{b^2k^2(k-1)^2}\Bigg[\sum_{s\geq 1}\mu_s^2 n_s r^2(k-1)^2 \\
&+\sum_{s,t,\gamma\geq 0}\sum_{\alpha,\beta\geq 1}\mu_s\mu_t\lambda_\alpha\lambda_\beta p_{\alpha\beta}^{\gamma}p_{st}^{\gamma}n_\gamma \\
&-2r(k-1)\sum_{s,t\geq 0}\sum_{\alpha\geq 1}\mu_s\mu_t\lambda_\alpha p_{st}^{\alpha}n_\alpha\Bigg].
\end{aligned} \tag{4.6}$$

This can be rearranged as follows :

$$\text{(4.7)} \qquad \operatorname{Var}\theta = \frac{2(v-1)}{b^2(k-1)^2} - \frac{2(v-1)^2}{b^3(k-1)^3}$$

$$+\frac{2v^2}{b^3k^2(k-1)^3}\sum_{s\neq t}(\mu_0-\mu_s)(\mu_s-\mu_t)\lambda_s\lambda_t n_s n_t$$

$$+\left(\frac{2(2k-3)}{k(k-1)}-\frac{2\Gamma_0}{\Delta_0^2}\right)\left[\frac{v}{b^2k(k-1)}\sum_{s\geqslant 1}\mu_s^2\lambda_s n_s\right.$$

$$-\frac{4}{b^2k(k-1)(k-2)}\sum_{s,t\geqslant 1}\mu_s\mu_t\lambda_{st}^{(1)}$$

$$\left.+\frac{3}{b^2k(k-1)(k-2)(k-3)}\sum_{s,t\geqslant 1}\mu_s\mu_t\lambda_{st}^{(2)}\right].$$

Now for cases where $k = 2, 3$, we have

$$2\Gamma_0 = \Delta_0^2,$$

and therefore the last term in (4.7) vanishes for those cases.

If we consider the limiting process such as $b\to\infty$ fixing v, n_1, n_2, p_{ik}^i as constants, then, since $vr = bk$ and $n_1\lambda_1+n_2\lambda_2 = r(k-1)$, r and at least one of λ_1 and λ_2 are of the same order as b, and

$$\mu_1\mu_2 = O\left(\frac{1}{b^2}\right).$$

It can be seen that

$$\lambda_{st}^{(1)} \leqslant bk(k-1)(k-2),$$

$$\lambda_{st}^{(2)} \leqslant bk(k-1)(k-2)(k-3),$$

and

$$\Gamma_0/\Delta_0^2 \leqslant 1.$$

Thus we get

$$\text{(4.8)} \qquad \operatorname{Var}\theta = \frac{2(v-1)}{b^2(k-1)^2}\left(1+O\left(\frac{1}{b}\right)\right) \quad \text{as } b\to\infty.$$

For a BIBD this is

$$\text{Var}\ \theta = \frac{2(v-1)}{b^2(k-1)^2}\left(1-\frac{v-1}{b(k-1)}\right),$$

which confirms Ogawa's result [5], and for a case when $\lambda_1 = 0$, $\lambda_2 \neq 0$ (4.6) reduces

$$\text{Var}\ \theta = \frac{2(v-1)}{b^2(k-1)^2}\left(1-\frac{v-1}{b(k-1)}\right),$$

which can be seen from (4.7). This confirms N. Giri's result [2].

5. THE APROXIMATE NULL-DISTRIBUTION OF THE F AFTER THE RANDOMIZATION

We take the Beta-distribution

$$\text{(5.1)} \qquad \frac{\Gamma\left(\frac{\nu_1+\nu_2}{2}\right)}{\Gamma\left(\frac{\nu_1}{2}\right)\Gamma\left(\frac{\nu_2}{2}\right)}\ \theta^{\frac{\nu_1}{2}-1}(1-\theta)^{\frac{\nu_2}{2}-1}\,d\theta, \quad 0 \leqslant \theta \leqslant 1,$$

as an approximation to the permutation distribution of θ due to the randomization under the uniformity conditions (4.1). Then

$$\text{(5.2)} \qquad \frac{\nu_1}{\nu_1+\nu_2} = \mathcal{E}(\theta), \ \frac{2\nu_1\nu_2}{(\nu_1+\nu_2)^2(\nu_1+\nu_2+2)} = \text{Var}\ \theta.$$

Hence we get

$$\nu_1 = \frac{2[\mathcal{E}(\theta)-\mathcal{E}^2(\theta)-\text{Var}\ \theta]}{\text{Var}\ \theta}\ \mathcal{E}(\theta),$$

$$\text{(5.3)} \qquad \nu_2 = \frac{2[\mathcal{E}(\theta)-\mathcal{E}^2(\theta)-\text{Var}\ \theta]}{\text{Var}\ \theta}\,(1-\mathcal{E}(\theta)).$$

Notice that

$$\mathcal{E}(\theta) = \mu_0-\mu_2-\frac{\mu_1-\mu_2}{bk(k-1)}\sum_p \lambda_{pp}^{(1)11} = \frac{v-1}{b(k-1)}$$

and hence

$$1-\mathcal{E}(\theta) = \frac{n-b-v+1}{b(k-1)}.$$

If we put

(5.4) $$\nu_1=(v-1)\varphi \text{ and } \nu_2=(n-b-v+1)\varphi,$$

then

(5.5) $$\varphi=\frac{1}{b(k-1)}\,\frac{2[\mathcal{E}(\theta)-\mathcal{E}^2(\theta)-\operatorname{Var}\theta]}{\operatorname{Var}\theta}$$

$$=\frac{2}{b(k-1)}\,\frac{\frac{v-1}{b(k-1)}\left(1-\frac{v-1}{b(k-1)}\right)}{\frac{2(v-1)}{b^2(k-1)^2}\left(1+O\left(\frac{1}{b}\right)\right)}-1$$

$$=1+O\left(\frac{1}{b}\right).$$

Therefore, for sufficiently large value of b, the permutation distribution of θ to the randomization can be approximated by the Beta-distribution

(5.6) $$\frac{\Gamma\left(\frac{n-b}{2}\right)}{\Gamma\left(\frac{v-1}{2}\right)\Gamma\left(\frac{n-b-v+1}{2}\right)}\,\theta^{\frac{v-1}{2}-1}(1-\theta)^{\frac{n-b-v+1}{2}-1}d\theta,\ 0\leqslant\theta\leqslant 1,$$

under the uniformity conditions (4.1). Consequently the null-distribution of the statistic F after the randomization can be approximated by the familiar central F-distribution

(5.7) $$\frac{\Gamma\left(\frac{n-b}{2}\right)}{\Gamma\left(\frac{v-1}{2}\right)\Gamma\left(\frac{n-b-v+1}{2}\right)}\left(\frac{v-1}{n-b-v+1}F\right)^{\frac{v-1}{2}-1}$$

$$\left(1+\frac{v-1}{n-b-v+1}F\right)^{-\frac{n-b}{2}}d\left(\frac{v-1}{n-b-v+1}F\right)$$

which is obtained by averaging (2.14) with respect to θ having the probability element given by (5.6).

References

[1] Bose, R. C. (1949). "Least Square Aspects of Analysis of Variance," Institute of Statistics, Mimeo Series No. 6, University of North Carolina.

[2] Giri, N. (1965). "The F-Test in the Intra-Block Analysis of a Class of Two Associate PBIB Designs," J. *Amer. Statist. Assoc.*, **60**, 1965 285-293.

[3] Ogawa, J. (1961). "The Effect of Randomization on the Analysis of Randomized Block Designs," *Ann. Inst. Statist. Math.*, **13**, 105–117.

[4] Ogawa, J. (1962). "On the Randomization in Latin-Square Designs under the Neyman Model," *Proc. Inst. Statist. Math.*, **10**, 1–16.

[5] Ogawa, J. (1963). "On the Null-Distribution of the F-Statistic in a Randomized Balanced Incomplete Block Design under the Neyman Model," *Ann. Math. Statist.*, **34**, 1558–1568.

[6] Ogawa, J. and Ishii, G. (1964). "On the Analysis of a Partially Balanced Incomplete Block Design in the Regular Case", Institute of Statistics, Mimeo Series No. 412, University of North Carolina.

[7] Ogawa, J. and Ishii, G. (1965). "The Relationship Algebra and the Analysis of Variance of a Partially Balanced Incomplete Block Design," *Ann. Math. Statist.*, **36**, 1815–1828.

[8] Ogawa, J., Ikeda, S. and Ogasawara, M. (1965). "On the Null-Distribution of the F-Statistic for Testing a 'Partial' Null-Hypothesis in a Randomized Partially Balanced Incomplete Block Design with m Associate Classes under the Neyman Model," presented at the 35th Session of the International Statistical Institute, Belgrade, Yugoslavia.

(*Received Jan. 1, 1966.*)

CHAPTER 28

Computation of the Probability Integral of the Non-Central Chi-Square

J. PACHARES, *Hughes Aircraft Company*

1. INTRODUCTION

Need for the c.d.f. of the non-central chi-square arises in a radar scintillation study. Assume that the radar return from a ground patch is the sum of a steady signal and a fluctuating signal. This is expressed by the model

$$\begin{aligned} y(t) &= P \cos(\omega_c t + \psi) + x(t) \\ &= V(t) \cos(\omega_c t + \phi(t)), \end{aligned}$$

where P (a constant) denotes the magnitude of the steady component, $x(t)$ is a stationary narrow-band Gaussian random process with variance σ^2. The random variable ψ is uniformly distributed $(0, 2\pi)$ and is independent of $x(t)$. The amplitude $V(t)$ of the total signal is called the envelope. The p.d.f. of $V(t)$ for a single pulse is given in [2] p. 166. $x(t)$ represents atmospheric disturbances and moving objects, such as leaves on a tree. P depends on the nature of stationary objects; e.g., if one is flying over a desert, then the value of P would be large since there is no movement. ω_c denotes the radar transmitter frequency.

Let $\bar{\lambda} = P^2/2\sigma^2$ denote the ratio of steady power to average random power (ratio of d.c. power to a.c. power). The parameter $\bar{\lambda}$ is a function of "target" characteristics.

Consider a square law device which squares the envelope value sampled, sums and averages the pulses received; i.e., if the ith sampled value of the envelope is denoted by V_i, then the p.d.f. of

$$\overline{V^2} = \sum_{i=1}^{M} V_i^2/M$$

is of interest.

$\overline{V^2}$ is proportional to the "target" corss-section. In particular, one is interested in the p.d.f. of the ratio

$$Z = \overline{V^2}/E(\overline{V^2}),$$

where $E(\overline{V^2})$ is the mean of $\overline{V^2}$.

Each V_i^2/σ^2 has a non-central chi-square distribution with two degrees of freedom and parameter $\lambda = 2\bar{\lambda}$. It follows that $E(\overline{V^2}) = 2\sigma^2+P^2$ and that Z is distributed as the ratio of a non-central chi-square (with $2M$ degrees of freedom and parameter $\lambda = 2M\bar{\lambda}$) to its mean.

It is of interest to determine how

$$P = \Pr(1/c \leqslant Z \leqslant c)$$

varies with M for given constants $\bar{\lambda}$ and c. For example, if $c = 2$ and $\bar{\lambda} = 1{\cdot}0$, how many pulses ($M$) must be sampled in order for $P \geqslant {\cdot}90$, say ? From the table it is seen that the answer is $M = 5$, since then $P = {\cdot}910$. For two pulses $P = {\cdot}711$, etc.

Fix [3] has tabled the values of λ corresponding to given Type I and Type II errors. Unfortunately, the above question cannot be answered by the tables in [3]. I am indebted to one of the editors for pointing out the important reference [5].

The purpose of this note is to suggest ways to compute the c.d.f. of the non-central chi-square. Useful approximations are given in [1] which are recommended for cases which are likely to be difficult to evaluate by conventional methods; e.g., if the ratio of the mean to the standard deviation exceeds six, then the p.d.f. hugs the axis for a long time before rising, resulting in excessive computation. Since these are the cases which are approaching normality, it is expected that the approximations of [1] would be quite accurate.

2. THE PROBLEM

Let X be a random variable having the non-central chi-square distribution with N degrees of freedom and non-centrality parameter λ. Let it be required to evaluate

(1) $$F(x) = \Pr(X \leqslant x) = \int_0^x p(t)dt$$

until F exceeds a given limit. We may write

(2) $$p(x) = f(x) \cdot H(x),$$

where

(3) $$f(x) = (2^{N/2}\Gamma(N/2))^{-1} e^{-x/2} x^{(N-2)/2}$$

(4) $$H(x) = \Gamma(N/2)e^{-\lambda/2} \sum_{i=0}^{\infty} (\lambda x/4)^i/i\,!\,\Gamma(i+N/2).$$

We intend to consider several alternative methods which could be used to evaluate $F(x)$. One such method was used in a preliminary study to obtain the results in the enclosed table; however, other methods suggested themselves during the course of the computations. These methods have not yet been evaluated. Time permitting, it is planned to do so. The methods are essentially based on performing numerical integration of ordinary differential equations which $H(x)$ and $F(x)$ satisfy.

If a versatile integration routine is available, then it is sometimes convenient to restate a given problem as a system of ordinary differential equations, if possible. If this approach is taken, there are several pitfalls one must guard against. Some of these will be discussed later. The differential equation approach was chosen to avoid the necessity for summing the infinite series repeatedly.

All methods suggested take advantage of the fact that if

$$P(y) = \sum_{i=0}^{\infty} y^i/i\,!\,\Gamma(i+k)$$

then $P(y)$ satisfies the differential equation

$$\ddot{P}+(k\dot{P}-P)/y = 0$$

The method used was based on solving the following system:

$$\dot{F} = \exp(\log f(x)+\log H(x)) \tag{5}$$

$$\ddot{H}+(\dot{H}(2N)-\lambda H)/4x = 0. \tag{6}$$

The differential equations were integrated using the subroutine RWDE2F from Share Distribution No. 450, which uses the fourth order Adams-Moulton predictor-corrector with a variable interval and error control. The independent variable was first transformed to u, where $x = u^2$.

The values of P listed in the table give the probability that the ratio of a non-central chi-square to its mean will be between 1/2 and 2, where $N = 2M$, $\lambda = N\bar{\lambda}$. Checks indicate that the maximum error is one unit in the third figure. These values were found by linear interpolation in the computed c.d.f. The values for the case $\bar{\lambda} = 0$, were found from [4], Table 7.

M \ $\bar{\lambda}$	0	1	4	9	16	25	36
1	.471	.531	.721	.865	.943	.979	.993
2	.644	.712	.876	.962	.991	.998	.999
3	.747	.810	.938	.988	.998	.999	
4	.815	.870	.968	.996	.999		
5	.862	.910	.982	.998			
6	.896	.936					

3. ALTERNATIVE PROCEDURES

In this section we propose two methods to be evaluated.

Method I.

If (6) is to be integrated, we see from (4) that the initial condition $H(0) = \exp(-\lambda/2)$, may result in underflow when λ is large. If $\lambda/2 > 88$, say transform to

$$Z = \log H,$$

getting the equation

$$\ddot{Z} + \dot{Z}^2 + (2N\dot{Z} - \lambda)/4x = 0. \tag{7}$$

On the other hand, if $\lambda/2 < 88$, and (6) is being integrated, there may be a loss of significance detected in computing $\ddot{H}$. If so, we should remove the first derivative by putting

$$J = X^{N/4}H,$$

getting the equation

$$\ddot{J} = J(N(3N-4)/4x + \lambda)/4x. \tag{8}$$

Since J may grow large, if it exceeds a specified value, put

$$W = \log J,$$

getting the equation

$$\ddot{W} + \dot{W}^2 = (N(3N-4)/4x + \lambda)/4x. \tag{9}$$

Clearly, when any transformation is made on H or $\log H$, the compensating change must be made on $f(x)$ in (5).

Method II.

Let $G(x) = 1 - F(x)$, and assume we may write

$$G(x) = \int_x^\infty f(t)H(t)dt = f(x)H(x)K(x). \tag{10}$$

It follows that $K(x)$ satisfies the equation

$$\dot{K} = -1 - K(\dot{H}/H + ((N-2)/x - 1)2), \tag{11}$$

and that $K(\infty) = 1$. However, $H(x)$ may grow large, and if it exceeds a specified value, put

$$Z = \log H$$

in (6) getting the Equation (7) and replacing $\dot{H}/H$ with $\dot{Z}$ in (11).

The switch from the pair (5), (6) to the pair (7), (11), should be made at a given signal, such as when $F(x) \geqslant \cdot 5$.

Subsequent experimentation indicates that it is better to replace (5), (6) with (5′), (6′), where

$$\dot{F} = \exp\left((N-1)\log u + Z\right) \tag{5′}$$

$$\ddot{Z} = \lambda - \dot{Z}(\dot{Z} + 2u + (N-1)/u) - (N + u^2) \tag{6′}$$

4. SUMMARY OF SOME RULES TO BE OBSERVED

It is useful to have a set of guidelines handy which are the results of numerical experience. A few are summarized below.

1. If a second order differential equation is to be solved numerically by a method such as Runge-Kutta, say, then it must have the ability to get started. For example, Bessel's equation for $I_N(x)$ if $N \geqslant 3$, will not go, but a transformation to $I_N(x)/X^N$ will get started.

2. Transform a set of differential equations such that all equations will require nearly the same increment, Δx. It is quite frustrating to use $\Delta x = 10^{-7}$ to integrate one equation while the second requires only $\Delta x = 10^{-2}$, say.

3. Do not allow the solution to tend to zero because the error may eventually mask it. Transform to a solution which grows gradually.

4. Guard against overflow or underflow by transforming to log at a given signal.

5. Group terms in order to preserve accuracy if possible. For example, let B_1 and B_2 be nearly equal and large and let S

be small, then the grouping $(B_1+S)-B_2$ is bad, while the grouping $(B_1-B_2)+S$ is better. As the ratio B_1/B_2 approaches unity, the answer should approach S, but if the first grouping is used, the answer will approach zero.

6. Loss of significance in the subtraction $A-B$ can be detected by the following approach :

If $\cdot 9 < A/B < 1\cdot 1$, compute

$$L = -\log_{10} | 1-(A/B) | ,$$

which gives the number of figures lost in the subtraction. If L exceeds a critical value when computing the second derivative, then we should transform to eliminate the first derivative.

7. Transform to a differential equation whose second derivative does not oscillate, if possible.

8. Avoid large values for the derivatives, since this leads to small Δx.

9. Since a single differential equation may not behave in the desired manner over the entire region, one should not hesitate to switch from one equation to another through appropriate transformations at given signals.

10. Since $F(x)$ is a monotonic function of x, one may find the percentage points in a convenient manner by making F the independent variable in the integration.

Acknowledgment

I wish to thank R. M. Hill for discussions concerning the radar problem mentioned in the note and E. S. Levitan for his unselfish programming support which led to the tabled results.

References

[1] Abdel-Aty, S. H. (1954). "Approximate Formulae for the Percentage Points and the Probability Integral of the Non-Central χ^2 Distribution," *Biometrika,* **41,** 538–540.

[2] Davenport, W. B., Jr., and Root, W. L. (1958). *Random Signals and Noise*, McGraw-Hill, New York.

[3] Fix, E. (1949). "Tables of Noncentral χ^2," *Univ. California Publ. Statist.*, 1, 15–19.

[4] Pearson, E. S., and Hartley, H. O., 1958. *Biometrika Tables for Statisticians*, 1, Cambridge University Press.

[5] Haynam, G. E., Govindarajulu, Z. and Leone, F. C., 1962. "Tables of the Cumulative Non-Central Chi-Square Distribution," Case Statistical Laboratory Publication No. 104.

(*Received Sept. 29, 1965.*)

CHAPTER 29

On the Non-Central Distributions of the Largest Roots of Two Matrices in Multivariate Analysis

K. C. SREEDHARAN PILLAI[1], *Purdue University*

1. INTRODUCTION AND SUMMARY

Tests of three multivariate hypotheses based on Roy's largest characteristic root criterion [18] have been shown to have monotonicity of power with respect to each population characteristic root ; namely, (i) that of equality of covariance matrices of two p-variate normal polulations, (ii) that of equality of p-dimensional mean vectors of l p-variate normal populations having a common covariance matrix, and (iii) that of independence between a p-set and a q-set of variates in a $(p+q)$-variate normal population [1], [2], [3], [19]. To facilitate these tests, the c.d.f. of the largest characteristic root criterion under the null hypothesis (i)-(iii) has been studied by Pillai [6], [8] [10] [11], [12], [15] with a view to obtaining an approximation to the c.d.f. useful

1 This research was supported by the National Science Foundation, Grant No. GP-4600.

for computing the upper percentage points. Further, for test (iii), the c.d.f. of the largest of two characteristic roots under a single non-zero population characteristic root has been obtained by Pillai [14] and the power function studied. In the present paper, this study is extended to the case of the c.d.f. of the largest of two roots when both population roots are non-zero and that of the largest of three roots under a single non-zero population root. In addition, similar non-central c.d.f.'s are obtained for test (ii) for the case of two roots with two non-null population roots, and that of three roots with single non-zero population root. Tabulations of the power functions have been made for both tests for various values of the parameters.

Further, Pillai's general expression [15] approximating at the upper end, the c.d.f. of the largest root under the null hypotheses, has been used to compute the upper 5 and 1 percent points of the largest of eleven and twelve roots.

2. NON-CENTRAL C.D.F. OF THE LARGEST ROOT FOR TEST (iii)

Let the columns of $\begin{bmatrix} \boldsymbol{X} \\ \boldsymbol{Y} \end{bmatrix}$ be ν independent normal $(p+q)$-variates ($p \leqslant q$, $p+q \leqslant \nu$, $\nu+1 = n'$, the sample size) with zero means and covariance matrix

$$\boldsymbol{\Sigma} = \begin{bmatrix} \boldsymbol{\Sigma}_{11} & \boldsymbol{\Sigma}_{12} \\ \boldsymbol{\Sigma}'_{12} & \boldsymbol{\Sigma}_{22} \end{bmatrix} \begin{matrix} p \\ q \end{matrix}$$
$$\qquad\qquad p \qquad q$$

Let $\boldsymbol{R} = \text{diag}(r_i)$, where $r_1^2, \ldots, r_p^2$ are the characteristic roots of the equation

$$|\boldsymbol{X}\boldsymbol{Y}'(\boldsymbol{Y}\boldsymbol{Y}')^{-1}\boldsymbol{Y}\boldsymbol{X}' - r^2\boldsymbol{X}\boldsymbol{X}'| = 0 \tag{2.1}$$

and $\boldsymbol{P} = \text{diag}(\rho_i)$, where $\rho_1^2, \ldots, \rho_p^2$ are the characteristic roots of the equation

$$|\boldsymbol{\Sigma}_{12}\boldsymbol{\Sigma}_{22}^{-1}\boldsymbol{\Sigma}'_{12} - \rho^2\boldsymbol{\Sigma}_{11}| = 0. \tag{2.2}$$

Then, the distribution of $r_1^2, \ldots, r_p^2$ is given by Constantine [4], in the following form [5] :

$$(2.3)\qquad |1-\boldsymbol{P}^2|^{\frac{\nu}{2}}\ {}_2F_1(\tfrac{1}{2}\nu, \tfrac{1}{2}\nu; \tfrac{1}{2}f_2; \boldsymbol{P}^2, \boldsymbol{R}^2)$$

$$\times C(p, m, n)\,|\boldsymbol{R}^2|^m\,|\boldsymbol{I}-\boldsymbol{R}^2|^n \prod_{i>j}(r_i^2-r_j^2)\prod_{i=1}^{p} dr_i^2,$$

$$0 < r_1^2 \leqslant \ldots \leqslant r_p^2 < 1,$$

where $f_2 = q$, $m = \frac{1}{2}(q-p-1)$, $n = \frac{1}{2}(\nu-q-p-1)$, (and if n is defined as $\frac{1}{2}(f_1-p-1)$, $\nu = f_1+f_2$),

$$C(p, m, n) = \pi^{\frac{1}{2}p} \prod_{i=1}^{p} \Gamma[\tfrac{1}{2}(2m+2n+p+i+2)]/$$

$$\{\Gamma[\tfrac{1}{2}(2m+i+1)]\ \Gamma[\tfrac{1}{2}(2n+i+1)]\ \Gamma(\tfrac{1}{2}i)\};$$

and the hypergeometric function of matrix argument is defined, as in [5], by

$$(2.4)\quad {}_sF_t(a_1, \ldots, a_s;\ b_1, \ldots, b_t\ ;\ \mathbf{S}, \boldsymbol{T})$$

$$= \sum_{k=0}^{\infty} \sum_{\mathcal{K}} \frac{(a_1)_{\mathcal{K}} \cdots (a_s)_{\mathcal{K}} C_{\mathcal{K}}(\mathbf{S}) C_{\mathcal{K}}(\boldsymbol{T})}{(b_1)_{\mathcal{K}} \cdots (b_t)_{\mathcal{K}} C_{\mathcal{K}}(\boldsymbol{I}_p) k!}$$

where $a_1, \ldots, a_s, b_1, \ldots, b_t$ are real or complex constants and the multivariate coefficient $(a)_{\mathcal{K}}$ is given by

$$(\mathrm{a})_{\mathcal{K}} = \prod_{i=1}^{p} (a-\tfrac{1}{2}(i-1))k_i,$$

where

$$(a)_k = a(a+1) \ldots (a+k-1),$$

and the partition $\mathcal{K}$ of k is such that

$$\mathcal{K} = (k_1, k_2, \ldots k_p), \quad k_1 \geqslant k_2 \geqslant \ldots \geqslant k_p \geqslant 0,$$

$k_1+\ldots+k_p=k$; and the zonal polynomials, $C_{\mathcal{K}}(\mathbf{S})$, are expressible in terms of elementary symmetric functions (e.s.f.) of the characteristic roots of $\mathbf{S}$, [5].

Now define by $V(x;\, q_p, \ldots, q_1;\, n)$, the determinant

$$(2.5)\qquad \begin{vmatrix} \int_0^x x_p^{q_p}(1-x_p)^n dx_p \ldots \int_0^x x_p^{q_1}(1-x_p)^n dx_p \\ \\ \int_0^{x_2} x_1^{q_p}(1-x_1)^n dx_1 \ldots \int_0^{x_2} x_1^{q_1}(1-x_1)^n dx_1 \end{vmatrix} .$$

It may be observed that the c.d.f. of the largest root from (2.3) under the null hypothesis (iii) can be thrown into the form (2.5) multiplied by $C(p, m, n)$,[6], [7], [9]. Further, in view of the fact that the zonal polynomials $C_{\mathcal{K}}(\mathbf{S})$ in (2.4) can be expressed in terms of e.s.f.'s of characteristic roots of $\mathbf{S}$, by the use of Pillai's lemma on the multiplication of the basic Vandermonde type determinant by powers of e.s.f.'s, [13], it is easy to see that the non-central distribution of the c.d.f. of r_p^2 in (2.3) can be expressed as a series whose terms are linear compounds of determinants of the type (2.5). Further, it has been shown [6], [7] that

$$(2.6)\quad V(x;\, q_s, q_{s-1}, \ldots, q_1;\, n) = (q_s+n+1)^{-1}(A^{(s)}+B^{(s)}+q\, C^{(s)}),$$

where

$$A^{(s)} = -I_0(x;\, q_s, n+1)\, V(x;\, q_{s-1}, \ldots, q_1;\, n)$$

$$B^{(s)} = 2 \sum_{j=s-1}^{1} (-1)^{s-j-1} I(x;\, q_s+q_j,\ 2n+1)\, V(x;\, q_{s-1}, \ldots, q_{j+1}, q_{j-1}, \ldots, q_1;\, n),$$

$$C^{(s)} = V(x;\, q_s-1, q_{s-1}, \ldots, q_1;\, n),$$

$$I_0(x;\, q_s, n+1) = x^{q_s}(1-y)^{n+1}\Big|_0^x,$$

and

$$I(x;\, q, r) = \int_0^x x_1^q(1-x_1)^r\, dx .$$

It may be noted that $C^{(s)}$ vanishes if $q_s = q_{s-1}+1$. Using (2.6) in each of the determinants of the linear compounds involved in the series obtainable from (2.3), after the necessary number of reductions, the c.d.f. of the largest canonical correlation coefficient, r_p^2, can be ultimately reduced in terms of simple incomplete beta functions.

3. NON-CENTRAL C.D.F. OF r_2^2.

Now putting $p = 2$ in (2.3) and using the method outlined in the preceding section, the c.d.f. of the largest of two canonical correlation coefficients is obtained in the following form :

$$(3.1)\quad \Pr\{r_2^2 \leqslant x\} = K\Big\{-I_0(x;\, m+1,\, n+1)\Big[\Big(\sum_{i=0}^{6} B_i x^i\Big) I(x;\, m,\, n)$$

$$+\Big(\sum_{i=2}^{6} C_i x^{i-1}\Big) I(x;\, m+1,\, n)$$

$$+\Big(\sum_{i=4}^{6} D_i x^{i-2}\Big) I(x;\, m+2,\, n)$$

$$+E_6 x^3 I(x;\, m+3,\, n)\Big] + 2[(B_6+C_6+D_6+E_6) I(x;\, 2m+7,\, 2n+1)$$
$$+(B_5+C_5+D_5) I(x;\, 2m+6,\, 2n+1)$$
$$+(B_4+C_4+D_4) I(x;\, 2m+5,\, 2n+1)+(B_3+C_3) I(2m+4,\, 2n+1)$$
$$+(B_2+C_2) I(x;\, 2m+3,\, 2n+1)+B_1 I(x;\, 2m+2,\, 2n+1)$$
$$+B_0 I(x;\, 2m+1,\, 2n+1)\Big\}],$$

where

$$K = [(1-\rho_1^2)(1-\rho_2^2)]^{\frac{1}{2}\nu}\ C(2,\, m,\, n);$$

$$B_6 = \frac{a_{61}}{m+n+8}, \qquad B_5 = \frac{a_{51}+(m+7)B_6}{m+n+7}, \quad B_4 = \frac{a_{41}+(m+6)B_5}{m+n+6}$$

$$B_3 = \frac{a_{31}+(m+5)B_4}{m+n+5}, \quad B_2 = \frac{a_{21}+(m+4)B_3}{m+n+4}, \quad B_1 = \frac{a_{11}+(m+3)B_2}{m+n+3}$$

$$B_0 = \frac{1+(m+2)B_1}{m+n+2}, \quad C_6 = \frac{a_{62}}{m+n+7}, \qquad C_5 = \frac{a_{52}+(m+6)C_6}{m+n+6}$$

$$C_4 = \frac{a_{42}+(m+5)C_5}{m+n+5}, \quad C_3 = \frac{a_{32}+(m+4)C_4}{m+n+4}, \quad C_2 = \frac{a_{22}+(m+3)C_3}{m+n+3}$$

$$D_6 = \frac{a_{63}}{m+n+6}, \qquad D_5 = \frac{a_{53}+(m+5)D_6}{m+n+5}, \; D_4 = \frac{a_{43}+(m+4)D_5}{m+n+4}$$

$$E_6 = \frac{a_{64}}{m+n+5}, \qquad a_{61} = 231A_{61}, \qquad a_{62} = 35(A_{62}-3A_{61})$$

$$a_{63} = 3(A_{63}-5A_{62}-7A_{61}), \quad a_{64} = A_{64}-A_{63}-2A_{62}-5A_{61},$$

$$a_{51} = 63A_{51}, \quad a_{52} = 5A_{52}-28A_{51}, \quad a_{53} = A_{53}-2A_{52}-5A_{51}$$

$$a_{41} = 35A_{41}, \quad a_{42} = 3(A_{42}-5A_{41}), \quad a_{43} = A_{43}-A_{42}-2A_{41}$$

$$a_{31} = 5A_{31}, \quad a_{32} = A_{32}-2A_{31}, \quad a_{21} = 3A_{21}, \quad a_{22} = A_{22}-A_{21},$$

$$a_{11} = A_{11} = \frac{\nu^2}{4f_2}\, b_1;$$

$$A_{61} = \frac{[\nu(\nu+2)\ldots(\nu+10)]^2[231b_1^6-1260b_1^4b_2+1680b_1^2b_2^2-320b_2^3]}{2^6f_2(f_2+2)\ldots(f_2+10) \quad 8!4!176}$$

$$A_{62} = \frac{[\nu(\nu+2)\ldots(\nu+8)(\nu-1)]^2}{2^6f_2(f_2+2)\ldots(f_2+8)(f_2-1)} \; \frac{3}{8!\,44}\,[35b_1^4b_2-120b_1^2b_2^2+48b_2^3],$$

$$A_{63} = \frac{[\nu\ldots(\nu+6)(\nu-1)(\nu+1)]^2}{2^6f_2\ldots(f_2+6)(f_2-1)(f_2+1)} \; \frac{10}{7!3}\,[3b_1^2b_2^2-4b_2^3],$$

$$A_{64} = \frac{[\nu(\nu+2)(\nu+4)(\nu-1)(\nu+1)(\nu+3)]^2}{2^6f_2(f_2+2)(f_2+4)(f_2-1)(f_2+1)(f_2+3)} \; \frac{4}{5.7.9}\, b_2^3,$$

$$A_{51} = \frac{[\nu(\nu+2)\ldots(\nu+8)]^2}{2^5f_2(f_2+2)\ldots(f_2+8)} \, \frac{1}{8!48}\,[63b_1^5-280b_1^3b_2+240b_1b_2^2],$$

$$A_{52} = \frac{[\nu(\nu+2)(\nu+4)(\nu+6)(\nu-1)]^2}{2^5f_2\ldots(f_2+6)(f_2-1)} \; \frac{1}{6!3}\,[5b_1^3b_2-12b_1b_2^2],$$

$$A_{53} = \frac{[\nu(\nu+2)(\nu+4)(\nu-1)(\nu+1)]^2}{2^5f_2(f_2+2)(f_2+4)(f_2-1)(f_2+1)} \; \frac{1}{7.5}\, b_1b_2^2,$$

$$A_{41} = \frac{[\nu(\nu+2)\ldots(\nu+6)]^2}{2^4f_2(f_2+2)\ldots(f_2+6)} \; \frac{3}{8!8}\,[35b_1^4-120b_1^2b_2+48b_2^2],$$

$$A_{42} = \frac{[\nu(\nu+2)(\nu+4)(\nu-1)]^2}{2^4f_2(f_2+2)(f_2+4)(f_2-1)} \; \frac{1}{4!7}\,[3b_1^2b_2-4b_2^2],$$

$$A_{43} = \frac{[\nu(\nu+2)(\nu-1)(\nu+1)]^2}{2^4 f_2(f_2+2)(f_2-1)(f_2+1)} \frac{2}{3.5} b_2^2,$$

$$A_{31} = \frac{[\nu(\nu+2)(\nu+4)]^2}{2^3 f_2(f_2+2)(f_2+4)} \frac{1}{5!4} [5b_1^3 - 12b_1b_2],$$

$$A_{32} = \frac{[\nu(\nu+2)(\nu-1)]^2}{2^3 f_2(f_2+2)(f_2-1)} \frac{1}{5} b_1b_2,$$

$$A_{21} = \frac{[\nu(\nu+2)]^2}{2^2 f_2(f_2+2)\ 4!2} [3b_1^2 - 4b_2]$$

$$A_{22} = \frac{[\nu(\nu-1)]^2}{2^2 f_2(f_2-1)} \frac{4b_2}{3!}, \quad b_1 = \rho_1^2+\rho_2^2, \text{ and } b_2 = \rho_1^2\rho_2^2.$$

In obtaining the c.d.f. of r_2^2 in (3.1), zonal polynomials of degrees 1 to 6 were included. The expression for the c.d.f. of r_2^2 in (3.1) has been used to compute the power of test (iii) for various pairs of values of (ρ_1^2, ρ_2^2) and the results are presented in Table 1.

4. NON-CENTRAL C.D.F. OF r_3^2 UNDER A SINGLE NON-ZERO ρ

When there is only a single non zero population root, ρ^2, the joint density function (2.3) can be expressed in the following simple form :

$$(4.1) \qquad (1-\rho^2)^{\frac{1}{2}\nu} C(p, m, n)$$

$$\sum_{i=0}^{\infty} \frac{[\nu(\nu+2)\ldots(\nu+2(i-1))\ \rho^{2i} Z_i(R^2)}{2^i f_2(f_2+2)\ldots(f_2+2(i-1))\, p(p+2)\ldots(p+2(i-1))i!}$$

where $\boldsymbol{Z}_i(\boldsymbol{R}^2)$ are ith degree zonal polynomials which are given a different normalizing constant than that of $C_i(\boldsymbol{R}^2)$, [5]. For $p = 3$, using methods as above, the c.d.f. of the largest root is obtained from (4.1) and given below.

$$(4.2) \quad \Pr\{r_3^2 \leqslant x\} = K_1\Big\{2I(x; m, n)\Big[\Big(B_6^{(3)} - C_6^{(3)} - E_6^{(3)} - \frac{225A_6}{m+n+6}\Big)$$

$$I(x;\,2m+9,\,2n+1)+(B_5^{(3)}-C_5^{(3)}-E_5^{(3)})I(x;\,2m+8,\,2n+1)$$
$$+(B_4^{(3)}-C_4^{(3)}-E_4^{(3)})\,I(x;\,2m+7,\,2n+1)$$
$$+(B_3^{(3)}-C_3^{(3)})I(x;\,2m+6,\,2n+1)$$
$$+(B_2^{(3)}-C_2^{(3)})\,I(x;\,2m+5,\,2n+1)+B_1^{(3)}\,I(x;\,2m+4,\,2n+1)$$
$$+B_0^{(3)}\,I(x;\,2m+3,\,2n+1)]$$
$$-2I(x;\,m+1,\,n)[(B_6^{(3)}-D_6^{(3)}-F_6^{(3)})\,I(x;\,2m+8,\,2n+1)$$
$$+(B_5^{(3)}-D_5^{(3)}-F_5^{(3)})\,I(x;\,2m+7,\,2n+1)$$
$$+(B_4^{(3)}-D_4^{(3)})\,I(x;\,2m+6,\,2n+1)$$
$$+(B_3^{(3)}-D_3^{(3)})\,I(x;\,2m+5,\,2n+1)$$
$$+B_2^{(3)}I(x;\,2m+4,\,2n+1)+B_1^{(3)}I(x;\,2m+3,\,2n+1)$$
$$+B_0^{(3)}I(x;\,2m+2,\,2n+1)]$$
$$+2I(x;\,m+2,\,n)\Big[\Big(C_6^{(3)}-D_6^{(3)}-\frac{45A_6}{m+n+5}\Big)\,I(x;\,2m+7,\,2n+1)$$
$$+(C_5^{(3)}-D_5^{(3)})I(x;\,2m+6,\,2n+1)$$
$$+(C_4^{(3)}-D_4^{(3)})I(x;\,2\text{m}+5,\,2n+1)+(C_3^{(3)}-D_3^{(3)})I(x;\,2m+4,$$
$$2n+1)+C_2^{(3)}I(x;\,2m+3,\,2n+1)\Big]$$
$$+2I(x;\,m+3,\,n)\Big[\Big(E_6^{(3)}-F_6^{(3)}+\frac{45A_6}{m+n+5}\Big)I(x;\,2m+6,\,2n+1)$$
$$+(E_5^{(3)}-F_5^{(3)})\,I(x;\,2m+5,\,2n+1)+E_4^{(3)}I(x;\,2m+4,\,2n+1)\Big]$$
$$+\frac{450A_6}{m+n+6}\,I(x;\,2m+5,\,2n+1)\,I(x;\,m+4,\,n)$$
$$-V(x;\,1,\,0;\,n)\,I_0(x;\,m+2,\,n+1)[B_6^{(3)}x^6+B_5^{(3)}x^5+\ldots+B_0^{(3)}]$$
$$+V(x;\,2,\,0;;\,n)\,I_0(x;\,m+3,\,n+1)[C_6^{(3)}x^4+C_5^{(3)}x^3+C_4^{(3)}x^2$$
$$+C_3^{(3)}x+C_2^{(3)}]$$
$$-V(x;\,2,\,1;\,n)\,I_0(x;\,m+3,\,n+1)[D_6^{(3)}x^3+D_5^{(3)}x^2+D_3^{(3)}x+D_3^{(3)}]$$
$$+V(x;\,3,\,0;\,n)I_0(x;\,m+4,\,n+1[E_6^{(3)}x^2+E_5^{(3)}x+E_4^{(3)}]$$
$$-V(x;\,3,\,1;\,n)\,I_0(x;\,m+4,\,n+1)[F_6^{(3)}x+F_5^{(3)}]$$
$$+\frac{225A_6}{m+n+6}\,I_0(x;\,m+5,\,n+1)+\frac{45A_6}{m+n+5}I_0(x;\,m+4,\,n+1)\Big\},$$

where

$$B_6^{(3)} = \frac{10395A_6}{m+n+9}, \qquad B_5^{(3)} = \frac{(m+8)\,B_6^{(3)}+945A_5}{m+n+8}$$

$$B_4^{(3)} = \frac{(m+7)B_5^{(3)}+105A_4}{m+n+7}, \quad B_3^{(3)} = \frac{(m+6)B_4^{(3)}+15A_3}{m+n+6},$$

$$B_2^{(3)} = \frac{(m+5)B_3^{(3)}+3A_2}{m+n+5}, \qquad B_1^{(3)} = \frac{(m+4)B_2^{(3)}+A_1}{m+n+4},$$

$$B_0^{(3)} = \frac{(m+3)B_1^{(3)}+A_0}{m+n+3};$$

$$C_6^{(3)} = \frac{4725A_6}{m+n+8}, \qquad C_5^{(3)} = \frac{(m+7)C_6^{(3)}+420A_5}{m+n+7},$$

$$C_4^{(3)} = \frac{(m+6)C_5^{(3)}+45A_4}{(m+n+6)}, \qquad C_3^{(3)} = \frac{(m+5)C_4^{(3)}+6A_3}{m+n+5},$$

$$C_2^{(3)} = \frac{(m+4)C_3^{(3)}+A_2}{m+n+4}$$

$$D_6^{(3)} = \frac{3150A_6}{m+n+6}, \qquad D_5^{(3)} = \frac{(m+6)D_6^{(3)}+270A_5}{m+n+5},$$

$$D_4^{(3)} = \frac{(m+5)D_5^{(3)}+27A_4}{m+n+4}, \quad D_3^{(3)} = \frac{(m+4)D_4^{(3)}+3A_3}{m+n+3},$$

$$E_6^{(3)} = \frac{945A_6}{m+n+7}, \qquad E_5^{(3)} = \frac{(m+6)E_6^{(3)}+75A_5}{m+n+6}$$

$$E_4^{(3)} = \frac{(m+5)E_5^{(3)}+6A_4}{m+n+5},$$

$$F_6^{(3)} = \frac{720A_6}{m+n+6}, \qquad F_5^{(3)} = \frac{(m+5)F_6^{(3)}+45A_5}{m+n+5};$$

$$K_1 = C(3, m, n)(1-\rho^2)^{\nu/2}, \qquad A_0 = 1,$$

$$A_i = \frac{[\nu\ldots(\nu+2(i-1))]^2\rho^{2i}}{2^i f_2\ldots(f_2+2(i-1))\ 3.5\ldots(3+2(i-1))i!},$$

$$i = 1, 2, \ldots .$$

The determinants of the $V(x; q_2, q_1; n)$ type occurring in (4.2) can further be reduced in terms of simple incomplete beta functions using (2.6). The c.d.f. (4.2) also has been obtained as before using the zonal polynomials of degrees one to six. Power function tabulations from (4.2) are presented in Table 2.

5. NON-CENTRAL C.D.F. OF THE LARGEST ROOT FOR TEST (ii).

Let $\boldsymbol{X}$ be a $p \times f_2$ matrix variate ($p \leqslant f_2$), $\boldsymbol{Y}$ be a $p \times f_1$ matrix variate ($p \leqslant f_1$), and the columns be all independently normally distributed with covariance matrix $\boldsymbol{\Sigma}$; $E(\boldsymbol{X}) = \boldsymbol{M}$ and $E(\boldsymbol{Y}) = \boldsymbol{0}$. Let $l_1, \ldots, l_p$ be the characteristic roots of

$$(5.1) \qquad |\boldsymbol{XX}' - l(\boldsymbol{YY}' + \boldsymbol{XX}')| = 0,$$

and $\omega_1, \ldots, \omega_p$, those of

$$(5.2) \qquad |\boldsymbol{MM}' - \omega\boldsymbol{\Sigma}| = 0.$$

Then the joint density function of $l_1, \ldots, l_p$ is given by Constantine [4], [5] in the form

$$(5.3) \quad e^{-\frac{1}{2}tr\boldsymbol{\Omega}}\,{}_1F_1(\tfrac{1}{2}\nu;\ \tfrac{1}{2}f_2;\ \tfrac{1}{2}\boldsymbol{\Omega}, \boldsymbol{L})C(p, m, n)\,|\boldsymbol{L}|^m\,|\boldsymbol{I} - \boldsymbol{L}|^n \prod_{i>j} (l_i - l_j)$$

where $\boldsymbol{L} = \boldsymbol{X}'(\boldsymbol{YY}' + \boldsymbol{XX}')^{-1}\boldsymbol{X}$, $\boldsymbol{\Omega} = \boldsymbol{M}'\boldsymbol{\Sigma}^{-1}\boldsymbol{M}$, (the determinants in (5.3) expressed as the products of the roots), $m = \frac{1}{2}(f_2 - p - 1)$, $n = \frac{1}{2}(f_1 - p - 1)$ and $\nu = f_1 + f_2$. In the context of test (ii) $f_2 = l - 1$ and $f_1 = N - l$, N being the pooled sample size of the samples from the l populations. It should be pointed out that the same symbols m and n are used in connection with both tests because of the reason that these definitions of m and n will leave the joint distribution of $l_1, \ldots l_p$ the same as that of $r_1^2, \ldots, r_p^2$ under the null hypotheses (ii) and (iii) respectively as may be seen from (5.3) and (2.3). Now the c.d.f. of the largest root, l_p, can be derived from (5.3) by the method outlined in Section 2.

6. NON-CENTRAL CDF OF l_2.

Now putting $p = 2$ in (5.3) and following the method described in Section 2, the c.d.f. of the largest of two roots can be obtained in the same form as (3.1) with the following changes:

$$r_2^2 \rightarrow l_2,$$

$$K \rightarrow K' = e^{-\frac{1}{2}(\omega_1+\omega_2)}C(2, m, n),$$

$A_{ij} \rightarrow A'_{ij}$ where A'_{ij} is obtained from A_{ij} by multiplying A_{ij} by 2^i and each linear factor involving ν in the numerator of A_{ij} is raised only to a single power instead of power two in A_{ij}, and defining b_1 as $\frac{1}{2}(\omega_1+\omega_2)$ and b_2 as $\omega_1\omega_2/4$. For example,

$$A'_{21} = \frac{2^2 A_{21}[3(\frac{1}{2}(\omega_1+\omega_2))^2-4(\omega_1\omega_2/4)]}{\nu(\nu+2)(3b_1^2-4b_2)}$$

Further, $a_{ij} \rightarrow a'_{ij}$ where a'_{ij} is the same function of A'_{ij}'s as a_{ij} is of A_{ij}'s. Similarly

$$(B_i, C_i, D_i, E_i) \rightarrow (B'_i, C'_i, D'_i, E'_i),$$

where the meaning of (′) is obvious. In other words, the basic changes are in the A_{ij} coefficients, the K coefficient and the different definitions of the b's. Now, tabulation of the power of test (ii) based on l_2 is presented in Table 3.

7. NON-CENTRAL C.D.F OF l_3 UNDER A SINGLE NON-ZERO ω.

When there is only one non-zero population root, ω, the joint density function (5.3) takes the simple form

$$(7.1) \quad e^{-\frac{1}{2}\omega}C(p, m, n) \sum_{i=0}^{\infty} \frac{[\nu(\nu+2)\ldots(\nu+2(i-1))]}{f_2(f_2+2)\ldots(f_2+2(i-1))}$$

$$\frac{(\frac{1}{2}\omega)^i Z_i(\boldsymbol{L})}{p(p+2)\ldots(p+2(i-1))i!}\,.$$

As before, the c.d.f. of the largest root l_3 can be obtained in the same form as (4.2) with the following changes :

$$r_3^2 \rightarrow l_3,$$

$$K_1 \rightarrow K_1' = e^{-\frac{1}{2}\omega} C(3, m, n).$$

$$A_i \rightarrow A_i' \text{ where } A_i' = 2^i(\omega/2\rho^2)^i A_i/[\nu(\nu+2) \ldots (\nu+2(i-1))],$$

$$i = 1, 2, \ldots,$$

$$(B_i^{(3)}, C_i^{(3)}, D_i^{(3)}, E_i^{(3)}, F_i^{(3)}) \rightarrow (B_i^{(3)\prime}, C_i^{(3)\prime}, D_i^{(3)\prime}, E_i^{(3)\prime}, F_i^{(3)\prime})$$

where (′) means that A_is are replaced by A_i's in in the corresponding original coefficients to get the primed ones. Now for the c.d.f. of l_3 as well as l_2, the degrees of the zonal polynomials used were as before. Numerical values of the power function from the c.d.f. of l_3 are presented in Table 4.

8. THE C.D.F. OF THE LARGEST ROOT UNDER THE THREE NULL HYPOTHESES

It has been shown [6], [8], that when each of the three null hypotheses is true, the c.d.f. of the largest of s non-null characteristic roots of a matrix is given by

$$C(s, m, n)\, V(x; m+s-1, m+s-2, \ldots, m; n). \tag{8.1}$$

A general expression approximating (8.1) for computing the upper percentage points has been obtained by Pillai [15] which is given below :

For *odd values of s*, (8.1) is approximated by

$$\frac{I(x; m, n)}{\beta(m+1, n+1)} + \sum_{i=1}^{s-1} (-1)^i k_{m+s-i} I_0(x; m+s-i, n+1) \tag{8.2}$$

where

$$k_{m+s-i} = (m+n+s-i+1)^{-1}[C(s, m, n)\, V(1; m+s-1, \ldots, m+s-i+1, m+s-i-1\ \ldots, m; n)$$
$$-(m+s-i+1)k_{m+s-i+1}] \tag{8.3}$$

where $k_{m+s} = 0$.

For *even values of s*, (8.1) is approximated by

$$(8.4)\qquad 1+\sum_{i=1}^{s-1}(-1)^i k_{m+s-i}\, I_0(x;\, m+s-i,\, n+1).$$

Further it has been shown [17] that

$$(8.5)\quad V(1;\, m+s-1, m+s-2, \ldots, m+s-i+1, m+s-i-1, \ldots m; n)$$

$$= \left[\binom{s-1}{i-1} \prod_{j=1}^{i-1} \frac{(2m+s-j+1)}{(2m+2n+2s-j+1)} \right] \Big/ C(s-1,\, m,\, n).$$

Using (8.2), upper 5 and 1 per cent points of the largest root for $s = 11$ have been computed and presented in Tables 5 and 6. Similarly (8.4) has been used to compute similar percentage points for $s = 12$, which are given in Tables 7 and 8. The computations were carried out on IBM 7094, but a trial value was first extrapolated from Pillai's tables [11], [12], [15] for each computed value in order to be fed into the machine. The error of approximation remains practically the same as discussed previously [16], [12], [15].

9. SOME REMARKS

The tabulation of power functions presented in Tables 1 to 4 reveal, in addition to monotonicity with respect to individual population roots, the following facts : a) the power is not monotonic with respect to either the sum of the population roots or their product; b) the power seems to be a minimum when the two population roots are equal and increases as they move apart.

Further, the tabulations of the power functions made in this paper are being used to compare the powers of the largest root with those of Hotelling's T_0^2 and Pillai's $V^{(s)}$ criterion and Wilks' Λ criterion and a report on this study will be soon forthcoming.

The suthor wishes to thank Mrs. Louise Mao Lui, Statistical Laboratory, Purdue University, for the excellent programming of the material for the computations in this paper carried out on IBM 7094 computer, Purdue University's Computer Science's Center.

References

[1] Anderson, T. W. and S. Das Gupta. "Monotonicity of the Power Functions of Some Tests of Independence Between Two Sets of Variates," *Ann. Math. Statist.*, **35** (1964), 206–208.

[2] Anderson, T. W. and S. Das Gupta. "A Monotonicity Property of the Power Functions of Some Tests of the Equality of Two Covariance Matrices," *Ann. Math. Statist.*, **35** (1964), 1059–1063.

[3] Das Gupta, S., T. W. Anderson and G. S. Mudholkar. "Monotonicity of Power Functions of Some Tests of the Multivariate Linear Hypothesis," *Ann. Math. Statist.*, **35** (1964), 200–205.

[4] Constantine, A. G. "Some Non-Central Distribution Problems in Multivariate Analysis," *Ann. Math. Statist.*, **34** (1963), 1270–1285.

[5] James, Alan T. "Distributions of Matrix Variates and Latent Roots Derived from Normal Samples," *Ann. Math. Statist.*, **35** (1964), 475–501.

[6] Pillai, K. C. S. "On Some Distribution Problems in Multivariate Analysis," Inst. of Statistics, Mimeo Series No. 88, Univ. of North Carolina, Chapel Hill, 1954.

[7] Pillai, K. C. S. "Some New Test Criteria in Multivariate Analysis," *Ann. Math. Statist.*, **26** (1955), 117–121.

[8] Pillai, K. C. S. "On the Distribution of the Largest or the Smallest Root of a Matrix in Multivariate Analysis," *Biometrika*, **43** (1956), 122–127.

[9] Pillai, K. C. S. "Some Results Useful in Multivariate Analysis," *Ann. Math. Statist.*, **27** (1956), 1106–1114.

[10] Pillai, K. C. S. *Concise Tables for Statisticians*, The Statistical Center, University of the Philippines, 1957.

[11] Pillai, K. C. S. *Statistical Tables for Tests of Multivariate Hypotheses*, The Statistical Center, University of the Philippines, 1960.

[12] Pillai, K. C. S. "On the Distribution of the Largest of Seven Roots of a Matrix in Multivariate Analysis," *Biometrika*, **51** (1964), 270–275.

[13] Pillai, K. C. S. "On the Moments of Elementary Symmetric Functions of the Roots of Two Matrices," *Ann. Math. Statist.*, **35** (1964), 1704–1712·

[14] Pillai, K. C. S. On the Non-Central Multivariate Beta Distribution and the Moments of Traces of Some Matrices, Mimeograph Series No. 11, Department of Statistics, Purdue University (Presented at the International Symposium on Multivariate Analysis, Dayton, Ohio, June, 1965).

[15] Pillai, K. C. S. "On the Distribution of the Largest Characteristic Root of a Matrix in Multivariate Analysis," *Biometrika*, **52** (1965), 405–411.

[16] Pillai, K. C. S. and C. G. Bantegui. "On the Distribution of the Largest of Six Roots of a Matrix in Multivariate Analysis," *Biometrika*, **46** (1959), 237–240.

[17] Pillai, K. C. S. and T. A. Mijares. "On the Moments of the Trace of a Matrix and Approximations to its Distribution," *Ann. Math. Statist.*, **30** (1959), 1135–1140.

[18] Roy, S. N. "The Individual Sampling Distribution of the Maximum, Minimum and any Intermediate of the p-Statistics on the Null Hypotheses," *Sankhyā*, **7** (1945), 133–158.

[19] Roy, S. N. and W. F. Mikail. "On the Monotonic Character of the Power Functions of Two Multivariate Tests," *Ann. Math. Statist.*, **32** (1961), 1145–1151.

(Received Oct. 1, 1965.)

TABLE 1

Powers of the r_2^2 test for testing: $\rho_1 = 0$, $\rho_2 = 0$ against different simple alternative hypotheses

$m = 0$, $\alpha = .05$

ρ_1^2	ρ_2^2	n = 5	10	15	20	25	30	40	60
0	$.0^51$	.0500004	.0500007	.0500011	.0500015	.0500018	.0500022	.050003	.050004
0	.0001	.0500394	.0500756	.0501120	.0501484	.0501848	.0502213	.050294	.050441
$\rho_1^2+\rho_2^2=$ .0025									
0	.0025	.0509921	.0519112	.0528426	.0537831	.0547320	.0556888	.057626	.061593
.0001	.0024	.0509916	.0519097	.0528395	.0537779	.0547240	.0556777	.057607	.061552
.0002	.0023	.0509912	.0519083	.0528366	.0537731	.0547168	.0556676	.057590	.061514
.0003	.0022	.0509907	.0519070	.0528340	.0537687	.0547102	.0556584	.057574	.061480
.0005	.0020	.0509900	.0519048	.0528296	.0537613	.0546991	.0556428	.057547	.061422
.001	.0015	.0509890	.0519016	.0528231	.0537504	.0546827	.0556198	.057508	.061337
.00125	.00125	.0509888	.0519012	.0528223	.0537490	.0546807	.0556170	.057503	.061326
$\rho_1^2+\rho_2^2=$.01									
0	.01	.0540580	.0579188	.0619218	.0660524	.0703052	.0746772	.083768	.103263
.0001	.0099	.0540558	.0579123	.0619088	.0660306	.0702725	.0746315	.083690	.103097
.0002	.0098	.0540537	.0579059	.0618960	.0660092	.0702405	.0745867	.083613	.102935
.0003	.0097	.0540517	.0578997	.0618834	.0659883	.0702091	.0745428	.083539	.102776
.0005	.0095	.0540477	.0578876	.0618592	.0659477	.0701483	.0744579	.083394	.102468
.001	.009	.0540385	.0578598	.0618031	.0658541	.0700079	.0742616	.083059	.101757
.002	.008	.0540234	.0578139	.0617108	.0656999	.0697765	.0739382	.082508	.100581
.003	.007	.0540126	.0577811	.0616448	.0655897	.0696112	.0737071	.082114	.099742
.005	.005	.0540040	.0577549	.0615920	.0655015	.0694790	.0735221	.081798	.099069
ρ_1^2 = constant (.0015)									
.0015	.0015	.0511876	.0522844	.0533928	.0545090	.0556321	.0567618	.059040	.063670
.0015	.0025	.0515865	.0530555	.0545433	.0560447	.0575585	.0590843	.062170	.068478
.0015	.0035	.0519879	.0538339	.0557083	.0576047	.0595215	.0614580	.065388	.073471
.0015	.0045	.0523916	.0546195	.0568880	.0591892	.0615211	.0638827	.068693	.078648
.0015	.0055	.0527977	.0554125	.0580824	.0607982	.0635572	.0663583	.072083	.084005
.0015	.0065	.0532062	.0562127	.0592914	.0624315	.0656298	.0688846	.075559	.089539
.0015	.0075	.0536171	.0570203	.0605152	.0640893	.0677388	.0714614	.079119	.095248
.0015	.0085	.0540304	.0578352	.0617536	.0657715	.0698840	.0740884	.082764	.101127
$\rho_1^2\rho_2^2$ = .00005									
.005	.01	.0560686	.0618237	.0677713	.0738876	.0801632	.0865914	.099884	.128065
.002	.025	.0604076	.0727491	.0849060	.097800	.111378	.12559	.15574	.2214
.001	.05	.0732128	.09792	.12552	.15553	.1876	.2211	.2915	.433

TABLE 1 (Contd.)

ρ_1^2	ρ_2^2	$m = 1$ $\alpha = .05$ n: 5	10	15	20	25	30	40	60
0	$.0^51$	.0500003	.0500005	.0500008	.0500011	.0500013	.0500016	.050002	.050003
0	.0001	.0500289	.0500549	.0500810	.0501073	.0501335	.0501598	.050212	.050318
$\rho_1^2+\rho_2^2 = .0025$									
0	.0025	.0507271	.0513874	.0520570	.0527333	.0534157	.0541039	.055497	.058350
.0001	.0024	.0507268	.0513863	.0520548	.0527297	.0534103	.0540963	.055484	.087322
.0002	.0023	.0507264	.0513854	.0520529	.0527264	.0534053	.0540894	.055472	.058296
.0003	.0022	.0507261	.0513844	.0520511	.0527234	.0534008	.0540831	.055461	.058273
.0005	.0020	.0507256	.0513829	.0520480	.0527183	.0533932	.0540723	.055443	.058233
.001	.0015	.0507249	.0513807	.0520435	.0527108	.0533819	.0540565	.055416	.058174
.00125	.00125	.0507248	.0513804	.0520430	.0527099	.0533805	.0540545	.055412	.058166
$\rho_1^2+\rho_2^2 = .01$									
0	.01	.0529726	.0557452	.0586222	.0615927	.0646534	.0678026	.074362	.088499
.0001	.0099	.0529711	.0557407	.0586131	.0615775	.0646305	.0677706	.074307	.088380
.0002	.0098	.0529696	.0557362	.0586041	.0615626	.0646081	.0677392	.074253	.088264
.0003	.0097	.0529681	.0557319	.0585954	.0615480	.0645862	.0677084	.074200	.088150
.0005	.0095	.0529653	.0557234	.0585785	.0615197	.0645437	.0676488	.074098	.087929
.001	.009	.0529588	.0557040	.0585394	.0614543	.0644455	.0675112	.073862	.087418
.002	.008	.0529481	.0556719	.0584750	.0613467	.0642837	.0672844	.073473	.086577
.003	.007	.0529405	.0556490	.0584290	.0612698	.0641681	.0671224	.073195	.085975
.005	.005	.0529344	.0556307	.0583922	.0612083	.0640756	.0669928	.072973	.085493
$\rho_1^2 =$ Constant (.0015)									
.0015	.0015	.0508704	.0516587	.0524560	.0532593	.0540677	.0548810	.056522	.059857
.0015	.0025	.0511628	.0522185	.0532886	.0543691	.0554589	.0565575	.058780	.063326
.0015	.0035	.0514569	.0527830	.0541317	.0554964	.0568762	.0582706	.061102	.066930
.0015	.0045	.0517527	.0533537	.0549852	.0566412	.0583199	.0600205	.063486	.070669
.0015	.0055	.0520502	.0539291	.0558493	.0578036	.0597898	.0618072	.065933	.074543
.0015	.0065	.0523494	.0545097	.0567239	.0589836	.0612863	.0636310	.068444	.078551
.0015	.0075	.0526504	.0550956	.0576091	.0601812	.0628092	.0654918	.071017	.082693
.0015	.0085	.0529531	.0556868	.0585049	.0613967	.0643588	.0673897	.073654	.086967
$\rho_1^2\rho_2^2 = .00005$									
.005	.01	.0544469	.0585846	.0628672	.0672776	.0718096	.0764603	.086105	.10670
.002	.025	.0583535	.0665110	.075286	.084633	.094531	.10496	.1274	.1770
.001	.05	.0669962	.084870	.10507	.1274	.1515	.1774	.2333	.355

TABLE 1 (Contd.)

		$m = 2$ $\alpha = .05$							
		n							
ρ_1^2	ρ_2^2	5	10	15	20	25	30	40	60
0	$.0^51$	.0500003	.0500005	.0500007	.0500009	.0500011	.0500013	.050002	.050003
0	.0001	.0500240	.0500451	.0500662	.0500874	.0501087	.0501299	.050172	.050258
$\rho_1^2+\rho_2^2 = .0025$									
0	.0025	.0506041	.0511385	.0516797	.0522259	.0527768	.0533322	.054456	.056755
.0001	.0024	.0506038	.0511377	.0516780	.0522231	.0527727	.0533264	.054446	.056733
.0002	.0023	.0506036	.0511369	.0516765	.0522206	.0527689	.0533211	.054437	.056713
.0003	.0022	.0506033	.0511362	.0516751	.0522183	.0527654	.0533163	.054429	.056696
.0005	.0020	.0506029	.0511350	.0516727	.0522144	.0527595	.0533080	.054415	.056665
.001	.0015	.0506023	.0511333	.0516692	.0522086	.0527509	.0532959	.054394	.056620
.00125	.00125	.0506022	.0511330	.0516688	.0522078	.0527498	.0532944	.054391	.056614
$\rho_1^2+\rho_2^2 = .01$									
0	.01	.0524780	.0547084	.0570277	.0594194	.0618813	.0644124	.069681	.081030
.0001	.0099	.0524668	.0547048	.0570206	.0594076	.0618636	.0643876	.069638	.080938
.0002	.0098	.0524656	.0547014	.0570137	.0593960	.0618462	.0643633	.069596	.080847
.0003	.0097	.0524645	.0546979	.0570069	.0593847	.0618292	.0643395	.069556	.080759
.0005	.0005	.0524622	.0546913	.0569937	.0593628	.0617963	.0642934	.069474	.080587
.001	.009	.0524570	.0546761	.0569633	.0593121	.0617203	.0641869	.069293	.080189
.002	.008	.0524485	.0546509	.0569132	.0592286	.0615951	.0640116	.068992	.079534
.003	.007	.0524424	.0546330	.0568774	.0591690	.0615057	.0638862	.068777	.079066
.005	.005	.0524375	.0546186	.0568488	.0591213	.0614341	.0637859	.068605	.078691
ρ_1^2 = Constant (.0015)									
.0015	.0015	.0507233	.0513614	.0520060	.0526552	.0533084	.0539654	.055290	.057982
.0015	.0025	.0519662	.0518206	.0526756	.0535586	.0544388	.0553258	.057120	.060786
.0015	.0035	.0512104	.0522840	.0533734	.0544756	.0555895	.0567147	.058998	.063695
.0015	.0045	.0514561	.0527514	.0540695	.0554063	.0567608	.0581323	.060926	.066710
.0015	.0055	.0517031	.0532230	.0547737	.0563507	.0579526	.0595788	.062902	.069831
.0015	.0065	.0519515	.0536987	.0554863	.0573090	.0591653	.0610542	.064929	.073060
.0015	.0075	.0522013	.0541785	.0562072	.0582812	.0603988	.0625589	.067006	.076396
.0015	.0085	.0524525	.0546625	.0569365	.0592674	.0616532	.0640930	.069132	.079839
$\rho_1^2\rho_2^2 = .00005$									
.005	.01	.0536927	.0570378	.0604935	.0640489	.0677007	.0714462	.079215	.09582
.002	.025	.0569281	.0635085	.070570	.078090	.086055	.09446	.11254	.1532
.001	.05	.0640729	.07848	.09477	.1128	.1325	.1538	.200	.305

TABLE 1 (Contd.)

ρ_1^2	ρ_2^2	$m = 5 \quad \alpha = .05$ n 5	10	15	20	25	30	40	60
0	$.0^51$	.0500002	.0500003	.0500005	.0500006	.0500008	.0500009	.050001	.050002
0	.0001	.0500180	.0500326	.0500471	.0500617	.0500762	.0500907	.050120	.050178
$\rho_1^2+\rho_2^2 = .0025$									
0	.0025	.0504521	.0509221	.0511932	.0515664	.0519420	.0523199	.053083	.054638
.0001	.0024	.0504519	.0508215	.0511922	.0515647	.0519394	.0523164	.053077	.054626
.0002	.0023	.0504517	.0508210	.0511912	.0515631	.0519371	.0523132	.053072	.054614
.0003	.0022	.0504515	.0508206	.0511903	.0515617	.0519350	.0523102	.053067	.054603
.0005	.0020	.0504513	.0508198	.0511888	.0515593	.0519314	.0523053	.053058	.054585
.001	.0015	.0504509	.0508187	.0511866	.0515557	.0519262	.0522980	.053046	.054559
.00125	.00125	.0504508	.0508185	.0511864	.0515553	.0519255	.0522970	.053044	.054556
$\rho_1^2+\rho_2^2 = .01$									
0	.01	.0518441	.0533889	.0549688	.0565876	.0582464	.0599453	.063465	.070999
.0001	.0099	.0518433	.0533866	.0549643	.0565803	.0582355	.0599302	.063440	.070948
.0002	.0098	.0518425	.0533843	.0549600	.0565731	.0582248	.0599154	.063414	.070888
.0003	.0097	.0518417	.0533821	.0549557	.0565661	.0582144	.0599008	.063390	.070835
.0005	.0095	.0518401	.0533778	.0549473	.0565525	.0581942	.0598727	.063342	.070730
.001	.009	.0518365	.0533679	.0549281	.0565210	.0581475	.0598078	.063231	.070491
.002	.008	.0518306	.0533515	.0548965	.0564692	.0580706	.0598008	.063049	.070096
.003	.007	.0518263	.0533398	.0548739	.0564322	.0580157	.0596244	.062918	.069813
.005	.005	.0518229	.0533305	.0548558	.0564026	.0579717	.0596322	.062814	.069587
ρ_1^2 = Constant (.0015)									
.0015	.0015	.0505414	.0509834	.0514258	.0518699	.0523160	.0527639	.053666	.055494
.0015	.0025	.0507231	.0513147	.0519081	.0525047	.0531049	.0537088	.054928	.057411
.0015	.0035	.0509057	.0516487	.0523995	.0531479	.0539064	.0546711	.056219	.059392
.0015	.0045	.0510893	.0519854	.0528881	.0537996	.0547206	.0556511	.057541	.061437
.0015	.0055	.0512739	.0523247	.0533859	.0544599	.0555476	.0566489	.058893	.063548
.0015	.0065	.0514594	.0526668	.0538890	.0551289	.0563874	.0657665	.060276	.065724
.0015	.0075	.0516459	.0530116	.0543974	.0558067	.0572403	.0586983	.061689	.067968
.0015	.0085	.0518333	.0533591	.0549112	.0564933	.0581063	.0597505	.063133	.070278
$\rho_1^2\rho_2^2 = .00005$									
.005	.01	.0527597	.0550674	.0574242	.0598350	.0623011	.0648223	.070034	.081121
.002	.025	.0551626	.0596736	.0644430	.069482	.074801	.08038	.09237	.1196
.001	.05	.06044	.07025	.08118	.0933	.1064	.1209	.152	.225

TABLE 2

Powers of the r_3^2 test for testing $\rho = 0$ against different simple alternative hypotheses

ρ	n = 5	10	15	20	25	30	40	60
				$m = 0 \quad \alpha = .05$				
.001	.0500003	.0500005	.0500007	.0500009	.0500011	.0500014	.0500019	.0500028
.005	.0500062	.0500118	.0500175	.0500232	.0500289	.0500346	.0500461	.0500690
.01	.0500246	.0500472	.0500700	.0500928	.0501157	.0501386	.0501845	.0502763
.02	.0500986	.0501891	.0502805	.0503722	.0504642	.0505565	.0507413	.0511127
.03	.0502223	.0504265	.0506332	.0508410	.0510597	.0512593	.0516805	.0525310
.05	.0506203	.0511937	.0517769	.0523667	.0529622	.0535630	.0547804	.0572566
				$m = 1 \quad \alpha = .05$				
.001	.0500002	.0500004	.0500006	.0500007	.0500009	.0500011	.0500014	.0500022
.005	.0500048	.0500091	.0500135	.0500178	.0500222	.0500266	.0500354	.0500530
.01	.0500192	.0500365	.0500539	.0500714	.0500889	.0501065	.0501416	.0502120
.02	.0500769	.0501460	.0502158	.0502858	.0503560	.0504274	.0505690	.0508535
.03	.0501729	.0503283	.0504849	.0506415	.0507977	.0509531	.0512604	.0518547
.05	.0504833	.0509216	.0513675	.0518184	.0522737	.0527329	.0536628	.0555669
				$m = 2 \quad \alpha = .05$				
.001	.0500002	.0500003	.0500004	.0500006	.0500007	.0500009	.0500012	.0500018
.005	.0500041	.0500077	.0500113	.0500149	.0500186	.0500222	.0500295	.0500441
.01	.0500163	.0500307	.0500452	.0500597	.0500743	.0500889	.0501181	.0501766
.02	.0500654	.0501130	.0501811	.0502394	.0502980	.0503566	.0504743	.0507105
.03	.0501474	.0502773	.0504086	.0505407	.0506734	.0508065	.0510742	.0516141
.05	.0504112	.0507754	.0511454	.0515195	.0518969	.0522775	.0530476	.0546224
				$m = 5 \quad \alpha = .05$				
.001	.0500001	.0500002	.0500003	.0500004	.0500005	.0500006	.0500008	.0500012
.005	.0500031	.0500057	.0500082	.0500108	.0500134	.0500159	.0500211	.0500313
.01	.0500125	.0500227	.0500330	.0500432	.0500535	.0500638	.0500843	.0501255
.02	.0500499	.0500909	.0501320	.0501731	.0502144	.0502557	.0503384	.0505055
.03	.0501125	.0502050	.0502978	.0503909	.0504842	.0505779	.0507658	.0511445
.05	.0503136	.0505729	.0508339	.0510968	.0513616	.0516281	.0521663	.0532631

TABLE 3

Powers of l_2 test for testing : $\omega_1 = 0$, $\omega_2 = 0$ against different simple alternative hypotheses

		$m = 0$ $\alpha = .05$							
		n							
ω_1	ω_2	5	10	15	20	25	30	40	60
0	$.0^4 1$	.0500003	.0500003	.0500003	.0500003	.0500003	.0500003	.0500003	.0500003
0	.0001	.0500025	.0500029	.0500031	.0500032	.0500033	.0500034	.0500034	.0500035
.0005	.0005	.0500246	.0500291	.0500311	.0500322	.0500330	.0500335	.0500342	.0500349
0	.01	.0502465	.0502909	.0503112	.0503228	.0503302	.0503355	.0503423	.0503495
ω_1 = Constant (.005)									
.005	.015	.0504931	.0505822	.0506228	.0506459	.0506609	.0506713	.0506850	.0506994
.005	.025	.0507404	.0508743	.0509354	.0509702	.0509927	.0510085	.0510290	.0510507
.005	.045	.0512367	.0514610	.0515633	.0510217	.0516595	.0516859	.0517204	.0517567
$\omega_1+\omega_2 = .1$									
0	.1	.0524889	.0529447	.0531527	.0532716	.0533485	.0534023	.0534726	.0535466
.01	.09	.0524846	.0529384	.0531455	.0532639	.0533404	.0533939	.0534639	.0535375
.02	.08	.0524813	.0529336	.0531400	.0532578	.0533341	.0533874	.0534571	.0535304
.03	.07	.0524789	.0529302	.0531360	.0532536	.0533296	.0533828	.0534523	.0535254
.05	.05	.0524770	.0529274	.0531328	.0532501	.0533260	.0533790	.0534484	.0535213
$\omega_1+\omega_2 = .5$									
0	.5	.0629750	.0654892	.0666565	.0673103	.0677405	.0680420	.0684364	.0688519
.01	.49	.0629519	.0654559	.0666080	.0672687	.0676969	.0679970	.0683896	.0688031
.02	.48	.0629297	.0654238	.0665710	.0672289	.0676552	.0679539	.0683446	.0687562
.03	.47	.0629085	.0653932	.0665356	.0671907	.0676152	.0679126	.0683016	.0687113
.05	.45	.0628689	.0653359	.0664696	.0671195	.0675405	.0678354	.0682212	.0686275
.10	.40	.0627864	.0652166	.0663320	.0669710	.0673849	.0676747	.0680538	.0684530
.15	.35	.0627274	.0651314	.0662337	.0668649	.0672737	.0675599	.0679342	.0683282
.20	.30	.0626920	.0650803	.0661727	.0668013	.0672069	.0674910	.0678054	.0682534
.25	.25	.0626802	.0650632	.0661550	.0667801	.0671847	.0674681	.0678624	.0682284
0	1	.077227	.082817	.085408	.086899	.087866	.088545	.089434	.0903708
0	2	.1092	.1224	.1286	.1321	.1344	.1361	.1381	.1404
1	1	.1048	.1160	.1212	.1242	.1262	.1275	.1293	.1312
0	3	.145	.168	.178	.184	.188	.191	.194	.198
$\omega_1\omega_2 = .01$									
.01	1	.077478	.083115	.085727	.087229	.088205	.088889	.089785	.090729
.02	.5	.0634740	.0660798	.0672788	.0679665	.0684122	.0687245	.0691330	.0695633
.03	.3333	.0592508	.0609990	.0618008	.0622600	.0625573	.0627656	.0630379	.0633246
.04	.25	.0573177	.0586838	.0593092	.0596670	.0598987	.0600609	.0602729	.0604960
.05	.2	.0562752	.0574380	.0579698	.0582740	.0584708	.0586086	.0587887	.0589782
.08	.125	.0551124	.0560512	.0564799	.0567296	.0568834	.0569944	.0571394	.0572919
.1	.1	.0549839	.0558980	.0563155	.0565540	.0567083	.0568163	.0569574	.0571059

TABLE 3 (Contd.)

		$m = 1$ $\alpha = .05$							
		n							
ω_1	ω_2	5	10	15	20	25	30	40	60
0	$.0^41$	.0500002	.0500002	.0500002	.0500002	.0500002	.0500002	.0500002	.0500002
0	.0001	.0500016	.0500019	.0500021	.0500022	.0500023	.0500023	.0500024	.0500025
.0005	.0005	.0500161	.0500196	.0500213	.0500223	.0500230	.0500235	.0500241	.0500248
0	.01	.0501606	.0501961	.0502134	.0502236	.0502303	.0502350	.0502413	.0502480
$\omega_1 =$ Constant (.005)									
.005	.015	.0503213	.0503924	.0504271	.0504474	.0504608	.0504703	.0504829	.0504968
.005	.025	.0504823	.0504893	.0506413	.0506720	.0506921	.0507064	.0507254	.0507456
.005	.045	.0508053	.0509846	.0510717	.0511230	.0511568	.0511808	.0512125	.0512464
$\omega_1 + \omega_2 = .1$									
0	.1	.0516194	.0519830	.0521598	.0522641	.0523329	.0523817	.0524463	.0525152
.01	.09	.0516170	.0519793	.0515544	.0522593	.0523278	.0523763	.0524406	.0525092
.02	.08	.0516152	.0519765	.0521520	.0522556	.0523238	.0523722	.0524362	.0525046
.03	.07	.0516139	.0519744	.0521150	.0522528	.0523209	.0523691	.0524331	.0525012
.05	.05	.0516128	.0519728	.0521476	.0522507	.0523186	.0523668	.0524305	.0524986
$\omega_1 + \omega_2 = .5$									
0	.5	.0584008	.0603901	.0613676	.0619473	.0623309	.0626034	.0629647	.0633512
.01	.49	.0583880	.0603702	.0613436	.0619209	.0623028	.0625741	.0629338	.0633185
.02	.48	.0583758	.0603520	.0613216	.0618956	.0622759	.0625459	.0629041	.0632871
.03	.47	.0583640	.0603326	.0612986	.0618713	.0622500	.0625190	.0628756	.0632571
.05	.45	.0583421	.0602983	.0612576	.0618260	.6022018	.0624687	.0628226	.0632009
.10	.40	.0582965	.0602983	.0611719	.0617315	.0621014	.0623640	.0627120	.0630840
.15	.35	.0582639	.0601758	.0611107	.0616641	.0620297	.0622892	.0626330	.0630005
.20	.30	.0582444	.0601452	.0610740	.0616236	.0619866	.0622443	.0625856	.0629503
.25	.25	.0582378	.0601349	.0610618	.0616101	.0619723	.0622293	.0625698	.0629336
0	1	.067551	.071952	.074137	.075349	.076303	.076918	.077734	.078609
0	2	.0880	.0984	.1037	.1068	.1089	.1104	.1124	.1145
1	1	.0854	.0944	.0988	.1015	.1033	.1045	.1062	.1079
0	3	.111	.129	.138	.144	.147	.150	.153	.157
$\omega_1 \omega_2 = .01$									
.01	1	.067715	.072154	.074357	.075670	.076541	.077161	.077985	.078867
.02	.5	.0587265	.0607895	.0618027	.0624037	.0628012	.0630836	.0634580	.0638585
.03	.3333	.0560032	.0573918	.0580708	.0584730	.0587383	.0588927	.0591765	.0594434
.04	.25	.0547540	.0558411	.0563716	.0566852	.0568922	.0570391	.0572337	.0574416
.05	.2	.0540795	.0550061	.0554575	.0557243	.0559004	.0560252	.0561906	.0563672
.08	.125	.0533264	.0540757	.0544401	.0546553	.0547972	.0548978	.0550311	.0551733
.1	.1	.0532431	.0539729	.0543278	.0545373	.0546755	.0547734	.0549031	.0550416

TABLE 3 (Contd.)

		$m = 2$ $\alpha = .05$ n							
ω_1	ω_2	5	10	15	20	25	30	40	60
0	$.0^41$	.0500001	.0500002	.0500002	.0500002	.0500002	.0500002	.0500002	.0500002
0	.0001	.0500012	.0500015	.0500017	.0500017	.0500018	.0500019	.0500019	.0500020
.0005	.0005	.0500120	.0500150	.0500166	.0500175	.0500181	.0500185	.0500191	.0500198
0	.01	.0501201	.0501503	.0501656	.0501749	.5001811	.0501856	.0501916	.0501980
ω_1 = Constant (.005)									
.005	.015	.0502402	.0503008	.0503314	.0503500	.0503625	.053714	.0503833	.0503963
.005	.025	.0503606	.0504515	.0504977	.0505256	.0505444	.0505578	.0505758	.0505952
.005	.045	.0506020	.0507540	.0508314	.0508782	.0509097	.0509325	.0509622	.0509948
$\omega_1+\omega_2 = .1$									
0	.1	.0512096	.0515174	.0516744	.0517695	.0518332	.0518789	.0519401	.0520063
.01	.09	.0512081	.0515150	.0516714	.0517661	.0518296	.0518751	.0519360	.0520019
.02	.08	.0512070	.0515131	.0516690	.0517634	.0518267	.0518721	.0519328	.0519985
.03	.07	.0512061	.0515117	.0516673	.0517615	.0518247	.0518700	.0519305	.0519960
.05	.05	.0512055	.0515106	.0516660	.0517600	.0518231	.0518682	.0519287	.0519941
$\omega_1+\omega_2 = .5$									
0	.5	.0562452	.0579126	.0587728	.0592988	.0596495	.0599029	.0602428	.0606115
.01	.49	.0562370	.0578991	.0587561	.0592781	.0596293	.0598817	.0602202	.0605873
.02	.48	.0562292	.0578862	.0587401	.0592601	.0596100	.0598614	.0601985	.0605641
.03	.47	.0562217	.0578738	.0587248	.0592430	.0595915	.0598420	.0601778	.0605418
.05	.45	.0562077	.0578506	.0586962	.0592109	.0595570	.0598057	.0601390	.0605004
.10	.40	.0561786	.0578023	.0586366	.0591441	.0594851	.0597300	.0600583	.0604140
.15	.35	.0561578	.0577678	.0585941	.0590963	.0594337	.0596760	.0600006	.0603522
.20	.30	.0561453	.0577471	.0585686	.0590677	.0594029	.0596436	.0599660	.0603152
.25	.25	.0561411	.0577402	.0585601	.0590581	.0593926	.0596328	.0599545	.0603028
0	1	.062981	.066637	.068546	.069716	.070506	.071075	.071840	.072672
0	2	.0779	.0865	.0911	.0939	.0958	.0972	.0990	.1011
1	1	.0762	.0837	.0876	.0900	.0916	.0928	.0943	.0960
0	3	.095	.109	.117	.125	.128	.131	.131	.134
$\omega_1\omega_2 = .01$									
.01	1	.0631040	.0667925	.0687178	.0698978	.0706945	.0712684	.072397	.0728782
.02	.5	.0564886	.0582185	.0591104	.0596385	.0600195	.0602822	.0606346	.0610167
.03	.3333	.0544726	.0556415	.0562414	.0566060	.0568509	.0570268	.0572624	.0575177
.04	.25	.0535456	.0544628	.0549322	.0552172	.0554086	.0555459	.0557297	.0559288
.05	.2	.0530444	.0538272	.0542273	.0544699	.0546328	.0547496	.0549060	.0550753
.08	.125	.0524843	.0531183	.0534418	.0536378	.0537693	.0538636	.0539897	.0541262
.1	.1	.0524223	.0530400	.0533550	.0535459	.0536740	.0537658	.0538886	.0540215

TABLE 3 (Contd.)

		$m = 5$ $\alpha = .05$							
		n							
ω_1	ω_2	5	10	15	20	25	30	40	60
0	$.0^41$	.0500001	.0500001	.0500001	.0500001	.0500001	.0500001	.0500001	.0500001
0	.0001	.0500007	.0500009	.0500010	.0500011	.0500012	.0500012	.0500012	.0500013
.0005	.0005	.0500069	.0500091	.0500102	.0500110	.0500116	.0500119	.0500125	.0500131
0	.01	.0500691	.0500905	.0501024	.0501101	.0501155	.0501194	.0501248	.0501308
ω_1 = Constant (.005)									
.005	.005	.0501383	.0501811	.0502050	.0502203	.0502310	.0502389	.0502497	.0502618
.005	.025	.0502076	.0502718	.0503078	.0503308	.0503469	.0503587	.0503749	.0503931
.005	.045	.0503464	.0504537	.0505138	.0505524	.0505793	.0505991	.0506263	.0506567
$\omega_1 + \omega_2 = .1$									
0	.1	.0506950	.0509115	.0510331	.0511111	.0501165	.0512056	.0512607	.0513223
.01	.09	.0506944	.0509104	.0510317	.0511905	.0511623	.0512036	.0512586	.0513200
.02	.08	.0506939	.0509096	.0510306	.0511082	.0511613	.0512021	.0512569	.0513182
.03	.07	.0506936	.0509090	.0510298	.0511073	.0511605	.0512011	.0512557	.0513169
.05	.05	.0506933	.0509085	.0510292	.0511066	.0523552	.0512002	.0512548	.0513158
$\omega_1 + \omega_2 = .5$									
0	.5	.0535538	.0547018	.0553540	.0557753	.0560702	.0562880	.0565885	.0569256
.01	.49	.0535506	.0546958	.0553462	.0557663	.0560602	.0562773	.0565768	.0569127
.02	.48	.0535475	.0546902	.0553388	.0557576	.0560506	.0562671	.0565655	.0569003
.03	.47	.0535446	.0546847	.0553317	.0557494	.0560415	.0562573	.0565547	.0568884
.05	.45	.0535391	.0546746	.0553184	.0557339	.0560244	.0562389	.0565346	.0568663
.10	.40	.0535276	.0546535	.0552908	.0557017	.0559888	.0562007	.0564927	.0568201
.15	.35	.0535194	.0546384	.0552711	.0556786	.0559633	.0561734	.0564628	.0567871
.20	.30	.0535145	.0546294	.0552592	.0556648	.0559480	.0561570	.0564449	.0567673
.25	.25	.0535128	.0546264	.0552553	.0556602	.0559430	.0561516	.0564389	.0567606
0	1	.057306	.059768	.061185	.062107	.062756	.063236	.063901	.064649
0	2	.0654	.0710	.0743	.0765	.0780	.0792	.0807	.0825
1	1	.0647	.0698	.0727	.0746	.0759	.0769	.0782	.0798
0	3	.074	.084	.089	.093	.096	.098	.101	.104
$\omega_1\omega_2 = .01$									
.01	1	.0573769	.0598613	.0612912	.0622215	.0628753	.0633599	.0640303	.0647852
.02	.5	.0536935	.0548838	.0555622	.0559994	.0563053	.0565313	.0568430	.0571927
.03	.3333	.0525563	.0533686	.0538278	.0541236	.0543302	.0544827	.0546927	.0549280
.04	.25	.0520306	.0526710	.0530320	.0532643	.0534264	.0535459	.0537105	.0538947
.05	.2	.0517457	.0522937	.0526022	.0528006	.0529389	.0530409	.0531812	.0538383
.08	.125	.0514266	.0518720	.0521224	.0522831	.0523952	.0524777	.0525913	.0527183
.1	.1	.0513913	.0518254	.0520693	.0522259	.0523350	.0524155	.0525261	.0526498

TABLE 4

Powers of the l_3 test for testing: $\omega = 0$ against different simple alternative hypotheses

ω	n = 5	10	15	20	25	30	40	60
				$m = 0 \quad \alpha = .05$				
$.0^41$	.0500001	.0500002	.0500002	.0500002	.0500002	.0500002	.0500002	.0500002
.0001	.0500014	.0500017	.0500019	.0500019	.0500020	.0500020	.0500021	.0500022
.001	.0500137	.0500169	.0500184	.0100193	.0100199	.0500204	.0500204	.0500216
.01	.0501370	.0501688	.0501843	.0101935	.0501996	.0502039	.0502096	.0502158
.02	.0502742	.0503379	.0503691	.0503876	.0503998	.0504085	.0504200	.0504323
.03	.0504117	.0505075	.0505544	.0505822	.0506006	.0506136	.0506309	.0506494
.05	.0506875	.0508479	.0509266	.0509732	.0501040	.0510259	.0510548	.0510859
.1	.0513816	.0517063	.0518657	.0519603	.0520228	.0520673	.0521262	.0521892
.2	.0527894	.0534543	.0537818	.0539763	.0541051	.0541967	.0543181	.0544481
.3	.0542234	.0552440	.0557483	.0560482	.0562470	.0563883	.0565759	.0567768
.5	.0571701	.0589486	.0598322	.0603593	.0607090	.0609580	.0612888	.0616434
1	.0649981	.0689415	.0709237	.0721124	.0729036	.0734680	.0742194	.0750266
2	.082720	.092108	.096905	.099800	.101736	.103118	.104964	.016949
3	.10345	.11965	.12798	.13303	.13641	.13882	.14203	.14551
				$m = 1 \quad \alpha = .05$				
$.0^41$	.0500001	.0500001	.0500001	.0500002	.0500002	.0500002	.0500002	.0500002
.0001	.0500010	.0500012	.0500014	.0500014	.0500015	.0500015	.0500016	.0500016
.001	.0500096	.0500122	.0500135	.0500142	.0500148	.0500152	.0500157	.0500163
.01	.0500960	.0501217	.0501349	.0501429	.0501483	.0501522	.0501574	.0501631
.02	.0501923	.0502436	.0502700	.0502861	.0502970	.0503048	.0503152	.0503266
.03	.0502887	.0503658	.0504055	.0504297	.0504461	.0504578	.0504735	.0504906
.05	.0504819	.0506109	.0506775	.0507181	.0507454	.0507651	.0507915	.0508200
.1	.0509677	.0512284	.0513632	.0514455	.0515009	.0515407	.0515942	.0516522
.2	.0519551	.0524831	.0527593	.0529281	.0530420	.0531238	.0532338	.0533532
.3	.0529499	.0537643	.0541885	.0544481	.0546233	.0547495	.0549190	.0551031
.5	.0549946	.0564065	.0571463	.0576006	.0579077	.0581291	.0584269	.0587510
1	.0103880	.0634791	.0651252	.0661426	.0668332	.0673325	.0680057	.0687402
2	.072320	.079631	.083596	.086070	.087760	.088985	.090643	.092459
3	.08579	.09841	.10537	.10973	.11274	.11498	.11787	.12110

TABLE 4 (Contd.)

ω	n = 5	10	15	20	25	30	40	60
$m = 2 \quad \alpha = .05$								
$.0^41$	.05000006	.05000008	.05000010	.05000011	.05000011	.05000013	.05000013	.05000013
.0001	.0500007	.0500010	.0500011	.0500012	.0500012	.0500012	.0500013	.0500013
.001	.0500074	.0500096	.0500108	.0500115	.0500120	.0500123	.0500128	.0500134
.01	.0500743	.0500960	.0501076	.0501149	.0501200	.0501235	.0501284	.0501337
.02	.0501487	.0501922	.0502155	.0502301	.0502400	.0502473	.0502471	.0502678
.03	.0502232	.0502885	.0506236	.0503455	.0503605	.0503714	.0503861	.0504023
.05	.0503726	.0504818	.0505406	.0505772	.0506023	.0506205	.0506452	.0506722
.1	.0507478	.0509682	.0510870	.0511611	.0512119	.0512487	.0512987	.0513537
.2	.0515058	.0519547	.0521974	.0523492	.0524532	.0525290	.0526313	.0527441
.3	.0522744	.0529597	.9533314	.0535644	.0537240	.0538403	.0539981	.0541717
.5	.0538427	.0550254	.0556708	.0560767	.0563554	.0565585	.0568347	.0571392
1	.0579464	.0605173	.0619402	.0628416	.0634634	.0639179	.0645378	.0652235
2	.068942	.072927	.076316	.078488	.079997	.081105	.082622	.084308
3	.07761	.08725	.09318	.09699	.09965	.01061	.10431	.10731
$m = 5 \quad \alpha = .05$								
$.0^41$	.05000003	.05000006	.05000007	.05000007	.05000008	.05000008	.05000008	.05000008
.0001	.0500004	.0500006	.0500007	.0500007	.0500008	.0500008	.0500009	.0500009
.001	.0500044	.0500060	.0500069	.0500074	.0500079	.0500082	.0500086	.0500091
.01	.0500445	.0500598	.0500687	.0500745	.0500787	.0500817	.0500860	.0500909
.02	.0500891	.0501196	.0501375	.0501492	.0501575	.0501636	.0501722	.0501820
.03	.0501338	.0501796	.0502064	.0502239	.0502364	.0502457	.0502586	.0502733
.05	.0502232	.0502997	.0503445	.0503739	.0503948	.0504103	.0504320	.0504565
.1	.0504475	.0506015	.0506918	.0507512	.0507933	.0508247	.0508684	.0509180
.2	.0508994	.0512114	.0513948	.0515157	.0516016	.0516656	.0517548	.0518562
.3	.0513555	.0518297	.0521092	.0522938	.0524249	.0525229	.0526595	.0528148
.5	.0522809	.0530916	.0535722	.0538098	.0541175	.0542872	.0545241	.0547940
1	.0546711	.0563965	.0574334	.0581261	.0586216	.0589936	.0595151	.0501118
2	.059784	.063666	.066059	.067674	.068842	.069235	.070968	.072407
3	.06534	.07183	.07586	.07871	.08074	.08228	.08446	.08697

TABLE 5

Upper 5% Points of the Largest Root for $s = 11$

n \ m	0	1	2	3	4	5	7	10	15
5	.91618	.92506	.93220	.93809	.94302	.94722	.95398	.96139	.96954
10	.80023	.81700	.83104	.84301	.85333	.86234	.87734	.89449	.91433
15	.69988	.72072	.73863	.75422	.76796	.78016	.80094	.82546	.85496
20	.61840	.64100	.66075	.67823	.69385	.70790	.73225	.76169	.79822
25	.55248	.57557	.59602	.61433	.63086	.64589	.67227	.70479	.74615
30	.49857	.52149	.54199	.56052	.57739	.59284	.62026	.65457	.69911
40	.41635	.43806	.45777	.47582	.49245	.50786	.53563	.57122	.61887
60	.31222	.33087	.34808	.36409	.37907	.39314	.41899	.45310	.50067
80	.24945	.26546	.28038	.29438	.30758	.32010	.34335	.37458	.41924
100	.20760	.22153	.23459	.24692	.25861	.26976	.29061	.31896	.36020
130	.16581	.17741	.18836	.19874	.20866	.21815	.23605	.26067	.29712
160	.13800	.14792	.15731	.16626	.17483	.18306	.19867	.22032	.25272
200	.11277	.12107	.12895	.13649	.14373	.15071	.16401	.18257	.21067
300	.077378	.083257	.088870	.094261	.099463	.10450	.11415	.12777	.14859*
500	.047533	.051239	.054792	.058217	.061535	.064759	.070972	.07983*	.09349*
1000	.024197	.026121	.027971	.029760	.031498	.033192	.03647*	.04118*	.04850*

* Value extrapolated

TABLE 6

Upper 1% Points of the Largest Root for $s = 11$

n \ m	0	1	2	3	4	5	7	10	15
5	.94070	.94704	.95213	.95632	.95982	.96280	.96760	.97284	.97860
10	.83835	.85213	.86365	.87344	.88187	.88922	.90143	.91535	.93140
15	.74297	.76119	.77680	.79036	.80228	.81285	.83081	.85193	.87725
20	.66254	.68299	.70083	.71657	.73061	.74322	.76501	.79127	.82370
25	.59590	.61729	.63618	.65307	.66828	.68208	.70625	.73593	.77351
30	.54051	.56209	.58135	.59873	.61450	.62893	.65447	.68631	.72746
40	.45466	.47557	.49449	.51179	.52770	.54242	.56887	.60266	.64769
60	.34384	.36223	.37916	.39488	.40956	.42333	.44858	.48178	.52790
80	.27602	.29200	.30687	.32080	.33392	.34633	.36934	.40017	.44407
100	.23042	.24443	.25755	.26991	.28163	.29277	.31359	.34182	.38272
130	.18457	.19634	.20741	.21792	.22792	.23749	.25551	.28023	.31669
160	.15391	.16402	.17357	.18266	.19136	.19970	.21550	.23734	.26994
200	.12599	.13447	.14253	.15022	.15760	.16471	.17822	.19705	.22546
300	.086653	.092700	.098468	.10400	.10934	.11450	.12437	.13827	.1594*
500	.053335	.057164	.060832	.064365	.067785	.071105	.077494	.08650*	.1004*
1000	.027190	.029186	.031103	.032956	.034754	.036505	.03990*	.04470*	.05216*

* Value extrapolated.

TABLE 7

Upper 5% Points of the Largest Root for $s = 12$

n \ m	0	1	2	3	4	5	7	10	15
5	.92561	.93309	.93918	.94424	.94852	.95218	.95812	.96469	.97199
10	.81794	.83260	.84499	.85560	.86481	.87289	.88640	.90194	.92008
15	.72182	.74049	.75664	.77077	.78326	.79439	.81343	.83602	.86335
20	.64214	.66273	.68082	.69690	.71130	.72430	.74689	.77432	.80851
25	.57671	.59800	.61696	.63398	.64940	.66344	.68816	.71872	.75775
30	.52260	.54394	.56311	.58049	.59635	.61091	.63678	.66927	.71157
40	.43909	.45959	.47826	.49540	.51123	.52592	.55243	.58648	.63218
60	.33172	.34962	.36618	.38162	.39608	.40969	.43472	.46779	.51399
80	.26618	.28168	.29616	.30978	.32265	.33485	.35755	.38808	.43178
100	.22215	.23572	.24847	.26053	.27199	.28291	.30339	.33124	.37180
130	.17792	.18929	.20004	.21026	.22002	.22938	.24705	.27138	.30741
160	.14835	.15811	.16736	.17620	.18467	.19282	.20828	.22974	.26189
200	.12143	.12962	.13741	.14488	.15206	.15900	.17221	.19067	.2187*
300	.083511	.089338	.094914	.10028	.10546	.11049	.12012	.13371	.1545*
500	.051398	.055086	.058629	.062051	.065370	.068599	.07481*	.08362*	.09726*
1000	.026203	.028123	.029974	.031767	.03350*	.03520*	.03846*	.04313*	.05041*

* Value extrapolated.

TABLE 8

Upper 1% Points of the Largest Root for $s = 12$

n \ m	0	1	2	3	4	5	7	10	15
5	.94743	.95276	.95709	.96069	.96372	.96632	.97053	.97517	.98033
10	.85289	.86491	.87504	.88371	.89121	.89779	.90876	.92136	.93602
15	.76213	.77839	.79242	.80468	.81549	.82512	.84154	.86096	.88438
20	.68401	.70257	.71886	.73330	.74621	.75785	.77802	.80244	.83276
25	.61832	.63798	.65543	.67109	.68523	.69809	.72068	.74853	.78393
30	.56311	.58314	.60110	.61734	.63213	.64569	.66974	.69983	.73885
40	.47652	.49619	.51407	.53045	.54555	.55955	.58475	.61701	.66013
60	.36304	.38062	.39687	.41199	.42613	.43942	.46381	.49594	.54065
80	.29270	.30813	.32252	.33603	.34878	.36086	.38328	.41335	.45624
100	.24503	.25865	.27142	.28349	.29494	.30584	.32624	.35292	.39408
130	.19684	.20833	.21918	.22949	.23932	.24874	.26649	.29086	.32685
160	.16444	.17435	.18375	.19270	.20128	.20952	.22514	.24676	.27905
200	.13483	.14319	.15113	.15874	.16604	.17308	.18649	.20518	.2334*
300	.092952	.098932	.10465	.11015	.11545	.12059	.13043	.14425	.1653*
500	.057323	.061125	.064776	.068299	.071712	.075030	.08141*	.09042*	.1043*
1000	.029268	.031255	.033169	.035022	.03682*	.03858*	.04197*	.04680*	05427*

* Value extrapolated.

CHAPTER 30

Inference on Discriminant Function Coefficients

C. Radhakrishna Rao, *Indian Statistical Institute*

1. INTRODUCTION

Some years ago, the author developed some tests for examining hypotheses such as "the coefficients of some specified characters in the linear discriminant function are zero" (Rao, 1946; 1948; 1949) and "the coefficients of two given characters are in a specified ratio" (Rao, 1952). These tests were meant to be generalizations of an earlier test by Fisher (1940) for a proposed (assigned) discriminant function (i.e., when the proportions of all the coefficients are specified). During the last few years, the author has received a number of queries regarding the theory and application of these tests. The present paper is an attempt to answer these queries and to propose tests for other hypotheses of interest in the use of discriminant functions.

It must be mentioned that the individual coefficients in the linear discriminant function are not definite population parameters, as only the ratios of the coefficients are unique. For this reason estimates of the individual coefficients and their standard errors are not meaningful. We can only draw inferences on the ratios of the coefficients.

All the tests considered in the paper are special cases of a test for examining the sufficiency of a given subset out of a larger set of variables, for purposes of discrimination between two populations. *The hypothesis of sufficiency* is explicitly defined as follows. Let $(X_1, \ldots, X_p)$ be a p dimensional random variable. Then the subset $(X_1, \ldots, X_q)$ is said to be sufficient for discrimination between two populations if the conditional distributions of $(X_1, \ldots, X_p)$ given $(X_1, \ldots, X_q)$ are the same for both the populations. We may also describe the hypothesis of sufficiency of $(X_1, \ldots, X_q)$ as the *absence of additional information* contained in $(X_{q+1}, \ldots, X_p)$ when the variables $(X_1, \ldots, X_q)$ are already available. We consider this general problem in Section 2.

The problem of inference on the coefficients of a genetic selection index as developed by Smith (1936) needs an entirely different approach. A test is described for examining the adequacy of a *straight selection function* or any proposed selection function. Other questions such as the adequacy of a subset of the phenotypic observations for assessing some well defined genetic worth of an individual need further study.

2. NOTATIONS AND PRELIMINARY RESULTS

Let $\boldsymbol{X}' = (X_1, \ldots, X_p)$ be a p dimensional random variable and consider the partition $\boldsymbol{X}' = (\boldsymbol{X}_1' : \boldsymbol{X}_2')$ where $\boldsymbol{X}_1$ consists of the first q components of $\boldsymbol{X}$ and $\boldsymbol{X}_2$, the rest of the components of $\boldsymbol{X}$. Assuming the first two moments of $\boldsymbol{X}$ to exist, we can write the corresponding partitions of $E(\boldsymbol{X})$, the expectation of $\boldsymbol{X}$, and $D(\boldsymbol{X})$, the dispersion matrix of $\boldsymbol{X}$, as

$$E(\boldsymbol{X}') = \boldsymbol{\mu}' = (E(\boldsymbol{X}_1') : E(\boldsymbol{X}_2')), \tag{1}$$

$$D(\boldsymbol{X}) = \boldsymbol{\Sigma} = \begin{pmatrix} \boldsymbol{\Sigma}_{11} & \boldsymbol{\Sigma}_{12} \\ \boldsymbol{\Sigma}_{21} & \boldsymbol{\Sigma}_{22} \end{pmatrix}. \tag{2}$$

Let H_1 and H_2 be two simple hypotheses specifying the means and dispersion matrices as

$$(3) \qquad E(\boldsymbol{X}' \mid H_1) = \boldsymbol{\mu}_1' = (\boldsymbol{\mu}_{11}' : \boldsymbol{\mu}_{12}'),\ D(\boldsymbol{X}) = \boldsymbol{\Sigma},$$

$$(4) \qquad E(\boldsymbol{X}' \mid H_2) = \boldsymbol{\mu}_2' = (\boldsymbol{\mu}_{21}' : \boldsymbol{\mu}_{22}'),\ D(\boldsymbol{X}) = \boldsymbol{\Sigma}.$$

If $\boldsymbol{\delta}' = \boldsymbol{\mu}_1' - \boldsymbol{\mu}_2' = (\boldsymbol{\delta}_1' : \boldsymbol{\delta}_2')$, then the linear discriminant function between H_1 and H_2 based on $\boldsymbol{X}$ is defined as $\boldsymbol{\delta}' \boldsymbol{\Sigma}^{-1}\boldsymbol{X}$, while that based on $\boldsymbol{X}_1$ alone is $\boldsymbol{\delta}_1' \boldsymbol{\Sigma}_{11}^{-1}\boldsymbol{X}_1$. We shall represent the linear discriminant function $\boldsymbol{\delta}' \boldsymbol{\Sigma}^{-1}\boldsymbol{X}$ by $\boldsymbol{\lambda}'\boldsymbol{X} = \lambda_1 X_1 + \ldots + \lambda_p X_p$ and consider some hypotheses on $\boldsymbol{\lambda}$.

Note that the Mahalanobis distance between H_1 and H_2 based on $\boldsymbol{X}$ is $\Delta_p^2 = \boldsymbol{\delta}'\boldsymbol{\Sigma}^{-1}\boldsymbol{\delta}$ while that based on $\boldsymbol{X}_1$ is $\Delta_q^2 = \boldsymbol{\delta}_1'\boldsymbol{\Sigma}_{11}^{-1}\boldsymbol{\delta}_1$.

With the above notations, the following statements are equivalent.

(a) $\boldsymbol{\delta}_2 - \boldsymbol{\Sigma}_{21}\boldsymbol{\Sigma}_{11}^{-1}\boldsymbol{\delta}_1 = \boldsymbol{0}$, i.e., the random variable $\boldsymbol{X}_2 - \boldsymbol{\Sigma}_{21}\boldsymbol{\Sigma}_{11}^{-1}\boldsymbol{X}_1$ obtained by subtracting from $\boldsymbol{X}_2$ its regression on $\boldsymbol{X}_1$ has the same expected value for both the populations.

(b) The coefficients $\lambda_{q+1}, \ldots, \lambda_p$ of the components $X_{q+1}, \ldots, X_p$, in the linear discriminant function based on $\boldsymbol{X}$ are all zero.

(c) $\Delta_p^2 = \Delta_q^2$, i.e., there is no additional distance contributed by the variables $X_{q+1}, \ldots, X_p$.

(d) Every linear function of $\boldsymbol{X}$ uncorrelated with $\boldsymbol{X}_1$ has the same expected value for both the populations.

(e) If $\boldsymbol{X} \sim \mathcal{N}_p$ (i.e., distributed as p-variate normal), then the conditional distribution of $\boldsymbol{X}$ given $\boldsymbol{X}_1$ is the same for both the populations, which is the same as saying that $\boldsymbol{X}_1$ is sufficient for discrimination between the populations.

The compounding vector $\boldsymbol{\lambda}$ of $\boldsymbol{X}$ in the linear discrimination function $\boldsymbol{\delta}\boldsymbol{\Sigma}^{-1}\boldsymbol{X}$ is

$$(5) \qquad (\boldsymbol{\lambda}_1' : \boldsymbol{\lambda}_2') = (\boldsymbol{\delta}_1' : \boldsymbol{\delta}_2') \cdot \begin{pmatrix} \boldsymbol{\Sigma}^{11} & \boldsymbol{\Sigma}^{12} \\ \boldsymbol{\Sigma}^{21} & \boldsymbol{\Sigma}^{22} \end{pmatrix}$$

where $\boldsymbol{\lambda}_2' = (\lambda_{q+1}, \ldots, \lambda_p)$ is the vector of coefficients of the components of $\boldsymbol{X}_2$. From (5) $\boldsymbol{\lambda}_2' = \boldsymbol{\delta}_1'\boldsymbol{\Sigma}^{12}+\boldsymbol{\delta}_2'\boldsymbol{\Sigma}^{22}$ and by the statement (b), $\boldsymbol{\lambda}_2 = \mathbf{0}$. By virtue of the algebraic equivalence

$$\boldsymbol{\delta}_1'\boldsymbol{\Sigma}^{12}+\boldsymbol{\delta}_2'\boldsymbol{\Sigma}^{22} = \mathbf{0} \Longleftrightarrow \boldsymbol{\delta}_2-\boldsymbol{\Sigma}_{21}\boldsymbol{\Sigma}_{11}^{-1}\boldsymbol{\delta}_1 = \mathbf{0}, \tag{6}$$

this proves the equivalence of statements (a) and (b).

It is easy to prove the identity

$$\begin{aligned}\Delta_p^2 &= \boldsymbol{\delta}\,\boldsymbol{\Sigma}^{-1}\boldsymbol{\delta} \\ &= \boldsymbol{\delta}_1'\boldsymbol{\Sigma}_{11}^{-1}\boldsymbol{\delta}_1+(\boldsymbol{\delta}_2-\boldsymbol{\Sigma}_{21}\boldsymbol{\Sigma}_{11}^{-1}\boldsymbol{\delta}_1)'(\boldsymbol{\Sigma}_{22}-\boldsymbol{\Sigma}_{21}\boldsymbol{\Sigma}_{11}^{-1}\boldsymbol{\Sigma}_{12})^{-1}(\boldsymbol{\delta}_2-\boldsymbol{\Sigma}_{21}\boldsymbol{\Sigma}_{11}^{-1}\boldsymbol{\delta}_1)\end{aligned} \tag{7}$$

If $\Delta_p^2 = \Delta_q^2 = \boldsymbol{\delta}_1'\boldsymbol{\Sigma}_{11}^{-1}\boldsymbol{\delta}_1$, the second term in (7), which is a positive definite quadratic form is zero. Hence $(\boldsymbol{\delta}_2-\boldsymbol{\Sigma}_{21}\boldsymbol{\Sigma}_{12}^{-1}\boldsymbol{\delta}_1) = \mathbf{0}$ and vice-versa, i.e., (a) $\Longleftrightarrow$ (c).

Let $\boldsymbol{L}_1'\boldsymbol{X}_1+\boldsymbol{L}_2'\boldsymbol{X}_2$ be a linear function of $\boldsymbol{X}$. Then the statement (d) implies $\boldsymbol{L}_1'\boldsymbol{\Sigma}_{11}+\boldsymbol{L}_2'\boldsymbol{\Sigma}_{21}=\mathbf{0}$, i.e., $\boldsymbol{L}_1=-\boldsymbol{\Sigma}_{11}^{-1}\boldsymbol{\Sigma}_{12}\boldsymbol{L}_2$. Consider $\boldsymbol{L}_1'\boldsymbol{\delta}_1+\boldsymbol{L}_2'\boldsymbol{\delta}_2=\boldsymbol{L}_2'(-\boldsymbol{\Sigma}_{21}\boldsymbol{\Sigma}_{11}^{-1}\boldsymbol{\delta}_1+\boldsymbol{\delta}_2)=\mathbf{0}$ for all $\boldsymbol{L}_2 \Longleftrightarrow \boldsymbol{\delta}_2-\boldsymbol{\Sigma}_{21}\boldsymbol{\Sigma}_{11}^{-1}\boldsymbol{\delta}_1 = \mathbf{0}$, i.e., (d) $\Longleftrightarrow$ (a).

To prove (c) $\Longleftrightarrow$ (a) we need only consider the conditional distribution of $\boldsymbol{X}_2$ given $\boldsymbol{X}_1$, which is $(p-q)$ variate normal with

$$E(\boldsymbol{X}_2|\boldsymbol{X}_1, H_1)-E(\boldsymbol{X}_2|\boldsymbol{X}_1, H_2) = \boldsymbol{\delta}_2-\boldsymbol{\Sigma}_{21}\boldsymbol{\Sigma}_{11}^{-1}\boldsymbol{\delta}_1 \tag{8}$$

$$D(\boldsymbol{X}_2|\boldsymbol{X}_1, H_1) = D(\boldsymbol{X}_2|\boldsymbol{X}_1, H_2) = \boldsymbol{\Sigma}_{22}-\boldsymbol{\Sigma}_{21}\boldsymbol{\Sigma}_{11}^{-1}\boldsymbol{\Sigma}_{12}. \tag{9}$$

Sufficiency of $\boldsymbol{X}_1$ is equivalent to $\boldsymbol{\delta}_2-\boldsymbol{\Sigma}_{21}\boldsymbol{\Sigma}_{11}^{-1}\boldsymbol{\delta}_1 = \mathbf{0}$, which proves the desired result.

3. TEST FOR SUFFICIENCY OF A SUBSET

We shall develop a test for the hypothesis that "$\boldsymbol{X}_1$ is sufficient" or "$\boldsymbol{X}_2$ has no additional information in the presence of $\boldsymbol{X}_1$" on the basis of samples of sizes n_1 and n_2 from the two populations. Let

$$\boldsymbol{d}' = (\boldsymbol{d}_1' \,\vdots\, \boldsymbol{d}_2') \tag{10}$$

represent the difference in sample means of $\boldsymbol{X}_1$ and $\boldsymbol{X}_2$ and

$$\text{(11)} \qquad \boldsymbol{S} = \begin{pmatrix} \boldsymbol{S}_{11} & \boldsymbol{S}_{12} \\ \boldsymbol{S}_{21} & \boldsymbol{S}_{22} \end{pmatrix}$$

the pooled sum of products (S. P. matrix) within samples on (n_1+n_2-2) degrees of freedom (d.f.).

Let us observe that under the assumption of normality

$$\text{(12)} \qquad E(\boldsymbol{X}_2 | \boldsymbol{X}_1,\ H_1) = \boldsymbol{\alpha}_1 + \boldsymbol{\Gamma X}_1,$$
$$E(\boldsymbol{X}_2 | \boldsymbol{X}_1,\ H_2) = \boldsymbol{\alpha}_2 + \boldsymbol{\Gamma X}_1,$$

where $\boldsymbol{\Gamma}$ represents the matrix of regression coefficients. If the dispersion matrix of $\boldsymbol{X}$ is the same under H_1 and H_2, then $D(\boldsymbol{X}_2 | \boldsymbol{X}_1, H_1) = D(\boldsymbol{X}_2 | \boldsymbol{X}_1, H_2)$. Hence the hypothesis under test is $\boldsymbol{\alpha}_1 = \boldsymbol{\alpha}_2$.

The formulation (12) may be recognized as the multivariate extension of the Gauss-Markoff set up, with $(p-k)$ variables (the components of $\boldsymbol{X}_2$). So, no new problem arises in the consideration of test criteria. One obtains, to begin with, the dispersion matrices due to deviation from hypothesis and due to error. We also notice that the set up (12) involves two sets of parameters, with the null hypothesis concerning only one of the sets. In such case the computations are simple, involving what is known as covariance adjustment (see the discussion on page 119 in Rao (1952) in the unvariate case and on pages 468-69 in Rao (1965) in the multivariate case).

The S.P. matrix within populations jointly for $\boldsymbol{X}_1$, $\boldsymbol{X}_2$ is

$$\text{(13)} \qquad \boldsymbol{W} = \begin{pmatrix} \boldsymbol{S}_{11} & \boldsymbol{S}_{12} \\ \boldsymbol{S}_{21} & \boldsymbol{S}_{22} \end{pmatrix} = \boldsymbol{S}$$

on (n_1+n_2-2) d.f. and that between populations is

$$\text{(14)} \qquad \boldsymbol{B} = \frac{n_1\, n_2}{n_1+n_2} \begin{pmatrix} \boldsymbol{d}_1\boldsymbol{d}_1' & \boldsymbol{d}_1\boldsymbol{d}_2' \\ \boldsymbol{d}_2\boldsymbol{d}_1' & \boldsymbol{d}_2\boldsymbol{d}_2' \end{pmatrix}$$

on 1 d.f., giving the total S.P. matrix

$$T = W + B = \begin{pmatrix} T_{11} & T_{12} \\ T_{21} & B_{22} \end{pmatrix}. \tag{15}$$

We now compute the residual S.P. matrices for X_2 making covariance adjustment for X_1, obtaining

$$S_{22} - S_{21} S_{11}^{-1} S_{12} \quad \text{with} \quad \text{d.f.} = n_1 + n_2 - 2 - q \tag{16}$$

for within, and

$$T_{22} - T_{21} T_{11}^{-1} T_{12} \text{ with d.f.} = n_1 + n_2 - 1 - q \tag{17}$$

for the total.

Now applying the Wilks Λ criterion which is the likelihood ratio test applied on the conditional distributions of X_2 given X_1 we have

$$\Lambda = \frac{|S_{22} - S_{21} S_{11}^{-1} S_{12}|}{|T_{22} - T_{21} T_{11}^{-1} T_{12}|} = \frac{|S|}{|S_{11}|} \frac{T}{|T_{11}|} \tag{18}$$

$$= \frac{|S|}{|T|} \cdot \frac{|S_{11}|}{|T_{11}|}$$

It is well known that in the special case when the matrix B, has 1 d.f., the statistic

$$\frac{n_1 + n_2 - p - 1}{p - q} \left(\frac{1}{\Lambda} - 1 \right) \tag{19}$$

has an F distribution on $(p-q)$ and (n_1+n_2-p-1) d.f.

Defining

$$D_p^2 = (n_1 + n_2 - 2) d' S^{-1} d, \; D_q^2 = (n_1 + n_2 - 2) d_1' S_{11}^{-1} d_1 \tag{20}$$

which are estimates of Δ_p^2 and Δ_q^2 and $c = n_1 n_2 / (n_1 + n_2)(n_1 + n_2 - 2)$ we can write the F statistic (19) as

$$\frac{n_1 + n_2 - p - 1}{p - q} \cdot \frac{c(D_p^2 - D_q^2)}{1 + c D_q^2} \tag{21}$$

the form in which the test was originally expressed (Rao, 1949; 1952).

Optimum properties of the test (21) have been recently investigated by Giri (1964).

4. INFERENCE ON DISCRIMINANT FUNCTION COEFFICIENTS

The test criterion (21) was developed for testing the hypothesis that the coefficients of specified variables in the linear discriminant function are zero. We can apply the same test by a suitable transformation of the variables in drawing inferences of various types on the coefficients of a linear discriminant function.

4.1. Test for a proposed discriminant function (Fisher, 1940)

Let $\boldsymbol{\lambda}'\boldsymbol{X}$ be the assigned discriminant function. Then the null hypotheses under test can be written as

$$\boldsymbol{\lambda}'\boldsymbol{X} \propto \boldsymbol{\delta}'\boldsymbol{\Sigma}^{-1}\boldsymbol{X} \rightarrow \boldsymbol{\delta} \propto \boldsymbol{\Sigma\lambda}, \ (\boldsymbol{\lambda} \text{ given}). \tag{22}$$

We make the transformation (assuming $\boldsymbol{\lambda}_1 \neq \mathbf{0}$ without loss of generality)

$$Y_1 = \boldsymbol{\lambda}'\boldsymbol{X}, \ \ Y_2 = X_2, \ldots, Y_p = X_q. \tag{23}$$

Then the null hypothesis says that in the discriminant based on $Y_1, \ldots, Y_p$, the coefficients of $Y_2, \ldots, Y_p$ are zero. Hence the test (21) applies with $q = 1$. We need the values of D_1^2 and D_p^2 based on the Y values. Since D^2 is invariant under a linear transformation

$$D_p^2(Y_1, \ldots, Y_p) = D_p^2(X_1, \ldots, X_p) = (n_1+n_2-2)d'S^{-1}d \tag{24}$$

$$D_1^2(Y_1) = (n_1+n_2-2)(\lambda'd)^2/\lambda'S\lambda \tag{25}$$

so that D_1^2 and D_p^2 are expressed in terms of the statistics d and S defined in (10) and (11) in terms of the original variables (X_1,

..., X_p). Substituting the values (24), (25) for D_p^2 and D_1^2 we obtain the statistic

$$\frac{n_1+n_2-p-1}{p-1}\,\frac{cD_p^2-cD_1^2}{1+cD_1^2} \tag{26}$$

which is an F statistic on $(p-1)$ and (n_1+n_2-p-1) d.f.

If we are using α probability level of significance, the test (26) can be written as

$$\frac{(\boldsymbol{\lambda}'\boldsymbol{d})^2}{\boldsymbol{\lambda}'S\boldsymbol{\lambda}} \geqslant \frac{c(n_1+n_2-p-1)D_p^2-(p-1)F_\alpha}{c(p-1)F_\alpha+c(n_1+n_2-p-1)} \tag{27}$$

where F_α is the upper α probability value of F on $(p-1)$ and (n_1+n_2-p-1) d.f. The inequality (27) provides a cone with vertex at the origin, within which the direction vector $\boldsymbol{\lambda}$ of the coefficients of the true discriminant function will lie with probability $1-\alpha$.

4.2. Test for a given ratio of the coefficients of two variables

Let the ratio of the coefficients of X_1 and X_2 be ρ, i.e. $\lambda_1/\lambda_2 = \rho$. For a given ρ, the discriminant function can be written

$$\lambda_2(\rho X_1+X_2)+\lambda_3 X_3+\ldots+\lambda_p X_p \tag{28}$$

where $\lambda_2, \ldots, \lambda_p$ are unknown. Equating (28) to $\boldsymbol{\delta}\Sigma^{-1}\boldsymbol{X}$ the null hypothesis can be written as

$$\boldsymbol{\delta} \propto \Sigma \begin{pmatrix} \boldsymbol{\rho} \\ \cdots \\ \boldsymbol{b} \end{pmatrix} \tag{29}$$

where $\boldsymbol{\rho}' = (\rho, 1)$ and the vector $\boldsymbol{b}$ is unknown.

Let us make the transformation

$$Y_1 = \rho X_1+X_2, \quad Y_2 = X_3, \ldots, Y_{p-1} = X_p, Y_p = X_1. \tag{30}$$

Then the null hypothesis (29) says that in the discriminant function based on $Y_1, \ldots, Y_p$, the coefficient of Y_p is zero. Hence the test (21) applies with $q = p-1$. We need the values of $D_p^2(Y_1, \ldots, Y_p) = D_p^2(X_1, \ldots, X_p)$ and $D_{p-1}^2(Y_1, \ldots, Y_{p-1})$. To compute D_{p-1}^2, consider the partition of the original variable $\boldsymbol{X}' = (\boldsymbol{X}_1' : \boldsymbol{X}_2')$ where $\boldsymbol{X}_1' = (X_1, X_2)$ and $\boldsymbol{X}_2' = (X_3, \ldots, X_p)$ with the corresponding partition

$$\boldsymbol{d}' = (\boldsymbol{d}_1' : \boldsymbol{d}_2') \tag{31}$$

of the sample mean differences and the partition

$$\boldsymbol{S} = \begin{pmatrix} \boldsymbol{S}_{11} & \boldsymbol{S}_{12} \\ \boldsymbol{S}_{21} & \boldsymbol{S}_{22} \end{pmatrix} \tag{32}$$

of the within pooled dispersion matrix. Then

$$\begin{aligned} &(n_1+n_2-2)^{-1}D_{p-1}^2(Y_1, \ldots, Y_{p-1}) \\ &= \boldsymbol{d}_2'\boldsymbol{S}_{22}^{-1}\boldsymbol{d}_2 + \frac{[\boldsymbol{\rho}'(\boldsymbol{d}_1 - \boldsymbol{S}_{12}\boldsymbol{S}_{22}^{-1}\boldsymbol{d}_2)]^2}{\boldsymbol{\rho}'(\boldsymbol{S}_{11} - \boldsymbol{S}_{12}\boldsymbol{S}_{22}^{-1}\boldsymbol{S}_{21})\boldsymbol{\rho}} \end{aligned} \tag{33}$$

and the variance ratio test on 1 and (n_1+n_2-p-1) d.f. for an assigned ρ is

$$\frac{n_1+n_2-p-1}{1} \quad \frac{c(D_p^2 - D_{p-1}^2)}{1+c\,D_{p-1}^2} \tag{34}$$

where $(n_1+n_2-2)^{-1}\ D_p^2 = \boldsymbol{d}'\boldsymbol{S}^{-1}\boldsymbol{d}$ and D_{p-1}^2 is as given in (33). To obtain a confidence interval for ρ, we consider the equation

$$\frac{n_1+n_2-p-1}{1} \quad \frac{cD_p^2 - cD_{p-1}^2}{1+cD_{p-1}^2} = F_\alpha \tag{35}$$

which is quadratic in ρ. The confidence bounds for ρ are obtained by determining the roots of the equation (35).

4.3. Test for assigned ratios of the coefficients of several variables

Let $\rho_1, \ldots, \rho_k$ be the assigned ratios of the coefficients of the first k variables $X_1, \ldots, X_k$, where some of the ρ_i may be zero. In such a case the discriminant function is of the form

$$\lambda_k(\rho_1 X_1 + \ldots + \rho_k X_k) + \lambda_{k+1} X_{k+1} + \ldots + \lambda_p X_p \tag{36}$$

where $\lambda_k, \ldots, \lambda_p$ are unknown. Equating (36) to $\boldsymbol{\delta}\boldsymbol{\Sigma}^{-1}\boldsymbol{X}$, the null hypothesis can be written as

$$\boldsymbol{\delta} \propto \boldsymbol{\Sigma} \begin{pmatrix} \boldsymbol{\rho} \\ \cdots \\ \boldsymbol{b} \end{pmatrix} \tag{37}$$

where $\boldsymbol{\rho}' = (\rho_1, \ldots, \rho_k)$ and $\boldsymbol{b}$ is unknown.

Let $\rho_k \neq 0$ without loss of generality. Then consider the transformation

$$\begin{aligned} Y_1 &= \rho_1 X_1 + \ldots + \rho_k X_k, \\ Y_2 &= X_{k+1}, \ldots, Y_{p-k+1} = X_p, \\ Y_{p-k+2} &= X_1, \ldots, Y_p = X_{k-1}. \end{aligned} \tag{38}$$

The hypothesis (37) says that in the discriminant function based on $Y_1, \ldots, Y_p$, the coefficients of the last $(k-1)$ variables are all zero. Hence the test (21) is applicable with the value of $q = p-k+1$. As in the earlier case, we shall express the values of D_p^2 and D_q^2 in terms of the original variables. Consider the partition $(\boldsymbol{X}_1', \boldsymbol{X}_2')$ of $\boldsymbol{X}'$ where $\boldsymbol{X}_1$ consists of the components $X_1, \ldots, X_k$ and $\boldsymbol{X}_2$ of the rest. Corresponding to such a partition of the random variable, we have the following partition of the sample mean difference and the within S.P. matrix.

$$\boldsymbol{d}' = (\boldsymbol{d}_1' \,\vdots\, \boldsymbol{d}_2'), \quad \boldsymbol{S} = \begin{pmatrix} \boldsymbol{S}_{11} & \boldsymbol{S}_{12} \\ \boldsymbol{S}_{21} & \boldsymbol{S}_{22} \end{pmatrix}. \tag{39}$$

Then

$$(n_1+n_2-2)^{-1} D_p^2 = \boldsymbol{d}'\boldsymbol{S}^{-1}\boldsymbol{d} \tag{40}$$

$$(41)\quad (n_1+n_2-2)^{-1} D_q^2 = d_2' S_{22}^{-1} d_2 + \frac{[\boldsymbol{\rho}'(d-S_{12}S_{22}^{-1}d_2)]^2}{\boldsymbol{\rho}'(S_{11}-S_{12}S_{22}^{-1}S_{21})\boldsymbol{\rho}}$$

substituting the expressions (40), (41) in (21), we find that the test criterion is an explicit function of $\boldsymbol{\rho}$. Hence we can test for any assigned value of $\boldsymbol{\rho}$ or determine the confidence zone of $\boldsymbol{\rho}$.

4.4. Test whether $\boldsymbol{\lambda}$ belongs to a given linear manifold

Let A be a $(p \times k)$ matrix providing a basis of the given manifold. Then the hypothesis $\lambda \epsilon M(A) \Longrightarrow \boldsymbol{\delta} = \boldsymbol{\Sigma} A \boldsymbol{\theta}$ where $\boldsymbol{\theta}$ is a $k \times 1$ vector. The values of $\boldsymbol{\delta}$, $\boldsymbol{\Sigma}$ and $\boldsymbol{\theta}$ are unknown but the hypothesis only specifies a relationship among them through the known matrix A. Consider the transformation

$$(42)\qquad Y_1 = A'X, \qquad Y_2 = B'X$$

where B is chosen such that $B'\boldsymbol{\Sigma}A = 0$, i.e., the linear functions in $B'X$ are uncorrelated with those in $A'X$. If $\boldsymbol{\delta} = \boldsymbol{\Sigma} A \boldsymbol{\theta}$, then

$$(43)\qquad E(B'X \mid H_1) - E(B'X \mid H_2) = B'\boldsymbol{\delta} = B'\boldsymbol{\Sigma} A \boldsymbol{\theta} = 0.$$

Thus, according to statement (e) in Section 2, $Y_1 = A'X$ is sufficient for discrimination between the hypotheses H_1 and H_2 or the coefficients of the components of Y_2 in the discriminant function expressed in terms of Y_1, Y_2 are all zero. Hence the test (21) applies with $q = k$. The matrix B in (42) is arbitrary subject to the condition $B'\boldsymbol{\Sigma} A = 0$, but we do not need to know B in order to evaluate the test criterion (21). We observe that

$$(44)\qquad D_p^2(Y_1, Y_2) = D_p^2(X) = (n_1+n_2-2) d'S^{-1}d$$

$$(45)\qquad D_k^2(Y_1) = D_k^2(A'X) = (n_1+n_2-2) d'A(A'SA)^{-1}Ad$$

which depend only on A, d and S, where d and S are as defined in (10) and (11).

5. DISCRIMINANT FUNCTION FOR GENETIC SELECTION

Consider an observable variable $\boldsymbol{X}$ which has the decomposition

$$\boldsymbol{X} = \boldsymbol{\gamma}+\boldsymbol{\epsilon} \tag{46}$$

in terms of two unobservable variables $\boldsymbol{\gamma}$ and $\boldsymbol{\epsilon}$ which are uncorrelated. The variable $\boldsymbol{\gamma}$ denotes the conceptual genotypic measurements and $\boldsymbol{\epsilon}$ denotes the environmental effects so that $\boldsymbol{X}$ may be considered as representing phenotypic measurements. Under the set up (46), Smith (1936) considered the problem of predicting a linear function $\boldsymbol{a}'\boldsymbol{\gamma}$ representing the genetic worth of an individual on the basis of phenotypic measurements $\boldsymbol{X}$. Let $D(\boldsymbol{X}) = \boldsymbol{\Sigma}$, $D(\boldsymbol{\gamma}) = \boldsymbol{\Gamma}$ and $D(\boldsymbol{\epsilon}) = \boldsymbol{E}$. Since $\boldsymbol{\gamma}$ and $\boldsymbol{\epsilon}$ are uncorrelated,

$$\boldsymbol{\Sigma} = \boldsymbol{\Gamma}+\boldsymbol{E} \tag{47}$$

$$\text{Cov}\,(\boldsymbol{X}, \boldsymbol{\gamma}) = \boldsymbol{\Gamma},\ \text{Cov}\,(\boldsymbol{X}, \boldsymbol{a}'\boldsymbol{\gamma}) = \boldsymbol{\Gamma a}$$

giving the regression of $\boldsymbol{a}'\boldsymbol{\gamma}$ on $\boldsymbol{X}$ as $\boldsymbol{\lambda}'\boldsymbol{X}$ where

$$\boldsymbol{\Sigma}\,\boldsymbol{\lambda} = \boldsymbol{\Gamma a} \quad \text{or} \quad \boldsymbol{\lambda} = \boldsymbol{\Sigma}^{-1}\boldsymbol{\Gamma a}. \tag{48}$$

If $\boldsymbol{\Sigma}$ and $\boldsymbol{\Gamma}$ are known, then the best predictor of $\boldsymbol{a}'\boldsymbol{\gamma}$ or the best selection index for $\boldsymbol{a}'\boldsymbol{\gamma}$ is the regression function of $\boldsymbol{a}'\boldsymbol{\gamma}$ on $\boldsymbol{X}$.

As an alternative to the regression function $\boldsymbol{a}'\boldsymbol{\Gamma}\,\boldsymbol{\Sigma}^{-1}\boldsymbol{X}$ we may consider the *straight selection function* $\boldsymbol{a}'\boldsymbol{X}$, which is simpler to compute and which does not involve $\boldsymbol{\Gamma}$ and $\boldsymbol{\Sigma}$. It is, therefore, of interest to find the conditions under which $\boldsymbol{a}'\boldsymbol{\Gamma\Sigma}^{-1}\boldsymbol{X}$ and $\boldsymbol{a}'\boldsymbol{X}$ are equivalent. Now, $\boldsymbol{a}'\boldsymbol{\Gamma\Sigma}^{-1}\boldsymbol{X} \propto \boldsymbol{a}'\boldsymbol{X}$ implies that there exists a constant μ such that

$$\boldsymbol{\Sigma}^{-1}\boldsymbol{\Gamma a} = \mu\,\boldsymbol{a} \ \text{or}\ \boldsymbol{\Gamma a} = \mu\,\boldsymbol{\Sigma a}, \tag{49}$$

or $\boldsymbol{a}$ is an eigen vector of the determinantal equation

$$|\boldsymbol{\Gamma}-\mu\,\boldsymbol{\Sigma}| = \boldsymbol{0}. \tag{50}$$

The condition (49) is the same as,

$$\boldsymbol{Ea} = \frac{\boldsymbol{a}'\boldsymbol{Ea}}{\boldsymbol{a}'\boldsymbol{\Sigma a}}\,\boldsymbol{\Sigma a}, \tag{51}$$

writing $\quad \boldsymbol{\Gamma} = \boldsymbol{\Sigma} - \boldsymbol{E}.$

Let a_1, the first component of $\boldsymbol{a}$, be non-zero. Then we consider the transformation from $\boldsymbol{X}$ to $\boldsymbol{Y}$,

$$Y_1 = \boldsymbol{a}'\boldsymbol{X},\ Y_2 = X_2,\ \ldots,\ Y_p = X_p \tag{52}$$

with the corresponding decomposition $\boldsymbol{Y} = \boldsymbol{\gamma}^* + \boldsymbol{\epsilon}^*$ and the dispersion matrices $\boldsymbol{\Sigma}^*$ and $\boldsymbol{E}^*$. In terms of $\boldsymbol{\Sigma}^*$ and $\boldsymbol{E}^*$, the condition (49) is equivalent to

$$\frac{\Sigma^*_{j1}}{\Sigma^*_{11}} = \frac{E^*_{j1}}{E^*_{11}},\quad j = 2, \ldots, p. \tag{53}$$

i.e., the regression coefficient of ϵ^*_j on ϵ^*_1 is the same as that of Y_j on Y_1. We shall consider a test of the hypothesis (53) in Section 5.1 on the basis of independent estimators of $\boldsymbol{\Sigma}^*$ and $\boldsymbol{E}^*$ having Wishart distributions.

A more general hypothesis of interest is the assignment of the ratios of the coefficients of a subset of the phenotypic measurements X_i in the selection index (regression function). In such a case the relationship between $\boldsymbol{\Sigma}$ and $\boldsymbol{\Gamma}$ can be written as

$$\boldsymbol{\Gamma a} = \mu\,\boldsymbol{\Sigma}\begin{pmatrix}\boldsymbol{\rho}\\ \cdots\\ \boldsymbol{b}\end{pmatrix} \tag{54}$$

where $\boldsymbol{\rho}$ is the vector of assigned ratios of the coefficients of the (say) first q variables and $\boldsymbol{b}$ is unknown. There does not seem to be a simple test of the hypothesis (54). However, one can determine the likelihood ratio test based on independent estimators of $\boldsymbol{\Sigma}$ and $\boldsymbol{E}$ and thus provide an asymptotic test with the usual chi-square approximation.

5.1. A test for the hypothesis (53)

Let $\boldsymbol{T}$ and $\boldsymbol{W}$ be independent random matrices such that

$$\boldsymbol{T} \sim \mathcal{W}_p(\boldsymbol{\Sigma}^*, k) \tag{55}$$

$$\boldsymbol{W} \sim \mathcal{W}_p(\boldsymbol{E}^*, m) \tag{56}$$

where $\mathcal{W}(\boldsymbol{A}, b)$ represents Wishart distribution of a $p \times p$ random matrix with the hypothetical matrix $\boldsymbol{A}$ and degrees of freedom b. Given T_{11}, the conditional distributions

$$\left(\frac{T_{12}}{T_{11}}, \ldots, \frac{T_{1p}}{T_{11}}\right) \sim \mathcal{N}_p\left[\nu_1, \frac{1}{T_{11}}(\Sigma^*_{ij.1})\right] \tag{57}$$

$$(T_{ij\cdot 1}) = \left(T_{ij} - \frac{T_{i1}\,T_{j1}}{T_{11}}\right) \sim \mathcal{W}_p[(\Sigma^*_{ij.1}), k-1)] \tag{58}$$

are independent. Similarly

$$\left(\frac{E_{12}}{E_{11}}, \ldots, \frac{E_{1p}}{E_{11}}\right) \sim \mathcal{N}_p\left[\nu_2, \frac{1}{E_{11}}(E^*_{ij.1})\right] \tag{59}$$

$$(E_{ij\cdot 1}) = \left(E_{ij} - \frac{E_{i1}E_{j1}}{E_{11}}\right) \sim \mathcal{W}_p[(E^*_{ij.1}), m-1] \tag{60}$$

are independent. The hypothesis (53) under test is the same as the hypothesis

$$\nu_1 = \nu_2 \tag{61}$$

i.e., the means of the two normal distributions (57) and (59) are equal. Since estimates of the dispersion matrices are available, it is possible to provide an appropriate test of the hypothesis (61).

If $(\Sigma^*_{ij\cdot 1})$, $(E^*_{ij\cdot 1})$, the true residual dispersion matrices, are equal, then we have the standard test based on the likelihood ratio

$$\Lambda = \frac{|(T_{ij}\cdot_1)+(W_{ij}\cdot_1)|(T_{11}+W_{11})}{|(T_{ij})+(W_{ij})|} \tag{62}$$

where

$$\frac{m+k-p}{p-1} \quad \frac{1-\Lambda}{\Lambda} \tag{63}$$

has an F distribution on $(p-1)$ and $(m+k-p)$ d.f.

If the residual dispersion matrices $(\Sigma^*_{ij\cdot 1})$ and $(E^*_{ij\cdot 1})$ are not equal, then we have the following approximate test. The differences

$$d_1 = \frac{T_{12}}{T_{11}} - \frac{W_{12}}{T_{11}}, \ldots, d_{p-1} = \frac{T_{1p}}{T_{11}} - \frac{W_{1p}}{W_{11}} \tag{64}$$

have the estimated dispersion matrix

$$(C_{ij}) = \frac{1}{(k-1)T_{11}}(T_{ij\cdot 1}) + \frac{1}{(m-1)W_{11}}(W_{ij\cdot 1}). \tag{65}$$

If (C^{ij}) is the reciprocal of (C_{ij}), then the statistic for testing the significance of the differences $d_1, \ldots, d_{p-1}$ is

$$\Sigma\Sigma\; C^{ij} d_i d_j \tag{66}$$

which can be used approximately as a chi square on $(p-1)$ d.f.

Note : The choice of a test of the hypothesis (53) or (61) depends on whether the residual dispersion matrices $(\Sigma^*_{ij\cdot 1})$ and $(E^*_{ij\cdot 1})$ are equal or not. Since we have the estimates $(T_{ij\cdot 1})$ and $(W_{ij\cdot 1})$ the hypothesis of equality $(\Sigma^*_{ij\cdot 1}) = (E^*_{ij\cdot 1})$ can be subjected to a suitable test.

In problems of genetic selection we have certain families (or individual lines) out of which some have to be selected on the basis of performance of individuals within family. We obtain observations on a certain number of characteristics from each of n individuals in a family. These observations (after transformation to the Y_i variables as in (51)) provide an analysis of dispersion (S.P. matrices) as between and within families, which correspond to the $\boldsymbol{T}$ and $\boldsymbol{W}$ matrices used in the test. For a numerical application of the tests (63) and (66) and further discussion, the reader is referred to a paper by the author (Rao, 1953).

References

Fisher, R. A. (1940). "The Precision of Discriminant Functions," *Ann. Eugenics*, **10**, 422-429.

Giri, N. (1964). "On the Likelihood Ratio Test of a Normal Multivariate Testing Problem," *Ann. Math. Statist.*, **35**, 181-189.

Rao, C. R. (1946). "Tests with Discriminant Functions in Multivariate Analysis," *Sankhyā*, **7**, 407-414.

Rao, C. R. (1948). "Tests of Significance in Multivariate Analysis," *Biometrika*, **35**, 58-79.

Rao, C. R. (1949). "On Some Problems Arising Out of Discrimination with Multiple Characters," *Sankhyā*, **9**, 343-364.

Rao, C. R. (1952). *Advanced Statistical Methods in Biometric Research*, John Wiley and Sons, New York.

Rao, C. R. (1953). "Discrimination Functions for Genetic Differentiation and Selection," *Sankhyā*, **12**, 229-246.

Rao, C. R. (1965). *Linear Statistical Inference with Applications*, John Wiley and Sons, New York.

Smith, Fairfield H. (1936). "A Discriminant Function for Plant Selection," *Ann. Eugenics*, **7**, 240.

(Received Jan. 18, 1966.)

CHAPTER 31

The Problem of the Three-Way Election[1]

RICHARD F. POTTHOFF,
The University of North Carolina at Chapel Hill[2]

1. INTRODUCTION

In an election among three candidates who are vying for a single position, how should the winner be determined? This question is of great practical importance, since a faulty election system may result in the selection of a political leader who is not "representative" of the electorate. Although the three-way election problem is an old one, and although a number of solutions to it have been offered, none of these solutions has successfully withstood criticism, and none has received general acceptance.

The basic notion which is advanced in the present paper is a rather simple one. We suggest that it may be helpful to treat the three-way election problem in the context of *game theory*. Such a treatment of the three-way election problem should at least serve the purpose of putting the problem into a clearer perspective. In addition, we will suggest a simple solution to the problem which is based on game theory considerations. However, this

[1]This research was supported by the Mathematics Division of the Air Force Office of Scientific Research.

[2]The author is now at Burlington Industries,

solution will require that the winner be chosen randomly (by rolling a die, say) in certain special instances; one would suspect that the possible use of such a random mechanism would be certain to generate objections, whether rational or otherwise.

2. THE DISTINCTION BETWEEN PREFERENCES AND STRATEGIES

Let the three candidates be designated as A, B, and C. We assume that each voter is able to rank the three candidates in the order of his *preference*. There would, of course, be $3! = 6$ possible preference orders : ABC, ACB, BAC, BCA, CAB, and CBA (where, e.g., BCA means that candidates B, C, and A would be the voter's first, second, and third preferences respectively). The electorate can thus be divided into six groups according to preference order; we will use G_{AB} to denote the group whose preference order is ABC, G_{AC} to denote the group with preference order ACB, and so forth. Let us use P_{AB} to denote the proportion of the electorate in preference group G_{AB}, P_{AC} to denote the proportion in G_{AC}, and so forth.

We will assume generally that, when the voter casts his ballot, he is required to indicate one of the three candidates as his first choice, and one of the two remaining candidates as his second choice; however, there will also be certain election systems which provide for the expression of a first choice only (see Section 3 below). In any event, the voter, in marking his ballot, must select a *strategy*. We will say (if both a first and a second choice are to be designated) that he adopts strategy s_{AB} if he marks A as his first choice and B as his second choice, strategy s_{AC} if he votes for A as his first choice and C as his second choice, and so forth for the four remaining possible voting strategies. Let us use p_{AB} to denote the proportion of the electorate which adopts strategy s_{AB}, p_{AC} to denote the proportion which adopts s_{AC}, and so forth.

The naive voter will adopt the strategy which conforms with his preference order. Thus, e.g., the naive voter in preference group G_{CB} will adopt the voting strategy s_{CB}. However, this strategy s_{CB} will not necessarily be the strategy which will best serve the voter's own interests; thus, e.g., s_{BC} might be the voting strategy which best serves his interests even though his preference order is CBA. The examples below will illustrate ways in which this sort of anomaly can occur.

In any event, the sophisticated voter will realize that he should not necessarily adopt the voting strategy which conforms with his preference order. Rather, in choosing his voting strategy, he will take into consideration (i) the method which is being used to determine the winner of the election, and (ii) his knowledge (perhaps gained from some form of public opinion poll) of the proportions of the electorate having different preferences among the three candidates.

It would seem to be most desirable to determine the winner of the election by a method under which no voter could gain anything by adopting a voting strategy other than the one which conforms with his preference order. Unfortunately, though, no such method has ever been found. However, a method which we will suggest later apparently comes closer to satisfying this criterion than any alternative method which is available.

3. METHODS OF DETERMINING THE WINNER

In this section we describe and examine certain methods which have been proposed for determining the winner of a three-way election. The *method of determining the winner* is essentially the same thing as the "rules of the game". In the usual game-theory problem, the rules of the game are specified, and one is interested in determining the (optimal) strategies which the players ought to follow. In the three-way election problem which we are considering here, however, the situation is somewhat reversed, in that we may be interested in trying to select the rules of the game (i.e., the method of determining the winner) in such a way that the players will adopt certain strategies; in particular, one might like the best strategy for each voter to be the one which conforms with the voter's preference order. More generally, we would like to choose the rules of the game so that the outcome of the election will be, in some sense, equitable. However, our problem is not an easy one; games other than the standard zero-sum two-person game generally are extremely difficult to analyze, and the three-way election problem seems to be no exception to this.

Some methods which have been proposed for determining the winner in a three-way election are as follows :

Method 1. The candidate who receives the highest number of first-choice votes is declared the winner. (Under this method, no attention is paid to the second-place votes, and, in fact, the voter normally does not even specify his second-choice vote. Thus the voter effectively has only three strategies to choose from, rather than six.)

Method 2. The candidate who receives the smallest number of first-place votes is eliminated, and the ballots which designate first-choice votes for him are distributed between the remaining two candidates on the basis of the second-choice votes which appear on these ballots. Thus the winner is whichever of these latter two candidates is ranked above the other by more than half the voters.

Method 2a. A variant of Method 2 is frequently used in practice. In an initial election, each voter casts a ballot which indicates only his first choice. The low man is eliminated, and a second election, or "run-off' election, is held between the remaining two candidates. Ordinarily the second election is carried out only if no candidate receives more than fifty percent of the votes in the first election.

Method 3. The candidate who receives the largest number of third-place votes is eliminated, and the ballots which designate first-choice votes for him are distributed between the remaining two candidates according to the second-choice votes which appear on these ballots. Thus the winner is whichever of these latter two candidates is ranked above the other by more than half the voters.

Method 4. The first-place vote on each ballot is given a score of two, and the second-place vote a score of one. The candidate with the highest total score is then declared the winner.

Method 5. If there is one candidate who (according to the ballots) would win a two-man election contest against each of the other two candidates individually, then this candidate is declared the winner. Thus, e.g., if A is placed above B by more than half the voters and if A is placed above C by more than half the voters, then A wins.

Extensive discussion of the historical background of some of the above methods is given elsewhere; Borda and Condorcet are credited with discovering Methods 4 and 5 respectively, and, later, the election problem was considered by Laplace and by C. L. Dodgson (Lewis Carroll), among others (see [1] and [4, Chapters 5-7], for example). Usiskin [5] dealt with a problem in probability theory which is related in a limited way to our election problem and to Method 5.

Methods 1 and 2a seem to be the ones most frequently used in practice.

Although it might appear that Methods 2 and 2a are equivalent except for the matter of timing, this is, in fact, not the case. Let us disregard the possibility that some voters may change their minds (i.e., change their preference orders) in the interim period between the two elections under Method 2a, and let us also assume that the same people vote in the run-off election as in the initial election under Method 2a. Then the two methods (2 and 2a) will indeed be equivalent if every voter votes in conformity with his true preferences. But if we allow (as we should) for the possibility that some voters may choose strategies not conforming to their preference orders, then Methods 2 and 2a are clearly not equivalent, because the voter is faced (respectively) with two different sets of strategies to choose from under the two methods. Under Method 2a, if a voter with preference order ABC decides to vote for B in the initial election, he would clearly vote for A in the run-off election if the run-off election should turn out to be between A and B; under Method 2, however, this voter would have no option of making such a switch from B to A at the second stage of the counting.

4. EVALUATION OF THE DIFFERENT METHODS

In evaluating the different methods of determining the winner of a three-way election, we will assume that it is desirable for a method to satisfy the following properties :

Property 1. If all voters vote their true preferences (i.e., if every voter's voting strategy conforms with his preference order) and if one candidate receives more than fifty percent of the first-place votes, then the method should always result in this candidate being declared the winner.

Property 2. If all voters vote their true preferences, and if there is one candidate who is ranked above the second candidate by more than half the voters and who is also rated above the third candidate by more than half the voters, then such a candidate should always be declared the winner. (Observe that (a) any method which satisfies Property 2 will automatically satisfy Property 1 as well, but not vice versa; and (b) Method 5 obviously satisfies Property 2.)

Property 3. No matter what the outcome of the voting, the method should come up with the definite selection of one of the three candidates as the winner.

Property 4. To the greatest extent possible, the method should be one under which every voter's "optimal" voting strategy will conform with his preference order. (We will later restate Property 4 in a more precise form.)

We will make no attempt to justify or "prove" the desirability of the above four properties, but rather we simply assume that our object is to find a method which satisfies these properties. Although it might be felt that the desirability of these properties is almost self-evident, we might point out, on the other hand, that Ross (who argues strongly for Method 4) not only refuses to accept Property 2 as a desirable one, but even refuses to acknowledge the desirability of Property 1[4, pp. 81, 84].

The effect of Property 4 is to discourage the voter from not voting according to his true preference, and to relieve him of the need to consider any voting strategy other than the one which conforms with his preference order. Suppose a voter decides to use a certain voting strategy (one which either does or does not conform with his preference order) because he believes that this strategy best serves his interests. He may learn to his sorrow after the election that he had misjudged the state of public opinion and/or the strategies of his opponents, and that an alternative strategy would have been a better one for him. If voters can gain something by using voting strategies which run counter to their preference orders, then, as Dodgson indicates, this "makes an election more of a game of skill than a real test of the wishes of the electors" [1, pp. 232-233]. This difficulty will be alleviated to the extent that we can find a method which satisfies Property 4.

Table 1 presents some examples which will be instructive with respect to determining the extent to which the different methods satisfy the different properties. In considering the examples of this table, we assume for the moment that every voter uses the voting strategy which conforms to his preference order. Then, with respect to Properties 1-3, these examples enable us to draw the following conclusions about the different methods :

(i) In Example I, almost 2/3 of the voters prefer B to A and almost 2/3 of them prefer C to A (with a slight majority preferring B to C). Nevertheless, A wins under Method 1. Hence, Method 1 does not satisfy Property 2. (However, Method 1 obviously does satisfy Property 1.)

(ii) In Example II, almost 2/3 of the voters prefer B to A, and over 2/3 of them prefer B to C. Nevertheless, C is the winner under Method 2 or 2a. Thus Methods 2 and 2a fail to satisfy Property 2. (However, they obviously do satisfy Property 1.)

(iii) In Example III, almost 2/3 of the first-place votes go to A, yet B is the winner under Method 3. Hence, Method 3 does not even satisfy Property 1.

(iv) In Example IV, candidate A receives almost 2/3 of the first-place votes, but B wins the election under Method 4. Thus we see that Method 4 also fails to satisfy Property 1.

(v) Example V demonstrates that Method 5 fails to satisfy Property 3. Method 5 produces no definite winner in Example V, because A is favored over B, B is favored over C, but C is favored over A. This sort of "paradox" is referred to by Dodgson and by Black as a situation of "cyclical majorities" (see, e.g., [1, p. 39ff., p. 46ff.]).

TABLE 1

Examples illustrating some possible effects of different methods of determining the winner of a three-way election, under different conditions on the P's

Example number	Value of						Winner (assuming that voting strategies conform to preference orders) under Method				
	P_{AB}	P_{AC}	P_{BA}	P_{BC}	P_{CA}	P_{CB}	1	2 or 2a	3	4	5
I	.168	.167	0	.333	0	.332	A	B	B	B	B
II	.335	0	.164	.168	0	.333	A	C	B	B	B
III	.333	.333	0	.168	0	.166	A	A	B	A	A
IV	.666	0	0	.334	0	0	A	A	A	B	A
V	.22	.10	.12	.21	.26	.09	C	B	C	A	—
VI	.25	0	.20	.20	.20	.15	B	B	B	B	B
VII	.40	0	.16	.16	0	.28	A	B	B	B	B
VIII	.45	0	.15	.10	0	.30	A	A	B	B	B
IX-A	0	.36	.13	.11	.30	.10	C	C	C	C	C
IXB	.30	.06	.24	0	.30	.10	C	A	A	A	A
IXC	.30	.06	.13	.11	0	.40	C	C	B	B	B

Thus we see that Methods 1, 2, and 2a are the only ones which satisfy both Property 1 and Property 3. This may partially explain why Methods 1 and 2a are used so frequently in practice. Methods 3 and 4 are the only ones which fail even to satisfy Property 1. Method 5 is the only one which does not satisfy Property 3, but it is also the only one which satisfies Property 2 as well as Property 1.

We now consider the extent to which voters might profit, under the different methods, from casting ballots which do not conform with their preference orders. From the examples of Table 1 we make the following observations which relate to Property 4 :

(vi) Under Method 1 in Example I, the members of G_{CB} would profit if some of them would vote for B rather than C, since then B rather than A would become the winner. (In this determination as well as in succeeding ones, we presume that the voters in all preference groups except the one we mention will vote in conformity with their preference orders.)

(vii) In Example II, if Method 2 or 2a is used, the voters in G_{AB} will gain if some of them vote for B, since then B rather than C will become the winner.

(viii) If Method 3 is used in Example III, the voters in G_{AB} can effect an outcome more favorable to them if some of them employ s_{AC} rather than s_{AB}; this will cause A rather than B to be elected.

(ix) Under Method 4 in Example IV, the members of G_{AB} can cause A rather than B to be elected if some of them adopt s_{AC} rather than s_{AB}.

(x) Under Method 1 in example VI, the members of G_{CA} can elect A rather than B if they all vote for A.

(xi) If either Method 2 or 2a is used in Example VII, the members of G_{AB} can effect the election of A rather than B if 1/8 of them vote for C instead of A (admittedly a risky and delicate strategy).

(xii) Under Method 3 in Example VIII, the voters in G_{AB} can cause A rather than B to be elected if they all use s_{AC} rather than s_{AB}.

(xiii) Suppose Method 4 is used in Example I. Then the voters of G_{CB} can bring about the election of C rather than B if a small fraction of them adopt s_{CA} instead of s_{CB}.

These cases (vi-xiii) thus present situations (in all of which there are no cyclical majorities with respect to the P's) where, under each of Methods 1, 2, 2a, 3, and 4, the voters in a particular preference group can bring about an outcome more favorable to them if some of them cast ballots which do not conform to their preference order. Note in particular that the new (different) outcome is a *more* desirable one in (vi-ix) but a *less* desirable one in (x-xiii), if one judges "desirability" on the basis of the criterion stated in Property 2.

In the next section, we propose a new method of determining the winner (i.e., a set of rules of the game) which represents an extension of Method 5. Under this method, if cyclical majorities do not exist with respect to the P's, then there will be no preference group whose members can derive any "gain" (as defined under a certain minimax assumption) by using a strategy other than the one conforming to their preference order, given that the voters in the other five preference groups all vote in conformity with their preference orders.

5. A SUGGESTED METHOD OF DETERMINING THE WINNER

As a possible method of determining the winner, consider the following modification of Method 5 :

Method 5*. If cyclical majorities do not exist (with respect to the p's), then the winner is determined in the same way as under Method 5. If cyclical majorities do exist (with respect to the p's), then the winner is chosen randomly from among the three candidates, with each candidate having probability 1/3 of being selected (for example, a die could be tossed, with the understanding that the winner would be A if a 1 or 2 turns up, B if a 3 or 4 is rolled, and C if a 5 or 6 turns up).

Method 5*, like Method 5, clearly satisfies Properties 1 and 2. Unlike Method 5, Method 5* produces a definite winner when cyclical majorities exist (with respect to the p's), and thus satisfies Property 3. We now show that, if cyclical majorities do not exist (with respect to the P's), then Method 5* satisfies Property 4 in a certain sense (see Property 4* below).

We define an *equilibrium point* to be a set of strategies of the voters such that, if the strategies of the members of any five of the preference groups are held fixed, then the voters in the remaining preference group can gain nothing if some or all of them change their strategies (the word "gain" will be defined more precisely in the next paragraph). It will be seen that this definition of "equilibrium point" is similar but not identical to that which appears elsewhere (see, e.g., [3, p. 106] or [2, p. 24]).

Under most methods of determining the winner, there are three possible outcomes of the voting (to be designated as outcomes O_A, O_B, and O_C, according as the voting causes candidate A, B, or C respectively to be elected), but under Method 5* there is a fourth possible outcome in addition to the three outcomes just indicated. This fourth outcome (to be designated by O_R) is the one where the winner is chosen randomly. Let us consider how the four outcomes O_A, O_B, O_C, and O_R would be ranked by a voter in a given preference group. To be specific, let us deal with G_{CA} as an example (an analogous treatment would hold for any other preference group). Clearly, the voter in G_{CA} would prefer O_C to O_A, O_C to O_B, O_A to O_B, O_C to O_R, and O_R to O_B. The only question, then, is what his choice would be as between O_A and O_R. We will assume that he uses the *minimax* principle in choosing between O_A and O_R; this will imply that he will prefer O_A to O_R, since the worst possible pay-off (result) under O_A (viz., A becoming the winner) is better than the worst possible pay-off under O_R (viz., B becoming the winner). Thus O_C will be his first-choice outcome, O_A his second-choice outcome, O_R his third choice, and O_B his fourth choice. We will then say that the voters in G_{CA} will be able to derive *gain* if, by some or all of them changing from one strategy to another, they can cause the outcome to change from a lower-choice to a higher-choice outcome.

It is not possible, of course, to provide a rigorous justification of our minimax assumption (as stated in the previous paragraph), since in practice some voters, in choosing between two outcomes, would probably take into account not only the worst possible pay-off but also the expectation of the pay-off (which would mean, e.g., that if a voter strongly prefers C to either A or B but has only a slight preference for A over B, then such a voter might prefer O_R to O_A). However, we will point out later how our conclusions can be modified in case the minimax assumption does not hold for all voters.

We are now ready to re-state Property 4 in a more precise form, as follows:

Property 4*. For all cases where cyclical majorities do not exist with respect to the P's, the method of determining the winner should be one under which there is no preference group whose members can derive any gain by some or all of them using a voting strategy which does not conform to their preference order, given that the voters in the other five preference groups all vote in conformity with their preference orders. In other words, whenever cyclical majorities do not exist with respect to the P's, the method should be one under which an equilibrium point is constituted by that set of strategies wherein each voter votes according to his preference order.

From the observations (vi-xiii) of the previous section it is apparent that Methods 1, 2, 2a, 3, and 4 do not satisfy Property 4*. We now show that Method 5* does satisfy Property 4*; what we will prove is similar to a theorem given by Black [1, pp. 43-44].

In order to prove that Method 5* satisfies Property 4*, let us suppose that the P's are such that over half the voters prefer A to B, over half the voters prefer A to C, and over half of them prefer C to B (we do not lose any generality by being this specific). We now consider each of the six preference groups individually, and determine if the voters in one preference group can derive any gain via a voting strategy which differs from their preference order, given that the voters in the other five preference groups all vote in conformity with their preference orders :

(1) Clearly, the members of G_{AB} would be able to derive on gain by using a strategy other than s_{AB}, since their first-choice candidate (viz., A) is being elected anyway.

(2) For the same reason, the members of G_{AC} can gain nothing by deviating from strategy s_{AC}.

(3) The members of G_{BC} could gain nothing by deviating from s_{BC} : there is no strategy switch they could make which could prevent A from being elected, inasmuch as they are already giving their last-place votes to A.

(4) For the same reason, the voters in G_{CB} would be able to derive no gain by switching away from strategy s_{CB}.

(5) There is only one change in outcome which the voters of G_{BA} might be able to effect by not using strategy s_{BA}, and this would be a change from O_A to O_C (which could be brought about in some cases if, e.g., the voters of G_{BA} were to use s_{BC} instead of s_{BA}). However, O_C would be a worse pay-off than O_A, and so the voters of G_{BA} would have much to gain and nothing to lose by adhering to s_{BA}.

(6) We consider finally the preference group G_{CA}. There is no way that the members of G_{CA} could bring about the outcome O_C. In some cases they would be able to bring about O_B, but they would not want to do so since B is their third-choice candidate. They will bring about O_A if they use s_{CA}. The only other real choice which might be open to them, then, would be to produce the outcome O_R : provided that groups G_{AB} and G_{AC} comprise less than half the electorate, the voters in G_{CA} could always change the outcome from O_A to O_R if they all use s_{CB} instead of s_{CA}. Under our minimax assumption, however, they would prefer O_A to O_R. Thus we conclude that the voters of G_{CA} could not gain by deviating from s_{CA}. (Although in this testing for an equilibrium point one assumes implicitly that the voters in G_{CA} know the values of the P's when considering s_{CA} versus s_{CB}, in practice they will not know the P's exactly. Thus, even if there are voters in G_{CA} who do not conform to our minimax assumption and who would prefer O_R to O_A, such voters might still adhere to s_{CA} rather

than switch to s_{CB}, because of uncertainty that the electorate really will rate C above B. If the electorate were actually to rate B above C, then any change in outcome caused by members of G_{CA} using s_{CB} instead of s_{CA} would, of course, be a change not from O_A to O_R but rather from O_A to O_B.)

Observe that it was only in (6) that we needed to use our minimax assumption (this assumption had led us to the proposition that a voter will prefer the definite election of his second-choice candidate to the outcome O_R). We now consider what modification needs to be made in our conclusions if this assumption is not fully satisfied. With particular reference to (6), let P'_{CA} denote the proportion of the electorate which has preference order CAB and which prefers O_A to O_R as well (note that P'_{CA} will be less than or equal to P_{CA}). So long as the sum of P_{AB}, P_{AC}, and P'_{CA} exceeds $\frac{1}{2}$, the essence of the argument given in (6) will still hold, and the set of strategies wherein each voter votes according to his preference order will still constitute an equilibrium point. Thus, if we should not wish to make the minimax assumption, then we can modify the condition stated in the initial clause of Property 4* by restricting this condition in the manner just indicated, and the equilibrium property of Method 5* will still hold under these less general conditions.

Even if we do make the minimax assumption, Method 5* of course will not produce the equilibrium point indicated in Property 4* if cyclical majorities do exist with respect to the P's. Consider Example V of Table 1. If, e.g., all the voters of G_{BC} utilize s_{CB} rather than s_{BC} (and if all other voters vote according to their preference orders), then the voters of G_{BC} will thereby change the outcome from O_R to O_C (the latter being the preferred outcome of the two, under the minimax assumption as applied to G_{BC}). A similar instability can be shown to exist in any other case where there are cyclical majorities with respect to the P's. It would not appear to be possible to find a method which would produce the equilibrium point of Property 4* whether or not cyclical majorities exist with respect to the P's, and which would satisfy Properties 1, 2, and 3 as well.

Black [1, p. 66] mentions a method whereby the winner is determined by Method 5 if cyclical majorities do not exist with respect to the p's, and by Method 4 if they do exist. However, Property 4* is satisfied neither by this method, nor by any other method which combines Method 5 with anything other than a random means for determining the winner in all those cases where cyclical majorities exist with respect to the p's. To show that this latter statement is true, we consider a specific set of p-values, viz.,

$$p_{AB} = .30,\ p_{AC} = .06,\ p_{BA} = .13,\ p_{BC} = .11,\ p_{CA} = .30,\ p_{CB} = .10,$$

but one may easily verify that our line of argument will apply also to any other set of six p-values for which there are cyclical majorities. We suppose that Method 5 would have been used to determine the winner if cyclical majorities (with respect to the p's) had not existed. We show that Property 4* will not be satisfied if, for any case of cyclical majorities with respect to the p's (we concentrate specifically on the case given above), any non-random means is used to select the winner. There are only three possible outcomes which such a non-random means could specify, viz., O_A, O_B, and O_C. We consider each in turn:

(i) Suppose the non-random means of selecting the winner specifies O_A to be the outcome if the p's are as given above. Then, if the P's are as in Example IXA of Table 1, the members of G_{AC} could change the outcome from O_C to O_A if 5/6 of them use s_{AB} instead of s_{AC} (we assume that all other voters vote according to their preference orders). Thus Property 4* is not satisfied if the non-random means of selecting the winner specifies O_A.

(ii) Suppose the non-random means specifies O_B. Then, if the P's are as in Example IXB, the voters in G_{BA} can switch the outcome from O_A to O_B if some of them use s_{BC} instead of s_{BA}. Again, Property 4* fails to hold.

(iii) Suppose O_C is specified. If the P's are as in Example IXC, the voters in G_{CB} can change the outcome from O_B to O_C

if some of them adopt s_{CA} instead of s_{CB}. As before, Property 4* is not satisfied; we will be forced to conclude that we can satisfy Property 4* only by specifying an indefinite or random outcome when cyclical majorities exist with respect to the p's.

In summary, we may say the following. Although Method 5* may not be especially palatable because of the fact that it sometimes utilizes a random mechanism to select the winner, there would seem to be no alternative method available which possesses all the desirable properties of Method 5*.

6. HOW PREVALENT WOULD CYCLICAL MAJORITIES BE ?

The degree to which Method 5* would be accepted might depend in large measure upon the prevalence of cyclical majorities in actual practice. That is, Method 5* might be readily accepted if cyclical majorities (with respect to either the P's or the p's) would be relatively infrequent, because then the random mechanism would not often have to be resorted to. It is therefore pertinent to consider live data from polls or elections in which the voters were asked to rank three or more candidates; such data may give some indication as to how prevalent cyclical majorities can be expected to be.

In March of 1952, a total of 562 voters (students) participated in a Presidential preference poll which was held at Swarthmore College. In this election, each voter was given a list of ten potential candidates for President of the United States, and was asked to rank these ten candidates in the order of his preference (however, most voters did not take the trouble to rank all ten of the candidates). It is probably safe to assume that the voters were voting their true preferences; thus, we will presume that the P's were the same as the p's. (The winner, incidentally, was actually determined by using a generalization of Method 2.) From among the ten candidates, one can choose a total of $\binom{10}{3}$ $= 120$ different groups of three candidates each. For each of these 120 trios of candidates, one can determine whether cyclical majorities exist.

Table 2 indicates, for every pair of candidates, the number of voters who preferred each candidate to the other; thus, e.g., Table 2 tells us that 173 voters ranked candidate D above candidate J, while 239 voters ranked J above D (the remaining 150 voters expressed no preference as between D and J). Although Table 2 simply identifies the ten candidates by letters (A, B, ..., J), their actual identities are indicated below the table. Candidate C was the winner of the election (as determined via a method which consisted of successively eliminating the candidate with the smallest number of ballots, and then re-distributing his ballots among the surviving candidates on the basis of the rankings on those ballots).

From Table 2 we find that cyclical majorities do not exist in a single one of the 120 trios of candidates (we see from the table that the ranking of the candidates is in the order C, B, A, J, D, G, I, H, F, E). If the results of this one election are at all indicative of what might happen in elections generally, then one must conclude that cyclical majorities will be a rare phenomenon indeed. To this extent, our analysis of the live election data of Table 2 enables us to be rather encouraged as to the potential practicability of Method 5*.

An obvious field for further investigation suggests itself at this point. If data from other appropriate elections could be analyzed in the same manner as indicated by Table 2, such analyses would provide us with more extensive evidence concerning the prevalence of cyclical majorities in actual practice.

Since the three-way election problem is itself a difficult problem, the present paper has been restricted to this problem and has not attempted to consider elections in which there are more than three candidates. However, since elections with four or more candidates are obviously of practical importance, such elections may provide a fruitful area for future investigation. Although it is of interest to consider (e.g.) the question of how the winner should be determined in a ten-way election such as the one which Table 2 is concerned with, our only immediate purpose in exhibiting the data of Table 2 was to obtain a rough indication of how frequently cyclical majorities might be expected in three-way elections.

TABLE 2

*Number of voters (out of a total of 562) who ranked Candidate * above Candidate ** in a Presidential preference poll held at Swarthmore College in March of* 1952

Candidate**	Candidate*									
	A	B	C	D	E	F	G	H	I	J
A		208	255	158	16	99	132	123	114	212
B	194		277	160	16	114	112	128	100	222
C	202	212		153	18	60	148	78	128	164
D	236	239	289		16	102	160	128	126	239
E	297	290	351	254		142	208	161	184	307
F	275	268	329	225	41		193	143	171	275
G	237	239	302	185	23	118		141	128	241
H	272	272	315	221	44	110	195		175	256
I	252	260	307	210	38	134	184	146		256
J	225	237	271	173	20	67	175	104	151	

The candidates were : A, Paul H. Douglas; B, William O. Douglas; C, Dwight D. Eisenhower; D, Estes Kefauver; E, Richard B. Russell; F, Harold E. Stassen; G, Adlai E. Stevenson; H, Robert A. Taft; I, Harry S. Truman; and J, Earl Warren.

Immediately after the election, the rankings on all 562 ballots were recorded. The figures in the above table, however, were tabulated at the time that this paper was written.

References

[1] Black, Duncan. (1958). *The Theory of Committees and Elections*, Cambridge University Press.

[2] Burger, Ewald. (1963). *Introduction to the Theory of Games* (translated by John E. Freund), Prentice Hall, Englewood Cliffs, New Jersey.

[3] Nash, J. F., and L. S. Shapley. "A Simple Three-Person Poker Game," *Contributions to the Theory of Games* (edited by H. W. Kuhn and A. W. Tucker), pp. 105-116, Princeton University Press, 1950.

[4] Ross, J. F. S. (1955). *Elections and Electors*, Eyre and Spottiswoode, London.

[5] Usiskin, Zalman. (1964). "Max-Min Probabilities in the Voting Paradox," *Ann. Math. Statist.*, **35**, 857-862.

(*Received Mar. 10, 1966.*)

CHAPTER 32

The Admissibility of the Largest Characteristic Root Test for the Normal Multivariate Linear Hypotheses

S. N. ROY AND W. F. MIKHAIL, *The University of North Carolina at Chapel Hill and IBM*

INTRODUCTION AND SUMMARY

The purpose of this paper is to show that the largest characteristic root test of multivariate analysis of variance is admissible among the class of similar tests that are invariant under the group of scale transformations [4].

The proof of the admissibility of the proposed test is established by showing that its power function is greater than that of any other similar and scale invariant test for some range of a particular class of alternative hypotheses. This is done in two steps. First, we shall select a particular class of alternative hypotheses and show that the acceptance region of the proposed test is the intersection of a family of Student t-regions defined in (2.1). Secondly, we shall show that for any other similar and scale invariant test, there exists a t-region that has a smaller probability of the second type of error. Proving the existence of such a region, that has the above property, completes the proof of the main result since the acceptance region of the proposed test is the intersection of all possible t-regions.

We note that while the largest characteristic root test is similar with respect to all free (nuisance) parameters, the above mentioned t-regions, individually, are not. Each of the t-regions is, however, similar with respect to the same proper subset of the totality of free parameters. They are used merely as a tool for obtaining the required result. We shall call any region that is similar with respect to a proper subset of the totality of free parameters a *quasi-similar* region to distinguish it from a similar region. It is to be noticed that similar regions are subclasses of quasi-similar regions.

1. MODEL AND HYPOTHESES

Let $\boldsymbol{X}(p \times n)$ be a random sample of size n from $N(E(\boldsymbol{x}_i), \boldsymbol{\Sigma})$, where $\boldsymbol{x}_i(p \times 1)$ is a p-column vector and $\boldsymbol{\Sigma}(p \times p)$ is symmetric positive definite (s.p.d.). Also, let

$$E(\boldsymbol{X}') = \boldsymbol{A}(n \times m)\boldsymbol{\xi}(m \times p)$$
$$= n\underset{r \qquad (m-r)}{[\boldsymbol{A}_1 \quad \boldsymbol{A}_2]} \underset{p}{\begin{bmatrix} \boldsymbol{\xi}_1 \\ \boldsymbol{\xi}_2 \end{bmatrix}} \begin{matrix} r \\ (m-r) \end{matrix}$$

where

(i) $\boldsymbol{A}$ is the design matrix with rank $r \leqslant m < n$ and $\boldsymbol{A}_1$ is a basis of $\boldsymbol{A}$ with a consequent partitioning of $\boldsymbol{\xi}$ into $\boldsymbol{\xi}_1$ and $\boldsymbol{\xi}_2$;

(ii) $\boldsymbol{\xi}$ is a matrix of unknown parameters.

For that model, we want to test

$$H_0 : \boldsymbol{C}(s \times m)\, \boldsymbol{\xi}(m \times p)\, \boldsymbol{M}(p \times u) = \boldsymbol{0}(s \times u) \tag{1.1}$$

or

$$s\underset{r \quad m-r}{[\boldsymbol{C}_1 \quad \boldsymbol{C}_2]} \begin{bmatrix} \boldsymbol{\xi}_1 \\ \boldsymbol{\xi}_2 \end{bmatrix} \boldsymbol{M} = 0$$

against

$$H_1 : \boldsymbol{C\xi M} = \boldsymbol{\eta}(s \times u) \neq \boldsymbol{0}, \tag{1.2}$$

where C and M are matrices given by the hypothesis such that $R(C) = s \leqslant r < n$ and $R(M) = u \leqslant p$ and η is an arbitrary unspecified non-null matrix.

The test of preassigned size α^*(say) is given by the following rule [2]. Accept H_0 if:

$$\mathcal{A} : Ch_M(S_1 S^{-1}) \leqslant \mu \tag{1.3}$$

and reject otherwise, where :

(i) $Ch_M(A)$ denotes the largest characteristic root of A;

(ii) μ is a constant depending on the preassigned level of significance α^* and the degrees of freedom u, s, and $(n-r)$;

(iii) $S_1(u \times u)$ is a symmetric and at least positive semi-definite (p.s.d.) matrix of rank t almost everywhere (a.e.), where $t = \min(u, s)$ and is given by

$$sS_1 = M'XA_1(A_1'A_1)^{-1}C_1'[C_1(A_1'A_1)^{-1}C_1']^{-1}C_1 \times (A_1'A_1)^{-1}A_1'X'M \,;$$

(iv) $S(u \times u)$ is symmetric p.d., a.e., and is given by

$$(n-r)S = M'X[I(n) - A_1(A_1'A_1)^{-1}A_1']X'M.$$

Now, X has the following frequency distribution:

$$f(X) = [(2\pi)^p |\Sigma|]^{-\frac{n}{2}} \exp[-1/2 \operatorname{tr} \Sigma^{-1}(X - E(X))(X' - E(X'))].$$

Put

$$X^*(u \times n) = M'X \,; \quad \Sigma^*(u \times u) = M'\Sigma M.$$

Hence, the frequency distribution of X^* is given by :

$$f(X^*) = [(2\pi)^u |\Sigma^*|]^{-\frac{n}{2}} \exp[-1/2 \operatorname{tr} \Sigma^{*-1}(X^* - E(X^*))(X^* - E(X^*))'] \tag{1.4}$$

Put

$$\begin{bmatrix} A_1' \\ A_2' \end{bmatrix} = \begin{bmatrix} T_1 \\ T_2 \end{bmatrix} L_1$$

where $T_1(r \times r)$ is a lower triangular matrix and $L_1 L_1' = I(r \times r)$.

Next, complete, L_1 into an orthogonal matrix $\underset{n}{\begin{bmatrix} L_1 \\ L \end{bmatrix}}\begin{matrix} r \\ (n-r) \end{matrix}$ and make the orthogonal transformation $\boldsymbol{X}^* = \boldsymbol{Y}^* \begin{bmatrix} \boldsymbol{L}_1 \\ \boldsymbol{L} \end{bmatrix}$ i.e., $\boldsymbol{Y}^* = \boldsymbol{X}^*$ $[\boldsymbol{L}_1' \boldsymbol{L}'] = [\boldsymbol{Y}_1 \ \boldsymbol{Y}]$ (say), so that $\boldsymbol{X}^* = \boldsymbol{Y}_1\boldsymbol{L}_1 + \boldsymbol{Y}\boldsymbol{L}$. Using (1.4), we find that the distribution of $(\boldsymbol{Y}_1, \boldsymbol{Y})$ is

$$(1.5) \quad f(\boldsymbol{Y}_1, \boldsymbol{Y}) = \text{const.}\exp.[-1/2 \operatorname{tr} \boldsymbol{\Sigma}^{*-1}\{(\boldsymbol{Y}_1 - (\boldsymbol{\xi}_1^* \boldsymbol{T}_1 + \boldsymbol{\xi}_2^* \boldsymbol{T}_2)) \times (\boldsymbol{Y}_1 - (\boldsymbol{\xi}_1^{*\prime} \boldsymbol{T}_1 + \boldsymbol{\xi}_2^{*\prime} \boldsymbol{T}_2))' + \boldsymbol{Y}\boldsymbol{Y}'\}],$$

where

$$\boldsymbol{\xi}^*(m \times u) = \begin{matrix} r \\ m-r \end{matrix} \begin{bmatrix} \boldsymbol{\xi}_1 \\ \boldsymbol{\xi}_2 \end{bmatrix} \boldsymbol{M}((p \times u) = \begin{bmatrix} \boldsymbol{\xi}_1^* \\ \boldsymbol{\xi}_2^* \end{bmatrix} \text{(say)}.$$

Next, put $\boldsymbol{C}_1 \boldsymbol{T}_1'^{-1} = \hat{\boldsymbol{V}}(s \times s)\boldsymbol{M}_1(s \times r)$ where $\hat{\boldsymbol{V}}$ is a lower triangular matrix and $\boldsymbol{M}_1\boldsymbol{M}_1' = \boldsymbol{I}(s \times s)$. Complete $\boldsymbol{M}_1$ into an orthogonal matrix $\begin{bmatrix} \boldsymbol{M}_1 \\ \boldsymbol{M}_2 \end{bmatrix}$ and make the orthogonal transformation

$$\boldsymbol{Y}_1(u \times r) = \boldsymbol{Y}^{**}(u \times r) \begin{bmatrix} \boldsymbol{M}_1 \\ \boldsymbol{M}_2 \end{bmatrix} \begin{matrix} s \\ r-s \end{matrix}, \text{ i.e.,}$$

$$\boldsymbol{Y}^{**} = \boldsymbol{Y}_1[\boldsymbol{M}_1' \boldsymbol{M}_2'] = [\boldsymbol{Y}_1^* \boldsymbol{Y}_2^*] \text{ (say).}$$

Then, from (1.5), the distribution of $(\boldsymbol{Y}_1^*, \boldsymbol{Y}_2^*, \boldsymbol{Y})$ is [2] :

$$f(\boldsymbol{Y}_1^*, \mathrm{Y}_2^*, \boldsymbol{Y}) = \text{const.}\exp[-1/2 \operatorname{tr} \boldsymbol{\Sigma}^{*-1}\{(\boldsymbol{Y}_1^* - \boldsymbol{\zeta}_1)(\boldsymbol{Y}_1^* - \boldsymbol{\zeta}_1)' +$$

$$(1.6) \quad (\boldsymbol{Y}_2^* - \boldsymbol{\zeta}_2)(\boldsymbol{Y}_2^* - \boldsymbol{\zeta}_2)' + \boldsymbol{Y}\boldsymbol{Y}'\}],$$

where

$$\boldsymbol{\zeta}_1(u \times s) = (\boldsymbol{\xi}_1^{*\prime} \boldsymbol{T}_1 + \boldsymbol{\xi}_2^{*\prime} \boldsymbol{T}_2)\boldsymbol{M}_1' \text{ and}$$

$$\boldsymbol{\zeta}_2(u \times \overline{r-s}) = (\boldsymbol{\xi}_1^{*\prime} \boldsymbol{T}_1 + \boldsymbol{\xi}_1^* \boldsymbol{T}_2)\boldsymbol{M}_2'.$$

Since $\boldsymbol{\Sigma}^*$ is s.p.d., then there exists the transformation

$$(1.7) \quad \boldsymbol{\Sigma}^* = \boldsymbol{B}\boldsymbol{B}'$$

where $\boldsymbol{B}(u \times u)$ is a non-singular matrix.

Let

$$(1.8) \quad \boldsymbol{Z}^*(u \times s) = \boldsymbol{B}^{-1}\boldsymbol{Y}_1^*, \ \boldsymbol{Z}^{**}(u \times \overline{r-s}) = \boldsymbol{B}^{-1}\boldsymbol{Y}_2^* \text{ and}$$

$$\boldsymbol{Z}(u \times \overline{n-r}) = \boldsymbol{B}^{-1}\boldsymbol{Y}.$$

Then, corresponding to (1.6), we get

$$(1.9) \quad f(\boldsymbol{Z}^*, \boldsymbol{Z}^{**}, \boldsymbol{Z}) = \text{const. exp.}[-1/2 \text{ tr } \{(\boldsymbol{Z}^*-\boldsymbol{\Lambda}_1)(\boldsymbol{Z}^*-\boldsymbol{\Lambda}_1)'+ +(\boldsymbol{Z}^{**}-\boldsymbol{\Lambda})(\boldsymbol{Z}^{**}-\boldsymbol{\Lambda})'+\boldsymbol{Z}\boldsymbol{Z}'\}]$$

where

$$(1.10) \quad \boldsymbol{\Lambda}_1 = \boldsymbol{B}^{-1}\boldsymbol{\zeta}_1; \ \boldsymbol{\Lambda} \equiv [\lambda_{ij}] = \boldsymbol{B}^{-1}\boldsymbol{\zeta}_2.$$

In [2], it has been shown that $\boldsymbol{\zeta}_1\boldsymbol{\zeta}_1' = \boldsymbol{\eta}'[\boldsymbol{C}_1(\boldsymbol{A}_1'\boldsymbol{A}_1)^{-1}\boldsymbol{C}_1']^{-1}\boldsymbol{\eta}$. Since $[\boldsymbol{C}_1(\boldsymbol{A}_1'\boldsymbol{A}_1)^{-1}\boldsymbol{C}_1']$ is s.p.d., it follows that $\eta(s\times u) = \boldsymbol{0}$ is equivalent to $\boldsymbol{\zeta}_1 \boldsymbol{\zeta}_1' = \boldsymbol{0}(u\times u)$ which implies that $\boldsymbol{\zeta}_1(u\times s) = \boldsymbol{0}$. Hence, the null and alternative hypotheses of (1.1) and (1.2) may be written in the form:

$$(1.11) \quad H_0 : \boldsymbol{\zeta}_1(u\times s) = \boldsymbol{0}; H_1 : \boldsymbol{\zeta}_1 \neq \boldsymbol{0}.$$

At this stage, we define the class of alternative hypothesis, referred to in the introduction, and which will be used in the course of the proof of the admissibility as:

$$(1.12) \quad H_1 : \boldsymbol{\zeta}_1 = \boldsymbol{B}\begin{bmatrix} \gamma_1 & \gamma_2 & \cdots & \gamma_s \\ \vdots & \boldsymbol{0} & & \\ 0 & & & \end{bmatrix}\begin{matrix} 1 \\ (u-1 \\ \\ \end{matrix} ; \gamma_j > 0 \quad (j = 1, 2, \dots s)$$
$$\qquad\qquad 1 \quad (s-1)$$

This implies that $\boldsymbol{\Lambda}_1 = \begin{bmatrix} \gamma_1 & \cdots & \gamma \\ : & \boldsymbol{0} & \\ 0 & & \end{bmatrix}$. Under this particular H_1, we can write (1.9) - after substituting from (1.10) - in the following form:

$$f(\boldsymbol{Z}^*,\boldsymbol{Z}^{**}, \boldsymbol{Z}) = \text{const. exp.}\Big[-1/2\Big\{ \sum_{j=1}^{s} (z_{ij}^*-\gamma_j)^2$$

$$(1.13) \quad + \sum_{i=2}^{u} \sum_{j=1}^{s} z_{ij}^{*2} + \sum_{i=1}^{u} \sum_{j=1}^{r-s} (z_{ij}^{**}-\lambda_{ij})^2 + \sum_{i=1}^{u} \sum_{j=1}^{n-r} z_{ij}^2\Big\}\Big]$$

Under H_0, $\gamma_j = 0 (j = 1, 2, \dots, s)$.

2. A MOST POWERFUL QUASI-SIMILAR TEST OF H_0 AGAINST H_1 IN THE CLASS OF NEYMAN-STRUCTURE TESTS.

Given $\boldsymbol{Z}^*$ and $\boldsymbol{Z}$, it is well known that the most powerful similar test of size α, for testing H_0 against H_1 of (1.12), is given by the rule:

Accept H_0 if

$$\mathscr{S}_0 : \sqrt{s-1+n+r}\ \cot\ \theta \leqslant t_\alpha \tag{2.1}$$

and reject otherwise, where

$$\cos\theta = \frac{\sum_{j=1}^{s} z_{1j}^{*}\gamma_j}{\left(\sum_{j=1}^{s} z_{1j}^{*2} + \sum_{j=1}^{n-r} z_{1j}^{2}\right)^{1/2}\left(\sum_{j=1}^{s}\gamma_j^2\right)^{1/2}}, \tag{2.2}$$

and t_α is the upper percentage point of the Student t-distribution with $(s-1+n-r)$ d.f. We shall denote $\mathscr{S}_0$ of (2.1) as a t-region and, as a matter of fact, $\mathscr{S}_0$ stands for such a region which we referred to in the introduction as a t-region.

We notice that $\mathscr{S}_0$ of (2.1) is invariant under the group of scale transformations $(z_{1j}^{*'} = cz_{1j}^{*}$ and $z_{1j}' = cz_{1j}$ (say); $c \neq 0)$. Also, $\mathscr{S}_0$ is similar with respect to the nuisance parameters $[\lambda_{ij}'^{s}]$ of (1.10). However, it depends on the unknown matrix $\boldsymbol{B}$ of (1.7) and cannot be determined without knowing $\boldsymbol{B}$. For this reason, it is called a quasi-similar region. The region $\mathscr{S}_0$ is used only as a tool for proving the admissibility of the proposed test.

Also, the proposed test $\mathscr{S}$ of (1.3) has the following properties [2], [3]: (i) it is similar with respect to all nuisance parameters; (ii) it is invariant under the group of scale transformations; (iii) it has size $\alpha^* > \alpha$ and (iv) it is the intersection of the class of t regions of the form (2.1) which is formed by letting the matrix $\boldsymbol{B}$ take all possible values. The last property can be proved as follows: For given values of z_{1j}^{*s} and $z_{1j}{}^{s}$, the intersection of the regions

$\mathscr{S}_0$ of (2.1) over different sets of $(\gamma_1, \gamma_2, \ldots, \gamma_j)$ is given by:

$$(2.3) \qquad t \equiv \sqrt{s-1+n-r}\,\frac{\left(\sum_{j=1}^{s} z_{1j}^{*2}\right)^{1/2}}{\left(\sum_{j=1}^{n-r} z_{1j}^{2}\right)^{1/2}} \leqslant t_\alpha$$

Now, it is easy to see that the intersection of the regions (2.3) over all values of the matrix $\boldsymbol{B}$ will give rise to the region $\mathscr{S}$ of (1.3)

3. FINAL STAGE IN THE PROOF OF ADMISSIBILITY

For testing H_0 of (1.11) against H_1 of (1.12), the region $\mathscr{S}_0$ of (2.1) is the acceptance region of the most powerful quasi-similar test of size α. Consider any test—other than $\mathscr{S}$ of (1.3)-of size $\alpha^* > \alpha$ which is similar and scale invariant. Let its acceptance region be $\mathscr{S}_1$. Since $\mathscr{S}_1$ is different from $\mathscr{S}$, which is the intersection of the class of t-regions, then there exists at least one t-region, $\mathscr{S}_0$ (say) such that $\mathscr{S}_2 = \mathscr{S}_1 \bigcap \mathscr{S}_0^*$ has a positive measure, where $\mathscr{S}_0^*$ denotes the complement of $\mathscr{S}_0$. We are going to show in the following lemma that —for some range of the alternative hypothesis H_1 of (1.12)—the probability of the second type of error of the test with acceptance region $\mathscr{S}_0$ is less than that of $\mathscr{S}_1$. This, in turn, implies that—for this particular class of alternatives—the probability of the second type of error of the proposed test $\mathscr{S}$ is less than that of the rival test $\mathscr{S}_1$ which proves the admissibility of the proposed test.

Lemma :

There exists a particular value of γ, *say* γ^* *such that*

$$(3.1) \qquad \int_{\mathscr{S}_0} f(\boldsymbol{Z}^*, \boldsymbol{Z}^{**}, \boldsymbol{Z})d\mu < \int_{\mathscr{S}_1} f(\boldsymbol{Z}^*, \boldsymbol{Z}^{**}, \boldsymbol{Z})d\mu$$

where $f(\boldsymbol{Z}^*, \boldsymbol{Z}^{**}, \boldsymbol{Z})$ *is given by* (1.13); $d\mu = d\boldsymbol{Z}\, d\boldsymbol{Z}^*\, d\boldsymbol{Z}^{**}$ *and*

$$(3.2) \qquad \gamma = \left(\sum_{j=1}^{s} \gamma_j^2\right)^{\frac{1}{2}}$$

Without loss of generality, we shall assume that $\gamma > 0$.

Proof :

Put

(3.3) $$r_i^2 = \sum_{j=1}^{s} z_{ij}^{*2} + \sum_{j=1}^{n-r} z_{ij}^2 \quad (i = 1, 2, \ldots, u)$$

Changing the variables into (1.13)—using (3.3) and (2.2)—the probability density function (1.13) can be expressed in the following form :

(3.4) $$f^*(\boldsymbol{R}, \theta, \boldsymbol{Z}^{**}) = \text{const. exp.}\Big[-1/2\Big\{r_1^2 - 2r_1\gamma\cos\theta + \gamma^2 + \sum_{i=2}^{u} r_1^2 + \sum_{i=1}^{u}\sum_{j=1}^{r-s}(z_{ij}^{**} - \lambda_{ij})^2\Big\}\Big] \times \prod_{i=1}^{u} r_i^{n^*}(\sin\theta)^{n^*-1}$$

where, $n^* = s-1+n-r$.

Since $\mathcal{D}_0$ is the acceptance test of the most powerful test and since $\mathcal{D}_2 = \mathcal{D}_1 \cap \mathcal{D}_0^*$ (where $\mathcal{D}_0^*$ is the complement of $\mathcal{D}_0$), then it follows from the Neyman-Pearson lemma that, in $\mathcal{D}_2$, f^* of (3.4) is no smaller than the expression (3.4) with $\cos\theta$ replaced by $\cos\theta_0$, where,

(3.5) $$\cos\theta_0 = \frac{t_\alpha/\sqrt{n^*}}{(1+t_\alpha^2/n^*)^{\frac{1}{2}}}$$

(t_α is the upper α-percentage point of the t-distribution and $n^* = s-1+n-r$).

Since $\mathcal{D}_1 \supset \mathcal{D}_2$, we conclude from the above argument that

$$\int_{\mathcal{D}_1} f^*(\boldsymbol{R}, \theta, \boldsymbol{Z}^{**})d\boldsymbol{\nu} \geqslant \int_{\mathcal{D}_2} f^*(\boldsymbol{R}, \theta, \boldsymbol{Z}^{**})d\boldsymbol{\nu} \geqslant \int_{\mathcal{D}_2} f_0^*(\boldsymbol{R}, \theta, \boldsymbol{Z}^{**})d\boldsymbol{\nu}$$

$$= \int_{\boldsymbol{T}} \text{const. exp. }[-1/2(\gamma^2 - 2r_1\gamma\cos\theta_0)]h(\boldsymbol{R}, \boldsymbol{Z}^{**})$$

(3.6) $$\times\Big\{\int_{\mathcal{D}_2(\boldsymbol{T})} (\sin\theta)^{n^*-1}d\theta\Big\}\, d\boldsymbol{R}\, d\boldsymbol{Z},$$

where :

(1) $f^*(\boldsymbol{R}, \theta, \boldsymbol{Z}^{**})$ is given by (3.4); $f_0^*(\boldsymbol{R}, \theta, \boldsymbol{Z}^{**})$ is the same expression (3.4) with $\cos\theta$ replaced by $\cos\theta_0$ given by (3.5) and $d\boldsymbol{\nu} = d\boldsymbol{R}d\theta\, d\boldsymbol{Z}^{**}$,

(2) $\boldsymbol{T} = [r_i^2, z_{ij}^{**},\ i = 1, 2, \ldots, u$ and $j = 1, 2, \ldots, (r-s)]$ and $\int\limits_{\boldsymbol{T}}$ stands for the integration over the sample space of $\boldsymbol{T}$, i.e., over the sample space of $\boldsymbol{R}$ and $\boldsymbol{Z}^{**}$.

(3) For each value $\boldsymbol{T}_0$ of $\boldsymbol{T}$, $\mathscr{W}_2(\boldsymbol{T}_0)$ denotes the intersection of $\mathscr{W}_2$ with the surface $\boldsymbol{T} = \boldsymbol{T}_0$.

(4) $h(\boldsymbol{R}, \boldsymbol{Z}^{**})$ is the marginal density of $(\boldsymbol{R}, \boldsymbol{Z}^{**})$.

At this stage, we notice the following :

(i) The integrand within the curly brackets of (3.6) does not depend on $\boldsymbol{T}$(i.e., does not depend on $\boldsymbol{R}$ and $\boldsymbol{Z}^{**}$) and is merely a function of θ.

(ii) The region $\mathscr{W}_0$-given by $\cos\theta \leqslant \cos\theta_0$ (where $\cos\theta_0$ is defined in (3.5))—does not depend on $\boldsymbol{T}$.

(iii) Also, the region $\mathscr{W}_1$, being invariant under the group of scale transformations must have a boundary surface of the form $F(\theta) = 0$ which is independent of $\boldsymbol{T}$.

We conclude from (ii) and (iii) above that the range of θ in $\mathscr{W}_2$ does not depend on $\boldsymbol{T}$. Hence, it follows from (i) above that the integral between curly brackets in (3.6) does not depend on $\boldsymbol{T}$ and is merely a pure constant $(1-\alpha^{**})$, say, where

$$1 > \alpha^{**} \geqslant \alpha^* > \alpha > 0$$

Hence, from (3.6), we get

$$\int\limits_{\mathscr{W}_1} f^*\, d\boldsymbol{v} \geqslant \text{const.} \int_0^\infty r_1^{n^*} \exp.[(-1/2)\{r_1^2 - 2r_1\gamma\cos\theta_0 + \gamma^2\}]dr_1$$

$$= \text{const.} \exp\left[(-1/2)\frac{n^*\gamma^2}{n^*+t_\alpha^2}\right] \int_0^\infty r_1^{n^*} \exp.\left[(-1/2)\left(r_1 - \frac{\gamma t_\alpha}{\sqrt{n^*+t_\alpha^2}}\right)^2\right] dr_1$$

$$= \beta^*(\gamma), \text{ say.} \tag{3.7}$$

(where γ is given by (3.2) and $n^* = (s-1+n-r)$)

On the other hand,

$\int\limits_{\mathscr{W}_0} f^*\, d\boldsymbol{v}$ = the probability of type II error of the t-distribution with n^* d.f. and non-centrality parameter γ.

This is given by [1]

$$\beta(\gamma) = \text{const} \int_0^\infty v^{n^*-1} \exp\left[(-1/2)(n^* v^2/t_\alpha^2)\right]$$

$$\times \left(1/\sqrt{2\pi} \int_{-\infty}^{v-\gamma} \exp\left[-1/2)y^2\right]dy\right) dv.$$

Now, we are going to show that $\lim_{\gamma\to\infty} \dfrac{\beta(\gamma)}{\beta^*(\gamma)} = 0$

As $\gamma \to \infty$, both $\beta(\gamma)$ and $\beta^*(\gamma)$ tend to zero. However using 'Hospital's rule and observing that

$$\int_0^\infty x/\delta)^n \exp\left[-1/2(x-\delta)^2\right]dx \to \sqrt{2\pi} \text{ as } \delta \to \infty,$$

we can easily show that

$$\lim_{\gamma\to\infty} \beta(\gamma)/\beta^*(\gamma) = 0.$$

This shows that there exists *a* γ^* such that for $\gamma \geqslant \gamma^*$, we have

$$\beta(\gamma) < \beta^*(\gamma).$$

Since,

$$\int_{\mathscr{D}_1} f^* \, d\mathbf{v} > \beta^*(\gamma),$$

then for all $\gamma \geqslant \gamma^*$, we have

$$\int_{\mathscr{D}_0} f^* \, d\mathbf{v} < \int_{\mathscr{D}_1} f^* d\mathbf{v}.$$

This proves the lemma and hence the admissibility property.

Acknowledgment

The second author wishes to thank the referee for his valuable comments and suggestions.

References

[1] Neyman, J. (1935), "Statistical Problems in Agricultural Experimentation," *J. Roy. Statist. Soc., Suppl.*, **2**, 107–180.

[2] Roy, S. N., "Multivariate Linear Hypothesis on Means," Unpublished notes, University of North Carolina.

[3] Roy, S. N. *Some Aspects of Multivariate Analysis*, John Wiley and Sons, New York, 1957.

[4] Roy, S. N. and Mikhail, W. F. (1960). "On the Admissibility of a Class of Tests in Normal Multivariate Analysis (Abstract). *Ann. Math. Statist.*, **31**, 536.

(Received Sept. 23, 1965. Revised Jan. 17, 1967.)

CHAPTER 33

*Rank Methods for Combination of Independent Experiments in Multivariate Analysis of Variance. Part One. Two Treatment Multiresponse Case**

PRANAB KUMAR SEN, *The University of North Carolina at Chapel Hill and University of Calcutta*

1. INTRODUCTION AND SUMMARY

In statistical experiments involving a set of multiresponse treatments and replicated under appreciably varied conditions, the assumptions underlying the usual parametric multivariate analysis of variance (MANOVA) procedures often appear to be quite stringent and dubious. The object of the present investigation is to propose and study a class of nonparametric MANOVA procedures, which not only remain valid for a broad class of parent distributions but also leave scope for the variation of the above distribution from one replicate to another in any arbitrary manner. The performance characteristics of the proposed

* Supported in part from Research Grant, GM-12868, from the National Institutes of Health, Public Health Services.

methods are also compared with that of the standard parametric methods.

Suppose, we have N subjects (plots) available for the experiment which are divided into r replicates (blocks) of sizes $N_1, \ldots, N_r$ respectively, so that $N_1+\ldots+N_r = N$. The ith replicate containing N_i subjects are further subdivided into two subgroups of sizes N_{i1} and N_{i2} respectively, and these two subgroups are treated with two different treatments A and B. Thus,

$$N_i = N_{i1}+N_{i2}, \quad i = 1, \ldots, r, \ r \geqslant 2. \tag{1.1}$$

The application of any treatment is followed by a quantitative p variate ($p \geqslant 2$) response, which is a stochastic variable $\boldsymbol{X}_{ik,\alpha}$ where

$$\boldsymbol{X}_{ik,\alpha} = (X^{(1)}_{ik,\alpha}, \ldots, X^{(p)}_{ik,\alpha}), \ \alpha = 1, \ldots, N_{ik}, \quad k = 1, 2, \quad i = 1, \ldots, r. \tag{1.2}$$

It is assumed that $\boldsymbol{X}_{ik,\alpha}$, $\alpha = 1, \ldots, N_{ik}$ are N_{ik} independent and identically distributed random variables (i.i.d.r.v.) distributed according to a continuous p-variate cumulative distribution function (c.d.f.) $F_{ik}(\boldsymbol{x})$, where it is given that

$$F_{i2}(\boldsymbol{x}) = F_i(\boldsymbol{x}+\boldsymbol{\theta}), \ F_{i1}(\boldsymbol{x}) = F_i(\boldsymbol{x}), \quad i = 1, \ldots, r; \tag{1.3}$$

$\boldsymbol{\theta}$ being a real p-vector. In the sequel, it will be assumed that $r, p \geqslant 2$.

In MANOVA, the standard parametric tools assume that F_i, in (1.3), is a p-variate normal c.d.f. for each $i = 1, \ldots, r$, and further all these r c.d.f.'s have a common dispersion matrix $\boldsymbol{\Sigma}$. In fact, these MANOVA procedures are appreciably sensitive to any departure from either of the two assumptions made above, and hence, may be regarded to have only a very limited scope of applicability. In many cases, the different batches of subjects in the different replicates may be quite heterogeneous, and may even be from appreciably different populations. Here, we shall be concerned with the following two problems, where we make no assumption regarding the specific forms of $F_1, \ldots, F_r$ in (1.3).

First, referred to (1.3) we desire to test the null hypothesis

$$H_0 : \boldsymbol{\theta} = \mathbf{0}, \tag{1.4}$$

against the set of alternatives that $\boldsymbol{\theta}$ is a non-null p-vector. Second, we may also desire to estimate $\boldsymbol{\theta}$ (by a point as well as region value), when we have the reasons to believe that $\boldsymbol{\theta}$ is non-null. For both purposes, we have proposed and studied appropriate nonparametric MANOVA procedures. The beauty of the proposed method is that it not only remains valid for any continuous p variate c.d.f., but also allows $F_1, \ldots, F_r$, in (1.3), to be arbitrarily different from each other. The findings of this paper generalize some of the single replicate nonparametric MANOVA tests of Chatterjee and Sen ([2], [3]), Sen ([13], [14]) and Puri and Sen [10] to the multireplicate case. The nonparametric estimation procedure considered here generalizes the technique of Hodges and Lehmann [9] and Sen [11] not only to the multivariate but also to the multireplicate case. This may also be regarded as a direct multivariate generalization of a similar univariate nonparametric procedure considered by the present author ([12], [16]).

2. REPLICATED NONPARAMETRIC MANOVA TESTS

We pool the N_i observations of the ith replicate into a pooled sample of size N_i. Then with respect to the jth variate values $X_{ik,\alpha}^{(j)}$, $\alpha = 1, \ldots, N_{ik}$, $k = 1, 2$, we arrange the N_i observations in order of magnitude and denote the rank of $X_{ik,\alpha}^{(j)}$, in this set, by $R_{ik,\alpha}^{(j)}$ for $\alpha = 1, \ldots, N_{ik}$, $k = 1, 2$. So that $R_{i1,1}^{(j)}, \ldots, R_{i2,N_{i2}}^{(j)}$ is a permutation of the N_i numbers $1, \ldots, N_i$. By virtue of the assumed continuity of F_i, in (1.3), the possibility of ties may be ignored, in probability. The above ranking is done separately for each $j = 1, \ldots, p$, within each of the r replicates. Thus, the observed variate values in (1.2) are mapped into r independent sets of p-tuplet rank values

$$\boldsymbol{R}_{ik,\alpha} = (R_{ik,\alpha}^{(1)}, \ldots, R_{ik,\alpha}^{(p)}), \alpha = 1, \ldots, N_{ik}, \ k = 1, 2; \quad i = 1, \ldots, r, \tag{2.1}$$

Let

(2.2) $$\boldsymbol{R}_i = (\boldsymbol{R}_{i1,1}, \ldots, \boldsymbol{R}_{i2N_{i2}}), \quad i = 1, \ldots, r;$$

(2.3) $$\boldsymbol{R} = (\boldsymbol{R}_1, \ldots, \boldsymbol{R}_r).$$

Then $\boldsymbol{R}$ is a $p \times N$ matrix which is partitioned into r submatrices of orders $p \times N_i$, $i = 1, \ldots, r$. Each of these submatrices is stochastic in nature and contains random rank p-tuplets. Thus $\boldsymbol{R}$ is a collection of r independent sets of random rank p-tuplets and will be termed the *compound collection matrix*, and $\boldsymbol{R}_1, \ldots, \boldsymbol{R}_r$ as the *component collection matrices.*

Now for each combination of $(i, j,\ i = 1, \ldots, r,\ j = 1, \ldots, p)$, and for every positive integer N, let us define a sequence of elements (which are known functions of N)

(2.4) $$\boldsymbol{E}_N^{(i,j)} = (E_{N1}^{(i,j)}, \ldots, E_{NN}^{(i,j)}), \quad j = 1, \ldots, p, \quad i = 1, \ldots, r;$$

the rp different sequences need not be identical. However, in the majority of cases, we would prefer having $\boldsymbol{E}_N^{(i,j)} = \boldsymbol{E}_N$ for all i, j. For the time being, we assume that

(2.5) $$\frac{1}{N} \sum_{\alpha=1}^{N} | E_{N\alpha}^{(i,j)} |^{2+\delta} < \infty, \text{ for some } \delta > 0, \text{ and all N},$$

and later, we shall impose some further conditions on $\boldsymbol{E}_N^{(i,j)}$. Then, let $\boldsymbol{Z}_{N_i\beta}^{(j)} = 1$, if the βth smallest observation on the jth variate values in the ith replicate is from the treatment A, and let $\boldsymbol{Z}_{N_i\beta}^{(j)} = 0$, otherwise; for $\beta = 1, \ldots, N_1, j = 1, \ldots, p, i = 1, \ldots, r$. We also denote by

(2.6) $$\boldsymbol{Z}_{N_i}^{(j)} = (Z_{N_i}^{(j)}, \ldots, Z_{N_iN_i}^{(j)}), \quad j = 1, \ldots, p, \quad i = 1, \ldots, r$$

and consider the rp random variables defined as

(2.7) $$T_{N_i}^{(j)} = (\boldsymbol{Z}_{N_i}^{(j)} \cdot \boldsymbol{E}_{N_i}^{(i,j)}) / (\boldsymbol{Z}_{N_i}^{(j)} \cdot \boldsymbol{Z}_{N_i}^{(j)})$$
$$= \frac{1}{N_{i1}} \sum_{\beta=1}^{N_i} Z_{N_i\beta}^{(j)} E_{N_i\beta}^{(j)}, \quad j = 1, \ldots, p, \quad i = 1, \ldots, r.$$

Further, let

(2.8) $$T^{(j)}_{N\cdot} = \sum_{i=1}^{r} [N_{i1}(N_i - 1)/N_i] T^{(j)}_{N_i},\ j = 1, \ldots, p$$

(2.9) $$\boldsymbol{T}_{N\cdot} = (T^{(1)}_{N\cdot}, \ldots, T^{(p)}_{N\cdot}).$$

Our proposed test is then based on $\boldsymbol{T}_{N\cdot}$. Before we present the test statistics and the test procedure, we consider the rationality of the test.

When the null hypothesis (1.4) is true, $\boldsymbol{X}_{ik,\alpha}$ $\alpha = 1, \ldots, N_{ik}$, $k = 1, 2$ are i.i.d.r.v. distributed according to the c.d.f. $F_i(\boldsymbol{x})$, defined in (1.3). Thus, by an adoptation of the rank-permutation argument of Chatterjee and Sen ([2], [3]) we may conclude that given the rank collection matrix $\boldsymbol{R}_i$, in (2.2), the conditional distribution over the N_i! permutations of the columns of $\boldsymbol{R}_i$ would be uniform under H_0 in (1.4), whatever the c.d.f. $F_i(\boldsymbol{x})$, may be. Consequently, given $\boldsymbol{R}_i$, all possible partitionings of the N_i rank p-tuplets into two subsets of sizes N_{i_1} and N_{i_2} respectively are equally likely (conditioned on the given $\boldsymbol{R}_i$) under H_0 in (1.4), and the permutational probability measure for each such partitioning is equal to $\binom{N_i}{N_{1i}}^{-1}$. We now consider the product permutational probability measure induced by the r independent sets of partitionings arising out of the r component collection matrices $\boldsymbol{R}_1, \ldots, \boldsymbol{R}_r$. Evidently, this is given by

(2.10) $$\prod_{i=1}^{r} \binom{N_i}{N_{i_1}}^{-1} = 1/N^* \text{ (say).}$$

Hence, conditioned on the component collection matrices of the compound collection matrix $\boldsymbol{R}$ to be all given, there are in all N^* possible partitionings and each such partitioning has (conditionally) a common permutational probability $1/N^*$, when H_0 in (1.4) is true. Thus, if we denote this permutational probability measure by $\mathcal{P}(\boldsymbol{R})$, and consider the permutation distribution of $\boldsymbol{T}_{N\cdot}$ (defined in (2.9),) induced by $\mathcal{P}(\boldsymbol{R})$, then any test function $\phi(\boldsymbol{T}_{N\cdot})$ based on this permutation distribution of $\boldsymbol{T}_{N\cdot}$ will have a strictly distribution-free structure when H_0 in (1.4) is true. The proposed test is thus a permutation test and is based on an extended rank-permutation argument.

To construct the test statistic S_N, we define first

$$\bar{E}_{N_i}^{(i,j)} = \frac{1}{N_i}\sum_{\beta=1}^{N_i} E_{N_i\beta}^{(i,j)}, \quad j = 1, \ldots, p, \quad i = 1, \ldots, r, \tag{2.11}$$

$$\bar{E}_N^{(j)} = \sum_{i=1}^{r} [N_{i1}(N_i - 1)/N_i]\bar{E}_{N_i}^{(i,j)}, \quad j = 1, \ldots, p; \tag{2.12}$$

$$\bar{\boldsymbol{E}}_N = (\bar{E}_N^{(1)}, \ldots, \bar{E}_N^{(p)}). \tag{2.13}$$

Also, let

$$v_{jl}^{(i)} = \frac{1}{N_i - 1}\left\{\sum_{k=1}^{2}\sum_{\alpha=1}^{N_{ik}} E_{ik,\alpha}^{(j)} E_{ik,\alpha}^{(l)} - N_i\, \bar{E}_{N_i}^{(i,j)}\bar{E}_{N_i}^{(i,l)}\right\}, \tag{2.14}$$

for $j, l = 1, \ldots, p$, $i = 1, \ldots, r$, where $E_{ik,\alpha}^{(j)}$ is the value of $E_{N_i}^{(i,j)}$ associated with the value of $\beta = R_{ik,\alpha}^{(j)}$ for $\alpha = 1, \ldots, N_{ik}$, $k = 1, 2$, $j = 1, \ldots, p$; $i = 1, \ldots, r$. Further, let

$$n_i = N_{i1}N_{i2}/N_i, \quad i = 1, \ldots, r \text{ and } n = \sum_{i=1}^{r} n_i; \tag{2.15}$$

$$v_{jl}(\boldsymbol{R}) = \sum_{i=1}^{r} n_i v_{jl}^{(i)}/n, \; j, l = 1, \ldots, p. \tag{2.16}$$

Then, it can be shown following a few simple steps that

$$E\{\boldsymbol{T}_{N\cdot} \mid \mathscr{P}(\boldsymbol{R})\} = \bar{\boldsymbol{E}}_N, \tag{2.17}$$

$$E\{\boldsymbol{T}_{N\cdot} - \bar{\boldsymbol{E}}_N)'(\boldsymbol{T}_{N\cdot} - \bar{\boldsymbol{E}}_N) \mid \mathscr{P}(\boldsymbol{R})\} = n\boldsymbol{V}(\boldsymbol{R}),$$

where

$$\boldsymbol{V}(\boldsymbol{R}) = ((v_{jl}(\boldsymbol{R})))_{j,l=1,\cdots,p}. \tag{2.18}$$

we now let $\boldsymbol{V}^{-1}(\boldsymbol{R}) = ((v^{jl}(\boldsymbol{R})))_{j,l=1,\cdots,p}$.

Following then essentially the same argument as in Puri and Sen [10], we may consider the following test-statistic

$$S_N = \frac{1}{n}(\boldsymbol{T}_{N\cdot} - \bar{\boldsymbol{E}}_N)\boldsymbol{V}^{-1}(\boldsymbol{R})(\boldsymbol{T}_{N\cdot} - \bar{\boldsymbol{E}}_N)', \tag{2.19}$$

which is a positive semidefinite quadratic form in $(\boldsymbol{T}_{N\cdot}-\overline{\boldsymbol{E}}_N)$. S_N will be small only when $(\boldsymbol{T}_{N\cdot}-\boldsymbol{E}_N)$ consists of elements of small magnitude. When H_0 in (1.4) is true, S_N will have have N^* possible (permuted) values (not necessarily all distinct), and the permutational probability measure attached to each of these points is $1/N^*$. On the other hand, if H_0 is not true i.e., $\boldsymbol{\theta} \neq 0$ (referred to (1.3), then it can be shown that by proper choice of $\boldsymbol{E}_{N_i}^{(i,j)}$, $i = 1, \ldots, r$, $j = 1, \ldots, p$, S_N can be made large, in probability. Thus, it seems reasonable to base our permutation test on S_N, using the right hand tail of the permutation c.d.f. of S_N as the appropriate critical region. Hence, we propose the following test function $\phi(S_N)$:

$$\phi(S_N) = \begin{cases} 1, & \text{, if } S_N > S_{N,\epsilon}, \\ a_{N,\epsilon}, & \text{if } S_N = S_{N,\epsilon}, \\ 0, & \text{, if } S_N < S_{N,\epsilon}, \end{cases} \tag{2.20}$$

where $S_{N,\epsilon}$ and $a_{N,\epsilon}$ are so chosen that

$$E\{\phi(S_N) \mid \mathscr{P}(\boldsymbol{R})\} = \epsilon : 0 < \epsilon < 1, \tag{2.21}$$

ϵ being the preassigned level of significance of the test. Note that (2.21) implies that $E\{\phi(S_N) \mid H_0\} = \epsilon$. So that $\phi(S_N)$ is a strictly size ϵ test. The values of $S_{N,\epsilon}$ and $a_{N,\epsilon}$ depend on the particular $\boldsymbol{R}$. In small samples, their values are to be evaluated from the exact permutation c.d.f. of S_N. The problem though appears to be deterministic, the labor involved in this numerical evaluation increases prohibitively with the increase in the sample sizes. In view of this, we present below the asymptotic permutation theory related to this problem and later, we shall see how the same can be used to simplify the large sample approach to this permutation test procedure.

Extending the ideas of Chernoff and Savage [4] (see also [5], [10]) to the multireplicate multivariate case, we write

$$E_{N_i\alpha}^{(i,j)} = J_{N_i}^{(i,j)}(\alpha/(N_i+1)), \quad 1 \leqslant \alpha \leqslant N_i, \quad j = 1, \ldots, p \quad i = 1, \ldots, r, \tag{2.22}$$

where $J_{N_i}^{(i,j)}$ need be defined only at $\alpha/(N_i+1)$, $\alpha = 1, \ldots, N_i$ and its domain of definition may be extended to the open interval (0, 1) in any conventional manner (cf. [10]). Let then

$$(2.23) \quad F_{N_i,k}^{(j)}(x) = \frac{1}{N_{ik}} \{\text{number of } X_{ik,\alpha}^{(j)} \leqslant x\}, \quad j = 1, \ldots, p,$$
$$i = 1, \ldots, r;\ k = 1, 2;$$

$$(2.24) \quad H_{N_i}^{(j)}(x) = \frac{1}{N_i} \{N_{i1}F_{N_i,1}^{(j)}(x) + N_{i2}F_{N_i,2}^{(j)}(x)\}, \quad j = 1, \ldots, p,$$
$$i = 1, \ldots, r.$$

The marginal c.d.f. of $X_{ik,\alpha}^{(j)}$ is denoted by $F_{i,k}^{(j)}(x)$ for $k = 1, 2$, and let

$$(2.25) \qquad H_i^{(j)}(x) = \frac{1}{N_i} \{N_{i1}\, F_{i,1}^{(j)}(x) + N_{i2}F_{1,2}^{(j)}(x)\},$$
$$i = 1, \ldots, r, \quad j = 1, \ldots, p.$$

Similarly, let

$$(2.26) \quad F_{N_i,k}^{(j,l)}(x, y) = \frac{1}{N_{ik}} \{\text{number of } (X_{ik,\alpha}^{(j)},\ X_{ik,\alpha}^{(l)}) \leqslant (x, y)\},$$
$$k = 1, 2,;$$

$$(2.27) \qquad H_{N_i}^{(j,l)}(x, y) = \frac{1}{N_i} \{N_{i1}F_{N_i,1}^{(i,l)}(x, y) + N_{i2}F_{N_i,2}^{(i,l)}(x, y),\},$$
$$j \neq l = 1, \ldots, p, \quad i = 1, \ldots, r.$$

Also, the joint (marginal) c.d.f. of $X_{ik,\alpha}^{(i)}$, $X_{ik,\alpha}^{(l)}$ is denoted by $F_{i,k}^{(j,l)}(x, y)$ for $k = 1, 2$, and we let

$$(2.28) \qquad H_i^{(j,l)}(x, y) = \frac{1}{N_i} \{N_{i1}F_{i,1}^{(j,l)}(x, y) + N_{i2}F_{1,2}^{(j,l)}(x, y)\},$$
$$j \neq l = 1, \ldots, p, \quad i = 1, \ldots, r.$$

It may be noted that if we define

$$(2.29) \qquad \lambda_{N_i}^{(i)} = N_{i1}/N_i, \qquad i = 1, \ldots, r,$$

then $H_i^{(j)}$ and $H_i^{(j,l)}$ both depend explicitly on $\lambda_{N_i}^{(i)}$, $i = 1, \ldots, r$. We now impose the following conditions :

(C.1) $\lambda_{N_1}^{(1)}, \ldots, \lambda_{N_r}^{(r)}$ are regarded as fixed and are all bounded away from zero and one.

(C.2) $J^{(i,j)}(H) = \lim_{N_i \to \infty} J_{N_i}^{(i,j)}(H)$ exists for all $0 < H < 1$ and is not a constant. For our purpose, we impose further that $J^{(i,j)}(H)$ is monotonic in H, for all $i = 1, \ldots, r$, $j = 1, \ldots, p$.

(C.3) If $I_{N_i}^{(j)} = \{x : 0 < H_{N_i}^{(j)}(x) < 1\}$, then

$$\int_{I_{N_i}^{(j)}} \left[J_{N_i}^{(i,j)}\left(\frac{N_i}{N_i+1} H_{N_i}^{(j)}\right) - J^{(i,j)}\left(\frac{N}{N_i+1} H_{N_i}^{(j)}\right)\right] dF_{N_j,1}^{(j)}(x)$$

$$= o_p(N_i^{-\frac{1}{2}}), \text{ for all } j = 1, \ldots, p \text{ and } i = 1, \ldots, r.$$

(C.4) $$\left| \frac{d^l}{dH^l} J^{(i,j)}(H) \right| \leqslant K[H(1-H)]^{-l-\frac{1}{2}+\delta},$$

for $l = 0, 1$ and some $\delta > 0$.

(C.5) $$\int_{I_{Ni}^{(j)} \times I_{N_i}^{(l)}} \int \left(J_{N_i}^{(i,j)}\left[\frac{N_i}{N_i+1} H_{N_i}^{(j)}\right) J_{N_i}^{(i,l)}\left(\frac{N_i}{N_i+1} H_{N_i}^{(l)}\right) -$$

$$J^{(i,j)}\left(\frac{N_i}{N_i+1} H_{N_i}^{(j)}\right) J^{(i,l)}\left(\frac{N_i}{N_i+1} H_{N_i}^{(l)}\right)\right] dH_{N_i}^{(j,l)}(x, y) = o_p(1),$$

for all $j, l = 1, \ldots, p$, $i = 1, \ldots, r$.

(C.6) Let us define

$$\nu_{jl}^{(i)} = \int_{-\infty}^{\infty} \int_{-\infty}^{\infty} [J^{(i,j)}(H_i^{(j)}) J^{(i,l)}(H_i^{(l)})] dH_i^{(j,l)}(x, y) \tag{2.30}$$

$$-\left[\int_{-\infty}^{\infty} J^{(i,j)}(H_i^{(j)}) dH_i^{(j)}(x)\right]\left[\int_{-\infty}^{\infty} J^{(i,l)}(H_i^{(l)}) dH_i^{(l)}(y)\right],$$

for $j, l = 1, \ldots, p$, and

$$\boldsymbol{\nu}^{(i)} = ((\nu_{jl}^{(i)}))_{j,l=1,\ldots,p} \quad i = 1, \ldots, r. \tag{2.31}$$

Then $\boldsymbol{\nu}^{(i)}$ is positive definite for all $i = 1, \ldots, r$. We shall see later on that conditions (C.1) through (C.6) are less restrictive than the ones required in the parametric case. Conditions (C.1) through (C.4) are the same as in the Chernoff-Savage theorem ([4], also [5]), while conditions (C.5) is required to study the convergence of the matrix $\boldsymbol{V}(\boldsymbol{R})$ (defined in (2.18)), and the condition (C.6) is required to assume that $\boldsymbol{V}(\boldsymbol{R})$ has a positive definite limit, in probability. Let us define now

$$\boldsymbol{\nu} = \sum_{i=1}^{r} n\, \boldsymbol{\nu}^{(i)}/n \tag{2.32}$$

where $n_1, \ldots, n_r$ and n are defined in (2.15). It then follows from (C.6), (2.31) and (2.32) that $\boldsymbol{\nu}$ is also positive definite.

Theorem 2.1. *For arbitrary continuous c.d.f.'s* $F_1, \ldots, F_r$ *and any real and finite* $\boldsymbol{\theta}$ *(in* (1.3),) *under the conditions* (C.1) *through* (C.5), $\boldsymbol{V}(\boldsymbol{R}) \xrightarrow{P} \boldsymbol{\nu}$.

Outline of proof. If follows from Theorem 4.2 of Puri and Sen [10] (which essentially relates to the unireplicate case) that under the stated regularity conditions

$$v_{jl}^{(i)} \xrightarrow{P} \nu_{jl}^{(i)} \text{ for all } j, l = 1, \ldots, p; \quad i = 1, \ldots, r. \tag{2.33}$$

Using then (2.16), (2.30) and (2.32), we get after some simple algebraic manipulations that

$$v_{jl}^{(i)} \xrightarrow{P} \nu_{jl} \text{ for all } j, l = 1, \ldots, p.$$

Hence, the theorem.

Theorem 2.2. *Under the conditions* (C.1), $i = 1, 2, 3, 4$ *and* $\boldsymbol{V}(\boldsymbol{R})$ *being positive definite, the permutation distribution of* S_N *(defined in* (2.19),) *asymptotically, reduces to a chi-square distribution with* p *degrees of freedom.*

Outline of proof. It follows from Theorem 5.1 of Puri and Sen [10] that under the stated regularity conditions, the joint permutation distribution of $N_i^{\frac{1}{2}}[T_{N_i}^{(j)} - \bar{E}_{N_i}^{(i,j)}]$, $j = 1, \ldots, p$ asymptotically reduces to a p-variate normal distribution. Now, the sets $N_i^{\frac{1}{2}}[T_{N_i}^{(j)} - \bar{E}_{N_i}^{(i,j)}]$, $j = 1, \ldots, p$ $(i = 1, \ldots, r)$ being all stochastically independent, it follows from (2.8) that the joint permutation distribution of $N^{-\frac{1}{2}}[\boldsymbol{T}_{N\cdot}-\boldsymbol{E}_N]$ can be obtained by convolution of r independent and asymptotically multinormal distributions. Hence, it can be easily shown that $n^{-\frac{1}{2}}\{\boldsymbol{T}_{N\cdot}-\boldsymbol{E}_N\}$ has a permutation distribution, which is asymptotically multinormal if $\boldsymbol{V}(\boldsymbol{R})$ is positive definite. Again, if $\boldsymbol{V}(\boldsymbol{R})$ is positive definite, the above permutation distribution is essentially non-singular and hence, applying a well-known result on the asymptotic distribution of quadratic forms associated with asymptotic normal distributions (cf. Sverdrup [17]), we arrive at the desired result that S_N has asymptotically a chi-square (permutation) distribution with degrees of freedom.

Hence, the theorem.

By virtue of Theorem 2.1 and condition (C. 6), we get that $\boldsymbol{V}(\boldsymbol{R}) \xrightarrow{P} \boldsymbol{\nu}$, which is positive definite. Hence, from the preceeding theorem, we arrive at the following.

Theorem 2.3. *Under conditions* (C.1), $i = 1, \ldots, 6$, *the permutation distribution of* S_N *reduces asymptotically in probability, to a* χ^2 *distribution with p.d.f.*

Thus, it follows from Theorem 2.3 that the test $\phi(S_N)$ in (2.20) asymptotically, in probability, reduces to

$$\text{(2.34)} \qquad \begin{aligned} \phi(S_N) &= 1, \text{ if } S_N \geqslant \chi^2_{p,\epsilon} \\ &= 0, \text{ if } S_N < \chi^2_{p,\epsilon}, \end{aligned}$$

where $\chi^2_{p.\epsilon}$, is the $100(1-\epsilon)\%$ point of a χ^2 distribution with p.d.f. Using condition (C.2), it can be shown with little difficulty that $\frac{1}{N}\{\boldsymbol{T}_{N\cdot}-\overline{\boldsymbol{E}}_N\}$ converges to a non-null vector if $\boldsymbol{\theta}$ (in (1.3),) is non-null. Consequently, using (2.19) and Theorem 2.3, it is easy to show that the test $\phi(S_N)$ in (2.20) or (2.34) is consistent against any $\boldsymbol{\theta} \neq \boldsymbol{0}$.

Now, as in most of the non-parametric tests in MANOVA, the study of the exact power of $\phi(S_N)$ seems to be considerably difficult and the same depends heavily on the parent c.d.f.'s $F_1, \ldots, F_r$ and the particular $\boldsymbol{\theta}$. No simple expression can be attached to such an exact power-function. However, we are in a position to study the asymptotic power properties of the test $\phi(S_N)$, and this we do for a sequence of alternatives $\{\boldsymbol{\theta}_N\}$, so chosen that $E\{\phi(S_N)\,|\,\theta_N\}$ tends to a finite limit $\gamma: \epsilon < \gamma < 1$ as $N \to \infty$. We now define n as in (2.15), and note that n is an increasing function of N_{ik}, $k = 1, 2$, $i = 1, \ldots, r$, and it tends to ∞ as $N \to \infty$, subject to (C.1). Then, we specify the sequence of alternative hypotheses $\{H_N\}$ by

$$H_N : \boldsymbol{\theta}_N = n^{-\frac{1}{2}} \,.\, \delta, \tag{2.35}$$

where $\boldsymbol{\delta}$ is any real and finite p-vector. We also define

$$a_{ij} = \int_{-\infty}^{\infty} \frac{d}{dx} J^{(i,j)}(H_i^{(j)}(x))dH_i^{(j)}(x), \quad i = 1, \ldots, r, \ j = 1, \ldots, p \tag{2.36}$$

$$\eta_{j,N} = \left(\sum_{i=1}^{r} n_i a_{ij}/n \right) \delta_j, \ j = 1, \ldots, p;$$

$$\eta_N = (\eta_{1,N}, \ldots, \eta_{p,N}) \tag{2.37}$$

At this stage, we require to put some restrictions on $n_1, \ldots, n_r$, in (2.15). We assume that as $n \to \infty$, $n_i/n \to p_i$, $i = 1, \ldots, r$ where $p_1, \ldots, p_r$ are all bounded away from zero and one and they add up to unity. Under this condition η_N, in (2.37), tends to a limit-vector, which is denoted by $\boldsymbol{\eta}$.

Theorem 2.4. *Under conditions* (C.1), $i = 1, \ldots, 6$ *and the one on* n_i*'s stated above* S_N *has asymptotically, under* $\{H_N\}$, *a non central* χ^2 *distribution with p.d.f. and the noncentrality parameter is*

$$\Delta_S = \boldsymbol{\eta}\boldsymbol{\nu}^{-1}\boldsymbol{\eta}'.$$

Proof: For arbitrary $F_1, \ldots, F_r$ and $\boldsymbol{\theta}$ in (1.3) (not necessarily the sequence $\{\theta_N\}$), the joint asymptotic normality of $[N_i^{\frac{1}{2}}\{T_{N_i}^{(j)} - \bar{E}_{N_i}^{(i,j)}\}, \ j = 1, \ldots, p]$ follows readily (under conditions (C.1), $i = 1, \ldots, 4$) from Theorem 5.1 of Puri and Sen [10]. Thus, again by simple convolution, we get that for arbitrary $F_1, \ldots, F_r$ and $\boldsymbol{\theta}$, $N^{-\frac{1}{2}}[\boldsymbol{T}_{N\cdot}-\bar{\boldsymbol{E}}_N]$ has asymptotically a multinormal distribution, under conditions (C.i), $i = 1, \ldots, 4$. Now, under $\{H_N\}$ in (2.35), it follows by more or less routine computation that

$$(2.38) \qquad E\{n^{-\frac{1}{2}}[\boldsymbol{T}_{N\cdot}-\boldsymbol{E}_N]/H_N\} \to \boldsymbol{\eta} \text{ as } N \to \infty,$$

and further it is also easily shown that

$$(2.39) \quad E\{n^{-1}[\boldsymbol{T}_{N\cdot}-\boldsymbol{E}_N]'[\boldsymbol{T}_{N\cdot}-\bar{\boldsymbol{E}}_N]/H_N\} \to \boldsymbol{\nu} \text{ as } N \to \infty,$$

where $\boldsymbol{\nu}$ is defined in (2.32). (Note that, in this case, in (2.30) and (2.36), $H_i^{(j)}$, $H_i^{(j,l)}$ are to be replaced by $F_i^{(j)}$ and $F_i^{(j,l)}$ respectively, for all $j, l = 1, \ldots, p$ and $i = 1, \ldots, r$). Consequently, by a well-known limit theorem, we get that under $\{H_N\}$

$$(2.40) \qquad S_N^* = \frac{1}{n}\,[(\boldsymbol{T}_{N\cdot}-\bar{\boldsymbol{E}}_N)\nu^{-1}(\boldsymbol{T}_{N\cdot}-\boldsymbol{E}_N)']$$

has asymptotically a noncentral χ^2 distribution with p.d.f. and the noncentrality parameter Δ_S, defined in the theorem. Finally, using Theorem 2.1 it is easy to show that under $\{H_N\}$ in (2.35), $\boldsymbol{V}(\boldsymbol{R}) \xrightarrow{P} \boldsymbol{\nu}$, where $\boldsymbol{V}(\boldsymbol{R})$ is defined in (2.18). Consequently, from (2.19), (2.35) and 2.40) we get after a few simple steps that under $\{\boldsymbol{H}_N\}$, $\boldsymbol{S}_N \overset{P}{\approx} S_N^*$.

Hence, the theorem.

Let us now consider the standard parametric MANOVA test which is based upon the assumption that the c.d.f.'s $F_1, \ldots, F_r$ (in (1.3),) are all p-variate normal with a common covariance matrix $\boldsymbol{\Sigma}$. The test-statistic (say R_N) is essentially a likelihood ratio test criterion (c.f. Wilks [18, p 561]) and it is easily seen that under H_0, in (1.4), R_N has asymptotically a χ^2 distribution with p d.f., while under $\{H_N\}$, in (2.35), R_N has asymptotically a noncentral χ^2 distribution with p d.f. and the noncentrality parameter

$$(2.41) \qquad \Delta_R = \boldsymbol{\delta}\boldsymbol{\Sigma}^{-1}\boldsymbol{\delta}'.$$

Thus the asymptotic efficiency (in the Pitman-sense) of the permutation MANOVA test based on S_N with respect to the standard parametric MANOVA test based on R_N comes out as

$$(2.42) \qquad e^{\{H_N\}}_{\{S_N\},\{R_N\}} = \boldsymbol{\eta}\boldsymbol{\nu}^{-1}\boldsymbol{\eta}'/\boldsymbol{\delta}\boldsymbol{\Sigma}^{-1}\,\boldsymbol{\delta}' = e(\boldsymbol{\delta}, \boldsymbol{J}, \boldsymbol{F}),$$

where $\boldsymbol{F} = (F_1, \ldots, F_r)$ and $\boldsymbol{J} = (J^{(1,1)}, \ldots, J^{(1,p)}, \ldots, J^{(r,1)}, \ldots, J^{(r,p)})$. Further, in the parametric case, if $F_1, \ldots, F_r$ are normal and have covariance matrices $\boldsymbol{\Sigma}_1, \ldots, \boldsymbol{\Sigma}_r$ respectively, which are not all identical, the statistic R_N has no simple distribution for small samples. However, for large samples, it can be shown that under $\{H_N\}$, in (2.35), R has asymptotically a noncentral χ^2 distribution with p d.f. and the noncentrality parameter

$$(2.43) \qquad \Delta_R^* = \boldsymbol{\delta}\bar{\boldsymbol{\Sigma}}^{-1}\boldsymbol{\delta}', \ \boldsymbol{\Gamma} = \sum_{i=1}^{r} n_i\Sigma_i/n;$$

(consequently, under H_0, R_N has asymptotically a χ^2 distribution with p d.f.). In this case, the asymptotic efficiency in (2.42) will have to be adjusted only by replacing $\boldsymbol{\Sigma}$ by $\bar{\boldsymbol{\Sigma}}$.

It may be noted in this connection that a second type of nonparametric MANOVA tests may also be constructed for the above purpose. This type of tests are somewhat restrictive in the sense that it allows $F_1, \ldots, F_r$ to be continuous p-variate c.d.f's but assumes that $F_1, \ldots, F_r$ have the same functional form apart from possible variation of the location vectors only. In a sense, it is thus essentially the parametric MANOVA test with the assumption of normality replaced by a broad family of continuous c.d.f's, but the assumption of identity of the covariance matrices being implicit. This type of tests is based upon rankings after alignment and may be regarded as a direct multivariate generalization of a similar class of univariate rank-tests considered by Hodges and Lehmann [8]. The study of such tests in the multivariate case poses some further problems, connected with rank-order tests, and because of the somewhat different type of approach, the solutions in this case and their lengthy deductions will be considered in a separate issue.

3. ESTIMATION OF $\boldsymbol{\theta}$ USING RANK-ORDER TESTS.

We shall now consider the problem of estimating $\boldsymbol{\theta}$ in (1.3), without assuming the functional forms of $F_1, \ldots, F_r$ or their covariance-matrices to be all identical. For this, we write

$$(3.1) \quad \boldsymbol{X}_{ik}^{(j)} = \left(X_{ik,1}^{(j)}, \ldots, X_{ik,N_{ik}}^{(j)}\right), \; k=1, 2, j = 1, \ldots, p, i = 1, \ldots r;$$

$$(3.2) \quad \boldsymbol{X}_{ik} = (\boldsymbol{X}_{ik}^{(1)}, \ldots, \boldsymbol{X}_{ik}^{(p)}), \; k = 1, 2, \; i = 1, \ldots, r;$$

$$(3.3) \quad \boldsymbol{X}_{.k}^{(j)} = (\boldsymbol{X}_{1k}^{(j)}, \ldots, \boldsymbol{X}_{rk}^{(j)}), \quad k = 1, 2, \quad j = 1, \ldots, p;$$

$$(3.4) \quad \boldsymbol{X}_{.k} = (\boldsymbol{X}_{.k}^{(1)}, \ldots, \boldsymbol{X}_{.k}^{(p)}), \quad k = 1, 2.$$

Let $\boldsymbol{I}_m$ be an m-vector with unit elements, and we denote

$$(3.5) \quad (x_1+a, \ldots, x_m+a) = \boldsymbol{X}_m+a\boldsymbol{I}_m$$

We then consider a rank-statistic $h(\boldsymbol{X}_m, \boldsymbol{Y}_n)$ which statisfies the following two conditions:

(a) $h(\boldsymbol{X}_m+a\boldsymbol{I}_m, \boldsymbol{Y}_n)$ is increasing in a,

(b) If the elements of $\boldsymbol{X}_m$ and $\boldsymbol{Y}_n$ are i.i.d.r.v., then $h(\boldsymbol{X}_m, \boldsymbol{Y}_n)$ has a strictly distribution-free structure, and the c.d.f of $h(\boldsymbol{X}_m, \boldsymbol{Y}_n)$ in this case is denoted by $G_0(h/m, n)$.

Let $\mu_{m,n}$ be any convenient measure of location of G_0, and as G_0 is distribution-free $\mu_{m,n}$ is a known function of (m, n). For the definition of $\mu_{m,n}$, we may adopt the convention in ([9], [16]). Let us then define

$$(3.6) \quad h_j^*(\boldsymbol{X}_{.1}^{(j)}+a_j\boldsymbol{I}_{N.1}, \boldsymbol{X}_{.2}^{(j)}) = \frac{1}{n} \sum_{i=1}^{r} n_i h_j(\boldsymbol{X}_{i1}^{(j)}+a_j\boldsymbol{I}_{N_{i1}}, \boldsymbol{X}_{i2}^{(j)}),$$

for $j = 1, \ldots, p$, where $n_1, \ldots, n_r$ and n are defined in (2.15). The appropriate location of the c.d.f. of $h_j^*(\boldsymbol{X}_{.1}^{(j)}+\theta_j\boldsymbol{I}_{N.1}, \boldsymbol{X}_{.2}^{(j)})$ can be easily derived from $\mu_{N_{i1}, N_{i2}}^{(j)}, N_2, \; i = 1, \ldots, r$ and is denoted by $\mu_N^{(j)}, j = 1, \ldots, p$. So that $\boldsymbol{\mu}_N = (\mu_N^{(1)}, \ldots, \mu_N^{(p)})$ is a known p-vector. Let us again define

$$(3.7) \quad \hat{\theta}_{j.1} = \text{Inf}\,\{\theta : h_j^*(\boldsymbol{X}_{.1}^{(j)}+\theta\boldsymbol{I}_{N.1}, \boldsymbol{X}_{.2}^{(j)}) > \mu_N^{(j)}\},$$

$$(3.8) \quad \hat{\theta}_{j.2} = \text{Sup}\,\{\theta : h_j^*(\boldsymbol{X}_{.1}^{(j)}+\theta\boldsymbol{I}_{N.1}, \boldsymbol{X}_{.2}^{(j)}) < \mu_N^{(j)}\}, \; j = 1, \ldots, p.$$

Then our proposed estimate of $\boldsymbol{\theta}$ in (1.3) is

$$(3.9) \quad \hat{\boldsymbol{\theta}} = (\hat{\theta}_1, \ldots, \hat{\theta}_p); \; \hat{\theta}_j = (\hat{\theta}_{j.1} + \hat{\theta}_{j.2})/2, \quad j = 1, \ldots, p.$$

The proposed method is a generalization of a similar univariate method considered by Hodges and Lehmann [9] and Sen [11] not only to the multivariate but also to the multireplicate case. Incidently, this also generalizes Bickel's [1] results not only to the replicated two sample case but also to a more general class of test statistics. Finally, in the univariate replicated case, Sen ([12], [16]) has considered a similar method and the present one is a direct generalization of the same to the multivariate case.

Following then essentially the same technique as in [9], [16]) it is easily shown that if $F_1, \ldots, F_r$ are all continuous (absolutely continuous) so also is the estimate $\hat{\boldsymbol{\theta}}$. Further $\hat{\boldsymbol{\theta}}$ is translation invariant, i.e.,

$$(3.10) \qquad \hat{\boldsymbol{\theta}}(\boldsymbol{X}_{.1} + \boldsymbol{a}\boldsymbol{I}_{N.1}, \boldsymbol{X}_{.2}) = \hat{\boldsymbol{\theta}}(\boldsymbol{X}_{.1}, \boldsymbol{X}_{.2}) + \boldsymbol{a},$$

for any real p-vector $\boldsymbol{a}$. Finally, if either $N_{i1} = N_{i2}$ for all $i = 1, \ldots, r$ or $F_1, \ldots, F_r$ are symmetric and if $\{h_j(\boldsymbol{X}_{i1}^{(j)}, \boldsymbol{X}_{i2}^{(j)}) + h_j(\boldsymbol{X}_{i2}^{(j)}, \boldsymbol{X}_{i1}^{(j)}\}$ = constant (which may depend on j), for all $j = 1, \ldots, p$, then the distribution of $(\hat{\boldsymbol{\theta}} - \boldsymbol{\theta})$ is symmetric about $\mathbf{0}$.

We shall now consider the asymptotic properties of the estimate $\boldsymbol{\theta}$. We then define $n_1, \ldots, n_r$ and n as in (2.15), and let

$$(3.11) \qquad n_i/n = p_i : 0 < p_i < 1 \text{ for all } i = 1, \ldots, r.$$

Further, we impose an asymptotic condition *on* $h(\;)$; namely, we assume that under (1.3), the joint distribution of

$$(3.12) \quad [n_i^{\frac{1}{2}}\{h_j(\boldsymbol{X}_{i1}^{(j)} + (\theta_j + n_i^{-\frac{1}{2}}a_j)\boldsymbol{I}_{N_{i1}}, \boldsymbol{X}_{i2}^{(j)}) - \mu_{N_{i1}, N_{i2}}^{(j)}\}, \; j = 1, \ldots, p]$$

is asymptotically p-variate normal with a mean vector $(a_1 B_1^{(i)}(F_i), \ldots, a_p B_p^{(i)}(F_i))$ and a dispersion matrix $\boldsymbol{\nu}^{(i)}$, $i = 1, \ldots, r$. Let us then define

$$(3.13) \quad \bar{B}_j(F) = \sum_{i=1}^{r} p_i B_j^{(i)}(F_i), \; j = 1, \ldots, p, \; F = (F_1, \ldots, F_r);$$

$$(3.14) \qquad \boldsymbol{\nu} = \sum_{i=1}^{r} p_i \boldsymbol{\nu}^{(i)}, \ \boldsymbol{\tau} = ((\tau_{jl})),$$

$$\tau_{jl} = \nu_{jl} / \bar{B}_j(E).\bar{B}_l(\boldsymbol{F}), \ l, j = 1, \ldots, p.$$

Then essentially by an adaptation of the same technique as in ([9], [16]) with more or less straight forward generalization to the multivariate case, we arrive at the following theorem, where $\hat{\theta}$ in (3.9) is replaced by a sequence $\{\hat{\boldsymbol{\theta}}_n\}$ for the sequence of n.

Theorem 3.1. *Under the conditions stated above,* $n^{\frac{1}{2}}(\hat{\boldsymbol{\theta}}_n - \boldsymbol{\theta})$ *has asymptotically a multinormal distribution with a null mean vector and a dispersion matrix* $\boldsymbol{\tau}$, *defined in* (3.14).

It may be noted here that if we now work with the class of rank-order tests $T_{N_i}^{(j)}$, $i = 1, \ldots, r$, $j = 1, \ldots, p$ and assume the conditions (C.i), $i = 1, 2, 3, 4$ of section 2 to hold, then the conditions (a) and (b) (of this section) imposed on $h(\ \ )$ are also satisfied. Further, in this case, the quantity $B_j^{(1)}(F_i)$ reduces to a_{ij}, defined in (2.36). Thus, the same class of rank-order tests may also be used in estimation problem.

In the parametric case, the conventional estimate of θ_j is $t_j = \sum_{i=1}^{r} p_i \bar{Z}_i^{(j)}$, $j = 1, \ldots, p$, where $\bar{Z}_i^{(j)}$ is the observed difference of means of $X_{i1,\alpha}^{(j)}$ and $X_{i2,\alpha}^{(j)}$, $j = 1, \ldots, p$, $i = 1, \ldots, r$. If $F_1, \ldots, F_r$ have a common covariance matrix Σ, it is easily seen that the covariance matrix of $n^{\frac{1}{2}}(\boldsymbol{t} - \boldsymbol{\theta})$ is also $\boldsymbol{\Sigma}$, (where $\boldsymbol{t} = (t_1, \ldots, t_p)$). If $F_1, \ldots, F_r$ have covariance matrices $\boldsymbol{\Sigma}_1, \ldots, \boldsymbol{\Sigma}_r$, then if we define $\bar{\boldsymbol{\Sigma}}$ as in (2.43), the covariance matrix of $n^{\frac{1}{2}}(\boldsymbol{t} - \boldsymbol{\theta})$ is $\bar{\boldsymbol{\Sigma}}$. The asymptotic multi-normality of this vector estimate follows readily from the Central Limit Theorems. Thus, if we employ the generalized variance as a measure of efficiency, the asymptotic efficiency of the estimate (3.9) with respect to the parametric estimate $\boldsymbol{t}$ comes out as

$$(3.15) \qquad |\bar{\boldsymbol{\Sigma}} \,.\, \boldsymbol{\tau}^{-1}|^{1/p} = \{|\bar{\boldsymbol{\Sigma}}| / |\boldsymbol{\tau}|\}^{1/p} = e(\boldsymbol{\tau}, \bar{\boldsymbol{\Sigma}}), \ \text{(say)}.$$

We can easily imagine the closeness of (2.42) and (3.15), and later we shall study these further.

Before that we consider the following problem of confidence region of $\boldsymbol{\theta}$ based on rank-order tests. In the univariate case, it has been shown by the present author (cf. [11], [16]) that distribution-free confidence interval for θ_j can be obtained from the distribution of $h_j^*(\boldsymbol{X}_{.1}^{(j)}+\theta_j\boldsymbol{I}_{N.1}, \boldsymbol{X}_{.2}^{(j)})$, for any $j = 1, \ldots, p$. However, there are certain difficulties associated with the simultaneous confidence region for $\boldsymbol{\theta}$. This is due to the fact that marginally each $h_j^*(\boldsymbol{X}_{.1}^{(j)}+\theta_j\,\boldsymbol{I}_{N.1}, \boldsymbol{X}_{.2}^{(j)})$ has a nonparametric distribution for $j = 1, \ldots, p$. But, jointly these p statistics has a distribution-free structure only under permutation model, and the permutation covariance-matrix in this case depends on the unknown $\boldsymbol{\theta}$ and the given $\boldsymbol{X}_{.1}$, $\boldsymbol{X}_{.2}$. Thus, the permutation distribution depends on the unknown $\boldsymbol{\theta}$ in a somewhat involved manner, and it seems to be fairly difficult to suggest any procedure for very small samples. The procedure sketched below remains valid for moderately large samples, and may therefore be regarded as an asymptotic procedure.

For the confidence region problem, we assume that the rank-order tests of Section 2 are used and borrow therefore the same notations. We can estimate $\nu_{jl}^{(i)}$ from the ith replicate in the following manner. Since, under (1.3), F_{i1} and F_{i2} have the same functional form and they differ only by a location vector, if we adopt the ranking procedure in Section 2 for each of the two samples (separately) within the ith replicate, and for each sample we use an estimate essentially similar to the one in (2.14), then by combining these two estimates within the ith replicate (with weights equal to $N_{i1}-1$ and $N_{i2}-1$ respectively), we get an estimate of $\nu_{jl}^{(i)}$, for $j,\ l = 1, \ldots, p,\ i = 1, \ldots, r$. It can be readily shown using Theorem 5.2 of Puri and Sen [10] that these are all consistent estimates. Once these are obtained, we estimate ν_{jl} in (2.32), by the same linear function (i.e. (2.32)) in the estimates.

We denote these $\hat{\nu}_{jl}$, $j, l = 1, \ldots, p$, and the matrix by $\hat{\boldsymbol{\nu}}$. Consequently, we replace (2.19) by

$$H_N(\boldsymbol{\theta}) = \frac{1}{n}\,\{(\boldsymbol{h}^*(\boldsymbol{\theta})-\boldsymbol{\mu}_N)\hat{\boldsymbol{\nu}}^{-1}(\boldsymbol{h}^*(\boldsymbol{\theta})-\boldsymbol{\mu}_N)'\}, \tag{3.16}$$

where

(3.17) $\boldsymbol{h}^*(\boldsymbol{\theta}) = (h_1^*(\boldsymbol{X}_{.1}^{(1)}+\theta_1 \boldsymbol{I}_{N.1}, \boldsymbol{X}_{.2}^{(1)}), \ldots, h_p^*(\boldsymbol{X}_{.1}^{(p)}+\theta\, I_{p.1}, \boldsymbol{X}_{.2}^{(p)}))$

Then, proceeding precisely on the same line as in Theorem 2.2, it can be shown that under (1.3) and asymptotically

$$P\{H_N(\boldsymbol{\theta}) \leqslant \chi^2_{p,\epsilon} \,|\, \boldsymbol{\theta}\} = 1-\epsilon,\ 0 < \epsilon < 10 \tag{3.18}$$

where $\chi^2_{p,\epsilon}$ is defined in (2.34). We now select $1-\epsilon$ as the desired confidence coefficient. Now $H_N(\boldsymbol{\theta}) \leqslant \chi^2_{p,\epsilon}$ describes an ellipsoid in $\boldsymbol{h}^*(\boldsymbol{\theta})$ with origin $\boldsymbol{\mu}_N$. For any point $\boldsymbol{a}$ on the boundary of this ellipsoid, we have

$$\boldsymbol{h}^*(\boldsymbol{\theta}) = \boldsymbol{a}. \tag{3.19}$$

We then solve the set of p equations in p unknowns in (3.19), and denote this solution as $\hat{\boldsymbol{\theta}}(\boldsymbol{a})$ where we adopt the following convention to achieve uniqueness of $\hat{\boldsymbol{\theta}}(\boldsymbol{a})$ for any $\boldsymbol{a}$.

$$\hat{\boldsymbol{\theta}}(\boldsymbol{a}) = \{\boldsymbol{h}^*(\hat{\boldsymbol{\theta}}) = \boldsymbol{a} : \|\boldsymbol{\theta}(\boldsymbol{a})-\hat{\boldsymbol{\theta}}\| \text{ is maximum}\}, \tag{3.20}$$

where $\hat{\boldsymbol{\theta}}$ is defined in (3.9), i.e., for any $\boldsymbol{a}$, we take the extreme (distant) value of $\boldsymbol{\theta}$ for which (3.19) holds. Now using the method of (3.19) and (3.20) and allowing $\boldsymbol{a}$ to assume all possible values on the boundary of the ellipsoid $H_N(\boldsymbol{\theta}) \leqslant \chi_{p,\epsilon}$, it is easily seen (using the monotonicity of $\boldsymbol{h}^*(\boldsymbol{\theta})$ with $\boldsymbol{\theta}$) that the set of points $\hat{\boldsymbol{\theta}}(\boldsymbol{a})$ obtained in this manner describes a closed convex set in $\boldsymbol{\theta}$ which contains $\hat{\boldsymbol{\theta}}$ as an inner point. This closed convex set is our desired confidence region for $\boldsymbol{\theta}$ and it is readily seen that this also remains translation invariant.

In actual practice, the problem of finding out this convex close set in $\boldsymbol{\theta}$ (say $C(\boldsymbol{\theta})$), is not very complicated. It can be shown (cf. Sen [15]) that under the regularity conditions of section 2,

$$n^{\frac{1}{2}}(h_j^*(\theta_j(\boldsymbol{a}))-\mu_N^{(j)}) \overset{\mathcal{P}}{\sim} \overline{B}_j(\boldsymbol{F})\,.n^{\frac{1}{2}}(\hat{\theta}_j(\boldsymbol{a})-\hat{\boldsymbol{\theta}}_j),\ j = 1, \ldots, p, \tag{3.21}$$

where $\overline{B}_j(F)$, $j = 1, \ldots, p$ are defined in (3.13). Thus, asymptotically $C(\boldsymbol{\theta})$ reduces to the following from

$$C(\boldsymbol{\theta}) = \{\boldsymbol{\theta} : n(\boldsymbol{\theta}-\hat{\boldsymbol{\theta}})\boldsymbol{\tau}^{-1}(\boldsymbol{\theta}-\hat{\boldsymbol{\theta}})' \leqslant \chi^2_{p,\epsilon}\}, \tag{3.22}$$

where $\boldsymbol{\tau}$ is defined in (3.14). If we take any point $\boldsymbol{a}$ on the boundary of the ellipsoid $H_N(\boldsymbol{\theta}) \leqslant \chi^2_{p,\varepsilon}$, then using (3.19), (3.20) and (3.21), we have

$$(3.23) \qquad \bar{B}_j(F) \overset{P}{\sim} (a_j - \mu_N^{(j)})/(\hat{\boldsymbol{\theta}}_j(\boldsymbol{a}) - \hat{\boldsymbol{\theta}}_j), \quad j = 1, \ldots, p.$$

Thus, on considering one or more points $\boldsymbol{a}$ on the boundary of the ellipsoid $H_N(\boldsymbol{\theta}) \leqslant \chi^2_{p,\varepsilon}$, we can estimate $\bar{B}_j(\boldsymbol{F})$ and then using (3.14) and $\hat{\boldsymbol{\nu}}$ (defined just before (3.16),), we get $\hat{\boldsymbol{\tau}} = ((\hat{\tau}_{jl}))$, where $\hat{\tau}_{jl} = \hat{\nu}_{jl}/\bar{B}_j^*(F)\, \bar{B}_l^*(\boldsymbol{F})$, $j, l = 1, \ldots, p$, and $\bar{B}_j^*(\boldsymbol{F})$ denote the estimate obtained by (3.23). Once $\boldsymbol{\tau}$ is obtained, if we define a set $\hat{C}(\theta)$ by

$$(3.24) \qquad \hat{C}(\boldsymbol{\theta}) = \{\boldsymbol{\theta} : n(\boldsymbol{\theta} - \hat{\boldsymbol{\theta}})\, \hat{\boldsymbol{\tau}}^{-1}(\boldsymbol{\theta} - \hat{\boldsymbol{\theta}})' \leqslant \chi^2_{p,\varepsilon}\},$$

then it is easily seen that $\hat{C}(\boldsymbol{\theta}) \overset{P}{\sim} C(\boldsymbol{\theta})$. Consequently, in actual practice, we may recommend (3.24) as the working confidence region for θ with confidence coefficient 1-ϵ. The simultaneous confidence region (3.22) or (3.24) remains valid for all $F_1, \ldots, F_r$ with covariance structures not necessarily identical. A word of clarification is necessary here. Though (3.24) is an asymptotic confidence region, in actual practice, if we select $\boldsymbol{h}^*$ some simple functions (e.g., rank-sum etc.), this remains valid for moderately large samples, where the usual parametric confidence-regions (under heterogeneity of dispersion matrices) may not be very simple.

It may be noted that if we borrow the idea of asymptotically smallest confidence region (cf. Wilks [18, pp 384-389]) then the asymptotic optimality or efficiency of the confidence region (3.22) or (3.24) with respect to the standard parametric confidence region for $\boldsymbol{\theta}$ (that may be derived with the help of likelihood ratio function) comes out to be

$$(3.25) \qquad \{|\bar{\boldsymbol{\Sigma}}|/|\boldsymbol{\tau}|\}^{\frac{1}{2}} = \{e(\boldsymbol{\tau}, \bar{\boldsymbol{\Sigma}})\}^{p/2},$$

where $\bar{\boldsymbol{\Sigma}}$, $\boldsymbol{\tau}$ and $e(\boldsymbol{\tau}, \bar{\boldsymbol{\Sigma}})$ are defined in (2.43), (3.14) and (3.15) respectively. Thus, the asymptotic optimality of the confidence region and the point estimate of $\boldsymbol{\theta}$ are functionally related to each other.

4. CHOICE OF RANK-ORDER TEST AND THE RELATED EFFICIENCY OF THE MANOVA PROCEDURES

We have so far considered a general class of rank-order statistics and developed some nonparametric MANOVA procedures. Now, we shall consider some specific rank-order statistics and study the related asymptotic efficiency factors. In particular, we shall consider the following three types of statistics, where,

$$\boldsymbol{E}_N = (E_{N1}, \ldots, E_{NN}), \quad E_{N\alpha} = J_N(\alpha/(N+1)) \; \alpha = 1, \ldots, N;$$

are characterized as follows.

(1) *Median procedure* : Here

$$E_{N\alpha} = 1, \text{ if } \alpha \leqslant [N/2], \qquad (4.1)$$

$$= 0, \text{ otherwise.}$$

(2) *Rank-sum procedure.* Here

$$E_{N\alpha} = \alpha \text{ for } 1 \leqslant \alpha \leqslant N. \qquad (4.2)$$

(3) *ψ-score procedure.* Let ψ be any specified c.d.f., and let $E_{N\alpha}$ be the expected value of the αth order statistic in a sample of size N drawn from a population with the c.d.f. ψ. In particular, if ψ is taken to be a standardized normal c.d.f, the procedure will be termed the *normal score procedure.*

In the univariate analysis of variance problem, it is known (cf. [6]), that against normal alternatives, the median procedure has an efficiency only $2/\pi$, though the same may be quite high for some non-normal c.d.f's. The asymptotic efficiency of rank-sum procedure for normal alternatives is $3/\pi$, and it has a lower bound .864 for any c.d.f. (cf [6]), though it can be arbitrarily large for some typical c.d.f. The normal score procedure has always an asymptotic efficiency greater than or equal to one, for all continuous c.d.f. (c.f. [7]).

In the multivariate case, it can be shown easily that all these procedures satisfy the regularity conditions (C.i), $i = 2, 3, 4, 5$. Further, if the c.d.f.'s $F_1, \ldots, F_r$ are non-singular in the sense that the cluster of the points is not confined in any $p-1$ dimensional subsequence of the p-dimensional Euclidean space, then it can be shown (cf. [10]) that (C.6) also holds. Regarding the expressions (2.42), (3.15) and (3.25), it can be shown that for normal $F_1, \ldots, F_r$ these are all equal to unity if we adopt the normal score procedure. Thus, the use of normal scores preserves the distribution-free property of the MANOVA procedure and at the same time, makes them asymptotically full efficient against normal alternatives. The efficiency factors can be shown to be greater than one for various non-normal c.d.f.'s but it can not be shown that they are greater than or equal to one for all $\boldsymbol{F} = F_1, \ldots, F_r)$ of the continuous type. Bickel [1] while considering the efficiency of Hodges-Lehmann [9] estimate of shift in the p-variate single sample case, came across a similar situation. He, however, considered only the rank-sum and median procedures. In the particular case of bivariate normal c.d.f., he deduced a lower bound (about .87) for the minimum efficiency of rank-sum procedure, while the minimum efficiency of the median procedure may be arbitrarily low. However, for more than two variates and/or for non-normal c.d.f.'s it is very difficult to prescribe any lower bound for the efficiency factors (2.42), (3.15) or (3.25), as they depend explicitly on the associated matrix $\boldsymbol{\tau}$, $\boldsymbol{\nu}$ and $\overline{\boldsymbol{\Sigma}}$, and nothing can be said, in general, about the magnitude or bounds for the characteristic roots of $\boldsymbol{\nu}\overline{\boldsymbol{\Sigma}}^{-1}$ or $\boldsymbol{\tau}\overline{\boldsymbol{\Sigma}}^{-1}$. However, if the p-variates are symmetrically dependent and have jointly a p-variate normal c.d.f. with a common correlation ρ, then it can be shown that the normal score procedure is better than the rank-sum procedure, which in turn is better than the median procedure.

In actual practice, the rank-sum procedure results in great simplification of the actual computation, while the normal score procedure is anticipated to have a better performance characteristic for nearly normal c.d.f.'s, and the choice will depend on the practical convenience and the degree of precision aimed at.

References

[1] Bickel, P. J. (1964). "On some Alternative Estimates for Shift in the p-Variate One Sample Problem," *Ann. Math. Statist.*, **35**, 1079–1090.

[2] Chatterjee, S. K. and Sen, P. K. (1964). "On Some Nonparametric Tests for the Bivariate Two Sample Location Problem," *Calcutta Statist. Asso. Bull.*, **13**, 18–58.

[3] Chatterjee, S. K. and Sen, P. K. (1969). "Nonparametric Tests for the Multisample Multivariate Location Problem," in *Contributions to Statistics and Probability, Essays in Memory of Samarendra Nath Roy*, University of North Carolina Press).

[4] Chernoff, H. and Savage, I. R. (1958). "Asymptotic Normality and Efficiency of Certain Nonparametric Test Statistics," *Ann. Math. Statist.*, **20**, 972–994.

[5] Govindarajulu, Z., Lecam, L. and Raghavachari, M. (1966). "Generalization of Chenoff-Savage Theorems on Asymptotic Normality of Nonparametric Test Statistics," *Proc. 5th Berkeley Symp. Math. Statist. Prob. Univ. of Calif.* 608-638.

[6] Hodges, J. L. Jr. and Lehmann, E. L. (1956). "The Efficiency of Some Nonparametric Competitors of the t-Test," *Ann. Math. Statist.*, **27**, 324–335.

[7] Hodges, J. L. Jr. and Lehmann, E. L. (1960). "Comparison of the Normal Scores and Wilcoxon Tests," *Proc. 4th Berkeley Symp. Math. Stat. Prob. Univ. of Calif.*, **1**, 83–92.

[8] Hodges, J. L. Jr. and Lehmann, E. L. (1962). "Rank Methods for Combination of Independent Experiments in Analysis of Variance," *Ann. Math. Statist.*, **33**, 482–497.

[9] Hodges, J. L. Jr. and Lehmann, E. L. (1963). "Estimates of Location Based on Rank Tests," *Ann. Math. Statist.*, **34**, 598–611.

[10] Puri, M. L. and Sen, P. K. (1966). On a Class of Multivariate Multisample Rank-Order Tests," *Sankhyā, Ser. A.*, **28**, 353–376.

[11] Sen, P. K. (1963). "On the Estimation of Relative Potency in Dilution (-Direct) Assays by Distribution-Free Methods," *Biometrics*, **19**, 532–552.

[12] Sen, P. K. (1965). "Some Further Applications of Non-parametric Methods in Dilution (-Direct) Assays," *Biometrics*, **21**, 799–810.

[13] Sen, P. K. (1966). "On a Class of Two Sample Bivariate Nonparametric Tests," *Proc. 5th Berkeley Symp. Math. Stat. Prob. Univ. of Calif.* 639-656.

[14] Sen, P. K. (1965). "On a Class of Multisample Multivariate Nonparametric Tests," (unpublished.)

[15] Sen, P. K. (1965). "On a Distribution-Free Method of Estimating Asymptotic Efficiency of a Class of Nonparametric Tests," *Ann. Math. Statist.*, **37**, 1759–1770.

[16] Sen, P. K. (1965). "On Pooled Estimation and Testing of Heterogeneity of Shift Parameters by Distributon-Free Methods. *Calcutta Statist. Assoc. Bull.* **16,** 139-157.

[17] Sverdrup, E. (1952). "The Limit Distribution of a Continuous Function of Random Variables," *Skand. Actdsfkt.*, **35**, 1–10.

[18] Wilks, S. S. (1962). *Mathematical Statistics*, John Wiley and Sons, New York.

(*Received Jan. 1, 1966.*)

CHAPTER 34

Probabilities of Deviations

J. SETHURAMAN[1], *Indian Statistical Institute*

1. INTRODUCTION

Limit theorems developed by probabilists have been widely used by statisticians in their problems. There are also occasions when probabilists have pointed out areas of interest to statisticians where limit theorems were not available. For instance, when the standard Central Limit Theorem and its modifications were available, nonparametric statisticians found that they needed a different sort of limit theorem for fixing significance points of their permutation tests and for computing Pitman efficiencies. From this need the Combinatorial Central Limit Theorem and the Chernoff Savage Theorem were born.

Again, there is now available a host of results developed by probabilists, pioneered by Cramér (1938), on the probabilities of deviations. These results give the rates of convergence to zero of the tails of various converging sequences of distributions met in practice. Chernoff (1952), Hodges and Lehman (1956), Bahadur (1960), etc., Rao (1962), have already used some of these results in

[1] Now at Florida State University.

their various concepts of efficiencies of tests. As statisticians start making more use of these results they will discover new areas where such results are not available. This has happened already: deviations of a certain order called *moderate* deviations have been introduced by Rubin and Sethuraman (1965a) while discussing Bayes' Risk Efficiency. It is with the hope that more such instances will be forthcoming that we summarize, *without proofs*, some of the important results in the theory of probabilities of deviations in this paper.

2. ASYMPTOTIC APPROXIMATIONS TO PROBABILITIES OF DEVIATIONS

Let $X_1, X_2, \ldots$ be independently and identically distributed in R_1, the real line, with common distribution function $F(x)$. Let

(1) $$\int x dF(x) = 0, \quad \int x^2 dF(x) = 1$$

and

(2) $$H_n(x) = P\left\{ \frac{X_1 + \ldots + X_n}{n} \leqslant x \right\}.$$

Then

(3) $$P\left\{ \frac{X_1 + \ldots + X_n}{n} > a/\sqrt{n} \right\} = 1 - H_n(a/\sqrt{n}) \to 1 - \Phi(a)$$

which is a number between 0 and 1, where

$$\Phi(a) = (1/\sqrt{2\pi}) \int_{-\infty}^{a} \exp(-t^2/2) dt.$$

We will call the deviation, $a/\sqrt{n}$, of $(X_1 + \ldots + X_n)/n$, the mean, as an *ordinary* deviation. A deviation, λ_n, of the mean will be called *excessive* if $n\lambda_n^2 \to \infty$. Probabilities of excessive deviations tend to zero or 1. An excessive deviation λ_n is called *large* if $\lambda_n = 0(1)$, *moderate* if $\lambda_n = c\sqrt{\log n/n}$. In this section we describe asymptotic expressions for the probabilities of excessive deviations of the mean of independent random variables.

The starting point is the famous result of Cramér (1938) which can be stated as follows.

Theorem 1. (*Cramér* (1938).)

Let $\int \exp(tx)dF(x) < \infty$ *for* t *in some open neighborhood of* 0. *Let* $a_n = o(1)$ *and* $na_n^2 > 1$. *Then*

$$\frac{1-H_n(a_n)}{1-\Phi(a_n\sqrt{n})} = \exp[na_n^3\,\lambda(a_n)]\,[1+0(a_n)] \tag{4}$$

$$\frac{H_n(-a_n)}{\Phi(-a_n\sqrt{n})} = \exp[-na_n^3\lambda(-a_n)][1+O(a_n)] \tag{5}$$

where $\lambda(z)$ *is a power series in* z, *called the Cramér series, and involves the cumulants of* X_1 *and converges for small* z.

From the above we note that if $a_n = o(n^{-1/3})$,

$$\frac{1-H_n(a_n)}{1-\Phi(a_n\sqrt{n})} \to 1 \tag{6}$$

and

$$\frac{H_n(-a_n)}{\Phi(-a_n\sqrt{n})} \to 1. \tag{7}$$

These therefore give asymptotic expressions to probabilities of deviations up to the order $o(n^{-1/3})$. However these results can be seen from another point of view due to Linnik (1960), who thus obtained some interesting generalizations. We first make the following definition. The interval $[0, \Psi(n)]$ is said to be a zone of normal convergence of the mean (*z.n.c.m*) if for any sequence $\{a_n\}$ with $a_n \epsilon [0, \psi(n)]$

$$\frac{1-H_n(a_n)}{1-\Phi(a_n\sqrt{n})} \to 1.$$

$[0, \psi(n)]$ is said to be a uniform z.n.c.m (*u.z.n.c.m*) if the above convergence is uniform w.r.t. $\{a_n\}$. $[-\psi(n), 0]$ is said to be a z.n.c.m. (u.z.n.c.m.) if the obvious analogous conditions hold. The remark at the end of Theorem 1 may be stated as follows.

Theorem 2. *With* $\psi(n) = o(n^{-1/3})$, $[0, \psi(n)]$ *and* $[-\psi(n), 0]$ *are z.n.c.m. if* $E(exp(tX_1)) < \infty$ *for* t *in* $(-\epsilon.\epsilon)$ *for some* $\epsilon > 0$.

[Here and elsewhere we use E to stand for expectation.]

Theorems 3 through 6 are due to Linnik (1960).

Theorem 3.

If for some $\beta < 0$, $[-n^{-\beta}, 0]$ *and* $[0, n^{-\beta}]$ *are z.n.c.m. then the distribution of* X_1 *is standard normal.*

In the following $\rho(n)$ will stand for a function of n tending to ∞ slowly as we please.

Remark : For Theorems 4 to 6 below, Linnik (1960) assumed that X_1 has a bounded density. It is not known if this condition can be relaxed.

Theorem 4.

Let $0 < \beta < 1/2$. *For* $\beta > 1/3$, *the condition*

$$E\left(\exp|X_1|^{\frac{1-2\beta}{1-\beta}}\right) < \infty \tag{8}$$

is necessary for $[-n^{-\beta}\rho(n), 0]$, $[0, n^{-\beta}\rho(n)]$ *to be z.n.c.m and conversely, condition* (8) *is sufficient for* $[-n^{-\beta}/\rho(n), 0]$, $[0, n^{-\beta}/\rho(n)]$ *to be u.z.n.c.m.*

For $0 < \beta \leqslant 1/3$, *consider the sequence of numbers*

$$\frac{1}{3}, \frac{1}{4}, \frac{1}{5} \ldots, \frac{1}{s+3} \ldots \to 0.$$

Let $\dfrac{1}{s+4} < \beta \leqslant \dfrac{1}{s+3}$. *Then the condition*

$$E\left(\exp|X_1|^{\frac{1-2\beta}{1-\beta}}\right) < \infty \tag{9}$$

and the condition that the first $s+3$ *moments of* X_1 *coincide with those of the standard normal are necessary for* $[-n^{-\beta}\rho(n), 0]$ *and* $[0, n^{-\beta}\rho(n)]$ *to be z.n.c.m. and are sufficient for* $[-n^{-\beta}/\rho(n), 0]$ *and* $[0, n^{-\beta}/\rho(n)]$ *to be u.z.n.c.m.*

Now, let $h(z)$ be a function on $(0, \infty)$ tending to ∞ monotonically as $z \to \infty$. Consider the two cases :

Case I.

$$(\log z)^{2+\partial} \leqslant h(z) \leqslant z^{\frac{1}{2}}, z \geqslant 1$$

where $\partial > 0$ is a fixed number. For simplicity we also assume that, with $h(z) = \exp(H(z))$, $H'(u) \leqslant 1$ and $H'(u)\exp(H(u)) \to \infty$ as $u \to \infty$.

Case II.

$$\rho_1(z)\log z \leqslant h(z) \leqslant (\log z)^{2+\partial}, z \geqslant 1$$

where $\rho_1(z)$ is some fixed function of z increasing to ∞ as slowly as we please as $z \to \infty$ and $\partial > 0$ is a fixed number.

In either case let $\Psi(n)$ be determined by

$$h(n\Psi(n)) = n\Psi^2(n)$$

Theorem 5.

Whenever $h(z)$ belongs to class I *or class* II, *the condition*

$$E[\exp(h(|X_1|))] < \infty \tag{10}$$

is necessary for $[-\Psi(n)\rho(n), 0]$ and $[0, -\Psi(n)\rho(n)]$ to be z.n.c.m. and is sufficient for $[-\Psi(n)/\rho(n), 0]$ and $[0, \Psi(n)/\rho(n)]$ to be u.z.n.c.m.

There is also a third case for $h(z)$ as defined below.

Case III.

$$3\log z \leqslant h(z) \leqslant M\log z$$

where $M \geqslant 3$ is a given constant.

Theorem 6.

Whenever $h(z)$ belongs to case III *the condition*

$$E[\exp(h(|X_1|))] < \infty \tag{11}$$

is necessary for $[-(\log n/n)^{\frac{1}{2}}\rho(n), 0]$ and $[0, (\log n/n)^{\frac{1}{2}}\rho(n)]$ to be z.n.c.m. and sufficient for $[-(\log n/n)^{\frac{1}{2}}/\rho(n), 0]$ and $[0, (\log n/n)^{\frac{1}{2}}/\rho(n)]$ to be u.z.n.c.m.

Continuing the work of Cramér (1938), Feller (1943) and Petrov (1954) have generalized Theorems 1 and 2 for the case of non-indentical summands. They were also able to use the Berry-Esseen remainder term approximation for the Central Limit Theorem which was not available in 1938. Let $X_1, X_2, \ldots$ be independent random variables on R_1 with distribution functions $F_1(x), F_2(x), \ldots$. Let $H_n(x) = P\{(X_1 + \ldots + X_n)/n \leqslant x\}$.

For $i = 1, 2, \ldots$, let

(12) $$\int x dF_i(x) = 0, \quad \int x^2 dF_i(x) = \sigma_i^2 > 0,$$

(13) $$0 < k \leqslant E(\exp(t X_i)) \leqslant K < \infty \text{ for } t \text{ in } (-\epsilon, \epsilon)$$

where $\epsilon > 0$ is independent of i. Also, let

(14) $$\frac{1}{n} \sum_{i=1}^{n} \sigma_i^2 = \sigma_n^{*2} \geqslant \partial > 0, \quad n = 1, 2, \ldots .$$

Assume that the Lindeberg-Lévy condition holds : for each $\theta > 0$

(15) $$\frac{1}{n\sigma_n^{*2}} \sum_{i=1}^{n} \int_{|y| > \theta\sqrt{n}\sigma_n^*} y^2 dF_i(y) \to 0.$$

Theorem 7.

Let conditions (12) *through* (15) *hold. Let* $a_n = o(1)$, $na_n^2 > 1$. *Then as* $n \to \infty$,

(16) $$\frac{1 - H_n(a_n)}{1 - \Phi(a_n\sqrt{n}/\sigma_n^*)} = \exp[na_n^3\lambda_n(a_n)][1 + O(a_n)]$$

(17) $$\frac{H_n(-a_n)}{\Phi(-a_n\sqrt{n}/\sigma_n^*)} = \exp[-na_n^3\lambda_n(-a_n)][1 + 0(a_n)]$$

where $\lambda_n(z)$ *is a series akin to the Cramér series and converges in the interval* $-c \leqslant z \leqslant c$ *where* c *is independent of* n.

Rubin and Sethuraman (1965b) investigated case III of Linnik (1960) in more detail. This case actually corresponds to a moderate deviation. Using the same notation as in Theorems 1—6 we have the following theorem.

Theorem 8.

The condition

(18) $$E(|X_1|^{c^2+2}) < \infty$$

with $c > 0$ *is necessary for* $[-a(\log n/n)^{\frac{1}{2}}, 0]$ *and* $[0, a(\log n/n)^{\frac{1}{2}}]$ *to be z.n.c.m. for all* $a > c$ *and is sufficient for* $[-a(\log n/n)^{\frac{1}{2}}, 0]$ *and* $[0, a(\log n/n)^{\frac{1}{2}}]$ *to be z.n.c.m. for all* $a < c$.

This theorem may also be stated as follows.

$$\frac{1-H_n(a(\log n/n)^{\frac{1}{2}})}{1-\Phi(a(\log n)^{\frac{1}{2}})} \to 1 \tag{19}$$

and

$$\frac{H_n(-a(\log n/n)^{\frac{1}{2}})}{\Phi(-a(\log n)^{\frac{1}{2}})} \to 1 \tag{20}$$

if

$$E(|X_1|^2) < \infty \tag{21}$$

for some $q > a^2+2$. Conversely, if (19) and (20) hold then (21) holds for all $q < a^2+2$.

Rubin and Sethuraman (1965b) generalized the above theorem on the lines of Petrov (1954) to means of independent random variables. They also considered probabilities of one-sided deviations which correspond to one-sided z.n.c.m. It is found that for results concerning positive deviations of the mean some conditions have to be imposed on the negative tails of the summands. This makes the conditions cumbersome but is interesting because it sheds light on the allowable magnitude of the negative tails of the summands.

Let $X_{n1}, X_{n2}, \ldots X_{nk_n}$ be independently distributed with distribution functions $F_{n1}(x), F_{n2}(x), \ldots F_{nk_n}(x)$, $n = 1, 2, \ldots$.

Let

$$\int x dF_{ni}(x) = 0, \quad \int x^2 dF_{ni}(x) = \sigma_{ni}^2 < 0. \tag{22}$$

We write N for k_n and assume that $N \to \infty$ as $n \to \infty$. Also let

$$\sigma_n^2 = \frac{1}{N} \sum_{i=1}^{N} \sigma_{ni}^2, \quad 0 < a \leqslant \sigma_n^2 \leqslant A < \infty. \tag{23}$$

Define

$$\overline{F}_n(x) = \frac{1}{N} \sum_{i=1}^{N} F_n(x) \tag{24}$$

We introduce a condition slightly stronger than the Lindeberg-Lévy condition. For each $\epsilon > 0$

$$(25) \qquad \frac{\log(N\sigma_n^2)}{\sigma_n^2} \int_{|x|>\epsilon\sqrt{N\sigma^2}/(\log(N\sigma_n^2))^{3/2}} x^2 d\bar{F}_n(x) \to 0$$

as $n \to \infty$

We can now state a main theorem of Rubin and Sethuraman (1965b).

Theorem 9.

Let (22) (23) *and* (25) *hold. Then the condition*

$$(26) \qquad \sup_n \int_0^\infty x^{c^2+2}\, d\bar{F}_n(x) < \infty.$$

is sufficient for

$$(27) \qquad \frac{P\left\{\frac{X_{n1}+\ldots+X_{nN}}{N} > a\sigma_n\sqrt{\frac{\log N}{N}}\right\}}{1-\Phi(a\sqrt{\log N})} \to 1$$

for all $a < c$. *Conversely* (27) *implies* (26) *for all* $c < a$.

3. ASYMPTOTIC EXPRESSIONS FOR THE LOGARITHM OF PROBABILITIES OF DEVIATIONS

This section deals with asymptotic approximations to the logarithm of the probability of an excessive deviation. The results are therefore weaker than those of the previous section. However one can deal with more general classes of statistics than means of independent random variables. Again, the class of statistics which have such approximations grows wider as the excessive deviation gets closer to an ordinary deviation. We now begin examining some of these results.

For large deviations of the mean of independent and identically distributed random variables Chernoff (1952) and Bahadur and Rao (1960) have established the following.

Theorem 10.

With the notation of Theorems 1-6, *let* $E(exp\ (tX_1)) < \infty$ *for some* $t > 0$. *Then*

$$(28) \qquad \frac{1}{n}\log P\left\{\frac{X_1+\ \ldots\ +X_n}{n} > a\right\} \to \log \rho^+(X_1, a)$$

where $0 \leqslant \rho^+(X_1, a) < 1$; $\rho^+(X_1, a)$ *is defined by*

$$(29) \qquad \rho^+(X_1, a) = \min_t \quad e^{-ta} E(\exp(tX_1)).$$

Again, if $E(exp(tX_1)) < \infty$ *for* t *in* $(-\epsilon, \epsilon)$ *with* $\epsilon > 0$,

$$(30) \qquad \frac{1}{n} \log P\left\{ \left| \frac{X_1+\ldots+X_n}{n} \right| > a \right\} \to \log \rho(X_1, a)$$

where

$$\rho(X_1, a) = \max(\rho^+(X_1, a), \quad \rho^-(X_1, a)),$$

$$\rho^-(X_1, a) = \rho^+(-X_1, a).$$

Theorem 10 has been extended to several general classes of statistics by Sethuraman (1964), (1965). We state them in Theorems 11 through 15. Let $X_1, X_2, \ldots$ be independent random variables in a separable complete metric space $\mathscr{X}$ distributed according to a common probability measure μ. Let μ_n denote the empirical probability measure of $X_1, \ldots X_n$.

For the following theorem we put $\mathscr{X} = R_k$, the Euclidean space of k dimensions. $F(x)$ and $G_n(x)$ are the distribution functions corresponding to μ and μ_n, respectively.

Theorem 11.

$$(32) \qquad \frac{1}{n} \log P\left\{ \sup_x |G_n(x) - F(x)| > a \right\} \to \log \rho(F, a).$$

$\rho(F, a)$ is defined in Sethuraman (1964). $\rho(F, a)$ depends on F and a. If F is a continuous distribution function then $\rho(F, a) = \rho(U, a)$ where U is the distribution function corresponding to the uniform distribution on [0, 1].

Theorem 12.

Let $\mathscr{G}$ *be a class of continuous functions on* $\mathscr{X}$ *such that (i)* $\mathscr{G}$ *is compact in the topology of uniform convergence on compacta, (ii) there is a continuous function* h *on* $\mathscr{X}$ *such that* $|g(x)| \leqslant h(x)$ *for each* g *in* $\mathscr{G}$ *and (ii)* $E[exp(th(X_1))] < \infty$ *for* t *in* $(-\epsilon, \epsilon)$ *for some* $\epsilon > 0$. *Then*

$$(33)\qquad \frac{1}{n}\log P\left\{\sup_{g\in\mathcal{G}}\left|\frac{g(X_1)+\ldots g(X_n)}{n}-E(g(X_1))\right|>a\right\}\rightarrow \log\rho(G, a)$$

where $\rho(\mathcal{G}, a)=\sup_{g\in\mathcal{G}}\rho(g(X_1), a)$

For any function h from $\mathcal{X}$ into R_k let $G_h(r)$ and $G_{n,h}(r)$ be the distribution function of $h(X_1)$ and the empirical distribution function of $h(X_1), \ldots, h(X_n)$, respectively.

Theorem 13.

Let $\mathcal{H}$ *be a class of continuous functions from* $\mathcal{X}$ *into* R_k *such that (i)* $\mathcal{H}$ *is compact in the topology of uniform convergence on compacta and (ii)* $G_h(r)$ *is a continuous distribution function for each* h *in* $\mathcal{H}$*. Then*

$$(34)\qquad \frac{1}{n}\log P\left\{\sup_{h\in\mathcal{H}}\sup_{r}|G_{n,h}(r)-G_h(r)|>a\right\}\rightarrow \log\rho(U, a)$$

where $\rho(U, a)$ *is as defined in the remark to Theorem* 11.

Several interesting and important theorems on large deviations follow from Theorems 10 and 11. As illustrations we present the following two theorems.

Theorem 14.

In the above notation let $\mathcal{X}=R_k$*. Let* μ *correspond to a continuous distribution. Let* $\mathfrak{C}$ *be the class of all measurable convex sets. Then*

$$(35)\qquad \frac{1}{n}\log P\left\{\sup_{C\in\mathfrak{C}}|\mu_n(C)-\mu(C)|>a\right\}\rightarrow \rho(U, a).$$

It is obvious that this theorem is true for suitable sub-classes of $\mathfrak{C}$, for instance, intersections of a fixed finite number of half-spaces.

Theorem 15.

In the above let $\mathcal{X}$ *be a separable Banach space with norm* $\|.\|$*. Let* $\mathcal{X}_1^*=\{x^*: x^*\epsilon\mathcal{X}^*, \|x^*\|=1\}$*, the boundary of the unit* *sphere in the adjoint of* $\mathscr{X}$. *Let* $E(exp(t\|X_1\|)) < \infty$ *for some* $t > 0$. *Then*

$$\frac{1}{n}\log P\left\{\left\|\frac{X_1+\dots+X_n}{n}\right\| > a\right\} \to \log \rho(\mathscr{X}_1^*, a) \tag{36}$$

where $\rho(\mathscr{X}_1^*, a)$ *is defined in a like manner as* $\rho(\mathscr{G}, a)$ *in Theorem* 12.

For the particular case $\mathscr{X} = D[0, 1]$, the space of functions on $[0, 1]$ with no discontinuities of the second kind with norm defined by $\|x\| = \sup_{t\in[0,1]} |x(t)|$, Theorem 15 is inapplicable since $\mathscr{X}$ is not a separable Banach space. However, in this case Sethuraman (1965) has established a large deviation result similar to Theorem 15.

We now state a moderate deviation result on U-statistics due to Rubin and Sethuraman (1965b). Let $\mathscr{X} = R_1$ in the above. Let $\psi(x_1, \dots, x_k)$ be a function symmetric in its arguments. then the U-statistic with kernal ψ is defined $U(X_1, \dots X_k) = \frac{1}{\binom{n}{k}} \Sigma\psi\left(X_{i_1}, \dots, X_{i_k}\right)$. Let $\psi(x_1) = E(\psi(x_1, X_2, \dots X_k)$ and where the summation is taken over all combinations $(i_1, \dots i_k)$. $E(\psi(X_1)) = 0$, $E(\psi^2(X_1)) = \sigma^2 > 0$.

Theorem 16.

Let

$$E(|\psi(X_1, \dots, X_k)|^q) < \infty \tag{37}$$

for all $q > 0$. *Then*

$$\frac{1}{\log n}\log P\left\{|U(X_1, \dots, X_n)| > k\sigma c\sqrt{\frac{\log n}{n}}\right\} \to -\frac{c^2}{2}. \tag{38}$$

Actually this theorem is true if (37) is true for some $q > c^2+2$. Again, this theorem is valid for generalized U-statistics.

The above results can be transformed to results concerning $P\{G_n \epsilon A_n\}$ where G_n is the empirical distribution function of $X_1, \dots X_n$ and A_n is some suitable subset of the space of distribution functions. Sanov (1957) was the pioneer in this approach but his paper contains some errors. Hoeffding (1965a), (1965b) and Hoadley (1967) have rectified the errors and extended Sanov's results. We now describe their results for only the case $\mathscr{X} = R_1$, for the sake of simplicity.

Let $\mathfrak{D}$ be the space of all distribution functions on R_1. For any two elements H_1 and H_2 in $\mathfrak{D}$ define

$$I(H_1, H_2) = \int h_1 \log (h_1/h_2) d\mu \tag{39}$$

where μ is a finite measure dominating μ_1 and μ_2 and $dH_1 = h_1 d\mu$, $dH_2 = h_2 d\mu$. It is easily seen that $0 \leqslant I(H_1, H_2) \leqslant \infty$, $I(H_1, H_2) = 0$ if only and if $H_1 = H_2$ and $I(H_1, H_2) < \infty$ only if H_2 dominates H_1. For any subset A of $\mathfrak{D}$ define

$$\begin{aligned} I(A, H_2) &= \inf_{H_1 \in A} I(H_1, H_2) \text{ if } A \text{ is nonempty} \\ &= +\infty \text{ if } A \text{ is empty.} \end{aligned}$$

Let $(a)_k = -\infty = a_0 < a_1 < \ldots < a_k = +\infty$ be a partition of R_1. Consider two sequences $\{\alpha_i\}$, $\{\beta_i\}$ $i = 1, \ldots, k-1$ with

$$0 \leqslant \beta_1 < \alpha_1 \leqslant \beta_2 < \alpha_2 \leqslant \ldots \leqslant \beta_{k-1} < \alpha_{k-1} \leqslant 1.$$

The set, S, of all distribution functions H with $\beta_i < H(a_i) \leqslant \alpha_i$ if $\beta_i > 0$ and $0 \leqslant H(a_i) \leqslant \alpha_i$ if $\beta_i = 0$, $i = 1, \ldots, k-1$ is called a *strip* of size k. Let $\{\gamma_i\}$, $i = 1, \ldots, k-1$ be a sequence of numbers with $0 \leqslant \gamma_1 \leqslant \gamma_2 \leqslant \ldots \leqslant \gamma_{k-1} \leqslant 1$. The set, C, of all distribution functions H with $H(a_i) = \gamma_i$ $i = 1, \ldots, k-1$ is called a *cylinder* set. If each γ_i is of the form n_i/n where n_i and n are integers then C is said to be an *n-cylinder*.

The following theorem on large deviations is due to Sanov (1957). It has been expressed in terms of multinomial distributions by Hoeffding (1965a), see Theorem 20. It is the starting point of this series of investigations. We state it in the form given in Hoadley (1967).

Theorem 17.

Let S be a strip in $\mathfrak{D}$ with $I(S, F) < \infty$. Then

$$P\{G_n \in S\} = \exp [-nI(S, F) + 0(\log n)] \tag{40}$$

Following Hoadley (1967) we now describe F-regular sets and several theorems concerning them rather than the more general and cumbersome F-distinguishable sets of Sanov (1957).

A set A in $\mathfrak{D}$ is said to be F-regular if

i) $I(A, F) < \infty$

ii) for each $\epsilon > 0$ there is a strip S such that $S \subset A$ and $I(S, F) < I(A, F)+\epsilon$

and

iii) for each $\epsilon > 0$ there is a finite number of strips $S_1, S_2, \ldots, S_m$ such that $A \subset \bigcup_{i=1}^{m} S_i$ and $I(A, F) < I(S_i, F)+\epsilon$ $i = 1, \ldots, m$.

The following theorem is immediate.

Theorem 18.

Let A be an F-regular set. Then

$$\frac{1}{n} \log P\{G_n \,\epsilon A\} \rightarrow -I(A, F). \tag{41}$$

A sequence A_n of subsets of $\mathfrak{D}$ is said to be F-regular if

i) $I(A_n, F) < \infty$ and $I(A_n, F) \rightarrow I < \infty$

ii) For each $\epsilon > 0$ there exist $k+1 = k(n, \epsilon)+1$ strips $S_0, S_{n1}, \ldots, S_{nk}$ such that $S_0 \subset A_n \subset \bigcup_{i=1}^{k} S_{ni}$, $I(S_0 F) -\epsilon < I(A_n, F) < I(S_i, F)+\epsilon$, $i = 1, \ldots, k$ and $k(n, \epsilon)$ $o(n/\log n)$.

Theorem 19.

Let A_n be an F-regular sequence.

Then

$$\frac{1}{n} \log P\{G_n \,\epsilon\, A_n\} \rightarrow -I. \tag{42}$$

Following Hoadley (1967) we give below a general class of subsets that are F-regular. $\mathfrak{D}$ can be considered as a subset of the space of functions of bounded variation, which is a Banach space with total variation as norm. Let T be a functional defined on $\mathcal{D}$ which is continuous in this norm. Let $B_r = \{H : T(H) \geqslant r\}$, $f_0(r) = I(B_r, F)$. Then B_r is F-regular if f_0 is continuous at r. With F-regular sets generated this way we can now deal with large deviations of many statistics like U-statistics with bounded kernels, Chernoff-Savage statistics with bounded J-functions etc. c-sample generalizations and multivariate random variables $X_1, X_2, \ldots$ are handled in a similar manner by Hoadley (1967).

Hoeffding (1965a) has obtained some very general results for the case $\mathcal{X} = \{\theta_1, \ldots, \theta_k\}$, a finite set. Let $P(X_1, = \theta_i) = p_i$, $i=1, \ldots, k$, $\boldsymbol{p} = (p_1, \ldots, p_k)$. The probability of any event E depends only on $\boldsymbol{p}$ and will therefore be denoted by $P(E \mid \boldsymbol{p})$. This is thus the familiar case of the multinomial. Let $\mathcal{D}^* = \{\boldsymbol{r} : \boldsymbol{r} = (r_1, \ldots, r_k),\ 0 \leqslant r_i \leqslant 1,\ \sum_1^k r_i = 1\}$. $\mathcal{D}^*$ is the space of all probability distributions on $\mathcal{X}$ and is also the space of all empirical probability distributions of $X_1, \ldots, X_n$. These latter distributions can be expressed in the following simple form. Let $G^*_{nj} =$ (no. of $X_i's = \theta_j$ among $X_1, \ldots, X_n)/n$, and $\boldsymbol{G}^*_n = (G^*_{n1}, \ldots, G^*_{nk})$. Then $\boldsymbol{G}^*_n \epsilon \mathcal{D}^*$. Let A^* be any subset of $\mathcal{D}^*$. Let $A^{*(n)}$ be the subset of A^* containing points of the form $(n_1/n, \ldots, n_k/n)$ where $n_1, \ldots, n_k$ are integers. Thus the events $\boldsymbol{G}^*_n \epsilon A^*$ and $\boldsymbol{G}^*_n \epsilon A^{*(n)}$ are equivalent. The I-function defined in (39) can now be re-interpreted in terms of points of $\mathcal{D}^*$. For h^*_1, $\boldsymbol{h}^*_2$ in $\mathcal{D}^*$ let

$$(43) \qquad I(\boldsymbol{h}^*_1, \boldsymbol{h}^*_2) = \sum_j h^*_{1j} \log (h^*_{1j}/h^*_{2j}).$$

For $A^* \subset D^*$, let

$$I(A^*, \boldsymbol{h}^*) = \inf_{\boldsymbol{h}^*_1 \epsilon A^n} I(\boldsymbol{h}^*_1, \boldsymbol{h}^*).$$

We can now state a theorem of Hoeffding (1965a) which is a slight generalization of Theorem 16.

Theorem 20.

For any subset A^* of $\mathcal{D}^*$ and $\boldsymbol{p}$ in $\mathcal{D}^*$

$$(44) \qquad c_0 n^{-(k-1)/2} \exp(-nI(A^{*(n)}, \boldsymbol{p})) \leqslant P\{\boldsymbol{G}^*_n \epsilon A^* \mid \boldsymbol{p}\}$$

$$\leqslant \binom{n+k-1}{k-1} \exp\left(-nI(A^{*(n)}, p)\right)$$

where c_0 *is an absolute constant*

A sequence $\{A_n^*\}$ of subsets of $\mathfrak{D}^*$ is said to be $\boldsymbol{p}$-regular if

$$I(A_n^{*(n)}, \boldsymbol{p}) = I(A_n^*, \boldsymbol{p}) + o\left(\frac{\log n}{n}\right).$$

Theorem 21.

Let $\{A_n^*\}$ *be a sequence of* $\boldsymbol{p}$*-regular sets in* $\mathfrak{D}^*$. *Then*

$$P\{G_n^* \epsilon A_n^* \mid \boldsymbol{p}\} = \exp\left(-I(A_n^*, \boldsymbol{p}) + o(\log n)\right). \tag{45}$$

Hoeffding (1965a) has given many examples of $\boldsymbol{p}$-regular sequences of subsets of $\mathfrak{D}^*$. They can be constructed from the general result embodied in the following theorem.

Theorem 22.

The sequence $\{A_n^*\}$ *of subsets of* $\mathfrak{D}^*$ *is* $\boldsymbol{p}$*-regular if there exist constants* n_0 *and* $c > 0$ *such that* $I(A_n^*, \boldsymbol{p}) < \infty$ *for* $n > n_0$ *and*

$$I(\boldsymbol{h}, \boldsymbol{p}) \leqslant I(A_n^*, \boldsymbol{p})$$

for some $\boldsymbol{h}$ *in* D^* *which is such that there is an* $\boldsymbol{r}$ *in* $A_n^{*(n)}$ *with* $|h_j - r_j| < c/n$ *if* $p_j > 0$ *and* $r_j = 0$ *if* $p_j = 0$.

As an important special case of this theorem we can deduce that complements of convex sets of $\mathfrak{D}^*$ are $\boldsymbol{p}$-regular for any $\boldsymbol{p}$. Hoeffding (1965b) has obtained more precise estimates for the probabilities in (44) for the present multinomial case and some other general situations, but since they get more complicated we shall not describe them.

4. OTHER APPROXIMATIONS

One can express the rate of convergence to zero of the probability of a deviation by showing that some series involving such probabilities converges. This is the approach taken by Baum and Katz (1963), (1964). We state their results in Theorems 23 through 26 in this section.

Let $X_1, X_2, \ldots$ be independent and identically distributed random variables in R_1. Let $S_n = X_1 + \ldots + X_n$.

Theorem 23.

Let $0 < t < 1$. *The following two statements are equivalent:*

$$E(|X_1|^t) < \infty; \tag{46}$$

$$\sum_{n=1}^{\infty} n^{-1} P\{|S_n| > \epsilon n^{1/t}\} < \infty \text{ for all } \epsilon > 0. \tag{47}$$

Let $1 \leqslant t < 2$. *The following two statements are equivalent:*

$$E(|X_1|^t) < \infty \text{ and } E(X_1) = \mu \tag{48}$$

$$\sum_{n=1}^{\infty} n^{-1} P\{|S_n - n\mu| > \epsilon n^{1/t}\} < \infty \text{ for all } \epsilon > 0 \tag{49}$$

Theorem 24.

Let $0 < t < 1$. *The following three statements are equivalent:*

$$E(|X_1|^t \log^+ |X_1|) < \infty; \tag{50}$$

$$\sum_{n=1}^{\infty} n^{-1} (\log n) P\{|S_n| > \epsilon n^{1/t}\} < \infty \text{ for all } \epsilon > 0; \tag{51}$$

$$\sum_{n=1}^{\infty} n^{-1} P\left\{\sup_{k \geqslant n} |S_k / k^{1/t}| > \epsilon\right\} < \infty \text{ for all } \epsilon > 0. \tag{52}$$

Let $1 \leqslant t < 2$. *The following three statements are equivalent:*

$$E(|X_1|^t \log^+ |X_1|) < \infty \text{ and } E(X_1) = \mu; \tag{53}$$

$$\sum_{n=1}^{\infty} n^{-1} (\log n) P\{|S_n - n\mu| < \epsilon n^{1/t}\} > \infty \text{ for all } \epsilon > 0; \tag{54}$$

$$\sum_{n=1}^{\infty} n^{-1} P\left\{\sup_{k \geqslant n} |(S_k - k\mu)/k^{1/t}| > \epsilon\right\} < \infty \text{ for all } \epsilon > 0. \tag{55}$$

Theorem 25.

Let $t > 1$, $r > 1$, $\frac{1}{2} < r/t < 1$. *The following three statements are equivalent:*

$$E(|X_1|^t) < \infty, \ E(X_1) = \mu; \tag{56}$$

(57) $$\sum_{n=1}^{\infty} n^{r-2}P\{|S_n - n\mu| > \epsilon n^{r/t}\} < \infty \text{ for all } \epsilon > 0;$$

(58) $$\sum_{n=1}^{\infty} n^{r-2}P\left\{\sup_{k\geqslant n}|(S_k - k\mu)/k^{r/t}| > \epsilon\right\} < \infty \text{ for all } \epsilon > 0.$$

Let $t > 0$, $r > 1$ and $r/t > 1$. The following three statements are equivalent:

(59) $$E(|X_1|^t) < \infty;$$

(60) $$\sum_{n=1}^{\infty} n^{r-2}P\{|S_n| > \epsilon n^{r/t}\} < \infty \text{ for all } \epsilon > 0;$$

(61) $$\sum_{n=1}^{\infty} n^{r-2}P\left\{\sup_{k\geqslant n}|S_k|k^{r/t}| > \epsilon\right\} < \infty \text{ for all } \epsilon > 0.$$

Finally we mention another result of Baum and Katz (1964) which gives conditions under which the probability of a large deviation goes to zero at the rate of a negative power of n. Compare the following theorem with Theorem 9.

Theorem 26.

Let $t \geqslant 0$. The following two statements are equivalent:

(62) $$n^{t+1}P\{|X_1| > n\} \to 0 \text{ and } \int_{|x|<n} x dF(x) \to 0;$$

(63) $$n^t P\{|S_n| > \epsilon n\} \to 0 \text{ for all } \epsilon > 0.$$

If $t > 0$, the above two statements are equivalent to:

(64) $$n^t P\left\{\sup_{k\geqslant n}|S_k/k| > \epsilon\right\} \to 0 \text{ for all } \epsilon > 0.$$

References

Bahadur, R. R. (1960). "Stochastic Comparison of Tests," *Ann. Math. Statist.*, **31**, 276–295.

Bahadur, R. R. and Rao, R. Ranga (1960). "On Deviations of the Sample Mean," *Ann. Math. Statist.*, **31**, 1015–1027.

Baum, L. E. and Katz, M. (1963). "Convergence Rates in the Law of Large Numbers," *Bull. Amer. Math. Soc.*, **69**, 771–772.

Baum, L. E. and Katz, M. (1964). "Convergence Rates in the Law of Large Numbers," *Trans. Amer. Math. Soc.*, **120**, 108–123.

Chernoff, H. (1952). "A Measure of the Asymptotic Efficiency for Tests of a Hypothesis Based on the Sum of Observations," *Ann. Math. Statist.*, **23**, 493–507.

Cramér, H. (1938). "Sur un nouveau théorem—limite de la théorie des probabilities," *Actualites Sci. Indust.*, **736**, 5–23.

Feller, W. (1943). "Generalization of a probability limit theorem of Cramér," *Trans. Amer. Math. Soc.*, **54**, 361–372.

Hoadley, A. B. (1967). "On the probability of large deviations of functions of several empirical c.d.f.'s," *Ann. Math. Statist.*, **38**, 360–381.

Hodges, J. L. and Lehmann, E. L. (1965). "The Efficiency of Some Nonparametric Competitors of the t-Test," *Ann. Math. Statist.*, **27**, 324–335.

Hoeffding, W. (1965a). "Asymptotically Optimal Tests for Multinomial Distributions,,' *Ann. Math. Statist.*, **36**, 369–401.

Hoeffding, W. (1965b). "On Probabilities of Large Deviations," *Proc. Fifth Berkeley Symp. Math. Stat. Prob.*, **1**, 203–219.

Linnik, Yu. V. (1960). "On the Probability of Large Deviations for Sums of Independent Variables," *Proc. Fourth Berkeley Symp. Math. Statist. Prob.*, **2**, 289–306.

Petrov, V. V. (1954). "Generalization of Cramér's Limit Theorem," (Russian) *Uspehi. Matem. Nauk* (N.S.), **9**, 195–202.

Rao, C. R. (1962). "Efficient Estimates and Optimum Inference Procedures in Large Samples," *J. Roy. Statist. Soc., Ser. B*, **24**, 46–72.

Rubin, H. and Sethuraman, J. (1965a). "Bayes Risk Efficiency," *Sankhyā, Ser. A.*, **27**, 347–356.

Rubin, H. and Sethuraman, J. (1965b). "Probabilities of Moderate Deviations," *Sankhyā, Ser. A.*, **27**, 325–346.

Sanov, I. N. (1957). "On the Probability of Large Deviations of Random Variables," (Russian) *Mat. Sbornik* (N.S.), **42**, 11–44, English translation in *Select. Transl. Math. Statist. Prob.*, **1** (1961), 213–244.

Sethuraman, J. (1964). "On the Probability of Large Deviations of Families of Sample Means," *Ann. Math. Statist.*, **35**, 1304–1316.

Sethuraman, J. (1965). "On the Probability of Large Deviations of the Mean for Random Variables in $D[0, 1]$," *Ann. Math. Statist.*, **36**, 280–285.

(Received Jan. 1, 1966.)

CHAPTER 35

A Note on Embedding for Hadamard Matrices

S. S. SHRIKHANDE AND BHAGWANDAS
University of Bombay

SUMMARY

A square matrix H_n of order n in symbols ± 1 is called a Hadamard matrix if the scalar product of any two rows (and hence of any two columns) is zero. It is known that H_n can exist if and only if $n = 2$ or $n = 4t$ where t is any positive integer. The main object of this note is to show that, for any t, any matrix of order $(4t-1, 4t)$ in symbols ± 1 which is orthogonal by rows can be embedded in a Hadamard matrix H_{4t}. Some further results are also obtained for the cases $t = 2$ and 3.

1. INTRODUCTION

An orthogonal array $A(N, p, 2, 2)$ is a matrix with p rows, N columns in two symbols which can be taken without loss of generality as 0 and 1 such that in any two rows each of the ordered pairs (0, 0), (0, 1), (1, 0), (1, 1) occur exactly t times. Obviously then $N = 4t$. The number p is called the number of constraints. It is known [3] that the maximum value of p with $N = 4t$ is $4t-1$. An orthogonal array $A(4t, 4t-1, 2, 2)$

will be called a maximal array. Without loss of generality any array $\boldsymbol{A}(4t, p, 2, 2)$ can be brought to the standard form in which the first column consists entirely of 1's. This can be done by interchanging 0 and 1 whenever necessary in any given row of the array. Since any row of the array contains each of the symbols 0 and 1 exactly $2t$ times, it is obvious that by omitting the initial column from a standard array $\boldsymbol{A}(4t, 4t-1, 2, 2)$ one obtains a balanced incomplete block design (BIBD) [2] with parameters $v = b = 4t-1$, $r = k = 2t-1$, $\lambda = t-1$. Conversely if $\boldsymbol{M} = (m_{ij})$ is the usual incidence matrix of the above BIBD where $m_{ij} = 1$ or 0 according as treatment i does or does not occur in block j, then it is obvious that the matrix $(\boldsymbol{J} : \boldsymbol{M})$ where $\boldsymbol{J}$ consists entirely of 1's is an orthogonal array $\boldsymbol{A}(4t, 4t-1, 2, 2)$. Again from any array $\boldsymbol{A}(4t, 4t-1, 2, 2)$, by changing 0 into -1 and adding an initial row consisting entirely of 1's, one obviously gets a matrix $\boldsymbol{H}_{4t}$. Conversely given a matrix $\boldsymbol{H}_{t4}$ one can interchange 1 and -1 in any given column without destroying its orthogonality property and hence obtain a Hadamard matrix in its standard form in which the initial row consists entirely of 1's. It is then obvious that every other row contains each of the symbols ± 1 exactly $2t$ times and that in any two rows of the standardized Hadamard matrix each of the ordered pairs (1, 1), (1, -1), (-1, 1), (-1, -1) occur exactly t times. It is then obvious that by omitting the initial row of the standardized matrix and changing -1 into 0 one obtains an array $\boldsymbol{A}(4t, 4t-1, 2, 2)$. Thus it is clear that the existence of any one of $\boldsymbol{H}_{4t}$, $\boldsymbol{A}(4t-1, 2, 2)$ and BIBD with $v = b = 4t-1$, $r = k = 2t-1$, $\lambda = t-1$ implies the existence of the remaining two.

Let $\boldsymbol{H}_{p,4t}$ denote a matrix of order $(p, 4t)$ in symbols ± 1 such that any two rows are orthogonal. Then using the same arguments as above it is easy to see that the existence of any one of $\boldsymbol{H}_{p,4t}$, an array $\boldsymbol{A}(4t, p-1, 2, 2)$ and a design with $v = p-1$, $b = 4t-1$, $r = 2t-1$, $\lambda = t-1$ implies the existence of the remaining two. It is then easy to see that the results in any one of the three systems can be interpreted in terms of the other two systems.

It may be mentioned that matrices $\boldsymbol{H}_{4t}$ have been constructed for an infinity of values of t [1] and it is actually conjectured that they exist for all values of t.

Two arrays $\boldsymbol{A}_1 = \boldsymbol{A}(4t, p, 2, 2)$ and $\boldsymbol{A}_2 = \boldsymbol{A}(4t, p, 2, 2)$ will be said to be isomorphic if $\boldsymbol{A}_1$ can be transferred into $\boldsymbol{A}_2$ by interchange of rows, or columns of $\boldsymbol{A}_1$ or the interchange of symbols 0, 1 in any given rows of $\boldsymbol{A}_1$. An orthogonal array will be said to be unique if it is unique up to isomorphism. Similarly by a unique solution of a BIBD we mean that the incidence matrices $\boldsymbol{M}_1$ and $\boldsymbol{M}_2$ of any two solutions of the BIBD are isomorphic, i.e., $\boldsymbol{M}_1$ can be transformed into $\boldsymbol{M}_2$ by a permutation of its rows or a permutations of its columns.

Now consider an array $\boldsymbol{A}(4t, p, 2, 2)$ in symbols 0 and 1. Two columns of the array will be said to have i coincidences if there are exactly i rows in which the symbols appearing in the two columns have the same value. Let n be the number of columns having i coincidences with the first column. Then from [3] we have

$$\sum_{i=0}^{p} n_i = 4t-1, \tag{1.1}$$

$$\sum_{i=0}^{p} in_i = p(2t-1), \tag{1.2}$$

$$\sum_{i=0}^{p} i(i-1)n_i = p(p-1)(t-1). \tag{1.3}$$

The above equations will be designated as necessary conditions for the existence of the array.

2. AN EMBEDDING THEOREM FOR ORTHOGONAL ARRAYS

Consider an orthogonal array $\boldsymbol{A}(4t, 4t-2, 2, 2)$ in symbols 0 and 1. Then the necessary conditions for existence of the array imply

$$\sum_{i=0}^{4t-2} n_i = 4t-1, \tag{2.1}$$

$$(2.2)\qquad \sum_{i=0}^{4t-2} in_i \qquad = (4t-2)(2t-1),$$

$$(2.3)\qquad \sum_{i=0}^{4t-2} i(i-1)n_i = (4t-2)(4t-3)(t-1).$$

Consider the function

$$(2.4)\qquad f(x) = \sum_{i=0}^{4t-2} (i-x)(i-1-x)n_i.$$

Then obviously $f(x) \geqslant 0$. Further using the above relations it is easy to verify that $f(2t-2) = 0$. Hence from (2.4) it follows that $n_i = 0$ for $i \neq 2t-2$ or $2t-1$, and then (2.1) and (2.2) give the unique solution

$$n_{2t-2} = 2t-1$$

$$(2.5)\qquad n_{2t-1} = 2t$$

$$n_i = 0 \text{ for } i \neq 2t-2,\ 2t-1.$$

As mentioned before without loss of generality we can consider the first column to consist entirely of 1's, then the array has the form

$$\boldsymbol{A} = (\boldsymbol{J}\ \boldsymbol{A}_1\ \boldsymbol{A}_2)$$

where $\boldsymbol{J}$ is a column vector with $4t-2$ components all equal to 1, $\boldsymbol{A}_1$ is a matrix with $2t-1$ columns each containing $2t-2$ unities and $\boldsymbol{A}_2$ is a matrix with $2t$ columns each containing $2t-1$ unities.

Let $\boldsymbol{x}_1$ and $\boldsymbol{x}_2$ be two vectors with n components in symbols 0 and 1. The distance $d(\boldsymbol{x}_1, \boldsymbol{x}_2)$ between them is defined as the number of places in which they differ. Then

$$(2.6)\qquad d(\boldsymbol{x}_1, \boldsymbol{x}_2) = w_1+w_2-2(\boldsymbol{x}_1, \boldsymbol{x}_2)$$

where w_1 and w_2 are respectively the number of unities in $\boldsymbol{x}_1$ and $\boldsymbol{x}_2$ and $\boldsymbol{x}_1 \cdot \boldsymbol{x}_2$ is their scalar product. If $i(\boldsymbol{x}_1, \boldsymbol{x}_2)$ is the number of coincidences then obviously

$$(2.7)\qquad i(\boldsymbol{x}_1, \boldsymbol{x}_2) = n-d(\boldsymbol{x}_1, \boldsymbol{x}_2)$$

Now let $\boldsymbol{x}_1, \boldsymbol{x}_2$ be two vectors from $\boldsymbol{A}_1$. Then from (2.7) it is clear that $i(\boldsymbol{x}_1, \boldsymbol{x}_2)$ is given by

$$i(\boldsymbol{x}_1, \boldsymbol{x}_2) = 2+2(\boldsymbol{x}_1 \cdot \boldsymbol{x}_2)$$

Since this is necessarily even, from the unique solution (2.5) we have

$$i(\boldsymbol{x}_1, \boldsymbol{x}_2) = 2t-2$$

and

$$d(\boldsymbol{x}_1, \boldsymbol{x}_2) = 2t$$

Hence the number of coincidences between any two vectors of $\boldsymbol{A}_1$ is $2t-2$ and the distance between them is $2t$. We now use the following lemma due to Schutzenberger [4].

Lemma : *Let* P *be a* (m, n) *matrix in symbols* 0 *and* 1. *Let* k_i *be the number of unities in the ith row, then*

$$m \operatorname{var}(k_i) = n\Sigma k_i - \frac{(\Sigma k_i)^2}{m} - \sum_{j<j'} d(\boldsymbol{x}_j, \boldsymbol{x}_{j'})$$

where $i = 1, 2, \ldots, m$; $j < j' = 1, 2, \ldots, n$ *and* $\boldsymbol{x}_j, \boldsymbol{x}_{j'}$ *are the column vectors of* $\boldsymbol{P}$.

Applying the lemma to the matrix $\boldsymbol{A}_1$, it is easy to verify that var $(k_i) = 0$ and hence each row of $\boldsymbol{A}_1$ contains the same number $t-1$ of 1's. Hence each row of $(\boldsymbol{J}\ \boldsymbol{A}_1)$ contains exactly t positions occupied by 0 and t positions occupied by 1 and hence the same is true for any row of $\boldsymbol{A}_2$. It is now obvious that if we add an additional row to $\boldsymbol{A}$ containing 1 in the first $2t$ positions and 0 in the remaining $2t$ positions, then we get an orthogonal array $\boldsymbol{A}(4t, 4t-1, 2, 2)$.

We can, therefore, state the following theorem.

Theorem : *An orthogonal array* $\boldsymbol{A}(4t, 4t-2, 2, 2)$ *can always be embedded in the array* $\boldsymbol{A}(4t, 4t-1, 2, 2)$.

As obvious consequences of the above theorem, we have the following corollaries.

Corollary 1 : A matrix $\boldsymbol{H}_{4t-1,4t}$ can always be embedded in a Hadamard matrix $\boldsymbol{H}_{4t}$.

Corollary 2 : An incomplete block design with parameters $v = 4t-2$, $b = 4t-1$, $r = 2t-1$, $\lambda = t-1$ can always be embedded in a symmetric BIBD with $v = b = 4t-1$, $r = k = 2t-1$, $\lambda = t-1$.

3. RESULTS FOR THE CASE $t = 2$

We prove in this section that any orthogonal array $\boldsymbol{P}(8, p, 2, 2)$ with $p < 7$ can always be embedded uniquely in the maximal array $\boldsymbol{A}(8, 7, 2, 2)$.

The array with $p = 1$ can be uniquely extended to the array $\boldsymbol{A}_2 = \boldsymbol{A}(8, 2, 2, 2)$,

$$\boldsymbol{A}_2 = \begin{pmatrix} 1 & 1 & 0 & 0 & 1 & 1 & 0 & 0 \\ 1 & 0 & 1 & 0 & 1 & 0 & 1 & 0 \end{pmatrix}$$

It is obvious that in any extension of $\boldsymbol{A}_2$ we can without loss of generality consider the case when only 1 is added in the initial column. If $\boldsymbol{x}_1'$ is any row vector with 1 in the first column such that

$$\begin{pmatrix} \boldsymbol{A}_2 \\ \boldsymbol{x}_1' \end{pmatrix}$$

is an array $\boldsymbol{A}(8, 3, 2, 2)$, then the four columns which contain 1 in $\boldsymbol{x}_1'$ must also contain two 1's and two 0's in each row of $\boldsymbol{A}_2$. It is then obvious from the structure of $\boldsymbol{A}_2$, that $\boldsymbol{x}_1'$ can be uniquely taken as

$$\boldsymbol{x}_1' = (1 \quad 1 \quad 1 \quad 1 \quad 0 \quad 0 \quad 0 \quad 0)$$

giving rise to $\boldsymbol{A}_3 = \boldsymbol{A}(8, 3, 2, 2)$

$$\boldsymbol{A}_3 = \begin{pmatrix} 1 & 1 & 0 & 0 & 1 & 1 & 0 & 0 \\ 1 & 0 & 1 & 0 & 1 & 0 & 1 & 0 \\ 1 & 1 & 1 & 1 & 0 & 0 & 0 & 0 \end{pmatrix}$$

If $\boldsymbol{A}_3$ is to be extended to an array with four constraints by adjoining a row vector $\boldsymbol{x}'$ then by considering the last row of $\boldsymbol{A}_3$ it is obvious that $\boldsymbol{x}'$ must contain 1 in the first position and one 1 and two 0's in the next three positions. We will say that a set of vectors of order $4t$ in symbols 0 and 1 are orthogonal if any two vectors form an orthogonal array $\boldsymbol{A}(4t, 2, 2, 2)$. Then by considering the orthogonality of $\boldsymbol{x}'$ with each of the rows of $\boldsymbol{A}_2$ we see that the only possible candidates for $\boldsymbol{x}'$ are

$$\begin{aligned}
\boldsymbol{x}_2' &= (1 \quad 1 \quad 0 \quad 0 \qquad 0 \quad 0 \quad 1 \quad 1) \\
\boldsymbol{x}_3' &= (1 \quad 0 \quad 1 \quad 0 \qquad 0 \quad 1 \quad 0 \quad 1) \\
\boldsymbol{x}_4' &= (1 \quad 0 \quad 0 \quad 1 \qquad 0 \quad 1 \quad 1 \quad 0) \\
\boldsymbol{x}_5' &= (1 \quad 0 \quad 0 \quad 1 \qquad 1 \quad 0 \quad 0 \quad 1)
\end{aligned}$$

Further we note that any two of the vectors $\boldsymbol{x}_i'$, $i = 2, \ldots, 5$ are orthogonal. Thus any array $\boldsymbol{A}(8, p, 2, 2)$ can be embedded in the unique maximal array $\boldsymbol{A}_8 = \boldsymbol{A}(8, 7, 2, 2)$

$$\boldsymbol{A}_8 = \begin{bmatrix} \boldsymbol{A}_2 \\ \boldsymbol{x}_1' \\ \boldsymbol{x}_2' \\ \boldsymbol{x}_3' \\ \boldsymbol{x}_4' \\ \boldsymbol{x}_5' \end{bmatrix}$$

The above results can also be interpreted as essentially unique extension of a matrix $\boldsymbol{H}_{p,8}$, $p < 8$ into a Hadamard matrix $\boldsymbol{H}_8$ as also the extension of $v = p$, $b = 7$, $r = 3$, $\lambda = 1$ to the unique BIBD design with $v = b = 7$, $r = k = 3$, $\lambda = 1$ which corresponds to the projective plane $PG(2, 2)$.

4. RESULTS FOR THE CASE $t = 3$

We show that an orthogonal array A$(12, p, 2, 2)$ can always be extended to an array $\boldsymbol{A}(12, p+1, 2, 2)$ except possibly for the case $p = 3$.

An array $\boldsymbol{A}(12, 2, 2, 2)$ can be uniquely written as $\boldsymbol{A}_2$ where

$$\boldsymbol{A}_2 = \begin{pmatrix} 1 & 1 & 0 & 0 & 1 & 1 & 0 & 0 & 1 & 1 & 0 & 0 \\ 1 & 0 & 1 & 0 & 1 & 0 & 1 & 0 & 1 & 0 & 1 & 0 \end{pmatrix}$$

which can obviously be extended to an array with three constraints by adjoining the row

$$(1 \quad 1 \quad 1 \quad 1 \quad 1 \quad 0 \quad 0 \quad 1 \quad 0 \quad 0 \quad 0 \quad 0).$$

If $\boldsymbol{A}_3 = \boldsymbol{A}(12, 3, 2, 2)$ is an array with three constraints it is easy to verify that the only solutions of equation (1.1), (1.2) and (1.3) with $0 \leqslant n_i \leqslant 11$ for $i = 0, 1, 2, 3$ are

$$(\alpha_1): \quad n_0 = 1, n_1 = 6, n_2 = 3, n_3 = 1,$$

$$(\alpha_2): \quad n_0 = 2, n_1 = 3, n_2 = 6, n_3 = 0,$$

and $$(\alpha_3): \quad n_0 = 0, n_1 = 9, n_2 = 0, n_3 = 2.$$

The solution (α_3) corresponds to a nonextendable case. For if the array with solution (α_3) was extendable to $\boldsymbol{A}(12, 4, 2, 2)$ then for this array with four constraints we have the necessary conditions

$$n_0 = 0, n_3+n_4 = 2, \tag{4.1}$$

besides the following equations obtained from (1.1), (1.2) and (1.3) :

$$n_0+n_1+n_2+n_3+n_4 = 11, \tag{4.2}$$

$$n_1+2n_2+3n_3+4n_4 = 20, \tag{4.3}$$

$$n_2+3n_3+6n_4 = 12. \tag{4.4}$$

It is easy to verify that the above four equations imply $2n_4 = 1$ which is impossible.

Now consider the case (α_1). Without loss of generality consider the solution (α_1) with respect to the initial column containing all 1's. Then from $n_3 = n_0 = 1$ the first three columns are

$$\begin{matrix} 1 & 1 & 0 \\ 1 & 1 & 0. \\ 1 & 1 & 0 \end{matrix}$$

Now we have to add six columns each containing one 1 and three columns each containing two 1's. We note that none of the three possible columns containing one 1 can be repeated more than twice for otherwise the pair (0, 0) will occur at least four times in two rows. Since $n_1 = 6$ it is obvious that each of the three columns with one 1 must be repeated twice. A similar argument shows that each of the three columns with two 1's must be taken exactly once. We thus have the unique array for the case (α_1) given by $\boldsymbol{A}_3(\alpha_1)$ when

$$\boldsymbol{A}_3(\alpha_1) = \begin{pmatrix} 1 & 1 & 0 & 0 & 0 & 0 & 0 & 1 & 1 & 1 & 1 & 0 \\ 1 & 1 & 0 & 0 & 0 & 1 & 1 & 0 & 0 & 1 & 0 & 1 \\ 1 & 1 & 0 & 1 & 1 & 0 & 0 & 0 & 0 & 0 & 1 & 1 \end{pmatrix}$$

A similar argument for the case (α_2) shows that the unique array $\boldsymbol{A}_3(\alpha_2)$ with the initial column of all 1's is given by

$$\boldsymbol{A}_3(\alpha_2) = \begin{pmatrix} 1 & 0 & 0 & 1 & 0 & 0 & 1 & 1 & 1 & 1 & 0 & 0 \\ 1 & 0 & 0 & 0 & 1 & 0 & 1 & 1 & 0 & 0 & 1 & 1 \\ 1 & 0 & 0 & 0 & 0 & 1 & 0 & 0 & 1 & 1 & 1 & 1 \end{pmatrix}$$

It is obvious that with respect to the second column of $\boldsymbol{A}_3(\alpha_2)$ the solution (α_1) holds and since there is a unique array for the case (α_1), the array $\boldsymbol{A}_3(\alpha_2)$ coincides with $\boldsymbol{A}_3(\alpha_1)$ when we interchange 0 and 1 in all the rows of $\boldsymbol{A}_3(\alpha_2)$ and suitably permute its columns. Hence without loss of generality we can take the unique array $\boldsymbol{A}_3$ with three constraints as given by $\boldsymbol{A}_3(\alpha_1)$ which is extendable to an array with four constraints by adding a row given by

$$(1 \quad 1 \quad 1 \quad 1 \quad 0 \quad 1 \quad 0 \quad 1 \quad 0 \quad 0 \quad 0 \quad 0).$$

We note that $A_3(\alpha_1)$ contains four pairs of columns such that each pair of columns gives 3 conicidences. Further if $A_3(\alpha_1)$ is to be extended to an array with four constraints by adjoining a row x', then it is easy to verify that x' must contain the same symbol in positions corresponding to at least one of the above four pairs of columns and hence in the array with four constraints there is at least one column with respect to which n_4 takes exactly the value 1.

Now consider an array $A(12, 4, 2, 2)$. We know from what has been proved above that there is a column in the array with respect to which $n_4 = 1$. With respect to this column it is easy to verify that equations (1.1), (1.2), (1.3) for $A(12, 4, 2, 2)$ together with $n_4 = 1$ imply

$$n_0 + n_3 = 0,$$

$$n_1 = 4 + 3n_3,$$

$$n_2 = 6 - 3n_3.$$

Since all the n_i's are $\geqslant 0$, it is easy to see that we have the unique solution

$$(\beta) : n_0 = n_3 = 0, \; n_1 = 4, \; n_2 = 6, \; n_4 = 1.$$

Taking the initial column to consist of all 1's and noting that $n_4 = 1$, it is easy to see that each of the 6 possible columns with two 1's must be taken exactly once. In the eight columns so obtained the ordered pair (1, 0) occurs twice in any two given rows and hence each of the four possible columns with exactly one 1 must be taken exactly once. We thus obtain the unique array A_4 given by

$$A_4 = \begin{bmatrix} 1 & 1 & 1 & 0 & 0 & 0 & 1 & 1 & 1 & 0 & 0 & 0 \\ 1 & 1 & 0 & 1 & 0 & 0 & 1 & 0 & 0 & 1 & 1 & 0 \\ 1 & 1 & 0 & 0 & 1 & 0 & 0 & 1 & 0 & 1 & 0 & 1 \\ 1 & 1 & 0 & 0 & 0 & 1 & 0 & 0 & 1 & 0 & 1 & 1 \end{bmatrix}$$

While trying to extend A_4 to an array with five constraints by adjoining a row x', we may, without loss of generality, take the first component of x' to be 1. It is then easy to verify that x' must contain either four or two 1's in positions 3, 4, 5 and 6. Hence the only possible candidates for x' are the following:

$$\begin{aligned}
x_0' &= (1\ 1\ 1\ 1\ 1\ 1\ 0\ 0\ 0\ 0\ 0\ 0) \\
y_1' &= (1\ 0\ 1\ 1\ 0\ 0\ 0\ 1\ 0\ 0\ 1\ 1) \\
y_2' &= (1\ 0\ 1\ 1\ 0\ 0\ 0\ 1\ 0\ 1\ 0\ 1) \\
y_3' &= (1\ 0\ 0\ 0\ 1\ 1\ 1\ 1\ 0\ 0\ 1\ 0) \\
y_4' &= (1\ 0\ 0\ 0\ 1\ 1\ 1\ 0\ 1\ 1\ 0\ 0) \\
y_5' &= (1\ 0\ 1\ 0\ 1\ 0\ 0\ 0\ 1\ 1\ 1\ 0) \\
y_6' &= (1\ 0\ 1\ 0\ 1\ 0\ 1\ 0\ 0\ 0\ 1\ 1) \\
y_7' &= (1\ 0\ 0\ 1\ 0\ 1\ 0\ 1\ 1\ 1\ 0\ 0) \\
y_8' &= (1\ 0\ 0\ 1\ 0\ 1\ 1\ 1\ 0\ 0\ 0\ 1) \\
y_9' &= (1\ 0\ 1\ 0\ 0\ 1\ 1\ 0\ 0\ 1\ 0\ 1) \\
y_{10}' &= (1\ 0\ 1\ 0\ 0\ 1\ 0\ 1\ 0\ 1\ 1\ 0) \\
y_{11}' &= (1\ 0\ 0\ 1\ 1\ 0\ 1\ 0\ 1\ 0\ 0\ 1) \\
y_{12}' &= (1\ 0\ 0\ 1\ 1\ 0\ 0\ 1\ 1\ 0\ 1\ 0).
\end{aligned}$$

Thus A_4 can be extended to an array with five constraints by adjoining either x_0' or y_i'.

We note that x_0' is orthogonal to any vector y_i and y_i' is orthogonal to y_j' if and only if i and j are of the same parity, i.e., either both even or both odd. It now follows that any orthogonal array with four or more constraints can always be extended to an array with one more constraint and hence to a maximal array with eleven constraints.

We now prove that the maximal array $\boldsymbol{A}(12, 11, 2, 2)$ is unique. To prove this it is sufficient to prove that there is up to an isomorphism a unique solution to the BIBD with $v = b = 11$, $r = k = 5$, $\lambda = 2$

Now suppose that there exists a BIBD D with

$$D: v = b = 11, \quad r = k = 5, \quad \lambda = 2.$$

Then utilizing the fact that any two blocks of D have two treatments in common, it follows that by omitting a block of D and omitting those treatments which occur in this block, the remaining blocks give a BIBD D_1 with

$$D_1 : v = 6, \ b = 10, \ r = 5, \ k = 3, \ \lambda = 2.$$

Conversely from [6] it follows that D_1 can be embedded uniquely up to an isomorphism in the design D. It is thus sufficient to prove that the design D_1 is unique up to an isomorphism. From [6] we know that the structure of the blocks of D_1 corresponds to the triangular association scheme T_5 [5]. This means that the blocks of D_1 can be numbered from 1, 2 up to 10 and can be exhibited in the scheme

$$(4.5) \qquad \begin{bmatrix} x & 1 & 2 & 3 & 4 \\ 1 & x & 5 & 6 & 7 \\ 2 & 5 & x & 8 & 9 \\ 3 & 6 & 8 & x & 10 \\ 4 & 7 & 9 & 10 & x \end{bmatrix}$$

(where x in the diagonal positions means a blank) and such that any block i has exactly one treatment in common with $n_1 = 6$ blocks which lie in the same row or same column containing i

and two treatments in common with remaining $n_2 = 3$ blocks. We will call two blocks of D_1 to be i-accociates if and only if they have i treatments in common, $i = 1, 2$. If B_m and B_n are two blocks of D_1 $(m \neq n)$ such that they are i-associates, we will denote this fact by writing $(B_m, B_n) = i$.

Let us denote the ten blocks numbered according to the association scheme (4.5) by $B_1, B_2, \ldots, B_{10}$. Then any two of the blocks B_1, B_2, B_3, B_4 are 1-associates and have exactly one treatment in common. Hence without loss of generality we can take B_1 and B_2 as follows:

B_1	B_2
1	1
1	0
1	0
0	1
0	1
0	0

Now if B_3 had 1 in the first position, then since B_1 and B_3 are 1-associates B_3 will have 0 in the next two positions and hence two 1's in the last three positions. This will imply that B_2 and B_3 are 2-associates which is a contradiction. Thus no three blocks which are mutually 1-associates have 1 in the same position. Thus we have either

(4.6)

B_1	B_2	B_3	B_4
1	1	0	0
1	0	1	0
1	0	0	1
0	1	1	0
0	1	0	1
0	0	1	1

or

B_1	B_2	B_3	B_4
1	1	0	0
1	0	1	0
1	0	0	1
0	1	0	1
0	1	1	0
0	0	1	1

The latter arrangement is however isomorphic to the previous one as can be seen by interchanging rows numbered 4 and 5. Hence without loss of generality we take the arrangement given by (4.6). Now since B_1, B_2, B_5 are mutually 1-associates B_5 cannot have 1 in the first position. If B_5 has (0 1 0) in the first three positions, then from $(B_2, B_5) = 1$ it must have 1 in the last position and exactly one 1 in positions numbered 4 and 5. If it had 1 in position numbered 4, then B_3 and B_5 will have three treatments in common which is impossible. Hence B_5 is given by (0 1 0 0 1 1). Similarly, if we start with (0 0 1) in the first three positions B_5 is given by (0 0 1 1 0 1). It is however easy to verify that these two ways of adjoining B_5 are isomorphic as can be seen by permuting the rows in the five blocks by the permutation $2 \to 3$, $3 \to 2$, $4 \to 5$, $5 \to 4$. Hence without loss of generality we have

B_1	B_2	B_3	B_4	B_5
1	1	0	0	0
1	0	1	0	1
1	0	0	1	0
0	1	1	0	0
0	1	0	1	1
0	0	1	1	1

Now since B_1, B_3, B_6 are mutually 1-associates B_6 must have 0 in the second position and exactly one 1 in the first three positions. The start (1 0 0) however cannot be completed to a B_6

satisfying the conditions implied by the association scheme. It is then easy to see that the unique B_6 is given by (0 0 1 1 1 0) giving

B_1	B_2	B_3	B_4	B_5	B_6
1	1	0	0	0	0
1	0	1	0	1	0
1	0	0	1	0	1
0	1	1	0	0	1
0	1	0	1	1	1
0	0	1	1	1	0

We note that (B_1, B_5, B_7), (B_1, B_6, B_7) and (B_5, B_6, B_7) form sets of mutual 1-associates and hence the unique blocks B_7 is given by (1 0 0 1 0 1). Similarly it is easy to verify that the blocks B_8, B_9 and B_{10} are uniquely determined giving the unique solution for D_1 given by

	B_1	B_2	B_3	B_4	B_5	B_6	B_7	B_8	B_9	B_{10}
	1	1	0	0	0	0	1	1	0	1
	1	0	1	0	1	0	0	0	1	1
	1	0	0	1	0	1	0	1	1	0
D_1 :	0	1	1	0	0	1	1	0	1	0
	0	1	0	1	1	1	0	0	0	1
	0	0	1	1	1	0	1	1	0	0

The above solution can be extended [6] to the unique (up to isomorphism) solution for $v = b = 11$, $r = k = 5$, $\lambda = 2$ where the blocks are given by (1, 2, 3, 7, 8), (1, 4, 5, 7, 9), (2, 4, 6, 7, 10), (3, 5, 6, 7, 11), (2, 5, 6, 8, 9), (3, 4, 5, 8, 10), (1, 4, 6, 8, 11), (1, 3, 6, 9, 10), (2, 3, 4, 9, 11), (1, 2, 5, 10, 11) and (7, 8, 9, 10, 11). It is then easy to verify that with the following permutations of the treatments

$$2 \to 11,\ \ 3 \to 2,\ \ 4 \to 9,\ \ 5 \to 6,\ \ 6 \to 8,$$
$$8 \to 4,\ \ 9 \to 5,\ \ 10 \to 3,\ \ 11 \to 10$$

the above solution reduces to the cyclic design developed (mod 11), from the difference set (1, 5, 6, 7, 9).

It is obvious that the above results imply the possibility of embedding a matrix $\boldsymbol{H}_{p,12}$ into a unique Hadamard matrix $\boldsymbol{H}_{12}$ if $p \geqslant 5$, Similarly if $p \geqslant 4$, a design with $v = p$, $b = 11$, $r = 5$, $\lambda = 2$ can be uniquely embedded in a BIBD with $v = b = 11$, $r = k = 5$, $\lambda = 2$.

5. CONCLUDING REMARKS

It is interesting to find how far the results of Section 2 can be improved for higher values of t. It is proposed to consider this problem in a subsequent communication.

References

[1] Baumert, L. D. and M. Hall, Jr. (1965). "A New Construction for Hadamard Matrices," *Bull. Amer. Math. Soc.*, **71**, 169–170.

[2] Bose, R. C. (1939). "On the Construction of Balanced Incomplete Block Designs," *Ann. Eugenics*, **1**, 353–399.

[3] Bose, R. C. and Bush, K. A. (1952). "Orthogonal Arrays of Strength Two and Three," *Ann. Math. Statist.*, **23**, 508–524.

[4] Schutzenberger, M. B. (1953). "Sur un problème de codage binaire," *Publ. Inst. Statist. Univ. Paris*, **2**, 125–127.

[5] Shrikhande, S. S. (1959). "On a Characterization of the Triangular Association Scheme," *Ann. Math. Statist.*, **30**, 39–47.

[6] Shrikhande, S. S. (1960). "Relations Between Certain Incomplete Block Designs," in *Contributions to Probability and Statistics, Essays in Honor of Harold Hotelling*, Stanford Univ. Press, pp. 388–395.

(*Received Jan. 1, 1966*)

CHAPTER 36

Optimal Balanced 2^m Fractional Factorial Designs

J. N. SRIVASTAVA, *University of Nebraska*

1. INTRODUCTION AND SUMMARY

It has been established by Bose [2] that the problem of construction of optimal orthogonal fractional factorial designs is mathematically equivalent to that of obtaining optimal linear error-correcting and error-detecting codes. Indeed, as he has shown, both problems resolve themselves to a certain problem in finite geometries, called *the packing problem.* Besides this link between coding theory and experimental design, a connection was found by Bose and Burton [3] (see also [4]) between the above coding theory problem and the theory of linear integer programing. These two connections together constitute an indirect link between the problem of construction of optimal orthogonal fractions and programing problems. However, orthogonal fractions are generally *uneconomic,* i.e. they involve more than the desirable number of *treatment combinations* or assemblies. Thus, one needs to consider non-orthogonal or irregular fractions as well (see for example [1, 5]). In this paper, *optimal balanced* irregular fractions of the 2^m type are investigated, and it is shown that for such fractions the problem of construction is reducible to an integer programing problem. This result reinforces the above indirect link between design theory and programing, and makes it wider.

It has been shown that a balanced factorial fraction is necessarily a partially balanced array (for definition, see Chakravarti [7], or [6]), and vice versa. Besides establishing the above-mentioned link, a main aim of this paper is to obtain parameters of certain classes of such arrays which are *optimal* in the sense that the number of assemblies N is of desirable order and, for given N, the covariance matrix of the estimates has certain optimal features. Another aim is to present a family of necessary conditions for the existence of certain classes of partially balanced arrays.

So far as the author is aware, the results in this paper are all new. They are dedicated to the memory of Professor S. N. Roy, whose life and work will always continue to inspire the author.

2. COMPLETELY BALANCED FRACTIONAL DESIGNS

Consider, for simplicity, a 2^m factorial, written $FE(2^m)$ for brevity. The treatment combinations or assemblies here are $a_1^{j_1} a_2^{j_2} \dots a_m^{j_m}$, where $j_r = 0$ or 1, a_r denotes rth factor, and j_r is the level of the rth factor. The various main effects and interactions will be written, as usual, as μ (the general mean) A_i (main effect of ith factor), A_iA_j or simply A_{ij} (the interaction between factors i and j), etc. We shall restrict our attention (throughout this paper) to effects involving up to two factors only, the number of which is obviously $\nu = 1+m+\binom{m}{2}$. We shall consider the problem of choosing a suitable set of assemblies (called a 'fraction') for estimating these ν parameters.

Let

$$L' = (\mu; A_1, A_2, \ldots, A_m; A_{12}, A_{13}, \ldots, A_{1m}, A_{23}, \ldots, A_{m-1, m})$$
$$= (\{\mu\}\ ;\ \{A_i\}\ ;\ \{A_{ij}\}),$$

say, be the vector of unknown parameters. Let T be any fraction, and suppose one observation is taken on each assembly in T using a completely randomized design, so that no block effects are involved. Suppose that under this design all parameters in $\boldsymbol{L}$ are estimatable, and $(\hat{\boldsymbol{L}})_T$ is the usual best linear unbiased estimate of T. Let $(\boldsymbol{V})_T$ be the variance covariance matrix of $(\hat{\boldsymbol{L}})_T$. Whenever no emphasis on T is needed, the suffix in these two symbols will be omitted.

A fractional factorial design is called completely balanced if $\boldsymbol{V}$ is symmetric with respect to the m factors, i.e. $\boldsymbol{V}$ should be invariant under any permutation of the factors. Thus if $\{A_{r_1}, \ldots, A_{r_m}\}$ is *any* permutation of $\{A_1, \ldots, A_m\}$, then the new $\boldsymbol{L}$ is $\boldsymbol{L}^* = (\mu\ ; A_{r_1}, \ldots, A_{r_m}; A_{r_1 r_2}, \ldots, A_{r_1 r_m}, A_{r_2 r_3}, \ldots, A_{r_{m-1} r_m})$. Let the rows and columns of $\boldsymbol{V}$ be permuted so as to correspond to $\boldsymbol{L}^*$, and let $\boldsymbol{V}^*$ be the matrix so obtained. Then the requirement for complete balance is that $\boldsymbol{V} = \boldsymbol{V}^*$.

We now obtain a necessary and sufficient condition for a fraction T to be completely balanced. For this purpose we use the λ-operator discussed in detail in [5]. Consider the symbols a_i^j $(i = 1, \ldots, m; j = 0, 1)$. Let θ be a product of r such symbols $(1 \leqslant r \leqslant m)$, say $\theta = a_{i_1}^{j_1} a_{i_2}^{j_2} \ldots a_{i_r}^{j_r}$ where the i's are all distinct. Then θ may be called a subtreatment or subassembly, since it clearly is a treatment combination in a 2^r factorial in which factors $i_1, \ldots, i_r$ are used. We defined

(1) $\lambda(\theta, T) =$ Number of assemblies in T which have θ as a subassembly.

If T is not to be emphasized, we just write $\lambda(\theta)$. Also we write

$$\lambda\left(a_{i_1}^{j_1} a_{i_2}^{j2} \ldots a_{i_r}^{j_r}\right) \equiv \lambda_{i_1 i_2 \ldots i_r}^{j_1 j_1 \ldots j_r} \tag{2}$$

Let $\theta_1, \theta_2, \ldots$ be any subassemblies each involving any number of factors whatsoever, and let $g_1, g_2, \ldots$ be scalars. Let $P = (g_1\theta_1 + g_2\theta_2 + \ldots)$ be a linear function of the subassemblies. Then the λ-operator is defined over the set of all possible P's by

$$\lambda(P) = g_1\lambda(\theta_1) + g_2\lambda(\theta_2) + \ldots . \tag{3}$$

Notice that P is just a polynomial in the symbols a_i^j. Thus for example, we could have $P = (a_i^1 - 2a_i^0)(a_j^1)$. Then

$$\lambda(P) = (\lambda_{ij}^{11}) - 2(\lambda_{ij}^{01}).$$

In [5], it is shown that the normal equations for estimating $\boldsymbol{L}$ are of the form $\boldsymbol{M}\hat{\boldsymbol{L}} = \boldsymbol{z}$, where $\boldsymbol{M}(\nu \times \nu)$ is a known matrix, and $\boldsymbol{z}$ is a known linear function of observations. If each element in $\boldsymbol{L}$ is estimable, then $\boldsymbol{M}$ is nonsingular. In this case we also call T to be nonsingular and we have $\boldsymbol{V}_T = \boldsymbol{M}^{-1}$. Consider $\boldsymbol{M}$. Its rows and columns correspond in order to the elements in $\boldsymbol{L}$. Let $\epsilon(e, e_2)$ denote the element in $\boldsymbol{M}$ which stands at the intersection of the row corresponding to e_1, and the column corresponding to e_2, where e_1, e_2 are any two not necessarily distinct elements of $\boldsymbol{L}$. Typical elements of $\boldsymbol{M}$ are presented below (see [5] for proof) in terms of the λ's:

$$\gamma_1 = \epsilon(\mu, \mu) = N = \epsilon\left(A_{i_1}, A_{i_1}\right) = \epsilon\left(A_{i_1i_2}, A_{i_1i_2}\right) \tag{4}$$

$$\gamma_2 = \epsilon\left(\mu, A_{i_1}\right) = \lambda\left(a_{i_1}^1 - a_{i_1}^0\right) = \epsilon\left(A_{i_2}, A_{i_1i_2}\right) \tag{5}$$

$$\gamma_3 = \epsilon\left(\mu, A_{i_1i_2}\right) = \lambda\left[\left(a_{i_1}^1 - a_{i_1}^0\right)\left(a_{i_2}^1 - a_{i_2}^0\right)\right] \tag{6}$$

$$= \epsilon\left(A_{i_1}, A_{i_2}\right) = \epsilon\left(A_{i_1i_3}, A_{i_2i_3}\right)$$

$$\gamma_4 = \epsilon\left(A_{i_1}, A_{i_2i_3}\right) = \lambda\left[\left(a_{i_1}^1 - a_{i_1}^0\right)\left(a_{i_2}^1 - a_{i_2}^0\right)\left(a_{i_3}^1 - a_{i_3}^0\right)\right] \tag{7}$$

$$\gamma_5 = \epsilon\left(A_{i_1i_2}, A_{i_3i_4}\right) = \lambda\left[\left(a_{i_1}^1 - a_{i_1}^0\right)\ldots\left(a_{i_4}^1 - a_{i_4}^0\right)\right] \tag{8}$$

where $i_1, \ldots, i_4$ are all distinct, and N is the total number of assemblies in T. For 'complete balance' of T we therefore require that each of the expressions in (4)—(8) is independent of $i_1, \ldots, i_4$. It can be shown that this is equivalent to the condition that for all possible $j_1, \ldots, j_4$,

$$\lambda_{i_1 i_2 i_3 i_4}^{j_1 j_2 j_3 j_4} = \lambda^{j_1 j_2 j_3 j_4}, \tag{9}$$

where the r.h.s. is a constant independent of the four chosen factors $i_1, \ldots, i_4$. A fraction T (belonging to $FE(2^m)$) which satisfies the conditions (8) is called a partially balanced array of strength 4, m constraints and 2 symbols. Thus T is completely balanced if and only if it is a partially balanced array of strength 4.

A close look at (9) shows that the 5 constants μ_i defined below occur naturally as parameters of the array:

$$\mu_0 = \lambda^{0000},\ \mu_1 = \lambda^{1000},\ \mu_2 = \lambda^{1100},\ \mu_3 = \lambda^{1110},\ \mu_4 = \lambda^{1111}. \tag{10}$$

It can be easily checked that (10) implies

$$\begin{aligned} \gamma_1 &= N = \mu_0+4\mu_1+6\mu_2+4\mu_3+\mu_4 \\ \gamma_2 &= (\mu_4-\mu_0)+2(\mu_3-\mu_1) \\ \gamma_3 &= \mu_4-2\mu_2+\mu_0 \\ \gamma_4 &= (\mu_4-\mu_0)-2(\mu_3-\mu_1) \\ \gamma_5 &= \mu_0-4\mu_1+6\mu_2-4\mu_3+\mu_4. \end{aligned} \tag{11}$$

Thus if T is a partially balanced array, $(V)_T$ is fully determined by the μ_i.

3. OPTIMALITY CRITERIA

We shall now consider different optimality criteria for selecting T. A basic criterion for screening undesirable fractions in most situations is that the selected T should be economic. This implies that $(N-\nu)$, which equals the number of degrees of freedom for error, should be desirably small. In this paper we generally take $5 \leqslant N-\nu \leqslant 30$. A fraction is called orthogonal, if $(\boldsymbol{V})_T$ is a diagonal matrix. As mentioned in the summary, the value of N (and hence $N-\nu$) is generally too large for orthogonal fractions.

The other criteria deal with the properties of $(\boldsymbol{V})_T$. Let T_1 and T_2 be two competing fractions. Then T_1 is said to be *better than* T_2 according to (a) the largest root criterion, if $\underset{\max}{\text{ch}}\ (\boldsymbol{V})_{T_1} < \underset{\max}{\text{ch}}\ (\boldsymbol{V})_{T_2}$, (b) the trace criterion, if $\text{tr}(\boldsymbol{V})_{T_1} < \text{tr}\ (\boldsymbol{V})_{T_2}$, (c) the generalized variance or determinant criterion, if $|(\boldsymbol{V})_{T_1}| < |(\boldsymbol{V})_{T_2}|$. For any T, the three criteria have the following physical interpretation.

(a) $\underset{\max}{\text{ch}}\ (\boldsymbol{V}) = \sup_{\boldsymbol{b}'\boldsymbol{b}=1} \{\text{var}\ (\boldsymbol{b}'\hat{\boldsymbol{L}})\}.$

Thus choosing T so as to minimize $\underset{\max}{\text{ch}}\ (\boldsymbol{V})_T$ means taking that T for which the maximum possible variance of any linear parametric function is a minimum. In this sense, the largest root is a minimax criterion.

(b) $\text{tr}(\boldsymbol{V}) \propto \int_{\boldsymbol{b}'\boldsymbol{b}=1} [\text{var}(\boldsymbol{b}'\hat{\boldsymbol{L}})]\, d\boldsymbol{b}.$

That is, tr $(\boldsymbol{V})$ is proportioned to the average of the variances of all normalised linear parametric functions. Thus in a sense, the trace criterion refers to the average variance.

(c) $|(\boldsymbol{V})|$ is proportioned to the volume of the ellipsoid of concentration. Thus, in a sense, it refers to the volume of the region within which the true parametric point may lie with a certain probability.

The author believes that the last criterion is not particularly suitable for comparison of fractional designs, since $|\boldsymbol{V}|$ could be small when variances are small and/or when the correlations between the different elements of $\hat{\boldsymbol{L}}$ are large. If e_1 and e_2 are two elements of $\boldsymbol{L}$, a large correlation between $\hat{e}_1$ and $\hat{e}_2$ indicates in a sense a partial confounding of e_1 and e_2. Thus large correlations are undesirable since confounding is, and hence $|\boldsymbol{V}|$ does not seem to be appropriate. Among the other two, the largest root is mathematically more tractable and convenient to study. However the mathematical approach in this paper is such that *all roots* of $\boldsymbol{V}$ are obtained first. Hence using them the interested reader can find the parameters of an optimal design under any of the criteria he likes.

4. SEMIORTHOGONAL COMPLETELY BALANCED (SOCB) FRACTIONS FOR 2^m DESIGNS

Consider now the matrix $\boldsymbol{V}$ for a fraction $T \in FE(2^m)$. Corresponding to the partition of $\boldsymbol{L}'$ given by $(\{\mu\}; \{A_i\}; \{A_{ij}\})$, we write following the notation of [5],

$$\boldsymbol{M} = \begin{bmatrix} \boldsymbol{M}_{00} & \boldsymbol{M}_{01} & \boldsymbol{M}_{02} \\ & \boldsymbol{M}_{11} & \boldsymbol{M}_{12} \\ \text{Sym.} & & \boldsymbol{M}_{22} \end{bmatrix} \tag{12}$$

where the $\boldsymbol{M}_{ij}$ are submatrices of appropriate order, and where the partitioning of $\boldsymbol{M}$ is induced by the partitioning of $\boldsymbol{L}$. Thus for example $\boldsymbol{M}_{01}$ is $(1 \times m)$, and its rows correspond to μ and columns to $\{A_i\}$ i.e. the m effects $A_1, \ldots, A_m$ respectively. Similarly $\boldsymbol{M}_{12}$ is $m \times \binom{m}{2}$ with rows and columns corresponding to the sets $\{A_i\}$ and $\{A_{ij}\}$ respectively, and so on.

A fraction is called semiorthogonal, if all the off-diagonal submatrices of $(\boldsymbol{M})$, here $\boldsymbol{M}_{01}$, $\boldsymbol{M}_{02}$ and $\boldsymbol{M}_{12}$, are zero. This means that the estimate of any element e belonging to *any* one out of the three sets $\{\mu\}$, $\{A_i\}$ and $\{A_{ij}\}$ is uncorrelated with the estimate of any element e^* belonging to a *different* set. Using (5)—(7) and (11), the condition for a fraction to be semiorthogonal completely balanced (SOCB) becomes

$$\gamma_2 = \gamma_3 = \gamma_4 = 0, \quad \text{or} \tag{13}$$

$$\mu_0 = \mu_2 = \mu_4; \ \mu_1 = \mu_3 \tag{14}$$

Consider the roots of $(\boldsymbol{M})_T$, where T is SOCB. The (scalar) matrix $\boldsymbol{M}_{00}$ equals N, and $\boldsymbol{M}_{11} = N\boldsymbol{I}_m$ (where $\boldsymbol{I}_v$ is a $v \times v$ identity matrix). Hence the problem is nontrivial only for $\boldsymbol{M}_{22}$, which using (4)-(14) is given by

$$\boldsymbol{M}_{22} = \sum_{\alpha=0}^{2} b_{22}^{\alpha} B_{22}^{\alpha}, \text{ where} \tag{15}$$

$$\begin{aligned} b_{22}^{0} &= \gamma_1 = 8(\mu_0+\mu_1) \\ b_{22}^{1} &= \gamma_3 = 0 \\ b_{22}^{2} &= \gamma_4 = 8(\mu_0-\mu_1); \end{aligned} \tag{16}$$

$$B_{22}^{0} = I_{m^0} \qquad m^0 = \binom{m}{2}; \tag{17}$$

(18) $B_{22}^{1}(A_{ij}, A_{kl}) = 1$, if exactly one element among the unordered pairs (i, j) and (k, l) is common
$= 0$, otherwise;

(19) $B_{22}^{2}(A_{ij}. A_{kl}) = 1$, if none among (i, j) and (k, l) are common
$= 0$, otherwise;

and where $B_{22}^{\alpha}(A_{ij}, A_{kl})$ denotes the element of B_{22}^{α} in the row corresponding to A_{ij} and the column corresponding to A_{kl}.

Let $\boldsymbol{\delta}_{22}^{\alpha}$ $(\alpha = 0, 1, 2)$ denote the vector containing the $\binom{m}{2}$ roots of B_{22}^{α}, arranged in a certain appropriate order, such that the roots of $\sum_{\alpha=0}^{2} g_{\alpha} B_{22}^{\alpha}$ are given by the vector $\Sigma g_{\alpha} \boldsymbol{\delta}_{22}^{\alpha}$, where g_{α} are any real numbers. In [6], it is shown that such arrangement of the roots of B_{22}^{α} is possible, and is exhibited in the matrix

$$\Delta' = \begin{bmatrix} \boldsymbol{\delta}_{22}^{0\prime} \\ \boldsymbol{\delta}_{22}^{1\prime} \\ \boldsymbol{\delta}_{22}^{2\prime} \end{bmatrix} = \begin{bmatrix} \mathrm{I} & \boldsymbol{J}'_{m-1} & \boldsymbol{J}'_{m'} \\ 2(m-2) & (m-4)\boldsymbol{J}'_{m-1} & (-2)\boldsymbol{J}'_{m'} \\ m'' & -(m-3)\boldsymbol{J}'_{m-1} & \boldsymbol{J}'_{m'} \end{bmatrix} \tag{20}$$

where

$$m' = \frac{1}{2}m(m-3), \; m'' = \frac{1}{2}(m-2)(m-3), \tag{21}$$

and J'_v is a vector with v coordinates each of whose elements is unity. Thus the (possibly) distinct roots of M_{22} are

(22) $$\pi_1 = 8(\mu_0+\mu_1)+\frac{1}{2}(m-2)(m-3)\,.\,8(\mu_0-\mu_1)$$

(23) $$\pi_2 = 8(\mu_0+\mu_1)-(m-3)\,.\,8(\mu_0-\mu_1)$$

(24) $$\pi_3 = 8(\mu_0+\mu_1)+8(\mu_0-\mu_1) = 16\mu_0.$$

Thus the matrix M for an SOCB fraction from $FE(2^m)$ has four possibly distinct roots N, π_1, π_2, and π_3, with respective multiplicities $m+1$, 1, $m-1$ and m'. In order that each element of L be estimable an obvious necessary and sufficient condition is that M be nonsingular. It can be easily checked that a necessary consequence of this is that $N \geqslant \nu_m$. Also, for any T, the matrix $(M)_T$ is always at least positive semidefinite. This implies that we must always have $\pi_i \geqslant 0$. Thus if we have a set of μ's such that *some* root $\pi_i < 0$, then a partially balanced array having these μ's as its parameters does not exist. The same is the case with any value of μ's for which $N < \nu_m$, even though $\pi_i > 0$, for all i. This leads us to

Theorem 1. *Consider a partially balanced array A (with m constraint ($m > 4$) and 2 symbols), such that the parameters satisfy the relations $\mu_0 = \mu_2 = \mu_4$ and $\mu_1 = \mu_3$. In order that A exists it is necessary that*

(25) $$N = 8(\mu_0+\mu_1) \geqslant \nu_m = \left[1+\frac{m(m+1)}{2}\right]$$

(26) $$\left[\frac{\left(\frac{m-2}{2}\right)-1}{\left(\frac{m-2}{2}\right)+1}\right]\mu_1 \leqslant \mu_0 \leqslant \mu_1\left[\frac{(m-2)}{(m-4)}\right],$$

where the condition (25) *is to be omitted if an equality is attained in* (26). *Furthermore for the nonsingularity of T (as a fraction), we must have inequality in* (26).

Consider now the question of optimal design, say with respect to the largest root criterion. It is clear that a fraction which minimizes the largest root of V maximizes the smallest root of $\boldsymbol{M}$. A look at the values of π_i's shows that if $(\mu_0-\mu_1) \geqslant 0$, $\underset{\max}{\text{ch}}\ (\boldsymbol{M})$ (the smallest root of $\boldsymbol{M}$) equals π_2; otherwise it equals π_1. It also follows that for $c > 0$ and fixed N, a fraction with $(\mu_0-\mu_1) = c$ is better than all fractions with $(\mu_0-\mu_1) \leqslant -2c(m-2)^{-1}$.

Coming to the trace criterion, we observe that the roots of $\boldsymbol{M}$ are N, π_1, π_2, and π_3 with respective multiplicities $m+1$, 1, $(m-1)$, and $m(m-3)/2$. Hence

$$\text{(27)} \qquad \text{tr}\ \boldsymbol{V} = \left[\frac{m+1}{N} + \frac{1}{N+(m-2)(m-3)N_1/2} + \frac{(m-1)}{N-(m-3)N_1} + \frac{m(m-3)}{2(N+N_1)}\right],$$

where $N_1 = 8(\mu_0-\mu_1)$. This formula throws light on the nature of the above two optimality criteria. Thus $\underset{\max}{\text{ch}}\ (\boldsymbol{V})$ is either π_1^{-1} or π_2^{-1} both of which have low multiplicities, while the root π_3^{-1} which generally has higher multiplicity does not come into the picture. The trace criterion on the other hand does take this into account. Since the multiplicities are widely different, it seems that at least for this class of 2^m factorial fractions, the trace criterion may give a better overall picture of the *goodness* of a fraction than the largest root.

In Table 1, we present the parameters of some arrays for which N is of the desirable order, and which for given N are optimal in the class of SOCB fractions from the viewpoint of trace criterion. The values of the π_i and trace ($\boldsymbol{V}$) are also given. The orthogonal cases (i.e. when $\mu_0 = \mu_1$) are not mentioned, since they are well known.

Consider now the design aspect. Since each root of $\boldsymbol{V}$ is the reciprocal of a root of $\boldsymbol{M}$ (granting $\boldsymbol{M}$ nonsingular) the roots of $\boldsymbol{V}$ are $\frac{1}{N}, \frac{1}{\pi_1}, \frac{1}{\pi_2}, \frac{1}{\pi_3}$ with multiplicities $(m+1)$, 1, $(m-1)$ and m' respectively. Thus any SOCB fraction could easily be evaluated in terms of any optimality criterion involving just the roots of the variance matrix $\boldsymbol{V}$.

For $5 \leqslant m \leqslant 10$, the following table gives the parameters of some arrays for which N is of the desirable order, and other useful information.

TABLE 1

Ser. No.	m	ν_m	μ_0	μ_1	N	π_1	π_2	π_3	(tr $\boldsymbol{V}$)
1	5	16	2	1	24	48	8	32	0.9271
2	6	22	2	1	24	72	0	16	
3	6	22	3	2	40	88	48	16	0.8513
4	7	29	3	2	40	120	8	48	1.2500
5	7	29	4	3	56	136	24	64	0.6191
6	8	37	4	3	56	176	16	64	0.9164
7	9	46	4	3	56	224	8	64	1.6050
8	9	46	5	4	72	240	24	80	0.8139
9	10	56	5	4	72	296	16	80	1.1565
10	10	56	6	5	88	312	32	96	0.7741

5. CONNECTION WITH LINEAR PROGRAMMING

We have seen that in the case of a SOCB fraction T, $\boldsymbol{M}$ has at most four distinct roots $N(=\pi_0$ say), π_1, π_2, π_3. Let these arranged in order be $\pi_{(0)} \leqslant \pi_{(1)} \leqslant \pi_{(2)} \leqslant \pi_{(3)}$.

Now define the 2^m variables $x_1, x_2, \ldots, x_{2^m}$.

(28) $$x_r = \lambda^{j_1 j_2 \ldots j_m}, \text{ if and only if}$$

(29) $$r = 1+j_1+2j_2+2^2j_3+\ldots+2^{m-1}j_m, \text{ where } j_s = 0, 1, \text{ all } s.$$

It is clear then that the μ_i are all known linear functions of the x_r, and that the $(2^m \times 1)$ vector $\boldsymbol{x}$ containing the x_r's completely specifies and is specified by T. This means that there exists a known matrix $\boldsymbol{A}$, such that the conditions for the fraction T to be SOCB could be written

(30) $$\boldsymbol{A}\boldsymbol{x} = \boldsymbol{\mu}$$

where $\boldsymbol{\mu}$ contains the quantities μ_i in some appropriate order, and where each μ_i may occur more than once in the vector $\boldsymbol{\mu}$. Also observe that

$$(31) \qquad \sum_r x_r = N,$$

and that there exist *known* vectors $\boldsymbol{b}_i$ such that $\pi_{(i)} = \boldsymbol{b}_i'\boldsymbol{x}$, so that for a nonsingular T, $\boldsymbol{x}$ obeys

$$(32) \qquad \boldsymbol{b}_i'\boldsymbol{x} > 0, \quad i = 0, \ldots, 3.$$

Also clearly,

$$(33) \qquad x_r \text{ is an integer and } x_r \geqslant 0, \; r = 1, \ldots, 2^m.$$

If we are interested in obtaining a nonsingular balanced fraction T for which N is a minimum, the corresponding $\boldsymbol{x}$ should satisfy the linear conditions (30), (32) and (33), and should be such that the *objective function* (31) is a minimum. This is a linear integer programing problem. Suppose now that aside from being nonsingulr, we want T to be optimal with respect to the largest root criterion. For this we may maximise $\underset{\min}{\text{ch}}(\boldsymbol{M})$ for a given N, or minimize N for a given value of $\underset{\min}{\text{ch}}(\boldsymbol{M})$. The best way seems to be to choose an appropriate number $c > 0$, and minimize N under the condition $\underset{\min}{\text{ch}}(\boldsymbol{M}) \geqslant c$. This in effect is the same programming problem as before except that the number zero on the r.h.s. of (32) replaced by c.

Note that the number of variables in our problem is 2^m. We might recall (see, e.g. [4]) that in the coding theory situation also the number is of the form 2^{m^*}. The important difference however is that while values of $m > 10$ are rarely of interest, m^* usually ranges between 30 to 200. This observation seems to justify the remark that programming methods are more likely to succeed in the theory of factorial fractions than in coding theory.

6. FRACTIONS WHICH ARE (1,0) SYMMETRIC WITH RESPECT TO TRIPLETS

It has been shown in [5] that a necessary and sufficient condition that $\boldsymbol{M}_{01}$ and $\boldsymbol{M}_{12}$ of (12) be zero matrices is that the fraction

T be (1, 0) symmetric with respect to triplets, i.e. for any three factors a_{i_1}, a_{i_2} and a_{i_3}, we have

$$\lambda\left(a_{i_1}^{j_1} a_{i_2}^{j_2} a_{i_3}^{j_3}\right) = \lambda\left(a_{i_1}^{j_1'} a_{i_2}^{j_2'} a_{i_3}^{j_3'}\right), \tag{34}$$

where the j's and j''s take values 0, 1 and 2, and the vector (j_1', j_2', j_3') is obtained from (j_1, j_2, j_3) by interchanging the symbols 2 and 0. Thus, in case T has this property, the matrix $\boldsymbol{M}$ is (apart from a permutation of rows and columns) a direct sum of $\boldsymbol{M}_{11}$ and the matrix

$$\boldsymbol{M}^* = \begin{bmatrix} \boldsymbol{M}_{00} & \boldsymbol{M}_{02} \\ \boldsymbol{M}_{20} & \boldsymbol{M}_{22} \end{bmatrix} \tag{35}$$

To obtain the roots of $\boldsymbol{M}^*$ when T is completely balanced, we observe that for any indeterminate θ,

$$\begin{aligned} |\boldsymbol{M}^*-\theta \boldsymbol{I}| &= |\boldsymbol{M}_{00}-\theta \boldsymbol{I}|\,|[\boldsymbol{M}_{22}-\theta \boldsymbol{I}]-\boldsymbol{M}_{20}[\boldsymbol{M}_{00}-\theta \boldsymbol{I}]^{-1}\boldsymbol{M}_{02}]| \\ &= |\boldsymbol{M}_{00}-\theta \boldsymbol{I}|\,|\psi(\theta)-\theta \boldsymbol{I}|, \quad \text{say,} \end{aligned} \tag{36}$$

where the matrices $\boldsymbol{I}$ are of appropriate orders. From (4)-(8), recall that $\boldsymbol{M}_{00}$ is just γ_1, and $\boldsymbol{M}_{02} = \gamma_3 \boldsymbol{J}'$, where $\boldsymbol{J}'$ is a row vector with $\binom{m}{2}$ coordinates. Hence

$$\begin{aligned} \psi(\theta) &= \boldsymbol{M}_{22}-\gamma_3^2(\gamma_1-\theta)^{-1}\boldsymbol{J}\,\boldsymbol{J}' \\ &= \sum_{\alpha=0}^{2} (b_{22}^{\alpha}-\gamma_3^2(\gamma_1-\theta)^{-1})B_{22}^{\alpha}, \end{aligned}$$

from (15). Thus the vector (θ say) of roots of $\psi(\theta)$ is given by $\sum_{\alpha=0}^{2} [b_{22}^{\alpha}-\gamma_3^2(\gamma_1-\theta)^{-1}]\delta_{22}^{\alpha}$. A look at (20) shows that $\psi(\theta)$ will have only three (possibly) distinct roots, say θ_1, θ_2, θ_3, with respective multiplicities 1, $m-1$ and m'. Thus

$$|\boldsymbol{M}^*-\theta \boldsymbol{I}| = (\gamma_1-\theta)(\theta_1-\theta)(\theta_2-\theta)^{m-1}(\theta_3-\theta)^{m'}. \tag{37}$$

Also, we note that

$$\boldsymbol{v} = \sum_{\alpha=0}^{2} b_{22}^{\alpha}\delta_{22}^{\alpha}-\gamma_3^2(\gamma_1-\theta)^{-1} \sum_{\alpha=0}^{2} \delta_{22}^{\alpha} \tag{38}$$

$$\sum_{\alpha=0}^{2} b_{22}^{\alpha}\delta_{22}^{\alpha} = \text{vector of roots of } \boldsymbol{M}_{22}$$

$$\sum_{\alpha=0}^{2} \delta_{22}^{\alpha} = \text{vector of roots of } \sum_{\alpha=0}^{2} \boldsymbol{M}_{22}^{\alpha}$$

Hence from (20) — (24), we get

$$(39) \qquad \theta_1 = \pi_1^* - \gamma_3^2(\gamma_1-\theta)^{-1}\binom{m}{2}, \quad \theta_2 = \pi_2^*, \quad \theta_3 = \pi_3^*,$$

where π_1^*, π_2^* and π_3^* are the three distinct roots of $\boldsymbol{M}_{22}$, with respective multiplicities 1, $m-1$ and m'. Hence

$$(40) \quad |\boldsymbol{M}^* - \theta \boldsymbol{I}| = \left\{(\gamma_1-\theta)(\pi_1^*-\theta) - \gamma_3^2\binom{m}{2}\right\}(\pi_2^*-\theta)^{m-1}(\pi_3^*-\theta)^{m'},$$

and the four possibly distinct roots of $\boldsymbol{M}^*$ are π_2^*, π_3^* and the roots π_4^* and π_5^* of the quadratic $(\gamma_1-\theta)(\pi_1^*-\theta) = \gamma_3^2\binom{m}{2}$ This gives

$$(41) \qquad \pi_4^*,\ \pi_5^* = \frac{1}{2}(\gamma+\pi_1^*) \pm \sqrt{\frac{1}{4}(\gamma_1-\pi_1^*)^2 + \gamma_3^2\binom{m}{2}}.$$

Since $\boldsymbol{M}_{01}$ and $\boldsymbol{M}_{12}$ are zero matrices, we have $\gamma_2 = \gamma_4 = 0$, or equivalently $\mu_0 = \mu_4$ and $\mu_1 = \mu_3$. The roots of $\boldsymbol{M}_{11}$ and $\boldsymbol{M}^*$ can then be expressed in terms of μ_0, μ_1 and μ_2. We thus have

Theorem 2. *Consider an array T with m constraints and 2 symbols under the restrictions that it is (1, 0) symmetric with respect to triplets, and is of strength 4. Then a set of necessary conditions that T exists is*

$$(42a) \qquad (m-2)\mu_1 \geqslant (m-4)\mu_2$$

$$(42b) \quad (m^2-m+2)\mu_0 - 4(m^2-5m+2)\mu_1 + (3m^2-19m+38)\mu_2 \geqslant 0$$

$$(42c) \quad (\mu_0+4\mu_1+3\mu_2)[(m^2-m)\mu_0 - 4(m-1)(m-4)\mu_1 + (3m^2-19m+32)\mu_2] \geqslant m(m-1)(\mu_0-\mu_2)^2.$$

The first condition is a consequence of $\pi_2^* \geqslant 0$, while the other two follow from : π_4^*, $\pi_5^* \geqslant 0$. From $\pi_3^* \geqslant 0$, we get a trivial one, viz. : $\mu_2 \geqslant 0$. Also, distinct roots of $\boldsymbol{M}_{11}$ are $m+4(\mu_1+2\mu_2+\mu_3)$ and

$4(\mu_1+2\mu_2+\mu_3)$, which are always non-negative. These trivial conditions are however important for estimation of $\boldsymbol{L}$, as indicated in

Theorem 3. *Consider the array T of Theorem 2. A necessary and sufficient condition that $\mathbf{L}$ is estimable from T (taken as a fraction), is that (42a, b, c) be satisfied with strict inequality in each case, and furthermore that $\mu_2 \geqslant 0$, and $N \geqslant \nu_m$.*

It can be shown that for any fraction T, a necessary condition for nonsingularity is that T contain at least ν_m *dinstinct* treatment combinations. Together with Theorem 3, it implies

Theorem 4. *A fraction T which satisfies the conditions of Theorem 3 has at least ν_m distinct elements.*

In other words, suppose that the existence of a fraction T with given parameters is under consideration. Suppose that on the one hand the parameters satisfy the conditions of Theorem 3, and on the other hand it is possible to show that T (if it exists) will have less than ν_m elements. Then from these facts one can infer that T does not exist.

From the foregoing results, it is clear that the roots of M are π_4^*, π_5^*, π_2^*, π_3^*, π_6^*, π_7^* with respective multiplicities 1, 1, $m-1$, $m(m-3)/2$, 1 and $m-1$, where π_4^* and π_5^* are given by (41), and

$$\pi_2^* = \gamma_1+(m-4)\gamma_3-(m-3)\gamma_5$$

$$\pi_3^* = \gamma_1-2\gamma_3+\gamma_5$$

$$\pi_6^* = \gamma_1+(m-1)\gamma_3$$

$$\pi_7^* = \gamma_1-\gamma_3$$

$$\pi_1^* = \gamma_1+2(m-2)\gamma_3+\binom{m-2}{2}\gamma_5,$$

where π_1^* occurs in (41). Since these roots can be easily computed, any two rival fractions can be quickly compared with respect to any criterion, once the values of μ_0, μ_1 and μ_2 are known. Unlike in Section 4, in this case there are quite a few competing sets (μ_0, μ_1, μ_2) for any m. Hence, because of lack of space, the analogue of Table 1 is omitted.

Coming to the programming aspect it is seen that unlike the case of SOCB fractions, we have to satisfy (42c) which is a nonlinear condition on the μ_i's (and hence on the x_r's of (28)). However, only one such condition is present, and the methods of Section 5 should be applicible here too without much added difficulty.

The author has extended the results in the above theorems to general arrays of strength four. However, the calculation of roots of $\boldsymbol{M}$ which requires some new methods, is more tedious and messy. The interested reader may look forward to [9], where the case 3^n and mixed factorials are also treated, and the methods are extended to existence problems for orthogonal arrays.

We close the paper with a concrete example of an optimal fraction.

Example. Consider $m = 7$. Then $\nu_m = 29$, and $N \geqslant 29$. Suppose we allow around 15 d.f. for error; this number is neither too large nor too small. We then try to obtain solutions $\boldsymbol{\mu}' = (\mu_0, \mu_1, \mu_2)$ of $N = 2(\mu_0+4\mu_1+3\mu_2) \approx 44$, the μ_i being non-negative integers. It can be checked that there are 26 such solutions, out of which more than 18 can be rejected out-right by using Theorems 3 and 4. Among the others, after a little inspection, two sets, viz. for which $\boldsymbol{\mu}'$ equals (4, 3, 2) and (5, 2, 3) respectively, stand out better.

In order to compare these, for $\boldsymbol{\mu}' = (4, 3, 2)$ we obtain: $\gamma_1 = 44$, $\gamma_2 = \gamma_4 = 0$, $\gamma_3 = 4$ and $\gamma_5 = -4$. Hence $\pi_1^* = 44$, $\pi_2^* = 72$, $\pi_3^* = 32$, $\pi_4^* = 44+4\sqrt{21}$, $\pi_5^* = 44-4\sqrt{21}$, $\pi_6^* = 68$ and $\pi_7^* = 40$. Thus $\underset{\max}{\text{ch}}\,(\boldsymbol{V}) = [44-4\sqrt{21}]^{-1}$, and

$$\text{tr}\ \boldsymbol{V} = tr\boldsymbol{M}^{-1} = \frac{6}{72}+\frac{14}{32}+\frac{11}{200}+\frac{1}{68}+\frac{6}{40} \approx \frac{3}{4}\,.$$

Similarly, when $\boldsymbol{\mu} = (5, 2, 3)$, we get: $\gamma_1 = 44$, $\gamma_2 = \gamma_4 = 0$, $\gamma_3 = 4$, $\gamma_5 = 12$, $\pi_1^* = 204$, $\pi_2^* = 20$, $\pi_3^* = 48$, $\pi_4^* = 208$, $\pi_5^* = 40$, $\pi_6^* = 68$ and $\pi_7^* = 40$. Hence $\underset{\max}{\text{ch}}\,(\boldsymbol{V}) = 20$, and $\text{tr}\,(\boldsymbol{V}) \approx 4/5$.

Thus the set (4, 3, 2) is *better* than (5, 2, 3) both with respect to the largest root and the trace criteria. If an orthogonal fraction T with N assemblies existed, we would have $V_T = \nu_m/N$. Let T^* be any fraction with N assemblies. We define *trace efficiency* of T^* as (trace $\mathbf{V}_T$)/(trace $\mathbf{V}_{T^*}$), i.e. $\left(\frac{\nu_m}{N}\right)$ (trace $\mathbf{V}_{T^*}$). In this sense the trace efficiencies of the above two fractions are respectively 0.88 and 0.82 nearly.

The set (4, 3, 2) is clearly optimal. The reader will be glad to know that an array with these parameters actually exsts. It is obtained by taking all the assemblies belonging to two arrays T_1 and T_2. T_1 is given below:

0	0 0 0	1 1 1	0 0 0	0 0 0	0 0 0	1 1 0 0 0 1
0	0 0 0	0 0 0	1 1 1	0 0 0	0 0 0	1 0 1 0 1 0
0	0 0 0	0 0 0	0 0 0	1 1 1	0 0 0	0 1 1 1 0 0
0	0 0 0	0 0 0	0 0 0	0 0 0	1 1 1	0 0 0 1 1 1
0	1 1 0	1 0 0	1 0 0	1 0 0	1 0 0	1 1 1 1 1 1
0	1 0 1	0 1 0	0 1 0	0 1 0	0 1 0	1 1 1 1 1 1
0	0 1 1	0 0 1	0 0 1	0 0 1	0 0 1	1 1 1 1 1 1

T_2 is obtained from T_1 by interchanging the symbols 0 and 1. Finally, it will be noted that one faces the problem of construction of only those fractions whose parameter set has first been shown to be optimal.

References

[1] Addelman, Sidney (1962). "Symmetrical and Asymmetrical Fractional Factorial Plans," *Technometrics*, **4**, 47–58.

[2] Bose, R. C. (1961). "On Some Connections Between the Design of Experiments and Information Theory," *Bull. Inst. Internat. Statist.*, **38**, 257–271.

[3] Bose, R. C. and Burton, R. C. (1957). "On a Problem in Abelian Groups and the Construction of Fractionally Replicated Designs," (Abstract), *Ann. Math. Statist.*, **28**, 533.

[4] Bose, R. C. and Srivastava, J. N. (1964a). "Analysis of Irregular Factorial Fractions," *Sankhyā, Ser. A.*, **26**, 117–144.

[5] Bose, R. C. and Srivastava, J. N. (1964b). "Multidimensional Partially Balanced Designs and Their Analysis With Applications to Partially Balanced Factorial Fractions," *Sankhya, Ser. A.*, **26**, 145–168.

[6] Burton, R. C. (1964). "An Application of Convex Sets to the Construction of Error Correcting Codes and Factorial Designs," Ph.D. Thesis, Univ. of North Carolina, Chapel Hill, N.C.

[7] Chakravarti, I. M. (1956). "Fractional Replication in Asymmetrical Factorial Designs and Partially Balanced Arrays," *Sankhyā*, **17**, 143–164.

[8] Srivastava, J. N. (1964). "On the Construction of a Class of Optimum Balanced Factorial Fractions by Linear Progaramming," (Abstract), *Ann. Math. Statist.*, **35**, 1389.

[9] Srivastava, J. N. (1965), "Some Necessary Conditions for the Existence of Partially Balanced Arrays," (Abstract), *Ann. Math. Statist.*, **36**, 1079.

(*Received Jan. 1, 1966*.)

CHAPTER 37

Some Remarks on a Distribution Occurring in Neural Studies

WALTER L. SMITH[1], *The University of North Carolina at Chapel Hill*

1. INTRODUCTION[2]

Suppose that X is a non-negative random variable (which does not vanish with probability one) such that $\mathcal{E} \log(1+X)$ is finite and, for real positive s, put $\varphi(s) = \mathcal{E}\, e^{-sX}$. For any $c > 0$ we shall show there exists a distribution function $G(x)$, of a non-negative random variable, such that

$$(1) \qquad G^*(s) = \int_0^\infty e^{-sx}\, dG(x) = e^{\left[-c \int_0^\infty \frac{-\varphi(z)}{z}\, dz\right]}$$

This distribution function $G(x)$ arises in a variety of contexts. The author obtained it many years ago in some unpublished

[1]This research was supported by the Office of Naval Research under contract No. Nonr-855(09) for research in probability and statistics at the University of North Carolina, Chapel Hill, N.C. Reproduction in whole or in part is permitted for any purpose of the United States Government.

[2]This paper was written in the fall of 1963 and has been delayed in publication. The appearence of Volume II of Feller's "Probability Theory and its Applications" in 1966 renders some of our remarks unnecessary.

work on the initiation of nerve pulses. It has also arisen in studies of a certain recording apparatus (Takacs, 1955) and of the "present value" of a renewal process (Dall'Aglio, 1964). More recently it was derived in a colloquium at University College, London, by Dr. J. Keilson, who raised the question of whether $G(x)$ is absolutely continuous and, if so, of how the corresponding probability density function behaves near the origin. It is the object of the present paper to prove the following theorem of several parts.

Theorem 1. *If $F(x)$ is the distribution function of X and we assume that*

$$\int_0^\infty \log(1+x)dF(x) < \infty, \tag{2}$$

then :

(1.1) *Equation* (1) *defines an absolutely continuous distribution function $G(x)$ with a probability density function $g(x)$, say, which is continuous on the open interval* $(0, \infty)$.

(1.2) *There is a strictly decreasing function $D_1(x)$ such that $G(x) = x^c D_1(x)$, and $D_1(0+)$ is finite if and only if (in addition to* (2))

$$\int_0^1 \frac{F(x)}{x}\, dx < \infty, \tag{3}$$

in which case

$$D_1(0+) = \frac{1}{\Gamma(1+c)} e^{-c \int_0^\infty \frac{1-e^{-x}-F(x)}{x} dx}$$

Furthermore, if $F(\tau) = 0$ for some $\tau > 0$ then $D_1(x)$ is constant in $(0, \tau)$.

(1.3) *There is a strictly decreasing convex function $D_2(x)$ such that $[1-G(x)] = x^c D_2(x)$. If, for some $0 \leqslant \gamma < 1$,*

$$\int_0^\infty [1-F(y)]dy \sim x^\gamma L(x), \quad \text{as } x \to \infty, \tag{4}$$

where $L(x)$ *is a function of slow growth, then*

(5) $$[1-G(x)] \sim \frac{c\gamma L(x),}{(1-\gamma)x^{(1-\gamma)}} \ as \rightarrow \infty.$$

If, however,

(6) $$\int_0^x [1-F(y)]dy \sim xL(x), \quad as\ x \rightarrow \infty,$$

then

(7) $$[1-G(x)] \sim cM(x), \quad as\ x \rightarrow \infty,$$

where

$$M(x) = \int_x^\infty \frac{L(z)}{z}\,dz$$

and $M(x)$ *is also a function of slow growth.*

(1.4) *The continuous probability density function* $g(x)$ *is such that* $x^{(1-c)}\,g(x) = d(x)$, *say, a strictly decreasing function of* x, *and* $xg(x)$ *is a function of bounded variation. Moreover* $d(0+)$ *is finite if and only if* (3) *holds, in which case* $d(0+) = cD_1(0+)$. *If* (4) *should hold, then*

(8) $$g(x) \sim \frac{c\gamma L(x)}{x^{2-\gamma}}, \quad as\ x \rightarrow \infty,$$

while, if (6) *should hold, then*

(9) $$g(x) \sim \frac{cL(x)}{x}, \qquad as\ x \rightarrow \infty.$$

(1.5) *If, for some* $A > 0$, $\lambda \geqslant 0$, $\nu \geqslant 0$, *and for all sufficiently large* x,

$$1-F(x) \leqslant \frac{Ae^{-\lambda x x}x^{\nu}}{\Gamma(\nu+1)}$$

then, as $x \rightarrow \infty$,

$$g(x) = 0\left\{\frac{\exp\left\{-\lambda x+\frac{\nu+1}{\nu}(A\ c)^{\frac{1}{\nu+1}}x^{\frac{\nu}{\nu+1}}\right\}}{x^{\frac{1}{2}\left(\frac{\nu+1}{\nu+2}\right)}}\right\}, \qquad \nu > 0,$$

$$= 0(e^{-\lambda x}x^{Ac-1}), \qquad \nu = 0.$$

To prove Theorem 1 we find it necessary to establish the following three theorems concerning a more general class of density functions.

Theorem 2. *If* $a(x) \geqslant 0$ *and*

$$\int_0^1 \frac{a(x)}{x}\,dx = \infty, \qquad \int_0^\infty \frac{a(x)}{1+x}\,dx < \infty,$$

and if we write, for $\mathcal{R}\, s \geqslant 0$,

$$a^0(s) = \int_0^\infty e^{-sx} a(x)dx,$$

then there is a probability density function $\Delta_a(x)$, *say, on* $(0, \infty)$ *such that*

$$\Delta_a^0(s) = \int_0^\infty e^{-sx}\Delta_a(x)dx = \exp\left[-\int_0^s a^0(z)dz\right]$$

where the contour integral in the exponent is taken along a straight line.

Theorem 3. *In the notation of Theorem* 2, *if* $a(x)$ *is continuous and of bounded variation then we may take* $\Delta_a(x)$ *as continuous and of bounded variation in any interval not containing the origin. Moreover, if* $a(x) < Ae^{-\eta x}$ *for some* $A > 0$, $\eta > 0$, *then* $\Delta_a(x) = O(e^{-\eta x}x^{A-1})$.

Theorem 4. *If* $a_1(x)$ *and* $a_2(x)$ *both satisfy the conditions of Theorem* 2 *and if* $a_1(x) \geqslant a_2(x)$ *for all* x *and*

$$\int_0^1 \frac{a_1(x)-a_2(x)}{x}\,dx < \infty,$$

then, in an obvious extension of notation,

$$\int_x^\infty \Delta_{a1}(y)dy \geqslant \int_x^\infty \Delta_{a2}(y)dy$$

for all $x \geqslant 0$, *and*

$$e^{\left[\int_0^\infty \frac{a_1(x)-a_2(x)}{x}dx\right]}\Delta_{a1}(x) \geqslant \Delta_{a2}(x)$$

for almost all x,

In part of our argument we make use of the continuity theorem for Laplace-Stieltjes transforms. There does not seem to be any convenient reference for this useful theorem (although its use occurs in the literature from time to time). We therefore append a short proof in an appendix.

Proof of Theorem 2. Write, for fixed $\delta > 0$,

$$I_\delta = \int_\delta^\infty \frac{a(x)}{x}\, dx$$

and define

$$g_\delta(x) = 0 \quad , \quad \text{for} \quad x < \delta,$$

$$= \frac{a(x)}{xI_\delta} , \quad \text{for } x \geqslant \delta.$$

Then $g_\delta(x)$ is a probability density function. Suppose that $Z_1, Z_2, Z_3, \ldots$ is an infinite sequence of independent random variables, each governed by the density function $g_\delta(x)$. Suppose M is an integer-valued random variable, independent of the $\{Z_n\}$, such that for $r = 0, 1, 2, \ldots$

$$P\{M = r\} = \frac{e^{-I_\delta} (I_\delta)^r}{r\,!} .$$

Define a random variable $Y = 0$ if $M = 0$, and $Y = Z_1 + Z_2 + \ldots + Z_M$ otherwise. Then it is an easy matter to see that Y has a distribution function $G_\delta(x)$, say, where

$$G_\delta^*(s) = \int_0^\infty e^{-sx} d\, G_\delta(x) = e^{-I_\delta + I_\delta g_\delta^0(s)}$$

(in the notation already suggested in the enunciation of Theorem 2 we have written $g_\delta^0(s)$ for the ordinary Laplace transform of $g_\delta(x)$). Thus we have

$$\text{(10)} \qquad \log G_\delta^*(s) = - \int_\delta^\infty \frac{(1-e^{-sx})}{x}\, a(x) dx.$$

As δ decreases to zero we see from (10) and Beppo Levi's theorem that $G^*_\delta(s) \to G^*_0(s)$, say, where

$$\log G^*_0(s) = -\int_0^\infty \frac{(1-e^{-sx})}{x} a(x)dx. \tag{11}$$

In view of our hypothesis about $a(x)$ it is clear that the integral on the right of (11) is absolutely convergent. Also, from Lebesgue's theorem on dominated convergence, we can deduce that $G^*_0(s) \to 1$ as s decreases through real values to zero. It follows therefore, from the continuity theorem for Laplace-Stieltjes transforms, that there is a distribution function $G_0(x)$ over $[0, \infty)$ such that

$$G^*_0(s) = \int_0^\infty e^{-sx} dG_0(x).$$

Furthermore, by Fubini's Theorem,

$$\int_0^\infty \frac{(1-e^{-sx})}{x} a(x)dx = \int_0^\infty \int_0^s e^{-zx} a(x)dz\, dx$$

$$= \int_0^s a^0(z)dz.$$

Thus the theorem will be proved if we show that $G_0(x)$ is absolutely continuous. To this end, we differentiate (11) and find that

$$-\frac{d}{ds} G^*_0(s) = a^0(s)\, G^*_0(s). \tag{12}$$

Hence, if $l(x)$ is defined by

$$l(x) = \int_0^x a(x-z)dG_0(z)$$

it is a consequence of (12) that

$$\int_0^\infty e^{-sx}\, x dG_0(x) = \int_0^\infty e^{-sx}\, l(x)dx. \tag{13}$$

From (13) we infer that, except for a possible discontinuity at the origin, $G_0(x)$ is absolutely continuous with a density function

$$\Delta_a(x) = \frac{d}{dx} G_0(x) = \frac{l(x)}{x}.$$

Finally, we rule out the possibility of a point mass of probability at the origin by observing that its weight must equal (taking the limit through real values)

$$\lim_{s\to\infty} e^{-\int_0^s a^o(z)\,dz} = e^{-\int_0^\infty \frac{a(x)}{x}\,dx}$$

which is zero, by our hypothesis that the integral

$$\int_0^1 \frac{a(x)}{x}\,dx$$

diverges. Thus the theorem is proved.

Proof of Theorem 3. The continuity and bounded variation properties claimed for $\Delta_a(x)$ are easy consequences of the representation

$$\Delta_a(x) = \frac{1}{x} \int_0^x a(x-z)dG_0(z).$$

From the equation we also see that if $a(x) < Ae^{-\eta x}$ then

$$e^{\eta x}\Delta_a(x) < \frac{A}{x} \int_0^x e^{\eta x}\Delta_a(z)dz. \tag{14}$$

Therefore

$$\frac{d}{dx} \log \int_0^x e^{\eta z}\Delta_a(z)dz < \frac{A}{x}$$

and so

$$\log \left\{ \frac{\int_0^x e^{\eta z}\Delta_a(z)dz}{\int_0^1 e^{\eta z}\Delta_a(z)dz} \right\} < A \log x.$$

Thus

$$\int_0^x e^{\eta z}\Delta_a(z)dz < x^A \int_0^1 e^{\eta z}\Delta_a(z)dz$$

and hence, from (14) again, we have

$$e^{\eta x}\Delta_a(x) < Ax^{A-1}\int_0^1 e^{\eta z}\Delta_a(z)dz,$$

which completes the proof of the theorem.

Proof of Theorem 4. We extend the notation used in the earlier proofs, with the aid of suffices, in an obvious way. Except for the convergence of the integral

$$\int_0^1 \frac{a_1(x)-a_2(x)}{x}\,dx \tag{15}$$

the non-negative function $a_1(x)-a_2(x)$ satisfies all the conditions imposed upon $a(x)$ in Theorem 2. Thus we can say there is a distribution function $H(x)$, say, of a non-negative random variable, such that

$$e^{-\int_0^s \{a_1^0(z)-a_2^0(z)\}dz} = H^*(s). \tag{16}$$

Because of the convergence of (15) it will be seen from the proof of Theorem 2 that the function $H(x)$ will have a discontinuity at the origin, but will otherwise be absolutely continuous. From (16) it follows that

$$G_{01}(x) = \int_0^x G_{02}(x-z)dH(z)$$

$$\leqslant G_{02}(x),$$

and therefore

$$\int_x^\infty \Delta_{a1}(y)dy \geqslant \int_x^\infty \Delta_{a2}(y)dy$$

as claimed.

Let us now write

$$g_\delta^{[2]}(x) = \int_0^x g_\delta(x-z)g_\delta(z)dz$$

and, for $n > 2$,

$$g_\delta^{[n]}(x) = \int_0^x g_\delta^{[n-1]}(x-z)g_\delta(z)dz.$$

Then, from the proof of Theorem 2, it is evident that $G_{\delta 1}(x)$ has a jump at the origin of amount $e^{-I_{\delta 1}}$, but is absolutely continuous otherwise, and, for almost all $x > 0$,

$$\frac{d}{dx} G_{\delta 1}(x) = \sum_{n=1}^{\infty} \frac{e^{-I_{\delta 1}}(I_{\delta 1})^n}{n\,!} g_{\delta 1}^{[n]}(x).$$

But, by our hypothesis,

$$(I_{\delta 1})^n g_{\delta 1}^{[n]}(x) \geqslant (I_{\delta 2})^n g_{\delta 2}^{[n]}(x)$$

for all x. Hence

$$\frac{d}{dx} G_{\delta 1}(x) \geqslant e^{-\{I_{\delta 1}-I_{\delta 2}\}} \frac{d}{dx} G_{\delta 2}(x).$$

Thus, if $0 < \alpha < \beta$,

$$G_{\delta 1}(\beta) - G_{\delta 1}(\alpha) \geqslant e^{-\{I_{\delta 1}-I_{\delta 2}\}} \{G_{\delta 2}(\beta) - G_{\delta 2}(\alpha)\}.$$

If we now let δ decrease to zero we find that

$$G_{01}(\beta) - G_{01}(\alpha) \geqslant e^{-\{I_{\delta 1}-I_{\delta 2}\}} \{G_{02}(\beta) - G_{02}(\alpha)\},$$

that is

$$\int_{\alpha}^{\beta} \Delta_{a_1}(x)dx \geqslant e^{-\int_0^{\infty} \frac{a_1(x)-a_2(x)}{x} dx} \int_{\alpha}^{\beta} \Delta_{a_2}(x)dx.$$

Since the last inequality holds for arbitrary α and β (> 0), the final contention of Theorem 4 is proved.

Proof of Theorem 1, Part (1.1). It is clear that the integral

$$\int_0^1 \frac{1-F(x)}{x} dx$$

diverges, and an integration by parts will show that

$$\int_0^{\infty} \log(1+x) dF(x) = \int_0^{\infty} \frac{1-F(x)}{1+x} dx.$$

Thus, if we put $a(x) = c\{1-F(x)\}$ then this function satisfies all the conditions of Theorem 2. Upon noting that

$$\int_0^\infty e^{-zx}\{1-F(x)\}dx = \frac{1-\varphi(z)}{z}$$

we can therefore infer that

$$e^{-c\int_0^\delta \frac{1-\varphi(z)}{z}dz}$$

is, indeed, the Laplace-Stieltjes transform of an absolutely continuous distribution function $G(x)$. Furthermore, if we write $g(x)$ for a density function corresponding to $G(x)$ then we may put

$$g(x) = \frac{c}{x}\int_0^x \{1-F(x-z)\}g(z)dz$$

$$= \frac{cG(x)}{x} - \frac{c}{x}\int_0^x G(x-z)dF(z). \tag{17}$$

The distribution function $G(x)$ is continuous and therefore

$$\int_0^x G(x-z)dF(z)$$

is also a continuous function of x. Equation (17) therefore shows $g(x)$ to be continuous as claimed.

Proof of Part (1.2). From (17) we see that

$$\frac{d}{dx}G(x) = \frac{c}{x}G(x) - \frac{c}{x}\int_0^x G(x-z)dF(z)$$

so that

$$\frac{d}{dx}[x^{-c}G(x)] = -\frac{c}{x^{(1+c)}}\int_0^x G(x-z)dF(z). \tag{18}$$

The right-hand side of (18) is negative; therefore $x^{-c}G(x) = D_1(x)$, say, is a decreasing function as was to be proved.

Suppose that $D_1(x)$ increases to a finite limit A, say, as x decreases to zero. Then, by a familiar Abelian theorem for

Laplace-Stieltjes transforms (Widder, 1941, p. 181), $s^c G^*(s) \to A\Gamma(1+c)$ as $s \to +\infty$ (through real values). Therefore

$$c \log s - c \int_0^s \frac{1-\varphi(z)}{z}\, dz \to \log [A\Gamma(1+c)],$$

that is

$$c \int_1^s \frac{\varphi(z)}{z}\, dz - c \int_0^s \frac{1-\varphi(z)}{z}\, dz \to \log [A\Gamma(1+c)]$$

as $s \to \infty$. But, by Fubini's theorem,

$$\int_1^s \frac{\varphi(z)}{z}\, dz = \int_1^s \int_0^\infty e^{-zx} F(x) dx\, dz$$

$$= \int_0^\infty \frac{e^{-x} - e^{-sx}}{x} F(x) dx.$$

We can thence deduce from Beppo Levi's theorem that

$$\int_1^\infty \frac{\varphi(z)}{z}\, dz = \int_0^\infty \frac{e^{-x} F(x)}{x}\, dx \leqslant \infty.$$

From all this we may conclude that A is finite if and only if $F(x)/x$ belongs to $L_1(0, 1)$, as is claimed in this part of the theorem. When A happens to be finite we see that

$$\log [A\Gamma(1+c)] = c \int_0^\infty \frac{e^{-x} F(x)}{x}\, dx - c \int_0^1 \frac{1-\varphi(z)}{z}\, dz.$$

But

$$\int_0^1 \frac{1-\varphi(z)}{z}\, dz = \int_0^\infty \frac{(1-e^{-x})}{x} [1-F(x)] dx,$$

and so

$$\log[A\Gamma(1+c)] = -c \int_0^\infty \frac{1-e^{-x}-F(x)}{x}\, dx,$$

which proves the value for $D_1(0+)$. We also note that should $F(x) = 0$ for all $x < \tau$ then, by (18), $x^{-c}G(x)$ is constant for all $x < \tau$. This completes the proof of Part (1.2).

Proof of Part (1.3). From (17) we have, for $x > 0$,

$$\frac{d}{dx}[x^{-c}\{1-G(x)\}] = -\frac{c}{x^{(1+c)}}\{1-K(x)\} \tag{19}$$

where $K(x)$ is the absolutely continuous distribution function

$$K(x) = \int_0^x F(x-z)dG(z). \tag{20}$$

From (19) it is apparent that $x^{-c}\{1-G(x)\} = D_2(x)$, say, is a decreasing function with an increasing derivative; in particular, $D_2(x)$ is convex.

Now suppose that for some $0 \leqslant \gamma \leqslant 1$ and some function of slow growth $L(x)$

$$\int_0^x [1-F(y)]dy \sim x^{\gamma} L(x), \qquad \text{as } x \to \infty.$$

Then, by a slightly more complicated Abelian theorem than the one we have already used (Doetsch, 1950), we have that

$$\frac{1-\varphi(s)}{s} \sim \frac{\Gamma(1+\gamma)L\left(\frac{1}{s}\right)}{s^{\gamma}}$$

as $s \to 0+$ through real values. However, we can discover from (1) that as $s \to 0+$

$$\frac{1-G^*(s)}{s} \sim \frac{c}{s}\int_0^s \frac{1-\varphi(z)}{z}dz$$

$$\sim \frac{c}{s}\Gamma(1+\gamma)\int_0^s \frac{L\left(\frac{1}{z}\right)}{z^{\gamma}}dz.$$

Before we can proceed we must discover the asymptotic behavior of the integral on the right. By an obvious change of variable we have

$$\frac{1}{s}\int_0^s \frac{L(z^{-1})}{z^{\gamma}}dz = \frac{1}{s^{\gamma}}\int_1^{\infty} \frac{L(u/s)}{u^{2-\gamma}}du.$$

Now Karamata (1930) has shown that for a given function of slow growth $L(x)$ there is necessarily a function $\rho(x)$ such that $\rho(x) \to 1$ as $x \to \infty$ and

$$L(x) = \frac{\rho(x)}{x} e^{\int_1^x \frac{\rho(v)}{v} dv}$$

From this fact it is an easy deduction that for arbitrary $\epsilon > 0$

$$0 < \frac{L(u/s)}{L(s^{-1})} < (1+\varepsilon)u^{\epsilon}$$

for all sufficiently large u and all sufficiently small s. Therefore, if $\gamma < 1$, we can appeal to dominated convergence to infer that

$$\lim_{s \to 0+} \int_1^\infty \frac{L(u/s)}{L(s^{-1})} \frac{du}{u^{2-\gamma}} = \frac{1}{(1-\gamma)}$$

and hence that

$$\frac{1}{s} \int_0^s \frac{L(z^{-1})}{z^{\gamma}} dz \sim \frac{L(s^{-1})}{s^{\gamma}(1-\gamma)}.$$

Hence

$$\frac{1-G^*(s)}{s} \sim \frac{c\Gamma(1+\gamma)L(s^{-1})}{(1-\gamma)s^{\gamma}},$$

as $s \to 0+$. From a Tauberian theorem for Laplace transforms (Doetsch, 1950, p. 511) we can then deduce that

$$\int_0^x \{1-G(y)\}dy \sim \frac{cL(x)x^{\gamma}}{(1-\gamma)}, \qquad \text{as } x \to \infty.$$

Furthermore, from (20),

$$K^*(s) = \varphi(s)G^*(s),$$

so that, as $s \to 0+$,

$$\frac{1-K^*(s)}{s} \sim \frac{1-\varphi(s)}{s} + \frac{1-G^*(s)}{s}$$

$$\sim \frac{\Gamma(1+\gamma)L(s^{-1})}{s^{\gamma}} + \frac{c\Gamma(1+\gamma)Ls(s^{-1})}{(1-\gamma)s^{\gamma}}$$

Hence, by another Tauberian argument,

$$\int_0^x \{1-K(y)\}dy \sim \frac{(1-\gamma+c)L(x)x^\gamma}{(1-\gamma)}, \quad \text{as } x \to \infty.$$

If we multiply (19) by $x^{(1+c)}$ and integrate by parts we find that

$$x\{1-G(x)\} = (1+c)\int_0^x \{1-G(y)\}dy - c\int_0^x \{1-K(y)\}dy. \tag{21}$$

From the asymptotic results we have obtained it follows from (21) that

$$1-G(x) \sim \frac{c\gamma L(x)}{(1-\gamma)x^{1-\gamma}}, \quad \text{as } x \to \infty.$$

In the case $\gamma = 1$ we cannot employ the dominated convergence argument and the results come out somewhat differently. Let us define

$$M(x) = \int_x^\infty \frac{L(z)}{z}\, dz.$$

Then for any fixed $\alpha > 0$

$$M(\alpha x) = \int_x^\infty \frac{L(\alpha z)}{z}\, dz$$

and hence, for an arbitrary $\epsilon > 0$ and all sufficiently large x,

$$(1-\epsilon)\int_x^\infty \frac{L(z)}{z}\, dz < M(\alpha x) < (1+\varepsilon)\int_x^\infty \frac{L(z)}{z}\, dz.$$

It is obvious therefore that $M(\alpha x) \sim M(x)$ as $x \to \infty$ and that $M(x)$ is consequently a function of slow growth. We thus obtain for this case

$$\frac{1-G^*(s)}{s} \sim \frac{cM(s^{-1})}{s}, \quad \text{as } s \to 0+,$$

and so, via the Tauberian theorem,

$$\int_0^x \{1-G(y)\}dy \sim cxM(x), \quad \text{as } x \to \infty.$$

We shall show in a moment that $L(x)/M(x) \to 0$ as $x \to \infty$. It then follows, as before, that

$$\frac{1-K^*(s)}{s} \sim \frac{L(s^{-1})}{s} + \frac{cM(s^{-1})}{s}$$

$$\sim \frac{cM(s^{-1})}{s}$$

and so,

$$\int_0^x \{1-K(y)\}dy \sim cxM(x).$$

From (21) we can then deduce that

$$\{1-G(x)\} \sim cM(x), \qquad \text{as } x \to \infty,$$

which was to be proved.

To see that $L(x)/M(x) \to 0$ as $x \to \infty$ we observe that for Δ arbitrarily large and positive

$$M(x) > \int_x^{x\Delta} \frac{L(z)}{z} dz$$

$$= \int_1^{\Delta} \frac{L(ux)}{u} du.$$

Thus

$$\frac{M(x)}{L(x)} > \int_1^{\Delta} \left\{ \frac{L(ux)}{L(u)} \right\} \frac{du}{u}$$

and so, by a dominated convergence which can be justified much as before,

$$\liminf_{x \to \infty} \frac{M(x)}{L(x)} \geqslant \int_1^{\Delta} \frac{du}{u}.$$

This establishes the correctness of our assertion.

Proof of Part (1.4). By (17) and (20) we have

(22) $$xg(x) = cG(x) - cK(x),$$

so that $xg(x)$ is of bounded variation as claimed. If we differentiate this last equation (and write $k(x)$ for the, necessarily continuous, density function associated with $K(x)$) we find that

$$xg'(x)+(1-c)g(x) = -ck(x)$$

which implies that

$$\frac{d}{dx}[x^{(1-c)}\,g(x)] = -\frac{ck(x)}{x^c}.$$

Therefore $x^{(1-c)}g(x) = d(x)$, say, where $d(x)$ is a strictly decreasing function. If $d(0+) = \infty$ then, given any large Δ we have $g(x) > \Delta x^{-(1-c)}$ and therefore $G(x) > x^c\Delta/c$, for all sufficiently small x. Hence $d(0+) = \infty$ only if $D_1(0+) = \infty$. On the other hand, if $d(0+)$ is finite it is clear that, for small x, $G(x) \sim d(0+)x^c/c$, so that $d(0+) = cD_1(0+)$ and, incidentally, $D_1(0+)$ is seen to be finite.

To complete the proof of this part we need the following:

Lemma 1. *If* $\int_0^x \{1-F(y)\}\,dy \sim x^\gamma L(x)$ *as* $x \to \infty$, *where* $0 \leqslant \gamma \leqslant 1$ *and* $L(x)$ *is a function of slow growth, then*

$$-\frac{s^\gamma \varphi'(s)}{L(s^{-1})} \to (1-\gamma)\Gamma(1+\gamma), \qquad \text{as } s \to 0+.$$

Proof. We note that, as $x \to \infty$,

$$\frac{1}{x^{1+\gamma}}\int_0^x\left\{\int_0^y [1-F(z)]dz\right\}dy \sim \frac{1}{x^{1+\gamma}}\int_0^x y^\gamma L(y)dy$$

$$\sim \frac{L(x)}{\gamma+1},$$

by Theorème 1 of Karamata (1930, p. 40). But an integration by parts shows

$$\frac{1}{x^{1+\gamma}}\int_0^x y\{1-F(y)\}dy$$

$$= \frac{1}{x^\gamma}\int_0^x \{1-F(y)\}dy - \frac{1}{x^{1+\gamma}}\int_0^x\left\{\int_0^y [1-F(z)]dz\right\}dy$$

and hence we have

$$\int_0^x y\{1-F(y)\}dy \sim \frac{\gamma x^{\gamma+1}L(x)}{\gamma+1}, \qquad \text{as } x \to \infty.$$

The Laplace transform of $x\{1-F(x)\}$ is

$$\frac{1-\varphi(s)}{s^2}+\frac{\varphi'(s)}{s}$$

and so, by the Abelian theorem for Laplace transforms, as $s \to 0+$,

$$\frac{1-\varphi(s)}{s^2}+\frac{\varphi'(s)}{s} \sim \frac{\gamma\Gamma(\gamma+2)L(s^{-1})}{(\gamma+1)s^{\gamma+1}}$$

But

$$\frac{1-\varphi(s)}{s} \sim \frac{\Gamma(1+\gamma)L(s^{-1})}{s^{\gamma}}$$

from the hypothesis $\int_0^x \{1-F(y)\}dy \sim x^{\gamma}L(x)$. Thus

$$-\frac{s^{\gamma}\varphi'(s)}{L(s^{-1})} \to (1-\gamma)\Gamma(1+\gamma)$$

as claimed.

Returning to the proof of Part (1.4), let us define

$$r(x) = \int_0^x g(x-z)z\,dF(z).$$

Then

$$r^0(s) = \int_0^\infty e^{-sx}r(x)dx = -\varphi'(s)G^*(s)$$

and so, under the conditions of Lemma 1,

$$\frac{s^{\gamma}r^0(s)}{L(s^{-1})} \to (1-\gamma)\Gamma(1+\gamma), \qquad \text{as } s \to 0+.$$

Therefore, by the Tauberian theorem we have been using,

$$\frac{1}{(x^{\gamma}L(x))}\int_0^x r(y)dy \to (1-\gamma), \qquad \text{as } x \to \infty.$$

If we convolute both sides of (22) with $F(x)$ we find

$$xk(x)-r(x) = cK(x)-cH(x) \tag{23}$$

where $H(x)$ is the distribution function

$$H(x) = \int_0^x K(x-z)dF(z).$$

On integrating (23) we obtain

$$-x\{1-K(x)\}+\int_0^x \{1-K(y)\}dy-\int_0^x r(y)dy \tag{24}$$

$$= c\int_0^x \{K(y)-H(y)\}dy.$$

From (21), (22), and (24) we then find

$$\frac{x^2g(x)}{c} = (1+c)\int_0^x \{G(y)-K(y)\}dy \tag{25}$$

$$-\int_0^x r(y)dy$$

$$-c\int_0^x \{K(y)-H(y)\}dy.$$

The function $\{G(y)-K(y)\}$ is non-negative and its Laplace transform is easily seen to be

$$\frac{1-\varphi(s)}{s}\,G^*(s) \sim \frac{\Gamma(1+\gamma)L(s^{-1})}{s^\gamma} \quad \text{as } s \to 0+.$$

Thus

$$\int_0^x \{G(y)-K(y)\}dy \sim x^\gamma L(x), \quad \text{as } x \to \infty.$$

Similarly, the non-negative function $\{K(y)-H(y)\}$ has Laplace transform

$$\frac{1-\varphi(s)}{s}\,\varphi(s)\,G^*(s)$$

and so

$$\int_3^x \{K(y)-H(y)\}dy \sim x^\gamma L(x), \quad \text{as } x \to \infty,$$

also.

We now have enough asymptotic results to deduce from (25) that

$$\frac{x^2 g(x)}{c} \sim \gamma x^\gamma L(x)$$

i.e.

$$g(x) \sim \frac{c\gamma L(x)}{x^{2-\gamma}}, \quad \text{as } x \to \infty.$$

This completes the proof of this part.

Proof of Part 1.5. We begin first with

Lemma 2. *If $l_1(x)$, $l_2(x)$, $l_3(x)$ are bounded integrable functions such that*

$$l_1(x) = \int_0^x l_2(x-z) l_3(z) dz$$

and if, for $x > 0$,

$$l_2(x) = O\left\{\frac{e^{-\gamma x + Ax^\alpha}}{x^\beta}\right\},$$

$$l_3(x) = O(e^{-\mu x}),$$

for some $\mu > \lambda > 0$, $A > 0$, $\beta \geqslant 0$, $1 > \alpha > 0$, then

$$l_1(x) = O\left\{\frac{e^{-\lambda x + Ax^\alpha}}{x}\right\}.$$

Proof. For any $m > 0$

$$\frac{d}{dx}\left\{\frac{e^{Ax^\alpha}}{(m+x)^\beta}\right\} = \frac{e^{Ax^\alpha}}{(m+x)^\beta}\left\{\frac{A\alpha}{x^{1-\alpha}} - \frac{\beta}{(m+x)}\right\}$$

so that we can always choose m large enough to make

$$f(x) = \frac{e^{Ax^\alpha}}{(m+x)^\beta}$$

an increasing function of $x > 0$. Having chosen m we can then find constants N_1, N_2, such that

$$l_2(x) \leqslant N_1 \frac{e^{-\lambda x + Ax^\alpha}}{(m+x)^\beta},$$

$$l_3(x) \leqslant N_2 e^{-\mu x},$$

for all x. Thus

$$l_1(x) \leqslant N_1 N_2 \int_0^x \frac{e^{-\mu x + \mu z - \lambda z + Az^\alpha}}{(m+z)^\beta}$$

$$\leqslant \frac{N_1 N_2 e^{-\lambda x + Ax^\alpha}}{(m+x)^\beta} \int_0^x e^{-(\mu-\gamma)(x-z)}\, dz,$$

in view of the fact that $f(x)$ increases. Thus the lemma is proved.

That part of (1.5) concerning the case $\nu = 0$ is already covered by Theorem 3. We shall therefore assume from here on that $\nu > 0$, and suppose there are constants $C > 0$, $\Delta > 0$ such that

$$1 - F(x) < \frac{Ce^{-\gamma x} x^\nu}{\Gamma(\nu+1)}, \quad x \geqslant \Delta.$$

Define

$$\theta(x) = c \operatorname{Max} \left\{ 1 - F(x), \quad \frac{Ce^{-\mu x} x^\nu}{\Gamma(\nu+1)} \right\}.$$

Then $\theta(x) \geqslant c\{1-F(x)\}$ for all x and $\theta(x) = c\{1-F(x)\}$ for all sufficiently small x. Therefore

$$\int_0^1 \frac{\theta(x) - c\{1-F(x)\}}{x}\, dx < \infty$$

and we can deduce from Theorem 4 that

(26) $$g(x) = O(\Delta_\theta(x)),$$

for almost all x.

Define

$$\sigma(x) = \frac{Ce^{-\lambda x} x^\nu}{\Gamma(\nu+1)}$$

and

$$\tau(x) = \theta(x) - \sigma(x).$$

Then $\tau(x) \geqslant 0$ and $\tau(x) = 0$ for all $x > \Delta$. Hence $\Delta_\tau(x)$ is defined. Moreover, since $\tau(x) = O(e^{-\eta x})$ for arbitrarily large η, it follows from Theorem 3 that

$$\Delta_\tau(x) = O(e^{-\eta x}) \tag{27}$$

for η arbitrarily large. We also note that

$$\Delta_\theta(x) = \int_0^x \Delta_\tau(x-z)\Delta_\sigma(z)\, dz. \tag{28}$$

For typographic case, let us write

$$\psi(x) = \frac{\exp\left\{-\lambda x + \frac{\nu+1}{\nu}(Cc)^{\frac{1}{\nu+1}} x^{\frac{\nu}{\nu+1}}\right\}}{x^{\frac{1}{2}\left(\frac{\nu+1}{\nu+2}\right)}}.$$

Then, in view of (26), (27), (28), and Lemma 2, we shall have proved Theorem (1.5) if we show that $\Delta_\sigma(x) = O(\psi(x))$. Our task thus becomes one of estimating $\Delta_\sigma(x)$. From all that we have proved so far we can say that $\Delta_\sigma(x)$ is continuous and locally of bounded variation; also, from Theorem 3, $\Delta_\sigma(x) = O(e^{-\gamma x})$ for any $\gamma < \lambda$. Thus we can deduce from Theorem 7.3 of Widder (1941, p. 66) that

$$\Delta_\sigma(x) = \lim_{T\to\infty} \frac{1}{2\pi i} \int_{-\gamma-iT}^{-\gamma+iT} e^{sx+\frac{Cc}{\nu(s+\gamma)^\nu}}\, ds. \tag{29}$$

Let us put

$$h(s) = sx + \frac{Cc}{\nu(s+\lambda)^\upsilon}.$$

Then

$$h'(s) = x - \frac{Cc}{(s+\lambda)^{\upsilon+1}}$$

so that $h'(s) = 0$ where

$$s = -\lambda + \left(\frac{Cc}{x}\right)^{\frac{1}{\nu+1}}$$

$$= -\lambda + \delta(x), \text{ say,}$$

and this is a point on the real axis a little to the right of the point $s = -\lambda$.

Choose an arbitrarily small $\epsilon < 0$.

As the real parameter t runs from $-\epsilon\delta(x)$ to $+\epsilon\delta(x)$ the point

$$s = -\lambda + \delta(x) - t^2 + it$$

runs along a small parabolic arc $\mathfrak{C}$, say. Now

$$h''(s) = \frac{Cc(\nu+1)}{(s+\lambda)^{(\nu+2)}}$$

$$h'''(s) = -\frac{Cc(\nu+1)(\nu+2)}{(s+\lambda)^{(\nu+3)}}\,.$$

Thus, for all s on $\mathfrak{C}$ we have

$$|h'''(s)| < \frac{K_1}{\delta^{(\nu+3)}}$$

where K_1 is some constant which does not depend on δ and ϵ, provided they are both small. Therefore, if s is any point on $\mathfrak{C}$,

$$h''(s) = \frac{Cc(\nu+1)}{\delta^{(\nu+2)}} - (t^2 - it)h'''(s^*)$$

where s^* is some point on $\mathfrak{C}$ between s and $-\lambda+\delta$. Hence, on $\mathfrak{C}$,

$$h''(s) = \frac{Cc(\nu+1)}{\delta^{(\nu+2)}}\{1+\rho_1(t)\}$$

where $|\rho_1(t)| < K_2\epsilon\delta$, K_2 being some further constant which does not depend on ϵ or δ. On $\mathfrak{C}$ we thus have

$$\mathcal{R}\alpha(s) = -\lambda x + x\delta + \frac{Cc}{\nu\delta} + \mathcal{R}\left\{\frac{1}{2}(-t^2 - 2it^3 + t^4)\frac{Cc(\nu+1)}{\delta^{(\nu+2)}}[1+\rho_1(t)]\right\}$$

$$= -\lambda x + x\delta + \frac{Cc}{\nu\delta^{\nu}} - \frac{Cc(\nu+1)t^2}{2\delta^{(\nu+2)}}\{1+\rho_2(t)\}$$

where $|\rho_2(t)| < K_3\epsilon\delta$, for some constant K_3 not depending on ϵ or δ. Hence, noting that $ds = (i-2t)dt$, we have

$$\left| \int_{\mathcal{C}} e^{h(s)}\, ds \right|$$

$$\leqslant (1+2\epsilon\delta)e^{-\lambda x+x\delta+\frac{Cc}{\nu\delta^\nu}} \int_{-\epsilon\delta}^{+\epsilon\delta} e^{-\frac{Cc(\nu+1)\{1-K_3\epsilon\delta\}}{2\delta^{(\nu+2)}}t^2} dt$$

$$\leqslant \frac{(1+2\epsilon\delta)(2\pi)^{\frac{1}{2}} e^{-\lambda x+x\delta+\frac{Cc}{\nu\delta^\nu}}}{\sqrt{\left\{\frac{Cc(\nu+1)\{1-K_3\epsilon\delta\}}{\delta^{(\nu+2)}}\right\}}}.$$

If we substitute for δ in terms of x in the last inequality, we discover

$$(30) \qquad \int_{\mathcal{C}} e^{h(s)}ds = O[\psi(x)], \qquad \text{as } x \to \infty.$$

Let T be a large positive number, η a small one, and let $\mathcal{L}(T)$ be the line mapped out by

$$s = (-\lambda+\delta-\epsilon^2\delta^2-\eta t)+i(\epsilon\delta+t)$$

as t runs from 0 to T. Notice that $\mathcal{L}(T)$ is a straight line segment sloping away from the imaginary axis and linking up with one end of $\mathcal{C}$. On the line $\mathcal{L}(T)$

$$\mathcal{R}\, h(s) < -\lambda x+x\delta-\epsilon^2\delta^2 x-\eta xt+\frac{Cc}{\nu r^\nu}$$

where

$$r^2 = \delta^2\{(1-\epsilon^2\delta^2)^2+\epsilon^2\}.$$

Thus

$$\mathcal{R}\, h(s) < -\lambda x+\frac{\nu+1}{\nu}\, x^{\frac{\nu}{\nu+1}}\, (Cc)^{\frac{1}{\nu+1}} -\eta xt+v, \text{ say,}$$

where

$$v = x\delta-s^2\delta^2x+\frac{Cc}{\nu r^\nu}-\frac{\nu+1}{\nu}\, x^{\frac{\nu}{\nu+1}}\, (Cc)^{\frac{1}{\nu+1}}.$$

On substituting for δ in terms of x we find

$$v = (Cc)^{\frac{1}{(\nu+1)}} x^{\frac{\nu}{(\nu+1)}} w, \text{ say,}$$

where

$$w = 1-\epsilon^2\delta+\frac{1}{\nu}\{(1-\epsilon^2\delta^2)^2+\epsilon^2\}^{-\frac{1}{2}\nu}-\frac{\nu+1}{\nu}$$

$$= -\left(\frac{1}{2}+\delta-\delta^2\right)\epsilon^2+O(\epsilon^4).$$

Hence there is a $k > 0$ such that $w > k-\epsilon^2$ and we see that on $\mathcal{L}(T)$

$$\mathcal{R}\, h(s) < -\lambda x+\left(\frac{\nu+1}{\nu}\right)x^{\frac{1}{\nu+1}} (Cc)^{\frac{1}{\nu+1}}\left(1-\frac{\kappa\nu\epsilon^2}{\nu+1}\right)-\eta xt.$$

Thus, noting that $ds = (i-\eta)dt$ on $\mathcal{L}(T)$, and that

$$\int_0^T e^{-\eta xt}\, dt < \frac{1}{\eta x},$$

we have

$$\left|\int_{\mathcal{L}} e^{h(s)}ds\right| < \frac{(1+\eta)}{\eta x}\exp\left\{-\lambda x+\frac{\nu+1}{\nu}\, x^{\frac{\nu}{\nu+1}} (Cc)^{\frac{1}{\nu+1}}\left(1-\frac{\kappa\nu\epsilon^2}{\nu+1}\right)\right\}.$$

Hence,

(31) $$\int_{\mathcal{L}} e^{h(s)}\, ds = o[\psi(x)], \qquad \text{as } x \to \infty,$$

and this result is uniform in T.

Lastly, consider the straight line segment $\mathcal{I}(T)$, say, which is parallel to the real axis and mapped out by

$$s = (-\lambda+\delta-\epsilon^2\delta^2-t)+i(\varepsilon\delta+T)$$

as t runs from 0 to ηT. On $\mathcal{I}(T)$ we have

$$|e^{h(s)}| < K_4\, e^{-\lambda x+x\delta-x\varepsilon^2\delta^2-xt}$$

for some K_4 which is independent of T provided it is sufficiently large, and of ϵ and δ provided they are both small. Thus

$$\left| \int_{\mathcal{J}(T)} e^{h(s)} ds \right| < \frac{K_4}{x} e^{-\lambda x + x\delta - \epsilon^2\delta^2 x}$$

$$= o[\psi(x)], \qquad \text{as } x \to \infty,$$

uniformly in T.

In (29) we may suppose that $\gamma = \lambda + \epsilon^2\delta^2 - \delta$, for then $\gamma < \lambda$ correctly if ϵ is small enough. Combining (29), (30,) and (31,), and noting especially the uniformity of (30) and (31) with respect to T, we can now easily prove

$$\Delta_\sigma(x) = O[\psi(x)], \qquad \text{as } x \to \infty.$$

This completes the proof of the theorem.

APPENDIX

Let $\{F_n(x)\}$ be an infinite sequence of distribution functions of non-negative random variables and, for real $s \geqslant 0$, let

$$F_n^*(s) = \int_{0^-}^{\infty} e^{-sx} \, dF_n(x), \qquad n = 1, 2, \ldots$$

be the corresponding Laplace-Stieltjes transforms. Suppose $F(x)$ is a further distribution function of a non-negative random variable and that $F_n(x) \to F(x)$ as $n \to \infty$, at every continuity point of $F(x)$. Then, by dominated convergence,

$$\int_0^\infty e^{-sx} F_n(x) dx \to \int_0^\infty e^{-sx} F(x) dx$$

as $n \to \infty$, for every fixed real $s > 0$. Hence $F_n^*(s) \to F^*(s)$ as $n \to \infty$, for every $s \geqslant 0$ (for $F_n^*(0) = F^*(0) = 1$ for all n).

On the other hand, suppose $F_n^*(s) \to \phi(s)$, for every real $s \geqslant 0$, as $n \to \infty$; suppose further that $\phi(s)$ is continuous to the right at the origin. By the usual Helly-Bray compactness argument there is a bounded non-decreasing function $M(x)$, say, and a subsequence $\left\{F_{n_m}(x)\right\}$ such that $F_{n_m}(x) \to M(x)$ at every continuity point $M(x)$. Moreover, we can take $M(x) = 0$ for $x < 0$. By the dominated convergence argument already used we see $F^*_{n_m}(s) \to M^*(s)$ and so $M^*(s) = \phi^*(s)$ for all real $s > 0$. But $F^*_n(0) = 1$ for all n and so $\phi^*(0) = 1$. However, $\phi^*(s)$ is continuous to the right at the origin and hence $M^*(0+) = 1$. This proves that $M(x)$ is a distribution function, and indeed, the unique distribution with Laplace-Stieltjes transform $M^*(s) = \phi(s)$. By familiar reasoning it now follows that $F_n(x) \to M(x)$ at every continuity point of $M(x)$.

References

Dall'Aglio, G. (1964). "Present Value of a Renewal Process," *Ann. Math. Statist.*, **35**, 1326–1331.

Doetsch, G. (1950). *Handbuch der Laplace-Transformation, Vol. I*, Verlag Birkhauser, Basel.

Karamata, J. (1930). "Sur un mode de croissance occents des fonctions," *Mathematica (luj)*, **4**, 38–53.

Takacs, L. (1955). "On Stochastic Processes Connected with Certain Physical Recording Apparatuses," *Acta. Math. acad. Sci., Hung.*, **6**, 363–380.

Widder, D. V. (1941). *The Laplace Transform*, Princeton University Press.

(*Received Jan. 1, 1966.*

CHAPTER 38

Component Tolerances Which Achieve a Specified System Tolerance

W. A. THOMPSON, JR. AND RICHARD K. TRASK,
The Florida State University

1. INTRODUCTION

This paper treats one specific aspect of a wider problem of considerable practical interest and significance: How should component variabilities (or tolerances) be chosen in order to stay within a specified system variability? A simple approximate tolerancing model is advanced for the multivariate normal case when the quantity of interest is Euclidean distance. It is demonstrated that tables of Grad and Solomon [1] yield the approximation error.

In order to fix ideas, two examples are provided. Many manufacturing processes result in a product whose critical dimensions may be subject to two or more sources of error. For example, let us assume a manufacturer making small plastic rods first cuts the rods from larger units and then must heat-treat the smaller rods for a fixed time at a given temperature. Hence, the length of the end product is subject to two sources of random error or variation; first, the error inherent in the cutting process and, second, that due to the heat treatment of the rods. Suppose now that we wish to know how small each of these errors must be in order that the length of the rods be within specified limits with some assigned probability.

Next consider an intercept problem; it is desired to design equipment which will engineer a collision between one object in space called the target, and a second object called the interceptor. For purposes of analyzing miss distance, this problem can usually be divided into four subsidiary parts: (1) intelligence: determining the present location of the target; (2) prediction: predicting the location of the target at the time the interceptor will arrive; (3) communication: "informing" the interceptor where to go; and (4) delivery: actually delivering the interceptor to the point in space to which it has been directed.

The errors generated in the intercept problem are such that errors accumulated at one stage are, on the average, equal to the error present at the previous stage. Thus, on the average, the error in determining future position will equal the present position error. Similarly, in the communication phase the interceptor will be directed to the point where it is predicted that the target will be and finally the delivery point in space will average out to the point where the interceptor was "told" to go.

As a simple example of an intercept problem consider a soldier practice-firing on a fixed paper target. Let us assume that the probability that a bullet hits the target is dependent only on the aiming error of the soldier and the round-to-round dispersion of the ammunition. The problem under consideration is that of determining how small these errors must be, in order that the bullet impact falls within a circle of specified radius with some assigned probability. Other intercept problems might involve electrons bombarding a screen, or a ship searching for a submarine.

2. A TOLERANCING MODEL

In this section we use the suggestive language of an intercept problem. The results for arbitrary multivariate random variables will be clear. Let $\boldsymbol{X}_i = (X_{i1}, X_{i2}, \ldots, X_{ip})$ be the p-dimensional error of an intercept problem accumulated up to and including the ith stage. The $\boldsymbol{X}_i$ are taken to be p-dimensional normal deviates for $i = 1, \ldots, n$. From the discussion of the previous section $\boldsymbol{X}_1$ is $N(\boldsymbol{0}, \boldsymbol{\Sigma}_1)$ and the conditional distribution of $\boldsymbol{X}_{i+1}$ given $\boldsymbol{X}_i$ is $N(\boldsymbol{X}_i, \boldsymbol{\Sigma}_{i+1})$ for $i = 1, \ldots, (n-1)$. [In accordance with the customary notation, $\boldsymbol{X}_1$ is $N(\boldsymbol{0}, \boldsymbol{\Sigma}_1)$ means that the random vector $\boldsymbol{X}_1$ is distributed according to the multivariate normal distribution with mean vector $\boldsymbol{0}$ and variance-covariance matrix $\boldsymbol{\Sigma}_1$.] Thus the final p-dimensional vector of miss distance is $\boldsymbol{X}_n$, and $\boldsymbol{X}_n$ is $N(\boldsymbol{0}, \boldsymbol{\Sigma})$ where $\boldsymbol{\Sigma} = (\sigma_{ij}) = \boldsymbol{\Sigma}_1 + \boldsymbol{\Sigma}_2 + \ldots + \boldsymbol{\Sigma}_n$. The radial miss distance for the intercept problem is $(X_{n1}^2 + X_{n2}^2 + \ldots + X_{np}^2)^{\frac{1}{2}} = r$, say. It is required to determine conditions on $\boldsymbol{\Sigma}_1, \ldots, \boldsymbol{\Sigma}_n$ such that

$$(1) \qquad \Pr(r < \delta) \geqslant P$$

is approximately true.

Given a specific system design (or more precisely, given the matrices $\boldsymbol{\Sigma}_1, \ldots, \boldsymbol{\Sigma}_n$) we may easily determine P, for given δ, by considering that r^2 is a quadratic form in p normal deviates. However, we are concerned with the more interesting reverse problem: what design systems may we consider which will yield a given desired δ and P?

The point of view of this paper is that some degree of approximation can be tolerated if the method yields explicit conditions on $\boldsymbol{\Sigma}_1, \ldots, \boldsymbol{\Sigma}_n$. We approximate the distribution of r^2 by assuming provisionally that $\Sigma = \bar{\sigma}^2 \mathrm{I}$ where $\bar{\sigma}^2 = E(r^2)/p$. Then $X_{n1}/\bar{\sigma}, X_{n2}/\bar{\sigma}, \ldots, X_{np}/\bar{\sigma}$ would be independently and identically $N(0, 1)$ and $r^2/\bar{\sigma}^2$ would be distributed as chi-square with p degrees of freedom. This $\bar{\sigma}$ approximation is exact when the characteristic roots of $\boldsymbol{\Sigma}$ are equal; hence, it is reasonable to believe that there will be a larger class of matrices $\boldsymbol{\Sigma}$ for which the approximation is not too bad. The extent to which this is true will be treated in the next section.

That this approximation does yield explicit conditions on $\boldsymbol{\Sigma}_1, \ldots, \boldsymbol{\Sigma}_n$ may be seen as follows:

$$p\bar{\sigma}^2 = E(r^2) = E\left(X_{n1}^2+X_{n2}^2+\ldots+X_{np}^2\right)$$
$$= (\sigma_{11}^2+\sigma_{22}^2+\ldots+\sigma_{pp}^2)$$
$$= t(\boldsymbol{\Sigma}) = t(\boldsymbol{\Sigma}_1)+t(\boldsymbol{\Sigma}_2)+\ldots+t(\boldsymbol{\Sigma}_n)$$

where $t(\boldsymbol{M})$, the trace of the square matrix $\boldsymbol{M}$, is the sum of the diagonal elements of $\boldsymbol{M}$. Thus $\bar{\sigma}^2 = \bar{\sigma}_1^2 + \bar{\sigma}_2^2+\ldots+\bar{\sigma}_p^2$ where $\bar{\sigma}_i^2$ is the average of the diagonal elements of $\boldsymbol{\Sigma}_i$. Furthermore, if δ is some assigned value of r and P is an assigned probability, then (1) holds for all $\bar{\sigma}^2 \leqslant \delta^2/\chi^2(P, p)$ or for all $\bar{\sigma}_1^2, \bar{\sigma}_2^2, \ldots, \bar{\sigma}_n^2$ such that

(2) $$\bar{\sigma}_1^2+ \bar{\sigma}_2^2 +\ldots+ \bar{\sigma}_n^2 \leqslant \delta^2/\chi^2(P, p).$$

It is worth noting that $t(\boldsymbol{\Sigma}_i)$ and hence $\bar{\sigma}_i^2$ is invariant under translations and rotations of axes. We may calculate $\bar{\sigma}_1^2$ and $\bar{\sigma}_2^2$ using whatever coordinate systems are respectively the most convenient; the two systems need not be the same.

3. WHY THE $\bar{\sigma}$ APPROXIMATION

The quantity being approximated may be written explicitly as

$$P = \text{const.} \underset{\sum_i x_i^2 < \delta^2}{\int \ldots \int} \exp\left(-\tfrac{1}{2} \sum_{i,j} \lambda_{ij}\, x_i x_j\right) dx_1 \ldots dx_p$$

where (λ_{ij}) is the inverse of the positive definite covariance matrix $\boldsymbol{\Sigma} = (\sigma_{ij})$. This integral can, of course, be transformed to the equivalent simpler expressions

$$P = \prod_{i=1}^{p} (2\pi a_i)^{-\frac{1}{2}} \underset{\sum_i x_i^2 < \delta^2}{\int \ldots \int} \exp\left(-\tfrac{1}{2} \sum_i x_i^2/a_i\right) dx_1 \ldots dx_p$$

and

$$P = (2\pi)^{-p/2} \underset{\sum_i a_i x_i^2 < \delta^2}{\int \ldots \int} \exp\left(-\tfrac{1}{2} \sum_i x_i^2\right) dx_1 \ldots dx_p$$

where $a_1, \ldots, a_p$ are the characteristic roots of $\boldsymbol{\Sigma}$. At this point it is important to notice that

$$\sigma_{11}^2+\sigma_{22}^2+\ldots+\sigma_{pp}^2 = t(\boldsymbol{\Sigma}) = a_1+\ldots+a_p.$$

and hence the $\bar{\sigma}$ approximation is exactly the assumption that $a_1 = \ldots = a_p$.

Alternative approximations are, of course, possible. For example, we might obtain a more accurate approximation by using the geometric rather than the arithmetic mean or perhaps the average of the maximum and minimum roots, but this is not solely a paper on approximating the distribution of quadratic forms and these two approximations are unacceptable on other grounds. Our purpose is two-fold; the approximation used must first of all yield simple explicit conditions on the component tolerance parameters $\mathbf{\Sigma}_1, \ldots, \mathbf{\Sigma}_n$ and secondly, it must be moderately accurate. As a case in point, consider the following conditions obtained by using the geometric mean:

$$\prod_{i=1}^{p} a_i \leqslant [\delta^2/\chi^2(P, p)]$$

where $a_1, \ldots, a_p$ are the roots of the matrix $\mathbf{\Sigma}_1+\mathbf{\Sigma}_2+\ldots+\mathbf{\Sigma}_n$. When compared with the simple sphere (2) given by the $\bar{\sigma}$ approximation, these conditions are very complex and, we believe, useless for the present purpose.

In two and three dimensions, the most likely case of practical interest, the error of the $\bar{\sigma}$ approximation can be determined from Grad and Solomon's [1] Tables I and II. For convenience (using our notation) we show several typical entries from their tables. Thus, for example, from Table I we see that if the $\bar{\sigma}$ approximation is used when $a_2 = \cdot 8$ and $a_1 = \cdot 2$ then $P(r < \delta) = \cdot 5464$ will be approximated as ·5034. Considering the exploratory nature of our problem we do not consider this error to be serious. At a later stage, when specific alternative systems have crystallized, then more exact probabilities may be calculated using Grad and Solomon's tables directly.

TABLE I

Pr $(r < \delta)$ in two dimensions

σ	$\bar{\sigma}$ approximation	a_2, a_1			
		.6, .4	.8, .2	.0, .1	.99, .01
$\sqrt{.7}$	.5034	.5080	.5464	.5780	.5962
$\sqrt{3.0}$	.9502	.9487	.9365	.9269	.9178

TABLE II

Pr $(r < \delta)$ in three dimensions

σ	$\bar{\sigma}$ approximation	a_3, a_2, a_1		
		.4, .4 ,.2	.6, .2, .6	.8, .1, .1
$\sqrt{.8}$	.50637	.5161	.5402	.5974
$\sqrt{3.0}$	.97071	.9668	.9577	.9378

4. COMPUTATIONAL EXAMPLES

Referring first to the example involving plastic rods, suppose σ_1^2 represents the error due to the cutting process and σ_2^2 represents the error introduced by the heat treatment. Then all values of σ_1 and σ_2 small enough so that the inaccuracy in the rods is less than, say 0·2 inches, with a probability of at least $\frac{1}{2}$ may be found by examining the positive quadrant of the circle defined by

$$\sigma_1^2+\sigma_2^2 \leqslant \frac{\delta^2}{\chi^2(\cdot 5,\ 1)} = \frac{\cdot 04}{\cdot 455} = \cdot 088 \text{ (inches)}^2.$$

Similarly, in the rifle marksmanship example mentioned earlier, suppose that σ_x^2, σ_y^2 and σ_{xy} are the variances and covariances of the horizontal and vertical components of error due to the soldier's aim or skill, and let τ_x^2, τ_y^2 and τ_{xy} be similarly defined for the dispersion error of the ammunition.

$$\mathbf{\Sigma}_1 = \begin{pmatrix} \sigma_x^2 & \sigma_{xy} \\ \sigma_{xy} & \sigma_y^2 \end{pmatrix} \quad \text{and} \quad \mathbf{\Sigma}_2 = \begin{pmatrix} \tau_x^2 & \tau_{xy} \\ \tau_{xy} & \tau_x^2 \end{pmatrix}$$

Then impacts will fall within a circle of radius, say two feet, with a probability of approximately $\frac{1}{2}$ for all values of σ_x, σ_y, τ_x, and τ_y such that $\sigma_x^2+\sigma_y^2+\tau_x^2+\tau_y^2 \leqslant 8/\chi^2(\cdot 5,\ 2) = 5{\cdot}77$ (feet)^2.

Reference

[1] Grad, A. and Solomon, H. "Distribution of Quadratic Forms and Some Applications," *Ann. Math. Statist.*, **26**, (1955), 464–477.

(*Received June. 7, 1965.*)

CHAPTER 39

Reflections on the Future of Mathematical Statistics

J. WOLFOWITZ[1], *Cornell University*

1. INTRODUCTION

During his stay at Columbia University in 1948-9, S. N. Roy often discussed with me the directions in which, we thought, (mathematical) statistics would or should develop. It is therefore not unfitting to discuss this subject in a volume dedicated to his memory. Nor is it entirely superfluous to do so, for the future of statistics, both as a field for research and as a discipline to be studied, is far from assured. Mathematical fashions change, and fields once regarded as of burning research interest, (e.g., projective geometry) are now considered uninteresting or played out. Some subjects, for example, differential equations, we will always have with us, at least as subjects to be studied; the needs of science and technology alone will suffice for this. However, one needs to examine whether present day statistics really

[1] The writing of this paper was supported by the Air Force Office of Scientific Research, Office of Aerospace Research, U.S. Air Force, under AFOSR Grant No. 396–63 to Cornell University.

meets scientific and technological needs. Statistics arose in response to such needs, and these are, or should be, intimately connected with the directions of research. It seems to me obvious that, were it not for these needs, the future of statistics as a discipline for research and its appeal to pure mathematicians would be limited indeed.

However, even at this point we must already introduce an obvious caution. Suppose a mathematical study does have its *raison d' être* in the solution of certain mathematical problems which arise in science and technology. It would be a grave mistake to construe the content of the science too narrowly. Even an applied subject needs a theory which can often facilitate the solution of "practical" problems. Rigorous justification of results obtained has many advantages: protection from error, a sense of psychological security, aesthetic grounds, the possibility of suggesting new solutions and interesting new problems. A colleague of mine once said that there are two kinds of applied mathematics, pure applied mathematics and applied applied mathematics. A branch of applied mathematics which is to endure needs to contain both.

2. THE NEO-BAYESIANS

The period after World War II saw the emergence of a school of neo-Bayesians, deeply critical of previous developments and filled with a proselytizing missionary zeal. Not their philosophical doctrines but their role in the future of statistics will be discussed here.

One of the leaders of this school has written (Savage [1]): "I hope that I have made you feel that statistics is going places today (i.e., is achieving considerable sucess, J. W.). There is a lot of exciting work to do that is relatively easy. Any of you who are inclined to work in statistical theory can get in on the ground floor now (i.e., be in at the founding of the subject, J. W.), for there are not any experts any more. Those who only use statistics without intending to do research in it must also remember that there are not any experts any more. Don't take any wooden nickels (i.e., don't be gullible, J. W.); do your own work honestly and thoughtfully without much reliance on rules; and don't believe in magic or powerful new tools". Even accepting this position for the sake of argument one cannot but raise certain obvious points. How can any scientific endeavor which is "relatively easy" be "exciting" or hold out much prospect as a research discipline? The role of the scientist is to push back the frontiers of knowledge in important directions. This is never easy, or the frontiers would not be where they are. If a certain field or certain directions have reached the state where they are sufficiently explored, the scientist worth his salt and not yet ready to retire, looks for challenging problems in other fields. According to the Bayesian point of view one has only to determine the a priori distribution and then compute the a posteriori distribution. Can the study of this engage for long the serious efforts of first-rate minds? What challenges can this offer to brilliant young students? It is obvious that, if one accepts the Bayesian view as the answer to the problems of statistics, then the answer is so sweeping as to deny the existence of enough challenging problems to constitute a research discipline with much attraction to first-rate minds.

Since the Bayesian answer is so sweeping and relatively so easy to give, and since (the Bayesians say) there are so few (if any) difficult problems to solve (when one adopts their point of view), it is a continual source of wonderment to me to see the growth of some Bayesian university statistics departments. What can one member of such a department teach that is so different from what his colleague is teaching at the same time?

3. DECISION THEORY

The rest of this essay is addressed only to those who believe that many, if not most, of the problems which actually rise in statistics can and should be solved in the framework of decision theory.

Decision theory arose to replace the theory of testing hypotheses. In the latter theory the formulation of the problems is inadequate for most of the problems which arise in actual statistical investigations. These inadequacies have often been discussed and need not be considered here.[2] (It is surely an irony of history that decision theory was founded by a theoretician like Wald, and that it was not those who use the results of statistical research, the experimental scientists, who first rebelled against fitting their problems into the Procrustean bed of testing hypotheses. No one is so impractical as the so-called practical men.) It is obvious that decision theory has failed the high hopes once held for it, and it is pertinent briefly to discuss here the reasons for this.

Overzealous converts to decision theory probably helped, by exaggerated claims, to launch the subject and Wald's book to a bad start. Even the jacket of the book contained advertising matter which claimed that the book explained a new theory which would enable the experimenter properly to design experiments and draw conclusions from them. It should have added "eventually, after the theory is properly developed". The practical statistician looked at the formidable notation and terminology, and at the theorems on the closure of classes of decision functions in various topologies, and drew the only conclusion he could draw from these, namely, that a statistical theory which needed such a formidable mathematical apparatus merely to explain its fundamental ideas was forever beyond his reach and useless to him[3]. Consequently it was certainly impossible for him to formulate his actual problems in decision theoretic terms and to ask the mathematicians for solutions which would be meaningful to him.

Thus the mathematician working in statistics was not presented with actual problems to which to apply the theory. Per-

[2] For a recent discussion see the paper listed as [7] below.

[3] The fundamental ideas can be explained, even rigorously, in much simpler and more readily accessible settings. See, for example, [2], Chapter 6, or [3], Section 1.

haps he had no feeling for, or much interest in, "applied" problems, except when the latter could be posed in attractive mathematical form. All his professional instincts always pushed him to generalize and deepen. When even the fundamentals of the theory were presented in such abstract terms, was he not justified in pushing further in the same direction? Also, Wald's proofs of his closure theorems and other developments could be made "cleaner" and much more elegant by use of the new abstract formulations which are so characteristic of the post-World War II mathematics. This was a challenge to the technician who, within varying limits, must be present in every professional mathematician. To meet this challenge required no new statistical ideas and really no basically new mathematical ideas, but simply technical mathematical knowledge and facility.

For these and perhaps other reasons the relatively few papers on decision theory written after Wald's death in 1950 are usually attempts at mathematical "cleanups", often *à la* Bourbaki, of his closure theorems. Such work obviously has a limited future, either on the basis of its mathematical interest or for its applicability to actual statistical problems. Only the emergence of a group of workers who will concern themselves with problems of greater mathematical novelty and greater statistical applicability will revivify the theory. On the basis of past history, it is highly unlikely that any stimulus in the latter direction will come from the applied statisticians. If there is to be such a revival it will have to come from the mathematical statisticians themselves.

4. IDEALIZATION

The role played by mathematical idealization in the "development" of decision theory and other branches of mathematical statistics is so important and so peculiarly different from its role in other sciences as to deserve special discussion. Mathematical idealization is a method of simplification and approximation and is employed in all sciences where mathematics is used, e.g., idealized gases in physics, etc. In mathematical statistics a crucial application occurs as follows: In actuality the observations which occur in any statistical experiment take values which belong to a (finite) set of integral multiples of some unit. Thus, the chance variables in the mathematical formulation of the problem should all take their values on a (finite) lattice. When the unit of measurement is small the problem can often be simplified and a good approximation obtained by replacing the distributions of the "actual" chance variables by continuous approximations. This is the idealization which occurs most frequently and which is of greatest importance in statistics.

In some problems involving chance variables with continuous distributions there arise measure theoretical difficulties not present when the chance variables are all discrete. (It is precisely to avoid these difficulties that the book [2] on decision theory deals almost entirely with discrete chance variables. Presumably its writers considered the measure theoretic problems as extraneous to the essence of decision theory.) For example, many of the difficult and delicate problems in the study of sufficient statistics arise *solely* for this reason. Some subtle and difficult papers have been written to solve these problems. The trouble is that most, if not all, of these problems are of no particular interest to the measure theorist and really have no enduring place in mathematics except possibly in mathematical statistics. In mathematical statistics they owe their existence to the idealization described above. Is there not something basically wrong with an idealization which *creates* difficult problems rather than serves to avoid them? And is there not something basically wrong with a subject if difficult problems arise from a supposedly simplifying idealization?

The reasons for studying such problems are easy to explain. When the problems are difficult they present a challenge to first-rate research workers which the latter find hard to resist. Once such a paper is written it is a challenge to improve it or carry it further. It would be wrong to say of any one paper that it should not have been written. But, unless a subject can produce many interesting and difficult papers which are not of this artificial character, its future would seem to be far from assured.

Of course our discussion has been of the difficult papers which it is an achievement to write. Trivial papers are written in every subject but do not determine its future, except when only trivial papers are being written.

It is also unnecessary to say that nothing in the above is to be construed as an argument against the need for rigorous proofs of statistical theorems. It seems difficult to believe that the value of, and need for, rigorous proof can seriously be questioned, but this has been done by Barnard [5]. In an earlier paper [4] this author had omitted to justify an inversion in the order of two limit processes, the most difficult part of an otherwise easy argument. This omission having been called to his attention, he supplied the missing argument in [5]. In the course of the latter he inveighed against pettifogging pedants who think that, even in mathematical statistics, one needs to justify the inversion of two limit processes. (What remarkable "theorems" one could prove without this vulgar constraint!) It is amusing to note that this author, in the very same paper [4], expects his reader to be familiar with Weill's *Integration dans les groupes topologiques* and seriously discusses the case where the statistical parameter to be estimated is an element of a general topological group (presumably not Euclidean space!)

5. THE THEORY OF TESTING HYPOTHESES

It has been said earlier that the theory of decision functions arose because of the inadequacies of the theory of testing hypotheses, and that the remainder of this essay would be addressed to those who believe that many, if not most, statistical problems are to be solved in the context of decision theory. Nevertheless, it is in order to make some remarks about work in testing hypotheses because a) many workers who teach the desirability of decision theory still write about testing hypotheses, and b) one of the most distinguished books on statistics is the book by Lehmann [6] on testing hypotheses.

Probably the most telling criticism of the theory of testing hypotheses is that it poses the wrong problems. For example, an experimenter is rarely, if ever, really interested in testing the hypothesis that eight varieties of wheat all have the same yield. (It is almost incredible that they should. What is really being tested is that the variation among their yields is "not large".) Yet this is what several famous tests are designed to test. In fact, as the tests are universally applied, it is concluded that all eight varieties have the same, or essentially the same, yield, unless the contrary is clearly demonstrated, i.e., unless a "significant"result is obtained! (Nor do these tests say what one is to do when a significant result is obtained.) Yet, first-rate statisticians, who freely acknowledge the validity of these criticisms, write papers about the admissibility, stringency, etc., of these tests. At their best, these papers deal with difficult problems whose solution requires considerable skill and ingenuity. However, they do not contribute to the essential development of statistics or to the solution of realistic problems, nor are they of permanent mathematical interest per se. What a pity to waste this talent and energy!

This is not the place to review Lehmann's book, but some comments on it are in order because of the great influence this book has and is bound to have. (It also contains material which does not fall under the theory of testing hypotheses and is not included in our comments.) The exercises in the book are an essential and important part of it. They carry the theory further, and many of them are extremely clever and ingenious. In fact, they are much more interesting than many of the formally stated theorems, which are relatively obvious or have proofs which are rather easy.

One comes away with a general impression of relatively few deep and difficult theorems, and of many clever and ingenious examples, mostly involving the binomial, normal, Poisson, and other distributions of the exponential family. So many ingenious tests about the latter have been studied, and so few problems of practical interest solved. Is this material likely to survive into the future, either because it meets the needs of experimental science or because of its enduring mathematical interest per se?

The author of [6], in the preface to his book and elsewhere, acknowledges the validity of some of the criticisms of the theory of testing hypotheses. In fact, from his remarks and other evidence, one concludes that he published the book because he had already expended so much ingenuity and effort on its contents. I agree enthusiastically that this book should have been published and that it is a distinguished book. I am delighted to have at hand in permanent form so many clever and ingenious examples. But I would consider it a disaster for statistics if this book should determine the direction of research for any appreciable period of time[4].

---

[4] Perhaps this is the place to raise a different and minor point, relevant to a discussion of what is one of the best and most influential books on statistical theory. A student of pure mathematics learns Poincaré's recurrence theorem or Hilbert's Nullstellensatz or about Finsler spaces or Lie groups. After all, how better to identify theorems than by the names of their discoverers? Incidentally, then, one also gets a little feeling for the grandeur of the subject and of its great men. My reading of Lehmann's work which, in this respect, is admittedly only casual, turns up only two theorems named after their discoverers. One is the "Neyman-Pearson fundamental lemma" which, no matter how "fundamental" it may be, is pretty trivial to prove and not difficult to discover. The other is the "Hunt-Stein" theorem, also not difficult to prove. Surely these distinguished writers have discovered deeper theorems to which their names could, with more luster, properly be attached. And what about other eminent writers, in particular such colossi as Fisher and Wald? Are there no theorems which should bear their names? The proof of what is perhaps the deepest theorem in the book displays most conspicuously not the name of him who brilliantly conjectured it, but of someone who later gave an alternate proof of one of its lemmas. What kind of feeling for the history and development of the subject is the student likely to get from such a treatment? Just as a taste for good literature is developed by reading good books, so good taste in science comes from knowing the great achievements of the past and how they fitted into the structure of the knowledge of their time.

6. CONCLUDING REMARKS

Just as it was necessary to hope that Lehmann's book does not determine the course of future research, although the book is so distinguished and eminently worth publishing, so it is necessary to understand the criticisms which have been made in this paper in the light of their context. Let us, therefore, recapitulate and summarize the principal of these.

Except perhaps for a few of the deepest theorems, and perhaps not even these, most of the theorems of statistics would not survive in mathematics if the subject of statistics itself were to die out. In order to survive the subject must be more responsive to the needs of application. On the basis of past experience it is too much to expect that the formulation of actual problems will come from "practical" experimenters; the change will have to come from within the ranks of the mathematical statisticians themselves. Some current work, e.g., in the design of experiments, does meet these needs as applied applied statistics and/or pure applied statistics.

Even when a branch of, or tendency in, statistics represents a sterile direction, it is a serious mistake to say of any one paper that, for this reason alone, it should not have been written. Rather one can only say of a series of papers that they do not develop the subject in a fruitful direction. It is very important to stress this. Obviously one does not want to dissuade anyone who is challenged by the difficulty of a problem from attempting to solve it. Nor should one dictate within strict bounds what should or should not be published. Dictatorship in science is as dangerous and repugnant as it is in other fields. (Of course, some choice and selection are absolutely unavoidable. Obviously everything cannot and should not be published. This has as an immediate consequence that some papers will be dismissed as "not interesting". Probably the best hope of avoiding error lies in the multiplicity of journals in many different countries.) What is to be hoped for is that the scientific concensus (to borrow a word now in vogue in another context) will be such that gifted workers will be induced to work in directions fruitful for statistics. It is for this reason that I have avoided discussing specific books or papers except in one or two egregious instances, e.g., Lehmann's book because of its great importance.

It goes without saying that "responsiveness to the needs of science and technology" should never be construed in any narrow sense. The subject obviously needs a theory (pure applied statistics) which should forge ahead and also interact with problems in application. What we must guard against is the development of a theory which, on the one hand, bears little or no relation to the actual problems of statistics, and which, on the other hand, when viewed as pure mathematics, is not interesting per se nor likely to survive.

Finally, let me anticipate the *tu quoque* criticism by pleading guilty to it in advance. Obviously my views on, say, testing hypotheses have changed and developed over the years. Criticism can be cogent even if the critic himself is not without blemish.

For any science, as in nuclear reactions, there is a critical mass, this time of scientific workers, who should be numerous enough to stimulate each other, gifted enough to see and to solve problems, and, by the esteem in which they hold some work and the disesteem in which they hold other work, successful in guiding the science into fertile and fruitful directions. Perhaps this essay may evoke a sympathetic response among others of similar but hitherto unvoiced views, who will be numerous enough and gifted enough to constitute such a critical mass.

References

[1] Savage, L. J. "The Subjective Basis of Statistical Practice," mimeographed notes, University of Michigan, July, 1961.

[2] Blackwell, D. and M. A. Girshick. *Theory of Games and Statistical Decisions*, John Wiley and Sons, New York, 1954.

[3] Wald, A. and J. Wolfowitz. "Characterization of the Minimal Complete Class of Decision Functions when the Number of Distributions and Decisions is Finite," *Proceedings of the Second Berkeley Symposium on Probability and Statistics*, pp. 149–158, University of California Press, 1951.

[4] Barnard, G. A. "The Frequency Justification of Sequential Tests," *Biometrika,* **39** (1952), 144–150.

[5] Barnard, G. A. "The Frequency Justification of Sequential Tests—Addendum," *Biometrika,* **40** (1953), 468–469.

[6] Lehmann, E. L. *Testing Statistical Hypotheses,* John Wiley and Sons, New York, 1959.

[7] Wolfowitz, J. "Remarks on the theory of testing hypotheses" *The New York Statistician,* **18** (1967), No. 7, 1–3.

(*Received Jan. 1, 1966.*)

www.ingramcontent.com/pod-product-compliance
Lightning Source LLC
Chambersburg PA
CBHW021021210326
41598CB00016B/880
* 9 7 8 0 8 0 7 8 7 9 7 9 5 *